Simon Haykin, McMaster University

Barry Van Veen, University of Wisconsin

Signals and Systems

2005 JustAsk Edition

JOHN WILEY & SONS, INC.

EXECUTIVE EDITOR   *Bill Zobrist*
EDITORIAL ASSISTANT   *Bridget Morrisey*
SENIOR MARKETING MANAGER   *Katherine Hepburn*
SENIOR PRODUCTION EDITOR   *Caroline Sieg*
SENIOR DESIGNER   *Maddy Lesure*
ILLUSTRATION COORDINATOR   *Gene Aiello*
COVER PHOTO   *Erich Ziller/Eastman's West*

This book was set in Sabon Roman by Prepare Inc. and printed and bound by Hamilton Printing Company. The cover was printed by Brady Palmer.

This book is printed on acid free paper. ⊗

To order books or for customer service please call 1-800-CALL WILEY (225-5945).

Library of Congress Cataloging-in-Publication Data
Haykin, Simon S., 1931-
    Signals and systems / Simon Haykin, Barry Van Veen. -2nd ed.
       p.   cm.
    Includes index.
       ISBN 978-0-471-70789-9 (cloth · alk. paper)
       1. Signal processing.   2. System analysis.   3. Linear time invariant systems.
    4. Telecommunication systems.   I. Van Veen, Barry.   II. Title.
    TK5102.5.H37   2002
    621.382′2—dc21   2002027040
                                                               CIP
ISBN (Domestic) 978-0-471-70789-9
ISBN (WIE) 978-0-471-37851-8

Printed in the United States of America

10 9 8 7 6 5 4 3 2

Preface

The "Signals and Systems" Course in the Electrical Engineering Undergraduate Curriculum

A course on "signals and systems" is fundamental to the study of the many fields that constitute the ever-expanding discipline of electrical engineering. Signals and systems serves as the prerequisite for additional coursework in the study of communications, signal processing, and control. Given the pervasive nature of computing, concepts from signals and systems, such as sampling, are an important component of almost every electrical engineering field. Although the signals and systems that arise across these diverse fields are naturally different in their physical make-up and application, the principles and tools of signals and systems are applicable to all of them. An introductory course on "signals and systems", commonly takes one of two forms:

- ▶ A one-semester course that focuses on the analysis of deterministic signals and an important class of systems known as linear time-invariant (LTI) systems, with practical examples drawn from communication and control systems.

- ▶ A two-semester course that expands on the one-semester course by including more detailed treatment of signal processing, communication and control systems.

This course is usually offered at the sophomore or junior level and assumes the student has a background in calculus and introductory physics.

How this Book Satisfies the Essential Needs of this Course

Given the introductory nature of the signals and systems course and diversity of applications for the topic, the textbook must be easy to read, accurate, and contain an abundance of insightful examples, problems, and computer experiments to expedite learning the fundamentals of signals and systems in an effective manner. This book has been written with all of these objectives in mind.

The second edition builds on the first edition's success at providing a balanced and integrated treatment of continuous- and discrete-time forms of signals and systems. This approach has the pedagogical advantage of helping the student see the fundamental similarities and differences between continuous- and discrete-time representations and reflects the integrated nature of continuous- and discrete-time concepts in modern engineering practice. One consistent comment from users of the first edition and reviewers of the second is that the compelling nature of our approach becomes very apparent in Chapter 4 with the coverage of sampling continuous-time signals, reconstruction of continuous-time signals from samples, and other applications involving mixtures of different signal classes. The integrated approach is also very efficient in covering the large range of topics that are typically required in a signals and systems course. For example, the properties of all four Fourier representations are covered side-by-side in Chapter 3. Great care has been taken in the presentation of the integrated approach to enhance understanding and avoid confusion. As an example of this, the four Fourier representations are treated in Chapter 3 as similar, yet distinct representations that apply to distinct signal classes. Only after the student has mastered them individually is the possibility of using Fourier representations to cross the boundaries between signal classes introduced in Chapter 4.

Given the mathematical nature of signal representation and system analysis, it is rather easy for the reader to lose sight of their practical application. Chapters 5, 8, and 9 deal with applications drawn from the fields of communication systems, design of filters, and control systems in order to provide motivation for the reader. In addition, considerable effort has been expended in the second edition to provide an application focus throughout the tool-oriented chapters by including an abundance of application-oriented examples. A set of six theme examples, introduced in Chapter 1 and revisited throughout the remaining chapters, is used to show how different signal representation and system analysis tools provide different perspectives on the same underlying problem. The theme examples have been selected to sample the broad range of applications for signals and systems concepts.

The text has been written with the aim of offering maximum teaching flexibility in both coverage and order of presentation, subject to our philosophy of truly integrating continuous- and discrete-time concepts. When continuous- and discrete-time concepts are introduced sequentially, such as with convolution in Chapter 2 and Fourier representations in Chapter 3, the corresponding sections have been written so that the instructor may present either the continuous- or discrete-time viewpoint first. Similarly, the order of Chapters 6 and 7 may be reversed. A two-semester course sequence would likely cover most, if not all, of the topics in the book. A one-semester course can be taught in a variety of ways, depending on the preference of the instructor, by selecting different topics.

Structure Designed to Facilitate and Reinforce Learning

A variety of features have been incorporated into the second edition to facilitate and reinforce the learning process. We have endeavored to write in a clear, easy to follow, yet precise manner. The layout and format has been chosen to emphasize important concepts. For example, key equations and procedures are enclosed in boxes and each example is titled. The choice and layout of figures has been designed to present key signals and systems concepts graphically, reinforcing the words and equations in the text.

A large number of examples are included in each chapter to illustrate application of the corresponding theory. Each concept in the text is demonstrated by examples that emphasize the sequence of mathematical steps needed to correctly apply the theory and by examples that illustrate application of the concepts to real-world problems.

An abundance of practice is required to master the tools of signals and systems. To this end, we have provided a large number of problems with answers immediately following introduction of significant concepts, and a large number of problems without answers at the end of each chapter. The problems within the chapters provide the student with immediate practice and allow them to verify their mastery of the concept. The end of the chapter problems offer additional practice and span a wide range of difficulty and nature, from drilling basic concepts to extending the theory in the text to new applications of the material presented. Each chapter also contains a section illustrating how MATLAB, acronym for MATrix LABoratory and product of The Math Works, Inc., may be used to explore concepts and test system designs within the context of a "Software Laboratory". A complementary set of computer-oriented end of chapter problems is also provided.

New to the Second Edition of the Book

In general terms, this new edition of the book follows the organization and philosophy of the first edition. Nevertheless, over and above new examples and additional problems, some important changes have been made to the book. In addition to the layout and format improvements noted above, long sections in the first edition have been broken up into smaller units. The significant changes to each chapter are summarized as follows:

- ▶ Chapter 1: Two new sections, one on Theme Examples and the other on electrical noise, have been added. The Theme Examples, six in number, illustrate the broad range of problems to which signals and systems concepts apply and provide a sense of continuity in subsequent chapters of the book by showing different perspectives on the same problem. Two new subsections, one on MicroElectroMechanical Systems (MEMS) and the other on derivatives of the unit-impulse function, have also been added.

- ▶ Chapter 2: The treatment of discrete- and continuous-time convolution has been reorganized into separate, yet parallel sections. The material introducing the frequency response of LTI systems has been removed and incorporated into Chapter 3. The treatment of differential and difference equations has been expanded to clarify several subtle issues.

- ▶ Chapter 3. The chapter has been written with increased emphasis on applications of Fourier representations for signals through the introduction of new examples, incorporation of filtering concepts contained in Chapter 4 of the first edition, and reordering the presentation of properties. For example, the convolution property is presented much earlier in the second edition because of its practical importance. Derivations of the discrete-time Fourier series, Fourier series, and discrete-time Fourier transform have been removed and incorporated as advanced problems.

- ▶ Chapter 4: The focus has been tightened as reflected by the new title. Material on frequency response of LTI systems has been moved to Chapter 3 and advanced material on interpolation, decimation, and fast convolution has been removed and incorporated as advanced problems.

- ▶ Chapter 5: A new section on the Costas receiver for demodulation of double sideband-suppressed carrier modulated signals has been added.

▶ Chapter 6: The definition of the unilateral Laplace transform has been modified to include impulses and discontinuities at $t = 0$ and the material on Bode diagrams in Chapter 9 of the first edition is now incorporated in the discussion of graphical evaluation of frequency response.

▶ Chapter 9: A new section on the fundamental notion of feedback and "why feedback?" has been introduced. Moreover, the treatment of feedback control systems has been shortened, focusing on the fundamental issue of stability and its different facets.

▶ Chapter 10: The epilogue has been completely rewritten. In particular, more detailed treatments of wavelets and the stability of nonlinear feedback systems have been introduced.

▶ Appendix F: This new appendix presents a tutorial introduction to MATLAB.

Supplements

The following supplements are available from the publishers website:

www.wiley.com/college/haykin

PowerPoint Slides: Every illustration from the text is available in PowerPoint format enabling instructors to easily prepare lesson plans.

Solutions Manual: An electronic Solutions Manual is available for download from the website. If a print version is required, it may be obtained by contacting your local Wiley representative. Your representative may be determined by finding your school on Wiley's CONTACT/Find a Rep webpages.

MATLAB resources: M-files for the computer-based examples and experiments are available.

About the Cover of the Book

The cover of the book is an actual photograph of Mount Shasta in California. This picture was chosen for the cover to imprint in the mind of the reader a sense of challenge, exemplified by the effort needed to reach the peak of the Mount, and a sense of the new vistas that result from climbing to the peak. We thus challenge the reader to master the fundamental concepts in the study of signals and systems presented in the book and promise that an unparalleled viewpoint of much of electrical engineering will be obtained by rising to the challenge.

In Chapter 1 we have included an image of Mount Shasta obtained using a synthetic aperture radar (SAR) system. A SAR image is produced using many concepts from the study of signals and systems. Although the SAR image corresponds to a different view of Mount Shasta, it embodies the power of signals and systems concepts for obtaining different perspectives of the same problem. We trust that motivation for the study of signals and systems begins with the cover.

Acknowledgments

In writing the second edition, we have benefited enormously from insightful suggestions and constructive input received many instructors and students that used the first edition, anonymous reviewers, and colleagues. We are deeply grateful to Professor Aziz Inan of University of Portland for carefully reading the entire manuscript for both accuracy and readability and making innumerable suggestions to improve the presentation. In addition, the following colleagues have generously offered detailed input on the second edition:

- ▶ Professor Yogesh Gianchandani, *University of Michigan*
- ▶ Professor Dan Cobb, *University of Wisconsin*
- ▶ Professor John Gubner, *University of Wisconsin*
- ▶ Professor Chris Demarco, *University of Wisconsin*
- ▶ Professor Leon Shohet, *University of Wisconsin*
- ▶ Mr. Jacob Eapen, *University of Wisconsin*
- ▶ Dr. Daniel Sebald

We are grateful to them all for helping us in their own individual ways shape the second edition into its final form.

Barry Van Veen is indebted to his colleagues at the University of Wisconsin for the opportunity to regularly teach the Signals and Systems class. Simon Haykin thanks his students, past and present, for the pleasure of teaching them and conducting research with them.

We thank the many students at both McMaster and Wisconsin, whose suggestions and questions have helped us over the years to refine and in some cases rethink the presentation of the material in the book. In particular, we thank Chris Swickhamer and Kris Huber for their invaluable help in preparing some of the computer experiments, the Introduction to MATLAB, the solutions manual, and in reviewing page proofs.

Bill Zobrist, Executive Editor of Electrical Engineering texts, has skillfully guided the second edition from conception to completion. We are grateful for his strong support, encouragement, constructive input, and persistence. We thank Caroline Sieg for dexterously managing the production process under a very tight schedule, and Katherine Hepburn (Senior Marketing Manager) for her creative promotion of the book.

We are indebted to Fran Daniele and her staff of Preparé Inc. for their magnificent job in the timely production of the book; it was a pleasure to work with them.

Lastly, Simon Haykin thanks his wife Nancy, and Barry Van Veen thanks his wife Kathy and children Emily, David, and Jonathan for their support and understanding throughout the long hours involved in writing this book.

Simon Haykin
Barry Van Veen

To God
who created the universe
and gives meaning to our lives
through His love

Contents

CHAPTER 5   *Application to Communication Systems*   **425**

CHAPTER 6   *Representing Signals by Using Continuous-Time Complex Exponentials: the Laplace Transform*   **482**

CHAPTER 7   *Representing Signals by Using Discrete-Time Complex Exponentials: the z-Transform*   **553**

CHAPTER 8   *Application to Filters and Equalizers*   614

CHAPTER 9   *Application to Linear Feedback Systems*   663

CHAPTER 10 *Epilogue* 737

APPENDIX A *Selected Mathematical Identities* 763

APPENDIX B *Partial-Fraction Expansions* 767

APPENDIX C *Tables of Fourier Representations and Properties* 773

APPENDIX D   *Tables of Laplace Transforms and Properties*   **781**

APPENDIX E   *Tables of z-Tansforms and Properties*   **784**

APPENDIX F   *Introduction to MATLAB*   **786**

INDEX   **793**

Notation

[·] indicates discrete valued independent variable, e.g. $x[n]$

(·) indicates continuous valued independent variable, e.g. $x(t)$

▶ Complex numbers

$|c|$ magnitude of complex quantity c

$\arg\{c\}$ phase angle of complex quantity c

$\text{Re}\{c\}$ real part of c

$\text{Im}\{c\}$ imaginary part of c

c^* complex conjugate of c

▶ Lower case functions denote time-domain quantities, e.g. $x(t)$, $w[n]$

▶ Upper-case functions denote frequency- or transform-domain quantities

$X[k]$ discrete-time Fourier series coefficients for $x[n]$

$X[k]$ Fourier series coefficients for $x(t)$

$X(e^{j\Omega})$ discrete-time Fourier transform of $x[n]$

$X(j\omega)$ Fourier transform of $x(t)$

$X(s)$ Laplace transform of $x(t)$

$X(z)$ z-transform of $x[n]$

▶ Boldface lower-case symbols denote vector quantities, e.g., $\mathbf{q}$

▶ Boldface upper-case symbols denote matrix quantities, e.g., $\mathbf{A}$

▶ Subscript δ indicates continuous-time representation of a discrete-time signal

$x_\delta(t)$ continuous-time representation for $x[n]$

$X_\delta(j\omega)$ Fourier transform of $x_\delta(t)$

▶ Sans serif type indicates MATLAB variables or commands, e.g., `X = fft(x,n)`

▶ 0^0 is defined as 1 for convenience

▶ arctan refers to the four quadrant inverse tangent function and produces a value between $-\pi$ and π radians

Principal Symbols

j square root of -1

i square root of -1 used by MATLAB

T_s sampling interval of T_s in seconds

T fundamental period for continuous-time signal in seconds

N	fundamental period for discrete-time signal in samples
ω	(angular) frequency for continuous-time signal in radians/second
Ω	(angular) frequency for discrete-time signal in radians
ω_o	fundamental (angular) frequency for continuous-time periodic signal in radians/second
Ω_o	fundamental (angular) frequency for discrete-time periodic signal in radians
$u(t), u[n]$	step function of unit amplitude
$\delta[n], \delta(t)$	unit impulse
$H\{\cdot\}$	representation of a system as an operator H
$S^\tau\{\cdot\}$	time shift of τ units
$H^{\mathrm{inv}}, h^{\mathrm{inv}}$	superscript inv denotes inverse system
$*$	denotes convolution operation
$\circledast$	periodic convolution of two periodic signals
$H(e^{j\Omega})$	discrete-time system frequency response
$H(j\omega)$	continuous-time system frequency response
$h[n]$	discrete-time system impulse response
$h(t)$	continuous-time system impulse response
$y^{(h)}$	superscript (h) denotes homogeneous solution
$y^{(n)}$	superscript (n) denotes natural response
$y^{(f)}$	superscript (f) denotes forced response
$y^{(p)}$	superscript (p) denotes particular solution
$\xleftrightarrow{DTFS;\ \Omega_o}$	discrete-time Fourier series pair with fundamental frequency Ω_o
$\xleftrightarrow{FS;\ \omega_o}$	Fourier series pair with fundamental frequency ω_o
$\xleftrightarrow{DTFT}$	discrete-time Fourier transform pair
$\xleftrightarrow{FT}$	Fourier transform pair
$\xleftrightarrow{\mathcal{L}}$	Laplace transform pair
$\xleftrightarrow{\mathcal{L}_u}$	unilateral Laplace transform pair
$\xleftrightarrow{z}$	z-transform pair
$\xleftrightarrow{z_u}$	unilateral z-transform pair
$\mathrm{sinc}(u)$	$\dfrac{\sin(\pi u)}{\pi u}$
$\cap$	intersection
$T(s)$	closed-loop transfer function
$F(s)$	return difference
$L(s)$	loop transfer function

Abbreviations

A	amperes (units for electric current)
A/D	analog-to-digital (converter)
AM	amplitude modulation
BIBO	bounded input-bounded output
BPSK	binary phase-shift keying
CD	compact disc
CW	continuous wave
D/A	digital-to-analog (converter)
dB	decibel
DSB-SC	double-sideband suppressed carrier
DTFS	discrete-time Fourier series
DTFT	discrete-time Fourier transform
ECG	electrocardiogram
F	Farads (units for capacitance)
FDM	frequency-division multiplexing
FFT	fast Fourier transform
FIR	finite-duration impulse response
FM	frequency modulation
FS	Fourier series
FT	Fourier transform
H	Henries (units for inductance)
Hz	Hertz
IIR	infinite-duration impulse response
LTI	linear time-invariant (system)
MEMS	microelectricalmechanical system
MSE	mean squared error
PAM	pulse-amplitude modulation
PCM	pulse-code modulation
PM	phase modulation
QAM	quadrature-amplitude modulation
RF	radio frequency
ROC	region of convergence
rad	radian(s)
s	second(s)
SSB	single sideband modulation
STFT	short-time Fourier transform
TDM	time-division multiplexing
V	volts (units for electric potential)
VLSI	very large scale integration
VSB	vestigial sideband modulation
WT	wavelet transform

1 Introduction

1.1 What Is a Signal?

Signals, in one form or another, constitute a basic ingredient of our daily lives. For example, a common form of human communication takes place through the use of speech signals, in a face-to-face conversation or over a telephone channel. Another common form of human communication is visual in nature, with the signals taking the form of images of people or objects around us.

Yet another form of human communication is electronic mail over the *Internet*. In addition to providing mail, the Internet serves as a powerful medium for searching for information of general interest, for advertising, for telecommuting, for education, and for playing games. All of these forms of communication over the Internet involve the use of information-bearing signals of one kind or another. Other real-life examples in which signals of interest arise are discussed subsequently.

By listening to the heartbeat of a patient and monitoring his or her blood pressure and temperature, a doctor is able to diagnose the presence or absence of an illness or disease. The patient's heartbeat and blood pressure represent signals that convey information to the doctor about the state of health of the patient.

In listening to a weather forecast over the radio, we hear references made to daily variations in temperature, humidity, and the speed and direction of prevailing winds. The signals represented by these quantities help us, for example, to form an opinion about whether to stay indoors or go out for a walk.

The daily fluctuations in the prices of stocks and commodities on world markets, in their own ways, represent signals that convey information on how the shares in a particular company or corporation are doing. On the basis of this information, decisions are made regarding whether to venture into new investments or sell off old ones.

A probe exploring outer space sends valuable information about a faraway planet back to a station on Earth. The information may take the form of radar images representing surface profiles of the planet, infrared images conveying information on how hot the planet is, or optical images revealing the presence of clouds around the planet. By studying these images, our knowledge of the unique characteristics of the planet in question is enhanced significantly.

Indeed, the list of what constitutes a signal is almost endless.

A signal is formally defined as a function of one or more variables that conveys information on the nature of a physical phenomenon. When the function depends on a single variable, the signal is said to be *one dimensional*. A speech signal is an example of a one-dimensional signal whose amplitude varies with time, depending on the spoken word and who speaks it. When the function depends on two or more variables, the signal is said to be *multidimensional*. An image is an example of a two-dimensional signal, with the horizontal and vertical coordinates of the image representing the two dimensions.

1.2 What Is a System?

In the examples of signals mentioned in the preceding section, there is always a system associated with the generation of each signal and another system associated with the extraction of information from the signal. For example, in speech communication, a sound or signal excites the vocal tract, which represents a system. The processing of speech signals usually relies on the use of our ears and auditory pathways in the brain. In this case, the systems responsible for the production and reception of signals are biological in nature. These systems could also be implemented using electronic systems that try to emulate or mimic their biological counterparts. For example, the processing of a speech signal may be performed by an automatic speech recognition system in the form of a computer program that recognizes words or phrases.

A system does not have a unique purpose. Rather, the purpose depends on the application of interest. In an automatic speaker recognition system, the function of the system is to extract information from an incoming speech signal for the purpose of *recognizing* or *identifying* the speaker. In a communication system, the function of the system is to *transport* the information contained in a message over a communication channel and deliver that information to a destination in a reliable fashion. In an *aircraft landing system*, the requirement is to keep the aircraft on the extended centerline of a runway.

A system is formally defined as an entity that manipulates one or more signals to accomplish a function, thereby yielding new signals. The interaction between a system and its associated signals is illustrated schematically in Fig. 1.1. Naturally, the descriptions of the input and output signals depend on the intended application of the system:

▶ In an automatic speaker recognition system, the input signal is a speech (voice) signal, the system is a computer, and the output signal is the identity of the speaker.

▶ In a communication system, the input signal could be a speech signal or computer data, the system itself is made up of the combination of a transmitter, channel, and receiver, and the output signal is an estimate of the information contained in the original message.

▶ In an aircraft landing system, the input signal is the desired position of the aircraft relative to the runway, the system is the aircraft, and the output signal is a correction to the lateral position of the aircraft.

1.3 Overview of Specific Systems

In describing what we mean by signals and systems in the previous two sections, we mentioned several applications. In this section, we will expand on six of those applications,

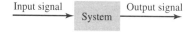

FIGURE 1.1 Block diagram representation of a system.

FIGURE 1.2   Elements of a communication system. The transmitter changes the message signal into a form suitable for transmission over the channel. The receiver processes the channel output (i.e., the received signal) to produce an estimate of the message signal.

namely, communication systems, control systems, microelectromechanical systems, remote sensing, biomedical signal processing, and auditory systems.

■ 1.3.1 COMMUNICATION SYSTEMS

As depicted in Fig. 1.2, there are three basic elements to every communication system: the *transmitter*, the *channel*, and the *receiver*. The transmitter is located at one point in space, the receiver is located at some other point separate from the transmitter, and the channel is the physical medium that connects the two together. Each of these three elements may be viewed as a system with associated signals of its own. The purpose of the transmitter is to convert the message signal produced by a source of information into a form suitable for transmission over the channel. The message signal could be a speech signal, a television (video) signal, or computer data. The channel may be an optical fiber, a coaxial cable, a satellite channel, or a mobile radio channel; each of these channels has its specific area of application.

As the transmitted signal propagates over the channel, it is distorted due to the physical characteristics of the channel. Moreover, noise and interfering signals (originating from other sources) contaminate the channel output, with the result that the received signal is a corrupted version of the transmitted signal. The function of the receiver is to operate on the received signal so as to reconstruct a recognizable form (i.e., produce an estimate) of the original message signal and deliver it to its destination. The signal-processing role of the receiver is thus the reverse of that of the transmitter; in addition, the receiver reverses the effects of the channel.

Details of the operations performed in the transmitter and receiver depend on the type of communication system being considered. The communication system can be of an analog or digital type. In signal-processing terms, the design of an *analog communication system* is relatively simple. Specifically, the transmitter consists of a *modulator* and the receiver consists of a *demodulator*. *Modulation* is the process of converting the message signal into a form that is compatible with the transmission characteristics of the channel. Ordinarily, the transmitted signal is represented as amplitude, phase, or frequency variations of a sinusoidal carrier wave. We thus speak of amplitude modulation, phase modulation, or frequency modulation, respectively. Correspondingly, through the use of amplitude demodulation, phase demodulation, or frequency demodulation, an estimate of the original message signal is produced at the receiver output. Each one of these analog modulation–demodulation techniques has its own advantages and disadvantages.

In contrast, a *digital communication system*, as described below, is considerably more complex. If the message signal is of analog form, as in speech and video signals, the transmitter performs the following operations to convert it into digital form:

▸ *Sampling*, which converts the message signal into a sequence of numbers, with each number representing the amplitude of the message signal at a particular instant of time.

   ▶ *Quantization*, which involves representing each number produced by the sampler to the nearest level selected from a finite number of discrete amplitude levels. For example, we may represent each sample as a 16-bit number, in which case there are 2^{16} amplitude levels. After the combination of sampling and quantization, we have a representation of the message signal that is *discrete* in both time and amplitude.

   ▶ *Coding*, the purpose of which is to represent each quantized sample by a code word made up of a finite number of symbols. For example, in a binary code, the symbols may be 1's or 0's.

Unlike the operations of sampling and coding, quantization is irreversible; that is, a loss of information is always incurred by its application. However, this loss can be made small, and nondiscernible for all practical purposes, by using a quantizer with a sufficiently large number of discrete amplitude levels. As the number of such levels increases, the length of the code word must increase correspondingly.

If the source of information is discrete to begin with, as in the case of a digital computer, none of the preceding operations is needed.

The transmitter may involve the additional operations of data compression and channel encoding. The purpose of data compression is to remove redundant information from the message signal and thereby provide for efficient utilization of the channel by reducing the number of bits per sample required for transmission. Channel encoding, on the other hand, involves the insertion of redundant elements (e.g., extra symbols) into the code word in a controlled manner in order to protect against noise and interfering signals picked up during the course of the signal's transmission through the channel. Finally, the coded signal is modulated onto a carrier wave (usually sinusoidal) for transmission over the channel.

At the receiver, the operations of coding and sampling are *reversed* (i.e., the roles of their individual input and output signals are interchanged) in that order, producing an estimate of the original message signal, which is then delivered to its intended destination. Because quantization is irreversible, it has no counterpart in the receiver.

It is apparent from this discussion that the use of digital communications may require a considerable amount of electronic circuitry. This is not a significant problem, since the electronics are relatively inexpensive, due to the ever-increasing availability of very large scale integrated (VLSI) circuits in the form of silicon chips. Indeed, with continuing improvements in the semiconductor industry, digital communications are often more cost effective than analog communications.

There are two basic modes of communication:

1. *Broadcasting*, which involves the use of a single powerful transmitter and numerous receivers that are relatively cheap to build. Here, information-bearing signals flow only in one direction.

2. *Point-to-point communication*, in which the communication process takes place over a link between a single transmitter and a single receiver. In this case, there is usually a bidirectional flow of information-bearing signals, with a transmitter and a receiver at each end of the link.

The broadcasting mode of communication is exemplified by radio and television, which have become integral parts of our daily lives. In contrast, the ubiquitous telephone provides the means for one form of point-to-point communication. Note, however, that in this case the link is part of a highly complex telephone network designed to accommodate a large number of users on demand.

Another example of point-to-point communication is the *deep-space communication* link between an Earth station and a robot navigating the surface of a distant planet. Unlike telephonic communication, the composition of the message signal depends on the direction of the communication process. The message signal may be in the form of computer-generated instructions transmitted from an Earth station that commands the robot to perform specific maneuvers, or it may contain valuable information about the chemical composition of the soil on the planet that is sent back to Earth for analysis. In order to communicate reliably over such great distances, it is necessary to use digital communications. Figure 1.3(a) shows a photograph of the *Pathfinder* robot, which landed on Mars on July 4, 1997, a historic day in the National Aeronautics and Space Administration's (NASA's) scientific investigation of the solar system. Figure 1.3(b) shows a photograph of the high-precision, 70-meter antenna located at Canberra, Australia. The antenna is an integral part

(a)

(b)

FIGURE 1.3   (a) Snapshot of *Pathfinder* exploring the surface of Mars. (b) The 70-meter (230-foot) diameter antenna is located at Canberra, Australia. The surface of the 70-meter reflector must remain accurate within a fraction of the signal's wavelength. (Courtesy of Jet Propulsion Laboratory.)

of NASA's worldwide Deep Space Network (DSN), which provides the vital two-way communications link that guides and controls robotic planetary explorers and brings back images and new scientific information collected by them. The successful use of DSN for planetary exploration represents a triumph of communication theory and technology over the challenges presented by the unavoidable presence of noise.

Unfortunately, every communication system suffers from the presence of *channel noise* in the received signal. Noise places severe limits on the quality of received messages. For example, owing to the enormous distance between our own planet Earth and Mars, the average power of the information-bearing component of the received signal, at either end of the link, is relatively small compared with the average power of the noise component. Reliable operation of the link is achieved through the combined use of (1) *large antennas* as part of the DSN and (2) *error control*. For a parabolic-reflector antenna (i.e., the type of antenna portrayed in Fig. 1.3(b)), the *effective area* of the antenna is generally between 50% and 65% of the physical area of the antenna. The received power available at the terminals of the antenna is equal to the effective area times the power per unit area carried by the incident electromagnetic wave. Clearly, the larger the antenna, the larger the received signal power will be and hence the use of large antennas in DSN.

Error control involves the use of a *channel encoder* at the transmitter and a *channel decoder* at the receiver. The channel encoder accepts message bits and adds *redundancy* according to a prescribed rule, thereby producing encoded data at a higher bit rate. The redundant bits are added for the purpose of protection against channel noise. The channel decoder exploits the redundancy to decide which message bits were actually sent. The combined goal of the channel encoder and decoder is to minimize the effect of channel noise; that is, the number of errors between the channel encoder input (derived from the source of information) and the encoder output (delivered to the user by the receiver) is minimized, on average.

■ 1.3.2 CONTROL SYSTEMS

The *control* of physical systems is widespread in our industrial society. Aircraft autopilots, mass-transit vehicles, automobile engines, machine tools, oil refineries, paper mills, nuclear reactors, power plants, and robots are all examples of the application of control. The object to be controlled is commonly referred to as a *plant*; in this context, an aircraft is a plant.

There are many reasons for using control systems. From an engineering viewpoint, the two most important ones are the attainment of a satisfactory response and robust performance:

1. *Response.* A plant is said to produce a satisfactory response if its output follows or tracks a specified reference input. The process of holding the plant output close to the reference input is called *regulation*.

2. *Robustness.* A control system is said to be robust if it regulates its objects well, despite the presence of external disturbances (e.g., turbulence affecting the flight of an aircraft) and in the face of changes in the plant parameters due to varying environmental conditions.

The attainment of these desirable properties usually requires the use of *feedback*, as illustrated in Fig. 1.4. The system shown is referred to as a *closed-loop control system* or *feedback control system*. For example, in an aircraft landing system, the plant is represented by the aircraft's body and actuator, the sensors are used by the pilot to determine the lateral position of the aircraft, and the controller is a digital computer.

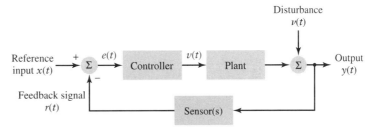

FIGURE 1.4   Block diagram of a feedback control system. The controller drives the plant, whose disturbed output drives the sensor(s). The resulting feedback signal is subtracted from the reference input to produce an error signal $e(t)$, which, in turn, drives the controller. The feedback loop is thereby closed.

In any event, the plant is described by mathematical operations that generate the output $y(t)$ in response to the plant input $v(t)$ and external disturbance $\nu(t)$. The sensor included in the feedback loop measures the plant output $y(t)$ and converts it into another form, usually electrical. The sensor output $r(t)$ constitutes the feedback signal and is compared with the reference input $x(t)$ to produce a difference or error signal $e(t)$. This latter signal is applied to a *controller*, which, in turn, generates the actuating signal $v(t)$ that performs the controlling action on the plant. A control system with a single input and single output, as illustrated in Fig. 1.4, is referred to as a *single-input, single-output (SISO) system*. When the number of plant inputs or outputs is more than one, the system is referred to as a *multiple-input, multiple-output (MIMO) system*.

In either case, the controller may be in the form of a digital computer or microprocessor, in which case we speak of a *digital control system*. The use of digital control systems is becoming increasingly common because of the flexibility and high degree of accuracy afforded by the use of a digital computer as the controller. By its very nature, a digital control system involves the operations of sampling, quantization, and coding described previously.

Figure 1.5 shows a photograph of a NASA space shuttle launch, which relies on the use of a digital computer for its control.

■ 1.3.3   MICROELECTROMECHANICAL SYSTEMS

Dramatic developments in microelectronics have made it possible to pack millions of transistors on a single silicon chip in a commercially viable manner. Thanks to silicon chips, today's computers are orders of magnitude cheaper, smaller, and more powerful than the computers of the 1960s. Digital signal processors, built on silicon chips, are integral parts of digital wireless communication systems and digital cameras, among many other applications. Microfabrication techniques have led to the creation of miniature silicon sensors such as optical detector arrays, which, in their own ways, are revolutionizing photography.

In addition to purely electrical circuits, it is now feasible to build *microelectromechanical systems* (MEMS) that merge mechanical systems with microelectronic control circuits on a silicon chip. The result is a new generation of smaller, more powerful, and less noisy "smart" sensors and actuators that have a broad range of applications, including health care, biotechnology, automotive, and navigation systems. MEMS are fabricated by means of surface micromachining techniques similar to those used in the fabrication of electrical silicon chips. From a manufacturing perspective, the rapid development of MEMS is due largely to two factors:

FIGURE 1.5 NASA space shuttle launch. (Courtesy of NASA.)

▶ An improved understanding of the mechanical properties of thin films—particularly polysilicon—that are basic to building devices with freely moving parts.

▶ The development and utilization of reactive ion-etching techniques to define features and spacing precisely in the thin films that are deposited.

Figure 1.6(a) shows the structure of a lateral capacitive accelerometer. The device has a number of moving sense fingers that are attached to the proof mass, which is suspended in a manner that allows it to move relative to the substrate. The moving sense fingers are interleaved with fixed fingers attached to the supporting structure. The interdigitization of

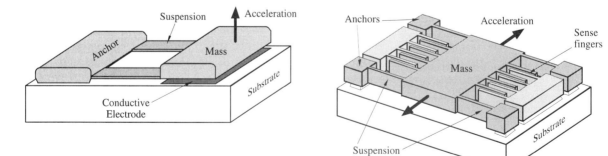

FIGURE 1.6 (a) Structure of lateral capacitive accelerometer. Part (b) of the figure is on the next page. (Courtesy of Navid Yazdi, Farroh Ayazi, and Khalil Najafi. *Micromachined Inertial Sensors*, Proc. IEEE, vol. 86, No. 8, August 1998. ©1998 IEEE.)

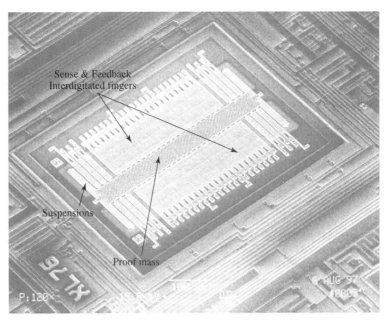

FIGURE 1.6 (Continued) (b) SEM view of Analog Device's ADXLO5 surface-micromachined poly-silicon accelerometer. (Courtesy of IEEE and Analog Devices.)

these fingers creates a *sense capacitance*, the value of which depends on the position of the proof mass. Acceleration displaces the proof mass, thereby changing the capacitance of the device. The change in capacitance is detected by the microelectronic control circuitry, which, in turn, is used to determine the value of acceleration. The sense direction of motion is in the proof-mass plane—hence the designation "lateral."

The accelerometer of Fig. 1.6(a) lends itself to *micromachining*, whereby the sensor and electronic control circuits are integrated on a single chip. Figure 1.6(b) shows a scanning electron microscope (SEM) view of the ADXLO5, a polysilicon accelerometer developed and produced by Analog Devices, Inc.

The basic micromechanical structure used to build an accelerometer can be employed to build a *gyroscope*—a device that senses the angular motion of a system. This property makes gyroscopes ideally suited for use in automatic flight control systems. The operation of a gyroscope follows the *law of conservation of angular momentum*, which states that if no external torques act upon a system made up of different pieces (particles), the angular momentum of the system remains constant. An insightful way of demonstrating the gyroscopic effect is to sit on a swivel chair and use both hands to hold a spinning wheel with the axis horizontal; in this situation, the wheel has an angular momentum about the horizontal axis. If, now, the axis of the spinning wheel is turned into the vertical axis, then, in order to balance the whole system in accordance with the law of conservation of angular momentum, a remarkable thing happens: Both the chair and the person sitting on it turn in the direction *opposite* that of the spin of the wheel.

In the MEMS version of a gyroscope, two adjacent proof masses are used. A voltage is applied across the interdigitized fingers, causing the proof masses to vibrate in antiphase at the resonant frequency of the structure, which may range from 1 kHz to 700 kHz. An external rotation due to motion introduces an apparent force called the *Coriolis force*, which causes the proof masses to be displaced vertically. The displacement is then measured by capacitive sensors located under the proof mass and is used to determine the motion of the object of interest.

■ 1.3.4 REMOTE SENSING

Remote sensing is defined as the process of acquiring information about an object of interest without being in physical contact with it. Basically, the acquisition of information is accomplished by *detecting and measuring the changes that the object imposes on the field surrounding it*. The field can be electromagnetic, acoustic, magnetic, or gravitational, depending on the application of interest. The acquisition of information can be performed in a *passive* manner, by listening to the field (signal) that is naturally emitted by the object and processing it, or in an *active* manner, by purposely illuminating the object with a well-defined field (signal) and processing the echo (i.e., signal returned) from the object.

This definition of remote sensing is rather broad, in that it applies to every possible field. In practice, however, the term "remote sensing" is commonly used in the context of electromagnetic fields, with the techniques used for information acquisition covering the whole electromagnetic spectrum. It is this specialized form of remote sensing that we are concerned with here.

The scope of remote sensing has expanded enormously since the 1960s, due to both the advent of satellites and planetary probes as space platforms for the sensors and the availability of sophisticated digital signal-processing techniques for extracting information from the data gathered by the sensors. In particular, sensors on Earth-orbiting satellites provide highly valuable information about global weather patterns, the dynamics of clouds, Earth's surface vegetation cover and seasonal variations, and ocean surface temperatures. Most importantly, they do so in a reliable way and on a continuing basis. In planetary studies, space-borne sensors have provided us with high-resolution images of the surfaces of various planets; the images, in turn, have uncovered new kinds of physical phenomena, some similar to, and others completely different from, what we are familiar with on planet Earth.

The electromagnetic spectrum extends from low-frequency radio waves through the microwave, submillimeter, infrared, visible, ultraviolet, X-ray, and gamma-ray regions of the spectrum. Unfortunately, a single sensor by itself can cover only a small part of the electromagnetic spectrum, with the mechanism responsible for wave–matter interaction being influenced by a limited number of physical properties of the object of interest. If, therefore, we are to undertake a detailed study of a planetary surface or atmosphere, then we must simultaneously use *multiple sensors* covering a large part of the electromagnetic spectrum. For example, to study a planetary surface, we may require a suite of sensors covering selected bands as follows:

- ▶ *Radar* sensors to provide information on the surface physical properties of the planet (e.g., its topography, roughness, moisture, and dielectric constant)
- ▶ *Infrared* sensors to measure the near-surface thermal properties of the planet
- ▶ *Visible and near-infrared* sensors to provide information about the surface chemical composition of the planet
- ▶ *X-ray* sensors to provide information on radioactive materials contained in the planet

The data gathered by these highly diverse sensors are processed on a computer to generate a set of images that can be used collectively to increase our scientific knowledge of the planet's surface.

Among electromagnetic sensors, a special type of radar known as *synthetic-aperture radar* (SAR) stands out as a unique imaging system in remote sensing. SAR offers the following attractive features:

- ▶ Satisfactory operation day and night and under all weather conditions
- ▶ A high-resolution imaging capability that is independent of the sensor's altitude or wavelength

FIGURE 1.7 Perspectival view of Mount Shasta (California), derived from a pair of stereo radar images acquired from orbit with the Shuttle Imaging Radar (SIR-B). (Courtesy of Jet Propulsion Laboratory.)

The realization of a high-resolution image with radar requires the use of an antenna with a large aperture. From a practical perspective, however, there is a physical limit on the size of an antenna that can be accommodated on an airborne or spaceborne platform. In a SAR system, a large aperture is synthesized by signal-processing means—hence the name "synthetic-aperture radar." The key idea behind SAR is that an array of antenna elements equally spaced along a straight line is equivalent to a single antenna moving along the array line at a uniform speed. This is true, provided that we satisfy the following requirement: The signals received by the single antenna at equally spaced points along the array line are *coherently* recorded; that is, amplitude and phase relationships among the received signals are maintained. Coherent recording ensures that signals received from the single antenna correspond to signals received from the individual elements of an equivalent array of antennas. In order to obtain a high-resolution image from the single-antenna signals, highly sophisticated signal-processing operations are necessary. A central operation in this signal processing is the *Fourier transform*, which is implemented efficiently on a digital computer using an algorithm known as the *fast Fourier transform (FFT) algorithm*. Fourier analysis of signals is one of the main focal points of this book.

The photograph in Fig. 1.7 shows a perspective view of Mt. Shasta (California), which was derived from a stereo pair of SAR images acquired from Earth orbit with the Shuttle Imaging Radar (SIR-B). The photograph on the front cover of the book presents the characteristics of the same mountain as seen from a different elevation in the visible portion of the electromagnetic spectrum.

■ **1.3.5 BIOMEDICAL SIGNAL PROCESSING**

The goal of biomedical signal processing is to extract information from a biological signal. The information then helps us to improve our understanding of basic mechanisms of biological functioning or aids us in the diagnosis or treatment of a medical condition. The generation of many *biological signals* found in the human body is traced to the electrical

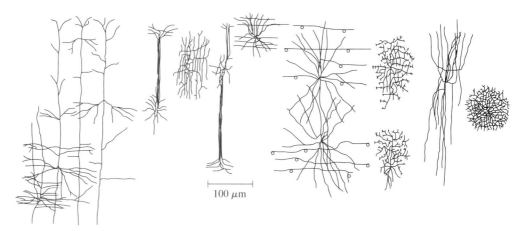

FIGURE 1.8 Morphological types of nerve cells (neurons) identifiable in a monkey's cerebral cortex, based on studies of primary somatic sensory and motor cortices. (Reproduced from E. R. Kandel, J. H. Schwartz, and T. M. Jessel, *Principles of Neural Science*, 3d ed., 1991; courtesy of Appleton and Lange.)

activity of large groups of nerve cells or muscle cells. Nerve cells in the brain are commonly referred to as *neurons*. Figure 1.8 shows morphological types of neurons that are identifiable in a monkey's cerebral cortex, based on studies of the monkey's primary somatic sensory and motor cortices. The figure illustrates the many different shapes and sizes of neurons.

Irrespective of the origin of the signal, biomedical signal processing begins with a temporal record of the biological event of interest. For example, the electrical activity of the heart is represented by a record called the *electrocardiogram* (ECG). The ECG represents changes in the potential (voltage) due to electrochemical processes involved in the formation and spatial spread of electrical excitations in heart cells. Accordingly, detailed inferences about the heart can be made from the ECG.

Another important example of a biological signal is the *electroencephalogram* (EEG). The EEG is a record of fluctuations in the electrical activity of large groups of neurons in the brain. Specifically, the EEG measures the electrical field associated with the current flowing through a group of neurons. To record an EEG (or an ECG, for that matter), at least two electrodes are needed. An active electrode is placed over the particular site of neuronal activity that is of interest, and a reference electrode is placed at some remote distance from this site; the EEG is measured as the voltage or potential difference between the active and reference electrodes. Figure 1.9 shows three examples of EEG signals recorded from the hippocampus of a rat.

A major issue of concern in biomedical signal processing—in the context of ECG, EEG, or some other biological signal—is the detection and suppression of artifacts. An *artifact* is that part of a signal produced by events that are extraneous to the biological event of interest. Artifacts arise in a biological signal at different stages of processing and in many different ways. Among the various kinds of artifacts are the following:

▸ *Instrumental artifacts*, generated by the use of an instrument. An example of an instrumental artifact is the 60-Hz interference picked up by recording instruments from an electrical main's power supply.

▸ *Biological artifacts*, in which one biological signal contaminates or interferes with another. An example of a biological artifact is the shift in electrical potential that may be observed in the EEG due to heart activity.

▸ *Analysis artifacts*, which may arise in the course of processing the biological signal to produce an estimate of the event of interest.

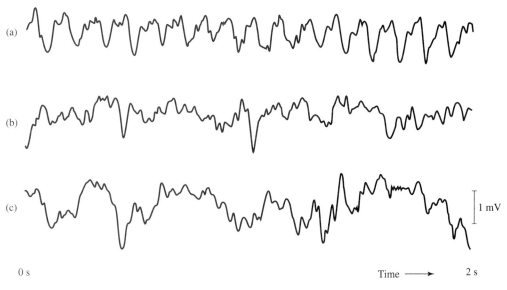

(a)

(b)

(c)

1 mV

0 s　　　　　　　　　　　　　　　　　　　　　Time ⟶　　2 s

FIGURE 1.9 The traces shown in (a), (b), and (c) are three examples of EEG signals recorded from the hippocampus of a rat. Neurobiological studies suggest that the hippocampus plays a key role in certain aspects of learning and memory.

Analysis artifacts are, in a way, controllable. For example, round-off errors due to the quantization of signal samples, which arise from the use of digital signal processing, can be made nondiscernible for all practical purposes by setting the number of discrete amplitude levels in the quantizer large enough.

What about instrumental and biological artifacts? A common method of reducing their effects is through the use of *filtering*. A *filter* is a *system* that passes signals containing frequencies in one frequency range, termed the filter passband, and removes signals containing frequencies in other frequency ranges. Assuming that we have a priori knowledge about the signal of interest, we may estimate the range of frequencies inside which the significant components of the desired signal are located. Then, by designing a filter whose passband corresponds to the frequencies of the desired signal, artifacts with frequency components outside this passband are removed by the filter. The assumption made here is that the desired signal and the artifacts contaminating it occupy essentially nonoverlapping frequency bands. If, however, the frequency bands overlap each other, then the filtering problem becomes more difficult and requires a solution that is beyond the scope of the present book.

■ 1.3.6 AUDITORY SYSTEM

For our last example of a system, we turn to the mammalian auditory system, the function of which is to discriminate and recognize complex sounds on the basis of their frequency content.

Sound is produced by vibrations such as the movements of vocal cords or violin strings. These vibrations result in the compression and rarefaction (i.e., increased or reduced pressure) of the surrounding air. The disturbance so produced radiates outward from the source of sound as an *acoustical wave* with alternating highs and lows of pressure.

The ear—the organ of hearing—responds to incoming acoustical waves. The ear has three main parts, each of which performs a particular function:

> ▶ The *outer ear* aids in the collection of sounds.

> ▶ The *middle ear* provides an acoustic impedance match between the air and the cochlear fluids, thereby conveying the vibrations of the *tympanic membrane* (eardrum) that are due to the incoming sounds to the inner ear in an efficient manner.

> ▶ The *inner ear* converts the mechanical vibrations from the middle ear to an "electrochemical" or "neural" signal for transmission to the brain.

The inner ear consists of a bony, spiral-shaped, fluid-filled tube called the *cochlea*. Sound-induced vibrations of the tympanic membrane are transmitted into the *oval window* of the cochlea by a chain of bones called *ossicles*. The lever action of the ossicles amplifies the mechanical vibrations of the tympanic membrane. The cochlea tapers in size like a cone toward a tip, so that there is a *base* at the oval window and an *apex* at the tip. Through the middle of the cochlea stretches the *basilar membrane*, which gets wider as the cochlea gets narrower.

The vibratory movement of the tympanic membrane is transmitted as a *traveling wave* along the length of the basilar membrane, starting from the oval window to the apex at the far end of the cochlea. The wave propagates along the basilar membrane, much as the snapping of a rope tied at one end causes a wave to propagate along the rope from the snapped end to the fixed end. As illustrated in Fig. 1.10, the wave attains its peak amplitude at a specific location along the basilar membrane that depends on the frequency of the incoming sound. Thus, although the wave itself travels along the basilar membrane, the envelope of the wave is "stationary" for a given frequency. The peak displacements for high frequencies occur toward the base (where the basilar membrane is narrowest and stiffest). The peak displacements for low frequencies occur toward the apex (where the basilar membrane is widest and most flexible). That is, as the wave propagates along the basilar

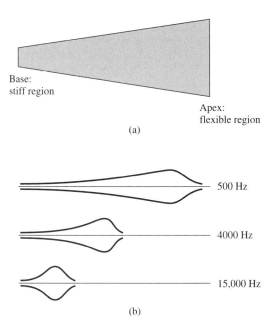

FIGURE 1.10   (a) In this diagram, the basilar membrane in the cochlea is depicted as if it were uncoiled and stretched out flat; the "base" and "apex" refer to the cochlea, but the remarks "stiff region" and "flexible region" refer to the basilar membrane. (b) This diagram illustrates the travelling waves along the basilar membrane, showing their envelopes induced by incoming sound at three different frequencies.

membrane, a *resonance* phenomenon takes place, with the end of the basilar membrane at the base of the cochlea resonating at about 20,000 Hz and its other end at the apex of the cochlea resonating at about 20 Hz; the resonance frequency of the basilar membrane *decreases* gradually with distance from base to apex. Consequently, the spatial axis of the cochlea is said to be *tonotopically ordered*, because each location is associated with a particular resonance frequency or tone.

The basilar membrane is a *dispersive medium*, in that higher frequencies propagate more slowly than do lower frequencies. In a dispersive medium, we distinguish two different velocities: *phase velocity* and *group velocity*. The phase velocity is the velocity at which a crest or valley of the wave propagates along the basilar membrane. The group velocity is the velocity at which the envelope of the wave and its energy propagate.

The mechanical vibrations of the basilar membrane are transduced into electrochemical signals by hair cells that rest on the basilar membrane in an orderly fashion. There are two main types of hair cells: *inner hair cells* and *outer hair cells*, the latter by far the more numerous. The outer hair cells are *motile* elements; that is, they are capable of altering their length, and perhaps other mechanical characteristics, a property that is believed to be responsible for the compressive *nonlinear* effect seen in the basilar membrane vibrations. There is also evidence that the outer hair cells contribute to the sharpening of tuning curves from the basilar membrane on up the system. However, the inner hair cells are the main sites of *auditory transduction*. By means of its terminals, a neuron transmits information about its own activity to the receptive surfaces of other neurons or cells in the brain; the point of contact is called a *synapse*. Thus, each auditory neuron synapses (i.e., establishes contact) with an inner hair cell at a particular location on the basilar membrane. The neurons that synapse with inner hair cells near the base of the basilar membrane are found in the periphery of the auditory nerve bundle, and there is an orderly progression toward synapsing at the apex end of the basilar membrane, with movement toward the center of the bundle. The tonotopic organization of the basilar membrane is therefore anatomically preserved in the auditory nerve. The inner hair cells also perform *rectification* and *compression*. The mechanical signal is approximately half-wave rectified, thereby responding to motion of the basilar membrane in one direction only. Moreover, the mechanical signal is compressed nonlinearly, such that a large range of incoming sound intensities is reduced to a manageable excursion of electrochemical potential. The electrochemical signals so produced are carried over to the brain, where they are further processed to become our hearing sensations.

In sum, in the cochlea we have a wonderful example of a biological system that operates as a *bank of filters* tuned to different frequencies and that uses nonlinear processing to reduce the dynamic range of sounds heard. The cochlea enables us to discriminate and recognize complex sounds, despite the enormous differences in intensity levels that can arise in practice.

■ 1.3.7 ANALOG VERSUS DIGITAL SIGNAL PROCESSING

The signal-processing operations involved in building communication systems, control systems, microelectromechanical systems, instruments for remote sensing, and instruments for the processing of biological signals, among the many applications of signal processing, can be implemented in two fundamentally different ways: (1) an analog, or continuous-time, approach and (2) a digital, or discrete-time, approach. The analog approach to signal processing was dominant for many years, and it remains a viable option for many applications. As the name implies, *analog signal processing* relies on the use of analog circuit elements such as resistors, capacitors, inductors, transistor amplifiers, and diodes. *Digital signal processing*, by contrast, relies on three basic digital computer elements: adders and multipliers (for arithmetic operations) and memory (for storage).

The main attribute of the analog approach is its inherent capability to solve differential equations that describe physical systems without having to resort to approximate solutions. Analog solutions are also obtained in *real time*, irrespective of the input signal's frequency range, since the underlying mechanisms responsible for the operations of the analog approach are all physical in nature. In contrast, the digital approach relies on numerical computations for its operation. The time required to perform these computations determines whether the digital approach is able to operate in real time (i.e., whether it can keep up with the changes in the input signal). In other words, the analog approach is assured of real-time operation, but there is no such guarantee for the digital approach.

However, the digital approach has the following important advantages over analog signal processing:

▶ *Flexibility*, whereby the same digital machine (hardware) can be used for implementing different versions of a signal-processing operation of interest (e.g., filtering) merely by making changes to the software (program) read into the machine. In the case of an analog machine, the system has to be redesigned every time the signal-processing specifications are changed.

▶ *Repeatability*, which refers to the fact that a prescribed signal-processing operation (e.g., control of a robot) can be repeated exactly over and over again when it is implemented by digital means. In contrast, analog systems suffer from parameter variations that can arise due to changes in the supply voltage or room temperature.

For a given signal-processing operation, however, we usually find that the use of a digital approach requires a more complex circuit than does an analog approach. This was an issue of major concern in years past, but it no longer is. As remarked earlier, the ever-increasing availability of VLSI circuits in the form of silicon chips has made digital electronics relatively cheap. Consequently, we are now able to build digital signal processors that are cost competitive with respect to their analog counterparts over a wide frequency range that includes both speech and video signals. In the final analysis, however, the choice of an analog or digital approach for the solution of a signal-processing problem can be determined only by the application of interest, the resources available, and the cost involved in building the system. Note that the vast majority of systems built in practice are mixed in nature, combining the desirable features of both analog and digital approaches to signal processing.

1.4 *Classification of Signals*

In this book, we will restrict our attention to one-dimensional signals defined as single-valued functions of time. "Single valued" means that for every instant of time, there is a unique value of the function. This value may be a real number, in which case we speak of a *real-valued signal*, or it may be a complex number, in which case we speak of a *complex-valued signal*. In either case, the independent variable, namely, time, is real valued.

The most useful method for representing a signal in a given situation hinges on the particular type of signal being considered. Five methods of classifying signals, based on different features, are common:

1. *Continuous-time and discrete-time signals.*
One way of classifying signals is on the basis of how they are defined as a function of time. In this context, a signal $x(t)$ is said to be a *continuous-time signal* if it is defined for all time t.

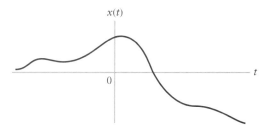

FIGURE 1.11   Continuous-time signal.

Figure 1.11 represents an example of a continuous-time signal whose amplitude or value varies continuously with time. Continuous-time signals arise naturally when a physical waveform such as an acoustic wave or a light wave is converted into an electrical signal. The conversion is effected by means of a *transducer*; examples include the microphone, which converts variations in sound pressure into corresponding variations in voltage or current, and the photocell, which does the same for variations in light intensity.

In contrast, a *discrete-time signal* is defined only at discrete instants of time. Thus, the independent variable has discrete values only, which are usually uniformly spaced. A discrete-time signal is often derived from a continuous-time signal by *sampling* it at a uniform rate. Let T_s denote the sampling period and n denote an integer that may assume positive and negative values. Then sampling a continuous-time signal $x(t)$ at time $t = nT_s$ yields a sample with the value $x(nT_s)$. For convenience of presentation, we write

$$x[n] = x(nT_s), \qquad n = 0, \pm 1, \pm 2, \ldots. \tag{1.1}$$

Consequently, a discrete-time signal is represented by the sequence of numbers $\ldots, x[-2]$, $x[-1], x[0], x[1], x[2], \ldots$, which can take on a continuum of values. Such a sequence of numbers is referred to as a *time series*, written as $\{x[n], n = 0, \pm 1, \pm 2, \ldots\}$, or simply $x[n]$. The latter notation is used throughout this book. Figure 1.12 illustrates the relationship between a continuous-time signal $x(t)$ and a discrete-time signal $x[n]$ derived from it as described by Eq. (1.1).

Throughout this book, we use the symbol t to denote time for a continuous-time signal and the symbol n to denote time for a discrete-time signal. Similarly, parentheses $(\cdot)$ are used to denote continuous-valued quantities, while brackets $[\cdot]$ are used to denote discrete-valued quantities.

2. *Even and odd signals*.
A continuous-time signal $x(t)$ is said to be an *even signal* if

$$x(-t) = x(t) \qquad \text{for all } t. \tag{1.2}$$

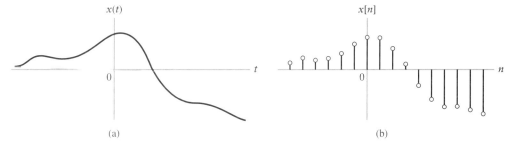

FIGURE 1.12   (a) Continuous-time signal $x(t)$. (b) Representation of $x(t)$ as a discrete-time signal $x[n]$.

The signal $x(t)$ is said to be an *odd signal* if

$$x(-t) = -x(t) \qquad \text{for all } t. \tag{1.3}$$

In other words, even signals are *symmetric* about the vertical axis, or time origin, whereas odd signals are *antisymmetric* about the time origin. Similar remarks apply to discrete-time signals.

EXAMPLE 1.1 EVEN AND ODD SIGNALS   Consider the signal

$$x(t) = \begin{cases} \sin\left(\dfrac{\pi t}{T}\right), & -T \le t \le T \\ 0, & \text{otherwise} \end{cases}.$$

Is the signal $x(t)$ an even or an odd function of time t?

Solution: Replacing t with $-t$ yields

$$x(-t) = \begin{cases} \sin\left(-\dfrac{\pi t}{T}\right), & -T \le t \le T \\ 0, & \text{otherwise} \end{cases}$$

$$= \begin{cases} -\sin\left(\dfrac{\pi t}{T}\right), & -T \le t \le T \\ 0, & \text{otherwise} \end{cases}$$

$$= -x(t) \qquad \text{for all } t,$$

which satisfies Eq. (1.3). Hence, $x(t)$ is an odd signal. ∎

Suppose we are given an arbitrary signal $x(t)$. We may develop an even–odd decomposition of $x(t)$ by applying the corresponding definitions. To that end, let $x(t)$ be expressed as the sum of two components $x_e(t)$ and $x_o(t)$ as follows:

$$x(t) = x_e(t) + x_o(t).$$

Define $x_e(t)$ to be even and $x_o(t)$ to be odd; that is,

$$x_e(-t) = x_e(t)$$

and

$$x_o(-t) = -x_o(t)$$

Putting $t = -t$ in the expression for $x(t)$, we may write

$$x(-t) = x_e(-t) + x_o(-t)$$
$$= x_e(t) - x_o(t).$$

Solving for $x_e(t)$ and $x_o(t)$, we thus obtain

$$\boxed{x_e(t) = \frac{1}{2}[x(t) + x(-t)]} \tag{1.4}$$

and

$$x_o(t) = \frac{1}{2}[x(t) - x(-t)]. \tag{1.5}$$

EXAMPLE 1.2 ANOTHER EXAMPLE OF EVEN AND ODD SIGNALS   Find the even and odd components of the signal

$$x(t) = e^{-2t} \cos t.$$

Solution: Replacing t with $-t$ in the expression for $x(t)$ yields

$$x(-t) = e^{2t} \cos(-t)$$
$$= e^{2t} \cos t.$$

Hence, applying Eqs. (1.4) and (1.5) to the problem at hand, we get

$$x_e(t) = \frac{1}{2}(e^{-2t} \cos t + e^{2t} \cos t)$$
$$= \cosh(2t) \cos t$$

and

$$x_o(t) = \frac{1}{2}(e^{-2t} \cos t - e^{2t} \cos t)$$
$$= -\sinh(2t) \cos t,$$

where $\cosh(2t)$ and $\sinh(2t)$ respectively denote the hyperbolic cosine and sine of time t.   ∎

▶ **Problem 1.1**   Find the even and odd components of each of the following signals:

(a) $x(t) = \cos(t) + \sin(t) + \sin(t) \cos(t)$
(b) $x(t) = 1 + t + 3t^2 + 5t^3 + 9t^4$
(c) $x(t) = 1 + t \cos(t) + t^2 \sin(t) + t^3 \sin(t) \cos(t)$
(d) $x(t) = (1 + t^3) \cos^3(10t)$

Answers:

(a) Even: $\cos(t)$
      Odd: $\sin(t)(1 + \cos(t))$
(b) Even: $1 + 3t^2 + 9t^4$
      Odd: $t + 5t^3$
(c) Even: $1 + t^3 \sin(t) \cos(t)$
      Odd: $t \cos(t) + t^2 \sin(t)$
(d) Even: $\cos^3(10t)$
      Odd: $t^3 \cos^3(10t)$ ◀

In the case of a complex-valued signal, we may speak of conjugate symmetry. A complex-valued signal $x(t)$ is said to be *conjugate symmetric* if

$$x(-t) = x^*(t), \tag{1.6}$$

where the asterisk denotes complex conjugation. Let

$$x(t) = a(t) + jb(t),$$

where $a(t)$ is the real part of $x(t)$, $b(t)$ is the imaginary part, and $j = \sqrt{-1}$. Then the complex conjugate of $x(t)$ is

$$x^*(t) = a(t) - jb(t).$$

Substituting $x(t)$ and $x^*(t)$ into Eq. (1.6) yields

$$a(-t) + jb(-t) = a(t) - jb(t).$$

Equating the real part on the left with that on the right, and similarly for the imaginary parts, we find that $a(-t) = a(t)$ and $b(-t) = -b(t)$. It follows that a complex-valued signal $x(t)$ is conjugate symmetric if its real part is even and its imaginary part is odd. (A similar remark applies to a discrete-time signal.)

▶ **Problem 1.2** The signals $x_1(t)$ and $x_2(t)$ shown in Figs. 1.13(a) and (b) constitute the real and imaginary parts, respectively, of a complex-valued signal $x(t)$. What form of symmetry does $x(t)$ have?

Answer: The signal $x(t)$ is conjugate symmetric.   ◀

3. *Periodic signals and nonperiodic signals.*
A *periodic signal* $x(t)$ is a function of time that satisfies the condition

$$x(t) = x(t + T) \qquad \text{for all } t, \tag{1.7}$$

where T is a positive constant. Clearly, if this condition is satisfied for $T = T_0$, say, then it is also satisfied for $T = 2T_0, 3T_0, 4T_0, \ldots$. The smallest value of T that satisfies Eq. (1.7) is called the *fundamental period* of $x(t)$. Accordingly, the fundamental period T defines the duration of one complete cycle of $x(t)$. The reciprocal of the fundamental period T is called the *fundamental frequency* of the periodic signal $x(t)$; it describes how frequently the periodic signal $x(t)$ repeats itself. We thus formally write

$$f = \frac{1}{T}. \tag{1.8}$$

The frequency f is measured in hertz (Hz), or cycles per second. The *angular frequency*, measured in radians per second, is defined by

$$\omega = 2\pi f = \frac{2\pi}{T}, \tag{1.9}$$

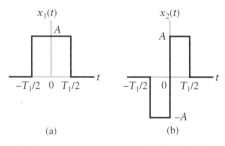

(a)   (b)

FIGURE 1.13 (a) One example of continuous-time signal. (b) Another example of a continuous-time signal.

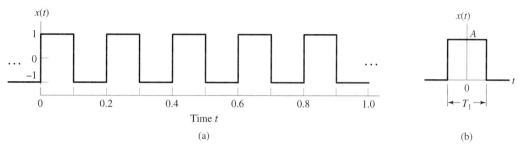

FIGURE 1.14   (a) Square wave with amplitude $A = 1$ and period $T = 0.2$ s. (b) Rectangular pulse of amplitude A and duration T_1.

since there are 2π radians in one complete cycle. To simplify terminology, ω is often referred to simply as the frequency.

Any signal $x(t)$ for which no value of T satisfies the condition of Eq. (1.7) is called an *aperiodic*, or *nonperiodic*, *signal*.

Figures 1.14(a) and (b) present examples of periodic and nonperiodic signals, respectively. The periodic signal represents a square wave of amplitude $A = 1$ and period $T = 0.2$ s, and the nonperiodic signal represents a single rectangular pulse of amplitude A and duration T_1.

▶ **Problem 1.3**   Figure 1.15 shows a triangular wave. What is the fundamental frequency of this wave? Express the fundamental frequency in units of Hz and rad/s.

Answer: 5 Hz, or 10π rad/s.   ◀

The classification of signals into periodic and nonperiodic signals presented thus far applies to continuous-time signals. We next consider the case of discrete-time signals. A discrete-time signal $x[n]$ is said to be periodic if

$$x[n] = x[n + N] \qquad \text{for integer } n, \tag{1.10}$$

where N is a positive integer. The smallest integer N for which Eq. (1.10) is satisfied is called the fundamental period of the discrete-time signal $x[n]$. The fundamental angular frequency or, simply, fundamental frequency of $x[n]$ is defined by

$$\Omega = \frac{2\pi}{N}, \tag{1.11}$$

which is measured in radians.

The differences between the defining equations (1.7) and (1.10) should be carefully noted. Equation (1.7) applies to a periodic continuous-time signal whose fundamental period T has any positive value. Equation (1.10) applies to a periodic discrete-time signal whose fundamental period N can assume only a positive integer value.

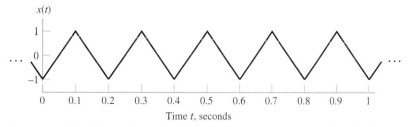

FIGURE 1.15   Triangular wave alternating between -1 and $+1$ for Problem 1.3.

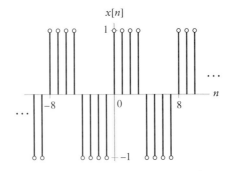

FIGURE 1.16   Discrete-time square wave alternating between -1 and $+1$.

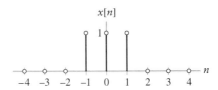

FIGURE 1.17   Nonperiodic discrete-time signal consisting of three nonzero samples.

Two examples of discrete-time signals are shown in Figs. 1.16 and 1.17; the signal of Fig. 1.16 is periodic, whereas that of Fig. 1.17 is nonperiodic.

▶ **Problem 1.4**   Determine the fundamental frequency of the discrete-time square wave shown in Fig. 1.16.

Answer: $\pi/4$ radians.   ◀

▶ **Problem 1.5**   For each of the following signals, determine whether it is periodic, and if it is, find the fundamental period:

(a) $x(t) = \cos^2(2\pi t)$
(b) $x(t) = \sin^3(2t)$
(c) $x(t) = e^{-2t}\cos(2\pi t)$
(d) $x[n] = (-1)^n$
(e) $x[n] = (-1)^{n^2}$
(f) $x[n] = \cos(2n)$
(g) $x[n] = \cos(2\pi n)$

Answers:

(a) Periodic, with a fundamental period of 0.5 s
(b) Periodic, with a fundamental period of $(1/\pi)$ s
(c) Nonperiodic

(d) Periodic, with a fundamental period of 2 samples

(e) Periodic, with a fundamental period of 2 samples

(f) Nonperiodic

(g) Periodic, with a fundamental period of 1 sample   ◀

4. *Deterministic signals and random signals.*

A *deterministic signal* is a signal about which there is no uncertainty with respect to its value at any time. Accordingly, we find that deterministic signals may be modeled as completely specified functions of time. The square wave shown in Fig. 1.13 and the rectangular pulse shown in Fig. 1.14 are examples of deterministic signals, and so are the signals shown in Figs. 1.16 and 1.17.

By contrast, a *random signal* is a signal about which there is uncertainty before it occurs. Such a signal may be viewed as belonging to an ensemble, or a group, of signals, with each signal in the ensemble having a different waveform. Moreover, each signal within the ensemble has a certain probability of occurrence. The ensemble of signals is referred to as a *random process*. The electrical *noise* generated in the amplifier of a radio or television receiver is an example of a random signal. Its amplitude fluctuates between positive and negative values in a completely random fashion. (Section 1.9 presents a typical waveform of electrical noise.) Another example of a random signal is the signal received in a radio communication system. This signal consists of an information-bearing component, an interference component, and unavoidable electrical noise generated at the front end of the radio receiver. The information-bearing component may represent, for example, a voice signal that typically consists of randomly spaced bursts of energy of random durations. The interference component may represent spurious electromagnetic signals produced by other communication systems operating in the vicinity of the radio receiver. The net result of all three components is a received signal that is completely random in nature. Yet another example of a random signal is the EEG signal, exemplified by the waveforms shown in Fig. 1.9.

5. *Energy signals and power signals.*

In electrical systems, a signal may represent a voltage or a current. Consider a voltage $v(t)$ developed across a resistor R, producing a current $i(t)$. The *instantaneous power* dissipated in this resistor is defined by

$$p(t) = \frac{v^2(t)}{R}, \tag{1.12}$$

or, equivalently,

$$p(t) = Ri^2(t). \tag{1.13}$$

In both cases, the instantaneous power $p(t)$ is proportional to the square of the amplitude of the signal. Furthermore, for a resistance R of 1 ohm, Eqs. (1.12) and (1.13) take on the same mathematical form. Accordingly, in signal analysis, it is customary to define power in terms of a 1-ohm resistor, so that, regardless of whether a given signal $x(t)$ represents a voltage or a current, we may express the instantaneous power of the signal as

$$p(t) = x^2(t). \tag{1.14}$$

On the basis of this convention, we define the *total energy* of the continuous-time signal $x(t)$ as

$$
\begin{aligned}
E &= \lim_{T \to \infty} \int_{-T/2}^{T/2} x^2(t)\, dt \\
&= \int_{-\infty}^{\infty} x^2(t)\, dt
\end{aligned}
\tag{1.15}
$$

and its *time-averaged*, or *average*, *power* as

$$P = \lim_{T \to \infty} \frac{1}{T} \int_{-T/2}^{T/2} x^2(t)\, dt. \tag{1.16}$$

From Eq. (1.16), we readily see that the time-averaged power of a periodic signal $x(t)$ of fundamental period T is given by

$$P = \frac{1}{T} \int_{-T/2}^{T/2} x^2(t)\, dt. \tag{1.17}$$

The square root of the average power P is called the *root mean-square* (rms) value of the periodic signal $x(t)$.

In the case of a discrete-time signal $x[n]$, the integrals in Eqs. (1.15) and (1.16) are replaced by corresponding sums. Thus, the total energy of $x[n]$ is defined by

$$E = \sum_{n=-\infty}^{\infty} x^2[n], \tag{1.18}$$

and its average power is defined by

$$P = \lim_{N \to \infty} \frac{1}{2N} \sum_{n=-N}^{N} x^2[n]. \tag{1.19}$$

Here again, from Eq. (1.19), the average power in a periodic signal $x[n]$ with fundamental period N is given by

$$P = \frac{1}{N} \sum_{n=0}^{N-1} x^2[n]. \tag{1.20}$$

A signal is referred to as an *energy signal* if and only if the total energy of the signal satisfies the condition

$$0 < E < \infty.$$

The signal is referred to as a *power signal* if and only if the average power of the signal satisfies the condition

$$0 < P < \infty.$$

The energy and power classifications of signals are mutually exclusive. In particular, an energy signal has zero time-averaged power, whereas a power signal has infinite energy. It is of interest to note that periodic signals and random signals are usually viewed as power signals, whereas signals that are both deterministic and nonperiodic are usually viewed as energy signals.

▶ **Problem 1.6**

(a) What is the total energy of the rectangular pulse shown in Fig. 1.14(b)?

(b) What is the average power of the square wave shown in Fig. 1.14(a)?

Answers: (a) $A^2 T_1$. (b) 1   ◀

▶ **Problem 1.7**   Determine the average power of the triangular wave shown in Fig. 1.15.

Answer: 1/3   ◀

▶ **Problem 1.8**   Determine the total energy of the discrete-time signal shown in Fig. 1.17.

Answer: 3   ◀

▶ **Problem 1.9**   Categorize each of the following signals as an energy signal or a power signal, and find the energy or time-averaged power of the signal:

(a) $x(t) = \begin{cases} t, & 0 \le t \le 1 \\ 2 - t, & 1 \le t \le 2 \\ 0, & \text{otherwise} \end{cases}$

(b) $x[n] = \begin{cases} n, & 0 \le n < 5 \\ 10 - n, & 5 \le n \le 10 \\ 0, & \text{otherwise} \end{cases}$

(c) $x(t) = 5\cos(\pi t) + \sin(5\pi t), \; -\infty < t < \infty$

(d) $x(t) = \begin{cases} 5\cos(\pi t), & -1 \le t \le 1 \\ 0, & \text{otherwise} \end{cases}$

(e) $x(t) = \begin{cases} 5\cos(\pi t), & -0.5 \le t \le 0.5 \\ 0, & \text{otherwise} \end{cases}$

(f) $x[n] = \begin{cases} \sin(\pi n), & -4 \le n \le 4 \\ 0, & \text{otherwise} \end{cases}$

(g) $x[n] = \begin{cases} \cos(\pi n), & -4 \le n \le 4 \\ 0, & \text{otherwise} \end{cases}$

(h) $x[n] = \begin{cases} \cos(\pi n), & n \ge 0 \\ 0, & \text{otherwise} \end{cases}$

Answers:

(a) Energy signal, energy $= \frac{2}{3}$

(b) Energy signal, energy $= 85$

(c) Power signal, power $= 13$

(d) Energy signal, energy $= 25$

(e) Energy signal, energy $= 12.5$

(f) Zero signal

(g) Energy signal, energy $= 9$

(h) Power signal, power $= \frac{1}{2}$ ◀

1.5 *Basic Operations on Signals*

An issue of fundamental importance in the study of signals and systems is the use of systems to process or manipulate signals. This issue usually involves a combination of some basic operations. In particular, we may identify two classes of operations.

■ 1.5.1 OPERATIONS PERFORMED ON DEPENDENT VARIABLES

Amplitude scaling. Let $x(t)$ denote a continuous-time signal. Then the signal $y(t)$ resulting from amplitude scaling applied to $x(t)$ is defined by

$$y(t) = cx(t), \tag{1.21}$$

where c is the scaling factor. According to Eq. (1.21), the value of $y(t)$ is obtained by multiplying the corresponding value of $x(t)$ by the scalar c for each instant of time t. A physical example of a device that performs amplitude scaling is an electronic *amplifier*. A resistor also performs amplitude scaling when $x(t)$ is a current, c is the resistance of the resistor, and $y(t)$ is the output voltage.

In a manner similar to Eq. (1.21), for discrete-time signals, we write

$$y[n] = cx[n].$$

Addition. Let $x_1(t)$ and $x_2(t)$ denote a pair of continuous-time signals. Then the signal $y(t)$ obtained by the addition of $x_1(t)$ and $x_2(t)$ is defined by

$$y(t) = x_1(t) + x_2(t). \tag{1.22}$$

A physical example of a device that adds signals is an audio *mixer*, which combines music and voice signals.

In a manner similar to Eq. (1.22), for discrete-time signals, we write

$$y[n] = x_1[n] + x_2[n].$$

Multiplication. Let $x_1(t)$ and $x_2(t)$ denote a pair of continuous-time signals. Then the signal $y(t)$ resulting from the multiplication of $x_1(t)$ by $x_2(t)$ is defined by

$$y(t) = x_1(t)x_2(t). \tag{1.23}$$

That is, for each prescribed time t, the value of $y(t)$ is given by the product of the corresponding values of $x_1(t)$ and $x_2(t)$. A physical example of $y(t)$ is an *AM radio signal*, in which $x_1(t)$ consists of an audio signal plus a dc component and $x_2(t)$ consists of a sinusoidal signal called a carrier wave.

In a manner similar to Eq. (1.23), for discrete-time signals, we write

$$y[n] = x_1[n]x_2[n].$$

Differentiation. Let $x(t)$ denote a continuous-time signal. Then the derivative of $x(t)$ with respect to time is defined by

$$y(t) = \frac{d}{dt}x(t). \tag{1.24}$$

For example, an *inductor* performs differentiation. Let $i(t)$ denote the current flowing through an inductor of inductance L, as shown in Fig. 1.18. Then the voltage $v(t)$ developed across the inductor is defined by

$$v(t) = L\frac{d}{dt}i(t). \tag{1.25}$$

FIGURE 1.18   Inductor with current $i(t)$, inducing voltage $v(t)$ across its terminals.

FIGURE 1.19 Capacitor with current $i(t)$ inducing, voltage $v(t)$.

Integration. Let $x(t)$ denote a continuous-time signal. Then the integral of $x(t)$ with respect to time t is defined by

$$y(t) = \int_{-\infty}^{t} x(\tau)\, d\tau, \tag{1.26}$$

where τ is the integration variable. For example, a capacitor performs integration. Let $i(t)$ denote the current flowing through a capacitor of capacitance C, as shown in Fig. 1.19. Then the voltage $v(t)$ developed across the capacitor is defined by

$$v(t) = \frac{1}{C} \int_{-\infty}^{t} i(\tau)\, d\tau. \tag{1.27}$$

■ 1.5.2 OPERATIONS PERFORMED ON THE INDEPENDENT VARIABLE

Time scaling. Let $x(t)$ denote a continuous-time signal. Then the signal $y(t)$ obtained by scaling the independent variable, time t, by a factor a is defined by

$$y(t) = x(at).$$

If $a > 1$, the signal $y(t)$ is a *compressed* version of $x(t)$. If $0 < a < 1$, the signal $y(t)$ is an *expanded* (stretched) version of $x(t)$. These two operations are illustrated in Fig. 1.20.

In the discrete-time case, we write

$$y[n] = x[kn], \qquad k > 0,$$

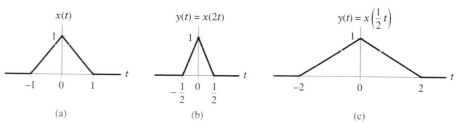

FIGURE 1.20 Time-scaling operation: (a) continuous-time signal $x(t)$, (b) version of $x(t)$ compressed by a factor of 2, and (c) version of $x(t)$ expanded by a factor of 2.

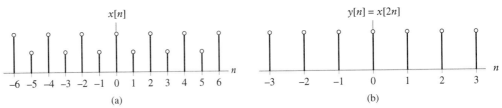

FIGURE 1.21   Effect of time scaling on a discrete-time signal: (a) discrete-time signal $x[n]$ and (b) version of $x[n]$ compressed by a factor of 2, with some values of the original $x[n]$ lost as a result of the compression.

which is defined only for integer values of k. If $k > 1$, then some values of the discrete-time signal $y[n]$ are lost, as illustrated in Fig. 1.21 for $k = 2$. The samples $x[n]$ for $n = \pm 1$, $\pm 3, \ldots$ are lost because putting $k = 2$ in $x[kn]$ causes these samples to be skipped.

▶ **Problem 1.10**   Let

$$x[n] = \begin{cases} n & \text{for } n \text{ odd} \\ 0 & \text{otherwise} \end{cases}.$$

Determine $y[n] = x[2n]$.

Answer: $y[n] = 0$ for all n.   ◀

Reflection. Let $x(t)$ denote a continuous-time signal. Let $y(t)$ denote the signal obtained by replacing time t with $-t$; that is,

$$y(t) = x(-t)$$

The signal $y(t)$ represents a reflected version of $x(t)$ about $t = 0$.
    The following two cases are of special interest:

▶ Even signals, for which we have $x(-t) = x(t)$ for all t; that is, an even signal is the same as its reflected version.

▶ Odd signals, for which we have $x(-t) = -x(t)$ for all t; that is, an odd signal is the negative of its reflected version.

Similar observations apply to discrete-time signals.

EXAMPLE 1.3   REFLECTION   Consider the triangular pulse $x(t)$ shown in Fig. 1.22(a). Find the reflected version of $x(t)$ about the amplitude axis (i.e., the origin).

Solution: Replacing the independent variable t in $x(t)$ with $-t$, we get $y(t) = x(-t)$, as shown in the figure.

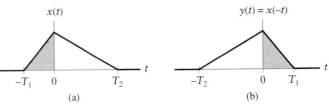

FIGURE 1.22   Operation of reflection: (a) continuous-time signal $x(t)$ and (b) reflected version of $x(t)$ about the origin.

Note that for this example, we have

$$x(t) = 0 \quad \text{for } t < -T_1 \text{ and } t > T_2.$$

Correspondingly, we find that

$$y(t) = 0 \quad \text{for } t > T_1 \text{ and } t < -T_2.$$ ∎

▶ **Problem 1.11** The discrete-time signal

$$x[n] = \begin{cases} 1, & n = 1 \\ -1, & n = -1 \\ 0, & n = 0 \text{ and } |n| > 1 \end{cases}.$$

Find the composite signal

$$y[n] = x[n] + x[-n].$$

Answer: $y[n] = 0$ for all integer values of n. ◀

▶ **Problem 1.12** Repeat Problem 1.11 for

$$x[n] = \begin{cases} 1, & n = -1 \text{ and } n = 1 \\ 0, & n = 0 \text{ and } |n| > 1 \end{cases}.$$

Answer: $y[n] = \begin{cases} 2, & n = -1 \text{ and } n = 1 \\ 0, & n = 0 \text{ and } |n| > 1 \end{cases}.$ ◀

Time shifting. Let $x(t)$ denote a continuous-time signal. Then the time-shifted version of $x(t)$ is defined by

$$y(t) = x(t - t_0),$$

where t_0 is the time shift. If $t_0 > 0$, the waveform of $y(t)$ is obtained by shifting $x(t)$ toward the right, relative to the time axis. If $t_0 < 0$, $x(t)$ is shifted to the left.

EXAMPLE 1.4 TIME SHIFTING Figure 1.23(a) shows a rectangular pulse $x(t)$ of unit amplitude and unit duration. Find $y(t) = x(t - 2)$.

Solution: In this example, the time shift t_0 equals 2 time units. Hence, by shifting $x(t)$ to the right by 2 time units, we get the rectangular pulse $y(t)$ shown in the figure. The pulse $y(t)$ has exactly the same shape as the original pulse $x(t)$; it is merely shifted along the time axis. ∎

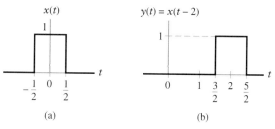

(a)   (b)

FIGURE 1.23 Time-shifting operation: (a) continuous-time signal in the form of a rectangular pulse of amplitude 1 and duration 1, symmetric about the origin; and (b) time-shifted version of $x(t)$ by 2 time units.

In the case of a discrete-time signal $x[n]$, we define its time-shifted version as

$$y[n] = x[n - m],$$

where the shift m must be a positive or negative integer.

▶ **Problem 1.13** The discrete-time signal

$$x[n] = \begin{cases} 1, & n = 1, 2 \\ -1, & n = -1, -2 \\ 0, & n = 0 \text{ and } |n| > 2 \end{cases}.$$

Find the time-shifted signal $y[n] = x[n + 3]$.

Answer: $y[n] = \begin{cases} 1, & n = -1, -2 \\ -1, & n = -4, -5 \\ 0, & n = -3, n < -5, \text{ and } n > -1 \end{cases}.$ ◀

■ **1.5.3 PRECEDENCE RULE FOR TIME SHIFTING AND TIME SCALING**

Let $y(t)$ denote a continuous-time signal that is derived from another continuous-time signal $x(t)$ through a combination of time shifting and time scaling; that is,

$$y(t) = x(at - b). \tag{1.28}$$

This relation between $y(t)$ and $x(t)$ satisfies the conditions

$$y(0) = x(-b) \tag{1.29}$$

and

$$y\left(\frac{b}{a}\right) = x(0), \tag{1.30}$$

which provide useful checks on $y(t)$ in terms of corresponding values of $x(t)$.

To obtain $y(t)$ from $x(t)$, the time-shifting and time-scaling operations must be performed in the correct order. The proper order is based on the fact that the scaling operation always replaces t by at, while the time-shifting operation always replaces t by $t - b$. Hence, the time-shifting operation is performed first on $x(t)$, resulting in an intermediate signal

$$v(t) = x(t - b).$$

The time shift has replaced t in $x(t)$ by $t - b$. Next, the time-scaling operation is performed on $v(t)$, replacing t by at and resulting in the desired output

$$y(t) = v(at)$$

$$= x(at - b).$$

To illustrate how the operation described in Eq. (1.28) can arise in a real-life situation, consider a voice signal recorded on a tape recorder. If the tape is played back at a rate faster than the original recording rate, we get compression (i.e., $a > 1$). If, however, the tape is played back at a rate slower than the original recording rate, we get expansion (i.e., $a < 1$). The constant b, assumed to be positive, accounts for a delay in playing back the tape.

EXAMPLE 1.5 PRECEDENCE RULE FOR CONTINUOUS-TIME SIGNAL   Consider the rectangular pulse $x(t)$ of unit amplitude and a duration of 2 time units, depicted in Fig. 1.24(a). Find $y(t) = x(2t + 3)$.

Solution: In this example, we have $a = 2$ and $b = -3$. Hence, shifting the given pulse $x(t)$ to the left by 3 time units relative to the amplitude axis gives the intermediate pulse $v(t)$ shown in Fig. 1.24(b). Finally, scaling the independent variable t in $v(t)$ by $a = 2$, we get the solution $y(t)$ shown in Fig. 1.24(c).

Note that the solution presented in Fig. 1.24(c) satisfies both of the conditions defined in Eqs. (1.29) and (1.30).

Suppose next that we purposely do not follow the precedence rule; that is, we first apply time scaling and then time shifting. For the given signal $x(t)$ shown in Fig. 1.25(a), the application of time scaling by factor of 2 produces the intermediate signal $v(t) = x(2t)$, which is shown in Fig. 1.25(b). Then shifting $v(t)$ to the left by 3 time units yields the signal shown in Fig. 1.25(c), which is defined by

$$y(t) = v(t + 3) = x(2(t + 3)) \neq x(2t + 3)$$

Hence, the signal $y(t)$ fails to satisfy Eq. (1.30).   ∎

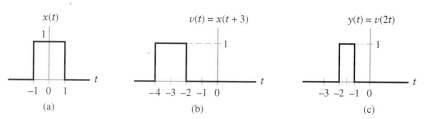

FIGURE 1.24   The proper order in which the operations of time scaling and time shifting should be applied in the case of the continuous-time signal of Example 1.5. (a) Rectangular pulse $x(t)$ of amplitude 1.0 and duration 2.0, symmetric about the origin. (b) Intermediate pulse $v(t)$, representing a time-shifted version of $x(t)$. (c) Desired signal $y(t)$, resulting from the compression of $v(t)$ by a factor of 2.

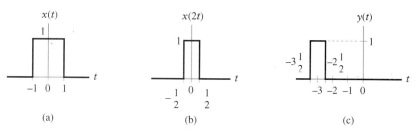

FIGURE 1.25   The incorrect way of applying the precedence rule. (a) Signal $x(t)$. (b) Time-scaled signal $v(t) = x(2t)$. (c) Signal $y(t)$ obtained by shifting $v(t) = x(2t)$ by 3 time units, which yields $y(t) = x(2(t + 3))$.

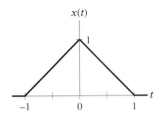

FIGURE 1.26   Triangular pulse for Problem 1.14.

▶ **Problem 1.14**   A triangular pulse signal $x(t)$ is depicted in Fig. 1.26. Sketch each of the following signals derived from $x(t)$:

(a) $x(3t)$

(b) $x(3t + 2)$

(c) $x(-2t - 1)$

(d) $x(2(t + 2))$

(e) $x(2(t - 2))$

(f) $x(3t) + x(3t + 2)$

Answers:

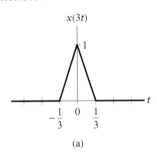

(a)

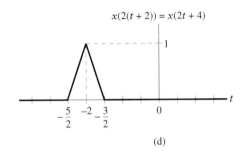

(d)

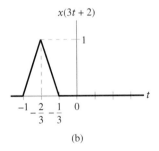

(b)

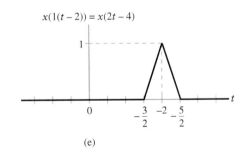

(e)

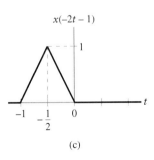

(c)

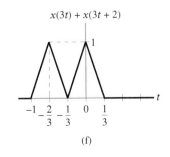

(f)

Example 1.5 clearly illustrates that if $y(t)$ is defined in terms of $x(t)$ by Eq. (1.28), then $y(t)$ can correctly be obtained from $x(t)$ only by adhering to the precedence rule for time shifting and time scaling. Similar remarks apply to the case of discrete-time signals, as illustrated next.

EXAMPLE 1.6   PRECEDENCE RULE FOR DISCRETE-TIME SIGNAL   A discrete-time signal is defined by

$$x[n] = \begin{cases} 1, & n = 1, 2 \\ -1, & n = -1, -2 \\ 0, & n = 0 \text{ and } |n| > 2 \end{cases}.$$

Find $y[n] = x[2n + 3]$.

Solution: The signal $x[n]$ is displayed in Fig. 1.27(a). Time shifting $x[n]$ to the left by 3 yields the intermediate signal $v[n]$ shown in Fig. 1.27(b). Finally, scaling n in $v[n]$ by 2, we obtain the solution $y[n]$ shown in Fig. 1.27(c).

Note that as a result of the compression performed in going from $v[n]$ to $y[n] = v[2n]$, the nonzero samples of $v[n]$ at $n = -5$ and $n = -1$ (i.e., those contained in the original signal at $n = -2$ and $n = 2$) are lost.   ∎

▶ **Problem 1.15**   Consider a discrete-time signal

$$x[n] = \begin{cases} 1, & -2 \leq n \leq 2 \\ 0, & |n| > 2 \end{cases}.$$

Find $y[n] = x[3n - 2]$.

Answer: $y[n] = \begin{cases} 1, & n = 0, 1 \\ 0, & \text{otherwise} \end{cases}.$ ◀

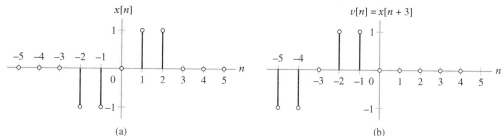

(a)

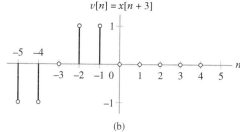

(b)

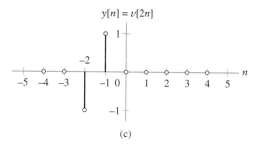

(c)

FIGURE 1.27   The proper order of applying the operations of time scaling and time shifting for the case of a discrete-time signal. (a) Discrete-time signal $x[n]$, antisymmetric about the origin. (b) Intermediate signal $v(n)$ obtained by shifting $x[n]$ to the left by 3 samples. (c) Discrete-time signal $y[n]$ resulting from the compression of $v[n]$ by a factor of 2, as a result of which two samples of the original $x[n]$, located at $n = -2, +2$, are lost.

▍1.6 *Elementary Signals*

Several elementary signals feature prominently in the study of signals and systems. Among these signals are exponential and sinusoidal signals, the step function, the impulse function, and the ramp function, all of which serve as building blocks for the construction of more complex signals. They are also important in their own right, in that they may be used to model many physical signals that occur in nature. In what follows, we will describe these elementary signals, one by one.

■ 1.6.1 EXPONENTIAL SIGNALS

A real exponential signal, in its most general form, is written as

$$x(t) = Be^{at}, \tag{1.31}$$

where both B and a are real parameters. The parameter B is the amplitude of the exponential signal measured at time $t = 0$. Depending on whether the other parameter a is positive or negative, we may identify two special cases:

▶ *Decaying exponential*, for which $a < 0$
▶ *Growing exponential*, for which $a > 0$

These two forms of an exponential signal are illustrated in Fig. 1.28. Part (a) of the figure was generated using $a = -6$ and $B = 5$. Part (b) of the figure was generated using $a = 5$ and $B = 1$. If $a = 0$, the signal $x(t)$ reduces to a dc signal equal to the constant B.

For a physical example of an exponential signal, consider a so-called lossy capacitor, as depicted in Fig. 1.29. The capacitor has capacitance C, and the loss is represented by shunt resistance R. The capacitor is charged by connecting a battery across it, and then the battery is removed at time $t = 0$. Let V_0 denote the initial value of the voltage developed across the capacitor. From the figure, we readily see that the operation of the capacitor for $t \geq 0$ is described by

$$RC\frac{d}{dt}v(t) + v(t) = 0, \tag{1.32}$$

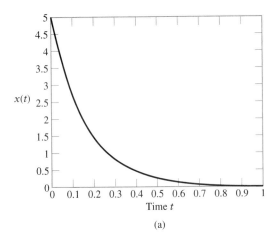

(a)

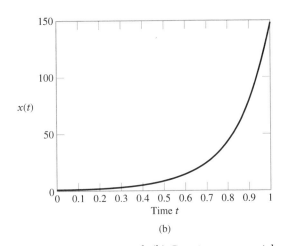

(b)

FIGURE 1.28   (a) Decaying exponential form of continuous-time signal. (b) Growing exponential form of continuous-time signal.

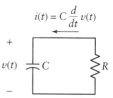

$$i(t) = C \frac{d}{dt} v(t)$$

FIGURE 1.29   Lossy capacitor, with the loss represented by shunt resistance R.

where $v(t)$ is the voltage measured across the capacitor at time t. Equation (1.32) is a *differential equation of order one*. Its solution is given by

$$v(t) = V_0 e^{-t/(RC)}, \tag{1.33}$$

where the product term RC plays the role of a *time constant*. Equation (1.33) shows that the voltage across the capacitor decays exponentially with time at a rate determined by the time constant RC. The larger the resistor R (i.e., the less lossy the capacitor), the slower will be the rate of decay of $v(t)$ with time.

The discussion thus far has been in the context of continuous time. In discrete time, it is common practice to write a real exponential signal as

$$x[n] = Br^n. \tag{1.34}$$

The exponential nature of this signal is readily confirmed by defining

$$r = e^{\alpha}$$

for some α. Figure 1.30 illustrates the decaying and growing forms of a discrete-time exponential signal corresponding to $0 < r < 1$ and $r > 1$, respectively. Note that when $r < 0$, the discrete-time exponential signal $x[n]$ assumes alternating signs for then r^n is positive for n even and negative for n odd.

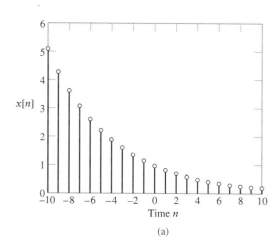

(a)

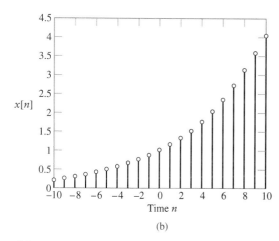

(b)

FIGURE 1.30   (a) Decaying exponential form of discrete-time signal. (b) Growing exponential form of discrete-time signal.

RODUCTION

onential signals shown in Figs. 1.28 and 1.30 are all real valued. It is possi-
xponential signal to be complex valued. The mathematical forms of complex
al signals are the same as those shown in Eqs. (1.31) and (1.34), with the following
ces: In the continuous-time case, in Eq. (1.30), the parameter B, the parameter a,
ch assume complex values. Similarly, in the discrete-time case, in Eq. (1.34), the pa-
eter B, the parameter r, or both assume complex values. Two commonly encountered
examples of complex exponential signals are $e^{j\omega t}$ and $e^{j\Omega n}$.

■ 1.6.2 SINUSOIDAL SIGNALS

The continuous-time version of a sinusoidal signal, in its most general form, may be written as

$$x(t) = A\cos(\omega t + \phi), \tag{1.35}$$

where A is the amplitude, ω is the frequency in radians per second, and ϕ is the phase angle
in radians. Figure 1.31(a) presents the waveform of a sinusoidal signal for $A = 4$ and
$\phi = +\pi/6$. A sinusoidal signal is an example of a periodic signal, the period of which is

$$T = \frac{2\pi}{\omega}.$$

We may readily show that this is the period for the sinusoidal signal of Eq. (1.35) by writing

$$\begin{aligned}
x(t + T) &= A\cos(\omega(t + T) + \phi) \\
&= A\cos(\omega t + \omega T + \phi) \\
&= A\cos(\omega t + 2\pi + \phi) \\
&= A\cos(\omega t + \phi) \\
&= x(t),
\end{aligned}$$

which satisfies the defining condition of Eq. (1.7) for a periodic signal.

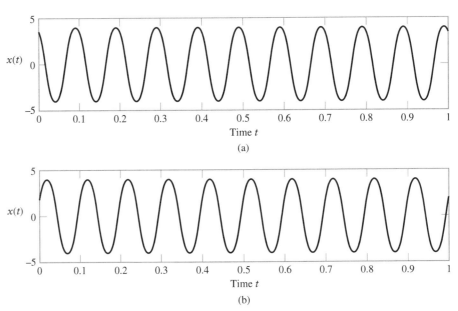

FIGURE 1.31 (a) Sinusoidal signal $A\cos(\omega t + \phi)$ with phase $\phi = +\pi/6$ radians. (b) Sinusoidal
signal $A\sin(\omega t + \phi)$ with phase $\phi = +\pi/6$ radians.

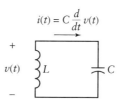

$$i(t) = C\frac{d}{dt}v(t)$$

FIGURE 1.32 Parallel LC circuit, assuming that the inductor L and capacitor C are both ideal.

To illustrate the generation of a sinusoidal signal, consider the circuit of Fig. 1.32, consisting of an inductor and a capacitor connected in parallel. It is assumed that the losses in both components of the circuit are small enough for them to be considered "ideal." The voltage developed across the capacitor at time $t = 0$ is equal to V_0. The operation of the circuit for $t \geq 0$ is described by

$$LC\frac{d^2}{dt^2}v(t) + v(t) = 0, \tag{1.36}$$

where $v(t)$ is the voltage across the capacitor at time t, C is the capacitance of the capacitor, and L is the inductance of the inductor. Equation (1.36) is a *differential equation of order two*. Its solution is given by

$$v(t) = V_0\cos(\omega_0 t), \qquad t \geq 0, \tag{1.37}$$

where

$$\omega_0 = \frac{1}{\sqrt{LC}} \tag{1.38}$$

is the *natural angular frequency of oscillation of the circuit*. Equation (1.37) describes a sinusoidal signal of amplitude $A = V_0$, frequency $\omega = \omega_0$, and zero phase angle.

Consider next the discrete-time version of a sinusoidal signal, written as

$$x[n] = A\cos(\Omega n + \phi). \tag{1.39}$$

This discrete-time signal may or may not be periodic. For it to be periodic with a period of, say, N samples, it must satisfy Eq. (1.10) for all integer n and some integer N. Substituting $n + N$ for n in Eq. (1.39) yields

$$x[n + N] = A\cos(\Omega n + \Omega N + \phi).$$

For Eq. (1.10) to be satisfied, in general, we require that

$$\Omega N = 2\pi m \qquad \text{radians,}$$

or

$$\Omega = \frac{2\pi m}{N} \text{ radians/cycle,} \qquad \text{integer } m, N. \tag{1.40}$$

The important point to note here is that, unlike continuous-time sinusoidal signals, not all discrete-time sinusoidal systems with arbitrary values of Ω are periodic. Specifically, for the discrete-time sinusoidal signal described in Eq. (1.39) to be periodic, the angular frequency

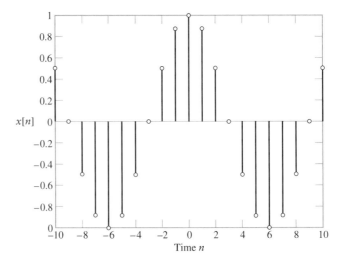

FIGURE 1.33   Discrete-time sinusoidal signal.

Ω must be a rational multiple of 2π, as indicated in Eq. (1.40). Figure 1.33 illustrates a discrete-time sinusoidal signal given by Eq. (1.39) for $A = 1$, $\phi = 0$, and $N = 12$.

Note also that, since ΩN represents an angle, it is measured in radians. Moreover, since N is the number of samples contained in a single cycle of $x[n]$, it follows that Ω is measured in radians per cycle, as stated in Eq. (1.40).

EXAMPLE 1.7   DISCRETE-TIME SINUSOIDAL SIGNALS   A pair of sinusoidal signals with a common angular frequency is defined by

$$x_1[n] = \sin[5\pi n]$$

and

$$x_2[n] = \sqrt{3}\cos[5\pi n].$$

(a) Both $x_1[n]$ and $x_2[n]$ are periodic. Find their common fundamental period.

(b) Express the composite sinusoidal signal

$$y[n] = x_1[n] + x_2[n]$$

in the form $y[n] = A\cos(\Omega n + \phi)$, and evaluate the amplitude A and phase ϕ.

Solution:

(a) The angular frequency of both $x_1[n]$ and $x_2[n]$ is

$$\Omega = 5\pi \text{ radians/cycle.}$$

Solving Eq. (1.40) for the period N, we get

$$N = \frac{2\pi m}{\Omega}$$
$$= \frac{2\pi m}{5\pi}$$
$$= \frac{2m}{5}.$$

For $x_1[n]$ and $x_2[n]$ to be periodic, N must be an integer. This can be so only for $m = 5, 10, 15, \ldots$, which results in $N = 2, 4, 6, \ldots$.

(b) Recall the trigonometric identity

$$A \cos(\Omega n + \phi) = A \cos(\Omega n) \cos(\phi) - A \sin(\Omega n) \sin(\phi).$$

Letting $\Omega = 5\pi$, we see that the right-hand side of this identity is of the same form as $x_1[n] + x_2[n]$. We may therefore write

$$A \sin(\phi) = -1 \quad \text{and} \quad A \cos(\phi) = \sqrt{3}.$$

Hence,

$$\tan(\phi) = \frac{\sin(\phi)}{\cos(\phi)} = \frac{\text{amplitude of } x_1[n]}{\text{amplitude of } x_2[n]}$$
$$= \frac{-1}{\sqrt{3}},$$

from which we find that $\phi = -\pi/3$ radians. Substituting this value into the equation

$$A \sin(\phi) = -1$$

and solving for the amplitude A, we get

$$A = -1/\sin\left(-\frac{\pi}{3}\right)$$
$$= 2.$$

Accordingly, we may express $y[n]$ as

$$y[n] = 2 \cos\left(5\pi n - \frac{\pi}{3}\right). \qquad \blacksquare$$

▶ **Problem 1.16**   Determine the fundamental period of the sinusoidal signal

$$x[n] = 10 \cos\left(\frac{4\pi}{31} n + \frac{\pi}{5}\right).$$

Answer: $N = 31$ samples.   ◀

▶ **Problem 1.17**   Consider the following sinusoidal signals:

(a) $x[n] = 5 \sin[2n]$
(b) $x[n] = 5 \cos[0.2\pi n]$
(c) $x[n] = 5 \cos[6\pi n]$
(d) $x[n] = 5 \sin[6\pi n/35]$

Determine whether each $x(n)$ is periodic, and if it is, find its fundamental period.

Answers: (a) Nonperiodic. (b) Periodic, fundamental period = 10. (c) Periodic, fundamental period = 1. (d) Periodic, fundamental period = 35.   ◀

▶ **Problem 1.18**   Find the smallest angular frequencies for which discrete-time sinusoidal signals with the following fundamental periods would be periodic: (a) $N = 8$, (b) $N = 32$, (c) $N = 64$, and (d) $N = 128$.

Answers: (a) $\Omega = \pi/4$. (b) $\Omega = \pi/16$. (c) $\Omega = \pi/32$. (d) $\Omega = \pi/64$.   ◀

■ **1.6.3 RELATION BETWEEN SINUSOIDAL AND COMPLEX EXPONENTIAL SIGNALS**

Consider the complex exponential $e^{j\theta}$. Using *Euler's identity*, we may expand this term as

$$e^{j\theta} = \cos\theta + j\sin\theta. \tag{1.41}$$

This result indicates that we may express the continuous-time sinusoidal signal of Eq. (1.35) as the real part of the complex exponential signal $Be^{j\omega t}$, where

$$B = Ae^{j\phi} \tag{1.42}$$

is itself a complex quantity. That is, we may write

$$A\cos(\omega t + \phi) = \text{Re}\{Be^{j\omega t}\} \tag{1.43}$$

where Re{ } denotes the real part of the complex quantity enclosed inside the braces. We may readily prove Eq. (1.43) by noting that

$$\begin{aligned} Be^{j\omega t} &= Ae^{j\phi}e^{j\omega t} \\ &= Ae^{j(\omega t + \phi)} \\ &= A\cos(\omega t + \phi) + jA\sin(\omega t + \phi). \end{aligned}$$

Equation (1.43) follows immediately. The sinusoidal signal of Eq. (1.35) is defined in terms of a cosine function. Of course, we may also define a continuous-time sinusoidal signal in terms of a sine function, such as

$$x(t) = A\sin(\omega t + \phi), \tag{1.44}$$

which is represented by the imaginary part of the complex exponential signal $Be^{j\omega t}$. That is, we may write

$$A\sin(\omega t + \phi) = \text{Im}\{Be^{j\omega t}\}, \tag{1.45}$$

where B is defined by Eq. (1.42) and Im{ } denotes the imaginary part of the complex quantity enclosed inside the braces. The sinusoidal signal of Eq. (1.44) differs from that of Eq. (1.35) by a phase angle of 90°. That is, the sinusoidal signal $A\cos(\omega t + \phi)$ lags behind the sinusoidal signal $A\sin(\omega t + \phi)$, as illustrated in Fig. 1.31 for $\phi = \pi/6$.

Similarly, in the discrete-time case, we may write

$$A\cos(\Omega n + \phi) = \text{Re}\{Be^{j\Omega n}\} \tag{1.46}$$

and

$$A\sin(\Omega n + \phi) = \text{Im}\{Be^{j\Omega n}\}, \tag{1.47}$$

where B is defined in terms of A and ϕ by Eq. (1.42). Figure 1.34 shows the two-dimensional representation of the complex exponential $e^{j\Omega n}$ for $\Omega = \pi/4$ and $n = 0, 1, \ldots, 7$. The projection of each value on the real axis is $\cos(\Omega n)$, while the projection on the imaginary axis is $\sin(\Omega n)$.

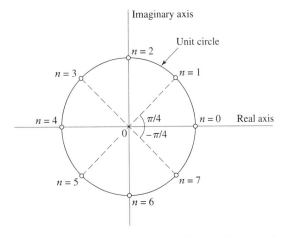

FIGURE 1.34 Complex plane, showing eight points uniformly distributed on the unit circle. The projection of the points on the real axis is $\cos(\pi/4n)$, while the projection on the imaginary axis is $\sin(\pi/4n); n = 0, 1, \ldots, 7$.

■ 1.6.4 EXPONENTIALLY DAMPED SINUSOIDAL SIGNALS

The multiplication of a sinusoidal signal by a real-valued decaying exponential signal results in a new signal referred to as an *exponentially damped sinusoidal signal*. Specifically, multiplying the continuous-time sinusoidal signal $A \sin(\omega t + \phi)$ by the exponential $e^{-\alpha t}$ results in the exponentially damped sinusoidal signal

$$x(t) = Ae^{-\alpha t} \sin(\omega t + \phi), \qquad \alpha > 0. \tag{1.48}$$

Figure 1.35 shows the waveform of this signal for $A = 60$, $\alpha = 6$, and $\phi = 0$. For increasing time t, the amplitude of the sinusoidal oscillations decreases in an exponential fashion, approaching zero for infinite time.

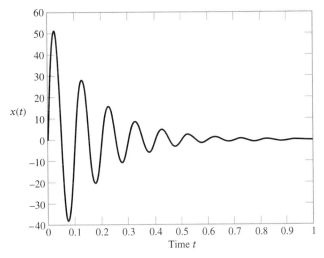

FIGURE 1.35 Exponentially damped sinusoidal signal $Ae^{-\alpha t} \sin(\omega t)$, with $A = 60$ and $\alpha = 6$.

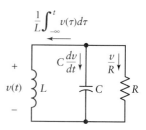

FIGURE 1.36   Parallel LRC circuit, with inductor L, capacitor C, and resistor R all assumed to be ideal.

To illustrate the generation of an exponentially damped sinusoidal signal, consider the parallel circuit of Fig. 1.36, consisting of a capacitor of capacitance C, an inductor of inductance L, and a resistor of resistance R. Let V_0 denote the initial voltage developed across the capacitor at time $t = 0$. Then the operation of the circuit is described by

$$C\frac{d}{dt}v(t) + \frac{1}{R}v(t) + \frac{1}{L}\int_{-\infty}^{t} v(\tau)\, d\tau = 0, \qquad (1.49)$$

where $v(t)$ is the voltage across the capacitor at time $t \geq 0$. Equation (1.49) is an *integro-differential equation*. Its solution is given by

$$v(t) = V_0 e^{-t/(2CR)} \cos(\omega_0 t), \qquad t \geq 0 \qquad (1.50)$$

where

$$\omega_0 = \sqrt{\frac{1}{LC} - \frac{1}{4C^2 R^2}}. \qquad (1.51)$$

In Eq. (1.51), it is assumed that $R > \sqrt{L/(4C)}$. Comparing Eq. (1.50) with (1.48), we have $A = V_0, \alpha = 1/(2CR), \omega = \omega_0$, and $\phi = \pi/2$.

The circuits of Figs. 1.29, 1.32, and 1.36 served as examples in which an exponential signal, a sinusoidal signal, and an exponentially damped sinusoidal signal, respectively, arose naturally as solutions to physical problems. The operations of these circuits are described by the differential equations (1.32), (1.36), and (1.49), whose solutions were merely stated. Methods for solving these differential equations are presented in subsequent chapters.

Returning to the subject matter at hand, we describe the discrete-time version of the exponentially damped sinusoidal signal of Eq. (1.48) by

$$x[n] = Br^n \sin[\Omega n + \phi]. \qquad (1.52)$$

For the signal of Eq. (1.52) to decay exponentially with time, the parameter r must lie in the range $0 < |r| < 1$.

S & S
Solutions

▶ **Problem 1.19**   Equation (1.51) assumes that the resistance $R > \sqrt{L/(4C)}$. What happens to the waveform $v(t)$ of Eq. (1.50) if this condition is not satisfied—that is, if $R < \sqrt{L/(4C)}$?

Answer: If $R < \sqrt{L/(4C)}$, then the signal $v(t)$ consists of the sum of two damped exponentials with different time constants, one equal to $2CR/\left(1 + \sqrt{1 - 4R^2 C/L}\right)$ and the other equal to $2CR/\left(1 - \sqrt{1 - 4R^2 C/L}\right)$.   ◀

▶ **Problem 1.20**   Consider the complex-valued exponential signal

$$x(t) = Ae^{\alpha t + j\omega t}, \qquad a > 0.$$

Evaluate the real and imaginary components of $x(t)$ for the following cases:

(a) α real, $\alpha = \alpha_1$
(b) α imaginary, $\alpha = j\omega_1$
(c) α complex, $\alpha = \alpha_1 + j\omega_1$

Answers:

(a) $\text{Re}\{x(t)\} = Ae^{\alpha_1 t}\cos(\omega t)$; $\text{Im}\{x(t)\} = Ae^{\alpha_1 t}\sin(\omega t)$
(b) $\text{Re}\{x(t)\} = A\cos(\omega_1 t + \omega t)$; $\text{Im}\{x(t)\} = A\sin(\omega_1 t + \omega t)$
(c) $\text{Re}\{x(t)\} = Ae^{\alpha_1 t}\cos(\omega_1 t + \omega t)$; $\text{Im}\{x(t)\} = Ae^{\alpha_1 t}\sin(\omega_1 t + \omega t)$ ◀

▶ **Problem 1.21**   Consider the pair of exponentially damped sinusoidal signals

$$x_1(t) = Ae^{\alpha t}\cos(\omega t), \qquad t \geq 0$$

and

$$x_2(t) = Ae^{\alpha t}\sin(\omega t), \qquad t \geq 0.$$

Assume that A, α, and ω are all real numbers; the exponential damping factor α is negative and the frequency of oscillation ω is positive; the amplitude A can be positive or negative.

(a) Derive the complex-valued signal $x(t)$ whose real part is $x_1(t)$ and imaginary part is $x_2(t)$.
(b) The formula

$$a(t) = \sqrt{x_1^2(t) + x_2^2(t)}$$

defines the *envelope* of the complex signal $x(t)$. Determine $a(t)$ for the $x(t)$ defined in part (a).
(c) How does the envelope $a(t)$ vary with time t?

Answers:

(a) $x(t) = Ae^{st}, \qquad t \geq 0$, where $s = \alpha + j\omega$
(b) $a(t) = |A|e^{\alpha t}, \qquad t \geq 0$
(c) At $t = 0$, $a(0) = |A|$, and then $a(t)$ decreases exponentially as time t increases; as t approaches infinity, $a(t)$ approaches zero   ◀

■ 1.6.5 STEP FUNCTION

The discrete-time version of the unit-step function is defined by

$$u[n] = \begin{cases} 1, & n \geq 0 \\ 0, & n < 0 \end{cases} \tag{1.53}$$

which is illustrated in Fig. 1.37.

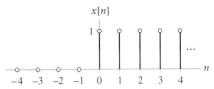

FIGURE 1.37   Discrete-time version of step function of unit amplitude.

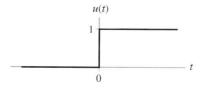

FIGURE 1.38   Continuous-time version of the unit-step function of unit amplitude.

The continuous-time version of the unit-step function is defined by

$$u(t) = \begin{cases} 1, & t > 0 \\ 0, & t < 0 \end{cases}. \qquad (1.54)$$

Figure 1.38 depicts the unit-step function $u(t)$. It is said to exhibit a discontinuity at $t = 0$, since the value of $u(t)$ changes instantaneously from 0 to 1 when $t = 0$. It is for this reason that we have left out the equals sign in Eq. (1.54); that is, $u(0)$ is undefined.

The unit-step function $u(t)$ is a particularly simple signal to apply. Electrically, a battery or dc source is applied at $t = 0$ by, for example, closing a switch. As a test signal, the unit-step function is useful because the output of a system due to a step input reveals a great deal about how quickly the system responds to an abrupt change in the input signal. A similar remark applies to $u[n]$ in the context of a discrete-time system.

The unit-step function $u(t)$ may also be used to construct other discontinuous waveforms, as illustrated in the next example.

▎**EXAMPLE 1.8 RECTANGULAR PULSE**   Consider the rectangular pulse $x(t)$ shown in Fig. 1.39(a). This pulse has an amplitude A and duration of 1 second. Express $x(t)$ as a weighted sum of two step functions.

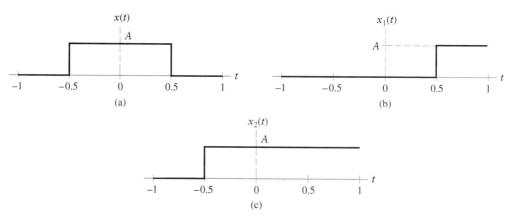

FIGURE 1.39   (a) Rectangular pulse $x(t)$ of amplitude A and duration of 1 s, symmetric about the origin. (b) Step function of amplitude A, shifted to the right by 0.5s. (c) Step function of amplitude A, shifted to the left by 0.5s. Note that $x(t) = x_2(t) - x_1(t)$.

Solution: The rectangular pulse $x(t)$ may be written in mathematical terms as

$$x(t) = \begin{cases} A, & 0 \le |t| < 0.5 \\ 0, & |t| > 0.5 \end{cases}, \tag{1.55}$$

where $|t|$ denotes the magnitude of time t. The rectangular pulse $x(t)$ is represented as the difference of two time-shifted step functions, $x_1(t)$ and $x_2(t)$, which are defined in Figs. 1.39(b) and 1.39(c), respectively. On the basis of this figure, we may express $x(t)$ as

$$x(t) = Au\left(t + \frac{1}{2}\right) - Au\left(t - \frac{1}{2}\right), \tag{1.56}$$

where $u(t)$ is the unit-step function.   ∎

EXAMPLE 1.9   *RC* CIRCUIT   Consider the simple RC circuit shown in Fig. 1.40(a). The capacitor C is assumed to be initially uncharged. At $t = 0$, the switch connecting the dc voltage source V_0 to the RC circuit is closed. Find the voltage $v(t)$ across the capacitor for $t \ge 0$.

Solution: The switching operation is represented by a step function $V_0 u(t)$, as shown in the equivalente circuit of Fig. 1.40(b). The capacitor cannot charge suddenly, so, with it being initially uncharged, we have

$$v(0) = 0.$$

For $t = \infty$, the capacitor becomes fully charged; hence,

$$v(\infty) = V_0.$$

Recognizing that the voltage across the capacitor increases exponentially with a time constant RC, we may thus express $v(t)$ as

$$v(t) = V_0(1 - e^{-t/(RC)})u(t). \tag{1.57}$$ ∎

▶ **Problem 1.22**   A discrete-time signal

$$x[n] = \begin{cases} 1, & 0 \le n \le 9 \\ 0, & \text{otherwise} \end{cases}.$$

Using $u[n]$, describe $x[n]$ as the superposition of two step functions.

Answer: $x[n] = u[n] - u[n - 10]$.   ◀

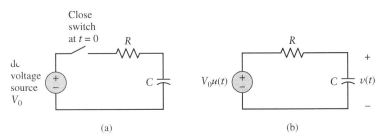

(a)   (b)

FIGURE 1.40   (a) Series RC circuit with a switch that is closed at time $t = 0$, thereby energizing the voltage source. (b) Equivalent circuit, using a step function to replace the action of the switch.

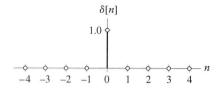

FIGURE 1.41   Discrete-time form of the unit impulse.

■ 1.6.6 IMPULSE FUNCTION

The discrete-time version of the *unit impulse* is defined by

$$\delta[n] = \begin{cases} 1, & n = 0 \\ 0, & n \neq 0 \end{cases}. \tag{1.58}$$

Equation (1.58) is illustrated in Fig. 1.41.

The continuous-time version of the unit impulse is defined by the following pair of relations:

$$\delta(t) = 0 \quad \text{for} \quad t \neq 0 \tag{1.59}$$

and

$$\int_{-\infty}^{\infty} \delta(t) \, dt = 1. \tag{1.60}$$

Equation (1.59) says that the impulse $\delta(t)$ is zero everywhere except at the origin. Equation (1.60) says that the total area under the unit impulse is unity. The impulse $\delta(t)$ is also referred to as the *Dirac delta function*.

A graphical description of the unit-impulse $\delta[n]$ for discrete time is straightforward, as shown in Fig. 1.41. In contrast, visualization of the unit impulse $\delta(t)$ for continuous time requires more detailed attention. One way to visualize $\delta(t)$ is to view it as the limiting form of a rectangular pulse of unit area, as illustrated in Fig. 1.42(a). Specifically, the duration of the pulse is decreased, and its amplitude is increased, such that the area under the pulse is maintained constant at unity. As the duration decreases, the rectangular pulse approximates the impulse more closely. Indeed, we may generalize this assertion by stating that

$$\delta(t) = \lim_{\Delta \to 0} x_\Delta(t), \tag{1.61}$$

where $x_\Delta(t)$ is any pulse that is an even function of time t with duration Δ and unit area. The area under the pulse defines the *strength* of the impulse. Thus, when we speak of the

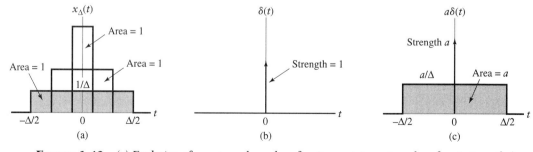

FIGURE 1.42   (a) Evolution of a rectangular pulse of unit area into an impulse of unit strength (i.e., unit impulse). (b) Graphical symbol for unit impulse. (c) Representation of an impulse of strength a that results from allowing the duration Δ of a rectangular pulse of area a to approach zero.

impulse function $\delta(t)$, in effect, we are saying that its strength is unity. The graphical symbol for the unit impulse is depicted in Fig. 1.42(b). An impulse of strength a is written as $a\delta(t)$; such an impulse results from allowing the duration Δ of a rectangular pulse of constant area a to approach zero, as illustrated in Fig. 1.42(c).

The impulse $\delta(t)$ and the unit-step function $u(t)$ are related to each other in that if we are given either one, we can uniquely determine the other. Specifically, $\delta(t)$ is the derivative of $u(t)$ with respect to time t, or

$$\delta(t) = \frac{d}{dt}u(t). \tag{1.62}$$

Conversely, the step function $u(t)$ is the integral of the impulse $\delta(t)$ with respect to time t:

$$u(t) = \int_{-\infty}^{t} \delta(\tau)\, d\tau \tag{1.63}$$

S & S
Solutions

EXAMPLE 1.10 RC CIRCUIT (CONTINUED)   Consider the simple circuit shown in Fig. 1.43, in which the capacitor is initially uncharged and the switch connecting it to the dc voltage source V_0 is suddenly closed at time $t = 0$. (This circuit is the same as that of the RC circuit in Fig. 1.40, except that we now have zero resistance.) Determine the current $i(t)$ that flows through the capacitor for $t \geq 0$.

Solution: The switching operation is equivalent to connecting the voltage source $V_0 u(t)$ across the capacitor, as shown in Fig. 1.43(b). We may thus express the voltage across the capacitor as

$$v(t) = V_0 u(t).$$

By definition, the current flowing through the capacitor is

$$i(t) = C\frac{dv(t)}{dt}.$$

Hence, for the problem at hand, we have

$$i(t) = CV_0\frac{du(t)}{dt}$$
$$= CV_0\delta(t),$$

where, in the second line, we have used Eq. (1.62). That is, the current that flows through the capacitor C in Fig. 1.43(b) is an impulsive current of strength CV_0.   ∎

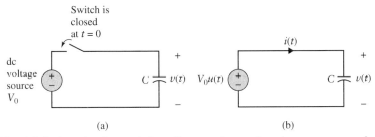

FIGURE 1.43   (a) Series circuit consisting of a capacitor, a dc voltage source, and a switch; the switch is closed at time $t = 0$. (b) Equivalent circuit, replacing the action of the switch with a step function $u(t)$.

From the defining equation (1.61), it immediately follows that the unit impulse $\delta(t)$ is an even function of time t; that is,

$$\delta(-t) = \delta(t). \tag{1.64}$$

For $\delta(t)$ to have mathematical meaning, however, it has to appear as a factor in the integrand of an integral with respect to time, and then, strictly speaking, only when the other factor in the integrand is a continuous function of time at the time at which the impulse occurs. Let $x(t)$ be such a function, and consider the product of $x(t)$ and the time-shifted delta function $\delta(t - t_0)$. In light of the two defining equations (1.59) and (1.60), we may express the integral of this product as

$$\boxed{\int_{-\infty}^{\infty} x(t)\delta(t - t_0)\,dt = x(t_0).} \tag{1.65}$$

It is assumed that $x(t)$ is continuous at time $t = t_0$, where the unit impulse is located.

The operation indicated on the left-hand side of Eq. (1.65) sifts out the value $x(t_0)$ of the function $x(t)$ at time $t = t_0$. Accordingly, Eq. (1.65) is referred to as the *sifting property* of the unit impulse. This property is sometimes used as the definition of a unit impulse; in effect, it incorporates Eqs. (1.59) and (1.60) into a single relation.

Another useful property of the unit impulse $\delta(t)$ is the *time-scaling property*, described by

$$\boxed{\delta(at) = \frac{1}{a}\delta(t), \qquad a > 0.} \tag{1.66}$$

To prove Eq. (1.66), we replace t in Eq. (1.61) with at and so write

$$\delta(at) = \lim_{\Delta \to 0} x_\Delta(at). \tag{1.67}$$

To represent the function $x_\Delta(t)$, we use the rectangular pulse shown in Fig. 1.44(a), which has duration Δ, amplitude $1/\Delta$, and therefore unit area. Correspondingly, the time-scaled function $x_\Delta(at)$ is shown in Fig. 1.44(b) for $a > 1$. The amplitude of $x_\Delta(at)$ is left unchanged by the time-scaling operation. Consequently, in order to restore the area under this pulse to unity, $x_\Delta(at)$ is scaled by the same factor a, as indicated in Fig. 1.44(c), in which the time function is thus denoted by $ax_\Delta(at)$. Using this new function in Eq. (1.67) yields

$$\lim_{\Delta \to 0} x_\Delta(at) = \frac{1}{a}\delta(t), \tag{1.68}$$

from which Eq. (1.66) follows.

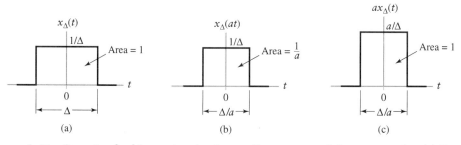

FIGURE 1.44   Steps involved in proving the time-scaling property of the unit impulse. (a) Rectangular pulse $x_\Delta(t)$ of amplitude $1/\Delta$ and duration Δ, symmetric about the origin. (b) Pulse $x_\Delta(t)$ compressed by factor a. (c) Amplitude scaling of the compressed pulse, restoring it to unit area.

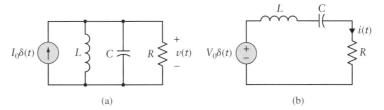

FIGURE 1.45   (a) Parallel LRC circuit driven by an impulsive current signal. (b) Series LRC circuit driven by an impulsive voltage signal.

Having defined what a unit impulse is and described its properties, we have one more question that needs to be addressed: What is the practical use of a unit impulse? We cannot generate a physical impulse function, since that would correspond to a signal of infinite amplitude at $t = 0$ and zero amplitude elsewhere. However, the impulse function serves a mathematical purpose by providing an approximation to a physical signal of extremely short duration and high amplitude. The response of a system to such an input reveals much about the character of the system. For example, consider the parallel LRC circuit of Fig. 1.36, assumed to be initially at rest. Suppose that a current signal approximating an impulse function is applied across the circuit at $t = 0$. Let $I_0\delta(t)$ denote the weighted representation of this impulsive current signal, as indicated in Fig. 1.45(a). At time $t = 0$, the inductor acts as an open circuit, whereas the capacitor acts as a short circuit. Accordingly, the entire impulsive current signal $I_0\delta(t)$ flows through the capacitor, thereby causing the voltage across the capacitor at time $t = 0^+$ to suddenly rise to the new value

$$V_0 = \frac{1}{C} \int_{0^-}^{0^+} I_0\delta(t)\,d(t)$$

$$= \frac{I_0}{C}.$$

(1.69)

Here, $t = 0^+$ and $t = 0^-$ refer to zero time approached from the positive and negative sides of the time axis. Thereafter, the circuit operates without additional input. The resulting value of the voltage $v(t)$ across the capacitor is defined by Eq. (1.50). The response $v(t)$ is called the *transient response* of the circuit, the evaluation of which is facilitated by the application of an impulse function as the test signal.

▶ **Problem 1.23**   The parallel LRC circuit of Fig. 1.45(a) and the series LRC circuit of Fig. 1.45(b) constitute a pair of *dual* circuits, in that their descriptions in terms of the voltage $v(t)$ in Fig. 1.45(a) and the current $i(t)$ in Fig. 1.45(b) are mathematically identical. Given what we already know about the parallel circuit, do the following for the series LRC circuit of Fig. 1.45(b), assuming that it is initially at rest:

(a) Find the value of the current $i(t)$ at time $t = 0^+$.

(b) Write the integro-differential equation defining the evolution of $i(t)$ for $t \geq 0^+$.

Answers:

(a) $I_0 = V_0/L$

(b) $L\dfrac{d}{dt}i(t) + Ri(t) + \dfrac{1}{C}\displaystyle\int_{0^+}^{t} i(\tau)\,d\tau = 0$ ◀

■ 1.6.7 DERIVATIVES OF THE IMPULSE

In systems analysis, we sometimes encounter the problem of having to determine the first and higher order derivatives of the impulse $\delta(t)$; this issue requires careful attention.

From Fig. 1.42(a), we recall that the impulse $\delta(t)$ is the limiting form of a rectangular pulse of duration Δ and amplitude $1/\Delta$. On this basis, we may view the first derivative of $\delta(t)$ as the limiting form of the first derivative of the same rectangular pulse.

Next, from Example 1.8, we recognize that this rectangular pulse is equal to the step function $(1/\Delta)u(t + \Delta/2)$ minus the step function $(1/\Delta)u(t - \Delta/2)$. Equation (1.62) indicates that the derivative of a unit-step function is a unit impulse, so differentiating the rectangular pulse with respect to time t yields a pair of impulses:

▶ One impulse of strength $1/\Delta$, located at $t = -\Delta/2$
▶ A second impulse of strength $-1/\Delta$, located at $t = \Delta/2$

As the duration Δ of the pulse is allowed to approach zero, two things happen. First, the two impulses resulting from the differentiation move toward each other; in the limit, they become practically coincident at the origin. Second, the strengths of the two impulses approach the limiting values of $+\infty$ and $-\infty$, respectively. We thus conclude that the first derivative of the impulse $\delta(t)$ consists of a pair of impulses, one of positive infinite strength at time $t = 0^-$ and a second of negative infinite strength at $t = 0^+$, where, as before, 0^- and 0^+ denote zero time approached from the negative and positive sides, respectively. The first derivative of the unit impulse is termed a *doublet*, which is denoted by $\delta^{(1)}(t)$. The doublet may be interpreted as the output of a system that performs differentiation, such as the inductor in Eq. (1.25), in response to a unit-impulse input.

As with the unit impulse, the doublet has mathematical meaning only as a factor in the integrand of an integral with respect to time, and then, strictly speaking, only when the other factor in the integrand has a continuous derivative at the time at which the doublet occurs. The properties of the doublet follow from its description as a limit of two impulses and the properties of the impulse. For example, writing

$$\delta^{(1)}(t) = \lim_{\Delta \to 0} \frac{1}{\Delta}\left(\delta(t + \Delta/2) - \delta(t - \Delta/2)\right), \tag{1.70}$$

we may show the following fundamental properties of the doublet:

$$\int_{-\infty}^{\infty} \delta^{(1)}(t)\, dt = 0; \tag{1.71}$$

$$\int_{-\infty}^{\infty} f(t)\delta^{(1)}(t - t_0)\, dt = \frac{d}{dt}f(t)\Big|_{t=t_0}. \tag{1.72}$$

In Eq. (1.72), $f(t)$ is a continuous function of time with a continuous derivative at $t = t_0$. The property exhibited by Eq. (1.72) is analogous to the sifting property of the impulse.

We may also use Eq. (1.70) to determine higher order derivatives of the unit impulse. In particular, the second derivative of the unit impulse is the first derivative of the doublet. That is,

$$\frac{d^2}{dt^2}\delta(t) = \frac{d}{dt}\delta^{(1)}(t)$$

$$= \lim_{\Delta \to 0} \frac{\delta^{(1)}(t + \Delta/2) - \delta^{(1)}(t - \Delta/2)}{\Delta}. \tag{1.73}$$

Equation (1.73) may be generalized to define the nth derivative of the unit impulse, which we denote by $\delta^{(n)}(t)$.

▶ **Problem 1.24**

(a) Evaluate the sifting property of $\delta^{(2)}(t)$.

(b) Generalize your result to describe the sifting property of the nth derivative of the unit impulse.

Answers:

(a) $\displaystyle\int_{-\infty}^{\infty} f(t)\delta^{(2)}(t - t_0)\, dt = \frac{d^2}{dt^2} f(t)\big|_{t=t_0}$

(b) $\displaystyle\int_{-\infty}^{\infty} f(t)\delta^{(n)}(t - t_0)\, dt = \frac{d^n}{dt^n} f(t)\big|_{t=t_0}$ ◀

■ 1.6.8 RAMP FUNCTION

The impulse function $\delta(t)$ is the derivative of the step function $u(t)$ with respect to time. By the same token, the integral of the step function $u(t)$ is a ramp function of unit slope. This latter test signal is formally defined as

$$r(t) = \begin{cases} t, & t \geq 0 \\ 0, & t < 0 \end{cases}. \tag{1.74}$$

Equivalently, we may write

$$r(t) = tu(t). \tag{1.75}$$

The ramp function $r(t)$ is shown graphically in Fig. 1.46.

In mechanical terms, a ramp function may be visualized as follows. If $f(t)$ represents the angular displacement of a shaft, then the constant-speed rotation of the shaft provides a representation of the ramp function. As a test signal, the ramp function enables us to evaluate how a continuous-time system would respond to a signal that increases linearly with time.

The discrete-time version of the ramp function is defined by

$$r[n] = \begin{cases} n, & n \geq 0 \\ 0, & n < 0 \end{cases}, \tag{1.76}$$

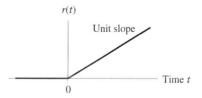

FIGURE 1.46   Ramp function of unit slope.

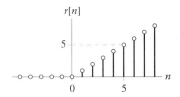

FIGURE 1.47   Discrete-time version of the ramp function.

or, equivalently,

$$r[n] = nu[n]. \tag{1.77}$$

The discrete-time ramp function is illustrated in Fig. 1.47.

EXAMPLE 1.11 PARALLEL CIRCUIT   Consider the parallel circuit of Fig. 1.48(a) involving a dc current source I_0 and an initially uncharged capacitor C. The switch across the capacitor is suddenly opened at time $t = 0$. Determine the current $i(t)$ flowing through the capacitor and the voltage $v(t)$ across it for $t \geq 0$.

Solution: Once the switch is opened, at time $t = 0$ the current $i(t)$ jumps from zero to I_0, and this behavior can be expressed in terms of the unit-step function as

$$i(t) = I_0 u(t).$$

We may thus replace the circuit of Fig. 1.48(a) with the equivalent circuit shown in Fig. 1.48(b). By definition, the capacitor voltage $v(t)$ is related to the current $i(t)$ by the formula

$$v(t) = \frac{1}{C} \int_{-\infty}^{t} i(\tau)\, d\tau.$$

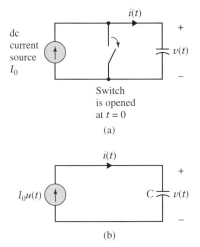

FIGURE 1.48   (a) Parallel circuit consisting of a current source, switch, and capacitor; the capacitor is initially assumed to be uncharged, and the switch is opened at time $t = 0$. (b) Equivalent circuit replacing the action of opening the switch with the step function $u(t)$.

Hence, using $i(t) = I_0 u(t)$ in this integral, we may write

$$v(t) = \frac{1}{C} \int_{-\infty}^{t} I_0 u(\tau)\, d\tau$$

$$= \begin{cases} 0 & \text{for } t < 0 \\ \dfrac{I_0}{C} t & \text{for } t \geq 0 \end{cases}$$

$$= \frac{I_0}{C} t u(t)$$

$$= \frac{I_0}{C} r(t).$$

That is, the voltage across the capacitor is a ramp function with slope I_0/C.   ∎

1.7 *Systems Viewed as Interconnections of Operations*

In mathematical terms, a system may be viewed as an *interconnection of operations* that transforms an input signal into an output signal with properties different from those of the input signal. The signals may be of the continuous-time or discrete-time variety or a mixture of both. Let the overall *operator H* denote the action of a system. Then the application of a continuous-time signal $x(t)$ to the input of the system yields the output signal

$$y(t) = H\{x(t)\}. \tag{1.78}$$

Figure 1.49(a) shows a block diagram representation of Eq. (1.78). Correspondingly, for the discrete-time case, we may write

$$y[n] = H\{x[n]\}, \tag{1.79}$$

where the discrete-time signals $x[n]$ and $y[n]$ denote the input and output signals, respectively, as depicted in Fig. 1.49(b).

S & S
Solutions

EXAMPLE 1.12 MOVING-AVERAGE SYSTEM   Consider a discrete-time system whose output signal $y[n]$ is the average of the three most recent values of the input signal $x[n]$; that is,

$$y[n] = \frac{1}{3}(x[n] + x[n-1] + x[n-2]).$$

Such a system is referred to as a *moving-average system*, for two reasons. First, $y[n]$ is the average of the sample values $x[n]$, $x[n-1]$, and $x[n-2]$. Second, the value of $y[n]$ changes as n moves along the discrete-time axis. Formulate the operator H for this system; hence, develop a block diagram representation for it.

(a)   (b)

FIGURE 1.49   Block diagram representation of operator H for (a) continuous time and (b) discrete time.

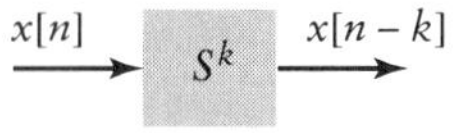

$$x[n] \longrightarrow \boxed{S^k} \xrightarrow{x[n-k]}$$

Figure 1.50   Discrete-time-shift operator S^k, operating on the discrete-time signal $x[n]$ to produce $x[n-k]$.

Solution: Let the operator S^k denote a system that shifts the input $x[n]$ by k time units to produce an output equal to $x[n-k]$, as depicted in Fig. 1.50. Accordingly, we may define the overall operator H for the moving-average system as

$$H = \frac{1}{3}(1 + S + S^2)$$

Two different implementations of H (i.e., the moving-average system) that suggest themselves are presented in Fig. 1.51. The implementation shown in part (a) of the figure uses the *cascade* connection of two identical unity time shifters, namely, $S^1 = S$. By contrast, the implementation shown in part (b) of the figure uses two different time shifters, S and S^2, connected in *parallel*. In both cases, the moving-average system is made up of an interconnection of three functional blocks, namely, two time shifters and an adder, connected by a scalar multiplication.   ■

▶ **Problem 1.25**   Express the operator that describes the input–output relation

$$y[n] = \frac{1}{3}(x[n+1] + x[n] + x[n-1])$$

in terms of the time-shift operator S.

Answer: $H = \frac{1}{3}(S^{-1} + 1 + S^1)$ ◀

In the interconnected systems shown in Figs. 1.51(a) and (b), the signal flows through each one of them in the forward direction only. Another way of combining systems is through the use of *feedback* connections. Figure 1.4 shows an example of a *feedback system* that is characterized by two paths. The forward path involves the cascade connection of the controller and plant. The feedback path is made possible through the use of a sensor connected to the output of the system at one end and the input at the other end. The use of feedback has many desirable benefits, but gives rise to problems of its own that require special attention; the subject of feedback is discussed in Chapter 9.

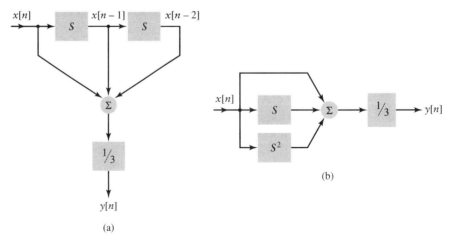

(a)

(b)

Figure 1.51   Two different (but equivalent) implementations of the moving-average system: (a) cascade form of implementation and (b) parallel form of implementation.

1.8 *Properties of Systems*

The properties of a system describe the characteristics of the operator H representing the system. In what follows, we study some of the most basic properties of systems.

■ 1.8.1 STABILITY

A system is said to be *bounded-input, bounded-output (BIBO) stable* if and only if every bounded input results in a bounded output. The output of such a system does not diverge if the input does not diverge.

To put the condition for BIBO stability on a formal basis, consider a continuous-time system whose input–output relation is as described in Eq. (1.78). The operator H is BIBO stable if the output signal $y(t)$ satisfies the condition

$$\boxed{|y(t)| \leq M_y < \infty \quad \text{for all } t} \tag{1.80}$$

whenever the input signals $x(t)$ satisfy the condition

$$\boxed{|x(t)| \leq M_x < \infty \quad \text{for all } t.} \tag{1.81}$$

Both M_x and M_y represent some finite positive numbers. We may describe the condition for the BIBO stability of a discrete-time system in a similar manner.

From an engineering perspective, it is important that a system of interest remain stable under all possible operating conditions. Only then is the system guaranteed to produce a bounded output for a bounded input. Unstable systems are usually to be avoided, unless some mechanism can be found to stabilize them.

One famous example of an unstable system is the first Tacoma Narrows suspension bridge, which collapsed on November 7, 1940, at approximately 11:00 A.M., due to wind-induced vibrations. Situated on the Tacoma Narrows in Puget Sound, near the city of Tacoma, Washington, the bridge had been open for traffic only a few months before it collapsed. (See Fig. 1.52 for photographs taken just prior to failure of the bridge and soon thereafter.)

EXAMPLE 1.13 MOVING-AVERAGE SYSTEM (CONTINUED)   Show that the moving-average system described in Example 1.12 is BIBO stable.

Solution: Assume that

$$|x[n]| < M_x < \infty \quad \text{for all } n.$$

Using the given input–output relation

$$y[n] = \frac{1}{3}(x[n] + x[n-1] + x[n-2]),$$

we may write

$$|y[n]| = \frac{1}{3}|x[n] + x[n-1] + x[n-2]|$$

$$\leq \frac{1}{3}(|x[n]| + |x[n-1]| + |x[n-2]|)$$

$$\leq \frac{1}{3}(M_x + M_x + M_x)$$

$$= M_x.$$

Hence, the absolute value of the output signal $y[n]$ is always less than the maximum absolute value of the input signal $x[n]$ for all n, which shows that the moving-average system is stable.   ∎

(a)

(b)

FIGURE 1.52 Dramatic photographs showing the collapse of the Tacoma Narrows suspension bridge on November 7, 1940. (a) Photograph showing the twisting motion of the bridge's center span just before failure. (b) A few minutes after the first piece of concrete fell, this second photograph shows a 600-ft section of the bridge breaking out of the suspension span and turning upside down as it crashed in Puget Sound, Washington. Note the car in the top right-hand corner of the photograph. (Courtesy of the Smithsonian Institution.)

EXAMPLE 1.14 UNSTABLE SYSTEM Consider a discrete-time system whose input–output relation is defined by

$$y[n] = r^n x[n],$$

where $r > 1$. Show that this system is unstable.

Solution: Assume that the input signal $x[n]$ satisfies the condition

$$|x[n]| \leq M_x < \infty \quad \text{for all } n.$$

We then find that

$$
\begin{aligned}
|y[n]| &= |r^n x[n]| \\
&= |r^n| \cdot |x[n]|.
\end{aligned}
$$

With $r > 1$, the multiplying factor r^n diverges for increasing n. Accordingly, the condition that the input signal is bounded is not sufficient to guarantee a bounded output signal, so the system is unstable. To prove stability, we need to establish that all bounded inputs produce a bounded output. ∎

▶ **Problem 1.26** The input-output relation of a discrete-time system is described by

$$y[n] = \sum_{k=0}^{\infty} \rho^k x[n - k].$$

Show that the system is BIBO unstable if $|\rho| \geq 1$. ◀

■ 1.8.2 MEMORY

A system is said to possess *memory* if its output signal depends on past or future values of the input signal. The temporal extent of past or future values on which the output depends defines how far the memory of the system extends into the past or future. In contrast, a system is said to be *memoryless* if its output signal depends only on the present value of the input signal.

For example, a resistor is memoryless, since the current $i(t)$ flowing through it in response to the applied voltage $v(t)$ is defined by

$$i(t) = \frac{1}{R} v(t),$$

where R is the resistance of the resistor. On the other hand, an inductor has memory, since the current $i(t)$ flowing through it is related to the applied voltage $v(t)$ by

$$i(t) = \frac{1}{L} \int_{-\infty}^{t} v(\tau) \, d\tau,$$

where L is the inductance of the inductor. That is, unlike the current through a resistor, that through an inductor at time t depends on all past values of the voltage $v(t)$; the memory of an inductor extends into the infinite past.

The moving-average system of Example 1.12 described by the input–output relation

$$y[n] = \frac{1}{3}(x[n] + x[n - 1] + x[n - 2])$$

has memory, since the value of the output signal $y[n]$ at time n depends on the present and on two past values of the input signal $x[n]$. In contrast, a system described by the input–output relation

$$y[n] = x^2[n]$$

is memoryless, since the value of the output signal $y[n]$ at time n depends only on the present value of the input signal $x[n]$.

▶ **Problem 1.27**   How far does the memory of the moving-average system described by the input–output relation

$$y[n] = \frac{1}{3}(x[n] + x[n-2] + x[n-4])$$

extend into the past?

Answer: Four time units.   ◀

▶ **Problem 1.28**   The input–output relation of a semiconductor diode is represented by

$$i(t) = a_0 + a_1 v(t) + a_2 v^2(t) + a_3 v^3(t) + \cdots,$$

where $v(t)$ is the applied voltage, $i(t)$ is the current flowing through the diode, and a_0, $a_1, a_3, \ldots$ are constants. Does this diode have memory?

Answer: No.   ◀

▶ **Problem 1.29**   The input–output relation of a capacitor is described by

$$v(t) = \frac{1}{C} \int_{-\infty}^{t} i(\tau)\, d\tau.$$

What is the extent of the capacitor's memory?

Answer: The capacitor's memory extends from time t back to the infinite past.   ◀

■ 1.8.3 CAUSALITY

A system is said to be *causal* if the present value of the output signal depends only on the present or past values of the input signal. In contrast, the output signal of a *noncausal* system depends on one or more future values of the input signal.

For example, the moving-average system described by

$$y[n] = \frac{1}{3}(x[n] + x[n-1] + x[n-2])$$

is causal. By contrast, the moving-average system described by

$$y[n] = \frac{1}{3}(x[n+1] + x[n] + x[n-1])$$

is noncausal, since the output signal $y[n]$ depends on a future value of the input signal, namely, $x[n+1]$.

The important point to note here is that causality is required for a system to be capable of operating in *real time*. Thus, in the first moving-average system just described, the output $y[n]$ is determined once the present sample $x[n]$ is received, thereby permitting real-time operation of the system for all n. By contrast, the second moving-average system has to wait for the future sample $x[n+1]$ to arrive before it can produce the output $y[n]$; thus, this second system can only operate in a non-real-time fashion.

▶ **Problem 1.30**   Consider the RC circuit shown in Fig. 1.53 with input voltage $v_1(t)$ and output voltage $v_2(t)$. Is this system causal or noncausal?

Answer: Causal.   ◀

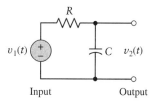

FIGURE 1.53   Series RC circuit driven from an ideal voltage source $v_1(t)$, producing output voltage $v_2(t)$.

▶ **Problem 1.31**   Suppose k in the operator of Fig. 1.50 is replaced by $-k$. Is the resulting system causal or noncausal for positive k?

Answer: Noncausal.   ◀

▪ 1.8.4 INVERTIBILITY

A system is said to be *invertible* if the input of the system can be recovered from the output. We may view the set of operations needed to recover the input as a second system connected in cascade with the given system, such that the output signal of the second system is equal to the input signal applied to the given system. To put the notion of invertibility on a formal basis, let the operator H represent a continuous-time system, with the input signal $x(t)$ producing the output signal $y(t)$. Let the output signal $y(t)$ be applied to a second continuous-time system represented by the operator H^{inv}, as illustrated in Fig. 1.54. Then the output signal of the second system is defined by

$$H^{inv}\{y(t)\} = H^{inv}\{H\{x(t)\}\}$$
$$= H^{inv}H\{x(t)\},$$

where we have made use of the fact that two operators H and H^{inv} connected in cascade are equivalent to a single operator $H^{inv}H$. For this output signal to equal the original input signal $x(t)$, we require that

$$H^{inv}H = I, \tag{1.82}$$

where I denotes the *identity operator*. The output of a system described by the identity operator is exactly equal to the input. Equation (1.82) expresses the condition the new operator H^{inv} must satisfy in relation to the given operator H in order for the original input signal $x(t)$ to be recovered from $y(t)$. The operator H^{inv} is called the *inverse operator*, and the associated system is called the *inverse system*. In general, the problem of finding the inverse of a given system is a difficult one. In any event, a system is not invertible unless distinct inputs applied to the system produce distinct outputs. That is, there must be a one-to-one mapping between input and output signals for a system to be invertible. Identical conditions must hold for a discrete-time system to be invertible.

The property of invertibility is of particular importance in the design of communication systems. As remarked in Section 1.3.1, when a transmitted signal propagates through a

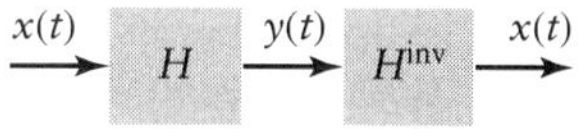

$$x(t) \xrightarrow{} \boxed{H} \xrightarrow{y(t)} \boxed{H^{inv}} \xrightarrow{x(t)}$$

FIGURE 1.54   The notion of system invertibility. The second operator H^{inv} is the inverse of the first operator H. Hence, the input $x(t)$ is passed through the cascade connection of H and H^{inv} completely unchanged.

communication channel, it becomes distorted due to the physical characteristics of the channel. A widely used method of compensating for this distortion is to include in the receiver a network called an *equalizer*, which is connected in cascade with the channel in a manner similar to that described in Fig. 1.54. By designing the equalizer to be the inverse of the channel, the transmitted signal is restored to its original form, assuming ideal (i.e., noiseless) conditions.

EXAMPLE 1.15 INVERSE OF SYSTEM   Consider the time-shift system described by the input–output relation

$$y(t) = x(t - t_0) = S^{t_0}\{x(t)\},$$

where the operator S^{t_0} represents a time shift of t_0 seconds. Find the inverse of this system.

Solution: For this example, the inverse of a time shift of t_0 seconds is a time shift of $-t_0$ seconds. We may represent the time shift of $-t_0$ by the operator S^{-t_0}, which is the inverse of S^{t_0}. Thus, applying S^{-t_0} to the output signal of the given time-shift system, we get

$$\begin{aligned} S^{-t_0}\{y(t)\} &= S^{-t_0}\{S^{t_0}\{x(t)\}\} \\ &= S^{-t_0}S^{t_0}\{x(t)\}. \end{aligned}$$

For this output signal to equal the original input signal $x(t)$, we require that

$$S^{-t_0}S^{t_0} = I,$$

which is in perfect accord with the condition for invertibility described in Eq. (1.82).   ■

▶ **Problem 1.32**   An inductor is described by the input–output relation

$$y(t) = \frac{1}{L} \int_{-\infty}^{t} x(\tau)\, d\tau.$$

Find the operation representing the inverse system.

Answer: $x(t) = L\dfrac{d}{dt}y(t)$ ◀

EXAMPLE 1.16 NON-INVERTIBLE SYSTEM   Show that a square-law system described by the input–output relation

$$y(t) = x^2(t)$$

is not invertible.

Solution: Note that the square-law system violates a necessary condition for invertibility, namely, that distinct inputs must produce distinct outputs. Specifically, the distinct inputs $x(t)$ and $-x(t)$ produce the same output $y(t)$. Accordingly, the square-law system is not invertible.   ■

■ 1.8.5 TIME INVARIANCE

A system is said to be *time invariant* if a time delay or time advance of the input signal leads to an identical time shift in the output signal. This implies that a time-invariant system responds identically no matter when the input signal is applied. Put another way, the char-

(a)   (b)

FIGURE 1.55 The notion of time invariance. (a) Time-shift operator S^{t_0} preceding operator H. (b) Time-shift operator S^{t_0} following operator H. These two situations equivalent, provided that H is time invariant.

acteristics of a time-invariant system do not change with time. Otherwise, the system is said to be *time variant*.

Consider a continuous-time system whose input–output relation is described by Eq. (1.78), reproduced here, in the form

$$y_1(t) = H\{x_1(t)\}.$$

Suppose the input signal $x_1(t)$ is shifted in time by t_0 seconds, resulting in the new input $x_1(t-t_0)$. Consistently with the notation introduced in Fig. 1.50, this operation may be described by writing

$$x_2(t) = x_1(t - t_0) = S^{t_0}\{x_1(t)\},$$

where the operator S^{t_0} represents a time shift equal to t_0 seconds for the situation at hand. Let $y_2(t)$ denote the output signal of the system H produced in response to the time-shifted input $x_1(t - t_0)$. We may then write

$$\begin{aligned} y_2(t) &= H\{x_1(t - t_0)\} \\ &= H\{S^{t_0}\{x_1(t)\}\} \\ &= HS^{t_0}\{x_1(t)\}, \end{aligned} \tag{1.83}$$

which is represented by the block diagram shown in Fig. 1.55(a). Now suppose $y_1(t - t_0)$ represents the output of the system H shifted in time by t_0 seconds, as shown by

$$\begin{aligned} y_1(t - t_0) &= S^{t_0}\{y_1(t)\} \\ &= S^{t_0}\{H\{x_1(t)\}\} \\ &= S^{t_0}H\{x_1(t)\} \end{aligned} \tag{1.84}$$

which is represented by the block diagram shown in Fig. 1.55(b). The system is time invariant if the outputs $y_2(t)$ and $y_1(t - t_0)$ defined in Eqs. (1.83) and (1.84), respectively, are equal for any identical input signal $x_1(t)$. Hence, we require that

$$\boxed{HS^{t_0} = S^{t_0}H.} \tag{1.85}$$

That is, for a system described by the operator H to be time invariant, the system operator H and the time-shift operator S^{t_0} must *commute* with each other for all t_0. A similar relation must hold for a discrete-time system to be time invariant.

EXAMPLE 1.17 INDUCTOR   Use the voltage $v(t)$ across an ordinary inductor to represent the input signal $x_1(t)$ and the current $i(t)$ flowing through the inductor to represent the output signal $y_1(t)$. Thus, the inductor is described by the input–output relation

$$y_1(t) = \frac{1}{L} \int_{-\infty}^{t} x_1(\tau)\, d\tau,$$

where L is the inductance. Show that the inductor so described is time invariant.

Solution: Let the input $x_1(t)$ be shifted by t_0 seconds, yielding $x_1(t - t_0)$. The response $y_2(t)$ of the inductor to $x_1(t - t_0)$ is

$$y_2(t) = \frac{1}{L} \int_{-\infty}^{t} x_1(\tau - t_0) \, d\tau.$$

Next, let $y_1(t - t_0)$ denote the original output of the inductor, shifted by t_0 seconds; that is,

$$y_1(t - t_0) = \frac{1}{L} \int_{-\infty}^{t - t_0} x_1(\tau) \, d\tau.$$

Although at first examination $y_2(t)$ and $y_1(t - t_0)$ look different, they are in fact equal, as shown by a simple change in the variable of integration. Let

$$\tau' = \tau - t_0.$$

Then for a constant t_0, we have $d\tau' = d\tau$. Hence, changing the limits of integration, the expression for $y_2(t)$ may be rewritten as

$$y_2(t) = \frac{1}{L} \int_{-\infty}^{t - t_0} x_1(\tau') \, d\tau',$$

which, in mathematical terms, is identical to $y_1(t - t_0)$. It follows that an ordinary inductor is time invariant. ▪

EXAMPLE 1.18 THERMISTOR A thermistor has a resistance that varies with time due to temperature changes. Let $R(t)$ denote the resistance of the thermistor, expressed as a function of time. Associating the input signal $x_1(t)$ with the voltage applied across the thermistor and the output signal $y_1(t)$ with the current flowing through the thermistor, we may express the input–output relation of the device as

$$y_1(t) = \frac{x_1(t)}{R(t)}.$$

Show that the thermistor so described is time variant.

Solution: Let $y_2(t)$ denote the response of the thermistor produced by a time-shifted version $x_1(t - t_0)$ of the original input signal. We may then write

$$y_2(t) = \frac{x_1(t - t_0)}{R(t)}.$$

Next, let $y_1(t - t_0)$ denote the original output of the thermistor due to the input $x_1(t)$, shifted in time by t_0; that is,

$$y_1(t - t_0) = \frac{x_1(t - t_0)}{R(t - t_0)}.$$

We now see that since, in general, $R(t) \neq R(t - t_0)$ for $t_0 \neq 0$, it follows that

$$y_1(t - t_0) \neq y_2(t) \quad \text{for } t_0 \neq 0.$$

Hence, a thermistor is time variant, which is intuitively satisfying. ▪

▶ **Problem 1.33**   Is a discrete-time system described by the input–output relation

$$y(n) = r^n x(n)$$

time invariant?

Answer: No   ◀

■ 1.8.6 LINEARITY

A system is said to be *linear* in terms of the system input (excitation) $x(t)$ and the system output (response) $y(t)$ if it satisfies the following two properties of superposition and homogeneity:

1. *Superposition.* Consider a system that is initially at rest. Let the system be subjected to an input $x(t) = x_1(t)$, producing an output $y(t) = y_1(t)$. Suppose next that the same system is subjected to a different input $x(t) = x_2(t)$, producing a corresponding output $y(t) = y_2(t)$. Then for the system to be linear, it is necessary that the composite input $x(t) = x_1(t) + x_2(t)$ produce the corresponding output $y(t) = y_1(t) + y_2(t)$. What we have described here is a statement of the *principle of superposition* in its simplest form.

2. *Homogeneity.* Consider again a system that is initially at rest, and suppose an input $x(t)$ results in an output $y(t)$. Then the system is said to exhibit the property of homogeneity if, whenever the input $x(t)$ is scaled by a constant factor a, the output $y(t)$ is scaled by exactly the same constant factor a.

When a system violates either the principle of superposition or the property of homogeneity, the system is said to be *nonlinear*.

Let the operator H represent a continuous-time system. Let the signal applied to the system input be defined by the weighted sum

$$x(t) = \sum_{i=1}^{N} a_i x_i(t), \tag{1.86}$$

where $x_1(t), x_2(t), \ldots, x_N(t)$ denote a set of input signals and $a_1, a_2, \ldots, a_N$ denote the corresponding weighting factors. The resulting output signal is written as

$$\begin{aligned} y(t) &= H\{x(t)\} \\ &= H\left\{ \sum_{i=1}^{N} a_i x_i(t) \right\}. \end{aligned} \tag{1.87}$$

If the system is linear, then, in accordance with the principle of superposition and the property of homogeneity, we may express the output signal of the system as

$$y(t) = \sum_{i=1}^{N} a_i y_i(t), \tag{1.88}$$

where $y_i(t)$ is the output of the system in response to the input $x_i(t)$ acting alone—that is, where

$$y_i(t) = H\{x_i(t)\}, \quad i = 1, 2, \ldots, N. \tag{1.89}$$

The weighted sum of Eq. (1.88), describing the output signal $y(t)$, is of the same

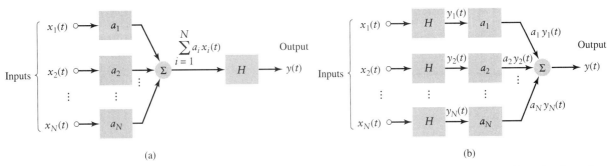

(a)   (b)

FIGURE 1.56   The linearity property of a system. (a) The combined operation of amplitude scaling and summation precedes the operator H for multiple inputs. (b) The operator H precedes amplitude scaling for each input; the resulting outputs are summed to produce the overall output $y(t)$. If these two configurations produce the same output $y(t)$, the operator H is linear.

mathematical form as that of Eq. (1.86), describing the input signal $x(t)$. For Eqs. (1.87) and (1.88) to yield exactly the same output signal $y(t)$, we require the following:

$$
\begin{aligned}
y(t) &= H\left\{ \sum_{i=1}^{N} a_i x_i(t) \right\} \\
&= \sum_{i=1}^{N} a_i H\{ x_i(t) \} \\
&= \sum_{i=1}^{N} a_i y_i(t).
\end{aligned}
\tag{1.90}
$$

In words, the system operation described by H must *commute* with the summation and amplitude scaling, as illustrated in Fig. 1.56. The commutation can only be justified if the operator H is *linear*.

For a linear discrete-time system, an equation similar to Eq. (1.90) holds, as illustrated in Example 1.19.

EXAMPLE 1.19 LINEAR DISCRETE-TIME SYSTEM   Consider a discrete-time system described by the input–output relation

$$ y[n] = nx[n]. $$

Show that this system is linear.

Solution: Let the input signal $x[n]$ be expressed as the weighted sum

$$ x[n] = \sum_{i=1}^{N} a_i x_i[n]. $$

We may then express the resulting output signal of the system as

$$
\begin{aligned}
y[n] &= n \sum_{i=1}^{N} a_i x_i[n] \\
&= \sum_{i=1}^{N} a_i n x_i[n] \\
&= \sum_{i=1}^{N} a_i y_i[n],
\end{aligned}
$$

where

$$y_i[n] = nx_i[n]$$

is the output due to each input acting independently. We thus see that the given system satisfies both superposition and homogeneity and is therefore linear.   ∎

EXAMPLE 1.20 NONLINEAR CONTINUOUS-TIME SYSTEM   Consider next the continuous-time system described by the input–output relation

$$y(t) = x(t)x(t - 1).$$

Show that this system is nonlinear.

Solution: Let the input signal $x(t)$ be expressed as the weighted sum

$$x(t) = \sum_{i=1}^{N} a_i x_i(t).$$

Correspondingly, the output signal of the system is given by the double summation

$$y(t) = \sum_{i=1}^{N} a_i x_i(t) \sum_{j=1}^{N} a_j x_j(t - 1)$$

$$= \sum_{i=1}^{N} \sum_{j=1}^{N} a_i a_j x_i(t) x_j(t - 1).$$

The form of this equation is radically different from that describing the input signal $x(t)$. That is, here we cannot write $y(t) = \sum_{i=1}^{N} a_i y_i(t)$. Thus, the system violates the principle of superposition and is therefore nonlinear.   ∎

▶ **Problem 1.34**   Show that the moving-average system described by

$$y[n] = \frac{1}{3}(x[n] + x[n - 1] + x[n - 2])$$

is a linear system.   ◀

▶ **Problem 1.35**   Is it possible for a linear system to be noncausal?

Answer: Yes.   ◀

▶ **Problem 1.36**   The hard limiter is a memoryless device whose output y is related to the input x by

$$y = \begin{cases} 1, & x \geq 0 \\ 0, & x < 0 \end{cases}.$$

Is the hard limiter linear?

Answer: No   ◀

S & S
Solutions

EXAMPLE 1.21 IMPULSE RESPONSE OF *RC* CIRCUIT   In this example, we use linearity, time invariance, and the representation of an impulse as the limiting form of a pulse to obtain the impulse response of the series circuit shown in Fig. 1.57. This circuit was discussed in Example 1.9, in light of which the step response of the circuit [i.e., the voltage $y(t)$ across the capacitor] is written as

$$y(t) = (1 - e^{-(t/RC)})u(t), \qquad x(t) = u(t). \tag{1.91}$$

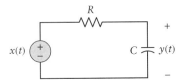

FIGURE 1.57 *RC* circuit for Example 1.20, in which we are given the capacitor voltage $y(t)$ in response to the step input $x(t) = u(t)$ and the requirement is to find $y(t)$ in response to the unit-impulse input $x(t) = \delta(t)$.

Equation (1.91) is a restatement of Eq. (1.57) with $V_0 = 1$ and $y(t)$ used in place of $v(t)$. Given this step response, the goal is to find the impulse response of the circuit, which relates the new input voltage $x(t) = \delta(t)$ to the corresponding voltage across the capacitor, $y(t)$.

Solution: To find the response $y(t)$ produced by the input $x(t) = \delta(t)$, we use four concepts: the properties of linearity and time invariance discussed in Section 1.8, the graphical definition of an impulse depicted in Fig. 1.42, and the definition of the derivative of a continuous function of time.

Following the discussion presented in Section 1.6.6, we proceed by expressing the rectangular pulse input $x(t) = x_\Delta(t)$ depicted in Fig. 1.58 as the difference between two weighted and time-shifted step functions:

$$x_1(t) = \frac{1}{\Delta} u\left(t + \frac{\Delta}{2} \right)$$

and

$$x_2(t) = \frac{1}{\Delta} u\left(t - \frac{\Delta}{2} \right).$$

Let $y_1(t)$ and $y_2(t)$ be the responses of the *RC* circuit to the step functions $x_1(t)$ and $x_2(t)$, respectively. Then, applying the time-invariance property to Eq. (1.91), we have

$$y_1(t) = \frac{1}{\Delta} \left(1 - e^{-(t+\Delta/2)/(RC)} \right) u\left(t + \frac{\Delta}{2} \right), \qquad x(t) = x_1(t),$$

and

$$y_2(t) = \frac{1}{\Delta} \left(1 - e^{-(t-\Delta/2)/(RC)} \right) u\left(t - \frac{\Delta}{2} \right), \qquad x(t) = x_2(t).$$

Next, recognizing that

$$x_\Delta(t) = x_1(t) - x_2(t),$$

we invoke the property of linearity to express the corresponding response of the *RC* circuit as

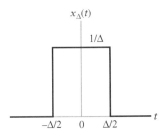

FIGURE 1.58   Rectangular pulse of unit area, which, in the limit, approaches a unit impulse as $\Delta \to 0$.

$$y_\Delta(t) = \frac{1}{\Delta}\left(1 - e^{-(t+\Delta/2)/(RC)}\right)u\left(t + \frac{\Delta}{2}\right) - \frac{1}{\Delta}\left(1 - e^{-(t-\Delta/2)/(RC)}\right)u\left(t - \frac{\Delta}{2}\right)$$

$$= \frac{1}{\Delta}\left(u\left(t + \frac{\Delta}{2}\right) - u\left(t - \frac{\Delta}{2}\right)\right) \tag{1.92}$$

$$- \frac{1}{\Delta}\left(e^{-(t+\Delta/2)/(RC)}u\left(t + \frac{\Delta}{2}\right) - e^{-(t-\Delta/2)/(RC)}u\left(t - \frac{\Delta}{2}\right)\right).$$

All that remains for us to do is to determine the limiting form of Eq. (1.92) as the duration Δ of the pulse approaches zero. Toward that end, we invoke the following two definitions:

1. Representation of an impulse as the limiting form of the pulse $x_\Delta(t)$:

$$\delta(t) = \lim_{\Delta \to 0} x_\Delta(t).$$

2. The derivative of a continuous function of time, say, $z(t)$:

$$\frac{d}{dt}z(t) = \lim_{\Delta \to 0}\left\{\frac{1}{\Delta}\left(z\left(t + \frac{\Delta}{2}\right) - z\left(t - \frac{\Delta}{2}\right)\right)\right\}.$$

Applying these two definitions to the last line of Eq. (1.92) with the duration Δ of the pulse approaching zero, we obtain the desired impulse response:

$$y(t) = \lim_{\Delta \to 0} y_\Delta(t)$$

$$= \delta(t) - \frac{d}{dt}\left(e^{-t/(RC)}u(t)\right)$$

$$= \delta(t) - e^{-t/(RC)}\frac{d}{dt}u(t) - u(t)\frac{d}{dt}\left(e^{-t/(RC)}\right)$$

$$= \delta(t) - e^{-t/(RC)}\delta(t) + \frac{1}{RC}e^{-t/(RC)}u(t), \qquad x(t) = \delta(t).$$

Note that in the second line we applied the rule for differentiating the product of two time functions $u(t)$ and $e^{-t/(RC)}$. Finally, since $\delta(t)$ is confined to the origin and $e^{-t/(RC)} = 1$ at $t = 0$, the terms $\delta(t)$ and $e^{-t/(RC)}\delta(t)$ cancel each other, and the expression for the impulse response of the RC circuit simplifies to

$$\boxed{y(t) = \frac{1}{RC}e^{-t/(RC)}u(t), \qquad x(t) = \delta(t).} \tag{1.93}$$

This is the required result.   ∎

S & S
Solutions

▶ **Problem 1.37   *LR* Circuit**   Figure 1.59 shows the circuit diagram of a series inductance–resistance (LR) circuit. Given the step response of the circuit,

$$y(t) = (1 - e^{-Rt/L})u(t), \qquad x(t) = u(t),$$

find the impulse response of the circuit—that is, the voltage across the resistor, $y(t)$, in response to the unit-impulse input voltage $x(t) = \delta(t)$.

Answer: $\dfrac{R}{L}e^{-Rt/L}u(t)$ ◀

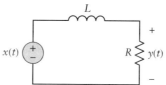

FIGURE 1.59   *LR* circuit for Problem 1.37.

1.9 Noise

The term *noise* is used customarily to designate unwanted signals that tend to disturb the operation of a system and over which we have incomplete control. The sources of noise that may arise in practice depend on the system of interest. For example, in a communication system, there are many potential sources of noise affecting the operation of the system. In particular, we have the following two broadly defined categories of noise:

▶ *External sources of noise*, examples of which include atmospheric noise, galactic noise, and human-made noise. The last of these may be an interfering signal picked up by the receiver of the communication system due to the spectral characteristics of the interference lying inside the operating frequency range for which the system is designed.

▶ *Internal sources of noise*, which include an important type of noise that arises from *spontaneous fluctuations* of the current or voltage signal in electrical circuits. For this reason, the latter type of noise is commonly referred to as *electrical noise*. The omnipresence and inevitability of electrical noise in all kinds of electronic systems impose a basic limitation on the transmission or detection of signals. Figure 1.60 shows a sample waveform of electrical noise generated by a thermionic diode noise generator, which consists of a vacuum-tube diode with a heated cathode and a plate (the anode) that collects the electrons emitted by the cathode.

Noiselike phenomena, irrespective of their origin, have a common property: Typically, it is not possible to specify their magnitudes as functions of time in precise terms. The inability to provide complete descriptions of noiselike phenomena may be attributed to one or more of the following reasons:

1. There is insufficient knowledge about the physical laws responsible for the generation of noise.

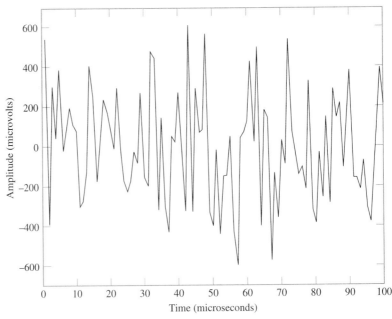

FIGURE 1.60   Sample waveform of electrical noise generated by a thermionic diode with a heated cathode. Note that the time-averaged value of the noise voltage displayed is approximately zero.

2. The mechanisms responsible for the generation of noise are so complicated that a complete description of the noise is impractical.

3. Insofar as system analysis is concerned, an *average* characterization of the noiselike phenomenon is adequate for the problem at hand.

■ **1.9.1 THERMAL NOISE**

A ubiquitous form of electrical noise is *thermal noise*, which arises from the random motion of electrons in a conductor. Let $v(t)$ denote the thermal noise voltage appearing across the terminals of a resistor. Then the noise so generated has the following two characteristics:

▶ *A time-averaged value*, defined by

$$\bar{v} = \lim_{T \to \infty} \frac{1}{2T} \int_{-T}^{T} v(t)\, dt, \tag{1.94}$$

where $2T$ is the total interval over which the noise voltage $v(t)$ is observed. In the limit, the time-averaged value $\bar{v}$ approaches zero as T approaches infinity. This result is justified on the grounds that the number of electrons in a resistor is typically very large and their random motions inside the resistor produce positive and negative values of the noise voltage $v(t)$ that average to zero in the course of time. (See, e.g., Fig. 1.60.)

▶ *A time-average-squared value*, defined by

$$\overline{v^2} = \lim_{T \to \infty} \frac{1}{2T} \int_{-T}^{T} v^2(t)\, dt. \tag{1.95}$$

In the limit, as T approaches infinity, we have

$$\overline{v^2} = 4kT_{\text{abs}}R\Delta f \quad \text{volts}^2, \tag{1.96}$$

where k is *Boltzmann's constant*, approximately equal to 1.38×10^{-23} joule per degree kelvin, T_{abs} is the *absolute temperature* in degrees kelvin, R is the resistance in ohms, and Δf is the width of the frequency band in hertz over which the noise voltage $v(t)$ is measured. We may thus model a noisy resistor by the *Thévenin equivalent circuit*, consisting of a noise voltage generator of time-average square value $\overline{v^2}$ in series with a noiseless resistor, as in Fig. 1.61(a). Alternatively, we may use the *Norton equivalent circuit*, consisting of a noise current generator in parallel with a noiseless conductor, as in Fig. 1.61(b). The time-average-squared value of the noise current generator is

$$\overline{i^2} = \lim_{T \to \infty} \frac{1}{2T} \int_{-T}^{T} \left(\frac{v(t)}{R} \right)^2 dt$$
$$= 4kT_{\text{abs}}G\Delta f \quad \text{amps}^2, \tag{1.97}$$

where $G = 1/R$ is the conductance in siemens.

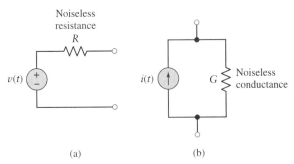

FIGURE 1.61   (a) Thévenin equivalent circuit of a noisy resistor. (b) Norton equivalent circuit of the same resistor.

Noise calculations involve the transfer of power, so we find that the use of the *maximum-power transfer theorem* is applicable to such calculations. This theorem states that the maximum possible power is transferred from a source of internal resistance R to a load of resistance R_l when $R_l = R$. Under this *matched* condition, the power produced by the source is divided equally between the internal resistance of the source and the load resistance, and the power delivered to the load is referred to as the *available power*. Applying the maximum-power transfer theorem to the Thévenin equivalent circuit of Fig. 1.61(a) or the Norton equivalent circuit of Fig. 1.61(b), we find that a noisy resistor produces an *available noise power* equal to $kT_{\text{abs}}\Delta f$ watts. There are therefore two operating factors that affect the available noise power:

1. The temperature at which the resistor is maintained.
2. The width of the frequency band over which the noise voltage across the resistor is measured.

Clearly, the available noise power increases with both of these parameters.

From the foregoing discussion, it is apparent that the time-averaged power is particularly important in the characterization of electrical noise—hence its wide use in practice.

■ 1.9.2 OTHER SOURCES OF ELECTRICAL NOISE

Another common source of electrical noise is *shot noise*, which arises in electronic devices such as diodes and transistors because of the discrete nature of current flow in those devices. For example, in a photodetector circuit, a pulse of current is generated every time an electron is emitted by the cathode due to incident light from a source of constant intensity. The electrons are naturally emitted at random times denoted by τ_k, where $-\infty < k < \infty$. Here, it is assumed that the random emissions of electrons have been going on for a long time; that is, the device is in a steady state. Thus, the total current flowing through the photodetector may be modeled as an infinite sum of current pulses as shown by

$$x(t) = \sum_{k=-\infty}^{\infty} h(t - \tau_k), \qquad (1.98)$$

where $h(t - \tau_k)$ is the current pulse generated at time τ_k. The random manner in which these pulses are generated causes the total current $x(t)$ to fluctuate randomly with time.

Finally, the type of electrical noise called $1/f$ *noise* is always present when an electric current is flowing. This noise is so named because the time-averaged power at a given frequency is inversely proportional to the frequency; $1/f$ noise is observed in all semiconductor devices that are used to amplify and detect signals of interest at low frequencies.

1.10 *Theme Examples*

In this section, we introduce six theme examples, which, as explained in the preface, run through several chapters of the book. The purpose of a theme example is twofold:

▶ To introduce a signal-processing operation or system application of *practical* importance and to explore how it can be implemented.

▶ To obtain different perspectives on the operation or application of interest, depending on the tools used to analyze it.

■ 1.10.1 DIFFERENTIATION AND INTEGRATION: *RC* CIRCUITS

The operations of differentiation and integration are basic to the study of linear time-invariant systems. The need for differentiation arises when the *sharpening* of a pulse is required. Let $x(t)$ and $y(t)$ denote the input and output signals of a differentiator, respectively. Ideally, the *differentiator* is defined by

$$y(t) - \frac{d}{dt}x(t). \tag{1.99}$$

Figure 1.62 shows a simple *RC* circuit for *approximating* this ideal operation. The input–output relation of this circuit is given by

$$v_2(t) + \frac{1}{RC}\int_{-\infty}^{t} v_2(\tau)\, d\tau = v_1(t).$$

Equivalently, we may write

$$\frac{d}{dt}v_2(t) + \frac{1}{RC}v_2(t) = \frac{d}{dt}v_1(t). \tag{1.100}$$

Provided that the time constant RC is small enough for the left-hand side of Eq. (1.100) to be dominated by the second term, $(1/(RC))v_2(t)$, over the time interval of interest [i.e., if RC is small relative to the rate of change of the input signal $v_1(t)$], we may approximate Eq. (1.100) as

$$\frac{1}{RC}v_2(t) \approx \frac{d}{dt}v_1(t),$$

or

$$v_2(t) \approx RC\frac{d}{dt}v_1(t) \quad \text{for } RC \text{ small.} \tag{1.101}$$

Comparing Eqs. (1.99) and (1.101), we see that the input $x(t) = RCv_1(t)$ and the output $y(t) = v_2(t)$.

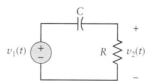

FIGURE 1.62   Simple *RC* circuit with small time constant, used to approximate a differentiator.

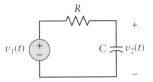

FIGURE 1.63   Simple RC circuit with large time constant used to approximate an integrator.

Next, consider the operation of integration whose purpose is the *smoothing* of an input signal. With $x(t)$ denoting the input and $y(t)$ denoting the output, an ideal *integrator* is defined by

$$y(t) = \int_{-\infty}^{t} x(\tau)\, d\tau. \tag{1.102}$$

For an approximate realization of the integrator, we may use the simple RC circuit rearranged as shown in Fig. 1.63. The input–output relation of this second RC circuit is given by

$$RC\frac{d}{dt}v_2(t) + v_2(t) = v_1(t),$$

or, equivalently,

$$RCv_2(t) + \int_{-\infty}^{t} v_2(\tau)\, d\tau = \int_{-\infty}^{t} v_1(\tau)\, d\tau. \tag{1.103}$$

Provided that, this time, the time constant RC is chosen large enough so that the integral on the left-hand side is dominated by the term $RCv_2(t)$ [i.e., if RC is large relative to the average value of the output signal $v_2(t)$ over the time interval of interest], we may approximate Eq. (1.103) as

$$RCv_2(t) \approx \int_{-\infty}^{t} v_1(\tau)\, d\tau,$$

or

$$v_2(t) \approx \frac{1}{RC}\int_{-\infty}^{t} v_1(\tau)\, d\tau \quad \text{for large } RC. \tag{1.104}$$

Comparing Eqs. (1.102) and (1.104), we see that the input $x(t) = [1/(RC)v_1(t)]$ and the output $y(t) = v_2(t)$.

From Eqs. (1.101) and (1.104), we also see that the more closely the RC circuits of Figs. 1.62 and 1.63 approximate an ideal differentiator and ideal integrator, respectively, the smaller will be their outputs. The RC circuits used to implement differentiators and integrators will be studied throughout Chapters 2 through 4. In later chapters of the book, we study more advanced methods of implementing differentiators and integrators.

■ 1.10.2 MEMS ACCELEROMETER

In Section 1.3.3, we described a microaccelerometer; see Fig. 1.6. We may model this device by a *second-order mass–damper–spring system* as shown in Fig. 1.64. As a result of external acceleration, the support frame is displaced relative to the proof mass. This displacement, in turn, produces a corresponding change in the internal stress in the suspension spring.

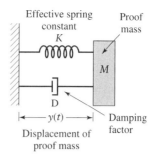

FIGURE 1.64   Mechanical lumped model of an accelerometer.

Let M denote the proof mass, K the effective spring constant, and D the damping factor affecting the dynamic movement of the proof mass. Let $x(t)$ denote the external acceleration due to motion and $y(t)$ denote the displacement of the proof mass. The net force on the proof mass must sum to zero. The inertial force of the proof mass is $Md^2y(t)/dt^2$, the damping force is $Ddy(t)/dt$, and the spring force is $Ky(t)$. Equating the sum of these three forces to the force due to external acceleration, $Mx(t)$, gives

$$Mx(t) = M\frac{d^2y(t)}{dt^2} + D\frac{dy(t)}{dt} + Ky(t),$$

or, equivalently,

$$\frac{d^2y(t)}{dt^2} + \frac{D}{M}\frac{dy(t)}{dt} + \frac{K}{M}y(t) = x(t). \tag{1.105}$$

We find it insightful to reformulate this second-order differential equation by defining two new quantities:

1. The *natural frequency* of the accelerometer:

$$\omega_n = \sqrt{\frac{K}{M}}. \tag{1.106}$$

   The mass M is measured in grams, and the spring constant K is measured in grams per second squared. Accordingly, the natural frequency ω_n is measured in radians per second.

2. The *quality factor* of the accelerometer:

$$Q = \frac{\sqrt{KM}}{D}. \tag{1.107}$$

   With the mass M measured in grams, the spring constant K in grams per second squared, and the damping factor D in grams per second, it follows that the quality factor Q is dimensionless.

Using the definitions of Eqs. (1.106) and (1.107) in Eq. (1.105), we may rewrite the second-order differential equation in terms of the two parameters ω_n and Q as

$$\frac{d^2y(t)}{dt^2} + \frac{\omega_n}{Q}\frac{dy(t)}{dt} + \omega_n^2 y(t) = x(t). \tag{1.108}$$

From Eq. (1.106), we see that the natural frequency ω_n can be increased by increasing the spring constant K and decreasing the proof mass M. From Eq. (1.107), we see that the quality factor Q can be increased by increasing the spring constant K and proof mass M and by reducing the damping factor D. In particular, a low value of Q—unity or less—permits the accelerometer to respond to a broad class of input signals.

The MEMS accelerometer is an example of a system described by a second-order differential equation. Electrical circuits containing two energy storage elements (capacitors or inductors) and other mechanical spring–mass–damper systems are also described by second-order differential equations of the same form as that of Eq. (1.108). Problem 1.79 discusses a series LRC circuit, which may be viewed as the electrical analog of the MEMS accelerometer.

■ 1.10.3 RADAR RANGE MEASUREMENT

In Section 1.3.4, we discussed the use of radar as an imaging system for remote sensing. For our third theme example, we consider another important application of radar: the measurement of how far away a target (e.g., an aircraft) is from the radar site.

Figure 1.65 shows a commonly used radar signal for measuring the range of a target. The signal consists of a periodic sequence of *radio frequency (RF) pulses*. Each pulse has a duration of T_0 in the order of microseconds and repeats regularly at a rate equal to $1/T$ pulses per second. More specifically, the RF pulse is made up of a sinusoidal signal whose frequency, denoted by f_c, is on the order of megahertz or gigahertz, depending on the application of interest. In effect, the sinusoidal signal acts as a *carrier*, facilitating the transmission of the radar signal and the reception of the echo from the radar target.

Suppose the radar target is at a range d, measured in meters from the radar. The *round-trip time* is equal to the time taken by a radar pulse to reach the target and for the echo from the target to come back to the radar. Thus, denoting the round-trip time by τ, we may write

$$\tau = \frac{2d}{c},\tag{1.109}$$

where c is the speed of light, measured in meters per second. Insofar as measuring the range is concerned, there are two issues of concern:

▶ *Range resolution.* The duration T_0 of the pulse places a lower limit on the shortest round-trip delay time that the radar can measure. Correspondingly, the smallest target range that the radar can measure reliably is $d_{\min} = cT_0/2$ meters. (Note that we have ignored the presence of electrical noise at the front end of the radar receiver.)

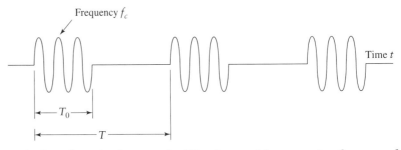

FIGURE 1.65   Periodic train of rectangular RF pulses used for measuring the range of a target.

▶ *Range ambiguity.* The interpulse period T places an upper limit on the largest target range that the radar can measure, since the echo from one pulse must return before the next pulse is transmitted, or else there will be ambiguity in the range estimate. Correspondingly, the largest target range that the radar can measure unambiguously is $d_{max} = cT/2$.

The radar signal of Fig. 1.65 provides an insightful setting for spectral analysis and its different facets, as demonstrated in subsequent chapters.

Similar range measurement methods are employed with sound (sonar), ultrasound (biomedical remote sensing), and infrared (automatic-focusing cameras) and at optical frequencies (laser range finders). In each case, the round-trip travel time of a pulse and its propagation velocity are used to determine the distance of an object.

■ 1.10.4 MOVING-AVERAGE SYSTEMS

An important application of discrete-time systems is the enhancement of some feature in a data set, such as identifying the underlying trend in data that are fluctuating. Moving-average systems, like the one introduced in Example 1.12, are often used for this purpose. Treating the data $x[n]$ as the input signal, we may express the output of an N-point moving-average system as

$$y[n] = \frac{1}{N} \sum_{k=0}^{N-1} x[n-k]. \tag{1.110}$$

The value N determines the degree to which the system smooths the input data. Consider, for example, the weekly closing stock price of Intel over a three-year period, as depicted in Fig. 1.66(a). The fluctuations in this data set highlight the volatile nature of Intel stock

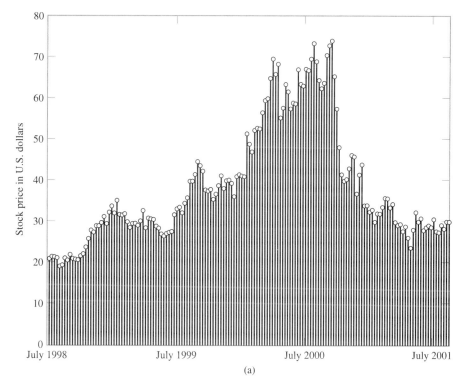

(a)

FIGURE 1.66   (a) Fluctuations in the closing stock price of Intel over a three-year period.

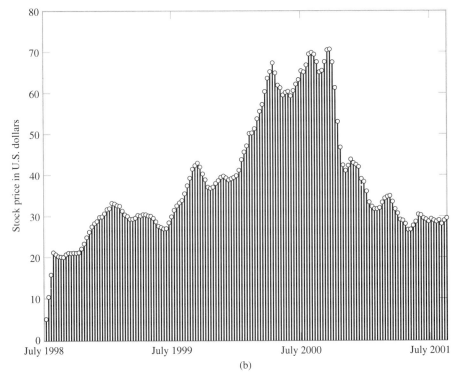

(b)

FIGURE 1.66   (b) Output of a four-point moving-average system.

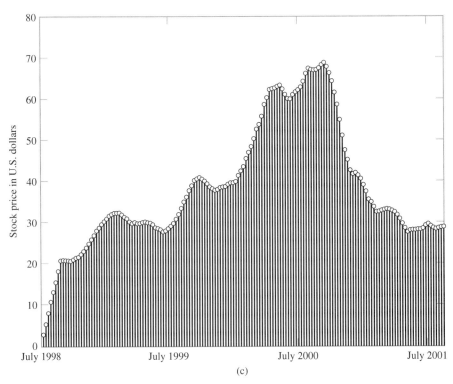

(c)

FIGURE 1.66   (c) Output of an eight-point moving-average system.

and the stock market in general. Figures 1.66(b) and (c) illustrate the effect of passing the data through moving-average systems with $N = 4$ and $N = 8$, respectively. Note that the moving-average systems significantly reduce the short-term fluctuations in the data, and the system with the larger value of N produces a smoother output. The challenge with smoothing applications of moving-average systems is how to choose the window length N so as to identify the underlying trend of the input data in the most informative manner.

In the most general form of a moving-average system application, unequal weighting is applied to past values of the input:

$$y[n] = \sum_{k=0}^{N-1} a_k x[n - k]. \tag{1.111}$$

In such a system, the weights a_k are chosen to extract a particular aspect of the data, such as fluctuations of a certain frequency, while eliminating other aspects, such as the time-averaged value. Specific methods for choosing the weights to accomplish such effects are described in Chapter 8; however, various aspects of moving-average systems will be explored throughout the chapters that follow.

■ 1.10.5 MULTIPATH COMMUNICATION CHANNELS

In Sections 1.3.1 and 1.9, we mentioned channel noise as a source that degrades the performance of a communication system. Another major source of degradation is the dispersive nature of the communication channel itself—that is, the fact that the channel has memory. In wireless communication systems, dispersive characteristics result from multipath propagation—the presence of more than one propagation path between the transmitter and receiver, as illustrated in Fig. 1.67. Multipath propagation results from the scattering of the transmitted signal from multiple objects. In the context of a digital communication system, multipath propagation manifests itself in the form of *intersymbol interference* (ISI), a term that refers to the residual effects at the receiver of symbols transmitted before and after the symbol of interest. A *symbol* is the waveform that represents a given set of bits in a binary representation

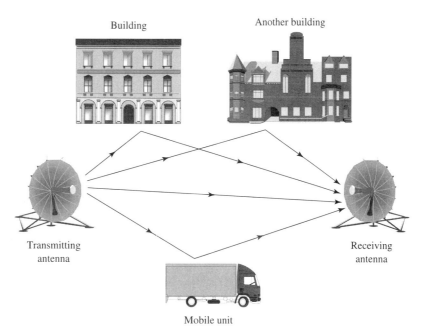

FIGURE 1.67 Example of multiple propagation paths in a wireless communication environment.

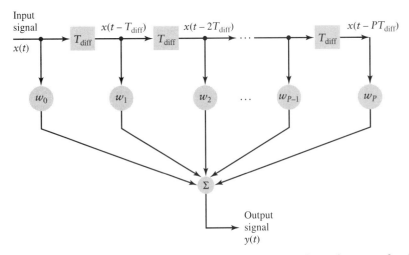

FIGURE 1.68   Tapped-delay-line model of a linear communication channel, assumed to be time-invariant.

of the message signal. To mitigate the ISI problem, the receiver incorporates an equalizer whose function is to compensate for dispersion in the channel.

It is desirable to have a *model* for the multipath propagation channel in order to understand and compensate for dispersive effects. The transmitted signal typically involves some form of modulation, the primary purpose of which is to shift the frequency band of the signal so that it coincides with the usable frequency band of the channel. For the purpose of modeling, it is helpful to work in terms of a *baseband model* that describes the effect of the channel on the original message signal, rather than the effect of the channel on the modulated signal. Depending on the form of modulation used, the baseband model can be real or complex valued.

One common baseband model is the tapped-delay line depicted in Fig. 1.68. The output of the model, which represents the received signal, is expressed in terms of the input as

$$y(t) = \sum_{i=0}^{P} w_i x(t - iT_{\text{diff}}), \qquad (1.112)$$

where T_{diff} represents the smallest detectable time difference between different paths. The value for T_{diff} depends on the characteristics of the transmitted symbols. [Equation (1.112) ignores the effect of noise at the channel output.] The quantity PT_{diff} represents the longest time delay of any significant path relative to the first arrival of the signal. The model coefficients w_i are used to approximate the gain of each path. For example, if $P = 1$, then

$$y(t) = w_0 x(t) + w_1 x(t - T_{\text{diff}}),$$

which could describe a propagation channel consisting of a direct path $w_0 x(t)$ and a single reflected path $w_1 x(t - T_{\text{diff}})$.

The signal processing in digital communication receivers is often performed by using discrete-time systems. We may obtain a discrete-time model for the multipath communication channel by sampling the baseband model of Eq. (1.112) at intervals of T_{diff} to obtain

$$y[n] = \sum_{k=0}^{P} w_k x[n - k]. \qquad (1.113)$$

Note that this model is an example of a linearly weighted moving-average system. A special case of the discrete-time multipath channel model that will be studied repeatedly in the chapters that follow is a normalized version of the case with $P = 1$, expressed as

$$y[n] = x[n] + ax[n - 1]. \qquad (1.114)$$

■ 1.10.6 RECURSIVE DISCRETE-TIME COMPUTATION

A particular form of computation known as recursive discrete-time computation is pervasive in its practical applications. In this form of computation, expressed in the most general terms, the current value of the output signal resulting from the computation depends on two sets of quantities: (1) the current and past values of the input signal and (2) the past values of the output signal itself. The term *recursive* signifies the dependence of the output signal on its own past values.

We illustrate this computation by introducing the simple and highly meaningful example of a *first-order recursive discrete-time filter*. Let $x[n]$ denote the input signal and $y[n]$ denote the output signal of the filter, both measured at time n. We may express the relationship between $y[n]$ and $x[n]$ for the filter by the linear constant-coefficient difference equation of order one, written as

$$y[n] = x[n] + \rho y[n - 1], \qquad (1.115)$$

where ρ is a constant. The recursive equation (1.115) is a special case of a linear difference equation in that it is void of past values of the input signal. Figure 1.69 shows a block diagram representation of the filter, where the block labeled S denotes the discrete-time time-shift operator. The structure described in the figure is an example of a *linear discrete-time feedback system*, with the coefficient ρ being responsible for the presence of feedback in the system. In other words, the use of feedback in a discrete-time system is related to recursive computation.

The solution of Eq. (1.115) is given by

$$y[n] = \sum_{k=0}^{\infty} \rho^k x[n - k]. \qquad (1.116)$$

Systematic procedures for arriving at this solution are presented in Chapters 2 and 7. For now, we may demonstrate the validity of Eq. (1.116) by proceeding as follows: Isolating the term corresponding to $k = 0$, we write

$$y[n] = x[n] + \sum_{k=1}^{\infty} \rho^k x[n - k]. \qquad (1.117)$$

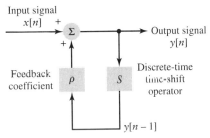

FIGURE 1.69   Block diagram of first-order recursive discrete-time filter. The operator S shifts the output signal $y[n]$ by one sampling interval, producing $y[n - 1]$. The feedback coefficient ρ determines the stability of the filter.

Next, setting $k - 1 = l$, or, equivalently, $k = l + 1$ in Eq. (1.117) yields

$$y[n] = x[n] + \sum_{l=0}^{\infty} \rho^{l+1}[n - 1 - l]$$
$$= x[n] + \rho \sum_{l=0}^{\infty} \rho^{l}[n - 1 - l].$$

(1.118)

In light of Eq. (1.116), we readily see that the summation term on the right-hand side of Eq. (1.118) is equal to $y[n - 1]$. Accordingly, we may rewrite Eq. (1.118) in the simplified form

$$y[n] = x[n] + \rho y[n - 1],$$

which we immediately recognize to be the original first-order recursive equation (1.115) of which Eq. (1.116) is the solution.

Depending on the value assigned to the constant ρ in the solution given by Eq. (1.116), we may identify three special cases:

1. $\rho = 1$, for which Eq. (1.116) reduces to

$$y[n] = \sum_{k=0}^{\infty} x[n - k].$$

(1.119)

   Equation (1.119) defines an *accumulator*, which represents the discrete-time equivalent of an ideal integrator.

2. $|\rho| < 1$, in which case successive contributions of past values of the input signal $x[n]$ to the output signal $y[n]$ are continually *attenuated* in absolute value. We may therefore refer to the resulting first-order recursive filter of Fig. 1.68 as a *leaky accumulator*, with the leakage becoming smaller as the magnitude of ρ approaches unity.

3. $|\rho| > 1$, in which case successive contributions of past values of the input signal $x[n]$ to the output signal $y[n]$ are *amplified* in absolute value as time goes on.

From this discussion, it is apparent that in case 2 the first-order recursive filter of Fig. 1.69 is stable in the BIBO sense, and it is unstable in the BIBO sense in both cases 1 and 3; see Problem 1.26.

In subsequent chapters, we discuss applications of the first-order recursive filter of Fig. 1.69 in such diverse fields as digital signal processing, financial computations, and digital control systems.

1.11 *Exploring Concepts with MATLAB*

The basic *object* used in MATLAB is a rectangular numerical matrix with possibly complex elements. The kinds of data objects encountered in the study of signals and systems are all well suited to matrix representations. In this section, we use MATLAB to explore the generation of some of the elementary signals described in previous sections. The exploration of systems and more advanced signals is deferred to subsequent chapters.

The MATLAB Signal Processing Toolbox has a large variety of functions for generating signals, most of which require that we begin with the vector representation of time t or n. To generate a vector t of time values with a *sampling interval* T_s of 1 ms on the interval from 0 to 1s, for example, we use the command

```
t = 0:.001:1;
```

This vector encompasses 1000 time samples each second, or a *sampling rate* of 1000 Hz. To generate a vector n of time values for discrete-time signals, say, from $n = 0$ to $n = 1000$, we use the command

```
n = 0:1000;
```

Given t or n, we may then proceed to generate the signal of interest.

In MATLAB, a discrete-time signal is represented *exactly*, because the values of the signal are described as the elements of a vector. On the other hand, MATLAB provides only an *approximation* to a continuous-time signal. The approximation consists of a vector whose individual elements are samples of the underlying continuous-time signal. When we use this approximate approach, it is important that we choose the sampling interval T_s sufficiently small so as to ensure that the samples capture all the details of the signal.

In this section, we consider the generation of both continuous-time and discrete-time signals of various kinds.

■ 1.11.1 PERIODIC SIGNALS

It is an easy matter to generate periodic signals such as square waves and triangular waves with MATLAB. Consider first the generation of a square wave of amplitude A, fundamental frequency w0 (measured in radians per second), and duty cycle rho. That is, rho is the fraction of each period for which the signal is positive. To generate such a signal, we use the basic command

```
A*square(w0*t , rho);
```

The square wave shown in Fig. 1.14(a) was generated with the following complete set of commands:

```
>> A = 1;
>> w0 = 10*pi;
>> rho = 0.5;
>> t = 0:.001:1;
>> sq = A*square(w0*t , rho);
>> plot(t, sq)
>> axis([0 1 -1.1 1.1])
```

In the second command, pi is a built-in MATLAB function that returns the floating-point number closest to π. The plot command is used to view the square wave. The command plot draws lines connecting the successive values of the signal and thus gives the appearance of a continuous-time signal.

Consider next the generation of a triangular wave of amplitude A, fundamental frequency w0 (measured in radians per second), and width W. Let the period of the wave be T, with the first maximum value occurring at t = WT. The basic command for generating this second periodic signal is

```
A*sawtooth(w0*t , W);
```

Thus, to generate the symmetric triangular wave shown in Fig. 1.15, we used the following commands:

```
>> A = 1;
>> w0 = 10*pi;
>> W = 0.5;
>> t = 0:0.001:1;
>> tri = A*sawtooth(w0*t , W);
>> plot(t, tri)
```

As mentioned previously, a signal generated in MATLAB is inherently of a discrete-time nature. To visualize a discrete-time signal, we may use the **stem** command. Specifically, **stem(n, x)** depicts the data contained in vector **x** as a discrete-time signal at the time values defined by **n**. The vectors **n** and **x** must, of course, have compatible dimensions.

Consider, for example, the discrete-time square wave shown in Fig. 1.16. This signal is generated by using the following commands:

```
>> A = 1;
>> omega = pi/4;
>> n = -10:10;
>> x = A*square(omega*n);
>> stem(n, x)
```

▶ **Problem 1.38**   Use the MATLAB code given at the top of this page to generate the triangular wave depicted in Fig. 1.15.   ◀

■ 1.11.2 EXPONENTIAL SIGNALS

Moving on to exponential signals, we have decaying exponentials and growing exponentials. The MATLAB command for generating a decaying exponential is

```
B*exp(-a*t);
```

To generate a growing exponential, we use the command

```
B*exp(a*t);
```

In both cases, the exponential parameter **a** is positive. The following commands were used to generate the decaying exponential signal shown in Fig. 1.28(a):

```
>> B = 5;
>> a = 6;
>> t = 0:.001:1;
>> x = B*exp(-a*t);   % decaying exponential
>> plot(t, x)
```

The growing exponential signal shown in Figure 1.28(b) was generated with these commands:

```
>> B = 1;
>> a = 5;
>> t = 0:0.001:1;
>> x = B*exp(a*t);   % growing exponential
>> plot(t, x)
```

Consider next the exponential sequence defined in Eq. (1.34). The decaying form of this exponential is shown in Fig. 1.30(a), generated with the following commands:

```
>> B = 1;
>> r = 0.85
>> n = -10:10;
>> x = B*r.^n;  % decaying exponential
>> stem(n, x)
```

Note that, in this example, the base r is a scalar, but the exponent is a vector—hence the use of the symbol .^ to denote *element-by-element powers*.

▶ **Problem 1.39** Use MATLAB to generate the growing exponential sequence depicted in Fig. 1.30(b). ◀

■ **1.11.3 Sinusoidal Signals**

MATLAB also contains trigonometric functions that can be used to generate sinusoidal signals. A cosine signal of amplitude A, frequency w0 (measured in radians per second), and phase angle phi (in radians) is obtained by using the command

```
A*cos(w0*t + phi);
```

Alternatively, we may use the sine function to generate a sinusoidal signal with the command

```
A*sin(w0*t + phi);
```

These two commands were used as the basis for generating the sinusoidal signals shown in Fig. 1.31. For example, for the cosine signal shown in Fig. 1.31(a), we used the following commands:

```
>> A = 4;
>> w0 = 20*pi;
>> phi = pi/6;
>> t = 0:.001:1;
>> cosine = A*cos(w0*t + phi);
>> plot(t, cosine)
```

▶ **Problem 1.40** Use MATLAB to generate the sine signal shown in Fig. 1.31(b). ◀

Consider next the discrete-time sinusoidal signal defined in Eq. (1.39). This periodic signal is plotted in Fig. 1.33, generated with the use of the following commands:

```
>> A = 1;
>> omega = 2*pi/12;  % angular frequency
>> n = -10:10;
>> y = A*cos(omega*n);
>> stem(n, y)
```

■ **1.11.4 Exponentially Damped Sinusoidal Signals**

In all of the MATLAB signal-generation commands, just described, we have generated the desired amplitude by multiplying a scalar A by a vector representing a unit-amplitude signal (e.g., sin(w0*t + phi)). This operation is described by using an asterisk. We next consider the generation of a signal that requires the *element-by-element multiplication* of two vectors.

Suppose we multiply a sinusoidal signal by an exponential signal to produce an exponentially damped sinusoidal signal. With each signal component represented by a vector, the generation of such a product signal requires the multiplication of one vector by another on an element-by-element basis. MATLAB represents element-by-element multiplication by a dot followed by an asterisk. Thus, the command for generating the exponentially damped sinusoidal signal

$$x(t) = A \sin(\omega_0 t + \phi)\, e^{-at}$$

is

```
A*sin(w0*t + phi).*exp(-a*t);
```

For a decaying exponential, a is positive. This command was used in the generation of the waveform shown in Fig. 1.35. The complete set of commands is as follows:

```
>> A = 60;
>> w0 = 20*pi;
>> phi = 0;
>> a = 6;
>> t = 0:.001:1;
>> expsin = A*sin(w0*t + phi).*exp(-a*t);
>> plot(t, expsin)
```

Consider next the exponentially damped sinusoidal sequence depicted in Fig. 1.70. This sequence is obtained by multiplying the sinusoidal sequence $x[n]$ of Fig. 1.33 by the decaying exponential sequence $y[n]$ of Fig. 1.30(a). Both sequences are defined for n = -10:10. Thus, letting $z[n]$ denote this product sequence, we may use the following commands to generate and visualize it:

```
>> z = x.*y;  % elementwise multiplication
>> stem(n, z)
```

Note that there is no need to include the definition of n in the generation of z, as it is already included in the commands for both x and y, which are defined on page 83.

▶ **Problem 1.41**   Use MATLAB to generate a signal defined as the product of the growing exponential of Fig. 1.30(b) and the sinusoidal signal of Fig. 1.33.   ◀

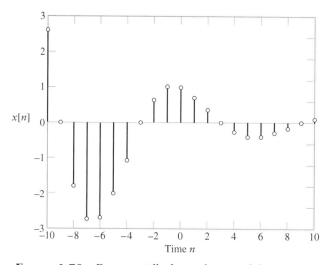

FIGURE 1.70   Exponentially damped sinusoidal sequence.

■ 1.11.5 Step, Impulse, and Ramp Functions

In MATLAB, `ones (M, N)` is an *M*-by-*N* matrix of ones, and `zeros (M, N)` is an *M*-by-*N* matrix of zeros. We may use these two matrices to generate two commonly used signals:

- ▶ *Step function.* A unit-amplitude step function is generated by writing

  ```
  u = [zeros(1, 50), ones(1, 50)];
  ```

- ▶ *Discrete-time impulse.* A unit-amplitude discrete-time impulse is generated by writing

  ```
  delta = [zeros(1, 49), 1, zeros(1, 49)];
  ```

  To generate a ramp sequence, we simply write

  ```
  ramp = 0:.1:10
  ```

In Fig. 1.39, we illustrated how a pair of step functions shifted in time relative to each other may be used to produce a rectangular pulse. In light of the procedure illustrated therein, we may formulate the following set of commands for generating a rectangular pulse centered on the origin:

```
>> t = -1:1/500:1;
>> u1 = [zeros(1, 250), ones(1, 751)];
>> u2 = [zeros(1, 751), ones(1, 250)];
>> u = u1 - u2;
```

The first command defines time running from -1 second to 1 second in increments of 2 milliseconds. The second command generates a step function `u1` of unit amplitude, beginning at time $t = -0.5$ second. The third command generates a second step function `u2`, beginning at time $t = 0.5$ second. The fourth command subtracts `u2` from `u1` to produce a rectangular pulse of unit amplitude and unit duration centered on the origin.

■ 1.11.6 User-Defined Function

An important feature of the MATLAB environment is that it permits us to create our own *M-files*, or subroutines. Two types of M-files exist: scripts and functions. *Scripts*, or *script files*, automate long sequences of commands; *functions*, or *function files*, provide extensibility to MATLAB by allowing us to add new functions. Any variables used in function files do not remain in memory. For this reason, input and output variables must be declared explicitly.

We may thus say that a function M-file is a separate entity characterized as follows:

1. It begins with a statement defining the function name, its input arguments, and its output arguments.
2. It also includes additional statements that compute the values to be returned.
3. The inputs may be scalars, vectors, or matrices.

Consider, for example, the generation of the rectangular pulse depicted in Fig. 1.39(a). Suppose we wish to generate the pulse with the use of an M-file. The pulse is to have unit amplitude and unit duration. To generate it, we create a file called `rect.m` containing the following statements:

```
>> function g = rect(x)
>> g = zeros(size(x));
>> set1 = find(abs(x)<= 0.5);
>> g(set1) = ones(size(set1));
```

In the last three statements of this M-file, we have introduced two useful functions:

1. The function `size` returns a two-element vector containing the row and column dimensions of a matrix.
2. The function `find` returns the indices of a vector or matrix that satisfy a prescribed relation. For the example at hand, `find(abs(x)<= T)` returns the indices of the vector `x`, where the absolute value of `x` is less than or equal to `T`.

The new function `rect.m` can be used like any other MATLAB function. In particular, we may use it to generate a rectangular pulse with the following command:

```
>> t = -1:1/500:1;
>> plot(t, rect(t));
```

1.12 Summary

In this chapter, we presented an overview of signals and systems, setting the stage for the rest of the book. A particular theme that stands out in the discussion presented herein is that signals may be of the continuous-time or discrete-time variety, and likewise for systems:

▶ A continuous-time signal is defined for all values of time. In contrast, a discrete-time signal is defined only for discrete instants of time.
▶ A continuous-time system is described by an operator that changes a continuous-time input signal into a continuous-time output signal. In contrast, a discrete-time system is described by an operator that changes a discrete-time input signal into a discrete-time output signal.

In practice, many systems mix continuous-time and discrete-time components. Analyzing *mixed* systems is an important part of the material presented in Chapters 4, 5, 8, and 9.

In discussing the various properties of signals and systems, we took special care in treating the two classes of signals and systems side by side. In so doing, much is gained by emphasizing their similarities and differences. This practice is followed in later chapters, too, as is appropriate.

Another noteworthy point is that, in the study of systems, particular attention is given to the analysis of *linear time-invariant systems*. A linear system obeys both the principle of superposition and the property of homogeneity. The characteristics of a time-invariant system do not change with time. By invoking these two properties, the analysis of systems becomes mathematically tractable. Indeed, a rich set of tools has been developed for analyzing linear time-invariant systems, providing direct motivation for much of the material on system analysis presented in the book.

In this chapter, we also explored the use of MATLAB for generating elementary waveforms representing continuous-time and discrete-time signals. MATLAB provides a powerful environment for exploring concepts and testing system designs, as will be illustrated in subsequent chapters.

FURTHER READING

1. For a readable account of signals, their representations, and their use in communication systems, see
   ▶ Pierce, J. R., and A. M. Noll, *Signals: The Science of Telecommunications* (Scientific American Library, 1990)

2. For examples of control systems, see Chapter 1 of

   ▶ Kuo, B. C., *Automatic Control Systems*, 7th ed. (Prentice Hall, 1995)

   and Chapters 1 and 2 of

   ▶ Phillips, C. L., and R. D. Harbor, *Feedback Control Systems*, 3rd ed. (Prentice Hall, 1996)

3. For a general discussion of remote sensing, see

   ▶ Hord, R. M., *Remote Sensing: Methods and Applications* (Wiley, 1986)

   For material on the use of spaceborne radar for remote sensing, see

   ▶ Elachi, C., *Introduction to the Physics and Techniques of Remote Sensing* (Wiley, 1987)

   For a detailed description of synthetic aperture radar and the role of signal processing in its implementation, see

   ▶ Curlander, J. C., and R. N. McDonough, *Synthetic Aperture Radar: Systems and Signal Processing* (Wiley, 1991)

   For an introductory treatment of radar, see

   ▶ Skolnik, M. I., *Introduction to Radar Systems*, 3rd ed. (McGraw-Hill, 2001)

4. Figure 1.6 is taken from

   ▶ Yazdi, D., F. Ayazi, and K. Najafi, "Micromachined Inertial Sensors," *Proceedings of the Institute of Electrical and Electronics Engineers*, vol. 86, pp. 1640–1659, August 1998

   This paper presents a review of silicon micromachined accelerometers and gyroscopes. It is part of a special issue devoted to integrated sensors, microactuators, and microsystems (MEMS). For additional papers on MEMS and their applications, see

   ▶ Wise, K. D., and K. Najafi, "Microfabrication Techniques for Integrated Sensors and Microsystems," *Science*, vol. 254, pp. 1335–1342, Novemeber 1991

   ▶ S. Cass, "MEMS in space," *IEEE Spectrum*, pp. 56–61, July 2001

5. For a collection of essays on biological signal processing, see

   ▶ Weitkunat, R., ed., *Digital Biosignal Processing* (Elsevier, 1991)

6. For a detailed discussion of the auditory system, see

   ▶ Dallos, P., A. N. Popper, and R. R. Fay, eds., *The Cochlea* (Springer-Verlag, 1996)

   ▶ Hawkins, H. L., and T. McMullen, eds., *Auditory Computation* (Springer-Verlag, 1996)

   ▶ Kelly, J. P., "Hearing." In E. R. Kandel, J. H. Schwartz, and T. M. Jessell, *Principles of Neural Science*, 3rd ed. (Elsevier, 1991)

   The cochlea has provided a source of motivation for building an electronic version of it, using silicon integrated circuits. Such an artificial implementation is sometimes referred to as a *silicon cochlea*. For a discussion of the silicon cochlea, see

   ▶ Lyon, R. F., and C. Mead, "Electronic Cochlea." In C. Mead, *Analog VLSI and Neural Systems* (Addison-Wesley, 1989)

7. For an account of the legendary story of the first Tacoma Narrows suspension bridge, see

   ▶ Smith, D., "A Case Study and Analysis of the Tacoma Narrows Bridge Failure," 99.497 Engineering Project, Department of Mechanical Engineering, Carleton University, March 29, 1974 (supervised by Professor G. Kardos)

8. For treatments of the different aspects of electrical noise, see

   ▶ Bennet, W. R., *Electrical Noise* (McGraw-Hill, 1960)

   ▶ Van der Ziel, A., *Noise: Sources, Characterization, Measurement* (Prentice-Hall, 1970)

   ▶ Gupta, M. S., ed., *Electrical Noise: Fundamentals and Sources* (IEEE Press, 1977)

The edited book by Gupta covers (1) the history of the subject of electrical noise, (2) an introduction to physical mechanisms, mathematical methods, and applications, (3) the principal noise-generating processes, (4) the types of electronic devices in which noise phenomena have been studied, and, finally, (5) noise generators.

9. For an introductory treatment of MATLAB, see Appendix F.

ADDITIONAL PROBLEMS

1.42 Determine whether the following signals are periodic. If they are periodic, find the fundamental period.

(a) $x(t) = (\cos(2\pi t))^2$

(b) $x(t) = \sum_{k=-5}^{5} w(t - 2k)$ for $w(t)$ depicted in Fig. P1.42b.

(c) $x(t) = \sum_{k=-\infty}^{\infty} w(t - 3k)$ for $w(t)$ depicted in Fig. P1.42b.

(d) $x[n] = (-1)^n$

(e) $x[n] = (-1)^{n^2}$

(f) $x[n]$ depicted in Fig. P1.42f.

(g) $x(t)$ depicted in Fig. P1.42g.

(h) $x[n] = \cos(2n)$

(i) $x[n] = \cos(2\pi n)$

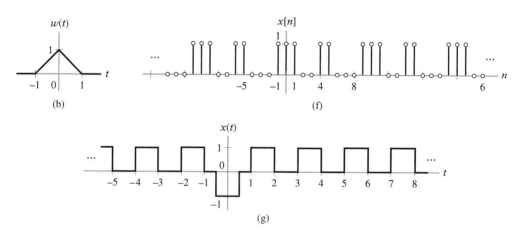

(b)

(f)

(g)

FIGURE P1.42

1.43 The sinusoidal signal

$$x(t) = 3\cos(200t + \pi/6)$$

is passed through a square-law device defined by the input–output relation

$$y(t) = x^2(t).$$

Using the trigonometric identity

$$\cos^2 \theta = \tfrac{1}{2}(\cos 2\theta + 1),$$

show that the output $y(t)$ consists of a dc component and a sinusoidal component.

(a) Specify the dc component.

(b) Specify the amplitude and fundamental frequency of the sinusoidal component in the output $y(t)$.

1.44 Consider the sinusoidal signal

$$x(t) = A\cos(\omega t + \phi).$$

Determine the average power of $x(t)$.

1.45 The angular frequency Ω of the sinusoidal signal

$$x[n] = A\cos(\Omega n + \phi)$$

satisfies the condition for $x[n]$ to be periodic. Determine the average power of $x[n]$.

1.46 The raised-cosine pulse $x(t)$ shown in Fig. P1.46 is defined as

$$x(t) = \begin{cases} \tfrac{1}{2}[\cos(\omega t) + 1], & -\pi/\omega \le t \le \pi/\omega \\ 0, & \text{otherwise} \end{cases}.$$

Determine the total energy of $x(t)$.

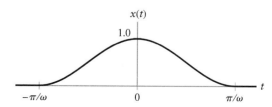

FIGURE P1.46

1.47 The trapezoidal pulse $x(t)$ shown in Fig. P1.47 is defined by

$$x(t) = \begin{cases} 5 - t, & 4 \le t \le 5 \\ 1, & -4 \le t \le 4 \\ t + 5 & -5 \le t \le -4 \\ 0, & \text{otherwise} \end{cases}.$$

Determine the total energy of $x(t)$.

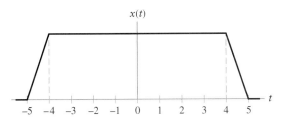

FIGURE P1.47

1.48 The trapezoidal pulse $x(t)$ of Fig. P1.47 is applied to a differentiator, defined by

$$y(t) = \frac{d}{dt}x(t).$$

(a) Determine the resulting output $y(t)$ of the differentiator.

(b) Determine the total energy of $y(t)$.

1.49 A rectangular pulse $x(t)$ is defined by

$$x(t) = \begin{cases} A, & 0 \le t \le T \\ 0, & \text{otherwise} \end{cases}.$$

The pulse $x(t)$ is applied to an integrator defined by

$$y(t) = \int_{0^-}^{t} x(\tau)\, d\tau.$$

Find the total energy of the output $y(t)$.

1.50 The trapezoidal pulse $x(t)$ of Fig. P1.47 is time scaled, producing the equation

$$y(t) = x(at).$$

Sketch $y(t)$ for (a) $a = 5$ and (b) $a = 0.2$.

1.51 Sketch the trapezoidal pulse $y(t)$ related to that of Fig. P1.47 as follows:

$$y(t) = x(10t - 5).$$

1.52 Let $x(t)$ and $y(t)$ be given in Figs. P1.52(a) and (b), respectively. Carefully sketch the following signals:

(a) $x(t)y(t - 1)$

(b) $x(t - 1)y(-t)$

(c) $x(t + 1)y(t - 2)$

(d) $x(t)y(-1 - t)$

(e) $x(t)y(2 - t)$

(f) $x(2t)y(\tfrac{1}{2}t + 1)$

(g) $x(4 - t)y(t)$

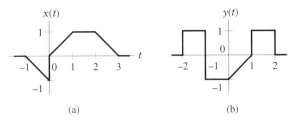

FIGURE P1.52

1.53 Figure P1.53(a) shows a staircaselike signal $x(t)$ that may be viewed as the superposition of four rectangular pulses. Starting with a compressed version of the rectangular pulse $g(t)$ shown in Fig. P1.53(b), construct the waveform of Fig. P1.53(a), and express $x(t)$ in terms of $g(t)$.

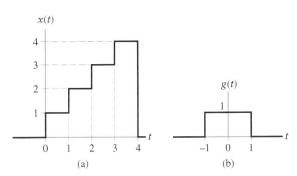

FIGURE P1.53

1.54 Sketch the waveforms of the following signals:

(a) $x(t) = u(t) - u(t - 2)$

(b) $x(t) = u(t + 1) - 2u(t) + u(t - 1)$

(c) $x(t) = -u(t + 3) + 2u(t + 1)$
$$- 2u(t - 1) + u(t - 3)$$

(d) $y(t) = r(t + 1) - r(t) + r(t - 2)$

(e) $y(t) = r(t + 2) - r(t + 1)$
$$- r(t - 1) + r(t - 2)$$

1.55 Figure P1.55(a) shows a pulse $x(t)$ that may be viewed as the superposition of three rectangular pulses. Starting with the rectangular pulse $g(t)$ of Fig. P1.55(b), construct the waveform of Fig. P1.55, and express $x(t)$ in terms of $g(t)$.

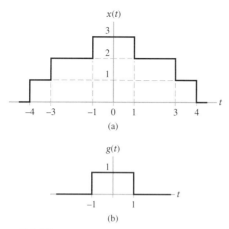

(a)

(b)

FIGURE P1.55

1.56 Let $x[n]$ and $y[n]$ be given in Figs. P1.56(a) and (b), respectively. Carefully sketch the following signals:

(a) $x[2n]$
(b) $x[3n - 1]$
(c) $y[1 - n]$
(d) $y[2 - 2n]$
(e) $x[n - 2] + y[n + 2]$
(f) $x[2n] + y[n - 4]$
(g) $x[n + 2]y[n - 2]$
(h) $x[3 - n]y[n]$
(i) $x[-n]y[-n]$
(j) $x[n]y[-2 - n]$
(k) $x[n + 2]y[6 - n]$

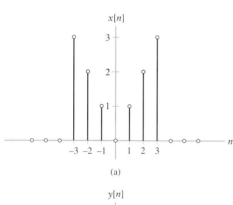

(a)

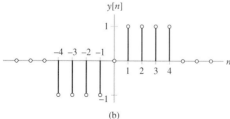

(b)

FIGURE P1.56

1.57 Determine whether the following signals are periodic, and for those which are, find the fundamental period:

(a) $x[n] = \cos\left(\frac{8}{15}\pi n\right)$
(b) $x[n] = \cos\left(\frac{7}{15}\pi n\right)$
(c) $x(t) = \cos(2t) + \sin(3t)$
(d) $x(t) = \sum_{k=-\infty}^{\infty}(-1)^k\delta(t - 2k)$
(e) $x[n] = \sum_{k=-\infty}^{\infty}\{\delta[n - 3k] + \delta[n - k^2]\}$
(f) $x(t) = \cos(t)u(t)$
(g) $x(t) = v(t) + v(-t)$, where
     $v(t) = \cos(t)u(t)$
(h) $x(t) = v(t) + v(-t)$, where
     $v(t) = \sin(t)u(t)$
(i) $x[n] = \cos\left(\frac{1}{5}\pi n\right)\sin\left(\frac{1}{3}\pi n\right)$

1.58 The sinusoidal signal $x[n]$ has fundamental period $N = 10$ samples. Determine the smallest angular frequency Ω for which $x[n]$ is periodic.

1.59 A complex sinusoidal signal $x(t)$ has the following components:

$$\text{Re}\{x(t)\} = x_R(t) = A\cos(\omega t + \phi);$$
$$\text{Im}\{x(t)\} = x_I(t) = A\sin(\omega t + \phi).$$

The amplitude of $x(t)$ is defined by the square root of $x_R^2(t) + x_I^2(t)$. Show that this amplitude equals A and is therefore independent of the phase angle ϕ.

1.60 Consider the complex-valued exponential signal

$$x(t) = Ae^{\alpha t + j\omega t}, \qquad \alpha > 0.$$

Evaluate the real and imaginary components of $x(t)$.

1.61 Consider the continuous-time signal

$$x(t) = \begin{cases} t/\Delta + 0.5, & -\Delta/2 \leq t \leq \Delta/2 \\ 1, & t > \Delta/2 \\ 0, & t < -\Delta/2 \end{cases},$$

which is applied to a differentiator. Show that the output of the differentiator approaches the unit impulse $\delta(t)$ as Δ approaches zero.

1.62 In this problem, we explore what happens when a unit impulse is applied to a differentiator. Consider a triangular pulse $x(t)$ of duration Δ and amplitude $2/\Delta$, as depicted in Fig. P1.62. The area under the pulse is unity. Hence, as the duration Δ approaches zero, the triangular pulse approaches a unit impulse.

(a) Suppose the triangular pulse $x(t)$ is applied to a differentiator. Determine the output $y(t)$ of the differentiator.

(b) What happens to the differentiator output $y(t)$ as Δ approaches zero? Use the definition of a unit impulse $\delta(t)$ to express your answer.

(c) What is the total area under the differentiator output $y(t)$ for all Δ? Justify your answer.

Based on your findings in parts (a) through (c), describe in succinct terms the result of differentiating a unit impulse.

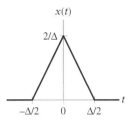

FIGURE P1.62

1.63 A system consists of several subsystems connected as shown in Fig. P1.63. Find the operator H relating $x(t)$ to $y(t)$ for the following subsystem operators:

$$H_1: y_1(t) = x_1(t)x_1(t-1);$$
$$H_2: y_2(t) = |x_2(t)|;$$
$$H_3: y_3(t) = 1 + 2x_3(t);$$
$$H_4: y_4(t) = \cos(x_4(t)).$$

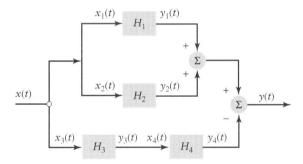

FIGURE P1.63

1.64 The systems that follow have input $x(t)$ or $x[n]$ and output $y(t)$ or $y[n]$. For each system, determine whether it is (i) memoryless, (ii) stable, (iii) causal, (iv) linear, and (v) time invariant.

(a) $y(t) = \cos(x(t))$

(b) $y[n] = 2x[n]u[n]$

(c) $y[n] = \log_{10}(|x[n]|)$

(d) $y(t) = \int_{-\infty}^{t/2} x(\tau)\,d\tau$

(e) $y[n] = \sum_{k=-\infty}^{n} x[k+2]$

(f) $y(t) = \dfrac{d}{dt}x(t)$

(g) $y[n] = \cos(2\pi x[n+1]) + x[n]$

(h) $y(t) = \dfrac{d}{dt}\{e^{-t}x(t)\}$

(i) $y(t) = x(2-t)$

(j) $y[n] = x[n]\sum_{k=-\infty}^{\infty}\delta[n-2k]$

(k) $y(t) = x(t/2)$

(l) $y[n] = 2x[2^n]$

1.65 The output of a discrete-time system is related to its input $x[n]$ as follows:

$$y[n] = a_0 x[n] + a_1 x[n-1]$$
$$+ a_2 x[n-2] + a_3 x[n-3].$$

Let the operator S^k denote a system that shifts the input $x[n]$ by k time units to produce $x[n-k]$. Formulate the operator H for the system relating $y[n]$ to $x[n]$. Then develop a block diagram representation for H, using (a) cascade implementation and (b) parallel implementation.

1.66 Show that the system described in Problem 1.65 is BIBO stable for all a_0, a_1, a_2, and a_3.

1.67 How far does the memory of the discrete-time system described in Problem 1.65 extend into the past?

1.68 Is it possible for a noncausal system to possess memory? Justify your answer.

1.69 The output signal $y[n]$ of a discrete-time system is related to its input signal $x[n]$ as follows:

$$y[n] = x[n] + x[n-1] + x[n-2].$$

Let the operator S denote a system that shifts its input by one time unit.

(a) Formulate the operator H for the system relating $y[n]$ to $x[n]$.

(b) The operator H^{inv} denotes a discrete-time system that is the inverse of the given system. How is H^{inv} defined?

1.70 Show that the discrete-time system described in Problem 1.65 is time invariant, independent of the coefficients a_0, a_1, a_2, and a_3.

1.71 (a) Is it possible for a time-variant system to be linear? Justify your anwer.

(b) Consider the RC circuit of Fig. P1.71, in which the resistive component $R(t)$ is time varying. For all time t, the time constant of the circuit is large enough to justify approximating the circuit as an integrator. Show that the circuit is indeed linear.

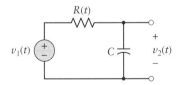

FIGURE P1.71

1.72 Show that a pth power-law device defined by the input–output relation

$$y(t) = x^p(t), \qquad p \text{ integer and } p \neq 0, 1,$$

is nonlinear.

1.73 A linear time-invariant system may be causal or non-causal. Give an example of each of these possibilities.

1.74 Figure 1.56 shows two equivalent system configurations on condition that the system operator H is linear. Which of these configurations is simpler to implement? Justify your answer.

1.75 A system H has its input–output pairs given. Determine whether the system could be memoryless, causal, linear, and time invariant for (a) signals depicted in Fig. P1.75(a), and (b) signals depicted in Fig. P1.75(b). For all cases, justify your answers.

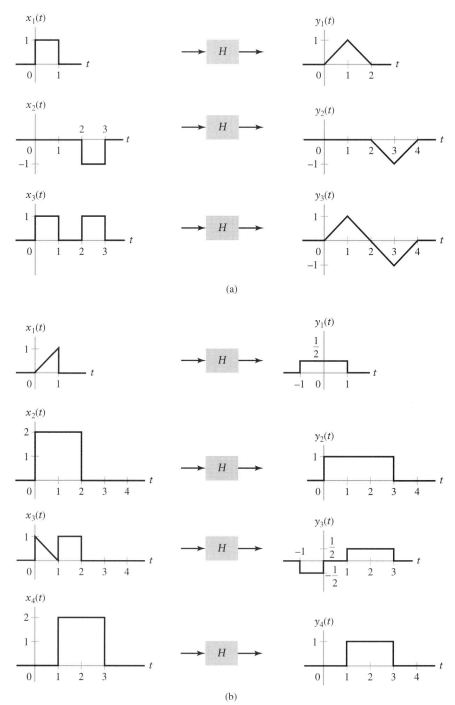

FIGURE P1.75

1.76 A linear system H has the input–output pairs depicted in Fig. P1.76(a). Answer the following questions, and explain your answers:

(a) Could this system be causal?

(b) Could this system be time invariant?

(c) Could this system be memoryless?

(d) What is the output for the input depicted in Fig. P1.76(b)?

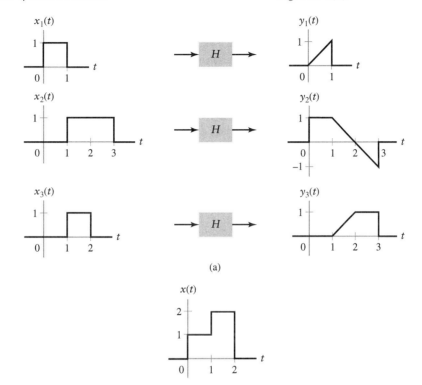

(a)

(b)

FIGURE P1.76

1.77 A discrete-time system is both linear and time invariant. Suppose the output due to an input $x[n] = \delta[n]$ is given in Fig. P1.77(a).

(a) Find the output due to an input $x[n] = \delta[n-1]$.

(b) Find the output due to an input $x[n] = 2\delta[n] - \delta[n-2]$.

(c) Find the output due to the input depicted in Fig. P1.77(b).

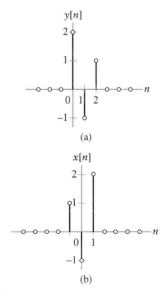

(a)

(b)

FIGURE P1.77

ADVANCED PROBLEMS

1.78 (a) An arbitrary real-valued continuous-time signal may be expressed as

$$x(t) = x_e(t) + x_o(t),$$

where $x_e(t)$ and $x_o(t)$ are, respectively, the even and odd components of $x(t)$. The signal $x(t)$ occupies the entire interval $-\infty < t < \infty$. Show that the energy of the signal $x(t)$ is equal to the sum of the energy of the even component $x_e(t)$ and the energy of the odd component $x_o(t)$. That is, show that

$$\int_{-\infty}^{\infty} x^2(t)\, dt = \int_{-\infty}^{\infty} x_e^2(t)\, dt + \int_{-\infty}^{\infty} x_o^2(t)\, dt.$$

(b) Show that an arbitrary real-valued discrete-time signal $x[n]$ satisfies a relationship similar to that satisfied by the continuous signal in Part (a). That is, show that

$$\sum_{n=-\infty}^{\infty} x^2[n] = \sum_{n=-\infty}^{\infty} x_e^2[n] + \sum_{n=-\infty}^{\infty} x_o^2[n],$$

where $x_e[n]$ and $x_o[n]$ are, respectively, the even and odd components of $x[n]$.

1.79 The *LRC* circuit of Fig. P1.79 may be viewed as an *analog* of the MEMS accelerometer represented by the lumped-circuit model of Fig. 1.64.

(a) Write the second-order differential equation defining the time-domain behavior of the circuit.

(b) Comparing the equation of Part (a) with Eq. (1.108), derive a table describing the analogies between the *LRC* circuit of Fig. P1.79 and the MEMS accelerometer.

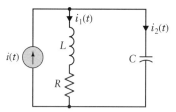

FIGURE P1.79

1.80 It may be argued that, for the limiting form of a pulse to approach a unit impulse, the pulse does not have to be an even function of time. All that the pulse has to satisfy is the unit-area requirement. To explore this matter, consider the asymmetric triangular pulse shown in Fig. P1.80. Then

(a) Explain what happens to this pulse as the duration Δ approaches zero.

(b) Determine whether the limiting form of the pulse so obtained satisfies all the properties of the unit impulse discussed in Section 1.6.6.

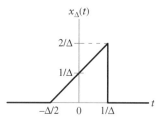

FIGURE P1.80

1.81 Consider a linear time-invariant system denoted by the operator H, as indicated in Fig. P1.81. The input signal $x(t)$ applied to the system is periodic with period T. Show that the corresponding response of the system, $y(t)$, is also periodic with the same period T.

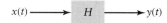

FIGURE P1.81

1.82 It is proposed that the unit impulse $\delta(t)$ be approximated by the symmetric double exponential pulse shown in Fig. P1.82, which is defined by

$$x_\Delta(t) = \frac{1}{\Delta}\left(e^{+t/\tau}u(-t) + e^{-t/\tau}u(t)\right).$$

(a) Determine the amplitude A attained by $x_\Delta(t)$ at $t = -\Delta/2$.

(b) Find the necessary condition that the time constant τ must satisfy for $x_\Delta(t)$ to approach $\delta(t)$ as the parameter Δ approaches zero, as given by

$$\delta(t) = \lim_{\Delta \to 0} x_\Delta(t).$$

(c) Illustrate graphically the nature of this approximation for $\Delta = 1, 0.5, 0.25,$ and 0.125.

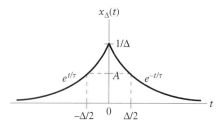

FIGURE P1.82

1.83 The operations of differentiation and integration are closely related. In light of this close relationship, it is tempting to state that they are the inverse of each other.

(a) Explain why it would be wrong, in a rigorous sense, to make this statement.

(b) The simple LR circuits of Figs. P1.83(a) and P1.83(b) may be used as approximators to differentiating and integrating circuits. Derive the conditions the elements of these two circuits would have to satisfy for them to fulfill their approximate functions.

(c) Use the examples of Figs. P1.83(a) and P1.83(b) in support of the explanation in part (a) of the problem.

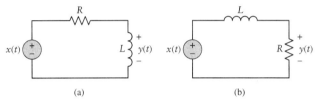

(a)   (b)

FIGURE P1.83

1.84 Figure P1.84 shows the block diagram of a linear time-varying system that consists simply of a multiplier that multiplies the input signal $x(t)$ by the output of an oscillator, $A_0 \cos(\omega_0 t + \phi)$, thereby producing the system output

$$y(t) = A_0 \cos(\omega_0 t + \phi)x(t).$$

Demonstrate the following:

(a) The system is linear; that is, it satisfies both the principle of superposition and the property of homogeneity.

(b) The system is time variant; that is, it violates the time-shift property. To show this, you may use the impulse input $x(t) = \delta(t)$.

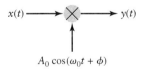

$A_0 \cos(\omega_0 t + \phi)$

FIGURE P1.84

1.85 In Problem 1.84 we considered an example of a linear time-varying system. In this problem, we consider a more complex nonlinear time-varying system. The output of the system, $y(t)$, is related to the input $x(t)$ as

$$y(t) = \cos\left(2\pi f_c t + k \int_{-\infty}^{t} x(\tau)\,d\tau\right),$$

where k is a constant parameter.

(a) Show that the system is nonlinear.

(b) Evaluate $y(t)$ for $x(t) = \delta(t)$ and its time-shifted version $x(t) = \delta(t - t_0)$, where $t_0 > 0$. Hence, demonstrate that the system is time variant.

1.86 In this problem, we explore a useful application of nonlinearity: *A nonlinear device provides a means for mixing two sinusoidal components.*

Consider a square-law device:

$$y(t) = x^2(t).$$

Let the input

$$x(t) = A_1 \cos(\omega_1 t + \phi_1) + A_2 \cos(\omega_2 t + \phi_2).$$

Determine the corresponding output $y(t)$. Show that $y(t)$ contains new components with the following frequencies: 0, $2\omega_1$, $2\omega_2$, $\omega_1 \mp \omega_2$. What are their respective amplitudes and phase shifts?

1.87 In this problem, we explore another application of nonlinearity: *A nonlinear device provides a basis for harmonic generation.*

Consider a cubic-law device:

$$y(t) = x^3(t).$$

Let the input

$$x(t) = A \cos(\omega t + \phi).$$

Determine the corresponding output $y(t)$. Show that $y(t)$ contains components with the frequencies ω and 3ω. What are their respective amplitudes and phase shifts?

What form of nonlinearity would you use to generate the pth harmonic of a sinusoidal component ω? Justify your answer.

1.88 (a) The step response of a second-order system produced by the input $x(t)$ is given by

$$y(t) = [1 - e^{-\alpha t} \cos(\omega_n t)]u(t), \qquad x(t) = u(t),$$

where the exponential parameter α and the frequency parameter ω_n are both real. Show that the impulse response of the system is given by

$$y(t) = [\alpha e^{-\alpha t} \cos(\omega_n t) + \omega_n e^{-\alpha t} \sin(\omega_n t)]u(t)$$

for $x(t) = \delta(t)$.

(b) Suppose next that the parameter ω_n is imaginary—say, $\omega_n - j\alpha_n$, where $\alpha_n \lesssim \alpha$. Show that the impulse response of the corresponding second-order system consists of the weighted sum of two decaying exponentials,

$$y(t) = \left[\frac{\alpha_1}{2}e^{-\alpha_1 t} + \frac{\alpha_2}{2}e^{-\alpha_2 t}\right]u(t), \qquad x(t) = \delta(t),$$

where $\alpha_1 = \alpha - \alpha_n$ and $\alpha_2 - \alpha + \alpha_n$.

shows the block diagram of a first-order screte-time filter. This filter differs from . 1.69 in that the output $y[n]$ also requires ge of the past input $x[n-1]$ for its evaluilding on the solution given in Eq. (1.116), an expression for the output $y[n]$ in terms of the input $x[n]$.

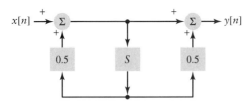

FIGURE P1.89

1.90 It is proposed that the MEMS acceleometer described in the block diagram of Fig. 1.64 be simulated by a second-order discrete-time system that would lend itself for use on a digital computer. Derive the difference equation that defines the input–output behavior of this simulator. *Hint:* Use the approximation of a derivative given by

$$\frac{d}{dt}z(t) \approx \frac{1}{T_s}\left[z\left(t+\frac{T_s}{2}\right) - z\left(t-\frac{T_s}{2}\right)\right],$$

where T_s denotes the sampling interval. For the second derivative d^2z/dt^2, apply the approximation twice.

1.91 Typically, the received signal of a radar or communication receiver is corrupted by additive noise. To combat the degrading effect of the noise, the signal-processing operation performed at the front end of the receiver usually involves some form of integration. Explain why, in such an application, integration is preferred over differentiation.

1.92 Consider the parallel RC circuit shown in Fig. P1.92. The source of current is denoted by $i(t)$, and the resulting currents through the capacitor C and resistor R are respectively denoted by $i_1(t)$ and $i_2(t)$. By formulating $i_1(t)$ in terms of $i(t)$ and the voltage across the resistor, $v(t)$, the circuit may be viewed as a feedback system. Develop the block diagram of this particular method of representation.

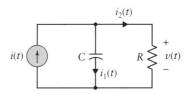

FIGURE P1.92

COMPUTER EXPERIMENTS

1.93 (a) The solution of a linear differential equation is given by

$$x(t) = 10e^{-t} - 5e^{-0.5t}.$$

Using MATLAB, plot $x(t)$ versus t for `t = 0:0.01:5`.

(b) Repeat Part (a) for
$$x(t) = 10e^{-t} + 5e^{-0.5t}.$$

1.94 An exponentially damped sinusoidal signal is defined by

$$x(t) = 20\sin(2\pi \times 1000t - \pi/3)\,e^{-at},$$

where the exponential parameter a is variable, taking on the set of values $a = 500, 750, 1000$. Using MATLAB, investigate the effect of varying a on the signal $x(t)$ for $-2 \leq t \leq 2$ milliseconds.

1.95 Write a set of MATLAB commands for approximating the following continuous-time periodic waveforms:

(a) Square wave of amplitude 5 volts, fundamental frequency 20 Hz, and duty cycle 0.6.

(b) Sawtooth wave of amplitude 5 volts and fundamental frequency 20 Hz.

Plot five cycles of each of these two waveforms.

1.96 A raised-cosine sequence is defined by

$$w[n] = \begin{cases} \cos(2\pi Fn), & -(1/2F) \leq n \leq (1/2F) \\ 0, & \text{otherwise} \end{cases}.$$

Use MATLAB to plot $w[n]$ versus n for $F = 0.1$.

1.97 A rectangular pulse $x(t)$ is defined by

$$x(t) = \begin{cases} 10, & 0 \leq t \leq 5 \\ 0, & \text{otherwise} \end{cases}.$$

Generate $x(t)$, using

(a) A pair of time-shifted step functions.

(b) An M-file.

2 Time-Domain Representations of Linear Time-Invariant Systems

2.1 Introduction

In this chapter, we examine several methods for describing the relationship between the input and output signals of linear time-invariant (LTI) systems. The focus here is on system descriptions that relate the output signal to the input signal when both are represented as functions of time—hence the terminology *time domain* in the chapter's title. Methods for relating system outputs and inputs in domains other than time are developed in later chapters. The descriptions developed herein are useful for analyzing and predicting the behavior of LTI systems and for implementing discrete-time systems on a computer.

We begin by characterizing an LTI system in terms of its *impulse response*, defined as the output of an LTI system due to a unit impulse signal input applied at time $t = 0$ or $n = 0$. The impulse response completely characterizes the behavior of any LTI system. This may seem surprising, but it is a basic property of all LTI systems. The impulse response is often determined from knowledge of the system configuration and dynamics or, in the case of an unknown system, it can be measured by applying an approximate impulse to the system input. The impulse response of a discrete-time system is usually easily obtained by setting the input equal to the impulse $\delta[n]$. In the continuous-time case, a true impulse signal having zero width and infinite amplitude cannot physically be generated and is usually approximated by a pulse of large amplitude and brief duration. Thus, the impulse response may be interpreted as the system behavior in response to a high-energy input of extremely brief duration. Given the impulse response, we determine the output due to an arbitrary input signal by expressing the input as a weighted superposition of time-shifted impulses. By linearity and time invariance, the output signal must be a weighted superposition of time-shifted impulse responses. This weighted superposition is termed the *convolution sum* for discrete-time systems and the *convolution integral* for continuous-time systems.

The second method we shall examine for characterizing the input–output behavior of LTI systems is the *linear constant-coefficient differential* or *difference equation*. Differential equations are used to represent continuous-time systems, while difference equations represent discrete-time systems. We focus on characterizing solutions of differential and difference equations with the goal of developing insight into the system's behavior.

The third system representation we discuss is the *block diagram*, which represents the system as an interconnection of three elementary operations: scalar multiplication, addition, and either a time shift for discrete-time systems or integration for continuous-time systems.

The final time-domain system representation discussed in this chapter is the *state-variable description*—a series of coupled first-order differential or difference equations that represent the behavior of the system's "state" and an equation that relates that state to the output of the system. The state is a set of variables associated with energy storage or memory devices in the system.

All four of these time-domain system representations are equivalent in the sense that identical outputs result from a given input. However, each representation relates the input to the output in a different manner. Different representations offer distinct views of the system, accompanied by different insights into the system's behavior. Each representation has advantages and disadvantages with respect to analyzing and implementing systems. Understanding how different representations are related and determining which offers the most insight and the most straightforward solution in a particular problem are important skills to develop.

2.2 *The Convolution Sum*

We begin by considering the discrete-time case. First, an arbitrary signal is expressed as a weighted superposition of shifted impulses. Then, the convolution sum is obtained by applying a signal represented in this manner to an LTI system. A similar procedure is used in Section 2.4 to obtain the convolution integral for continuous-time systems.

Let a signal $x[n]$ be multiplied by the impulse sequence $\delta[n]$; that is,

$$x[n]\delta[n] = x[0]\delta[n].$$

This relationship may be generalized to the product of $x[n]$ and a time-shifted impulse sequence, to obtain

$$x[n]\delta[n - k] = x[k]\delta[n - k],$$

where n represents the time index; hence, $x[n]$ denotes the entire signal, while $x[k]$ represents a specific value of the signal $x[n]$ at time k. We see that multiplication of a signal by a time-shifted impulse results in a time-shifted impulse with amplitude given by the value of the signal at the time the impulse occurs. This property allows us to express $x[n]$ as the following weighted sum of time-shifted impulses:

$$
\begin{aligned}
x[n] = \cdots &+ x[-2]\delta[n + 2] + x[-1]\delta[n + 1] + x[0]\delta[n] \\
&+ x[1]\delta[n - 1] + x[2]\delta[n - 2] + \cdots .
\end{aligned}
$$

We may rewrite this representation for $x[n]$ in the concise form

$$x[n] = \sum_{k=-\infty}^{\infty} x[k]\delta[n - k]. \tag{2.1}$$

A graphical illustration of Eq. (2.1) is given in Fig. 2.1. Equation (2.1) represents the signal as a weighted sum of basis functions, which are time-shifted versions of the unit impulse signal. The weights are the values of the signal at the corresponding time shifts.

Let the operator H denote the system to which the input $x[n]$ is applied. Then, using Eq. (2.1) to represent the input $x[n]$ to the system results in the output

$$
\begin{aligned}
y[n] &= H\{x[n]\} \\
&= H\left\{ \sum_{k=-\infty}^{\infty} x[k]\delta[n - k] \right\}.
\end{aligned}
$$

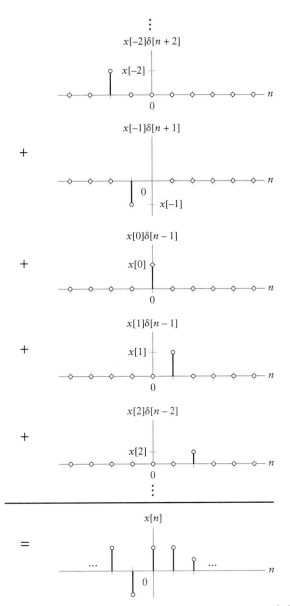

FIGURE 2.1   Graphical example illustrating the representation of a signal $x[n]$ as a weighted sum of time-shifted impulses.

Now we use the linearity property to interchange the system operator H with the summation and obtain

$$y[n] = \sum_{k=-\infty}^{\infty} H\{x[k]\delta[n-k]\}.$$

Since n is the time index, the quantity $x[k]$ is a constant with respect to the system operator H. Using linearity again, we interchange H with $x[k]$ to obtain

$$y[n] = \sum_{k=-\infty}^{\infty} x[k]H\{\delta[n-k]\}. \tag{2.2}$$

Equation (2.2) indicates that the system output is a weighted sum of the response of the system to time-shifted impulses. This response completely characterizes the system's input–output behavior and is a fundamental property of linear systems.

If we further assume that the system is time invariant, then a time shift in the input results in a time shift in the output. This relationship implies that the output due to a time-shifted impulse is a time-shifted version of the output due to an impulse; that is,

$$H\{\delta[n-k]\} = h[n-k], \tag{2.3}$$

where $h[n] = H\{\delta[n]\}$ is the impulse response of the LTI system H. The response of the system to each basis function in Eq. (2.1) is determined by the system impulse response. Substituting Eq. (2.3) into Eq. (2.2), we may rewrite the output as

$$y[n] = \sum_{k=-\infty}^{\infty} x[k]h[n-k]. \tag{2.4}$$

Thus, the output of an LTI system is given by a weighted sum of time-shifted impulse responses. This is a direct consequence of expressing the input as a weighted sum of time-shifted impulse basis functions. The sum in Eq. (2.4) is termed the *convolution sum* and is denoted by the symbol $*$; that is,

$$x[n] * h[n] = \sum_{k=-\infty}^{\infty} x[k]h[n-k].$$

The convolution process is illustrated in Fig. 2.2. In Fig. 2.2(a) an LTI system with impulse response $h[n]$ and input $x[n]$ is shown, while in Fig. 2.2(b) the input is decomposed as a sum of weighted and time-shifted unit impulses, with the kth input component given by $x[k]\delta[n-k]$. The output of the system associated with the kth impulse input is represented in the right half of the figure as

$$H\{x[k]\delta[n-k]\} = x[k]h[n-k].$$

This output component is obtained by shifting the impulse response k units in time and multiplying by $x[k]$. The total output $y[n]$ in response to the input $x[n]$ is obtained by summing all the individual outputs:

$$y[n] = \sum_{k=-\infty}^{\infty} x[k]h[n-k].$$

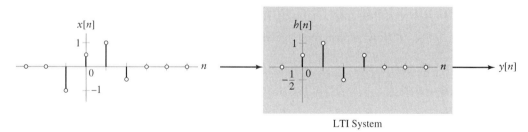

(a)

FIGURE 2.2   Illustration of the convolution sum. (a) LTI system with impulse response $h[n]$ and input $x[n]$, producing the output $y[n]$ to be determined.

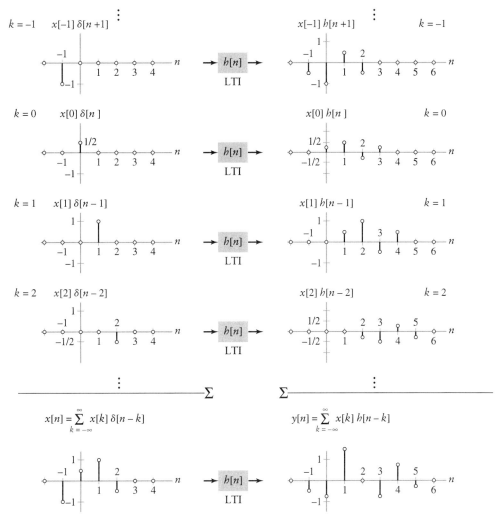

Figure 2.2   (b) The decomposition of the input $x[n]$ into a weighted sum of time-shifted impulses results in an output $y[n]$ given by a weighted sum of time-shifted impulse responses.

That is, for each value of n, we sum the outputs associated with each weighted and time-shifted impulse input from $k = -\infty$ to $k = \infty$. The following example illustrates this process.

**S & S
Solutions**

Example 2.1   Multipath Communication Channel: Direct Evaluation of the Convolution Sum   Consider the discrete-time LTI system model representing a two-path propagation channel described in Section 1.10. If the strength of the indirect path is $a = 1/2$, then

$$y[n] = x[n] + \frac{1}{2}x[n-1].$$

Letting $x[n] = \delta[n]$, we find that the impulse response is

$$h[n] = \begin{cases} 1, & n = 0 \\ \frac{1}{2}, & n = 1 \\ 0, & \text{otherwise} \end{cases}.$$

Determine the output of this system in response to the input

$$x[n] = \begin{cases} 2, & n = 0 \\ 4, & n = 1 \\ -2, & n = 2 \\ 0, & \text{otherwise} \end{cases}.$$

Solution: First, write $x[n]$ as the weighted sum of time-shifted impulses:

$$x[n] = 2\delta[n] + 4\delta[n - 1] - 2\delta[n - 2].$$

Here, the input is decomposed as a weighted sum of three time-shifted impulses because the input is zero for $n < 0$ and $n > 2$. Since a weighted, time-shifted impulse input, $\gamma\delta[n - k]$, results in a weighted, time-shifted impulse response output, $\gamma h[n - k]$, Eq. (2.4) indicates that the system output may be written as

$$y[n] = 2h[n] + 4h[n - 1] - 2h[n - 2].$$

Summing the weighted and shifted impulse responses over k gives

$$y[n] = \begin{cases} 0, & n < 0 \\ 2, & n = 0 \\ 5, & n = 1 \\ 0, & n = 2 \\ -1, & n = 3 \\ 0, & n \geq 4 \end{cases}.$$

▪

2.3 *Convolution Sum Evaluation Procedure*

In Example 2.1, we found the output corresponding to each time-shifted impulse and then summed each weighted, time-shifted impulse response to determine $y[n]$. This approach illustrates the principles that underlie convolution and is effective when the input is of brief duration so that only a small number of time-shifted impulse responses need to be summed. When the input is of long duration, the procedure can be cumbersome, so we use a slight change in perspective to obtain an alternative approach to evaluating the convolution sum in Eq. (2.4).

Recall that the convolution sum is expressed as

$$y[n] = \sum_{k=-\infty}^{\infty} x[k]h[n - k].$$

Suppose we define the intermediate signal

$$w_n[k] = x[k]h[n - k] \tag{2.5}$$

as the product of $x[k]$ and $h[n - k]$. In this definition, k is the independent variable and we explicitly indicate that n is treated as a constant by writing n as a subscript on w. Now, $h[n - k] = h[-(k - n)]$ is a reflected (because of $-k$) and time-shifted (by $-n$) version of $h[k]$. Hence, if n is negative, then $h[n - k]$ is obtained by time shifting $h[-k]$ to the left, while if n is positive, we time shift $h[-k]$ to the right. The time shift n determines the time at which we evaluate the output of the system, since

$$y[n] = \sum_{k=-\infty}^{\infty} w_n[k]. \tag{2.6}$$

Note that now we need only determine one signal, $w_n[k]$, for each time n at which we desire to evaluate the output.

EXAMPLE 2.2 CONVOLUTION SUM EVALUATION BY USING AN INTERMEDIATE SIGNAL
Consider a system with impulse response

$$h[n] = \left(\frac{3}{4}\right)^n u[n].$$

Use Eq. (2.6) to determine the output of the system at times $n = -5$, $n = 5$, and $n = 10$ when the input is $x[n] = u[n]$.

Solution: Here, the impulse response and input are of infinite duration, so the procedure followed in Example 2.1 would require summing a large number of time-shifted impulse responses to determine $y[n]$ for each n. By using Eq. (2.6), we form only one signal, $w_n[k]$, for each n of interest. Figure 2.3(a) depicts $x[k]$ superimposed on the reflected and time-shifted impulse response $h[n - k]$. We see that

$$h[n - k] = \begin{cases} \left(\frac{3}{4}\right)^{n-k}, & k \le n \\ 0, & \text{otherwise} \end{cases}.$$

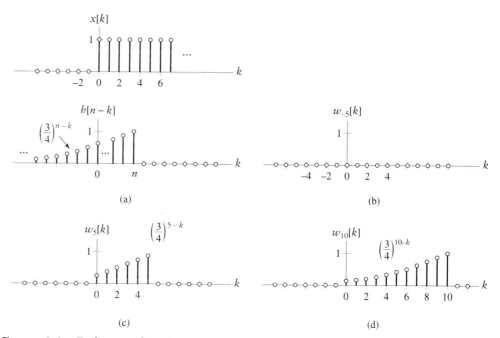

(a)

(b)

(c)

(d)

FIGURE 2.3 Evaluation of Eq. (2.6) in Example 2.2. (a) The input signal $x[k]$ above the reflected and time-shifted impulse response $h[n - k]$, depicted as a function of k. (b) The product signal $w_{-5}[k]$ used to evaluate $y[-5]$. (c) The product signal $w_5[k]$ used to evaluate $y[5]$. (d) The product signal $w_{10}[k]$ used to evaluate $y[10]$.

The intermediate signal $w_n[k]$ is now easily obtained by means of Eq. (2.5). Figures 2.3(b), (c), and (d) depict $w_n[k]$ for $n = -5$, $n = 5$, and $n = 10$, respectively. We have

$$w_{-5}[k] = 0,$$

and thus Eq. (2.6) gives $y[-5] = 0$. For $n = 5$, we have

$$w_5[k] = \begin{cases} \left(\frac{3}{4}\right)^{5-k}, & 0 \leq k \leq 5 \\ 0, & \text{otherwise} \end{cases},$$

so Eq. (2.6) gives

$$y[5] = \sum_{k=0}^{5} \left(\frac{3}{4}\right)^{5-k},$$

which represents the sum of the nonzero values of the intermediate signal $w_5[k]$ shown in Fig. 2.3(c). We then factor $\left(\frac{3}{4}\right)^5$ from the sum and apply the formula for the sum of a finite geometric series (see Appendix A.3) to obtain

$$y[5] = \left(\frac{3}{4}\right)^5 \sum_{k=0}^{5} \left(\frac{4}{3}\right)^k$$

$$= \left(\frac{3}{4}\right)^5 \frac{1 - \left(\frac{4}{3}\right)^6}{1 - \left(\frac{4}{3}\right)} = 3.288.$$

Last, for $n = 10$, we see that

$$w_{10}[k] = \begin{cases} \left(\frac{3}{4}\right)^{10-k}, & 0 \leq k \leq 10 \\ 0, & \text{otherwise} \end{cases},$$

and Eq. (2.6) gives

$$y[10] = \sum_{k=0}^{10} \left(\frac{3}{4}\right)^{10-k}$$

$$= \left(\frac{3}{4}\right)^{10} \sum_{k=0}^{10} \left(\frac{4}{3}\right)^k$$

$$= \left(\frac{3}{4}\right)^{10} \frac{1 - \left(\frac{4}{3}\right)^{11}}{1 - \left(\frac{4}{3}\right)} = 3.831.$$

Note that in this example $w_n[k]$ has only two different mathematical representations. For $n < 0$, we have $w_n[k] = 0$, since there is no overlap between the nonzero portions of $x[k]$ and $h[n - k]$. When $n \geq 0$, the nonzero portions of $x[k]$ and $h[n - k]$ overlap on the interval $0 \leq k \leq n$, and we may write

$$w_n[k] = \begin{cases} \left(\frac{3}{4}\right)^{n-k}, & 0 \leq k \leq n \\ 0, & \text{otherwise} \end{cases}.$$

Hence, we may determine the output for an arbitrary n by using the appropriate mathematical representation for $w_n[k]$ in Eq. (2.6). ■

The preceding example suggests that, in general, we may determine $y[n]$ for all n without evaluating Eq. (2.6) at an infinite number of distinct shifts n. This is accomplished by identifying intervals of n on which $w_n[k]$ has the same mathematical representation. We then need only to evaluate Eq. (2.6) using the $w_n[k]$ associated with each interval. Often, it is helpful to graph both $x[k]$ and $h[n - k]$ in determining $w_n[k]$ and identifying the appropriate intervals of shifts. We now summarize this procedure:

Procedure 2.1: Reflect and Shift Convolution Sum Evaluation

1. Graph both $x[k]$ and $h[n - k]$ as a function of the independent variable k. To determine $h[n - k]$, first reflect $h[k]$ about $k = 0$ to obtain $h[-k]$. Then shift by $-n$.
2. Begin with n large and negative. That is, shift $h[-k]$ to the far left on the time axis.
3. Write the mathematical representation for the intermediate signal $w_n[k]$.
4. Increase the shift n (i.e., move $h[n - k]$ toward the right) until the mathematical representation for $w_n[k]$ changes. The value of n at which the change occurs defines the end of the current interval and the beginning of a new interval.
5. Let n be in the new interval. Repeat steps 3 and 4 until all intervals of time shifts and the corresponding mathematical representations for $w_n[k]$ are identified. This usually implies increasing n to a very large positive number.
6. For each interval of time shifts, sum all the values of the corresponding $w_n[k]$ to obtain $y[n]$ on that interval.

The effect of varying n from $-\infty$ to ∞ is to first shift the reflected impulse response $h[-k]$ far to the left in time and then slide it past $x[k]$ by shifting it towards the right. Transitions in the intervals identified in Step 4 generally occur when a transition in the representation for $h[-k]$ slides through a transition in the representation for $x[k]$. Identification of these intervals is simplified by placing the graph of $h[n - k]$ beneath the graph of $x[k]$. Note that we can sum all the values in $w_n[k]$ as each interval of time shifts is identified (i.e., after Step 4), rather than waiting until all the intervals are identified.

We may interpret the interaction between the system and the input signal in the procedure just described as a moving assembly line acting on a stationary signal. The operations in the assembly line are represented by the values of the impulse response, and the order in which these operations are to be performed corresponds to the time index of each value. The values of the input signal are ordered from left (smaller time indices) to right (larger time indices), so the assembly line must move along the signal from left to right to process the input values in the correct order. The position of the assembly line on the signal is indicated by n. Since the assembly line is moving from left to right, the sequence of operations must be ordered from right to left so that the operations represented by the impulse response are applied in the correct order. This illustrates why the impulse response is reflected in the reflect-and-shift convolution sum evaluation procedure. The output of the assembly line at each position n is the sum of the products of the impulse response values and the corresponding input signal values.

EXAMPLE 2.3 MOVING-AVERAGE SYSTEM: REFLECT-AND-SHIFT CONVOLUTION SUM EVALUATION The output $y[n]$ of the four-point moving-average system introduced in Section 1.10 is related to the input $x[n]$ according to the formula

$$y[n] = \frac{1}{4} \sum_{k=0}^{3} x[n - k].$$

The impulse response $h[n]$ of this system is obtained by letting $x[n] = \delta[n]$, which yields

$$h[n] = \frac{1}{4} (u[n] - u[n - 4]),$$

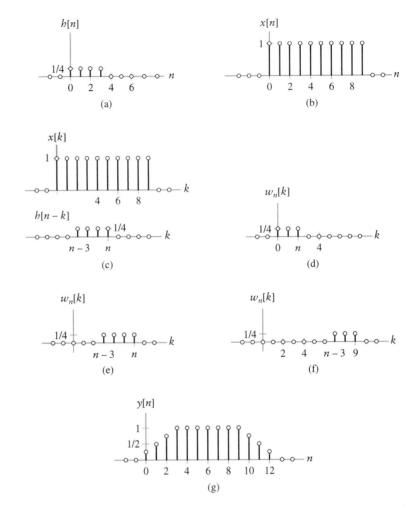

FIGURE 2.4 Evaluation of the convolution sum for Example 2.3. (a) The system impulse response $h[n]$. (b) The input signal $x[n]$. (c) The input above the reflected and time-shifted impulse response $h[n - k]$, depicted as a function of k. (d) The product signal $w_n[k]$ for the interval of shifts $0 \leq n \leq 3$. (e) The product signal $w_n[k]$ for the interval of shifts $3 < n \leq 9$. (f) The product signal $w_n[k]$ for the interval of shifts $9 < n \leq 12$. (g) The output $y[n]$.

as depicted in Fig. 2.4(a). Determine the output of the system when the input is the rectangular pulse defined as

$$x[n] = u[n] - u[n - 10]$$

and shown in Fig. 2.4(b).

Solution: First, we graph $x[k]$ and $h[n - k]$, treating n as a constant and k as the independent variable, as depicted in Fig. 2.4(c). Next, we identify intervals of time shifts on which the intermediate signal $w_n[k] = x[k]h[n - k]$ does not change its mathematical representation. We begin with n large and negative, in which case $w_n[k] = 0$, because there is no overlap in the nonzero portions of $x[k]$ and $h[n - k]$. By increasing n, we see that $w_n[k] = 0$, provided that $n < 0$. Hence, the first interval of shifts is $n < 0$.

When $n = 0$, the right edge of $h[n - k]$ slides past the left edge of $x[k]$, and a transition occurs in the mathematical representation for $w_n[k]$. For $n = 0$,

$$w_0[k] = \begin{cases} 1/4, & k = 0 \\ 0, & \text{otherwise} \end{cases}.$$

For $n = 1$,

$$w_1[k] = \begin{cases} 1/4, & k = 0, 1 \\ 0, & \text{otherwise} \end{cases}.$$

In general, for $n \geq 0$, we may write the mathematical representation for $w_n[k]$ as

$$w_n[k] = \begin{cases} 1/4, & 0 \leq k \leq n \\ 0, & \text{otherwise} \end{cases}.$$

This mathematical representation is depicted in Fig. 2.4(d) and is applicable until $n > 3$. When $n > 3$, the left edge of $h[n - k]$ slides past the left edge of $x[k]$, so the representation of $w_n[k]$ changes. Hence, the second interval of shifts is $0 \leq n \leq 3$.

For $n > 3$, the mathematical representation of $w_n[k]$ is given by

$$w_n[k] = \begin{cases} 1/4, & n - 3 \leq k \leq n \\ 0, & \text{otherwise} \end{cases},$$

as depicted in Fig. 2.4(e). This representation holds until $n = 9$, since at that value of n, the right edge of $h[n - k]$ slides past the right edge of $x[k]$. Thus, our third interval of shifts is $3 < n \leq 9$.

Next, for $n > 9$, the mathematical representation of $w_n[k]$ is given by

$$w_n[k] = \begin{cases} 1/4, & n - 3 \leq k \leq 9 \\ 0, & \text{otherwise} \end{cases},$$

as depicted in Fig. 2.4(f). This representation holds until $n - 3 = 9$, or $n = 12$, since, for $n > 12$, the left edge of $h[n - k]$ lies to the right of $x[k]$, and the mathematical representation for $w_n[k]$ again changes. Hence, the fourth interval of shifts is $9 < n \leq 12$.

For all values of $n > 12$, we see that $w_n[k] = 0$. Thus, the last interval of time shifts in this problem is $n > 12$.

The output of the system on each interval n is obtained by summing the values of the corresponding $w_n[k]$ according to Eq. (2.6). Evaluation of the sums is simplified by noting that

$$\sum_{k=M}^{N} c = c(N - M + 1).$$

Beginning with $n < 0$, we have $y[n] = 0$. Next, for $0 \leq n \leq 3$,

$$y[n] = \sum_{k=0}^{n} 1/4$$

$$= \frac{n + 1}{4}.$$

On the third interval, $3 < n \leq 9$, Eq. (2.6) gives

$$y[n] = \sum_{k=n-3}^{n} 1/4$$
$$= \frac{1}{4}(n - (n-3) + 1)$$
$$= 1.$$

For $9 < n \leq 12$, Eq. (2.6) yields

$$y[n] = \sum_{k=n-3}^{9} 1/4$$
$$= \frac{1}{4}(9 - (n-3) + 1)$$
$$= \frac{13 - n}{4}.$$

Last, for $n > 12$, we see that $y[n] = 0$. Figure 2.4(g) depicts the output $y[n]$ obtained by combining the results on each interval. ■

S & S

Solutions

EXAMPLE 2.4 FIRST-ORDER RECURSIVE SYSTEM: REFLECT-AND-SHIFT CONVOLUTION SUM EVALUATION   The input–output relationship for the first-order recursive system introduced in Section 1.10 is given by

$$y[n] - \rho y[n-1] = x[n].$$

Let the input be given by

$$x[n] = b^n u[n+4].$$

We use convolution to find the output of this system, assuming that $b \neq \rho$ and that the system is causal.

Solution: First we find the impulse response of this system by setting $x[n] = \delta[n]$ so that $y[n]$ corresponds to the impulse response $h[n]$. Thus, we may write

$$h[n] = \rho h[n-1] + \delta[n]. \tag{2.7}$$

Since the system is causal, the impulse response cannot begin before the impulse is applied, and we have $h[n] = 0$ for $n < 0$. Evaluating Eq. (2.7) for $n = 0, 1, 2, \ldots$, we find that $h[0] = 1, h[1] = \rho, h[2] = \rho^2, \ldots$ or

$$h[n] = \rho^n u[n]. \tag{2.8}$$

Next, we graph $x[k]$ and $h[n-k]$, treating time n as a constant and k as the independent variable, as depicted in Fig. 2.5(a). We see that

$$x[k] = \begin{cases} b^k, & -4 \leq k \\ 0, & \text{otherwise} \end{cases}$$

and

$$h[n-k] = \begin{cases} \rho^{n-k}, & k \leq n \\ 0, & \text{otherwise} \end{cases}.$$

Now we identify intervals of time shifts for which the mathematical representation of $w_n[k]$ is the same. We begin by considering n large and negative. We see that, for $n < -4$, $w_n[k] = 0$, since there are no values k such that $x[k]$ and $h[n-k]$ are both nonzero. Hence, the first interval is $n < -4$.

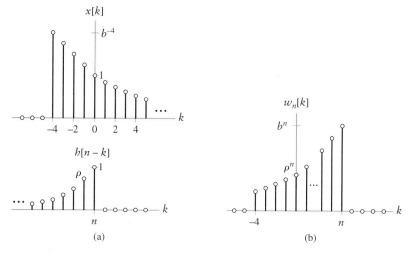

(a)   (b)

S & S
Solutions

FIGURE 2.5 Evaluation of the convolution sum for Example 2.4. (a) The input signal $x[k]$ depicted above the reflected and time-shifted impulse response $h[n - k]$. (b) The product signal $w_n[k]$ for $-4 \leq n$.

When $n = -4$, the right edge of $h[n - k]$ slides past the left edge of $x[k]$, and a transition occurs in the representation of $w_n[k]$. For $n \geq -4$,

$$w_n[k] = \begin{cases} b^k \rho^{n-k}, & -4 \leq k \leq n \\ 0, & \text{otherwise} \end{cases}.$$

This representation is correct, provided that $-4 \leq n$ and is depicted in Fig. 2.5(b).

We next determine the output $y[n]$ for each of these sets of time shifts by summing $w_n[k]$ over all k. Starting with the first interval, $n < -4$, we have $w_n[k] = 0$, and thus, $y[n] = 0$. For the second interval, $-4 \leq n$, we have

$$y[n] = \sum_{k=-4}^{n} b^k \rho^{n-k}.$$

Here, the index of summation is limited from $k = -4$ to n because these are the only times k for which $w_n[k]$ is nonzero. Combining terms raised to the kth power yields

$$y[n] = \rho^n \sum_{k=-4}^{n} \left(\frac{b}{\rho}\right)^k.$$

We may write the sum in a standard form by changing the variable of summation. Let $m = k + 4$; then

$$y[n] = \rho^n \sum_{m=0}^{n+4} \left(\frac{b}{\rho}\right)^{m-4}$$

$$= \rho^n \left(\frac{\rho}{b}\right)^4 \sum_{m=0}^{n+4} \left(\frac{b}{\rho}\right)^m.$$

Next, we apply the formula for summing a geometric series of $n + 5$ terms to obtain

$$y[n] = \rho^n \left(\frac{\rho}{b}\right)^4 \frac{1 - \left(\frac{b}{\rho}\right)^{n+5}}{1 - \frac{b}{\rho}}$$

$$= b^{-4} \left(\frac{\rho^{n+5} - b^{n+5}}{\rho - b}\right).$$

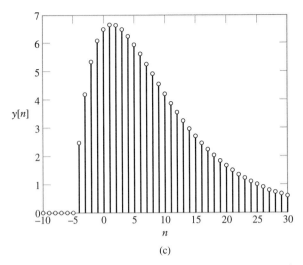

(c)

FIGURE 2.5   (c) The output $y[n]$ assuming that $\rho = 0.9$ and $b = 0.8$.

Combining the solutions for each interval of time shifts gives the system output:

$$y[n] = \begin{cases} 0, & n < -4, \\ b^{-4}\left(\dfrac{\rho^{n+5} - b^{n+5}}{\rho - b}\right), & -4 \le n \end{cases}.$$

Figure 2.5(c) depicts the output $y[n]$, assuming that $\rho = 0.9$ and $b = 0.8$.   ■

EXAMPLE 2.5 INVESTMENT COMPUTATION   The first-order recursive system of Example 2.4 may be used to describe the value of an investment earning compound interest at a fixed rate of $r\%$ per period if we set $\rho = 1 + \frac{r}{100}$. Let $y[n]$ be the value of the investment at the start of period n. If there are no deposits or withdrawals, then the value at time n is expressed in terms of the value at the previous time as $y[n] = \rho y[n-1]$. Now, suppose $x[n]$ is the amount deposited ($x[n] > 0$) or withdrawn ($x[n] < 0$) at the start of period n. In this case, the value of the account is expressed by the first-order recursive equation

$$y[n] = \rho y[n-1] + x[n].$$

We use convolution to find the value of an investment earning 8% per year if \$1000 is deposited at the start of each year for 10 years and then \$1500 is withdrawn at the start of each year for 7 years.

Solution: We expect the account balance to grow for the first 10 years, due to the deposits and accumulated interest. The value of the account will likely decrease during the next 7 years, however, because of the withdrawals, and afterwards the value will continue growing due to interest accrued. We may quantify these predictions by using the reflect-and-shift convolution sum evaluation procedure to evaluate $y[n] = x[n] * h[n]$ where $x[n]$ is depicted in Figure 2.6 and $h[n] = \rho^n u[n]$ is as shown in Example 2.4 with $\rho = 1.08$.

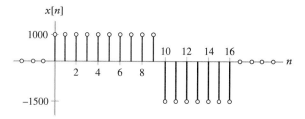

FIGURE 2.6   Cash flow into an investment. Deposits of $1000 are made at the start of each of the first 10 years, while withdrawals of $1500 are made at the start of each of the next 7 years.

First, we graph $x[k]$ above $h[n - k]$, as depicted in Figure 2.7(a). Beginning with n large and negative, we see that $w_n[k] = 0$, provided that $n < 0$. When $n = 0$, the right edge of $h[n - k]$ slides past the left edge of $x[k]$, and a transition occurs in the mathematical representation of $w_n[k]$.

For $n \geq 0$, we may write the mathematical representation of $w_n[k]$ as

$$w_n[k] = \begin{cases} 1000(1.08)^{n-k}, & 0 \leq k \leq n \\ 0, & \text{otherwise} \end{cases}.$$

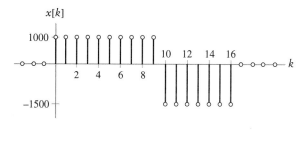

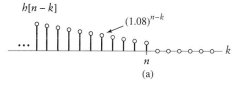

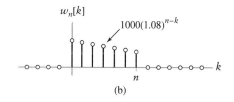

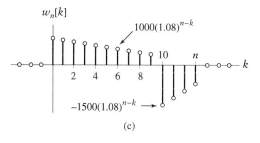

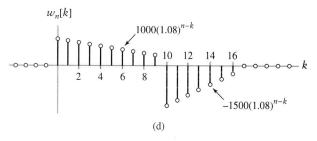

FIGURE 2.7   Evaluation of the convolution sum for Example 2.5. (a) The input signal $x[k]$ depicted above the reflected and time-shifted impulse response $h[n - k]$. (b) The product signal $w_n[k]$ for $0 \leq n \leq 9$. (c) The product signal $w_n[k]$ for $10 \leq n \leq 16$. (d) The product signal $w_n[k]$ for $17 \leq n$.

This mathematical representation is depicted in Figure 2.7(b) and is applicable until $n > 9$, at which point the right edge of $h[n - k]$ begins to slide past the transition from deposits to withdrawals, so the representation of $w_n[k]$ changes. Hence, our second interval of shifts is $0 \leq n \leq 9$. On this range, we obtain

$$y[n] = \sum_{k=0}^{n} 1000(1.08)^{n-k}$$

$$= 1000(1.08)^n \sum_{k=0}^{n} \left(\frac{1}{1.08}\right)^k.$$

Now we apply the formula for summing a geometric series and write

$$y[n] = 1000(1.08)^n \frac{1 - \left(\frac{1}{1.08}\right)^{n+1}}{1 - \frac{1}{1.08}}$$

$$= 12{,}500((1.08)^{n+1} - 1), \qquad 0 \leq n \leq 9.$$

For $n \geq 9$, we may write

$$w_n[k] = \begin{cases} 1000(1.08)^{n-k}, & 0 \leq k \leq 9 \\ -1500(1.08)^{n-k}, & 10 \leq k \leq n, \\ 0, & \text{otherwise} \end{cases}$$

as depicted in Figure 2.7(c). This mathematical representation is applicable until $n > 16$. Hence, our third interval of shifts is $10 \leq n \leq 16$. On this range, we obtain

$$y[n] = \sum_{k=0}^{9} 1000(1.08)^{n-k} - \sum_{k=10}^{n} 1500(1.08)^{n-k}$$

$$= 1000(1.08)^n \sum_{k=0}^{9} \left(\frac{1}{1.08}\right)^k - 1500(1.08)^{n-10} \sum_{m=0}^{n-10} \left(\frac{1}{1.08}\right)^m.$$

Note that in the second sum we changed the index of summation to $m = k - 10$ in order to write the sum in the standard form. Now we apply the formula for summing geometric series to both sums and write

$$y[n] = 1000(1.08)^n \left(\frac{1 - \left(\frac{1}{1.08}\right)^{10}}{1 - \frac{1}{1.08}}\right) - 1500(1.08)^{n-10} \left(\frac{1 - \left(\frac{1}{1.08}\right)^{n-9}}{1 - \frac{1}{1.08}}\right)$$

$$= 7246.89(1.08)^n - 18{,}750((1.08)^{n-9} - 1), \qquad 10 \leq n \leq 16.$$

The last interval of shifts is $17 \leq n$. On this interval, we may write

$$w_n[k] = \begin{cases} 1000(1.08)^{n-k}, & 0 \leq k \leq 9 \\ -1500(1.08)^{n-k}, & 10 \leq k \leq 16, \\ 0, & \text{otherwise} \end{cases}$$

as depicted in Figure 2.7(d). Thus, we obtain

$$y[n] = \sum_{k=0}^{9} 1000(1.08)^{n-k} - \sum_{k=10}^{16} 1500(1.08)^{n-k}.$$

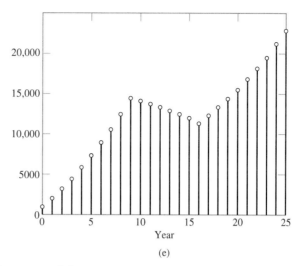

(e)

FIGURE 2.7 (e) The output $y[n]$ representing the value of the investment in U.S. dollars immediately after the deposit or withdrawal at the start of year n.

Using the same techniques as before to evaluate the sums, we get

$$y[n] = 1000(1.08)^{n-9}\frac{(1.08)^{10} - 1}{1.08 - 1} - 1500(1.08)^{n-16}\frac{(1.08)^{7} - 1}{1.08 - 1}$$

$$= 3{,}340.17(1.08)^{n}, \qquad 17 \leq n.$$

Figure 2.7(e) depicts $y[n]$, the value of the investment at the start of each period, by combining the results for each of the four intervals. ∎

▶ **Problem 2.1** Repeat the convolution in Example 2.1, using the convolution sum evaluation procedure.

Answer: See Example 2.1. ◀

S & S
Solutions

▶ **Problem 2.2** Evaluate the following discrete-time convolution sums:

(a) $y[n] = u[n] * u[n - 3]$

(b) $y[n] = (1/2)^{n}u[n - 2] * u[n]$

(c) $y[n] = \alpha^{n}\{u[n - 2] - u[n - 13]\} * 2\{u[n + 2] - u[n - 12]\}$

(d) $y[n] = (-u[n] + 2u[n - 3] - u[n - 6]) * (u[n + 1] - u[n - 10])$

(e) $y[n] = u[n - 2] * h[n]$, where

$$h[n] = \begin{cases} \gamma^{n}, & n < 0, |\gamma| > 1 \\ \eta^{n}, & n \geq 0, |\eta| < 1 \end{cases}$$

(f) $y[n] = x[n] * h[n]$, where $x[n]$ and $h[n]$ are shown in Fig. 2.8.

Answers:

(a)

$$y[n] = \begin{cases} 0, & n < 3 \\ n - 2, & n \geq 3 \end{cases}$$

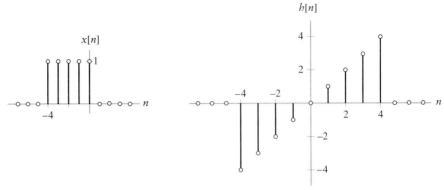

FIGURE 2.8   Signals for Problem 2.2(f).

S & S
Solutions

(b)

$$y[n] = \begin{cases} 0, & n < 2 \\ 1/2 - (1/2)^n, & n \geq 2 \end{cases}$$

(c)

$$y[n] = \begin{cases} 0, & n < 0 \\ 2\alpha^{n+2}\dfrac{1 - (\alpha)^{-1-n}}{1 - \alpha^{-1}}, & 0 \leq n \leq 10 \\ 2\alpha^{12}\dfrac{1 - (\alpha)^{-11}}{1 - \alpha^{-1}}, & 11 \leq n \leq 13 \\ 2\alpha^{12}\dfrac{1 - (\alpha)^{n-24}}{1 - \alpha^{-1}}, & 14 \leq n \leq 23 \\ 0, & n \leq 24 \end{cases}$$

(d)

$$y[n] = \begin{cases} 0, & n < -1 \\ -(n + 2), & -1 \leq n \leq 1 \\ n - 4, & 2 \leq n \leq 4 \\ 0, & 5 \leq n \leq 9 \\ n - 9, & 10 \leq n \leq 11 \\ 15 - n, & 12 \leq n \leq 14 \\ 0, & n > 14 \end{cases}$$

(e)

$$y[n] = \begin{cases} \dfrac{\gamma^{n-1}}{\gamma - 1}, & n < 2 \\ \dfrac{1}{\gamma - 1} + \dfrac{1 - \eta^{n-1}}{1 - \eta}, & n \geq 2 \end{cases}$$

(f)

$$y[n] = \begin{cases} 0, & n < -8, n > 4 \\ -10 + (n + 5)(n + 4)/2, & -8 \leq n \leq -5 \\ 5(n + 2), & -4 \leq n \leq 0 \\ 10 - n(n - 1)/2, & 1 \leq n \leq 4 \end{cases}$$

◀

2.4 *The Convolution Integral*

The output of a continuous-time LTI system may also be determined solely from knowledge of the input and the system's impulse response. The approach and result are analogous to those in the discrete-time case. We first express a continuous-time signal as the weighted superposition of time-shifted impulses:

$$x(t) = \int_{-\infty}^{\infty} x(\tau)\delta(t - \tau)\, d\tau. \tag{2.9}$$

Here, the superposition is an integral instead of a sum, and the time shifts are given by the continuous variable τ. The weights $x(\tau)\, d\tau$ are derived from the value of the signal $x(t)$ at the time τ at which each impulse occurs. Recall that Eq. (2.9) is a statement of the sifting property of the impulse [see Eq. (1.65)].

   Let the operator H denote the system to which the input $x(t)$ is applied. We consider the system output in response to a general input expressed as the weighted superposition in Eq. (2.9):

$$y(t) = H\{x(t)\}$$
$$= H\left\{ \int_{-\infty}^{\infty} x(\tau)\delta(t - \tau)\, d\tau \right\}.$$

Using the linearity property of the system, we may interchange the order of the operator H and integration to obtain

$$y(t) = \int_{-\infty}^{\infty} x(\tau)H\{\delta(t - \tau)\}\, d\tau. \tag{2.10}$$

As in the discrete-time case, the response of a continuous-time linear system to time-shifted impulses completely describes the input–output characteristics of the system.

   Next, we define the impulse response $h(t) = H\{\delta(t)\}$ as the output of the system in response to a unit impulse input. If the system is also time invariant, then

$$H\{\delta(t - \tau)\} = h(t - \tau). \tag{2.11}$$

That is, time invariance implies that a time-shifted impulse input generates a time-shifted impulse response output, as shown in Fig. 2.9. Hence, substituting this result into Eq. (2.10),

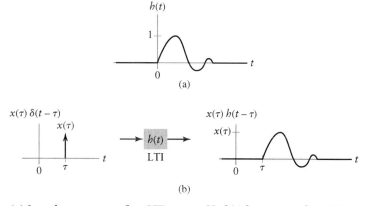

FIGURE 2.9   (a) Impulse response of an LTI system H. (b) The output of an LTI system to a time-shifted and amplitude-scaled impulse is a time-shifted and amplitude-scaled impulse response.

we see that the output of an LTI system in response to an input of the form of Eq. (2.9) may be expressed as

$$y(t) = \int_{-\infty}^{\infty} x(\tau)h(t - \tau)\,d\tau.$$

(2.12)

The output $y(t)$ is given as a weighted superposition of impulse responses time shifted by τ. Equation (2.12) is termed the *convolution integral* and is also denoted by the symbol $*$; that is,

$$x(t) * h(t) = \int_{-\infty}^{\infty} x(\tau)h(t - \tau)\,d\tau.$$

2.5 Convolution Integral Evaluation Procedure

As with the convolution sum, the procedure for evaluating the convolution integral is based on defining an intermediate signal that simplifies the evaluation of the integral. The convolution integral of Eq. (2.12) is expressed as

$$y(t) = \int_{-\infty}^{\infty} x(\tau)h(t - \tau)\,d\tau.$$

(2.13)

We redefine the integrand as the intermediate signal

$$w_t(\tau) = x(\tau)h(t - \tau).$$

In this definition, τ is the independent variable and time t is treated as a constant. This is explicitly indicated by writing t as a subscript and τ within the parentheses of $w_t(\tau)$. Hence, $h(t - \tau) = h(-(\tau - t))$ is a reflected and shifted (by $-t$) version of $h(\tau)$. If $t < 0$, then $h(-\tau)$ is time shifted to the left, while if $t > 0$, then $h(-\tau)$ is shifted to the right. The time shift t determines the time at which we evaluate the output of the system, since Eq. (2.13) becomes

$$y(t) = \int_{-\infty}^{\infty} w_t(\tau)\,d\tau.$$

(2.14)

Thus, the system output at any time t is the area under the signal $w_t(\tau)$.

In general, the mathematical representation of $w_t(\tau)$ depends on the value of t. As in the discrete-time case, we avoid evaluating Eq. (2.14) at an infinite number of values of t by identifying intervals of t on which $w_t(\tau)$ does not change its mathematical representation. We then need only to evaluate Eq. (2.14), using the $w_t(\tau)$ associated with each interval. Often, it is helpful to graph both $x(\tau)$ and $h(t - \tau)$ in determining $w_t(\tau)$ and identifying the appropriate set of shifts. This procedure is summarized as follows:

Procedure 2.2: Reflect-and-Shift Convolution Integral Evaluation

1. Graph $x(\tau)$ and $h(t - \tau)$ as a function of the independent variable τ. To obtain $h(t - \tau)$, reflect $h(\tau)$ about $\tau = 0$ to obtain $h(-\tau)$, and then shift $h(-\tau)$, by $-t$.

2. Begin with the shift t large and negative, that is, shift $h(-\tau)$ to the far left on the time axis.

3. Write the mathematical representation of $w_t(\tau)$.

4. Increase the shift t by moving $h(t - \tau)$ towards the right until the mathematical representation of $w_t(\tau)$ changes. The value t at which the change occurs defines the end of the current set of shifts and the beginning of a new set.

5. Let t be in the new set. Repeat steps 3 and 4 until all sets of shifts t and the corresponding representations of $w_t(\tau)$ are identified. This usually implies increasing t to a large positive value.

6. For each set of shifts t, integrate $w_t(\tau)$ from $\tau = -\infty$ to $\tau = \infty$ to obtain $y(t)$.

The effect of increasing t from a large negative value to a large positive value is to slide $h(-\tau)$ past $x(\tau)$ from left to right. Transitions in the sets of t associated with the same form of $w_t(\tau)$ generally occur when a transition in $h(-\tau)$ slides through a transition in $x(\tau)$. Identification of these intervals is simplified by graphing $h(t - \tau)$ beneath $x(\tau)$. Note that we can integrate $w_t(\tau)$ as each set of shifts is identified (i.e., after Step 4) rather than waiting until all sets are identified. Integration of $w_t(\tau)$ corresponds to finding the signed area under $w_t(\tau)$. The next three examples illustrate this procedure for evaluating the convolution integral.

EXAMPLE 2.6 REFLECT-AND-SHIFT CONVOLUTION EVALUATION   Evaluate the convolution integral for a system with input $x(t)$ and impulse response $h(t)$, respectively, given by

$$x(t) = u(t - 1) - u(t - 3)$$

and

$$h(t) = u(t) - u(t - 2),$$

as depicted in Fig. 2.10.

Solution: To evaluate the convolution integral, we first graph $h(t - \tau)$ beneath the graph of $x(\tau)$, as shown in Fig. 2.11(a). Next, we identify the intervals of time shifts for which the mathematical representation of $w_t(\tau)$ does not change, beginning with t large and negative. Provided that $t < 1$, we have $w_t(\tau) = 0$, since there are no values τ for which both $x(\tau)$ and $h(t - \tau)$ are nonzero. Hence, the first interval of time shifts is $t < 1$.

FIGURE 2.10   Input signal and LTI system impulse response for Example 2.6.

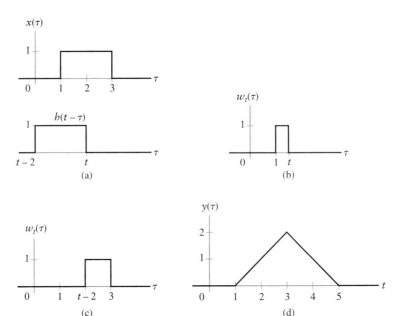

FIGURE 2.11   Evaluation of the convolution integral for Example 2.6. (a) The input $x(\tau)$ depicted above the reflected and time-shifted impulse response $h(t - \tau)$, depicted as a function of τ. (b) The product signal $w_t(\tau)$ for $1 \leq t < 3$. (c) The product signal $w_t(\tau)$ for $3 \leq t < 5$. (d) The system output $y(t)$.

Note that at $t = 1$ the right edge of $h(t - \tau)$ coincides with the left edge of $x(\tau)$. Therefore, as we increase the time shift t beyond 1, we have

$$w_t(\tau) = \begin{cases} 1, & 1 < \tau < t \\ 0, & \text{otherwise} \end{cases}.$$

This representation for $w_t(\tau)$ is depicted in Fig. 2.11(b). It does not change until $t > 3$, at which point both edges of $h(t - \tau)$ pass through the edges of $x(\tau)$. The second interval of time shifts t is thus $1 \leq t < 3$.

As we increase the time shift t beyond 3, we have

$$w_t(\tau) = \begin{cases} 1, & t - 2 < \tau < 3 \\ 0, & \text{otherwise} \end{cases},$$

as depicted in Fig. 2.11(c). This mathematical representation for $w_t(\tau)$ does not change until $t = 5$; thus, the third interval of time shifts is $3 \leq t < 5$.

At $t = 5$, the left edge of $h(t - \tau)$ passes through the right edge of $x(\tau)$, and $w_t(\tau)$ becomes zero. As we continue to increase t beyond 5, $w_t(\tau)$ remains zero, since there are no values τ for which both $x(\tau)$ and $h(t - \tau)$ are nonzero. Hence, the final interval of shifts is $t \geq 5$.

We now determine the output $y(t)$ for each of these four intervals of time shifts by integrating $w_t(\tau)$ over τ (i.e., finding the area under $w_t(\tau)$):

▶ For $t < 1$ and $t > 5$, we have $y(t) = 0$, since $w_t(\tau)$ is zero.

▶ For the second interval, $1 \leq t < 3$, the area under $w_t(\tau)$ shown in Fig. 2.11(b) is $y(t) = t - 1$.

▶ For $3 \leq t < 5$, the area under $w_t(\tau)$ shown in Fig. 2.11(c) is $y(t) = 3 - (t - 2)$.

Combining the solutions for each interval of time shifts gives the output

$$y(t) = \begin{cases} 0, & t < 1 \\ t - 1, & 1 \le t < 3 \\ 5 - t, & 3 \le t < 5 \\ 0, & t \ge 5 \end{cases},$$

as shown in Fig. 2.11(d).   ∎

EXAMPLE 2.7   *RC* CIRCUIT OUTPUT   Consider the *RC* circuit depicted in Fig. 2.12, and assume that the circuit's time constant is $RC = 1$ s. Example 1.21 shows that the impulse response of this circuit is

$$h(t) = e^{-t}u(t).$$

Use convolution to determine the voltage across the capacitor, $y(t)$, resulting from an input voltage $x(t) = u(t) - u(t - 2)$.

Solution: The circuit is linear and time invariant, so the output is the convolution of the input and the impulse response. That is, $y(t) = x(t) * h(t)$. Our intuition from circuit analysis indicates that the capacitor should charge toward the supply voltage in an exponential fashion beginning at time $t = 0$, when the voltage source is turned on, and then, at time $t = 2$, when the voltage source is turned off, start to discharge exponentially.

To verify our intuition using the convolution integral, we first graph $x(\tau)$ and $h(t - \tau)$ as functions of the independent variable τ. We see from Fig. 2.13(a) that

$$x(\tau) = \begin{cases} 1, & 0 < \tau < 2 \\ 0, & \text{otherwise} \end{cases}$$

and

$$h(t - \tau) = e^{-(t-\tau)}u(t - \tau) = \begin{cases} e^{-(t-\tau)}, & \tau < t \\ 0, & \text{otherwise} \end{cases}.$$

Now we identify the intervals of time shifts t for which the mathematical representation of $w_t(\tau)$ does not change. Begin with t large and negative. Provided that $t < 0$, we have $w_t(\tau) = 0$, since there are no values τ for which $x(\tau)$ and $h(t - \tau)$ are both nonzero. Hence, the first interval of shifts is $t < 0$.

Note that at $t = 0$ the right edge of $h(t - \tau)$ intersects the left edge of $x(\tau)$. For $t > 0$,

$$w_t(\tau) = \begin{cases} e^{-(t-\tau)}, & 0 < \tau < t \\ 0, & \text{otherwise} \end{cases}.$$

This representation for $w_t(\tau)$ is depicted in Fig. 2.13(b). It does not change its mathematical representation until $t > 2$, at which point the right edge of $h(t - \tau)$ passes through the right edge of $x(\tau)$. The second interval of shifts t is thus $0 \le t < 2$.

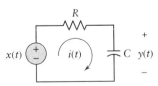

FIGURE 2.12   *RC* circuit system with the voltage source $x(t)$ as input and the voltage measured across the capacitor, $y(t)$, as output.

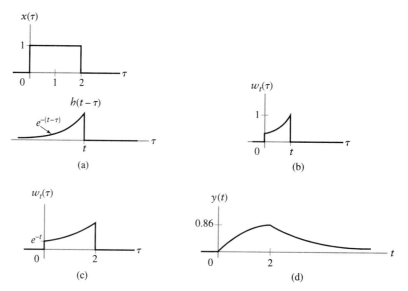

FIGURE 2.13 Evaluation of the convolution integral for Example 2.7. (a) The input $x(\tau)$ super-imposed over the reflected and time-shifted impulse response $h(t - \tau)$, depicted as a function of τ. (b) The product signal $w_t(\tau)$ for $0 \leq t < 2$. (c) The product signal $w_t(\tau)$ for $t \geq 2$. (d) The system output $y(t)$.

For $t \geq 2$, we have a third representation of $w_t(\tau)$, which is written as

$$w_t(\tau) = \begin{cases} e^{-(t-\tau)}, & 0 < \tau < 2 \\ 0, & \text{otherwise} \end{cases}.$$

Figure 2.13(c) depicts $w_t(\tau)$ for this third interval of time shifts, $t \geq 2$.

We now determine the output $y(t)$ for each of the three intervals of time shifts by integrating $w_t(\tau)$ from $\tau = -\infty$ to $\tau = \infty$. Starting with the first interval, $t < 0$, we have $w_t(\tau) = 0$, and thus, $y(t) = 0$. For the second interval, $0 \leq t < 2$,

$$y(t) = \int_0^t e^{-(t-\tau)} d\tau$$
$$= e^{-t}(e^\tau|_0^t)$$
$$= 1 - e^{-t}.$$

For the third interval, $t \geq 2$, we have

$$y(t) = \int_0^2 e^{-(t-\tau)} d\tau$$
$$= e^{-t}(e^\tau|_0^2)$$
$$= (e^2 - 1)e^{-t}.$$

Combining the solutions for the three intervals of time shifts gives the output

$$y(t) = \begin{cases} 0, & t < 0 \\ 1 - e^{-t}, & 0 \leq t < 2, \\ (e^2 - 1)e^{-t}, & t \geq 2 \end{cases}$$

as depicted in Fig. 2.13(d). This result agrees with our intuition from circuit analysis. ■

EXAMPLE 2.8 ANOTHER REFLECT-AND-SHIFT CONVOLUTION EVALUATION   Suppose the input $x(t)$ and impulse response $h(t)$ of an LTI system are, respectively, given by

$$x(t) = (t - 1)[u(t - 1) - u(t - 3)]$$

and

$$h(t) = u(t + 1) - 2u(t - 2).$$

Find the output of this system.

Solution: Graph $x(\tau)$ and $h(t - \tau)$ as shown in Fig. 2.14(a). From these graphical representations, we determine the intervals of time shifts, t, on which the mathematical representation of $w_t(\tau)$ does not change. We begin with t large and negative. For $t + 1 < 1$ or $t < 0$, the right edge of $h(t - \tau)$ is to the left of the nonzero portion of $x(\tau)$, and consequently, $w_t(\tau) = 0$.

  For $t > 0$, the right edge of $h(t - \tau)$ overlaps with the nonzero portion of $x(\tau)$, and we have

$$w_t(\tau) = \begin{cases} \tau - 1, & 1 < \tau < t + 1 \\ 0, & \text{otherwise} \end{cases}.$$

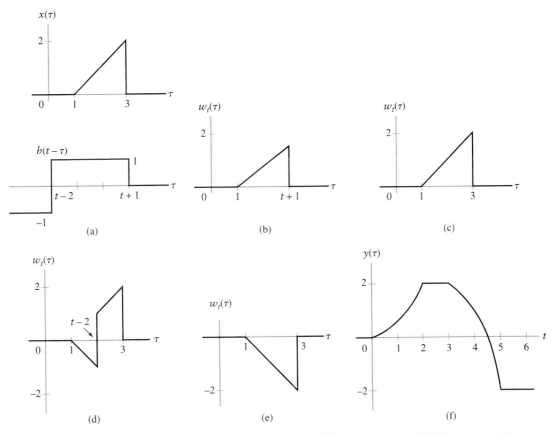

FIGURE 2.14   Evaluation of the convolution integral for Example 2.8. (a) The input $x(\tau)$ superimposed on the reflected and time-shifted impulse response $h(t - \tau)$, depicted as a function of τ. (b) The product signal $w_t(\tau)$ for $0 \leq t < 2$. (c) The product signal $w_t(\tau)$ for $2 \leq t < 3$. (d) The product signal $w_t(\tau)$ for $3 \leq t < 5$. (e) The product signal $w_t(\tau)$ for $t \geq 5$. (f) The system output $y(t)$.

This representation for $w_t(\tau)$ holds provided that $t + 1 < 3$, or $t < 2$, and is depicted in Fig. 2.14(b).

For $t > 2$, the right edge of $h(t - \tau)$ is to the right of the nonzero portion of $x(\tau)$. In this case, we have

$$w_t(\tau) = \begin{cases} \tau - 1, & 1 < \tau < 3 \\ 0, & \text{otherwise} \end{cases}.$$

This representation for $w_t(\tau)$ holds provided that $t - 2 < 1$, or $t < 3$, and is depicted in Fig. 2.14(c).

For $t \geq 3$, the edge of $h(t - \tau)$ at $\tau = t - 2$ is within the nonzero portion of $x(\tau)$, and we have

$$w_t(\tau) = \begin{cases} -(\tau - 1), & 1 < \tau < t - 2 \\ \tau - 1, & t - 2 < \tau < 3 \\ 0, & \text{otherwise} \end{cases}.$$

This representation for $w_t(\tau)$ is depicted in Fig. 2.14(d) and holds provided that $t - 2 < 3$, or $t < 5$.

For $t \geq 5$, we have

$$w_t(\tau) = \begin{cases} -(\tau - 1), & 1 < \tau < 3 \\ 0, & \text{otherwise} \end{cases},$$

as depicted in Fig. 2.14(e).

The system output $y(t)$ is obtained by integrating $w_t(\tau)$ from $\tau = -\infty$ to $\tau = \infty$ for each interval of time shifts just identified. Beginning with $t < 0$, we have $y(t) = 0$, since $w_t(\tau) = 0$. For $0 \leq t < 2$,

$$y(t) = \int_1^{t+1} (\tau - 1)\, d\tau$$
$$= \left(\frac{\tau^2}{2} - \tau \Big|_1^{t+1} \right)$$
$$= \frac{t^2}{2}.$$

For $2 \leq t < 3$, the area under $w_t(\tau)$ is $y(t) = 2$. On the next interval, $3 \leq t < 5$, we have

$$y(t) = -\int_1^{t-2} (\tau - 1)\, d\tau + \int_{t-2}^3 (\tau - 1)\, d\tau$$
$$= -t^2 + 6t - 7.$$

Finally, for $t \geq 5$, the area under $w_t(\tau)$ is $y(t) = -2$. Combining the outputs for the different intervals of time shifts gives the result

$$y(t) = \begin{cases} 0, & t < 0 \\ \frac{t^2}{2}, & 0 \leq t < 2 \\ 2, & 2 \leq t < 3 \\ -t^2 + 6t - 7, & 3 \leq t < 5 \\ -2, & t \geq 5 \end{cases}$$

as depicted in Fig. 2.14(f). ■

▶ **Problem 2.3**   Let the impulse response of an LTI system be $h(t) = e^{-t}u(t)$. Find the output $y(t)$ if the input is $x(t) = u(t)$.

Answer:

$$y(t) = (1 - e^{-t})u(t).$$ ◀

▶ **Problem 2.4**   Let the impulse response of an LTI system be $h(t) = e^{-2(t+1)}u(t + 1)$. Find the output $y(t)$ if the input is $x(t) = e^{-|t|}$.

Answer: For $t < -1$,

$$w_t(\tau) = \begin{cases} e^{-2(t+1)}e^{3\tau}, & -\infty < \tau < t + 1 \\ 0, & \text{otherwise} \end{cases},$$

so

$$y(t) = \frac{1}{3}e^{t+1}.$$

For $t > -1$,

$$w_t(\tau) = \begin{cases} e^{-2(t+1)}e^{3\tau}, & -\infty < \tau < 0 \\ e^{-2(t+1)}e^{\tau}, & 0 < \tau < t + 1 \\ 0, & \text{otherwise} \end{cases}$$

and

$$y(t) = e^{-(t+1)} - \frac{2}{3}e^{-2(t+1)}.$$ ◀

▶ **Problem 2.5**   Let the input $x(t)$ to an LTI system with impulse response $h(t)$ be given in Fig. 2.15. Find the output $y(t)$.

Answer:

$$y(t) = \begin{cases} 0, & t < -4, t > 2 \\ (1/2)t^2 + 4t + 8, & -4 \le t < -3 \\ t + 7/2, & -3 \le t < -2 \\ (-1/2)t^2 - t + 3/2, & -2 \le t < -1 \\ (-1/2)t^2 - t + 3/2, & -1 \le t < 0 \\ 3/2 - t, & 0 \le t < 1 \\ (1/2)t^2 - 2t + 2, & 1 \le t < 2 \end{cases}.$$ ◀

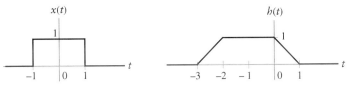

FIGURE 2.15   Signals for Problem 2.5.

▶ **Problem 2.6** Let the impulse response of an LTI system be given by $h(t) = u(t - 1) - u(t - 4)$. Find the output of this system in response to the input $x(t) = u(t) + u(t - 1) - 2u(t - 2)$.

Answer:

$$y(t) = \begin{cases} 0, & t < 1 \\ t - 1, & 1 \leq t < 2 \\ 2t - 3, & 2 \leq t < 3 \\ 3, & 3 \leq t < 4 \\ 7 - t, & 4 \leq t < 5 \\ 12 - 2t, & 5 \leq t < 6 \\ 0, & t \geq 6 \end{cases}.$$

◀

The convolution integral describes the behavior of a continuous-time system. The system impulse response provides insight into the operation of the system. We shall develop this insight in the next section and in subsequent chapters. To pave the way for our development, consider the following example.

S & S
Solutions

Example 2.9 Radar Range Measurement: Propagation Model   In Section 1.10, we introduced the problem of measuring the radar range to an object by transmitting a radio-frequency (RF) pulse and determining the round-trip time delay for the echo of the pulse to return to the radar. In this example, we identify an LTI system describing the propagation of the pulse. Let the transmitted RF pulse be given by

$$x(t) = \begin{cases} \sin(\omega_c t), & 0 \leq t \leq T_o \\ 0, & \text{otherwise} \end{cases},$$

as shown in Fig. 2.16(a).

Suppose we transmit an impulse from the radar to determine the impulse response of the round-trip propagation to the target. The impulse is delayed in time and attenuated in amplitude, which results in the impulse response $h(t) = a\delta(t - \beta)$, where a represents the attenuation factor and β the round-trip time delay. Use the convolution of $x(t)$ with $h(t)$ to verify this result.

Solution: First, find $h(t - \tau)$. Reflecting $h(\tau) = a\delta(\tau - \beta)$ about $\tau = 0$ gives $h(-\tau) = a\delta(\tau + \beta)$, since the impulse has even symmetry. Next, shift the independent variable τ by $-t$ to obtain $h(t - \tau) = a\delta(\tau - (t - \beta))$. Substitute this expression for

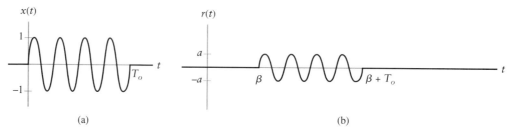

Figure 2.16 Radar range measurement. (a) Transmitted RF pulse. (b) The received echo is an attenuated and delayed version of the transmitted pulse.

$h(t - \tau)$ into the convolution integral of Eq. (2.12), and use the sifting property of the impulse to obtain the received signal as

$$r(t) = \int_{-\infty}^{\infty} x(\tau)a\delta(\tau - (t - \beta))\,d\tau$$

$$= ax(t - \beta).$$

Thus, the received signal is an attenuated and delayed version of the transmitted signal, as shown in Fig. 2.16(b). ∎

The preceding example establishes a useful result for convolution with impulses: The convolution of an arbitrary signal with a time-shifted impulse simply applies the same time shift to the input signal. The analogous result holds for convolution with discrete-time impulses.

▶ **Problem 2.7**   Determine $y(t) = e^{-t}u(t) * \{\delta(t + 1) - \delta(t) + 2\delta(t - 2)\}$.

Answer:

$$y(t) = e^{-(t+1)}u(t + 1) - e^{-t}u(t) + 2e^{-(t-2)}u(t - 2).$$ ◀

S & S
Solutions

EXAMPLE 2.10 RADAR RANGE MEASUREMENT (CONTINUED): THE MATCHED FILTER
In the previous example, the target range is determined by estimating the time delay β from the received signal $r(t)$. In principle, this may be accomplished by measuring the onset time of the received pulse. However, in practice, the received signal is contaminated with noise (e.g., thermal noise, discussed in Section 1.9) and may be weak. For these reasons, the time delay is determined by passing the received signal through an LTI system commonly referred to as a *matched filter*. An important property of this system is that it optimally discriminates against certain types of noise in the received waveform. The impulse response of the matched filter is a reflected, or time-reversed, version of the transmitted signal $x(t)$. That is, $h_m(t) = x(-t)$, so

$$h_m(t) = \begin{cases} -\sin(\omega_c t), & -T_o \le t \le 0, \\ 0, & \text{otherwise} \end{cases},$$

as shown in Fig. 2.17(a). The terminology "matched filter" refers to the fact that the impulse response of the radar receiver is "matched" to the transmitted signal.
    To estimate the time delay from the matched filter output, we evaluate the convolution $y(t) = r(t) * h_m(t)$.

Solution: First, we form $w_t(\tau) = r(\tau)h_m(t - \tau)$. The received signal $r(\tau)$ and the reflected, time-shifted impulse response $h_m(t - \tau)$ are shown in Fig. 2.17(b). Note that since

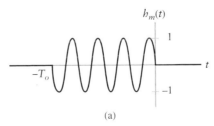

(a)

FIGURE 2.17   (a) Impulse response of the matched filter for processing the received signal.

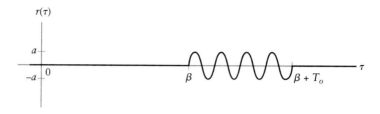

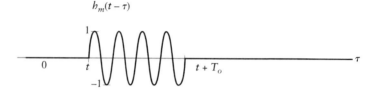

(b)

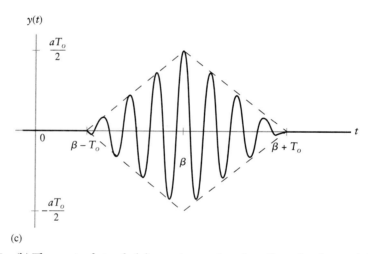

(c)

FIGURE 2.17 (b) The received signal $r(\tau)$ superimposed on the reflected and time-shifted matched filter impulse response $h_m(t - \tau)$, depicted as functions of τ. (c) Matched filter output $y(t)$.

$h_m(\tau)$ is a reflected version of $x(t)$, we have $h_m(t - \tau) = x(\tau - t)$. If $t + T_o < \beta$, then $w_t(\tau) = 0$, and thus, $y(t) = 0$ for $t < \beta - T_o$. When $\beta - T_o < t < \beta$,

$$w_t(\tau) = \begin{cases} a \sin(\omega_c(\tau - \beta)) \sin(\omega_c(\tau - t)), & \beta < \tau < t + T_o \\ 0, & \text{otherwise} \end{cases}.$$

Next, we apply the identity for the product of sine functions to redefine $w_t(\tau)$ and write $y(t)$ for $\beta - T_o < t \leq \beta$ as

$$y(t) = \int_{\beta}^{t+T_o} [(a/2) \cos(\omega_c(t - \beta)) + (a/2) \cos(\omega_c(2\tau - \beta - t))] \, d\tau$$

$$= (a/2) \cos(\omega_c(t - \beta))[t + T_o - \beta] + (a/4\omega_c) \sin(\omega_c(2\tau - \beta - t))|_{\beta}^{t+T_o}$$

$$= (a/2) \cos(\omega_c(t - \beta))[t - (\beta - T_o)] + (a/4\omega_c)[\sin(\omega_c(t + 2T_o - \beta))$$

$$- \sin(\omega_c(\beta - t))].$$

In practice we typically have $\omega_c > 10^6$ rad/s; thus, the se̶c̶
tions makes only a negligible contribution to the output, beca̶
$a/4\omega_c$. Similarly, when $\beta < t < \beta + T_o$, we have

$$w_t(\tau) = \begin{cases} a \sin(\omega_c(\tau - \beta)) \sin(\omega_c(\tau - t)), & t < \tau < \beta + \\ 0, & \text{otherwise} \end{cases}$$

and

$$y(t) = \int_t^{\beta + T_o} \left[(a/2) \cos(\omega_c(t - \beta)) + (a/2) \cos(\omega_c(2\tau - \beta - t)) \right] d\tau$$

$$= (a/2) \cos(\omega_c(t - \beta))[\beta + T_o - t] + (a/4\omega_c) \sin(\omega_c(2\tau - \beta - t))|_t^{\beta + T_o}$$

$$= (a/2) \cos(\omega_c(t - \beta))[\beta - t + T_o] + (a/4\omega_c)[\sin(\omega_c(\beta + 2T_o - t)) - \sin(\omega_c(t - \beta))].$$

Here again, division by ω_c renders the second term involving the sine functions negligible. The last interval is $\beta + T_o < t$. On this interval, $w_t(\tau) = 0$, so $y(t) = 0$. Combining the solutions for all three intervals and ignoring the negligibly small terms gives the output of the matched filter:

$$y(t) = \begin{cases} (a/2)[t - (\beta - T_o)] \cos(\omega_c(t - \beta)), & \beta - T_o < t \leq \beta \\ (a/2)[\beta - t + T_o] \cos(\omega_c(t - \beta)), & \beta < t < \beta + T_o. \\ 0, & \text{otherwise} \end{cases}$$

A sketch of the matched filter output is shown in Fig. 2.17(b). The envelope of $y(t)$ is a triangular waveform, as shown by the dashed lines. The peak value occurs at the round-trip time delay of interest, $t = \beta$. Thus, β is estimated by finding the time at which the matched filter output reaches its peak value. Estimating the round-trip time delay from the peak of the matched filter output gives much more accurate results in the presence of noise than finding the time at which the echo starts in $r(t)$ hence the common use of matched filtering in practice.   ∎

2.6   *Interconnections of LTI Systems*

In this section, we develop the relationships between the impulse response of an interconnection of LTI systems and the impulse responses of the constituent systems. The results for continuous- and discrete-time systems are obtained by using nearly identical approaches, so we derive the continuous-time results and then simply state the discrete-time results.

■ 2.6.1   PARALLEL CONNECTION OF LTI SYSTEMS

Consider two LTI systems with impulse responses $h_1(t)$ and $h_2(t)$ connected in parallel, as illustrated in Fig. 2.18(a). The output of this connection of systems, $y(t)$, is the sum of the outputs of the two systems:

$$y(t) = y_1(t) + y_2(t)$$
$$= x(t) * h_1(t) + x(t) * h_2(t).$$

We substitute the integral representation of each convolution:

$$y(t) = \int_{-\infty}^{\infty} x(\tau)h_1(t - \tau) \, d\tau + \int_{-\infty}^{\infty} x(\tau)h_2(t - \tau) \, d\tau.$$

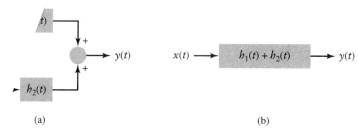

(a)   (b)

.terconnection of two LTI systems. (a) Parallel connection of two systems. (b) Equiv-

, a common input, the two integrals are combined to obtain

$$y(t) = \int_{-\infty}^{\infty} x(\tau)\{h_1(t - \tau) + h_2(t - \tau)\}\, d\tau$$

$$= \int_{-\infty}^{\infty} x(\tau)h(t - \tau)\, d\tau$$

$$= x(t) * h(t),$$

where $h(t) = h_1(t) + h_2(t)$. We identify $h(t)$ as the impulse response of the equivalent system representing the parallel connection of the two systems. This equivalent system is depicted in Fig. 2.18(b). The impulse response of the overall system represented by the two LTI systems connected in parallel is the sum of their individual impulse responses.

Mathematically, the preceding result implies that convolution possesses the *distributive property*:

$$x(t) * h_1(t) + x(t) * h_2(t) = x(t) * \{h_1(t) + h_2(t)\}. \tag{2.15}$$

Identical results hold for the discrete-time case:

$$x[n] * h_1[n] + x[n] * h_2[n] = x[n] * \{h_1[n] + h_2[n]\}. \tag{2.16}$$

■ 2.6.2 CASCADE CONNECTION OF SYSTEMS

Consider next the cascade connection of two LTI systems, as illustrated in Fig. 2.19(a). Let $z(t)$ be the output of the first system and therefore the input to the second system in the cascade. The output is expressed in terms of $z(t)$ as

$$y(t) = z(t) * h_2(t), \tag{2.17}$$

FIGURE 2.19 Interconnection of two LTI systems. (a) Cascade connection of two systems. (b) Equivalent system. (c) Equivalent system: Interchange system order.

or

$$y(t) = \int_{-\infty}^{\infty} z(\tau)h_2(t - \tau)\, d\tau. \tag{2.18}$$

Since $z(\tau)$ is the output of the first system, it is expressed in terms of the input $x(\tau)$ as

$$z(\tau) = x(\tau) * h_1(\tau)$$
$$= \int_{-\infty}^{\infty} x(\nu)h_1(\tau - \nu)\, d\nu, \tag{2.19}$$

where ν is used as the variable of integration in the convolution integral. Substituting Eq. (2.19) for $z(\tau)$ into Eq. (2.18) gives

$$y(t) = \int_{-\infty}^{\infty} \int_{-\infty}^{\infty} x(\nu)h_1(\tau - \nu)h_2(t - \tau)\, d\nu d\tau.$$

Now we perform the change of variable $\eta = \tau - \nu$ and interchange the order of integration to obtain

$$y(t) = \int_{-\infty}^{\infty} x(\nu) \left[\int_{-\infty}^{\infty} h_1(\eta)h_2(t - \nu - \eta)\, d\eta \right] d\nu. \tag{2.20}$$

The inner integral is identified as the convolution of $h_1(t)$ with $h_2(t)$, evaluated at $t - \nu$. That is, if we define $h(t) = h_1(t) * h_2(t)$, then

$$h(t - \nu) = \int_{-\infty}^{\infty} h_1(\eta)h_2(t - \nu - \eta)\, d\eta.$$

Substituting this relationship into Eq. (2.20) yields

$$y(t) = \int_{-\infty}^{\infty} x(\nu)h(t - \nu)\, d\nu$$
$$= x(t) * h(t). \tag{2.21}$$

Hence, the impulse response of an equivalent system representing two LTI systems connected in cascade is the convolution of their individual impulse responses. The cascade connection is input–output equivalent to the single system represented by the impulse response $h(t)$, as shown in Fig. 2.19(b).

Substituting $z(t) = x(t) * h_1(t)$ into the expression for $y(t)$ given in Eq. (2.17) and $h(t) = h_1(t) * h_2(t)$ into the alternative expression for $y(t)$ given in Eq. (2.21) establishes the fact that convolution possesses the *associative property*; that is,

$$\boxed{\{x(t) * h_1(t)\} * h_2(t) - x(t) * \{h_1(t) * h_2(t)\}.} \tag{2.22}$$

A second important property for the cascade connection of LTI systems concerns the ordering of the systems. We write $h(t) = h_1(t) * h_2(t)$ as the integral

$$h(t) = \int_{-\infty}^{\infty} h_1(\tau)h_2(t - \tau)\, d\tau$$

and perform the change of variable $\nu = t - \tau$ to obtain

$$h(t) = \int_{-\infty}^{\infty} h_1(t - \nu)h_2(\nu)\, d\nu \tag{2.23}$$
$$= h_2(t) * h_1(t).$$

Hence, the convolution of $h_1(t)$ and $h_2(t)$ can be performed in either order. This corresponds to interchanging the order of the LTI systems in the cascade without affecting the result, as shown in Fig. 2.19(c). Since

$$x(t) * \{h_1(t) * h_2(t)\} = x(t) * \{h_2(t) * h_1(t)\},$$

we conclude that the output of a cascade combination of LTI systems is independent of the order in which the systems are connected. Mathematically, we say that the convolution operation possesses the *commutative property*, or

$$\boxed{h_1(t) * h_2(t) = h_2(t) * h_1(t).} \tag{2.24}$$

The commutative property is often used to simplify the evaluation or interpretation of the convolution integral.

   Discrete-time LTI systems and convolutions have properties that are identical to their continuous-time counterparts. For example, the impulse response of a cascade connection of LTI systems is given by the convolution of the individual impulse responses, and the output of a cascade combination of LTI systems is independent of the order in which the systems are connected. Also, discrete-time convolution is associative, so that

$$\boxed{\{x[n] * h_1[n]\} * h_2[n] = x[n] * \{h_1[n] * h_2[n]\},} \tag{2.25}$$

and commutative, or

$$\boxed{h_1[n] * h_2[n] = h_2[n] * h_1[n].} \tag{2.26}$$

The next example demonstrates the use of convolution properties in finding a single system that is input–output equivalent to an interconnected system.

EXAMPLE 2.11 EQUIVALENT SYSTEM TO FOUR INTERCONNECTED SYSTEMS   Consider the interconnection of four LTI systems, as depicted in Fig. 2.20. The impulse responses of the systems are

$$h_1[n] = u[n],$$
$$h_2[n] = u[n + 2] - u[n],$$
$$h_3[n] = \delta[n - 2],$$

and

$$h_4[n] = \alpha^n u[n].$$

Find the impulse response $h[n]$ of the overall system.

Solution: We first derive an expression for the overall impulse response in terms of the impulse response of each system. We begin with the parallel combination of $h_1[n]$ and $h_2[n]$. The distributive property implies that the equivalent system has the impulse response $h_{12}[n] = h_1[n] + h_2[n]$, as illustrated in Fig. 2.21(a). This system is in series with $h_3[n]$, so the associative property implies that the equivalent system for the upper branch has the impulse response $h_{123}[n] = h_{12}[n] * h_3[n]$. Substituting for $h_{12}[n]$ in this expression, we have $h_{123}[n] = (h_1[n] + h_2[n]) * h_3[n]$, as depicted in Fig. 2.21(b). Last, the upper branch

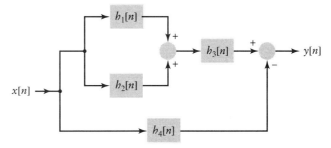

FIGURE 2.20 Interconnection of systems for Example 2.11.

is in parallel with the lower branch, characterized by $h_4[n]$; hence, application of the distributive property gives the overall system impulse response as $h[n] = h_{123}[n] - h_4[n]$. Substituting for $h_{123}[n]$ in this expression yields

$$h[n] = (h_1[n] + h_2[n]) * h_3[n] - h_4[n],$$

as shown in Fig. 2.21(c).

Now substitute the specified forms of $h_1[n]$ and $h_2[n]$ to obtain

$$h_{12}[n] = u[n] + u[n + 2] - u[n]$$
$$= u[n + 2].$$

Convolving $h_{12}[n]$ with $h_3[n]$ gives

$$h_{123}[n] = u[n + 2] * \delta[n - 2]$$
$$= u[n].$$

Finally, we sum $h_{123}[n]$ and $-h_4[n]$ to obtain the overall impulse response:

$$h[n] = \{1 - \alpha^n\}u[n]. \qquad \blacksquare$$

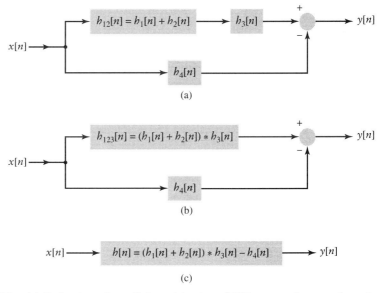

FIGURE 2.21 (a) Reduction of parallel combination of LTI systems in upper branch of Fig. 2.20. (b) Reduction of cascade of systems in upper branch of Fig. 2.21(a). (c) Reduction of parallel combination of systems in Fig. 2.21(b) to obtain an equivalent system for Fig. 2.20.

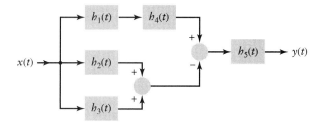

FIGURE 2.22   Interconnection of LTI systems for Problem 2.8.

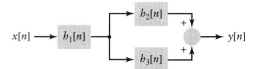

FIGURE 2.23   Interconnection of LTI systems for Problem 2.9.

▶ **Problem 2.8**   Find the expression for the impulse response relating the input $x(t)$ to the output $y(t)$ for the system depicted in Fig. 2.22.

Answer:

$$h(t) = [h_1(t) * h_4(t) - h_2(t) - h_3(t)] * h_5(t).$$ ◀

▶ **Problem 2.9**   An interconnection of LTI systems is depicted in Fig. 2.23. The impulse responses are $h_1[n] = (1/2)^n u[n + 2]$, $h_2[n] = \delta[n]$, and $h_3[n] = u[n - 1]$. Let the overall impulse response of the system relating $y[n]$ to $x[n]$ be denoted as $h[n]$.

(a) Express $h[n]$ in terms of $h_1[n]$, $h_2[n]$, and $h_3[n]$.

(b) Evaluate $h[n]$, using the results of Part (a).

Answers:

(a) $h[n] = h_1[n] * h_2[n] + h_1[n] * h_3[n]$

(b) $h[n] = (1/2)^n u[n + 2] + (8 - (1/2)^{n-1})u[n + 1]$ ◀

Interconnections among systems arise naturally out of the process of analyzing the systems. Often, it is easier to break a complex system into simpler subsystems, analyze each subsystem, and then study the entire system as an interconnection of subsystems than it is to analyze the overall system directly. This is an example of the "divide-and-conquer" approach to problem solving and is possible due to the assumptions of linearity and time invariance.

Table 2.1 summarizes the interconnection properties presented in this section.

Property	Continuous-time system	Discrete-time system
Distributive	$x(t) * h_1(t) + x(t) * h_2(t) =$ $x(t) * \{h_1(t) + h_2(t)\}$	$x[n] * h_1[n] + x[n] * h_2[n] =$ $x[n] * \{h_1[n] + h_2[n]\}$
Associative	$\{x(t) * h_1(t)\} * h_2(t) = x(t) * \{h_1(t) * h_2(t)\}$	$\{x[n] * h_1[n]\} * h_2[n] = x[n] * \{h_1[n] * h_2[n]\}$
Commutative	$h_1(t) * h_2(t) = h_2(t) * h_1(t)$	$h_1[n] * h_2[n] = h_2[n] * h_1[n]$

TABLE 2.1 *Interconnection Properties for LTI Systems.*

2.7 Relations between LTI System Properties and the Impulse Response

The impulse response completely characterizes the input–output behavior of an LTI system. Hence, properties of the system, such as memory, causality, and stability, are related to the system's impulse response. In this section, we explore the relationships involved.

■ 2.7.1 MEMORYLESS LTI SYSTEMS

We recall from Section 1.8.2 that the output of a memoryless LTI system depends only on the current input. Exploiting the commutative property of convolution, we may express the output of a discrete-time LTI system as

$$y[n] = h[n] * x[n]$$
$$= \sum_{k=-\infty}^{\infty} h[k]x[n-k].$$

It is instructive to expand the sum term by term:

$$y[n] = \cdots + h[-2]x[n+2] + h[-1]x[n+1] + h[0]x[n]$$
$$+ h[1]x[n-1] + h[2]x[n-2] + \cdots \tag{2.27}$$

For this system to be memoryless, $y[n]$ must depend only on $x[n]$ and therefore cannot depend on $x[n-k]$ for $k \neq 0$. Hence, every term in Eq. (2.27) must be zero, except $h[0]x[n]$. This condition implies that $h[k] = 0$ for $k \neq 0$; thus, a discrete-time LTI system is memoryless if and only if

$$\boxed{h[k] = c\delta[k],}$$

where c is an arbitrary constant.

Writing the output of a continuous-time system as

$$y(t) = \int_{-\infty}^{\infty} h(\tau)x(t-\tau)\,d\tau,$$

we see that, analogously to the discrete-time case, a continuous-time LTI system is memoryless if and only if

$$\boxed{h(\tau) = c\delta(\tau),}$$

for c an arbitrary constant.

The memoryless condition places severe restrictions on the form of the impulse response: All memoryless LTI systems simply perform scalar multiplication on the input.

■ 2.7.2 CAUSAL LTI SYSTEMS

The output of a causal LTI system depends only on past or present values of the input. Again, we write the convolution sum as

$$y[n] = \cdots + h[-2]x[n+2] + h[-1]x[n+1] + h[0]x[n]$$
$$+ h[1]x[n-1] + h[2]x[n-2] + \cdots.$$

We see that past and present values of the input, $x[n], x[n-1], x[n-2], \ldots$, are associated with indices $k \geq 0$ in the impulse response $h[k]$, while future values of the input, $x[n+1], x[n+2], \ldots$, are associated with indices $k < 0$. In order, then, for $y[n]$ to depend only on past or present values of the input, we require that $h[k] = 0$ for $k < 0$. Hence, for a discrete-time causal LTI system,

$$\boxed{h[k] = 0 \quad \text{for} \quad k < 0,}$$

and the convolution sum takes the new form

$$y[n] = \sum_{k=0}^{\infty} h[k]x[n-k].$$

The causality condition for a continuous-time system follows in an analogous manner from the convolution integral

$$y(t) = \int_{-\infty}^{\infty} h(\tau)x(t-\tau)\, d\tau.$$

A causal continuous-time LTI system has an impulse response that satisfies the condition

$$\boxed{h(\tau) = 0 \quad \text{for} \quad \tau < 0.}$$

The output of a continuous-time causal LTI system is thus expressed as the convolution integral

$$y(t) = \int_{0}^{\infty} h(\tau)x(t-\tau)\, d\tau.$$

The causality condition is intuitively satisfying. Recall that the impulse response is the output of a system in response to a unit-strength impulse input applied at time $t = 0$. Note that causal systems are nonanticipatory; that is, they cannot generate an output before the input is applied. Requiring the impulse response to be zero for negative time is equivalent to saying that the system cannot respond with an output prior to application of the impulse.

■ 2.7.3 STABLE LTI SYSTEMS

We recall from Section 1.8.1 that a system is bounded input–bounded output (BIBO) stable if the output is guaranteed to be bounded for every bounded input. Formally, if the input to a stable discrete-time system satisfies $|x[n]| \leq M_x \leq \infty$, then the output must satisfy $|y[n]| \leq M_y \leq \infty$. We shall now derive conditions on $h[n]$ that guarantee stability of the system by bounding the convolution sum. The magnitude of the output is given by

$$|y[n]| = |h[n] * x[n]|$$
$$= \left| \sum_{k=-\infty}^{\infty} h[k]x[n-k] \right|.$$

We seek an upper bound on $|y[n]|$ that is a function of the upper bound on $|x[n]|$ and the impulse response. The magnitude of a sum of terms is less than or equal to the sum of their magnitudes; for example, $|a + b| \leq |a| + |b|$. Accordingly, we may write

$$|y[n]| \leq \sum_{k=-\infty}^{\infty} |h[k]x[n-k]|.$$

Furthermore, the magnitude of a product of terms is equal to the product of their magnitudes; for example, $|ab| = |a||b|$. Thus, we have

$$|y[n]| \leq \sum_{k=-\infty}^{\infty} |h[k]||x[n-k]|.$$

If we assume that the input is bounded, or $|x[n]| \leq M_x < \infty$, then $|x[n-k]| \leq M_x$, and it follows that

$$|y[n]| \leq M_x \sum_{k=\infty}^{\infty} |h[k]|. \tag{2.28}$$

Hence, the output is bounded, or $|y[n]| \leq \infty$ for all n, provided that the impulse response of the system is absolutely summable. We conclude that the impulse response of a stable discrete-time LTI system satisfies the bound

$$\boxed{\sum_{k=-\infty}^{\infty} |h[k]| < \infty.}$$

Our derivation so far has established absolute summability of the impulse response as a sufficient condition for BIBO stability. In Problem 2.79, the reader is asked to show that this is also a necessary condition for BIBO stability.

A similar set of steps may be used to establish the fact that a continuous-time LTI system is BIBO stable if and only if the impulse response is absolutely integrable—that is, if and only if

$$\boxed{\int_{-\infty}^{\infty} |h(\tau)| \, d\tau < \infty.}$$

S & S
Solutions

EXAMPLE 2.12 PROPERTIES OF THE FIRST-ORDER RECURSIVE SYSTEM The first-order system introduced in Section 1.10 is described by the difference equation

$$y[n] = \rho y[n-1] + x[n]$$

and has the impulse response

$$h[n] = \rho^n u[n].$$

Is this system causal, memoryless, and BIBO stable?

Solution: The system is causal, since the impulse response $h[n]$ is zero for $n < 0$. The system is not memoryless, because $h[n]$ is nonzero for all values $n > 0$. The stability of the system is determined by checking whether the impulse response is absolutely summable, or, mathematically, whether

$$\sum_{k=-\infty}^{\infty} |h[k]| = \sum_{k=0}^{\infty} |\rho^k|$$

$$= \sum_{k=0}^{\infty} |\rho|^k < \infty.$$

The infinite geometric sum in the second line converges if and only if $|\rho| < 1$. Hence, the system is stable, provided that $|\rho| < 1$. Recall from Example 2.5 that the first-order recursive equation may be used to describe the value of an investment or loan by setting $\rho = 1 + \frac{r}{100}$, where $r > 0$ is used to represent the percentage interest rate per period. Thus, we find that interest calculations involve an unstable system. This is consistent with our intuition: When payments are not made on a loan, the balance outstanding continues to grow.   ■

▶ **Problem 2.10**   For each of the following impulse responses, determine whether the corresponding system is (i) memoryless, (ii) causal, and (iii) stable. Justify your answers.

(a) $h(t) = u(t + 1) - u(t - 1)$
(b) $h(t) = u(t) - 2u(t - 1)$
(c) $h(t) = e^{-2|t|}$
(d) $h(t) = e^{at}u(t)$
(e) $h[n] = 2^n u[-n]$
(f) $h[n] = e^{2n}u[n - 1]$
(g) $h[n] = (1/2)^n u[n]$

Answers:

(a) not memoryless, not causal, stable.
(b) not memoryless, causal, not stable.
(c) not memoryless, not causal, stable.
(d) not memoryless, causal, stable provided that $a < 0$.
(e) not memoryless, not causal, stable.
(f) not memoryless, causal, not stable.
(g) not memoryless, causal, stable.   ◀

We emphasize that a system can be unstable even though the impulse response has a finite value. For example, consider the ideal integrator, defined by the input–output relationship

$$y(t) = \int_{-\infty}^{t} x(\tau)\, d\tau. \tag{2.29}$$

Recall from Eq. (1.63) that the integral of an impulse is a step. Hence, the application of an impulse input $x(\tau) = \delta(\tau)$ shows that the impulse response of the ideal integrator is given by $h(t) = u(t)$. This impulse response is never greater than unity, but is not absolutely integrable, and thus, the system is unstable. Although the output of the system, as defined in Eq. (2.29), is bounded for some bounded inputs $x(t)$, it is not bounded for *every* bounded input. In particular, the constant input $x(t) = c$ clearly results in an unbounded output. A similar observation applies to the discrete-time ideal accumulator introduced in Section 1.10. The input–output equation of the ideal accumulator is

$$y[n] = \sum_{k=-\infty}^{n} x[k].$$

Thus, the impulse response is $h[n] = u[n]$, which is not absolutely summable, so the ideal accumulator is not stable. Note that the constant input $x[n] = c$ results in an unbounded output.

▶ **Problem 2.11**   A discrete-time system has impulse response $h[n] = \cos\left(\frac{\pi}{2}n\right)u[n + 3]$. Is the system stable, causal, or memoryless? Justify your answers.

Answer: The system is not stable, not causal, and not memoryless.   ◀

■ 2.7.4 INVERTIBLE SYSTEMS AND DECONVOLUTION

A system is *invertible* if the input to the system can be recovered from the output except for a constant scale factor. This requirement implies the existence of an inverse system that takes the output of the original system as its input and produces the input of the original system. We shall limit ourselves here to a consideration of inverse systems that are LTI. Figure 2.24 depicts the cascade of an LTI system having impulse response $h(t)$ with an LTI inverse system whose impulse response is denoted as $h^{\text{inv}}(t)$.

The process of recovering $x(t)$ from $h(t) * x(t)$ is termed *deconvolution*, since it corresponds to reversing or undoing the convolution operation. An inverse system performs deconvolution. Deconvolution problems and inverse systems play an important role in many signal-processing and systems applications. A common problem is that of reversing or "equalizing" the distortion introduced by a nonideal system. For example, consider the use of a high-speed modem to communicate over telephone lines. Distortion introduced by the telephone channel places severe restrictions on the rate at which information can be transmitted, so an equalizer is incorporated into the modem. The equalizer reverses the distortion and permits much higher data rates to be achieved. In this case, the equalizer represents an inverse system for the telephone channel. In practice, the presence of noise complicates the equalization problem. (We shall discuss equalization in more detail in Chapters 5 and 8.)

The relationship between the impulse response of an LTI system, $h(t)$, and that of the corresponding inverse system, $h^{\text{inv}}(t)$, is easily derived. The impulse response of the cascade connection in Fig. 2.24 is the convolution of $h(t)$ and $h^{\text{inv}}(t)$. We require the output of the cascade to equal the input, or

$$x(t) * (h(t) * h^{\text{inv}}(t)) = x(t).$$

This requirement implies that

$$\boxed{h(t) * h^{\text{inv}}(t) = \delta(t).} \tag{2.30}$$

Similarly, the impulse response of a discrete-time LTI inverse system, $h^{\text{inv}}[n]$, must satisfy

$$\boxed{h[n] * h^{\text{inv}}[n] = \delta[n].} \tag{2.31}$$

In many equalization applications, an exact inverse system may be difficult to find or implement. An approximate solution of Eq. (2.30) or Eq. (2.31) is often sufficient in such cases. The next example illustrates a case where an exact inverse system is obtained by directly solving Eq. (2.31).

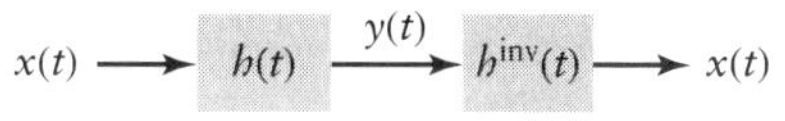

$$x(t) \longrightarrow \boxed{h(t)} \xrightarrow{y(t)} \boxed{h^{\text{inv}}(t)} \longrightarrow x(t)$$

FIGURE 2.24   Cascade of LTI system with impulse response $h(t)$ and inverse system with impulse response $h^{\text{inv}}(t)$.

EXAMPLE 2.13 MULTIPATH COMMUNICATION CHANNELS: COMPENSATION BY MEANS OF AN INVERSE SYSTEM Consider designing a discrete-time inverse system to eliminate the distortion associated with multipath propagation in a data transmission problem. Recall from Section 1.10 that a discrete-time model for a two-path communication channel is

$$y[n] = x[n] + ax[n - 1].$$

Find a causal inverse system that recovers $x[n]$ from $y[n]$. Check whether this inverse system is stable.

Solution: First we identify the impulse response of the system relating $y[n]$ and $x[n]$. We apply an impulse input $x[n] = \delta[n]$ to obtain the impulse response

$$h[n] = \begin{cases} 1, & n = 0 \\ a, & n = 1 \\ 0, & \text{otherwise} \end{cases}$$

as the impulse response of the multipath channel. The inverse system $h^{\text{inv}}[n]$ must satisfy $h[n] * h^{\text{inv}}[n] = \delta[n]$. Substituting for $h[n]$, we see that $h^{\text{inv}}[n]$ must satisfy the equation

$$h^{\text{inv}}[n] + ah^{\text{inv}}[n - 1] = \delta[n]. \tag{2.32}$$

Let us solve this equation for several different values of n. For $n < 0$, we must have $h^{\text{inv}}[n] = 0$ in order to obtain a causal inverse system. For $n = 0$, $\delta[n] = 1$, and Eq. (2.32) gives

$$h^{\text{inv}}[0] + ah^{\text{inv}}[-1] = 1.$$

Since causality implies that $h^{\text{inv}}[-1] = 0$, we find that $h^{\text{inv}}[0] = 1$. For $n > 0$, $\delta[n] = 0$, and Eq. (2.32) implies that

$$h^{\text{inv}}[n] + ah^{\text{inv}}[n - 1] = 0,$$

which may be rewritten as

$$h^{\text{inv}}[n] = -ah^{\text{inv}}[n - 1]. \tag{2.33}$$

Since $h^{\text{inv}}[0] = 1$, Eq. (2.33) implies that $h^{\text{inv}}[1] = -a$, $h^{\text{inv}}[2] = a^2$, $h^{\text{inv}}[3] = -a^3$, and so on. Hence, the inverse system has the impulse response

$$h^{\text{inv}}[n] = (-a)^n u[n].$$

To check for stability, we determine whether $h^{\text{inv}}[n]$ is absolutely summable, which will be the case if

$$\sum_{k=-\infty}^{\infty} |h^{\text{inv}}[k]| = \sum_{k=0}^{\infty} |a|^k$$

is finite. This geometric series converges; hence, the system is stable, provided that $|a| < 1$. This implies that the inverse system is stable if the multipath component $ax[n - 1]$ is weaker than the first component $x[n]$; otherwise the system is unstable. ∎

Obtaining an inverse system by directly solving Eq. (2.30) or Eq. (2.31) is difficult, in general. Furthermore, not every LTI system has a stable and causal inverse. The effect of the inverse system on noise also is an important consideration in many problems. Methods developed in later chapters provide additional insight into the existence and determination of inverse systems.

Table 2.2 summarizes the relationship between LTI system properties and impulse response characteristics.

TABLE 2.2   *Properties of the Impulse Response Representation for LTI Systems.*

Property	Continuous-time system	Discrete-time system
Memoryless	$h(t) = c\delta(t)$	$h[n] = c\delta[n]$
Causal	$h(t) = 0$ for $t < 0$	$h[n] = 0$ for $n < 0$
Stability	$\int_{-\infty}^{\infty}|h(t)|\,dt < \infty$	$\sum_{n=-\infty}^{\infty}|h[n]| < \infty$
Invertibility	$h(t) * h^{\text{inv}}(t) = \delta(t)$	$h[n] * h^{\text{inv}}[n] = \delta[n]$

2.8 *Step Response*

Step input signals are often used to characterize the response of an LTI system to sudden changes in the input. The *step response* is defined as the output due to a unit step input signal. Let $h[n]$ be the impulse response of a discrete-time LTI system, and denote the step response as $s[n]$. We thus write

$$s[n] = h[n] * u[n]$$
$$= \sum_{k=-\infty}^{\infty} h[k]u[n-k].$$

Now, since $u[n-k] = 0$ for $k > n$ and $u[n-k] = 1$ for $k \le n$, we have

$$s[n] = \sum_{k=-\infty}^{n} h[k].$$

That is, the step response is the running sum of the impulse response. Similarly, the step response $s(t)$ of a continuous-time system is expressed as the running integral of the impulse response:

$$s(t) = \int_{-\infty}^{t} h(\tau)\,d\tau. \tag{2.34}$$

Note that we may invert these relationships to express the impulse response in terms of the step response as

$$h[n] = s[n] - s[n-1]$$

and

$$h(t) = \frac{d}{dt}s(t).$$

S & S
Solutions

EXAMPLE 2.14 RC CIRCUIT: STEP RESPONSE   As shown in Example 1.21, the impulse response of the *RC* circuit depicted in Fig. 2.12 is

$$h(t) = \frac{1}{RC}e^{-\frac{t}{RC}}u(t)$$

Find the step response of the circuit.

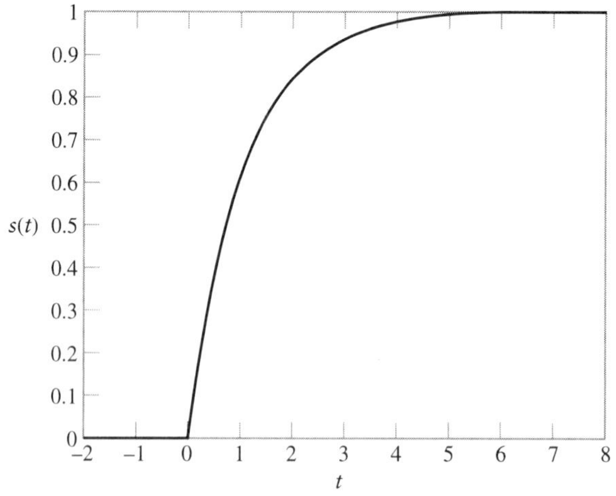

FIGURE 2.25 RC circuit step response for $RC = 1$ s.

Solution: The step represents a switch that turns on a constant voltage source at time $t = 0$. We expect the capacitor voltage to increase toward the value of the source in an exponential manner. Applying Eq. (2.34), we obtain

$$s(t) = \int_{-\infty}^{t} \frac{1}{RC} e^{-\frac{\tau}{RC}} u(\tau)\, d\tau.$$

Now we simplify the integral to get

$$s(t) = \begin{cases} 0, & t < 0 \\ \frac{1}{RC}\int_0^t e^{-\frac{\tau}{RC}}\, d\tau, & t \geq 0 \end{cases}$$

$$= \begin{cases} 0, & t < 0 \\ 1 - e^{-\frac{t}{RC}}, & t \geq 0 \end{cases}.$$

Figure 2.25 depicts the RC circuit step response for $RC = 1$ s. The result derived here confirms the premise used in Example 1.21 to derive the impulse response $h(t)$ for this circuit. ■

► **Problem 2.12** Find the step response of the first-order recursive system with impulse response

$$h[n] = \rho^n u[n],$$

assuming that $|\rho| < 1$.

Answer:

$$s[n] = \frac{1 - \rho^{n+1}}{1 - \rho} u[n].$$ ◄

► **Problem 2.13** Evaluate the step responses for the LTI systems represented by the following impulse responses:

(a) $h[n] = (1/2)^n u[n]$

(b) $h(t) = e^{-|t|}$

(c) $h(t) = \delta(t) - \delta(t - 1)$

Answers:

(a) $s[n] = (2 - (1/2)^n)u[n]$

(b) $s(t) = e^t u(-t) + (2 - e^{-t})u(t)$

(c) $s(t) = u(t) - u(t - 1)$ ◀

2.9 Differential and Difference Equation Representations of LTI Systems

Linear constant-coefficient difference and differential equations provide another representation for the input–output characteristics of LTI systems. Difference equations are used to represent discrete-time systems, while differential equations represent continuous-time systems. The general form of a linear constant-coefficient differential equation is

$$\sum_{k=0}^{N} a_k \frac{d^k}{dt^k} y(t) = \sum_{k=0}^{M} b_k \frac{d^k}{dt^k} x(t), \tag{2.35}$$

where the a_k and the b_k are constant coefficients of the system, $x(t)$ is the input applied to the system, and $y(t)$ is the resulting output. A linear constant-coefficient difference equation has a similar form, with the derivatives replaced by delayed values of the input $x[n]$ and output $y[n]$:

$$\sum_{k=0}^{N} a_k y[n - k] = \sum_{k=0}^{M} b_k x[n - k]. \tag{2.36}$$

The *order* of the differential or difference equation is (N, M), representing the number of energy storage devices in the system. Often, $N \geq M$, and the order is described using only N.

As an example of a differential equation that describes the behavior of a physical system, consider the *RLC* circuit depicted in Fig. 2.26. Suppose the input is the voltage source $x(t)$ and the output is the current around the loop, $y(t)$. Then summing the voltage drops around the loop gives

$$Ry(t) + L\frac{d}{dt}y(t) + \frac{1}{C}\int_{-\infty}^{t} y(\tau)\, d\tau = x(t).$$

Differentiating both sides of this equation with respect to t results in

$$\frac{1}{C}y(t) + R\frac{d}{dt}y(t) + L\frac{d^2}{dt^2}y(t) = \frac{d}{dt}x(t).$$

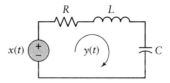

FIGURE 2.26 Example of an *RLC* circuit described by a differential equation.

This differential equation describes the relationship between the current $y(t)$ and the voltage $x(t)$ in the circuit. In this example, the order is $N = 2$, and we note that the circuit contains two energy storage devices: a capacitor and an inductor.

Mechanical systems also may be described in terms of differential equations that make use of Newton's laws. Recall that the behavior of the MEMS accelerometer modeled in Section 1.10 was given by the differential equation

$$\omega_n^2 y(t) + \frac{\omega_n}{Q}\frac{d}{dt}y(t) + \frac{d^2}{dt^2}y(t) = x(t),$$

where $y(t)$ is the position of the proof mass and $x(t)$ is the external acceleration. This system contains two energy storage mechanisms—a spring and a mass—and the order is again $N = 2$.

An example of a second-order difference equation is

$$y[n] + y[n - 1] + \frac{1}{4}y[n - 2] = x[n] + 2x[n - 1], \tag{2.37}$$

which may represent the relationship between the input and output signals of a system that processes data in a computer. Here, the order is $N = 2$, because the difference equation involves $y[n - 2]$, implying a maximum memory of 2 in the system output. Memory in a discrete-time system is analogous to energy storage in a continuous-time system.

Difference equations are easily rearranged to obtain recursive formulas for computing the current output of the system from the input signal and past outputs. We rewrite Eq. (2.36) so that $y[n]$ is alone on the left-hand side:

$$\boxed{y[n] = \frac{1}{a_0}\sum_{k=0}^{M} b_k x[n - k] - \frac{1}{a_0}\sum_{k=1}^{N} a_k y[n - k].}$$

This equation indicates how to obtain $y[n]$ from the present and past values of the input and the past values of the output. Such equations are often used to implement discrete-time systems in a computer. Consider computing $y[n]$ for $n \geq 0$ from $x[n]$ for the second-order difference equation (2.37) , rewritten in the form

$$y[n] = x[n] + 2x[n - 1] - y[n - 1] - \frac{1}{4}y[n - 2]. \tag{2.38}$$

Beginning with $n = 0$, we may determine the output by evaluating the sequence of equations

$$y[0] = x[0] + 2x[-1] - y[-1] - \frac{1}{4}y[-2], \tag{2.39}$$

$$y[1] = x[1] + 2x[0] - y[0] - \frac{1}{4}y[-1], \tag{2.40}$$

$$y[2] = x[2] + 2x[1] - y[1] - \frac{1}{4}y[0],$$

$$y[3] = x[3] + 2x[2] - y[2] - \frac{1}{4}y[1],$$

$$\vdots$$

In each equation, the current output is computed from the input and past values of the output. In order to begin this process at time $n = 0$, we must know the two most recent past values of the output, namely, $y[-1]$ and $y[-2]$. These values are known as *initial conditions*.

The initial conditions summarize all the information about the system's past that is needed to determine future outputs. No additional information about the past output is necessary. Note that, in general, the number of initial conditions required to determine the output is equal to the maximum memory of the system. It is common to choose $n = 0$ or $t = 0$ as the starting time for solving a difference or differential equation, respectively. In this case, the initial conditions for an Nth-order difference equation are the N values

$$y[-N], y[-N + 1], \ldots, y[-1],$$

and the initial conditions for an Nth-order differential equation are the values of the first N derivatives of the output—that is,

$$y(t)\big|_{t=0^-}, \frac{d}{dt}y(t)\bigg|_{t=0^-}, \frac{d^2}{dt^2}y(t)\bigg|_{t=0^-}, \ldots, \frac{d^{N-1}}{dt^{N-1}}y(t)\bigg|_{t=0^-}.$$

The initial conditions in a differential-equation description of an LTI system are directly related to the initial values of the energy storage devices in the system, such as initial voltages on capacitors and initial currents through inductors. As in the discrete-time case, the initial conditions summarize all information about the past history of the system that can affect future outputs. Hence, initial conditions also represent the "memory" of continuous-time systems.

EXAMPLE 2.15 RECURSIVE EVALUATION OF A DIFFERENCE EQUATION   Find the first two output values $y[0]$ and $y[1]$ for the system described by Eq. (2.38), assuming that the input is $x[n] = (1/2)^n u[n]$ and the initial conditions are $y[-1] = 1$ and $y[-2] = -2$.

Solution: Substitute the appropriate values into Eq. (2.39) to obtain

$$y[0] = 1 + 2 \times 0 - 1 - \frac{1}{4} \times (-2) = \frac{1}{2}.$$

Now substitute for $y[0]$ in Eq. (2.40) to find

$$y[1] = \frac{1}{2} + 2 \times 1 - \frac{1}{2} - \frac{1}{4} \times (1) = 1\frac{3}{4}. \qquad \blacksquare$$

EXAMPLE 2.16 EVALUATION OF A DIFFERENCE EQUATION BY MEANS OF A COMPUTER
A system is described by the difference equation

$$y[n] - 1.143y[n - 1] + 0.4128y[n - 2] =$$
$$0.0675x[n] + 0.1349x[n - 1] + 0.675x[n - 2].$$

Write a recursive formula that computes the present output from the past outputs and the current inputs. Use a computer to determine the step response of the system, the system output when the input is zero and the initial conditions are $y[-1] = 1$ and $y[-2] = 2$, and the output in response to the sinusoidal inputs $x_1[n] = \cos\left(\frac{\pi}{10}n\right)$, $x_2[n] = \cos\left(\frac{\pi}{5}n\right)$, and $x_3[n] = \cos\left(\frac{7\pi}{10}n\right)$, assuming zero initial conditions. Last, find the output of the system if the input is the weekly closing price of Intel stock depicted in Fig. 2.27, assuming zero initial conditions.

Solution: We rewrite the difference equation as

$$y[n] = 1.143y[n - 1] - 0.4128y[n - 2]$$
$$+ 0.0675x[n] + 0.1349x[n - 1] + 0.675x[n - 2].$$

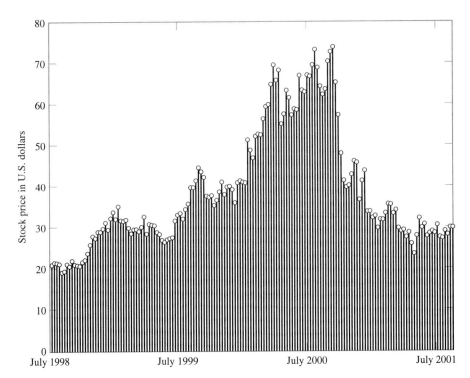

FIGURE 2.27   Weekly closing price of Intel stock.

This equation is evaluated in a recursive manner to determine the system output from the system input and the initial conditions $y[-1]$ and $y[-2]$.

The step response of the system is evaluated by assuming that the input is a step, $x[n] = u[n]$, and that the system is initially at rest, so that the initial conditions are zero. Figure 2.28(a) depicts the first 50 values of the step response. This system responds to a step by initially rising to a value slightly greater than the amplitude of the input and then decreasing to the value of the input at about $n = 13$. For n sufficiently large, we may consider the step to be a dc, or constant, input. Since the output amplitude is equal to the input amplitude, we see that this system has unit gain to constant inputs.

The response of the system to the initial conditions $y[-1] = 1$, $y[-2] = 2$, and zero input is shown in Fig. 2.28(b). Although the recursive nature of the difference equation suggests that the initial conditions affect all future values of the output, we see that the significant portion of the output due to the initial conditions lasts until about $n = 13$.

The outputs due to the sinusoidal inputs $x_1[n]$, $x_2[n]$, and $x_3[n]$ are depicted in Figs. 2.28(c), (d), and (e), respectively. Once the behavior of the system is distant from the initial conditions and the system enters a steady state, we see that the rapid fluctuations associated with the high-frequency sinusoidal input are attenuated. Figure 2.28(f) shows the system output for the Intel stock price input. We see that the output initially increases gradually in the same manner as the step response. This is a consequence of assuming that the input is zero prior to July 31, 1998. After about six weeks, the system has a smoothing effect on the stock price, since it attenuates rapid fluctuations while passing constant terms with unity gain. A careful comparison of the peaks in Figs. 2.27 and 2.28(f) shows that the system also introduces a slight delay, because the system computes the present output using past outputs and the present and past inputs.   ■

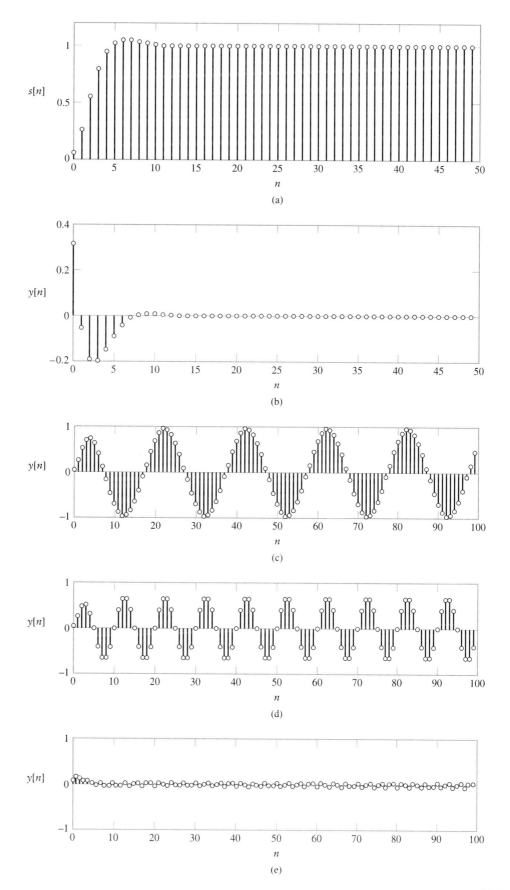

FIGURE 2.28
Illustration of the solution to Example 2.16. (a) Step response of system. (b) Output due to nonzero initial conditions with zero input. (c) Output due to $x_1[n] = \cos\left(\frac{1}{10}\pi n\right)$. (d) Output due to $x_2[n] = \cos\left(\frac{1}{5}\pi n\right)$. (e) Output due to $x_3[n] = \cos\left(\frac{7}{10}\pi n\right)$.

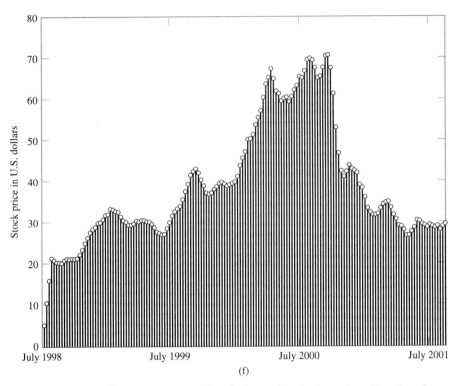

FIGURE 2.28 (f) Output associated with the weekly closing price of Intel stock.

▶ **Problem 2.14** Write a differential equation describing the relationship between the input voltage $x(t)$ and current $y(t)$ through the inductor in Fig. 2.29.

Answer:

$$Ry(t) + L\frac{d}{dt}y(t) = x(t). \qquad ◀$$

▶ **Problem 2.15** Calculate $y[n], n = 0, 1, 2, 3$ for the first-order recursive system

$$y[n] - (1/2)y[n - 1] = x[n]$$

if the input is $x[n] = u[n]$ and the initial condition is $y[-1] = -2$.

Answer:

$$y[0] = 0, \quad y[1] = 1, \quad y[2] = 3/2, \quad y[3] = 7/4. \qquad ◀$$

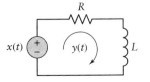

FIGURE 2.29 RL circuit.

2.10 *Solving Differential and Difference Equations*

In this section, we briefly review o⌐
tions. Our analysis offers a gener⌐
LTI system behavior.

   The output of a system c⌐
pressed as the sum of two con⌐
ferential or difference equati⌐
$y^{(h)}$. The second componer⌐
particular solution and den⌐
that we omit the arguments (t) o⌐

■ 2.10.1 THE HOMOGENEOUS SOLU⌐

The homogeneous form of a differential or differe⌐
terms involving the input to zero. Hence, for a continuou⌐
tion of the homogeneous equation

$$\sum_{k=0}^{N} a_k \frac{d^k}{dt^k} y^{(h)}(t) = 0.$$

The homogeneous solution for a continuous-time system is of the form

$$y^{(h)}(t) = \sum_{i=1}^{N} c_i e^{r_i t}, \tag{2.41}$$

where the r_i are the N roots of the system's *characteristic equation*

$$\sum_{k=0}^{N} a_k r^k = 0. \tag{2.42}$$

Substitution of Eq. (2.41) into the homogeneous equation establishes the fact that $y^{(h)}(t)$ is a solution for any set of constants c_i.

   In discrete time, the solution of the homogeneous equation

$$\sum_{k=0}^{N} a_k y^{(h)}[n - k] = 0$$

is

$$y^{(h)}[n] = \sum_{i=1}^{N} c_i r_i^n, \tag{2.43}$$

where the r_i are the N roots of the discrete-time system's characteristic equation

$$\sum_{k=0}^{N} a_k r^{N-k} = 0. \tag{2.44}$$

Again, substitution of Eq. (2.43) into the homogeneous equation establishes the fact that $y^{(h)}[n]$ is a solution for any set of constants c_i. In both cases, the c_i are determined later, in order that the complete solution satisfy the initial conditions. Note that the continuous- and discrete-time characteristic equations are different.

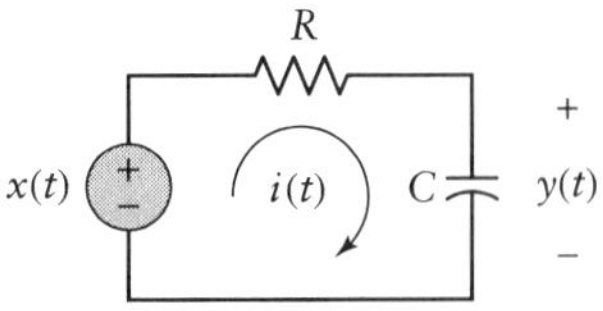

FIGURE 2.30 RC circuit.

The form of the homogeneous solution changes slightly when the characteristic equation described by Eq. (2.42) or Eq. (2.44) has repeated roots. If a root r_j is repeated p times, then there are p distinct terms in the solution of Eqs. (2.41) and (2.43) associated with r_j. These terms respectively involve the p functions

$$e^{r_j t}, te^{r_j t}, \ldots, t^{p-1}e^{r_j t}$$

and

$$r_j^n, nr_j^n, \ldots, n^{p-1}r_j^n.$$

The nature of each term in the homogeneous solution depends on whether the roots r_i are real, imaginary, or complex. Real roots lead to real exponentials, imaginary roots to sinusoids, and complex roots to exponentially damped sinusoids.

S & S
Solutions

EXAMPLE 2.17 RC CIRCUIT: HOMOGENEOUS SOLUTION The RC circuit depicted in Fig. 2.30 is described by the differential equation

$$y(t) + RC\frac{d}{dt}y(t) = x(t).$$

Determine the homogeneous solution of this equation.

Solution: The homogeneous equation is

$$y(t) + RC\frac{d}{dt}y(t) = 0.$$

The solution is given by Eq. (2.41), using $N = 1$ to obtain

$$y^{(h)}(t) = c_1 e^{r_1 t}\,\mathrm{V},$$

where r_1 is the root of the characteristic equation

$$1 + RCr_1 = 0.$$

Hence, $r_1 = -\frac{1}{RC}$, and the homogeneous solution for this system is

$$y^{(h)}(t) = c_1 e^{-\frac{t}{RC}}\,\mathrm{V}.$$

■

S & S
Solutions

EXAMPLE 2.18 FIRST-ORDER RECURSIVE SYSTEM: HOMOGENEOUS SOLUTION Find the homogeneous solution for the first-order recursive system described by the difference equation

$$y[n] - \rho y[n-1] = x[n].$$

Solution: The homogeneous equation is

$$y[n] - \rho y[n-1] = 0,$$

and its solution is given by Eq. (2.43) for $N = 1$:

$$y^{(h)}[n] = c_1 r_1^n.$$

The parameter r_1 is obtained from the root of the characteristic equation given by Eq. (2.44) with $N = 1$:

$$r_1 - \rho = 0.$$

Hence, $r_1 = \rho$, and the homogeneous solution is

$$y^{(b)}[n] = c_1\rho^n.$$ ■

▶ **Problem 2.16** Determine the homogeneous solution for the systems described by the following differential or difference equations:

(a)
$$\frac{d^2}{dt^2}y(t) + 5\frac{d}{dt}y(t) + 6y(t) = 2x(t) + \frac{d}{dt}x(t)$$

(b)
$$\frac{d^2}{dt^2}y(t) + 3\frac{d}{dt}y(t) + 2y(t) = x(t) + \frac{d}{dt}x(t)$$

(c)
$$y[n] - (9/16)y[n-2] = x[n-1]$$

(d)
$$y[n] + (1/4)y[n-2] = x[n] + 2x[n-2]$$

Answers:

(a)
$$y^{(b)}(t) = c_1 e^{-3t} + c_2 e^{-2t}$$

(b)
$$y^{(b)}(t) = c_1 e^{-t} + c_2 e^{-2t}$$

(c)
$$y^{(b)}[n] = c_1(3/4)^n + c_2(-3/4)^n$$

(d)
$$y^{(b)}[n] = c_1(1/2e^{j\pi/2})^n + c_2(1/2e^{-j\pi/2})^n$$ ◀

▶ **Problem 2.17** Determine the homogeneous solution for the *RLC* circuit depicted in Fig. 2.26 as a function of R, L, and C. Indicate the conditions on R, L, and C so that the homogeneous solution consists of real exponentials, complex sinusoids, and exponentially damped sinusoids.

Answers: For $R^2 \neq \frac{4L}{C}$,

$$y^{(b)}(t) = c_1 e^{r_1 t} + c_2 e^{r_2 t},$$

where

$$r_1 = \frac{-R + \sqrt{R^2 - \frac{4L}{C}}}{2L} \quad \text{and} \quad r_2 = \frac{-R - \sqrt{R^2 - \frac{4L}{C}}}{2L}.$$

For $R^2 = \frac{4L}{C}$,

$$y^{(n)}(t) = c_1 e^{-\frac{R}{2L}t} + c_2 t e^{-\frac{R}{2L}t}.$$

The solution consists of real exponentials for $R^2 \geq \frac{4L}{C}$, complex sinusoids for $R = 0$, and exponentially damped sinusoids for $R^2 \leq \frac{4L}{C}$. ◀

TABLE 2.3 *Form of Particular Solutions Corresponding to Commonly Used Inputs.*

Continuous Time		Discrete Time	
Input	*Particular Solution*	*Input*	*Particular Solution*
1	c	1	c
t	$c_1 t + c_2$	n	$c_1 n + c_2$
e^{-at}	ce^{-at}	α^n	$c\alpha^n$
$\cos(\omega t + \phi)$	$c_1 \cos(\omega t) + c_2 \sin(\omega t)$	$\cos(\Omega n + \phi)$	$c_1 \cos(\Omega n) + c_2 \sin(\Omega n)$

■ 2.10.2 THE PARTICULAR SOLUTION

The particular solution $y^{(p)}$ represents any solution of the differential or difference equation for the given input. Thus, $y^{(p)}$ is not unique. A particular solution is usually obtained by assuming an output of the same general form as the input. For example, if the input to a discrete-time system is $x[n] = \alpha^n$, then we assume that the output is of the form $y^{(p)}[n] = c\alpha^n$ and find the constant c so that $y^{(p)}[n]$ is a solution of the system's difference equation. If the input is $x[n] = A\cos(\Omega n + \phi)$, then we assume a general sinusoidal response of the form $y^{(p)}[n] = c_1 \cos(\Omega n) + c_2 \sin(\Omega n)$, where c_1 and c_2 are determined so that $y^{(p)}[n]$ satisfies the system's difference equation. Assuming an output of the same form as the input is consistent with our expectation that the output of the system be directly related to the input.

This approach for finding a particular solution is modified when the input is of the same form as one of the components of the homogeneous solution. In that case, we must assume a particular solution that is independent of all terms in the homogeneous solution. This is accomplished analogously to the procedure for generating independent natural-response components when there are repeated roots in the characteristic equation. Specifically, we multiply the form of the particular solution by the lowest power of t or n that will give a response component not included in the natural response, and then we solve for the coefficient by substituting the assumed particular solution into the differential or difference equation.

The forms of the particular solutions associated with common input signals are given in Table 2.3. More extensive tables are given in books devoted to solving difference and differential equations, such as those listed under Further Reading at the end of this chapter. The particular solutions given in Table 2.3 assume that the inputs exist for all time. If the input is specified after a starting time $t = 0$ or $n = 0$ [e.g., $x(t) = e^{-at}u(t)$], as is common in solving differential or difference equations subject to initial conditions, then the particular solution is valid only for $t > 0$ or $n \geq 0$.

EXAMPLE 2.19 FIRST-ORDER RECURSIVE SYSTEM (CONTINUED): PARTICULAR SOLUTION
Find a particular solution for the first-order recursive system described by the difference equation

$$y[n] - \rho y[n-1] = x[n]$$

if the input is $x[n] = (1/2)^n$.

Solution: We assume a particular solution of the form $y^{(p)}[n] = c_p \left(\frac{1}{2}\right)^n$. Substituting $y^{(p)}[n]$ and $x[n]$ into the given difference equation yields

$$c_p \left(\frac{1}{2}\right)^n - \rho c_p \left(\frac{1}{2}\right)^{n-1} = \left(\frac{1}{2}\right)^n.$$

We multiply both sides of the equation by $(1/2)^{-n}$ to obtain

$$c_p(1 - 2\rho) = 1. \tag{2.45}$$

Solving this equation for c_p gives the particular solution

$$y^{(p)}[n] = \frac{1}{1 - 2\rho}\left(\frac{1}{2}\right)^n.$$

If $\rho = \left(\frac{1}{2}\right)$, then the particular solution has the same form as the homogeneous solution found in Example 2.18. Note that in this case no coefficient c_p satisfies Eq. (2.45), and we must assume a particular solution of the form $y^{(p)}[n] = c_p n(1/2)^n$. Substituting this particular solution into the difference equation gives $c_p n(1 - 2\rho) + 2\rho c_p = 1$. Using $\rho = (1/2)$ we find that $c_p = 1$. ∎

EXAMPLE 2.20 RC CIRCUIT (CONTINUED): PARTICULAR SOLUTION   Consider the RC circuit of Example 2.17 and depicted in Fig. 2.30. Find a particular solution for this system with an input $x(t) = \cos(\omega_0 t)$.

Solution: From Example 2.17, the differential equation describing the system is

$$y(t) + RC\frac{d}{dt}y(t) = x(t).$$

We assume a particular solution of the form $y^{(p)}(t) = c_1 \cos(\omega_0 t) + c_2 \sin(\omega_0 t)$. Replacing $y(t)$ in the differential equation by $y^{(p)}(t)$ and $x(t)$ by $\cos(\omega_0 t)$ gives

$$c_1 \cos(\omega_0 t) + c_2 \sin(\omega_0 t) - RC\omega_0 c_1 \sin(\omega_0 t) + RC\omega_0 c_2 \cos(\omega_0 t) = \cos(\omega_0 t).$$

The coefficients c_1 and c_2 are obtained by separately equating the coefficients of $\cos(\omega_0 t)$ and $\sin(\omega_0 t)$. This gives the following system of two equations in two unknowns:

$$c_1 + RC\omega_0 c_2 = 1;$$
$$-RC\omega_0 c_1 + c_2 = 0.$$

Solving these equations for c_1 and c_2 gives

$$c_1 = \frac{1}{1 + (RC\omega_0)^2}$$

and

$$c_2 = \frac{RC\omega_0}{1 + (RC\omega_0)^2}.$$

Hence, the particular solution is

$$y^{(p)}(t) = \frac{1}{1 + (RC\omega_0)^2} \cos(\omega_0 t) + \frac{RC\omega_0}{1 + (RC\omega_0)^2} \sin(\omega_0 t) \text{ V}.$$

∎

▶ **Problem 2.18**   Determine the particular solution associated with the specified input for the systems described by the following differential or difference equations:

(a) $x(t) = e^{-t}$:

$$\frac{d^2}{dt^2}y(t) + 5\frac{d}{dt}y(t) + 6y(t) = 2x(t) + \frac{d}{dt}x(t)$$

(b) $x(t) = \cos(2t)$:

$$\frac{d^2}{dt^2}y(t) + 3\frac{d}{dt}y(t) + 2y(t) = x(t) + \frac{d}{dt}x(t)$$

(c) $x[n] = 2$:

$$y[n] - (9/16)y[n - 2] = x[n - 1]$$

(d) $x[n] = (1/2)^n$:

$$y[n] + (1/4)y[n - 2] = x[n] + 2x[n - 2]$$

Answers:

(a) $y^{(p)}(t) = (1/2)e^{-t}$

(b) $y^{(p)}(t) = (1/4)\cos(2t) + (1/4)\sin(2t)$

(c) $y^{(p)}[n] = 32/7$

(d) $y^{(p)}[n] = (9/2)(1/2)^n$ ◀

■ 2.10.3 THE COMPLETE SOLUTION

The complete solution of the differential or difference equation is obtained by summing the particular solution and the homogeneous solution and finding the unspecified coefficients in the homogeneous solution so that the complete solution satisfies the prescribed initial conditions. This procedure is summarized as follows:

Procedure 2.3: Solving a Differential or Difference Equation

1. Find the form of the homogeneous solution $y^{(h)}$ from the roots of the characteristic equation.

2. Find a particular solution $y^{(p)}$ by assuming that it is of the same form as the input, yet is independent of all terms in the homogeneous solution.

3. Determine the coefficients in the homogeneous solution so that the complete solution $y = y^{(p)} + y^{(h)}$ satisfies the initial conditions.

We assume that the input is applied at time $t = 0$ or $n = 0$, so the particular solution applies only to times $t > 0$ or $n \geq 0$, respectively. If so, then the complete solution is valid only for just those times. Therefore, in the discrete-time case, the initial conditions $y[-N], \ldots, y[-1]$ must be translated to new initial conditions $y[0], \ldots, y[N - 1]$ before Step 3 is performed. Translation of the initial conditions is accomplished by using the recursive form of the difference equation, as shown in Example 2.15 and in the examples that follow.

In the continuous-time case, the initial conditions at $t = 0^-$ must be translated to $t = 0^+$ to reflect the effect of applying the input at $t = 0$. While this process is often straightforward in problems involving capacitors and inductors, translating initial conditions for the most general differential equation is complicated and will not be discussed further. Rather, we shall only solve differential equations for which application of the input at $t = 0$ does not cause discontinuities in the initial conditions. A necessary and sufficient condition for the initial conditions at $t = 0^+$ to equal the initial conditions at $t = 0^-$ for a given input is that the right-hand side of the differential equation in Eq. (2.35), $\sum_{k=0}^{M} b_k \frac{d^k}{dt^k} x(t)$, contain no impulses or derivatives of impulses. For example, if $M = 0$, then the initial conditions do not need to be translated as long as there are no impulses in $x(t)$, but if $M = 1$, then any input involving a step discontinuity at $t = 0$ generates an impulse term due to the $\frac{d}{dt}x(t)$ term on the right-hand side of the differential equation, and the initial conditions at $t = 0^+$ are no longer equal to the initial conditions at $t = 0^-$. The Laplace transform method, described in Chapter 6, circumvents these difficulties.

S & S
Solutions

EXAMPLE 2.21 FIRST-ORDER RECURSIVE SYSTEM (CONTINUED): COMPLETE SOLUTION
Find the solution for the first-order recursive system described by the difference equation

$$y[n] - \frac{1}{4}y[n-1] = x[n] \tag{2.46}$$

if the input is $x[n] = (1/2)^n u[n]$ and the initial condition is $y[-1] = 8$.

Solution: The form of the solution is obtained by summing the homogeneous solution determined in Example 2.18 with the particular solution determined in Example 2.19 after setting $\rho = 1/4$:

$$y[n] = 2\left(\frac{1}{2}\right)^n + c_1\left(\frac{1}{4}\right)^n, \quad \text{for } n \geq 0. \tag{2.47}$$

The coefficient c_1 is obtained from the initial condition. First, we translate the initial condition to time $n = 0$ by rewriting Eq. (2.46) in recursive form and substituting $n = 0$ to obtain

$$y[0] = x[0] + (1/4)y[-1],$$

which implies that $y[0] = 1 + (1/4) \times 8 = 3$. Then we substitute $y[0] = 3$ into Eq. (2.47), yielding

$$3 = 2\left(\frac{1}{2}\right)^0 + c_1\left(\frac{1}{4}\right)^0,$$

from which we find that $c_1 = 1$. Thus, we may write the complete solution as

$$y[n] = 2\left(\frac{1}{2}\right)^n + \left(\frac{1}{4}\right)^n, \quad \text{for } n \geq 0. \quad \blacksquare$$

S & S
Solutions

EXAMPLE 2.22 RC CIRCUIT (CONTINUED): COMPLETE RESPONSE Find the complete response of the *RC* circuit depicted in Fig. 2.30 to an input $x(t) = \cos(t)u(t)$ V, assuming normalized values $R = 1\ \Omega$ and $C = 1$ F and assuming that the initial voltage across the capacitor is $y(0^-) = 2$ V.

Solution: The homogeneous solution was obtained in Example 2.17:

$$y^{(h)}(t) = ce^{-\frac{t}{RC}} \text{ V}.$$

A particular solution was obtained for this input in Example 2.20, namely,

$$y^{(p)}(t) = \frac{1}{1 + (RC)^2}\cos(t) + \frac{RC}{1 + (RC)^2}\sin(t) \text{ V},$$

where we have used $\omega_0 = 1$. Substituting $R = 1\ \Omega$ and $C = 1$ F, we find that the complete solution is

$$y(t) = ce^{-t} + \frac{1}{2}\cos t + \frac{1}{2}\sin t \text{ V} \qquad t > 0.$$

The input does not introduce impulses into the right-hand side of the differential equation, so the coefficient c is determined from the initial condition $y(0^-) = y(0^+) = 2$. We have

$$2 = ce^{-0^+} + \frac{1}{2}\cos 0^+ + \frac{1}{2}\sin 0^+$$

$$= c + \frac{1}{2},$$

so that $c = 3/2$, which gives

$$y(t) = \frac{3}{2}e^{-t} + \frac{1}{2}\cos t + \frac{1}{2}\sin t \text{ V} \qquad t > 0. \qquad ■$$

EXAMPLE 2.23 FINANCIAL COMPUTATIONS: LOAN REPAYMENT Example 2.5 showed that the first-order difference equation introduced in Section 1.10 and studied in Examples 2.18, 2.19, and 2.21 may be used to describe the value of an investment earning a fixed rate of interest. The same equation also describes the balance of a loan if $x[n] < 0$ represents the principal and interest payment made at the beginning of each period and $y[n]$ is the balance after the principal and interest payment is credited. As before, if $r\%$ is the interest rate per period, then $\rho = 1 + r/100$.

Use the complete response of the first-order difference equation to find the payment required to pay off a \$20,000 loan in 10 periods. Assume equal payments and a 10% interest rate.

Solution: We have $\rho = 1.1$ and $y[-1] = 20,000$, and we assume that $x[n] = b$ is the payment each period. Note that the first payment is made when $n = 0$. Since the loan balance is to be zero after 10 payments, we seek the payment b for which $y[9] = 0$.

The homogeneous solution is of the form

$$y^{(h)}[n] = c_h(1.1)^n,$$

while the particular solution is of the form

$$y^{(p)}[n] = c_p,$$

since the input (the payment) is constant. Solving for c_p by substituting $y^{(p)}[n] = c_p$ and $x[n] = b$ into the difference equation $y[n] - 1.1y[n-1] = x[n]$, we obtain

$$c_p = -10b.$$

Therefore, the complete solution is of the form

$$y[n] = c_h(1.1)^n - 10b, \qquad n \geq 0. \qquad (2.48)$$

We solve for c_h by first translating the initial condition forward one period to obtain

$$y[0] = 1.1y[-1] + x[0]$$
$$= 22,000 + b.$$

Next, we substitute $y[0]$ into Eq. (2.48) to obtain the equation for c_h:

$$22,000 + b = c_h(1.1)^0 - 10b.$$

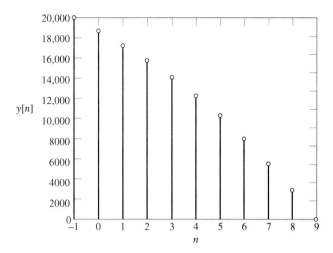

FIGURE 2.31 Balance on a $20,000 loan for Example 2.23 in U.S. dollars. Assuming 10% interest per period, the loan is paid off with 10 payments of $3,254.91.

Thus, $c_h = 22,000 + 11b$. This implies that the solution of the difference equation is given by

$$y[n] = (22,000 + 11b)(1.1)^n - 10b.$$

We now solve for the required payment b by setting $y[9] = 0$. That is,

$$0 = (22,000 + 11b)(1.1)^9 - 10b,$$

which implies that

$$b = \frac{-22,000(1.1)^9}{11(1.1)^9 - 10}$$

$$= -3,254.91.$$

Hence, a payment of $3,254.91 each period is required to pay off the loan in 10 payments. Figure 2.31 depicts the loan balance $y[n]$. ∎

S & S
Solutions

▶ **Problem 2.19** Find the output, given the input and initial conditions, for the systems described by the following differential or difference equations:

(a) $x(t) = e^{-t}u(t), y(0) = -\frac{1}{2}, \frac{d}{dt}y(t)|_{t=0} = \frac{1}{2}$:

$$\frac{d^2}{dt^2}y(t) + 5\frac{d}{dt}y(t) + 6y(t) = x(t)$$

(b) $x(t) = \cos(t)u(t), y(0) = -\frac{4}{5}, \frac{d}{dt}y(t)|_{t=0} = \frac{3}{5}$:

$$\frac{d^2}{dt^2}y(t) + 3\frac{d}{dt}y(t) + 2y(t) = 2x(t)$$

(c) $x[n] = u[n], y[-2] = 8, y[-1] = 0$:

$$y[n] - \frac{1}{4}y[n-2] = 2x[n] + x[n-1]$$

(d) $x[n] = 2^n u[n], y[-2] = 26, y[-1] = -1$:

$$y[n] - \left(\frac{1}{4}\right)y[n-1] - \left(\frac{1}{8}\right)y[n-2] = x[n] + \left(\frac{11}{8}\right)x[n-1]$$

Answers:

(a)

$$y(t) = \left(\left(\frac{1}{2} \right) e^{-t} + e^{-3t} - 2e^{-2t} \right) u(t)$$

(b)

$$y(t) = \left(\left(\frac{1}{5} \right) \cos(t) + \left(\frac{3}{5} \right) \sin(t) - 2e^{-t} + e^{-2t} \right) u(t)$$

(c)

$$y[n] = \left(-\left(\frac{1}{2} \right)^n + \left(-\frac{1}{2} \right)^n + 4 \right) u[n]$$

(d)

$$y[n] = \left(2(2)^n + \left(-\frac{1}{4} \right)^n + \left(\frac{1}{2} \right)^n \right) u[n] \qquad \blacktriangleleft$$

▶ **Problem 2.20** Find the response of the *RL* circuit depicted in Fig. 2.29 to the following input voltages, assuming that the initial current through the inductor is $y(0) = -1$ A:

(a) $x(t) = u(t)$

(b) $x(t) = tu(t)$

Answers:

(a)

$$y(t) = \left(\frac{1}{R} - \left(1 + \frac{1}{R} \right) e^{-\frac{R}{L}t} \right) \text{A}, \quad t \geq 0$$

(b)

$$y(t) = \left[\frac{1}{R}t - \frac{L}{R^2} + \left(\frac{L}{R^2} - 1 \right) e^{-\frac{R}{L}t} \right] \text{A}, \quad t \geq 0 \qquad \blacktriangleleft$$

2.11 *Characteristics of Systems Described by Differential and Difference Equations*

It is informative to express the output of a system described by a differential or difference equation as the sum of two components: one associated only with the initial conditions, the other due only to the input signal. We will term the component of the output associated with the initial conditions the *natural response* of the system and denote it as $y^{(n)}$. The component of the output due only to the input is termed the *forced response* of the system and is denoted as $y^{(f)}$. Thus, the complete output is $y = y^{(n)} + y^{(f)}$.

▪ 2.11.1 THE NATURAL RESPONSE

The natural response is the system output for zero input and thus describes the manner in which the system dissipates any stored energy or memory of the past represented by non-zero initial conditions. Since the natural response assumes zero input, it is obtained from the homogeneous solution given in Eq. (2.41) or Eq. (2.43) by choosing the coefficients c_i so that the initial conditions are satisfied. The natural response assumes zero input and thus does not involve a particular solution. Since the homogeneous solutions apply for all time, the natural response is determined without translating initial conditions forward in time.

S & S
Solutions

EXAMPLE 2.24 RC CIRCUIT (CONTINUED): NATURAL RESPONSE   The system in Examples 2.17, 2.20, and 2.22 is described by the differential equation

$$y(t) + RC\frac{d}{dt}y(t) = x(t).$$

Find the natural response of this system, assuming that $y(0) = 2$ V, $R = 1\,\Omega$, and $C = 1$ F.

Solution: The homogeneous solution, derived in Example 2.17, is

$$y^{(h)}(t) = c_1 e^{-t}\,\text{V}.$$

Hence, the natural response is obtained by choosing c_1 so that the initial condition $y^{(n)}(0) = 2$ is satisfied. The initial condition implies that $c_1 = 2$, so the natural response is

$$y^{(n)}(t) = 2e^{-t}\,\text{V for } t \ge 0.$$ ∎

S & S
Solutions

EXAMPLE 2.25 FIRST-ORDER RECURSIVE SYSTEM (CONTINUED): NATURAL RESPONSE
The system in Example 2.21 is described by the difference equation

$$y[n] - \frac{1}{4}y[n-1] = x[n].$$

Find the natural response of this system.

Solution: Recall from Example 2.21 that the homogeneous solution is

$$y^{(h)}[n] = c_1\left(\frac{1}{4}\right)^n.$$

Satisfaction of the initial condition $y[-1] = 8$ implies that

$$8 = c_1\left(\frac{1}{4}\right)^{-1},$$

or $c_1 = 2$. Thus, the natural response is

$$y^{(n)}[n] = 2\left(\frac{1}{4}\right)^n, \quad n \ge -1.$$ ∎

▶ **Problem 2.21**   Determine the natural response for the systems described by the following differential or difference equations and the specified initial conditions:

(a) $y(0) = 3, \frac{d}{dt}y(t)|_{t=0} = -7$:

$$\frac{d^2}{dt^2}y(t) + 5\frac{d}{dt}y(t) + 6y(t) = 2x(t) + \frac{d}{dt}x(t)$$

(b) $y(0) = 0, \frac{d}{dt}y(t)|_{t=0} = -1$:

$$\frac{d^2}{dt^2}y(t) + 3\frac{d}{dt}y(t) + 2y(t) = x(t) + \frac{d}{dt}x(t)$$

(c) $y[-1] = -4/3, y[-2] = 16/3$:
$$y[n] - (9/16)y[n-2] = x[n-1]$$

(d) $y[0] = 2, y[1] = 0$:
$$y[n] + (1/4)y[n-2] = x[n] + 2x[n-2]$$

Answers:

(a)

$$y^{(n)}(t) = e^{-3t} + 2e^{-2t}, \quad \text{for} \quad t \geq 0$$

(b)

$$y^{(n)}(t) = -e^{-t} + e^{-2t}, \quad \text{for} \quad t \geq 0$$

(c)

$$y^{(n)}[n] = (3/4)^n + 2(-3/4)^n, \quad \text{for} \quad n \geq -2$$

(d)

$$y^{(n)}[n] = (1/2e^{j\pi/2})^n + (1/2e^{-j\pi/2})^n, \quad \text{for} \quad n \geq 0 \qquad \blacktriangleleft$$

■ 2.11.2 THE FORCED RESPONSE

The forced response is the system output due to the input signal assuming zero initial conditions. Thus, the forced response is of the same form as the complete solution of the differential or difference equation. A system with zero initial conditions is said to be "at rest," since there is no stored energy or memory in the system. The forced response describes the system behavior that is "forced" by the input when the system is at rest.

The forced response depends on the particular solution, which is valid only for times $t > 0$ or $n \geq 0$. Accordingly, the at-rest conditions for a discrete-time system, $y[-N] = 0, \ldots, y[-1] = 0$, must be translated forward to times $n = 0, 1, \ldots, N - 1$ before solving for the undetermined coefficients, such as when one is determining the complete solution. As before, we shall consider finding the forced response only for continuous-time systems and inputs that do not result in impulses on the right-hand side of the differential equation. This ensures that the initial conditions at $t = 0^+$ are equal to the zero initial conditions at $t = 0^-$.

S & S
Solutions

EXAMPLE 2.26 FIRST-ORDER RECURSIVE SYSTEM (CONTINUED): FORCED RESPONSE
The system in Example 2.21 is described by the first-order difference equation

$$y[n] - \frac{1}{4}y[n-1] = x[n].$$

Find the forced response of this system if the input is $x[n] = (1/2)^n u[n]$.

Solution: The difference between this example and Example 2.21 is the initial condition. Recall that the complete solution is of the form

$$y[n] = 2\left(\frac{1}{2}\right)^n + c_1\left(\frac{1}{4}\right)^n, \quad n \geq 0.$$

To obtain c_1, we translate the at-rest condition $y[-1] = 0$ to time $n = 0$ by noting that

$$y[0] = x[0] + \frac{1}{4}y[-1],$$

which implies that $y[0] = 1 + (1/4) \times 0$. Now we use $y[0] = 1$ to solve for c_1 from the equation

$$1 = 2\left(\frac{1}{2}\right)^0 + c_1\left(\frac{1}{4}\right)^0,$$

which implies that $c_1 = -1$. Thus, the forced response of the system is

$$y^{(f)}[n] = 2\left(\frac{1}{2}\right)^n - \left(\frac{1}{4}\right)^n, \quad n \geq 0.$$

∎

S & S Solutions

EXAMPLE 2.27 RC CIRCUIT (CONTINUED): FORCED RESPONSE The system in Examples 2.17, 2.20, and 2.22 is described by the differential equation

$$y(t) + RC\frac{d}{dt}y(t) = x(t).$$

Find the forced response of this system, assuming that $x(t) = \cos(t)u(t)$ V, $R = 1\,\Omega$, and $C = 1$ F.

Solution: Example 2.22 established that the complete response is of the form

$$y(t) = ce^{-t} + \frac{1}{2}\cos t + \frac{1}{2}\sin t \text{ V}, \quad t > 0.$$

The forced response is obtained by choosing c under the assumption that the system is initially at rest—that is, assuming that $y(0^-) = y(0^+) = 0$. Thus, we obtain $c = -1/2$, and the forced response is given by

$$y^{(f)}(t) = -\frac{1}{2}e^{-t} + \frac{1}{2}\cos t + \frac{1}{2}\sin t \text{ V}.$$

Note that the sum of the forced response and the natural response in Example 2.24 is equal to the complete system response determined in Example 2.22. ∎

▶ **Problem 2.22** Determine the forced response for the systems described by the following differential or difference equations and the specified inputs:

(a) $x(t) = e^{-t}u(t)$

$$\frac{d^2}{dt^2}y(t) + 5\frac{d}{dt}y(t) + 6y(t) = x(t)$$

(b) $x(t) = \sin(2t)u(t)$

$$\frac{d^2}{dt^2}y(t) + 3\frac{d}{dt}y(t) + 2y(t) = x(t) + \frac{d}{dt}x(t)$$

(c) $x[n] = 2u[n]$

$$y[n] - (9/16)y[n-2] = x[n-1]$$

Answers:

(a) $y^{(f)}(t) = ((1/2)e^{-t} - e^{-2t} + (1/2)e^{-3t})u(t)$

(b) $y^{(f)}(t) = ((-1/4)\cos(2t) + (1/4)\sin(2t) + (1/4)e^{-2t})u(t))$

(c) $y^{(f)}[n] = (32/7 - 4(3/4)^n - (4/7)(-3/4)^n)u[n]$

◀

■ 2.11.3 THE IMPULSE RESPONSE

The method described in Section 2.10 for solving differential and difference equations cannot be used to find the impulse response directly. However, given the step response, the impulse response may be determined by exploiting the relationship between the two responses. The definition of the step response assumes that the system is at rest, so it represents the response of the system to a step input with zero initial conditions. For a continuous-time system, the impulse response $h(t)$ is related to the step response $s(t)$ via the formula $h(t) = \frac{d}{dt}s(t)$. For a discrete-time system, $h[n] = s[n] - s[n-1]$. Thus, the impulse response is obtained by differentiating or differencing the step response.

Note the basic difference between impulse-response descriptions and differential- or difference-equation system descriptions: There is no provision for initial conditions when one is using the impulse response; it applies only to systems that are initially at rest or when the input is known for all time. Differential- and difference-equation system descriptions are more flexible in this respect, since they apply to systems either at rest or with nonzero initial conditions.

■ 2.11.4 LINEARITY AND TIME INVARIANCE

The forced response of an LTI system described by a differential or difference equation is linear with respect to the input. That is, if $y_1^{(f)}$ is the forced response associated with an input x_1 and $y_2^{(f)}$ is the forced response associated with an input x_2, then the input $\alpha x_1 + \beta x_2$ generates the forced response $\alpha y_1^{(f)} + \beta y_2^{(f)}$. Similarly, the natural response is linear with respect to the initial conditions: If $y_1^{(n)}$ is the natural response associated with initial conditions I_1 and $y_2^{(n)}$ is the natural response associated with initial conditions I_2, then the composite initial conditions $\alpha I_1 + \beta I_2$ results in the natural response $\alpha y_1^{(n)} + \beta y_2^{(n)}$. The forced response is also time invariant: A time shift in the input results in a time shift in the output, since the system is initially at rest. By contrast, in general, the complete response of an LTI system described by a differential or difference equation is *not* time invariant, since the initial conditions will result in an output term that does not shift with a time shift of the input. Finally, we observe that the forced response is also causal: Since the system is initially at rest, the output does not begin prior to the time at which the input is applied to the system.

■ 2.11.5 ROOTS OF THE CHARACTERISTIC EQUATION

The forced response depends on both the input and the roots of the characteristic equation, since it involves both the homogeneous solution and a particular solution of the differential or difference equation. The basic form of the natural response is dependent entirely on the roots of the characteristic equation. The impulse response of an LTI system also depends on the roots of the characteristic equation, since it contains the same terms as the natural response. Thus, the roots of the characteristic equation afford considerable information about LTI system behavior.

For example, the stability characteristics of an LTI system are directly related to the roots of the system's characteristic equation. To see this, note that the output of a stable system in response to zero input must be bounded for any set of initial conditions. This follows from the definition of BIBO stability and implies that the natural response of the system must be bounded. Thus, each term in the natural response must be bounded. In the discrete-time case, we must have $|r_i^n|$ bounded, or $|r_i| < 1$ for all i. When $|r_i| = 1$, the natural response does not decay, and the system is said to be on the verge of instability. For continuous-time LTI systems, we require that $|e^{r_i t}|$ be bounded, which implies that

$\mathrm{Re}\{r_i\} < 0$. Here again, when $\mathrm{Re}\{r_i\} = 0$, the system is said to be on the verge of instability. These results imply that a discrete-time LTI system is unstable if any root of the characteristic equation has a magnitude greater than unity and a continuous-time LTI system is unstable if the real part of any root of the characteristic equation is positive.

This discussion leads to the idea that the roots of the characteristic equation indicate when an LTI system is unstable. In later chapters, we will prove that a discrete-time causal LTI system is stable if and only if all roots of the characteristic equation have magnitude less than unity, and a continuous-time causal LTI system is stable if and only if the real parts of all roots of the characteristic equation are negative. These stability conditions imply that the natural response of an LTI system goes to zero as time approaches infinity, since each term in the natural response is a decaying exponential. This "decay to zero" is consistent with our intuitive concept of an LTI system's zero input behavior. We expect a zero output when the input is zero if all the stored energy in the system has dissipated. The initial conditions represent energy that is present in the system: in a stable LTI system with zero input, the stored energy eventually dissipates and the output approaches zero.

The response time of an LTI system is also determined by the roots of the characteristic equation. Once the natural response has decayed to zero, the system behavior is governed only by the particular solution, which is of the same form as the input. Thus, the natural-response component describes the transient behavior of the system; that is, it describes the transition of the system from its initial condition to an equilibrium condition determined by the input. Hence, the time it takes an LTI system to respond to a transient is determined by the time it takes for the natural response to decay to zero. Recall that the natural response contains terms of the form r_i^n for a discrete-time LTI system and $e^{r_i t}$ for a continuous-time LTI system. The response time of a discrete-time LTI system to a transient is therefore proportional to the root of the characteristic equation with the largest magnitude, while that of a continuous-time LTI system is determined by the root with the largest real component. For a continuous-time LTI system to have a fast response time, all the roots of the characteristic equation must have large negative real parts.

2.12 *Block Diagram Representations*

In this section, we examine block diagram representations of LTI systems described by differential and difference equations. A *block diagram* is an interconnection of elementary operations that act on the input signal. The block diagram is a more detailed representation of a system than the impulse response or difference and differential equation descriptions, since it describes how the system's internal computations or operations are ordered. The impulse response and difference or differential equation descriptions represent only the input–output behavior of a system. We shall show that a system with a given input–output characteristic can be represented by different block diagrams. Each block diagram representation describes a different set of internal computations used to determine the system output.

Block diagram representations consist of an interconnection of three elementary operations on signals:

1. Scalar multiplication: $y(t) = cx(t)$ or $y[n] = cx[n]$, where c is a scalar.
2. Addition: $y(t) = x(t) + w(t)$ or $y[n] = x[n] + w[n]$.
3. Integration for continuous-time LTI systems: $y(t) = \int_{-\infty}^{t} x(\tau)\, d\tau$; and a time shift for discrete-time LTI systems: $y[n] = x[n-1]$.

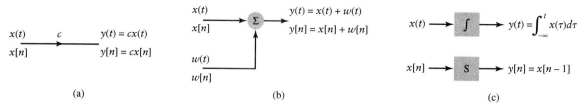

(a)   (b)   (c)

FIGURE 2.32   Symbols for elementary operations in block diagram descriptions of systems. (a) Scalar multiplication. (b) Addition. (c) Integration for continuous-time systems and time shifting for discrete-time systems.

Figure 2.32 depicts the block diagram symbols used to represent each of these operations. In order to express a continuous-time LTI system in terms of integration, we convert the differential equation into an integral equation. The operation of integration is usually used in block diagrams for continuous-time LTI systems instead of differentiation, because integrators are more easily built from analog components than are differentiators. Moreover, integrators smooth out noise in the system, while differentiators accentuate noise.

The integral or difference equation corresponding to the system behavior is obtained by expressing the sequence of operations represented by the block diagram in equation form. We begin with the discrete-time case. A discrete-time LTI system is depicted in Fig. 2.33. Let us write an equation corresponding to the portion of the system within the dashed box. The output of the first time shift is $x[n - 1]$. The second time shift has output $x[n - 2]$. The scalar multiplications and summations imply that

$$w[n] = b_0 x[n] + b_1 x[n - 1] + b_2 x[n - 2]. \tag{2.49}$$

Now we may write an expression for $y[n]$ in terms of $w[n]$. The block diagram indicates that

$$y[n] = w[n] - a_1 y[n - 1] - a_2 y[n - 2]. \tag{2.50}$$

The output of this system may be expressed as a function of the input $x[n]$ by substituting Eq. (2.49) for $w[n]$ into Eq. (2.50). We thus have

$$y[n] = -a_1 y[n - 1] - a_2 y[n - 2] + b_0 x[n] + b_1 x[n - 1] + b_2 x[n - 2],$$

or

$$y[n] + a_1 y[n - 1] + a_2 y[n - 2] = b_0 x[n] + b_1 x[n - 1] + b_2 x[n - 2]. \tag{2.51}$$

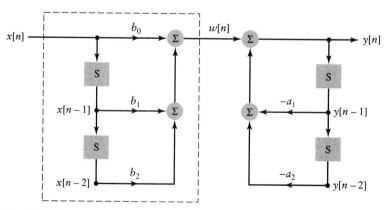

FIGURE 2.33   Block diagram representation of a discrete-time LTI system described by a second-order difference equation.

Therefore, the block diagram in Fig. 2.33 describes an LTI system whose input–output characteristic is represented by a second-order difference equation.

Note that the block diagram explicitly represents the operations involved in computing the output from the input and tells us how to simulate the system on a computer. The operations of scalar multiplication and addition are easily evaluated with a computer. The outputs of the time-shift operations correspond to memory locations in a computer. To compute the current output from the current input, we must have saved the past values of the input and output in memory. To begin a computer simulation at a specified time, we must know the input and past values of the output. The past values of the output are the initial conditions required to solve the difference equation directly.

▶ **Problem 2.23** Determine the difference equation corresponding to the block diagram description of the systems depicted in Fig. 2.34(a) and (b).

Answers:

(a)

$$y[n] + \frac{1}{2}y[n-1] - \frac{1}{3}y[n-3] = x[n] + 2x[n-2]$$

(b)

$$y[n] + (1/2)y[n-1] + (1/4)y[n-2] = x[n-1] \qquad ◀$$

The block diagram description of a system is not unique. We illustrate this fact by developing a second block diagram description of the system described by the second-order difference equation given by Eq. (2.51). We may view the system in Fig. 2.33 as a cascade of two systems: one with input $x[n]$ and output $w[n]$ described by Eq. (2.49) and a second with input $w[n]$ and output $y[n]$ described by Eq. (2.50). Since these are LTI systems, we may interchange their order without changing the input–output behavior of the cascade.

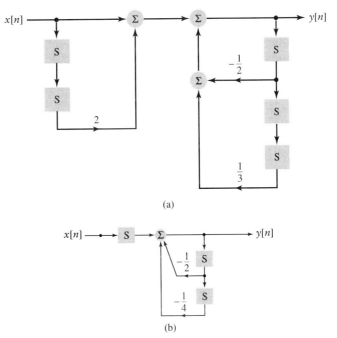

FIGURE 2.34 Block diagram representations for Problem 2.23.

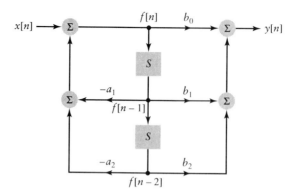

FIGURE 2.35   Direct form II representation of an LTI system described by a second-order difference equation.

Accordingly, let us interchange their order and denote the output of the new first system as $f[n]$. This output is obtained from Eq. (2.50) and the input $x[n]$ and is given by

$$f[n] = -a_1 f[n-1] - a_2 f[n-2] + x[n]. \tag{2.52}$$

The signal $f[n]$ is also the input to the second system. The output of the second system, obtained from Eq. (2.49), is

$$y[n] = b_0 f[n] + b_1 f[n-1] + b_2 f[n-2]. \tag{2.53}$$

Both systems involve time-shifted versions of $f[n]$. Hence, only one set of time shifts is needed in the block diagram for this second description of the system. We may represent the system described by Eqs. (2.52) and (2.53) by the block diagram of Fig. 2.35.

The block diagrams in Figs. 2.33 and 2.35 represent different implementations of a system with input-output behavior described by Eq. (2.51). The diagram in Fig. 2.33 is termed a "direct form I" implementation; that in Fig. 2.35 is termed a "direct form II" implementation. The direct form II implementation uses memory more efficiently. In this example, it requires only two memory locations, compared with the four required for the direct form I implementation.

▶ **Problem 2.24**   Draw direct form I and direct form II implementations of the systems described by the difference equation

$$y[n] + (1/4)y[n-1] + (1/8)y[n-2] = x[n] + x[n-1].$$

Answer: See Fig. 2.36.   ◀

There are many different implementations of an LTI system whose input–output behavior is described by a difference equation. All are obtained by manipulating either the difference equation or the elements in a block diagram representation. While these different systems are equivalent from an input–output perspective, they generally differ with respect to other criteria, such as memory requirements, the number of computations required per output value, and numerical accuracy.

Analogous results hold for continuous-time LTI systems. We may simply replace the time-shift operations in Figs. 2.33 and 2.35 with differentiation to obtain block diagram representations of LTI systems described by differential equations. However, in order to

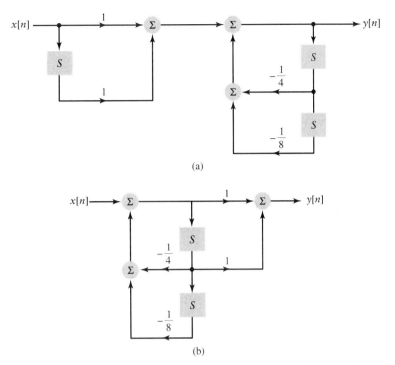

(a)

(b)

FIGURE 2.36   Solution to Problem 2.24. (a) Direct form I. (b) Direct form II.

depict the continuous-time LTI system in terms of the more easily implemented integration operation, we must first rewrite the differential equation, which takes the form

$$\sum_{k=0}^{N} a_k \frac{d^k}{dt^k} y(t) = \sum_{k=0}^{M} b_k \frac{d^k}{dt^k} x(t),$$ (2.54)

as an integral equation. To do so, we define the integration operation in a recursive manner to simplify the notation. Let $v^{(0)}(t) = v(t)$ be an arbitrary signal, and set

$$v^{(n)}(t) = \int_{-\infty}^{t} v^{(n-1)}(\tau)\, d\tau, \qquad n = 1, 2, 3, \ldots.$$

Hence, $v^{(n)}(t)$ is the n-fold integral of $v(t)$ with respect to time. This definition integrates over all past values of time. We may rewrite it in terms of an initial condition on the integrator as

$$v^{(n)}(t) = \int_{0}^{t} v^{(n-1)}(\tau)\, d\tau + v^{(n)}(0), \qquad n = 1, 2, 3, \ldots.$$

If we assume zero initial conditions, then integration and differentiation are inverse operations. That is,

$$\frac{d}{dt} v^{(n)}(t) = v^{(n-1)}(t), \qquad t > 0 \quad \text{and} \quad n = 1, 2, 3, \ldots.$$

Thus, if $N \geq M$ and we integrate Eq. (2.54) N times, we obtain the integral equation description of the system:

$$\sum_{k=0}^{N} a_k y^{(N-k)}(t) = \sum_{k=0}^{M} b_k x^{(N-k)}(t).$$ (2.55)

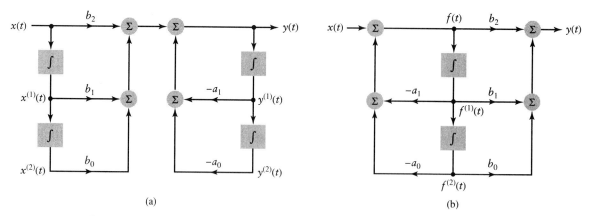

(a)   (b)

FIGURE 2.37 Block diagram representations of a continuous-time LTI system described by a second-order integral equation. (a) Direct form I. (b) Direct form II.

For a second-order system with $a_0 = 1$, Eq. (2.55) may be written as

$$y(t) = -a_1 y^{(1)}(t) - a_0 y^{(2)}(t) + b_2 x(t) + b_1 x^{(1)}(t) + b_0 x^{(2)}(t). \qquad (2.56)$$

Direct form I and direct form II implementations of this system are depicted in Figs. 2.37(a) and (b), respectively. Note that the direct form II implementation uses fewer integrators than the direct form I implementation.

▶ **Problem 2.25**   Find the differential equation description of the system depicted in Fig. 2.38.

Answer:

$$\frac{d^2}{dt^2} y(t) + 3y(t) = \frac{d}{dt} x(t) + 2\frac{d^2}{dt^2} x(t) \qquad\qquad ◀$$

Block diagram representations of continuous-time LTI systems may be used to specify analog computer simulations of systems. In such a simulation, signals are represented as voltages, resistors are used to implement scalar multiplication, and the integrators are constructed out of resistors, capacitors, and operational amplifiers. (Operational amplifiers are discussed in Chapter 9.) Initial conditions are specified as initial voltages on integrators. Analog computer simulations are much more cumbersome than digital computer sim-

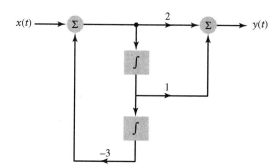

FIGURE 2.38   Block diagram representation for Problem 2.25.

ulations and suffer from drift. These serious practical problems are avoided by simulating continuous-time systems on digital computers, using numerical approximations to either integration or differentiation operations. However, care must be exercised with digital computer simulations to balance the complexity of computation against accuracy.

2.13 *State-Variable Descriptions of LTI Systems*

The state-variable description of an LTI system consists of a series of coupled first-order differential or difference equations that describe how the state of the system evolves and an equation that relates the output of the system to the current state variables and input. These equations are written in matrix form. Since the state-variable description is expressed in terms of matrices, powerful tools from linear algebra may be used to systematically study and design the behavior of the system.

The *state* of a system may be defined as a minimal set of signals that represent the system's entire memory of the past. That is, given only the value of the state at an initial point in time, n_i (or t_i), and the input for times $n \geq n_i$ (or $t \geq t_i$), we can determine the output for all times $n \geq n_i$ (or $t \geq t_i$). We shall see that the selection of signals indicating the state of a system is not unique and that there are many possible state-variable descriptions corresponding to a system with a given input–output characteristic. The ability to represent a system with different state-variable descriptions is a powerful attribute that finds application in advanced methods for control system analysis and discrete-time system implementation.

■ 2.13.1 THE STATE-VARIABLE DESCRIPTION

We shall develop the general state-variable description by starting with the direct form II implementation of a second-order LTI system, depicted in Fig. 2.39. In order to determine the output of the system for $n \geq n_i$, we must know the input for $n \geq n_i$ and the outputs of the time-shift operations labeled $q_1[n]$ and $q_2[n]$ at time $n = n_i$. This suggests that we may choose $q_1[n]$ and $q_2[n]$ as the state of the system. Note that since $q_1[n]$ and $q_2[n]$ are the outputs of the time-shift operations, the next value of the state, $q_1[n + 1]$ and $q_2[n + 1]$, must correspond to the variables at the input to the time-shift operations. The block diagram indicates that the next value of the state is obtained from the current state and the input via the two equations

$$q_1[n + 1] = -a_1 q_1[n] - a_2 q_2[n] + x[n] \qquad (2.57)$$

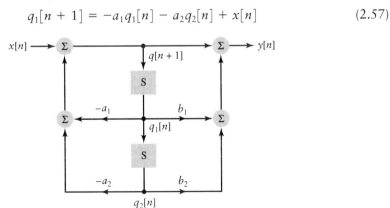

FIGURE 2.39   Direct form II representation of a second-order discrete-time LTI system depicting state variables $q_1[n]$ and $q_2[n]$.

and

$$q_2[n + 1] = q_1[n]. \tag{2.58}$$

The block diagram also indicates that the system output is expressed in terms of the input and the state of the system as

$$y[n] = x[n] - a_1 q_1[n] - a_2 q_2[n] + b_1 q_1[n] + b_2 q_2[n],$$

or

$$y[n] = (b_1 - a_1)q_1[n] + (b_2 - a_2)q_2[n] + x[n]. \tag{2.59}$$

We write Eqs. (2.57) and (2.58) in matrix form as

$$\begin{bmatrix} q_1[n + 1] \\ q_2[n + 1] \end{bmatrix} = \begin{bmatrix} -a_1 & -a_2 \\ 1 & 0 \end{bmatrix} \begin{bmatrix} q_1[n] \\ q_2[n] \end{bmatrix} + \begin{bmatrix} 1 \\ 0 \end{bmatrix} x[n], \tag{2.60}$$

while Eq. (2.59) is expressed as

$$y[n] = \begin{bmatrix} b_1 - a_1 & b_2 - a_2 \end{bmatrix} \begin{bmatrix} q_1[n] \\ q_2[n] \end{bmatrix} + [1]x[n]. \tag{2.61}$$

If we define the state vector as the column vector

$$\mathbf{q}[n] = \begin{bmatrix} q_1[n] \\ q_2[n] \end{bmatrix},$$

then we can rewrite Eqs. (2.60) and (2.61) as

$$\mathbf{q}[n + 1] = \mathbf{A}\mathbf{q}[n] + \mathbf{b}x[n] \tag{2.62}$$

and

$$y[n] = \mathbf{c}\mathbf{q}[n] + Dx[n], \tag{2.63}$$

where matrix $\mathbf{A}$, vectors $\mathbf{b}$ and $\mathbf{c}$, and scalar D are given by

$$\mathbf{A} = \begin{bmatrix} -a_1 & -a_2 \\ 1 & 0 \end{bmatrix}, \qquad \mathbf{b} = \begin{bmatrix} 1 \\ 0 \end{bmatrix},$$
$$\mathbf{c} = \begin{bmatrix} b_1 - a_1 & b_2 - a_2 \end{bmatrix}, \quad \text{and} \quad D = 1.$$

Equations (2.62) and (2.63) are the general form of a state-variable description corresponding to a discrete-time system. Previously, we studied impulse-response, difference equation, and block diagram representations of systems. Matrix $\mathbf{A}$, vectors $\mathbf{b}$ and $\mathbf{c}$, and scalar D represent another description of the system. Systems having different internal structures will be represented by different $\mathbf{A}$'s, $\mathbf{b}$'s, $\mathbf{c}$'s, and D's. The state-variable description is the only analytic system representation capable of specifying the internal structure of the system. Thus, the state-variable description is used in any problem in which the internal system structure needs to be considered.

If the input–output characteristics of the system are described by an Nth-order difference equation, then the state vector $\mathbf{q}[n]$ is N by 1, $\mathbf{A}$ is N by N, $\mathbf{b}$ is N by 1, and $\mathbf{c}$ is 1 by N. Recall that solving of the difference equation requires N initial conditions, which represent the system's memory of the past, as does the N-dimensional state vector. Also, an Nth-order system contains at least N time-shift operations in its block diagram representation. If the block diagram of a system has a minimal number of time shifts, then a natural choice for the states are the outputs of the unit delays, since the unit delays embody the memory of the system. This choice is illustrated in the next example.

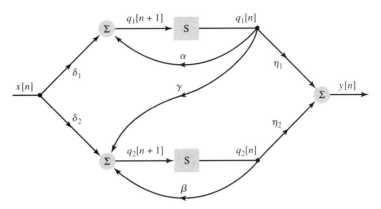

FIGURE 2.40 Block diagram of LTI system for Example 2.28.

EXAMPLE 2.28 STATE-VARIABLE DESCRIPTION OF A SECOND-ORDER SYSTEM Find the state-variable description corresponding to the system depicted in Fig. 2.40 by choosing the state variables to be the outputs of the unit delays.

Solution: The block diagram indicates that the states are updated according to the equations

$$q_1[n + 1] = \alpha q_1[n] + \delta_1 x[n]$$

and

$$q_2[n + 1] = \gamma q_1[n] + \beta q_2[n] + \delta_2 x[n]$$

and the output is given by

$$y[n] = \eta_1 q_1[n] + \eta_2 q_2[n].$$

These equations may be expressed in the state-variable forms of Eqs. (2.62) and (2.63) if we define

$$\mathbf{q}[n] = \begin{bmatrix} q_1[n] \\ q_2[n] \end{bmatrix},$$

$$\mathbf{A} = \begin{bmatrix} \alpha & 0 \\ \gamma & \beta \end{bmatrix}, \qquad \mathbf{b} = \begin{bmatrix} \delta_1 \\ \delta_2 \end{bmatrix},$$

$$\mathbf{c} = [\eta_1 \quad \eta_2], \quad \text{and} \quad D = [0]. \qquad \blacksquare$$

▶ **Problem 2.26** Find the state-variable description corresponding to the block diagram representations in Figs. 2.41(a) and (b). Choose the state variables to be the outputs of the unit delays, $q_1[n]$ and $q_2[n]$, as indicated in the figure.

Answers:

(a)

$$\mathbf{A} = \begin{bmatrix} -\frac{1}{2} & 0 \\ 1 & \frac{1}{3} \end{bmatrix}; \qquad \mathbf{b} = \begin{bmatrix} 1 \\ 3 \end{bmatrix};$$

$$\mathbf{c} = [0 \quad 1]; \qquad D = [2].$$

(b)

$$\mathbf{A} = \begin{bmatrix} 0 & -\frac{1}{3} \\ \frac{1}{4} & 0 \end{bmatrix}; \qquad \mathbf{b} = \begin{bmatrix} 2 \\ -1 \end{bmatrix};$$

$$\mathbf{c} = [1 \quad -2]; \qquad D = [0]. \qquad \blacktriangleleft$$

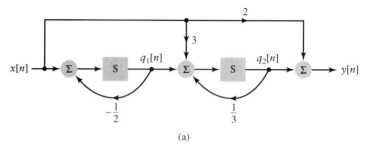

(a)

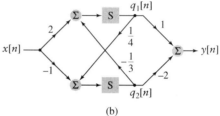

(b)

FIGURE 2.41   Block diagram of LTI systems for Problem 2.26.

The state-variable description of continuous-time systems is analogous to that of discrete-time systems, with the exception that the state equation given by Eq. (2.62) is expressed in terms of a derivative. We thus write

$$\frac{d}{dt}\mathbf{q}(t) = \mathbf{A}\mathbf{q}(t) + \mathbf{b}x(t) \tag{2.64}$$

and

$$y(t) = \mathbf{c}\mathbf{q}(t) + Dx(t). \tag{2.65}$$

Once again, matrix $\mathbf{A}$, vectors $\mathbf{b}$ and $\mathbf{c}$, and scalar D describe the internal structure of the system.

The memory of a continuous-time system is contained within the system's energy storage devices. Hence, state variables are usually chosen as the physical quantities associated with such devices. For example, in electrical systems, the energy storage devices are capacitors and inductors. Accordingly, we may choose state variables to correspond to the voltages across capacitors or the currents through inductors. In a mechanical system, the energy storage devices are springs and masses; thus, displacements of springs or velocities of masses may be chosen as state variables. The state-variable equations represented by Eqs. (2.64) and (2.65) are obtained from the equations that relate the behavior of the energy storage devices to the input and output, as the next example demonstrates.

EXAMPLE 2.29 STATE-VARIABLE DESCRIPTION OF AN ELECTRICAL CIRCUIT   Consider the electrical circuit depicted in Fig. 2.42. Derive a state-variable description of this system if the input is the applied voltage $x(t)$ and the output is the current $y(t)$ through the resistor.

Solution: Choose the state variables as the voltage across each capacitor. Summing the voltage drops around the loop involving $x(t)$, R_1, and C_1 gives

$$x(t) = y(t)R_1 + q_1(t),$$

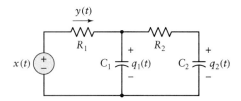

FIGURE 2.42   Circuit diagram of LTI system for Example 2.29.

or

$$y(t) = -\frac{1}{R_1}q_1(t) + \frac{1}{R_1}x(t). \tag{2.66}$$

This equation expresses the output as a function of the state variables and the input $x(t)$. Let $i_2(t)$ be the current through R_2. Summing the voltage drops around the loop involving C_1, R_2, and C_2, we obtain

$$q_1(t) = R_2 i_2(t) + q_2(t),$$

or

$$i_2(t) = \frac{1}{R_2}q_1(t) - \frac{1}{R_2}q_2(t). \tag{2.67}$$

However, we also know that

$$i_2(t) = C_2\frac{d}{dt}q_2(t).$$

We use Eq. (2.67) to eliminate $i_2(t)$ and obtain

$$\frac{d}{dt}q_2(t) = \frac{1}{C_2 R_2}q_1(t) - \frac{1}{C_2 R_2}q_2(t). \tag{2.68}$$

To conclude our derivation, we need a state equation for $q_1(t)$. This is obtained by applying Kirchhoff's current law to the node between R_1 and R_2. Letting $i_1(t)$ be the current through C_1, we have

$$y(t) = i_1(t) + i_2(t).$$

Now we use Eq. (2.66) for $y(t)$, Eq. (2.67) for $i_2(t)$, the relation

$$i_1(t) = C_1\frac{d}{dt}q_1(t)$$

and rearrange terms to obtain

$$\frac{d}{dt}q_1(t) = -\left(\frac{1}{C_1 R_1} + \frac{1}{C_1 R_2}\right)q_1(t) + \frac{1}{C_1 R_2}q_2(t) + \frac{1}{C_1 R_1}x(t). \tag{2.69}$$

The state-variable description, from Eqs. (2.66), (2.68), and (2.69), is

$$\mathbf{A} = \begin{bmatrix} -\left(\dfrac{1}{C_1 R_1} + \dfrac{1}{C_1 R_2}\right) & \dfrac{1}{C_1 R_2} \\[2ex] \dfrac{1}{C_2 R_2} & -\dfrac{1}{C_2 R_2} \end{bmatrix}, \quad \mathbf{b} = \begin{bmatrix} \dfrac{1}{C_1 R_1} \\[1ex] 0 \end{bmatrix}$$

$$\mathbf{c} = \begin{bmatrix} -\dfrac{1}{R_1} & 0 \end{bmatrix}, \quad \text{and} \quad D = \dfrac{1}{R_1}. \qquad\blacksquare$$

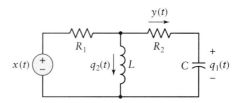

FIGURE 2.43   Circuit diagram of LTI system for Problem 2.27.

S & S
Solutions

▶ **Problem 2.27**   Find the state-variable description of the circuit depicted in Fig. 2.43. Choose the state variables $q_1(t)$ and $q_2(t)$ as the voltage across the capacitor and the current through the inductor, respectively.

Answer:

$$\mathbf{A} = \begin{bmatrix} \dfrac{-1}{(R_1 + R_2)C} & \dfrac{-R_1}{(R_1 + R_2)C} \\[2mm] \dfrac{R_1}{(R_1 + R_2)L} & \dfrac{-R_1 R_2}{(R_1 + R_2)L} \end{bmatrix}, \qquad \mathbf{b} = \begin{bmatrix} \dfrac{1}{(R_1 + R_2)C} \\[2mm] \dfrac{R_2}{(R_1 + R_2)L} \end{bmatrix},$$

$$\mathbf{c} = \begin{bmatrix} \dfrac{-1}{R_1 + R_2} & \dfrac{-R_1}{R_1 + R_2} \end{bmatrix}, \qquad D = \begin{bmatrix} \dfrac{1}{R_1 + R_2} \end{bmatrix}.$$ ◀

In a block diagram representation of a continuous-time system, the state variables correspond to the outputs of the integrators. Thus, the input to the integrator is the derivative of the corresponding state variable. The state-variable description is obtained by writing equations that correspond to the operations in the block diagram. The procedure is illustrated in the next example.

EXAMPLE 2.30 STATE-VARIABLE DESCRIPTION FROM A BLOCK DIAGRAM   Determine the state-variable description corresponding to the block diagram in Fig. 2.44. The choice of state variables is indicated on the diagram.

Solution: The block diagram indicates that

$$\frac{d}{dt} q_1(t) = 2q_1(t) - q_2(t) + x(t),$$

$$\frac{d}{dt} q_2(t) = q_1(t),$$

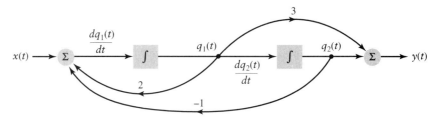

FIGURE 2.44   Block diagram of LTI system for Example 2.30.

and

$$y(t) = 3q_1(t) + q_2(t).$$

Hence, the state-variable description is

$$\mathbf{A} = \begin{bmatrix} 2 & -1 \\ 1 & 0 \end{bmatrix}, \qquad \mathbf{b} = \begin{bmatrix} 1 \\ 0 \end{bmatrix},$$

$$\mathbf{c} = [3 \quad 1], \quad \text{and} \quad D = [0].$$ ∎

■ 2.13.2 TRANSFORMATIONS OF THE STATE

We have claimed that there is no unique state-variable description of a system with a given input–output characteristic. Different state-variable descriptions may be obtained by transforming the state variables. The transformation is accomplished by defining a new set of state variables that are a weighted sum of the original ones. This changes the form of $\mathbf{A}$, $\mathbf{b}$, $\mathbf{c}$, and D, but does not change the input–output characteristics of the system. To illustrate the procedure, consider Example 2.30 again. Let us define new states $q_2'(t) = q_1(t)$ and $q_1'(t) = q_2(t)$. Here, we simply have interchanged the state variables: $q_2'(t)$ is the output of the first integrator and $q_1'(t)$ is the output of the second integrator. We have not changed the structure of the block diagram, so, clearly, the input–output characteristic of the system remains the same. The state-variable description is different, however, since we now have

$$\mathbf{A}' = \begin{bmatrix} 0 & 1 \\ -1 & 2 \end{bmatrix}, \qquad \mathbf{b}' = \begin{bmatrix} 0 \\ 1 \end{bmatrix},$$

$$\mathbf{c}' = [1 \quad 3], \quad \text{and} \quad D' = [0].$$

The example in the previous paragraph employs a particularly simple transformation of the original state. In general, we may define a new state vector as a transformation of the original state vector, or $\mathbf{q}' = \mathbf{T}\mathbf{q}$. We define $\mathbf{T}$ as the state-transformation matrix. Note that we have dropped the time index (t) or $[n]$ in order to treat both continuous- and discrete-time cases simultaneously. In order for the new state to represent the entire system's memory, the relationship between $\mathbf{q}'$ and $\mathbf{q}$ must be one to one. This implies that $\mathbf{T}$ must be a nonsingular matrix, that is, the inverse matrix $\mathbf{T}^{-1}$ exists. Hence, $\mathbf{q} = \mathbf{T}^{-1}\mathbf{q}'$. The original state-variable description is given by

$$\dot{\mathbf{q}} = \mathbf{A}\mathbf{q} + \mathbf{b}x \tag{2.70}$$

and

$$y = \mathbf{c}\mathbf{q} + Dx, \tag{2.71}$$

where the dot over the $\mathbf{q}$ denotes differentiation in continuous time or time advance ($[n + 1]$) in discrete time. The new state-variable description involving $\mathbf{A}'$, $\mathbf{b}'$, $\mathbf{c}'$, and D' is obtained from the relationship $\dot{\mathbf{q}}' = \mathbf{T}\dot{\mathbf{q}}$ by first substituting Eq. (2.70) for $\dot{\mathbf{q}}$ to obtain

$$\dot{\mathbf{q}}' = \mathbf{T}\mathbf{A}\mathbf{q} + \mathbf{T}\mathbf{b}x.$$

Now we use $\mathbf{q} = \mathbf{T}^{-1}\mathbf{q}'$ to write

$$\dot{\mathbf{q}}' = \mathbf{T}\mathbf{A}\mathbf{T}^{-1}\mathbf{q}' + \mathbf{T}\mathbf{b}x.$$

Next, we again use $\mathbf{q} = \mathbf{T}^{-1}\mathbf{q}'$, this time in Eq. (2.71) to obtain the output equation

$$y = \mathbf{c}\mathbf{T}^{-1}\mathbf{q} + Dx.$$

Hence, if we set

$$\mathbf{A}' = \mathbf{TAT}^{-1}, \qquad \mathbf{b}' = \mathbf{Tb},$$
$$\mathbf{c}' = \mathbf{cT}^{-1}, \quad \text{and} \quad D' = D, \tag{2.72}$$

then

$$\dot{\mathbf{q}}' = \mathbf{A}'\mathbf{q} + \mathbf{b}'x$$

and

$$y = \mathbf{c}'\mathbf{q} + D'x$$

together make up the new state-variable description.

EXAMPLE 2.31 TRANSFORMING THE STATE   A discrete-time system has the state-variable description

$$\mathbf{A} = \frac{1}{10}\begin{bmatrix} -1 & 4 \\ 4 & -1 \end{bmatrix}, \qquad \mathbf{b} = \begin{bmatrix} 2 \\ 4 \end{bmatrix},$$
$$\mathbf{c} = \frac{1}{2}[1 \quad 1], \quad \text{and} \quad D = 2.$$

Find the state-variable description $\mathbf{A}'$, $\mathbf{b}'$, $\mathbf{c}'$, and D' corresponding to the new states $q_1'[n] = -\frac{1}{2}q_1[n] + \frac{1}{2}q_2[n]$ and $q_2'[n] = \frac{1}{2}q_1[n] + \frac{1}{2}q_2[n]$.

Solution: We write the new state vector as $\mathbf{q}' = \mathbf{Tq}$, where

$$\mathbf{T} = \frac{1}{2}\begin{bmatrix} -1 & 1 \\ 1 & 1 \end{bmatrix}.$$

This matrix is nonsingular, and its inverse is

$$\mathbf{T}^{-1} = \begin{bmatrix} -1 & 1 \\ 1 & 1 \end{bmatrix}.$$

Hence, using these values of $\mathbf{T}$ and $\mathbf{T}^{-1}$ in Eq. (2.72) gives

$$\mathbf{A}' = \begin{bmatrix} -\frac{1}{2} & 0 \\ 0 & \frac{3}{10} \end{bmatrix}, \qquad \mathbf{b}' = \begin{bmatrix} 1 \\ 3 \end{bmatrix},$$
$$\mathbf{c}' = [0 \quad 1], \quad \text{and} \quad D' = 2.$$

Note that this choice for $\mathbf{T}$ results in $\mathbf{A}'$ being a diagonal matrix and thus separates the state update into the two decoupled first-order difference equations

$$q_1[n+1] = -\frac{1}{2}q_1[n] + x[n]$$

and

$$q_2[n+1] = \frac{3}{10}q_2[n] + 3x[n].$$

Because of its simple structure, the decoupled form of the state-variable description is particularly useful in analyzing systems.   ■

▶ **Problem 2.28**   A continuous-time system has the state-variable description

$$\mathbf{A} = \begin{bmatrix} -2 & 0 \\ 1 & -1 \end{bmatrix}, \qquad \mathbf{b} = \begin{bmatrix} 1 \\ 1 \end{bmatrix},$$

$$\mathbf{c} = \begin{bmatrix} 0 & 2 \end{bmatrix}, \quad \text{and} \quad D = 1.$$

Find the state-variable description $\mathbf{A}'$, $\mathbf{b}'$, $\mathbf{c}'$, and D' corresponding to the new states $q_1'(t) = 2q_1(t) + q_2(t)$ and $q_2'(t) = q_1(t) - q_2(t)$.

Answers:

$$\mathbf{A}' = \frac{1}{3}\begin{bmatrix} -4 & -1 \\ -2 & -5 \end{bmatrix}; \qquad \mathbf{b}' = \begin{bmatrix} 3 \\ 0 \end{bmatrix};$$

$$\mathbf{c}' = \frac{1}{3}\begin{bmatrix} 2 & -4 \end{bmatrix}; \qquad D' = 1. \qquad \blacktriangleleft$$

Note that each nonsingular transformation $\mathbf{T}$ generates a different state-variable description of an LTI system with a given input–output behavior. Different state-variable descriptions correspond to different ways of determining the LTI system output from the input. Both the block diagram and state-variable descriptions represent the internal structure of an LTI system. The state-variable description is advantageous because powerful tools from linear algebra may be used to systematically study and design the internal structure of the system. The ability to transform the internal structure without changing the input–output characteristics of the system is used to analyze LTI systems and identify implementations of such systems that optimize some performance criteria not directly related to input–output behavior, such as the numerical effects of round-off in a computer-based system implementation.

2.14 *Exploring Concepts with MATLAB*

Digital computers are ideally suited to implementing time-domain descriptions of discrete-time systems, because computers naturally store and manipulate sequences of numbers. For example, the convolution sum describes the relationship between the input and output of a discrete-time system and is easily evaluated with a computer as a sum of products of numbers. In contrast, continuous-time systems are described in terms of continuous functions, which are not easily represented or manipulated in a digital computer. For instance, the output of a continuous-time system is described by the convolution integral, the computer evaluation of which requires the use of either numerical integration or symbolic manipulation techniques, both of which are beyond the scope of this book. Hence, our exploration with MATLAB focuses on discrete-time systems.

A second limitation on exploring signals and systems is imposed by the finite memory or storage capacity and nonzero computation times inherent in all digital computers. Consequently, we can manipulate only finite-duration signals. For example, if the impulse response of a system has infinite duration and the input is of infinite duration, then the convolution sum involves summing an infinite number of products. Of course, even if we could store the infinite-length signals in the computer, the infinite sum could not be computed in a finite amount of time. In spite of this limitation, the behavior of a system in response to an infinite-length signal may often be inferred from its response to a carefully chosen finite-length signal. Furthermore, the impulse response of stable LTI systems decays to zero at infinite time and thus may often be well approximated by a truncated version.

Both the MATLAB Signal Processing Toolbox and Control System Toolbox are used in this section.

■ 2.14.1 CONVOLUTION

We recall from Section 2.2 that the convolution sum expresses the output of a discrete-time system in terms of the input and the impulse response of the system. MATLAB has a function named `conv` that evaluates the convolution of finite-duration discrete-time signals. If x and h are vectors representing signals, then the MATLAB command `y = conv(x,h)` generates a vector y representing the convolution of the signals represented by x and h. The number of elements in y is given by the sum of the number of elements in x and h, minus one. Note that we must know the point in time at which the signals represented by x and h originated in order to determine the origin of their convolution. In general, suppose the first and last elements of x correspond to times $n = k_x$ and $n = l_x$, respectively, while the first and last elements of h correspond to times $n = k_h$ and $n = l_h$. Then the first and last elements of y correspond to times $n = k_y = k_x + k_h$ and $n = l_y = l_x + l_h$. Observe that the lengths of $x[n]$ and $h[n]$ are $L_x = l_x - k_x + 1$ and $L_h = l_h - k_h + 1$. Thus, the length of $y[n]$ is $L_y = l_y - k_y + 1 = L_x + L_h - 1$.

To illustrate all this, let us repeat Example 2.1, this time using MATLAB. Here, the first nonzero value in the impulse response and input occurs at time $n = k_h = k_x = 0$. The last elements of the impulse response and input occur at times $n = l_h = 1$ and $n = l_x = 2$. Thus, the convolution y starts at time $n = k_y = k_x + k_h = 0$, ends at time $n = l_y = l_x + l_h = 3$, and has length $L_y = l_y - k_y + 1 = 4$. We evaluate this convolution in MATLAB as follows:

```
>>h=[-1, 0.5];
>>x=[2, 4, -2];
>>y=conv(x, h)
y =
        2  5  0  -1
```

In Example 2.3, we used hand calculation to determine the output of a system with impulse response given by

$$h[n] = (1/4)(u[n] - u[n - 4])$$

and input

$$x[n] = u[n] - u[n - 10].$$

We may use the MATLAB command `conv` to perform the convolution as follows: In this case, $k_h = 0, l_h = 3, k_x = 0$, and $l_x = 9$, so y starts at time $n = k_y = 0$, ends at time $n = l_y = 12$, and has length $L_y = 13$. The impulse response consists of four consecutive values of 0.25, while the input consists of 10 consecutive ones. These signals may be defined in MATLAB with the following commands:

```
>>h=0.25*ones(1, 4);
>>x=ones(1, 10);
```

The output is obtained and graphed using these commands:

```
>>n=0:12;
>>y=conv(x, h);
>>stem(n, y); xlabel('n'); ylabel('y[n]')
```

The result is depicted in Fig. 2.45.

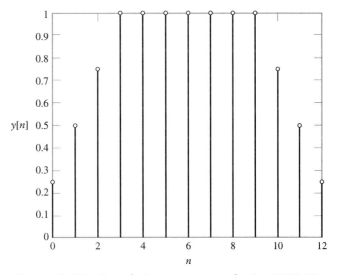

FIGURE 2.45 Convolution sum computed using MATLAB.

▶ **Problem 2.29** Use MATLAB to solve Problem 2.2(c) for $\alpha = 0.9$. That is, find the ouput of the system with input $x[n] = 2\{u[n + 2] - u[n - 12]\}$ and impulse response $h[n] = 0.9^n\{u[n - 2] - u[n - 13]\}$.

Answer: See Fig. 2.46.

■ 2.14.2 STEP RESPONSE

The step response is the ouput of a system in response to a step input and is infinite in duration, in general. However, we can evaluate the first p values of the step response using the `conv` function if the system impulse response is zero for times $n < k_h$ by convolving the first p values of $h[n]$ with a finite-duration step of length p. That is, we construct

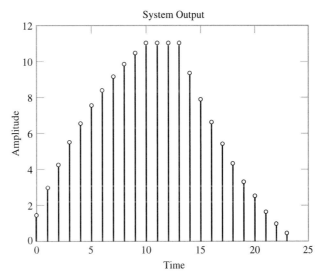

FIGURE 2.46 Solution to Problem 2.29.

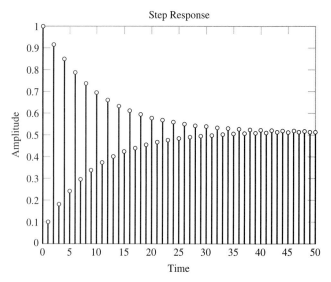

FIGURE 2.47   Step response computed using MATLAB.

a vector **h** from the first p nonzero values of the impulse response, define the step **u=ones(1, p)**, and evaluate **s=conv(u,h)**. The first element of **s** corresponds to time k_b, and the first p values of **s** represent the first p values of the step response. The remaining values of **s** do not correspond to the step response, but are an artifact of convolving finite-duration signals.

For example, we may determine the first 50 values of the step response of the system with impulse response given in Problem 2.12, namely,

$$h[n] = (\rho)^n u[n],$$

with $\rho = -0.9$, by using the following MATLAB commands:

```
>>h=(-0.9).^[0:49];
>>u=ones(1, 50);
>>s=conv(u, h);
```

The vector **s** has 99 values, the first 50 of which represent the step response and are depicted in Fig. 2.47. This figure is obtained using the MATLAB command **stem([0:49], s(1:50))**.

■ 2.14.3 SIMULATING DIFFERENCE EQUATIONS

In Section 2.9, we expressed the difference equation description of a system in a recursive form that allowed the system output to be computed from the input signal and past outputs. The **filter** command performs a similar function. We define vectors **a** = $[a_0, a_1, \ldots, a_N]$ and **b** = $[b_0, b_1, \ldots, b_M]$ representing the coefficients of the difference equation (2.36). If **x** is a vector representing the input signal, then the command **y=filter(b, a, x)** results in a vector **y** representing the output of the system for zero initial conditions. The number of output values in **y** corresponds to the number of input values in **x**. Nonzero initial conditions are incorporated by using the alternative command syntax **y=filter(b, a, x, zi)**, where **zi** represents the initial conditions required by **filter**. The initial conditions used by **filter** are not the past values of the output, since `filter` employs a modified form of the difference equation to determine the output. Rather, these initial conditions are obtained from knowledge of the past outputs, using the command `zi=filtic(b,a,yi)`, where `yi` is a vector containing the initial conditions in the order $[y[-1], y[-2], \ldots, y[-N]]$.

We illustrate the use of the `filter` command by revisiting Example 2.16. The system of interest is described by the difference equation

$$y[n] - 1.143y[n-1] + 0.4128y[n-2] =$$
$$0.0675x[n] + 0.1349x[n-1] + 0.675x[n-2]. \quad (2.73)$$

We determine the output in response to zero input and initial conditions $y[-1] = 1$ and $y[-2] = 2$ by using the following commands:

```
>>a=[1, -1.143, 0.4128];   b=[0.0675, 0.1349, 0.675];
>>x=zeros(1, 50);
>>zi=filtic(b, a, [1, 2]);
>>y=filter(b, a, x,zi);
```

The result is depicted in Fig. 2.28(b). We may determine the system response to an input consisting of the Intel stock price data with the following commands:

```
>>load Intc;
>>filtintc=filter(b, a, Intc);
```

Here, we have assumed that the Intel stock price data are in the file `Intc.mat`. The result is depicted in Fig. 2.28(g).

▶ **Problem 2.30** Use `filter` to determine the first 50 values of the step response of the system described by Eq. (2.73) and the first 100 values of the response to the input $x[n] = \cos\left(\frac{\pi}{5}n\right)$, assuming zero initial conditions.

Answer: See Figs. 2.28(a) and (d). ◀

The command `[h, t]=impz(b, a, n)` evaluates n values of the impulse response of a system described by a difference equation. The coefficients of the equation are contained in the vectors b and a, as they are in `filter`. The vector h contains the values of the impulse response, and t contains the corresponding time indices.

■ 2.14.4 STATE-VARIABLE DESCRIPTIONS

The MATLAB Control System Toolbox contains numerous routines for manipulating state-variable descriptions. A key feature of the Control System Toolbox is the use of LTI objects, which are customized data structures that enable manipulation of LTI system descriptions as single MATLAB variables. If `a,b,c` and `d` are MATLAB arrays representing the matrices $\mathbf{A}$, $\mathbf{b}$, $\mathbf{c}$, and D, respectively, in the state-variable description, then the command `sys=ss(a,b,c,d,-1)` produces an LTI object `sys` that represents the discrete-time system in state-variable form. Note that a continuous-time system is obtained by omitting the `-1`—that is, by using `sys=ss(a,b,c,d)`. LTI objects corresponding to other system representations are discussed in Sections 6.14 and 7.11.

Systems are manipulated in MATLAB by operations on their LTI objects. For example, if `sys1` and `sys2` are objects representing two systems in state-variable form, then `sys=sys1+sys2` produces the state-variable description for the parallel combination of `sys1` and `sys2`, while `sys=sys1*sys2` represents the cascade combination.

The function `lsim` simulates the output of an LTI system in response to a specified input. For a discrete-time system, the command has the form `y=lsim(sys,x)` where x is a vector containing the input and y represents the output. The command `h=impulse(sys,N)` places the first N values of the impulse response in h. Both of these may also be used for continuous-time LTI systems, although the command syntax changes slightly. In the continuous-time case, numerical methods are used to approximate the system response.

Recall that there is no unique state-variable description for a given LTI system. Different state-variable descriptions for the same system are obtained by transforming the state. Transformations of the state may be computed in MATLAB using the routine `ss2ss`. The state transformation is identical for both continuous- and discrete-time systems, so the same command is used for transforming either type of system. The command is of the form `sysT=ss2ss(sys, T)`, where `sys` represents the original state-variable description, T is the state-transformation matrix, and `sysT` represents the transformed state-variable description.

Let us use `ss2ss` to transform the state-variable description of Example 2.31, namely,

$$\mathbf{A} = \frac{1}{10}\begin{bmatrix} -1 & 4 \\ 4 & -1 \end{bmatrix}, \qquad \mathbf{b} = \begin{bmatrix} 2 \\ 4 \end{bmatrix},$$

$$\mathbf{c} = \frac{1}{2}[1 \quad 1], \quad \text{and} \quad D = 2,$$

using the state-transformation matrix

$$\mathbf{T} = \frac{1}{2}\begin{bmatrix} -1 & 1 \\ 1 & 1 \end{bmatrix}.$$

The following commands produce the desired result:

```
>>a=[-0.1, 0.4; 0.4, -0.1];    b=[2; 4];
>>c=[0.5, 0.5];     d=2;
>>sys=ss(a,b,c,d,-1);          % define the state-space
                                 object sys
>>T=0.5*[-1, 1; 1, 1];
>>sysT=ss2ss(sys, T)
a =
                   x1          x2
       x1    -0.50000           0
       x2           0     0.30000

b =
                   u1
       x1     1.00000
       x2     3.00000

c =
                   x1          x2
       y1            0     1.00000

d =
                   u1
       y1     2.00000

Sampling time: unspecified
Discrete-time system.
```

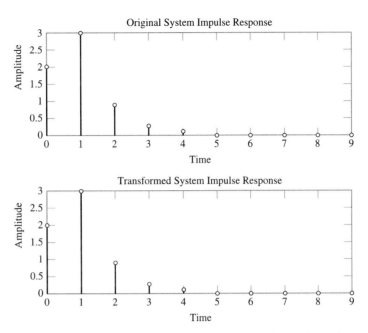

FIGURE 2.48 Impulse responses associated with the original and transformed state-variable descriptions computed using MATLAB.

This result agrees with that obtained in Example 2.31. We may verify that the two systems represented by `sys` and `sysT` have identical input–output characteristics by comparing their impulse responses via the following commands:

```
>>h=impulse(sys,10);    hT=impulse(sysT,10);
>>subplot(2, 1, 1)
>>stem([0:9], h)
>>title('Original System Impulse Response');
>>xlabel('Time'); ylabel('Amplitude')
>>subplot(2, 1, 2)
>>stem([0:9], hT)
>>title('Transformed System Impulse Response');
>>xlabel('Time'); ylabel('Amplitude')
```

Figure 2.48 depicts the first 10 values of the impulse responses of the original and transformed systems produced by this sequence of commands. We may verify that the original and transformed systems have the (numerically) identical impulse response by computing the error, `err=h-hT`.

▶ **Problem 2.31** Solve Problem 2.28 using MATLAB. ◀

2.15 *Summary*

There are many different methods for describing the action of an LTI system on an input signal. In this chapter, we have examined four different descriptions of LTI systems: the impulse response, difference and differential equation, block diagram, and state-variable descriptions. All four descriptions are equivalent in the input–output sense; that is, for a given input, each description will produce the identical output. However, different descriptions offer different insights into system characteristics and use different techniques for obtaining the output from the input. Thus, each description has its own advantages and disadvantages that come into play in solving a particular system problem.

The impulse response is the output of a system when the input is an impulse. The output of an LTI system in response to an arbitrary input is expressed in terms of the impulse response as a convolution operation. System properties, such as causality and stability, are directly related to the impulse response, which also offers a convenient framework for analyzing interconnections among systems. The input must be known for all time in order to determine the output of a system by using the impulse response and convolution.

The input and output of an LTI system may also be related by either a differential or difference equation. Differential equations often follow directly from the physical principles that define the behavior and interaction of continuous-time system components. The order of a differential equation reflects the maximum number of energy storage devices in the system, while the order of a difference equation represents the system's maximum memory of past outputs. In contrast to impulse response descriptions, the output of a system from a given point in time forward can be determined without knowledge of all past inputs, provided that the initial conditions are known. Initial conditions are the initial values of energy storage or system memory, and they summarize the effect of all past inputs up to the starting time of interest. The solution of a differential or difference equation can be separated into a natural and a forced response. The natural response describes the behavior of the system due to the initial conditions; the forced response describes the behavior of the system in response to the input acting alone.

A block diagram represents the system as an interconnection of elementary operations on signals. The manner in which these operations are interconnected defines the internal structure of the system. Different block diagrams can represent systems with identical input–output characteristics.

The state-variable description is yet another description of LTI systems that is used in controlling such systems and in advanced studies of structures for implementing difference equations. The state-variable description consists of a set of coupled first-order differential or difference equations representing the system's behavior. Written in matrix form, the description consists of two equations, one describing how the state of the system evolves, the other relating the state to the output. The state represents the system's entire memory of the past. The number of states corresponds to the number of energy storage devices or the maximum memory of past outputs present in the system. The choice of state is not unique: An infinite number of different state-variable descriptions can be used to represent LTI systems with the same input–output characteristic. Thus, state-variable descriptions are used to represent the internal structure of a physical system and provide a more detailed characterization of LTI systems than the impulse response or differential (difference) equations can.

❚ FURTHER READING

1. A concise summary and many worked-out problems for much of the material presented in this and later chapters is found in
   ▶ Hsu, H. P., *Signals and Systems*, Schaum's Outline Series (McGraw-Hill, 1995)

2. A general treatment of techniques for solving differential equations is given in
   ▶ Boyce, W. E., and R. C. DiPrima, *Elementary Differential Equations*, 6th ed. (Wiley, 1997)

3. Applications of difference equations to signal-processing problems and block diagram descriptions of discrete-time systems are described in the following texts:

   ▶ Proakis, J. G., and D. G. Manolakis, *Digital Signal Processing: Principles, Algorithms and Applications*, 3rd ed. (Prentice Hall, 1995)

   ▶ Oppenheim, A. V., R. W. Schafer, and J. R. Buck, *Discrete Time Signal Processing*, 2nd ed. (Prentice Hall, 1999)

   Both of the foregoing texts address numerical issues related to implementing discrete-time LTI systems in digital computers. Signal flow graph representations are often used to describe implementations of continuous- and discrete-time systems. They are essentially the same as a block diagram representation, except for a few differences in notation.

4. In this chapter, we determined the input–output characteristics of block diagrams by manipulating the equations representing the block diagram. *Mason's gain formula* provides a direct method for evaluating the input–output characteristic of any block diagram representation of an LTI system. The formula is described in detail in the following two texts:

   ▶ Dorf, R. C., and R. H. Bishop, *Modern Control Systems*, 7th ed. (Addison-Wesley, 1995)

   ▶ Phillips, C. L., and R. D. Harbor, *Feedback Control Systems*, 3rd ed. (Prentice Hall, 1996)

5. The role of differential equations and block diagram and state-variable descriptions in the analysis and design of feedback control systems is described in Dorf and Bishop and in Phillips and Harbor, both just mentioned.

6. More advanced treatments of state-variable-description-based methods for the analysis and design of control systems are discussed in

   ▶ Chen, C. T., *Linear System Theory and Design* (Holt, Rinehart, and Winston, 1984)

   ▶ Friedland, B., *Control System Design: An Introduction to State-Space Methods* (McGraw-Hill, 1986)

   A thorough, yet advanced application of state-variable descriptions for implementing discrete-time LTI systems and analyzing the effects of numerical round-off is given in

   ▶ Roberts, R. A., and C. T. Mullis, *Digital Signal Processing* (Addison-Wesley, 1987)

▎ ADDITIONAL PROBLEMS

2.32 A discrete-time LTI system has the impulse response $h[n]$ depicted in Fig. P2.32(a). Use linearity and time invariance to determine the system output $y[n]$ if the input is

(a) $x[n] = 3\delta[n] - 2\delta[n-1]$

(b) $x[n] = u[n+1] - u[n-3]$

(c) $x[n]$ as given in Fig. P2.32(b).

2.33 Evaluate the following discrete-time convolution sums:

(a) $y[n] = u[n+3] * u[n-3]$

(b) $y[n] = 3^n u[-n+3] * u[n-2]$

(c) $y[n] = \left(\frac{1}{4}\right)^n u[n] * u[n+2]$

(d) $y[n] = \cos\left(\frac{\pi}{2}n\right)u[n] * u[n-1]$

(e) $y[n] = (-1)^n * 2^n u[-n+2]$

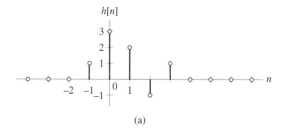

(a)

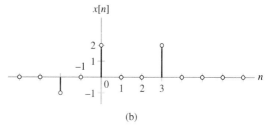

(b)

FIGURE P2.32

(f) $y[n] = \cos\left(\frac{\pi}{2}n\right) * \left(\frac{1}{2}\right)^n u[n-2]$

(g) $y[n] = \beta^n u[n] * u[n-3], \qquad |\beta| < 1$

(h) $y[n] = \beta^n u[n] * \alpha^n u[n-10], \qquad |\beta| < 1,$
    $|\alpha| < 1$

(i) $y[n] = (u[n+10] - 2u[n]$
    $+ u[n-4]) * u[n-2]$

(j) $y[n] = (u[n+10] - 2u[n]$
    $+ u[n-4]) * \beta^n u[n], \qquad |\beta| < 1$

(k) $y[n] = (u[n+10] - 2u[n+5]$
    $+ u[n-6]) * \cos\left(\frac{\pi}{2}n\right)$

(l) $y[n] = u[n] * \sum_{p=0}^{\infty} \delta[n-4p]$

(m) $y[n] = \beta^n u[n] * \sum_{p=0}^{\infty} \delta[n-4p], \quad |\beta| < 1$

(n) $y[n] = \left(\frac{1}{2}\right)^n u[n+2] * \gamma^{|n|}$

2.34 Consider the discrete-time signals depicted in Fig. P2.34. Evaluate the following convolution sums:

(a) $m[n] = x[n] * z[n]$

(b) $m[n] = x[n] * y[n]$

(c) $m[n] = x[n] * f[n]$

(d) $m[n] = x[n] * g[n]$

(e) $m[n] = y[n] * z[n]$

(f) $m[n] = y[n] * g[n]$

(g) $m[n] = y[n] * w[n]$

(h) $m[n] = y[n] * f[n]$

(i) $m[n] = z[n] * g[n]$

(j) $m[n] = w[n] * g[n]$

(k) $m[n] = f[n] * g[n]$

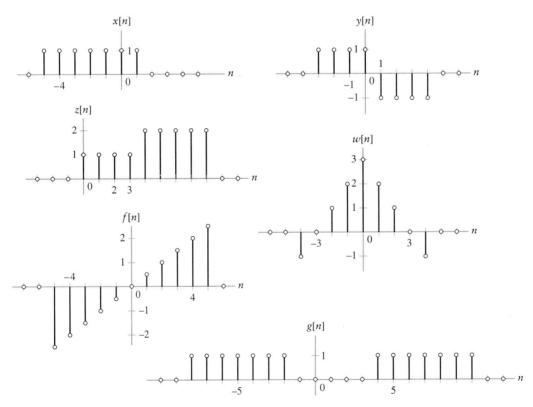

FIGURE P2.34

2.35 At the start of a certain year, $10,000 is deposited in a bank account earning 5% interest per year. At the start of each succeeding year, $1000 is deposited. Use convolution to determine the balance at the start of each year (after the deposit).

2.36 The initial balance of a loan is $20,000 and the interest rate is 1% per month (12% per year). A monthly payment of $200 is applied to the loan at the start of each month. Use convolution to calculate the loan balance after each monthly payment.

2.37 The convolution sum evaluation procedure actually corresponds to a formal statement of the well-known procedure for multiplying polynomials. To see this, we interpret polynomials as signals by setting the value of a signal at time n equal to the polynomial coefficient associated with monomial z^n. For example, the polynomial $x(z) = 2 + 3z^2 - z^3$ corresponds to the signal $x[n] = 2\delta[n] + 3\delta[n-2] - \delta[n-3]$. The procedure for multiplying polynomials involves forming the product of all polynomial coefficients that result in an nth-order monomial and then summing them to obtain the polynomial coefficient of the nth-order monomial in the product. This corresponds to determining $w_n[k]$ and summing over k to obtain $y[n]$.

   Evaluate the convolutions $y[n] = x[n] * h[n]$, both using the convolution sum evaluation procedure and taking a product of polynomials.

   (a) $x[n] = \delta[n] - 2\delta[n-1] + \delta[n-2]$,

       $h[n] = u[n] - u[n-3]$

   (b) $x[n] = u[n-1] - u[n-5]$,

       $h[n] = u[n-1] - u[n-5]$

2.38 An LTI system has the impulse response $h(t)$ depicted in Fig. P2.38. Use linearity and time invariance to determine the system output $y(t)$ if the input $x(t)$ is

   (a) $x(t) = 2\delta(t+2) + \delta(t-2)$

   (b) $x(t) = \delta(t-1) + \delta(t-2) + \delta(t-3)$

   (c) $x(t) = \sum_{p=0}^{\infty}(-1)^p\delta(t-2p)$

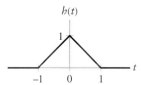

$h(t)$

FIGURE P2.38

2.39 Evaluate the following continuous-time convolution integrals:

   (a) $y(t) = (u(t) - u(t-2)) * u(t)$

   (b) $y(t) = e^{-3t}u(t) * u(t+3)$

   (c) $y(t) = \cos(\pi t)(u(t+1) - u(t-1)) * u(t)$

   (d) $y(t) = (u(t+3) - u(t-1)) * u(-t+4)$

   (e) $y(t) = (tu(t) + (10-2t)u(t-5)$
       $- (10-t)u(t-10)) * u(t)$

   (f) $y(t) = 2t^2(u(t+1) - u(t-1)) * 2u(t+2)$

   (g) $y(t) = \cos(\pi t)(u(t+1)$
       $- u(t-1)) * (u(t+1) - u(t-1))$

   (h) $y(t) = \cos(2\pi t)(u(t+1)$
       $- u(t-1)) * e^{-t}u(t)$

   (i) $y(t) = (2\delta(t+1) + \delta(t-5)) * u(t-1)$

   (j) $y(t) = (\delta(t+2) + \delta(t-2)) * (tu(t)$
       $+ (10-2t)u(t-5)$
       $- (10-t)u(t-10))$

   (k) $y(t) = e^{-\gamma t}u(t) * (u(t+2) - u(t))$

   (l) $y(t) = e^{-\gamma t}u(t) * \sum_{p=0}^{\infty}\left(\frac{1}{4}\right)^p\delta(t-2p)$

   (m) $y(t) = (2\delta(t)$
       $+ \delta(t-2)) * \sum_{p=0}^{\infty}\left(\frac{1}{2}\right)^p\delta(t-p)$

   (n) $y(t) = e^{-\gamma t}u(t) * e^{\beta t}u(-t)$

   (o) $y(t) = u(t) * h(t)$, where $h(t) = \begin{cases} e^{2t} & t < 0 \\ e^{-3t} & t \geq 0 \end{cases}$

2.40 Consider the continuous-time signals depicted in Fig. P2.40. Evaluate the following convolution integrals:

   (a) $m(t) = x(t) * y(t)$

   (b) $m(t) = x(t) * z(t)$

   (c) $m(t) = x(t) * f(t)$

   (d) $m(t) = x(t) * a(t)$

   (e) $m(t) = y(t) * z(t)$

   (f) $m(t) = y(t) * w(t)$

   (g) $m(t) = y(t) * g(t)$

   (h) $m(t) = y(t) * c(t)$

   (i) $m(t) = z(t) * f(t)$

   (j) $m(t) = z(t) * g(t)$

   (k) $m(t) = z(t) * b(t)$

   (l) $m(t) = w(t) * g(t)$

   (m) $m(t) = w(t) * a(t)$

   (n) $m(t) = f(t) * g(t)$

   (o) $m(t) = f(t) * d(t)$

   (p) $m(t) = z(t) * d(t)$

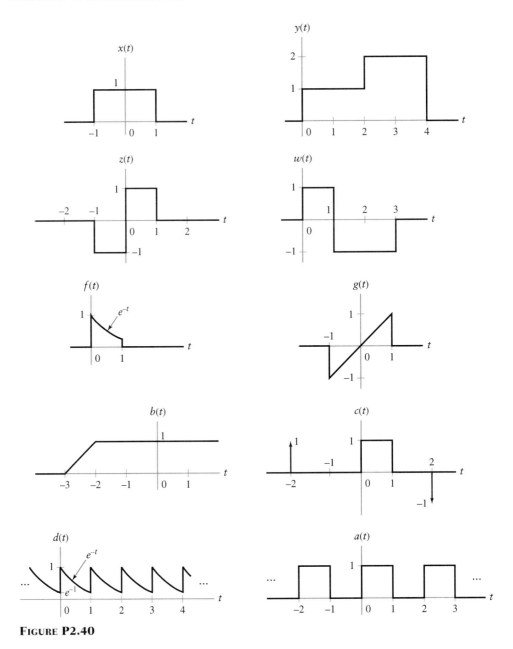

FIGURE P2.40

2.41 Suppose we model the effect of imperfections in a communication channel as the RC circuit depicted in Fig. P2.41(a). Here, the input $x(t)$ is the transmitted signal and the output $y(t)$ is the received signal. Suppose that the message is represented in binary format, that a "1" is communicated in an interval of length T by transmitting the waveform or symbol $p(t)$ depicted in Fig. P2.41(b) in the pertinent interval, and that a "0" is communicated by transmitting $-p(t)$ in the pertinent interval. Figure P2.41(c) illustrates

the waveform transmitted in communicating the sequence "1101001".

(a) Use convolution to calculate the received signal due to the transmission of a single "1" at time $t = 0$. Note that the received waveform extends beyond time T and into the interval allocated for the next bit, $T < t < 2T$. This contamination is called intersymbol interference (ISI), since the received waveform is interfered with by previous symbols. Assume that $T = 1/(RC)$.

(b) Use convolution to calculate the received signal due to the transmission of the sequences "1110" and "1000". Compare the received waveforms with the output of an ideal channel $(h(t) = \delta(t))$, to evaluate the effect of ISI for the following choices of RC:

(i) $RC = 1/T$

(ii) $RC = 5/T$

(iii) $RC = 1/(5T)$

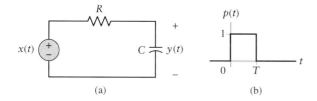

(a)   (b)

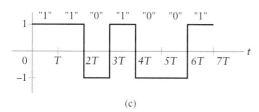

(c)

FIGURE P2.41

2.42 Use the definition of the convolution sum to derive the following properties:

(a) Distributive: $x[n] * (h[n] + g[n]) = x[n] * h[n] + x[n] * g[n]$

(b) Associative: $x[n] * (h[n] * g[n]) = (x[n] * h[n]) * g[n]$

(c) Commutative: $x[n] * h[n] = h[n] * x[n]$

2.43 An LTI system has the impulse response depicted in Fig. P2.43.

(a) Express the system output $y(t)$ as a function of the input $x(t)$.

(b) Identify the mathematical operation performed by this system in the limit as $\Delta \to 0$.

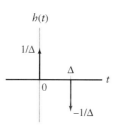

FIGURE P2.43

(c) Let $g(t) = \lim_{\Delta \to 0} h(t)$. Use the results of Part (b) to express the output of an LTI system with impulse response $h^n(t) = \underbrace{g(t) * g(t) * \cdots * g(t)}_{n \text{ times}}$ as a function of the input $x(t)$.

2.44 Show that if $y(t) = x(t) * h(t)$ is the output of an LTI system with input $x(t)$ and impulse response $h(t)$, then

$$\frac{d}{dt}y(t) = x(t) * \left(\frac{d}{dt}h(t)\right)$$

and

$$\frac{d}{dt}y(t) = \left(\frac{d}{dt}x(t)\right) * h(t).$$

2.45 If $h(t) = H\{\delta(t)\}$ is the impulse response of an LTI system, express $H\{\delta^{(2)}(t)\}$ in terms of $h(t)$.

2.46 Find the expression for the impulse response relating the input $x[n]$ or $x(t)$ to the output $y[n]$ or $y(t)$ in terms of the impulse response of each subsystem for the LTI systems depicted in

(a) Fig. P2.46(a)

(b) Fig. P2.46(b)

(c) Fig. P2.46(c)

2.47 Let $h_1(t), h_2(t), h_3(t)$, and $h_4(t)$ be impulse responses of LTI systems. Construct a system with impulse response $h(t)$, using $h_1(t), h_2(t), h_3(t)$, and $h_4(t)$ as subsystems. Draw the interconnection of systems required to obtain the following impulse responses:

(a) $h(t) = \{h_1(t) + h_2(t)\} * h_3(t) * h_4(t)$

(b) $h(t) = h_1(t) * h_2(t) + h_3(t) * h_4(t)$

(c) $h(t) = h_1(t) * \{h_2(t) + h_3(t) * h_4(t)\}$

2.48 For the interconnection of LTI systems depicted in Fig. P2.46(c), the impulse responses are $h_1(t) = \delta(t-1)$, $h_2(t) = e^{-2t}u(t)$, $h_3(t) = \delta(t-1)$ and $h_4(t) = e^{-3(t+2)}u(t+2)$. Evaluate $h(t)$, the impulse response of the overall system from $x(t)$ to $y(t)$.

2.49 For each of the following impulse responses, determine whether the corresponding system is (i) memoryless, (ii) causal, and (iii) stable.

(a) $h(t) = \cos(\pi t)$

(b) $h(t) = e^{-2t}u(t-1)$

(c) $h(t) = u(t+1)$

(d) $h(t) = 3\delta(t)$

(e) $h(t) = \cos(\pi t)u(t)$

(f) $h[n] = (-1)^n u[-n]$

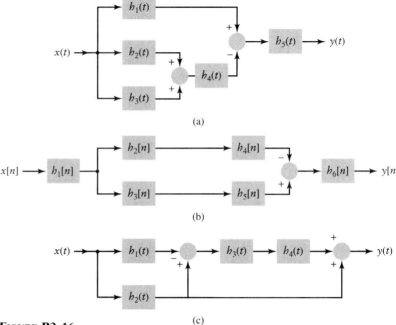

FIGURE P2.46

(g) $h[n] = (1/2)^{|n|}$

(h) $h[n] = \cos\left(\frac{\pi}{8}n\right)\{u[n] - u[n - 10]\}$

(i) $h[n] = 2u[n] - 2u[n - 5]$

(j) $h[n] = \sin\left(\frac{\pi}{2}n\right)$

(k) $h[n] = \sum_{p=-1}^{\infty} \delta[n - 2p]$

2.50 Evaluate the step response for the LTI systems represented by the following impulse responses:

(a) $h[n] = (-1/2)^n u[n]$

(b) $h[n] = \delta[n] - \delta[n - 2]$

(c) $h[n] = (-1)^n \{u[n + 2] - u[n - 3]\}$

(d) $h[n] = nu[n]$

(e) $h(t) = e^{-|t|}$

(f) $h(t) = \delta^{(2)}(t)$

(g) $h(t) = (1/4)(u(t) - u(t - 4))$

(h) $h(t) = u(t)$

2.51 Suppose the multipath propagation model is generalized to a k-step delay between the direct and indirect paths, as given by the input–output equation

$$y[n] = x[n] + ax[n - k].$$

Find the impulse response of the inverse system.

2.52 Write a differential equation description relating the output to the input of the electrical circuit shown in.

(a) Fig. P2.52(a)

(b) Fig. P2.52(b)

2.53 Determine the homogeneous solution for the systems described by the following differential equations:

(a) $5\dfrac{d}{dt}y(t) + 10y(t) = 2x(t)$

(b) $\dfrac{d^2}{dt^2}y(t) + 6\dfrac{d}{dt}y(t) + 8y(t) = \dfrac{d}{dt}x(t)$

(c) $\dfrac{d^2}{dt^2}y(t) + 4y(t) = 3\dfrac{d}{dt}x(t)$

(d) $\dfrac{d^2}{dt^2}y(t) + 2\dfrac{d}{dt}y(t) + 2y(t) = x(t)$

(e) $\dfrac{d^2}{dt^2}y(t) + 2\dfrac{d}{dt}y(t) + y(t) = \dfrac{d}{dt}x(t)$

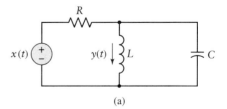

(a)

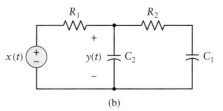

(b)

FIGURE P2.52

2.54 Determine the homogeneous solution for the systems described by the following difference equations:

(a) $y[n] - \alpha y[n-1] = 2x[n]$

(b) $y[n] - \frac{1}{4}y[n-1] - \frac{1}{8}y[n-2] = x[n] + x[n-1]$

(c) $y[n] + \frac{9}{16}y[n-2] = x[n-1]$

(d) $y[n] + y[n-1] + \frac{1}{4}y[n-2] = x[n] + 2x[n-1]$

2.55 Determine a particular solution for the systems described by the following differential equations, for the given inputs:

(a) $5\frac{d}{dt}y(t) + 10y(t) = 2x(t)$

   (i) $x(t) = 2$
   (ii) $x(t) = e^{-t}$
   (iii) $x(t) = \cos(3t)$

(b) $\frac{d^2}{dt^2}y(t) + 4y(t) = 3\frac{d}{dt}x(t)$

   (i) $x(t) = t$
   (ii) $x(t) = e^{-t}$
   (iii) $x(t) = (\cos(t) + \sin(t))$

(c) $\frac{d^2}{dt^2}y(t) + 2\frac{d}{dt}y(t) + y(t) = \frac{d}{dt}x(t)$

   (i) $x(t) = e^{-3t}u(t)$
   (ii) $x(t) = 2e^{-t}u(t)$
   (iii) $x(t) = 2\sin(t)$

2.56 Determine a particular solution for the systems described by the following difference equations, for the given inputs:

(a) $y[n] - \frac{2}{5}y[n-1] = 2x[n]$

   (i) $x[n] = 2u[n]$
   (ii) $x[n] = -\left(\frac{1}{2}\right)^n u[n]$
   (iii) $x[n] = \cos\left(\frac{\pi}{5}n\right)$

(b) $y[n] - \frac{1}{4}y[n-1] - \frac{1}{8}y[n-2] = x[n] + x[n-1]$

   (i) $x[n] = nu[n]$
   (ii) $x[n] = \left(\frac{1}{8}\right)^n u[n]$
   (iii) $x[n] = e^{j\frac{\pi}{4}n}u[n]$
   (iv) $x[n] = \left(\frac{1}{2}\right)^n u[n]$

(c) $y[n] + y[n-1] + \frac{1}{2}y[n-2] = x[n] + 2x[n-1]$

   (i) $x[n] = u[n]$
   (ii) $x[n] = \left(\frac{-1}{2}\right)^n u[n]$

2.57 Determine the output of the systems described by the following differential equations with input and initial conditions as specified:

(a) $\frac{d}{dt}y(t) + 10y(t) = 2x(t)$,
$$y(0^-) = 1, x(t) = u(t)$$

(b) $\frac{d^2}{dt^2}y(t) + 5\frac{d}{dt}y(t) + 4y(t) = \frac{d}{dt}x(t)$,
$$y(0^-) = 0, \frac{d}{dt}y(t)|_{t=0^-} = 1, x(t) = \sin(t)u(t)$$

(c) $\frac{d^2}{dt^2}y(t) + 6\frac{d}{dt}y(t) + 8y(t) = 2x(t)$,
$$y(0^-) = -1, \frac{d}{dt}y(t)|_{t=0^-} = 1, x(t) = e^{-t}u(t)$$

(d) $\frac{d^2}{dt^2}y(t) + y(t) = 3\frac{d}{dt}x(t)$,
$$y(0^-) = -1, \frac{d}{dt}y(t)|_{t=0^-} = 1, x(t) = 2te^{-t}u(t)$$

2.58 Identify the natural and forced responses for the systems in Problem 2.57.

2.59 Determine the output of the systems described by the following difference equations with input and initial conditions as specified:

(a) $y[n] - \frac{1}{2}y[n-1] = 2x[n]$,
$$y[-1] = 3, x[n] = \left(\frac{-1}{2}\right)^n u[n]$$

(b) $y[n] - \frac{1}{9}y[n-2] = x[n-1]$,
$$y[-1] = 1, y[-2] = 0, x[n] = u[n]$$

(c) $y[n] + \frac{1}{4}y[n-1] - \frac{1}{8}y[n-2] = x[n] + x[n-1]$,
$$y[-1] = 4, y[-2] = -2, x[n] = (-1)^n u[n]$$

(d) $y[n] - \frac{3}{4}y[n-1] + \frac{1}{8}y[n-2] = 2x[n]$,
$$y[-1] = 1, y[-2] = -1, x[n] = 2u[n]$$

2.60 Identify the natural and forced responses for the systems in Problem 2.59.

2.61 Write a differential equation relating the output $y(t)$ to the circuit in Fig. P2.61, and find the step response by applying an input $x(t) = u(t)$. Then, use the step response to obtain the impulse response. *Hint:* Use principles of circuit analysis to translate the $t = 0^-$ initial conditions to $t = 0^+$ before solving for the undetermined coefficients in the homogeneous component of the complete solution.

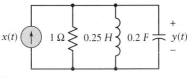

FIGURE P2.61

2.62 Use a first-order difference equation to calculate the monthly balance on a $100,000 loan at 1% per month interest, assuming monthly payments of $1200. Identify the natural and forced responses. In this case, the natural response represents the balance of the loan, assuming that no payments are made. How many payments are required to pay off the loan?

2.63 Determine the monthly payments required to pay off the loan in Problem 2.62 in 30 years (360 payments) and in 15 years (180 payments).

2.64 The portion of a loan payment attributed to interest is given by multiplying the balance after the previous payment was credited by $\frac{r}{100}$, where r is the rate per period, expressed in percent. Thus, if $y[n]$ is the loan balance after the nth payment, then the portion of the nth payment required to cover the interest cost is $y[n-1](r/100)$. The cumulative interest paid over payments for period n_1 through n_2 is thus

$$I = (r/100) \sum_{n=n_1}^{n_2} y[n-1].$$

Calculate the total interest paid over the life of the 30-year and 15-year loans described in Problem 2.63.

2.65 Find difference-equation descriptions for the three systems depicted in Fig. P2.65.

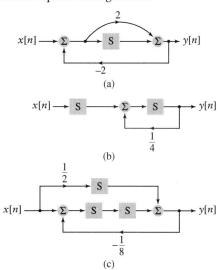

(a)

(b)

(c)

FIGURE P2.65

2.66 Draw direct form I and direct form II implementations for the following difference equations:

(a) $y[n] - \frac{1}{4}y[n-1] = 6x[n]$

(b) $y[n] + \frac{1}{2}y[n-1] - \frac{1}{8}y[n-2] = x[n] + 2x[n-1]$

(c) $y[n] - \frac{1}{9}y[n-2] = x[n-1]$

(d) $y[n] + \frac{1}{2}y[n-1] - y[n-3] = 3x[n-1] + 2x[n-2]$

2.67 Convert the following differential equations to integral equations, and draw direct form I and direct form II implementations of the corresponding systems:

(a) $\frac{d}{dt}y(t) + 10y(t) = 2x(t)$

(b) $\frac{d^2}{dt^2}y(t) + 5\frac{d}{dt}y(t) + 4y(t) = \frac{d}{dt}x(t)$

(c) $\frac{d^2}{dt^2}y(t) + y(t) = 3\frac{d}{dt}x(t)$

(d) $\frac{d^3}{dt^3}y(t) + 2\frac{d}{dt}y(t) + 3y(t) = x(t) + 3\frac{d}{dt}x(t)$

2.68 Find differential-equation descriptions for the two systems depicted in Fig. P2.68.

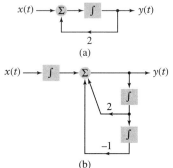

(a)

(b)

FIGURE P2.68

2.69 Determine a state-variable description for the four discrete-time systems depicted in Fig. P2.69.

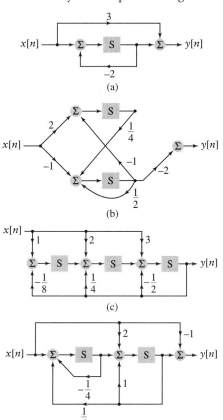

(a)

(b)

(c)

(d)

FIGURE P2.69

2.70 Draw block diagram representations corresponding to the discrete-time state-variable descriptions of the following LTI systems:

(a) $\mathbf{A} = \begin{bmatrix} 1 & -\frac{1}{2} \\ \frac{1}{3} & 0 \end{bmatrix}$; $\quad \mathbf{b} = \begin{bmatrix} 1 \\ 2 \end{bmatrix}$;
$\quad \mathbf{c} = [1 \quad 1]$; $\quad \mathbf{D} = [0]$

(b) $\mathbf{A} = \begin{bmatrix} 1 & -\frac{1}{2} \\ \frac{1}{3} & 0 \end{bmatrix}$; $\quad \mathbf{b} = \begin{bmatrix} 1 \\ 2 \end{bmatrix}$;
$\quad \mathbf{c} = [1 \quad -1]$; $\quad \mathbf{D} = [0]$

(c) $\mathbf{A} = \begin{bmatrix} 0 & -\frac{1}{2} \\ \frac{1}{3} & -1 \end{bmatrix}$; $\quad \mathbf{b} = \begin{bmatrix} 0 \\ 1 \end{bmatrix}$;
$\quad \mathbf{c} = [1 \quad 0]$; $\quad \mathbf{D} = [1]$

(d) $\mathbf{A} = \begin{bmatrix} 0 & 0 \\ 0 & 1 \end{bmatrix}$; $\quad \mathbf{b} = \begin{bmatrix} 2 \\ 3 \end{bmatrix}$;
$\quad \mathbf{c} = [1 \quad -1]$; $\quad \mathbf{D} = [0]$

2.71 Determine a state-variable description for the five continuous-time LTI systems depicted in Fig. P2.71.

2.72 Draw block diagram representations corresponding to the continuous-time state-variable descriptions of the following LTI systems:

(a) $\mathbf{A} = \begin{bmatrix} \frac{1}{3} & 0 \\ 0 & -\frac{1}{2} \end{bmatrix}$; $\quad \mathbf{b} = \begin{bmatrix} -1 \\ 2 \end{bmatrix}$;
$\quad \mathbf{c} = [1 \quad 1]$; $\quad \mathbf{D} = [0]$

(b) $\mathbf{A} = \begin{bmatrix} 1 & 1 \\ 1 & 0 \end{bmatrix}$; $\quad \mathbf{b} = \begin{bmatrix} -1 \\ 2 \end{bmatrix}$;
$\quad \mathbf{c} = [0 \quad -1]$; $\quad \mathbf{D} = [0]$

(c) $\mathbf{A} = \begin{bmatrix} 1 & -1 \\ 0 & -1 \end{bmatrix}$; $\quad \mathbf{b} = \begin{bmatrix} 0 \\ 5 \end{bmatrix}$;
$\quad \mathbf{c} = [1 \quad 0]$; $\quad \mathbf{D} = [0]$

(d) $\mathbf{A} = \begin{bmatrix} 1 & -2 \\ 1 & 1 \end{bmatrix}$; $\quad \mathbf{b} = \begin{bmatrix} 2 \\ 3 \end{bmatrix}$;
$\quad \mathbf{c} = [1 \quad 1]$; $\quad \mathbf{D} = [0]$

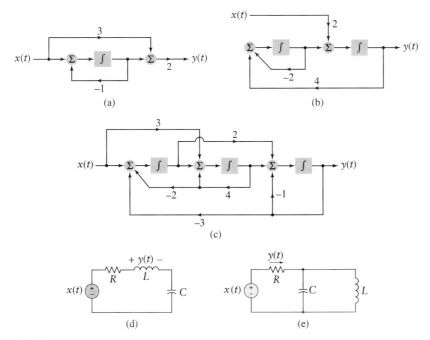

FIGURE P2.71

2.73 Let a discrete-time system have the state-variable description given by

$$\mathbf{A} = \begin{bmatrix} 1 & -\frac{1}{2} \\ \frac{1}{3} & 0 \end{bmatrix}; \quad \mathbf{b} = \begin{bmatrix} 1 \\ 2 \end{bmatrix};$$

$$\mathbf{c} = [1 \quad -1]; \quad \text{and} \quad D = [0].$$

(a) Define new states $q_1'[n] = 2q_1[n], q_2'[n] = 3q_2[n]$. Find the new state-variable description given by $\mathbf{A}', \mathbf{b}', \mathbf{c}'$, and D'.

(b) Define new states $q_1'[n] = 3q_2[n], q_2'[n] = 2q_1[n]$. Find the new state-variable description given by $\mathbf{A}', \mathbf{b}', \mathbf{c}'$, and D'.

(c) Define new states $q_1'[n] = q_1[n] + q_2[n], q_2'[n] = q_1[n] - q_2[n]$. Find the new state-variable description given by $\mathbf{A}', \mathbf{b}', \mathbf{c}'$, and D'.

2.74 Consider the continuous-time system depicted in Fig. P2.74.

(a) Find the state-variable description for this system, assuming that the states $q_1(t)$ and $q_2(t)$ are as labeled.

(b) Define new states $q_1'(t) = q_1(t) - q_2(t)$, $q_2'(t) = 2q_1(t)$. Find the new state-variable description given by $\mathbf{A}', \mathbf{b}', \mathbf{c}'$, and D'.

(c) Draw a block diagram corresponding to the new state-variable description in (b).

(d) Define new states $q_1'(t) = \frac{1}{b_1}q_1(t), q_2'(t) = b_2q_1(t) - b_1q_2(t)$. Find the new state-variable description given by $\mathbf{A}', \mathbf{b}', \mathbf{c}'$, and D'.

(e) Draw a block diagram corresponding to the new state-variable description in (d).

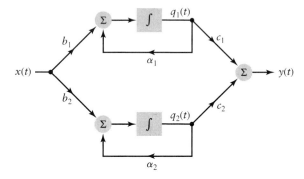

FIGURE P2.74

ADVANCED PROBLEMS

2.75 In this problem, we develop the convolution integral using linearity, time invariance, and the limiting form of a stair-step approximation to the input signal. Toward that end, we define $g_\Delta(t)$ as the unit area rectangular pulse depicted in Fig. P2.75(a).

(a) A stair-step approximation to a signal $x(t)$ is depicted in Fig. P2.75(b). Express $\tilde{x}(t)$ as a weighted sum of shifted pulses $g_\Delta(t)$. Does the quality of the approximation improve as Δ decreases?

(b) Let the response of an LTI system to an input $g_\Delta(t)$ be $h_\Delta(t)$. If the input to this system is $\tilde{x}(t)$, find an expression for the output of the system in terms of $h_\Delta(t)$.

(c) In the limit as Δ goes to zero, $g_\Delta(t)$ satisfies the properties of an impulse, and we may interpret $h(t) = \lim_{\Delta \to 0} h_\Delta(t)$ as the impulse response of the system. Show that the expression for the system output derived in (b) reduces to $x(t) * h(t)$ in the limit as Δ goes to zero.

2.76 The convolution of finite-duration discrete-time signals may be expressed as the product of a matrix and a vector. Let the input $x[n]$ be zero outside of $n = 0, 1, \ldots L - 1$ and the impulse response $h[n]$ zero outside of $n = 0, 1, \ldots M - 1$. The output $y[n]$ is then zero outside of $n = 0, 1, \ldots,$

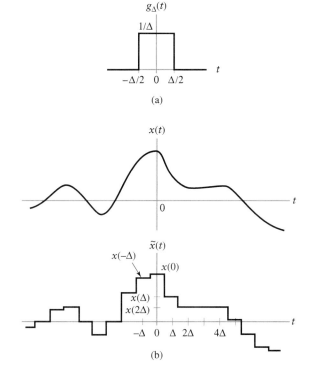

FIGURE P2.75

$L + M - 1$. Define column vectors $\mathbf{x} = [x[0], x[1],$ $\ldots x[L-1]]^T$ and $\mathbf{y} = [y[0], y[1], \ldots y[M-1]]^T$. Use the definition of the convolution sum to find a matrix $\mathbf{H}$ such that $\mathbf{y} = \mathbf{Hx}$.

2.77 Assume that the impulse response of a continous-time system is zero outside the interval $0 < t < T_o$. Use a Riemann sum approximation to the convolution integral to convert the integral to a convolution sum that relates uniformly spaced samples of the output signal to uniformly spaced samples of the input signal.

2.78 The cross-correlation between two real signals $x(t)$ and $y(t)$ is defined as

$$r_{xy}(t) = \int_{-\infty}^{\infty} x(\tau)y(\tau - t)d\tau.$$

This integral is the area under the product of $x(t)$ and a shifted version of $y(t)$. Note that the independent variable $\tau - t$ is the negative of that found in the definition of convolution. The autocorrelation, $r_{xx}(t)$, of a signal $x(t)$ is obtained by replacing $y(t)$ with $x(t)$.

(a) Show that $r_{xy}(t) = x(t) * y(-t)$

(b) Derive a step-by-step procedure for evaluating the cross-correlation that is analogous to the procedure for evaluating the convolution integral given in Section 2.5.

(c) Evaluate the cross-correlation between the following signals:

  (i) $x(t) = e^{-t}u(t), y(t) = e^{-3t}u(t)$

  (ii) $x(t) = \cos(\pi t)[u(t + 2) - u(t - 2)],$
     $y(t) = \cos(2\pi t)[u(t + 2) - u(t - 2)]$

  (iii) $x(t) = u(t) - 2u(t - 1) + u(t - 2),$
     $y(t) = u(t + 1) - u(t)$

  (iv) $x(t) = u(t - a) - u(t - a - 1),$
     $y(t) = u(t) - u(t - 1)$

(d) Evaluate the autocorrelation of the following signals:

  (i) $x(t) = e^{-t}u(t)$

  (ii) $x(t) = \cos(\pi t)[u(t + 2) - u(t - 2)]$

  (iii) $x(t) = u(t) - 2u(t - 1) + u(t - 2)$

  (iv) $x(t) = u(t - a) - u(t - a - 1)$

(e) Show that $r_{xy}(t) = r_{yx}(-t)$.

(f) Show that $r_{xx}(t) = r_{xx}(-t)$.

2.79 Prove that absolute summability of the impulse response is a necessary condition for the stability of a discrete-time system. (*Hint:* find a bounded input $x[n]$ such that the output at some time n_o satisfies $|y[n_o]| = \sum_{k=-\infty}^{\infty} |h[k]|$.)

2.80 Light with a complex amplitude $f(x, y)$ in the xy-plane propagating over a distance d along the z-axis in free space generates a complex amplitude

$$g(x, y) = \int_{-\infty}^{\infty} \int_{-\infty}^{\infty} f(x', y')h(x - x', y - y') \, dx' \, dy',$$

where

$$h(x, y) = h_0 e^{-jk(x^2 + y^2)/2d},$$

Where $k = 2\pi/\lambda$ is the wavenumber, λ is the wavelength, and $h_0 = j/(\lambda d)e^{-jkd}$. (We used the Fresnel approximation in deriving the expression for g.)

(a) Determine whether free-space propagation represents a linear system.

(b) Is this system space invariant? That is, does a spatial shift of the input, $f(x - x_0, y - y_0)$, lead to the identical spatial shift in the output?

(c) Evaluate the result of a point source located at (x_1, y_1) propagating a distance d. In this case, $f(x, y) = \delta(x - x_1, y - y_1)$, where $\delta(x, y)$ is the two-dimensional version of the impulse. Find the corresponding two-dimensional impulse response of this system.

(d) Evaluate the result of two point sources located at (x_1, y_1) and (x_2, y_2) and propagating a distance d.

2.81 The motion of a vibrating string depicted in Fig. P2.81 may be described by the partial differential equation

$$\frac{\partial^2}{\partial l^2} y(l, t) = \frac{1}{c^2} \frac{\partial^2}{\partial t^2} y(l, t),$$

where $y(l, t)$ is the displacement expressed as a function of position l and time t and c is a constant determined by the material properties of the string. The initial conditions may be specified as follows:

$$y(0, t) = 0, \qquad y(a, t) = 0, \qquad t > 0;$$
$$y(l, 0) = x(l), \qquad 0 < l < a;$$

$$\frac{\partial}{\partial t}y(l, t)\bigg|_{t=0} = g(l), \qquad 0 < l < a.$$

Here, $x(l)$ is the displacement of the string at $t = 0$, while $g(l)$ describes the velocity at $t = 0$. One approach to solving this equation is by separation of variables—that is, $y(l, t) = \phi(l)f(t)$, in which case the partial differential equation becomes

$$f(t)\frac{d^2}{dl^2}\phi(l) = \phi(l)\frac{1}{c^2}\frac{d^2}{dt^2}f(t).$$

This implies that

$$\frac{\frac{d^2}{dl^2}\phi(l)}{\phi(l)} = \frac{\frac{d^2}{dt^2}f(t)}{c^2 f(t)}, \qquad 0 < l < a, \qquad 0 < t.$$

For this equality to hold, both sides of the equation must be constant. Let the constant be $-\omega^2$, and separate the partial differential equation into two ordinary second-order differential equations linked by the common parameter ω^2:

$$\frac{d^2}{dt^2}f(t) + \omega^2 c^2 f(t) = 0, \qquad 0 < t;$$

$$\frac{d^2}{dl^2}\phi(l) + \omega^2\phi(l) = 0 \qquad 0 < l < a.$$

(a) Find the form of the solution for $f(t)$ and $\phi(l)$.

(b) The boundary conditions at the endpoints of the string are

$$\phi(0)f(t) = 0 \quad \text{and} \quad \phi(a)f(t) = 0.$$

Also, since $f(t) = 0$ gives a trivial solution for $y(l, t)$, we must have $\phi(0) = 0$ and $\phi(a) = 0$. Determine how these constraints restrict the permissible values for ω and the form of the solution for $\phi(l)$.

(c) Apply the boundary conditions in (b) to show that constant $(-\omega^2)$ used to separate the partial differential equation into two ordinary second-order differential equations must be negative.

(d) Assume that the initial position of the string is $y(l, 0) = x(l) = \sin(\pi l/a)$ and that the initial velocity is $g(l) = 0$. Find $y(l, t)$.

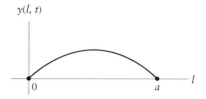

FIGURE P2.81

2.82 Suppose the N-by-N matrix $\mathbf{A}$ in a state-variable description has N linearly independent eigenvectors $\mathbf{e}_i$, $i = 1, 2, \ldots, N$ and corresponding distinct eigenvalues λ_i. Thus, $\mathbf{A}\mathbf{e}_i = \lambda_i \mathbf{e}_i$, $i = 1, 2, \ldots, N$.

(a) Show that we may decompose $\mathbf{A}$ as $\mathbf{A} = \mathbf{E}\Lambda\mathbf{E}^{-1}$, where Λ is a diagonal matrix with ith diagonal element λ_i.

(b) Find a transformation of the state that will diagonalize $\mathbf{A}$.

(c) Assume that

$$\mathbf{A} = \begin{bmatrix} 0 & -1 \\ 2 & -3 \end{bmatrix}, \quad \mathbf{b} = \begin{bmatrix} 2 \\ 3 \end{bmatrix},$$

$$\mathbf{c} = [1 \quad 0], \quad \text{and} \quad D = [0].$$

Find a transformation that converts this system to diagonal form.

(d) Sketch the block diagram representation of a discrete-time system corresponding to the system in part (c).

COMPUTER EXPERIMENTS

2.83 Repeat Problem 2.34, using the MATLAB `conv` command.

2.84 Use MATLAB to repeat Example 2.5.

2.85 Use MATLAB to evaluate the first 20 values of the step response for the systems described in Problem 2.50(a)–(d).

2.86 Two systems have impulse responses

$$h_1[n] = \begin{cases} \frac{1}{4}, & 0 \le n \le 3 \\ 0, & \text{otherwise} \end{cases}$$

and

$$h_2[n] = \begin{cases} \frac{1}{4}, & n = 0, 2 \\ -\frac{1}{4}, & n = 1, 3 \\ 0, & \text{otherwise} \end{cases}.$$

Use the MATLAB command `conv` to plot the first 20 values of the step response.

2.87 Use the MATLAB commands `filter` and `filtic` to repeat Example 2.16.

2.88 Use the MATLAB commands `filter` and `filtic` to verify the loan balance in Example 2.23.

2.89 Use the MATLAB commands `filter` and `filtic` to determine the first 50 output values in Problem 2.59.

2.90 Use the MATLAB command `impz` to determine the first 30 values of the impulse response for the systems described in Problem 2.59.

2.91 Use MATLAB to solve Problem 2.62.

2.92 Use MATLAB to solve Problem 2.63.

2.93 Use the MATLAB command `ss2ss` to solve Problem 2.73.

2.94 A system has the state-variable description

$$\mathbf{A} = \begin{bmatrix} \frac{1}{2} & -\frac{1}{2} \\ \frac{1}{3} & 0 \end{bmatrix}, \quad \mathbf{b} = \begin{bmatrix} 1 \\ 2 \end{bmatrix},$$

$$\mathbf{c} = [1 \quad -1], \quad \text{and} \quad D = [0].$$

(a) Use the MATLAB commands `lsim` and `impulse` to determine the first 30 values of the step and impulse responses of this system.

(b) Define new states $q_1[n] = q_1[n] + q_2[n]$ and $q_2[n] = 2q_1[n] - q_2[n]$. Repeat Part (a) for the transformed system.

3 Fourier Representations of Signals and Linear Time-Invariant Systems

3.1 Introduction

In this chapter, we represent a signal as a weighted superposition of complex sinusoids. If such a signal is applied to an LTI system, then the system output is a weighted superposition of the system response to each complex sinusoid. A similar application of the linearity property was exploited in the previous chapter in order to develop the convolution integral and convolution sum. There, the input signal was expressed as a weighted superposition of delayed impulses; the output was then given by a weighted superposition of delayed versions of the system's impulse response. The expression for the output that resulted from expressing signals in terms of impulses was termed "convolution." By representing signals in terms of sinusoids, we will obtain an alternative expression for the input–output behavior of an LTI system.

Representing signals as superpositions of complex sinusoids not only leads to a useful expression for the system output, but also provides an insightful characterization of signals and systems. The general notion of describing complicated signals as a function of frequency is commonly encountered in music. For example, the musical score for an orchestra contains parts for instruments having different frequency ranges, such as a string bass, which produces very low frequency sound, and a piccolo, which produces very high frequency sound. The sound that we hear when listening to an orchestra is a superposition of sounds generated by different instruments. Similarly, the score for a choir contains bass, tenor, alto, and soprano parts, each of which contributes to a different frequency range in the overall sound. In this chapter, the representations we develop of signals can be viewed analogously: The weight associated with a sinusoid of a given frequency represents the contribution of that sinusoid to the overall signal.

The study of signals and systems using sinusoidal representations is termed *Fourier analysis*, after Joseph Fourier (1768–1830) for his development of the theory. Fourier methods have widespread application beyond signals and systems, being used in every branch of engineering and science.

There are four distinct Fourier representations, each applicable to a different class of signals, determined by the periodicity properties of the signal and whether the signal is discrete or continuous in time. The focus of this chapter is a parallel study of these four Fourier representations and their properties. Applications involving mixtures of the signals from the four classes, such as sampling a continuous-time signal, are considered in the next chapter.

3.2 Complex Sinusoids and Frequency Response of LTI Systems

The response of an LTI system to a sinusoidal input leads to a characterization of system behavior that is termed the *frequency response* of the system. This characterization is obtained in terms of the impulse response by using convolution and a complex sinusoidal input signal. Consider the output of a discrete-time LTI system with impulse response $h[n]$ and unit amplitude complex sinusoidal input $x[n] = e^{j\Omega n}$. This output is given by

$$y[n] = \sum_{k=-\infty}^{\infty} h[k]x[n-k]$$

$$= \sum_{k=-\infty}^{\infty} h[k]e^{j\Omega(n-k)}.$$

We factor $e^{j\Omega n}$ from the sum to obtain

$$y[n] = e^{j\Omega n} \sum_{k=-\infty}^{\infty} h[k]e^{-j\Omega k}$$

$$= H(e^{j\Omega})e^{j\Omega n},$$

where we have defined

$$\boxed{H(e^{j\Omega}) = \sum_{k=-\infty}^{\infty} h[k]e^{-j\Omega k}.} \tag{3.1}$$

Hence, the output of the system is a complex sinusoid of the same frequency as the input, multiplied by the complex number $H(e^{j\Omega})$. This relationship is depicted in Fig. 3.1. The complex scaling factor $H(e^{j\Omega})$ is not a function of time n, but is only a function of frequency Ω and is termed the *frequency response* of the discrete-time system.

Similar results are obtained for continuous-time LTI systems. Let the impulse response of such a system be $h(t)$ and the input be $x(t) = e^{j\omega t}$. Then the convolution integral gives the output as

$$y(t) = \int_{-\infty}^{\infty} h(\tau)e^{j\omega(t-\tau)}\, d\tau$$

$$= e^{j\omega t} \int_{-\infty}^{\infty} h(\tau)e^{-j\omega \tau}\, d\tau \tag{3.2}$$

$$= H(j\omega)e^{j\omega t},$$

where we define

$$\boxed{H(j\omega) = \int_{-\infty}^{\infty} h(\tau)e^{-j\omega \tau}\, d\tau.} \tag{3.3}$$

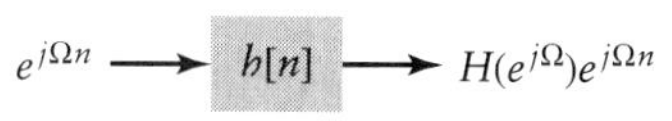

$$e^{j\Omega n} \longrightarrow \boxed{h[n]} \longrightarrow H(e^{j\Omega})e^{j\Omega n}$$

FIGURE 3.1   The output of a complex sinusoidal input to an LTI system is a complex sinusoid of the same frequency as the input, multiplied by the frequency response of the system.

The output of the system is thus a complex sinusoid of the same frequency as the input, multiplied by the complex number $H(j\omega)$. Note that $H(j\omega)$ is a function of only the frequency ω and not the time t and is termed the *frequency response* of the continuous-time system.

An intuitive interpretation of the sinusoidal steady-state response is obtained by writing the complex-valued frequency response $H(j\omega)$ in polar form. Recall that if $c = a + jb$ is a complex number, then we may write c in polar form as $c = |c|e^{j\,\arg\{c\}}$, where $|c| = \sqrt{a^2 + b^2}$ and $\arg\{c\} = \arctan\left(\frac{b}{a}\right)$. Hence, we have $H(j\omega) = |H(j\omega)|e^{j\,\arg\{H(j\omega)\}}$, where $|H(j\omega)|$ is now termed the *magnitude response* and $\arg\{H(j\omega)\}$ is termed the *phase response* of the system. Substituting this polar form into Eq. (3.2), we may express the output as

$$y(t) = |H(j\omega)|e^{j(\omega t + \arg\{H(j\omega)\})}.$$

The system thus modifies the amplitude of the input by $|H(j\omega)|$ and the phase by $\arg\{H(j\omega)\}$.

EXAMPLE 3.1 RC CIRCUIT: FREQUENCY RESPONSE   The impulse response of the system relating the input voltage to the voltage across the capacitor in Fig. 3.2 is derived in Example 1.21 as

$$h(t) = \frac{1}{RC}e^{-\frac{t}{RC}}u(t).$$

Find an expression for the frequency response, and plot the magnitude and phase response.

Solution: Substituting $h(t)$ into Eq. (3.3) gives

$$
\begin{aligned}
H(j\omega) &= \frac{1}{RC}\int_{-\infty}^{\infty} e^{-\frac{\tau}{RC}}u(\tau)e^{-j\omega\tau}\,d\tau \\[2mm]
&= \frac{1}{RC}\int_{0}^{\infty} e^{-\left(j\omega + \frac{1}{RC}\right)\tau}\,d\tau \\[2mm]
&= \frac{1}{RC}\frac{-1}{\left(j\omega + \frac{1}{RC}\right)}e^{-\left(j\omega + \frac{1}{RC}\right)\tau}\bigg|_{0}^{\infty} \\[2mm]
&= \frac{1}{RC}\frac{-1}{\left(j\omega + \frac{1}{RC}\right)}(0 - 1) \\[2mm]
&= \frac{\frac{1}{RC}}{j\omega + \frac{1}{RC}}.
\end{aligned}
$$

The magnitude response is

$$|H(j\omega)| = \frac{\frac{1}{RC}}{\sqrt{\omega^2 + \left(\frac{1}{RC}\right)^2}},$$

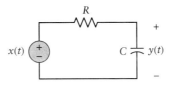

FIGURE 3.2   *RC* circuit for Example 3.1.

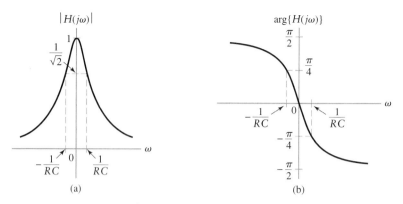

FIGURE 3.3   Frequency response of the RC circuit in Fig. 3.2. (a) Magnitude response. (b) Phase response.

while the phase response is

$$\arg\{H(j\omega)\} = -\arctan(\omega RC).$$

The magnitude response and phase response are presented in Figs. 3.3(a) and (b), respectively. The magnitude response indicates that the RC circuit tends to attenuate high-frequency $\left(\omega > \frac{1}{RC}\right)$ sinusoids. This agrees with our intuition from circuit analysis. The circuit cannot respond to rapid changes in the input voltage. High-frequency sinusoids also experience a phase shift of $-\frac{\pi}{2}$ radians. Low-frequency $\left(\omega < \frac{1}{RC}\right)$ sinusoids are passed by the circuit with much higher gain and acquire relatively little phase shift.   ■

We say that the complex sinusoid $\psi(t) = e^{j\omega t}$ is an *eigenfunction* of the LTI system H associated with the *eigenvalue* $\lambda = H(j\omega)$, because ψ satisfies an eigenvalue problem described by

$$H\{\psi(t)\} = \lambda\psi(t).$$

This eigenrepresentation is illustrated in Fig. 3.4. The effect of the system on an eigenfunction input signal is scalar multiplication: The output is given by the product of the input and a complex number. This eigenrepresentation is analogous to the more familiar matrix eigenproblem. If $\mathbf{e}_k$ is an eigenvector of a matrix $\mathbf{A}$ with eigenvalue λ_k, then

$$\mathbf{A}\mathbf{e}_k = \lambda_k\mathbf{e}_k.$$

In words, pre-multiplying $\mathbf{e}_k$ by the matrix $\mathbf{A}$ is equivalent to multiplying $\mathbf{e}_k$ by the scalar λ_k.

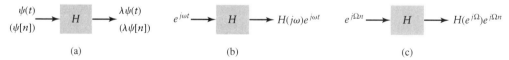

FIGURE 3.4   Illustration of the eigenfunction property of linear systems. The action of the system on an eigenfunction input is multiplication by the corresponding eigenvalue. (a) General eigenfunction $\psi(t)$ or $\psi[n]$ and eigenvalue λ. (b) Complex sinusoidal eigenfunction $e^{j\omega t}$ and eigenvalue $H(j\omega)$. (c) Complex sinusoidal eigenfunction $e^{j\Omega n}$ and eigenvalue $H(e^{j\Omega})$.

Signals that are eigenfunctions of systems play an important role in LTI systems theory. By representing arbitrary signals as weighted superpositions of eigenfunctions, we transform the operation of convolution to multiplication. To see this, consider expressing the input to an LTI system as the weighted sum of M complex sinusoids

$$x(t) = \sum_{k=1}^{M} a_k e^{j\omega_k t}.$$

If $e^{j\omega_k t}$ is an eigenfunction of the system with eigenvalue $H(j\omega_k)$, then each term in the input, $a_k e^{j\omega_k t}$, produces an output term $a_k H(j\omega_k) e^{j\omega_k t}$. Hence, we express the output of the system as

$$y(t) = \sum_{k=1}^{M} a_k H(j\omega_k) e^{j\omega_k t}.$$

The output is a weighted sum of M complex sinusoids, with the input weights a_k modified by the system frequency response $H(j\omega_k)$. The operation of convolution, $h(t) * x(t)$, becomes multiplication, $a_k H(j\omega_k)$, because $x(t)$ is expressed as a sum of eigenfunctions. An analogous relationship holds in the discrete-time case.

This property is a powerful motivation for representing signals as weighted superpositions of complex sinusoids. In addition, the weights provide an alternative interpretation of the signal: Rather than describing the signal's behavior as a function of time, the weights describe it as a function of frequency. This alternative view is highly informative, as we shall see in what follows.

3.3 *Fourier Representations for Four Classes of Signals*

There are four distinct Fourier representations, each applicable to a different class of signals. The four classes are defined by the periodicity properties of a signal and whether the signal is continuous or discrete in time. The Fourier series (FS) applies to continuous-time periodic signals, and the discrete-time Fourier series (DTFS) applies to discrete-time periodic signals. Nonperiodic signals have Fourier transform representations. The Fourier transform (FT) applies to a signal that is continuous in time and nonperiodic. The discrete-time Fourier transform (DTFT) applies to a signal that is discrete in time and nonperiodic. Table 3.1 illustrates the relationship between the temporal properties of a signal and the appropriate Fourier representation.

▪ 3.3.1 PERIODIC SIGNALS: FOURIER SERIES REPRESENTATIONS

Consider representing a periodic signal as a weighted superposition of complex sinusoids. Since the weighted superposition must have the same period as the signal, each sinusoid in the superposition must have the same period as the signal. This implies that the frequency of each sinusoid must be an integer multiple of the signal's fundamental frequency.

TABLE 3.1   *Relationship between Time Properties of a Signal and the Appropriate Fourier Representation.*

Time Property	Periodic	Nonperiodic
Continuous (t)	Fourier Series (FS)	Fourier Transform (FT)
Discrete [n]	Discrete-Time Fourier Series (DTFS)	Discrete-Time Fourier Transform (DTFT)

If $x[n]$ is a discrete-time signal with fundamental period N, then we seek to represent $x[n]$ by the DTFS

$$\hat{x}[n] = \sum_k A[k]e^{jk\Omega_o n}, \tag{3.4}$$

where $\Omega_o = 2\pi/N$ is the fundamental frequency of $x[n]$. The frequency of the kth sinusoid in the superposition is $k\Omega_o$. Each of these sinusoids has a common period N. Similarly, if $x(t)$ is a continuous-time signal of fundamental period T, we represent $x(t)$ by the FS

$$\hat{x}(t) = \sum_k A[k]e^{jk\omega_o t}, \tag{3.5}$$

where $\omega_o = 2\pi/T$ is the fundamental frequency of $x(t)$. Here, the frequency of the kth sinusoid is $k\omega_o$, and each sinusoid has a common period T. A sinusoid whose frequency is an integer multiple of a fundamental frequency is said to be a *harmonic* of the sinusoid at the fundamental frequency. Thus, $e^{jk\omega_o t}$ is the kth harmonic of $e^{j\omega_o t}$. In both Eqs. (3.4) and (3.5), $A[k]$ is the weight applied to the kth harmonic, and the hat (^) denotes approximate value, since we do not yet assume that either $x[n]$ or $x(t)$ can be represented exactly by a series of the form shown. The variable k indexes the frequency of the sinusoids, so we say that $A[k]$ is a function of frequency.

How many terms and weights should we use in each sum? The answer to this question becomes apparent for the DTFS described in Eq. (3.4) if we recall that complex sinusoids with distinct frequencies are not always distinct. In particular, the complex sinusoids $e^{jk\Omega_o n}$ are N-periodic in the frequency index k, as shown by the relationship

$$e^{j(N+k)\Omega_o n} = e^{jN\Omega_o n}e^{jk\Omega_o n}$$
$$= e^{j2\pi n}e^{jk\Omega_o n}$$
$$= e^{jk\Omega_o n}.$$

Hence, there are only N distinct complex sinusoids of the form $e^{jk\Omega_o n}$. A unique set of N distinct complex sinusoids is obtained by letting the frequency index k vary from $k = 0$ to $k = N - 1$. Accordingly, we may rewrite Eq. (3.4) as

$$\hat{x}[n] = \sum_{k=0}^{N-1} A[k]e^{jk\Omega_o n}. \tag{3.6}$$

The set of N consecutive values over which k varies is arbitrary and may be chosen to simplify the problem by exploiting symmetries in the signal $x[n]$. For example, if $x[n]$ is an even or odd signal, it may be simpler to use $k = -(N-1)/2$ to $(N-1)/2$ if N is odd.

In contrast to the discrete-time case, continuous-time complex sinusoids $e^{jk\omega_o t}$ with distinct frequencies $k\omega_o$ are always distinct. Hence, there are potentially an infinite number of distinct terms in the series of Eq. (3.5), and we express $x(t)$ as

$$\hat{x}(t) = \sum_{k=-\infty}^{\infty} A[k] e^{jk\omega_o t}. \tag{3.7}$$

We seek weights or coefficients $A[k]$ such that $\hat{x}[n]$ and $\hat{x}(t)$ are good approximations to $x[n]$ and $x(t)$, respectively. This is accomplished by minimizing the mean-square error (MSE) between the signal and its series representation. Since the series representations have the same period as the signals, the MSE is the average squared difference over any one period, or the average power in the error. In the discrete-time case, we have

$$MSE = \frac{1}{N} \sum_{n=0}^{N-1} |x[n] - \hat{x}[n]|^2. \tag{3.8}$$

Similarly, in the continuous-time case,

$$MSE = \frac{1}{T} \int_0^T |x(t) - \hat{x}(t)|^2 \, dt. \tag{3.9}$$

The DTFS and FS coefficients to be given in Sections 3.4 and 3.5 minimize the MSE. Determination of these coefficients is simplified by the properties of harmonically related complex sinusoids.

■ 3.3.2 NONPERIODIC SIGNALS: FOURIER-TRANSFORM REPRESENTATIONS

In contrast to the case of the periodic signal, there are no restrictions on the period of the sinusoids used to represent nonperiodic signals. Hence, the Fourier transform representations employ complex sinusoids having a continuum of frequencies. The signal is represented as a weighted integral of complex sinusoids where the variable of integration is the sinusoid's frequency. Discrete-time sinusoids are used to represent discrete-time signals in the DTFT, while continuous-time sinusoids are used to represent continuous-time signals in the FT. Continuous-time sinusoids with distinct frequencies are distinct, so the FT involves frequencies from $-\infty$ to ∞, as shown by the equation

$$\hat{x}(t) = \frac{1}{2\pi} \int_{-\infty}^{\infty} X(j\omega) e^{j\omega t} \, d\omega.$$

Here, $X(j\omega)/(2\pi)$ represents the "weight" or coefficient applied to a sinusoid of frequency ω in the FT representation.

Discrete-time sinusoids are unique only over a 2π interval of frequency, since discrete-time sinusoids with frequencies separated by an integer multiple of 2π are identical.

Consequently, the DTFT involves sinusoidal frequencies within a 2π interval, as shown by the relationship

$$\hat{x}[n] = \frac{1}{2\pi} \int_{-\pi}^{\pi} X(e^{j\Omega}) e^{j\Omega n} \, d\omega.$$

Thus the "weighting" applied to the sinusoid $e^{j\Omega n}$ in the DTFT representation is $X(e^{j\Omega})/(2\pi)$. The next four sections of this chapter present, in sequence, the DTFS, FS, DTFT, and FT.

▶ **Problem 3.1**   Identify the appropriate Fourier representation for each of the following signals:

(a) $x[n] = (1/2)^n u[n]$
(b) $x(t) = 1 - \cos(2\pi t) + \sin(3\pi t)$
(c) $x(t) = e^{-t} \cos(2\pi t) u(t)$
(d) $x[n] = \sum_{m=-\infty}^{\infty} \delta[n - 20m] - 2\delta[n - 2 - 20m]$

Answers:

(a) DTFT
(b) FS
(c) FT
(d) DTFS   ◀

3.4 Discrete-Time Periodic Signals: The Discrete-Time Fourier Series

The DTFS representation of a periodic signal $x[n]$ with fundamental period N and fundamental frequency $\Omega_o = 2\pi/N$ is given by

$$x[n] = \sum_{k=0}^{N-1} X[k] e^{jk\Omega_o n}, \tag{3.10}$$

where

$$X[k] = \frac{1}{N} \sum_{n=0}^{N-1} x[n] e^{-jk\Omega_o n} \tag{3.11}$$

are the DTFS coefficients of the signal $x[n]$. We say that $x[n]$ and $X[k]$ are a *DTFS pair* and denote this relationship as

$$x[n] \xleftrightarrow{\quad DTFS;\, \Omega_o \quad} X[k].$$

From N values of $X[k]$, we may determine $x[n]$ by using Eq. (3.10), and from N values of $x[n]$, we may determine $X[k]$ by using Eq. (3.11). Either $X[k]$ or $x[n]$ provides a complete description of the signal. We shall see that in some problems it is advantageous to represent the signal using its time-domain values $x[n]$, while in others the DTFS coefficients $X[k]$ offer a more convenient description of the signal. The DTFS coefficients $X[k]$ are

termed a *frequency-domain* representation for $x[n]$, because each coe
with a complex sinusoid of a different frequency. The variable k determines
of the sinusoid associated with $X[k]$, so we say that $X[k]$ is a function of freque
DTFS representation is exact; any periodic discrete-time signal may be described in ter
of Eq. (3.10).

The DTFS is the only Fourier representation that can be numerically evaluated and manipulated in a computer. This is because both the time-domain, $x[n]$, and frequency-domain, $X[k]$, representations of the signal are exactly characterized by a finite set of N numbers. The computational tractability of the DTFS is of great practical significance. The series finds extensive use in numerical signal analysis and system implementation and is often used to numerically approximate the other three Fourier representations. These issues are explored in the next chapter.

Before presenting several examples illustrating the DTFS, we remind the reader that the limits on the sums in Eqs. (3.10) and (3.11) may be chosen to be different from 0 to $N - 1$ because $x[n]$ is N periodic in n while $X[k]$ is N periodic in k. The range of the indices may thus be chosen to simplify the problem at hand.

EXAMPLE 3.2 DETERMINING DTFS COEFFICIENTS   Find the frequency-domain representation of the signal depicted in Fig. 3.5

Solution: The signal has period $N = 5$, so $\Omega_o = 2\pi/5$. Also, the signal has odd symmetry, so we sum over $n = -2$ to $n = 2$ in Eq. (3.11) to obtain

$$X[k] = \frac{1}{5} \sum_{n=-2}^{2} x[n]e^{-jk2\pi n/5}$$

$$= \frac{1}{5}\{x[-2]e^{jk4\pi/5} + x[-1]e^{jk2\pi/5} + x[0]e^{j0} + x[1]e^{-jk2\pi/5} + x[2]e^{-jk4\pi/5}\}.$$

Using the values of $x[n]$, we get

$$X[k] = \frac{1}{5}\left\{1 + \frac{1}{2}e^{jk2\pi/5} - \frac{1}{2}e^{-jk2\pi/5}\right\}$$

$$= \frac{1}{5}\{1 + j\sin(k2\pi/5)\}. \tag{3.12}$$

From this equation, we identify one period of the DTFS coefficients $X[k]$, $k = -2$ to $k = 2$, in rectangular and polar coordinates as

$$X[-2] = \frac{1}{5} - j\frac{\sin(4\pi/5)}{5} = 0.232e^{-j0.531}$$

$$X[-1] = \frac{1}{5} - j\frac{\sin(2\pi/5)}{5} = 0.276e^{-j0.760}$$

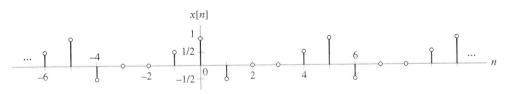

FIGURE 3.5   Time-domain signal for Example 3.2.

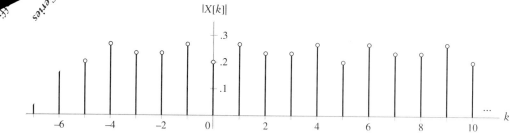

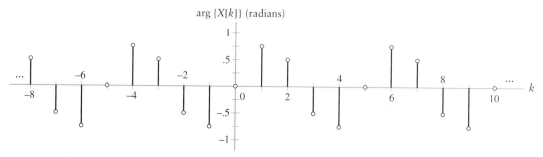

FIGURE 3.6   Magnitude and phase of the DTFS coefficients for the signal in Fig. 3.5.

$$X[0] = \frac{1}{5} = 0.2e^{j0}$$

$$X[1] = \frac{1}{5} + j\frac{\sin(2\pi/5)}{5} = 0.276e^{j0.760}$$

$$X[2] = \frac{1}{5} + j\frac{\sin(4\pi/5)}{5} = 0.232e^{j0.531}.$$

Figure 3.6 depicts the magnitude and phase of $X[k]$ as functions of the frequency index k.

Now suppose we calculate $X[k]$ using $n = 0$ to $n = 4$ for the limits on the sum in Eq. (3.11), to obtain

$$X[k] = \frac{1}{5}\{x[0]e^{j0} + x[1]e^{-j2\pi/5} + x[2]e^{-jk4\pi/5} + x[3]e^{-jk6\pi/5} + x[4]e^{-j8\pi/5}\}$$

$$= \frac{1}{5}\left\{1 - \frac{1}{2}e^{-jk2\pi/5} + \frac{1}{2}e^{-jk8\pi/5}\right\}.$$

This expression appears to differ from Eq. (3.12), which was obtained using $n = -2$ to $n = 2$. However, noting that

$$e^{-jk8\pi/5} = e^{-jk2\pi}e^{jk2\pi/5}$$
$$= e^{jk2\pi/5},$$

we see that both intervals, $n = -2$ to $n = 2$ and $n = 0$ to $n = 4$, yield equivalent expressions for the DTFS coefficients. ∎

The magnitude of $X[k]$, denoted $|X[k]|$ and plotted against the frequency index k, is known as the *magnitude spectrum* of $x[n]$. Similarly, the phase of $X[k]$, termed $\arg\{X[k]\}$, is known as the *phase spectrum* of $x[n]$. Note that in Example 3.2 $|X[k]|$ is even while $\arg\{X[k]\}$ is odd.

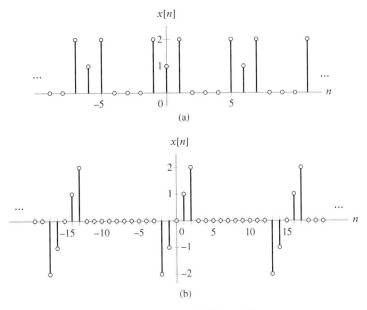

FIGURE 3.7   Signals $x[n]$ for Problem 3.2.

▶ **Problem 3.2**   Determine the DTFS coefficients of the periodic signals depicted in Figs. 3.7(a) and (b).

Answers:

Fig. 3.7(a):

$$x[n] \xleftrightarrow{\;DTFS;\,\pi/3\;} X[k] = \frac{1}{6} + \frac{2}{3}\cos(k\pi/3)$$

Fig. 3.7(b):

$$x[n] \xleftrightarrow{\;DTFS;\,2\pi/15\;} X[k] = \frac{-2j}{15}\big(\sin(k2\pi/15) + 2\sin(k4\pi/15)\big) \qquad ◀$$

If $x[n]$ is composed of real or complex sinusoids, then it is often easier to determine $X[k]$ by inspection than by evaluating Eq. (3.11). The method of inspection is based on expanding all real sinusoids in terms of complex sinusoids and comparing each term in the result with each term of Eq. (3.10), as illustrated by the next example.

EXAMPLE 3.3 COMPUTATION OF DTFS COEFFICIENTS BY INSPECTION   Determine the DTFS coefficients of $x[n] = \cos(\pi n/3 + \phi)$, using the method of inspection.

Solution: The period of $x[n]$ is $N = 6$. We expand the cosine by using Euler's formula and move any phase shifts in front of the complex sinusoids. The result is

$$x[n] = \frac{e^{j\left(\frac{\pi}{3}n+\phi\right)} + e^{-j\left(\frac{\pi}{3}n+\phi\right)}}{2} \tag{3.13}$$

$$= \frac{1}{2}e^{-j\phi}e^{-j\frac{\pi}{3}n} + \frac{1}{2}e^{j\phi}e^{j\frac{\pi}{3}n}.$$

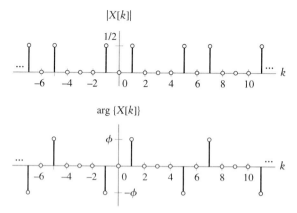

FIGURE 3.8   Magnitude and phase of DTFS coefficients for Example 3.3.

Now we compare Eq. (3.13) with the DTFS of Eq. (3.10) with $\Omega_o = 2\pi/6 = \pi/3$, written by summing from $k = -2$ to $k = 3$:

$$x[n] = \sum_{k=-2}^{3} X[k]e^{jk\pi n/3}$$
$$= X[-2]e^{-j2\pi n/3} + X[-1]e^{-j\pi n/3} + X[0] + X[1]e^{j\pi n/3} + X[2]e^{j2\pi n/3} + X[3]e^{j\pi n}. \tag{3.14}$$

Equating terms in Eq. (3.13) with those in Eq. (3.14) having equal frequencies, $k\pi/3$, gives

$$x[n] \xleftrightarrow{\;DTFS;\frac{\pi}{3}\;} X[k] = \begin{cases} e^{-j\phi}/2, & k = -1 \\ e^{j\phi}/2, & k = 1 \\ 0, & \text{otherwise on } -2 \le k \le 3 \end{cases}.$$

The magnitude spectrum, $|X[k]|$, and phase spectrum, $\arg\{X[k]\}$, are depicted in Fig. 3.8. ▪

▶ **Problem 3.3**   Use the method of inspection to determine the DTFS coefficients for the following signals:

(a) $x[n] = 1 + \sin(n\pi/12 + 3\pi/8)$

(b) $x[n] = \cos(n\pi/30) + 2\sin(n\pi/90)$

Answers:

$$\text{(a)}\quad x[n] \xleftrightarrow{\;DTFS;2\pi/24\;} X[k] = \begin{cases} -e^{-j3\pi/8}/(2j), & k = -1 \\ 1, & k = 0 \\ e^{j3\pi/8}/(2j), & k = 1 \\ 0, & \text{otherwise on } -11 \le k \le 12 \end{cases}$$

$$\text{(b)}\quad x[n] \xleftrightarrow{\;DTFS;2\pi/180\;} X[k] = \begin{cases} -1/j, & k = -1 \\ 1/j, & k = 1 \\ 1/2, & k = \pm 3 \\ 0, & \text{otherwise on } -89 \le k \le 90 \end{cases}$$ ◀

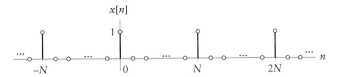

$x[n]$

FIGURE 3.9   A discrete-time impulse train with period N.

EXAMPLE 3.4 DTFS REPRESENTATION OF AN IMPULSE TRAIN   Find the DTFS coefficients of the N-periodic impulse train

$$x[n] = \sum_{l=-\infty}^{\infty} \delta[n - lN],$$

as shown in Fig. 3.9.

Solution: Since there is only one nonzero value in $x[n]$ per period, it is convenient to evaluate Eq. (3.11) over the interval $n = 0$ to $n = N - 1$ to obtain

$$X[k] = \frac{1}{N} \sum_{n=0}^{N-1} \delta[n] e^{-jkn2\pi/N}$$

$$= \frac{1}{N}.$$ ∎

Although we have focused on evaluating the DTFS coefficients, the similarity between Eqs. (3.11) and (3.10) indicates that the same mathematical methods can be used to find the time-domain signal corresponding to a set of DTFS coefficients. Note that in cases where some of the values of $x[n]$ are zero, such as the previous example, $X[k]$ may be periodic in k with period less than N. In this case, it is not possible to determine N from $X[k]$, so N must be known in order to find the proper time signal.

EXAMPLE 3.5 THE INVERSE DTFS   Use Eq. (3.10) to determine the time-domain signal $x[n]$ from the DTFS coefficients depicted in Fig. 3.10.

Solution: The DTFS coefficients have period 9, so $\Omega_o = 2\pi/9$. It is convenient to evaluate Eq. (3.10) over the interval $k = -4$ to $k = 4$ to obtain

$$x[n] = \sum_{k=-4}^{4} X[k] e^{jk2\pi n/9}$$

$$= e^{j2\pi/3} e^{-j6\pi n/9} + 2e^{j\pi/3} e^{-j4\pi n/9} - 1 + 2e^{-j\pi/3} e^{j4\pi n/9} + e^{-j2\pi/3} e^{j6\pi n/9}$$

$$= 2\cos(6\pi n/9 - 2\pi/3) + 4\cos(4\pi n/9 - \pi/3) - 1.$$ ∎

▶ **Problem 3.4**   One period of the DTFS coefficients of a signal is given by

$$X[k] = (1/2)^k, \quad \text{on } 0 \le k \le 9.$$

Find the time-domain signal $x[n]$ assuming $N = 10$.

Answer:

$$x[n] = \frac{1 - (1/2)^{10}}{1 - (1/2) e^{j(\pi/5)n}}$$ ◀

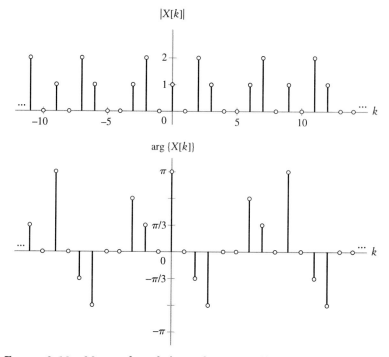

FIGURE 3.10   Magnitude and phase of DTFS coefficients for Example 3.5.

▶ **Problem 3.5**   Use the method of inspection to find the time-domain signal corresponding to the DTFS coefficients

$$X[k] = \cos(k4\pi/11) + 2j\sin(k6\pi/11).$$

Answer:

$$x[n] = \begin{cases} 1/2, & n = \pm 2 \\ 1, & n = 3 \\ -1, & n = -3 \\ 0, & \text{otherwise on } -5 \le n \le 5 \end{cases}$$

◀

EXAMPLE 3.6 DTFS REPRESENTATION OF A SQUARE WAVE   Find the DTFS coefficients for the N-periodic square wave given by

$$x[n] = \begin{cases} 1, & -M \le n \le M \\ 0, & M < n < N - M \end{cases}.$$

That is, each period contains $2M + 1$ consecutive ones and the remaining $N - (2M + 1)$ values are zero, as depicted in Fig. 3.11. Note that this definition requires that $N > 2M + 1$.

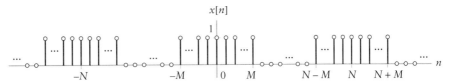

FIGURE 3.11 Discrete-time square wave for Example 3.6.

Solution: The period is N, so $\Omega_o = 2\pi/N$. It is convenient in this case to evaluate Eq. (3.11) over indices $n = -M$ to $n = N - M - 1$. We thus have

$$X[k] = \frac{1}{N}\sum_{n=-M}^{N-M-1} x[n]e^{-jk\Omega_o n}$$

$$= \frac{1}{N}\sum_{n=-M}^{M} e^{-jk\Omega_o n}.$$

We perform the change of variable on the index of summation by letting $m = n + M$ to obtain

$$X[k] = \frac{1}{N}\sum_{m=0}^{2M} e^{-jk\Omega_o(m-M)}$$

$$= \frac{1}{N}e^{jk\Omega_o M}\sum_{m=0}^{2M} e^{-jk\Omega_o m}. \qquad (3.15)$$

Now, for $k = 0, \pm N, \pm 2N, \ldots$ we have $e^{jk\Omega_o} = e^{-jk\Omega_o} = 1$, and Eq. (3.15) becomes

$$X[k] = \frac{1}{N}\sum_{m=0}^{2M} 1$$

$$= \frac{2M+1}{N}, \qquad k = 0, \pm N, \pm 2N, \ldots.$$

For $k \neq 0, \pm N, \pm 2N, \ldots$, we may sum the geometric series in Eq. (3.15) to obtain

$$X[k] = \frac{e^{jk\Omega_o M}}{N}\left(\frac{1 - e^{-jk\Omega_o(2M+1)}}{1 - e^{-jk\Omega_o}}\right), \qquad k \neq 0, \pm N, \pm 2N, \ldots, \qquad (3.16)$$

which may be rewritten as

$$X[k] = \frac{1}{N}\left(\frac{e^{jk\Omega_o(2M+1)/2}}{e^{jk\Omega_o/2}}\right)\left(\frac{1 - e^{-jk\Omega_o(2M+1)}}{1 - e^{-jk\Omega_o}}\right),$$

$$= \frac{1}{N}\left(\frac{e^{jk\Omega_o(2M+1)/2} - e^{-jk\Omega_o(2M+1)/2}}{e^{jk\Omega_o/2} - e^{-jk\Omega_o/2}}\right), \qquad k \neq 0, \pm N, \pm 2N, \ldots.$$

At this point, we divide the numerator and denominator by $2j$ to express $X[k]$ as a ratio of two sine functions:

$$X[k] = \frac{1}{N}\frac{\sin(k\Omega_o(2M+1)/2)}{\sin(k\Omega_o/2)}, \qquad k \neq 0, \pm N, \pm 2N, \ldots.$$

The technique used here to write the finite geometric-sum expression for $X[k]$ as a ratio of sine functions involves symmetrizing both the numerator, $1 - e^{-jk\Omega_o(2M+1)}$, and denominator, $1 - e^{-jk\Omega_o}$, in Eq. (3.16) with the appropriate power of $e^{jk\Omega_o}$. An alternative expression for $X[k]$ is obtained by substituting $\Omega_o = \frac{2\pi}{N}$, yielding

$$X[k] = \begin{cases} \dfrac{1}{N} \dfrac{\sin(k\pi(2M+1)/N)}{\sin(k\pi/N)}, & k \neq 0, \pm N, \pm 2N, \ldots \\ (2M+1)/N, & k = 0, \pm N, \pm 2N, \ldots \end{cases}.$$

Using L'Hôpital's rule by treating k as a real number, it is easy to show that

$$\lim_{k \to 0, \pm N, \pm 2N, \ldots} \left(\frac{1}{N} \frac{\sin(k\pi(2M+1)/N)}{\sin(k\pi/N)} \right) = \frac{2M+1}{N}.$$

For this reason, the expression for $X[k]$ is commonly written as

$$X[k] = \frac{1}{N} \frac{\sin(k\pi(2M+1)/N)}{\sin(k\pi/N)}.$$

In this form, it is understood that the value of $X[k]$ for $k = 0, \pm N, \pm 2N, \ldots$ is obtained from the limit as $k \to 0$. A plot of two periods of $X[k]$ as a function of k is depicted in Fig. 3.12 for both $M = 4$ and $M = 12$, assuming $N = 50$. Note that in this example $X[k]$ is real; hence, the magnitude spectrum is the absolute value of $X[k]$, and the phase spectrum is 0 for $X[k]$ positive and π for $X[k]$ negative. ■

S & S Solutions

▶ **Problem 3.6**   Find the DTFS coefficients of the signals depicted in Figs. 3.13(a) and (b).

Answers:

(a)

$$X[k] = \frac{8}{125} e^{jk2\pi/5} \frac{1 - \left(\frac{5}{4}e^{-jk\pi/5}\right)^7}{1 - \frac{5}{4}e^{-jk\pi/5}}$$

(b)

$$X[k] = -\frac{j}{5} \sin(k\pi/2) \frac{\sin(k2\pi/5)}{\sin(k\pi/10)}$$ ◀

It is instructive to consider the contribution of each term in the DTFS of Eq. (3.10) to the representation of the signal. We do so by examining the series representation of the square wave in Example 3.6. Evaluating the contribution of each term is particularly simple for this waveform because the DTFS coefficients have even symmetry (i.e., $X[k] = X[-k]$). Therefore, we may rewrite the DTFS of Eq. (3.10) as a series involving harmonically related cosines. Assume for convenience that N is even, so that $N/2$ is integer, and let k range from $-N/2 + 1$ to $N/2$. We thus write

$$x[n] = \sum_{k=-N/2+1}^{N/2} X[k]e^{jk\Omega_o n}.$$

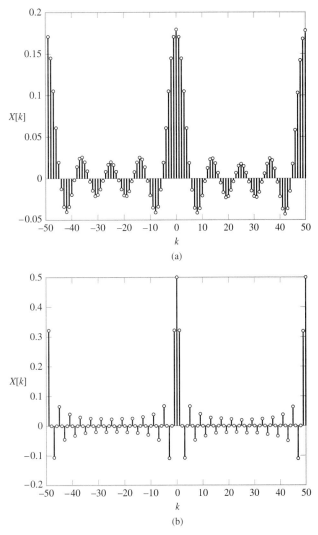

FIGURE 3.12 The DTFS coefficients for the square wave shown in Fig. 3.11, assuming a period $N = 50$: (a) $M = 4$. (b) $M = 12$.

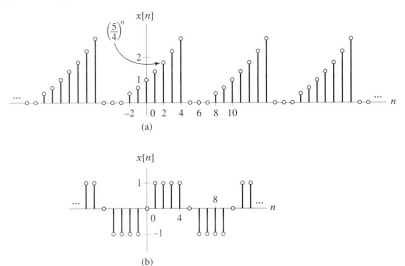

FIGURE 3.13 Signals $x[n]$ for Problem 3.6.

In order to exploit the symmetry in the DTFS coefficients, we pull the $k = 0$ and $k = N/2$ terms out of the sum and express the remaining terms using the positive index m:

$$x[n] = X[0] + X[N/2]e^{jN\Omega_o n/2} + \sum_{m=1}^{N/2-1} (X[m]e^{jm\Omega_o n} + X[-m]e^{-jm\Omega_o n}).$$

Now we use $X[m] = X[-m]$ and the identity $N\Omega_o = 2\pi$ to obtain

$$x[n] = X[0] + X[N/2]e^{j\pi n} + \sum_{m=1}^{N/2-1} 2X[m]\left(\frac{e^{jm\Omega_o n} + e^{-jm\Omega_o n}}{2}\right)$$

$$= X[0] + X[N/2]\cos(\pi n) + \sum_{m=1}^{N/2-1} 2X[m]\cos(m\Omega_o n),$$

where we have also used $e^{j\pi n} = \cos(\pi n)$ since $\sin(\pi n) = 0$ for integer n.
   Finally, we define the new set of coefficients

$$B[k] = \begin{cases} X[k], & k = 0, N/2 \\ 2X[k], & k = 1, 2, \ldots, N/2 - 1 \end{cases}$$

and write the DTFS for the square wave in terms of a series of harmonically related cosines as

$$x[n] = \sum_{k=0}^{N/2} B[k]\cos(k\Omega_o n). \tag{3.17}$$

A similar expression may be derived for N odd.

EXAMPLE 3.7 BUILDING A SQUARE WAVE FROM DTFS COEFFICIENTS   The contribution of each term to the square wave may be illustrated by defining the partial-sum approximation to $x[n]$ in Eq. (3.17) as

$$\hat{x}_J[n] = \sum_{k=0}^{J} B[k]\cos(k\Omega_o n), \tag{3.18}$$

where $J \leq N/2$. This approximation contains the first $2J + 1$ terms centered on $k = 0$ in Eq. (3.10). Assume a square wave has period $N = 50$ and $M = 12$. Evaluate one period of the Jth term in Eq. (3.18) and the $2J + 1$ term approximation $\hat{x}_J[n]$ for $J = 1, 3, 5, 23$, and 25.

Solution: Figure 3.14 depicts the Jth term in the sum, $B[J]\cos(J\Omega_o n)$, and one period of $\hat{x}_J[n]$ for the specified values of J. Only odd values for J are considered, because the even-indexed coefficients $B[k]$ are zero when $N = 25$ and $M = 12$. Note that the approximation improves as J increases, with $x[n]$ represented exactly when $J = N/2 = 25$. In general, the coefficients $B[k]$ associated with values of k near zero represent the low-frequency or slowly varying features in the signal, while the coefficients associated with values of k near $\pm\frac{N}{2}$ represent the high-frequency or rapidly varying features in the signal.   ■

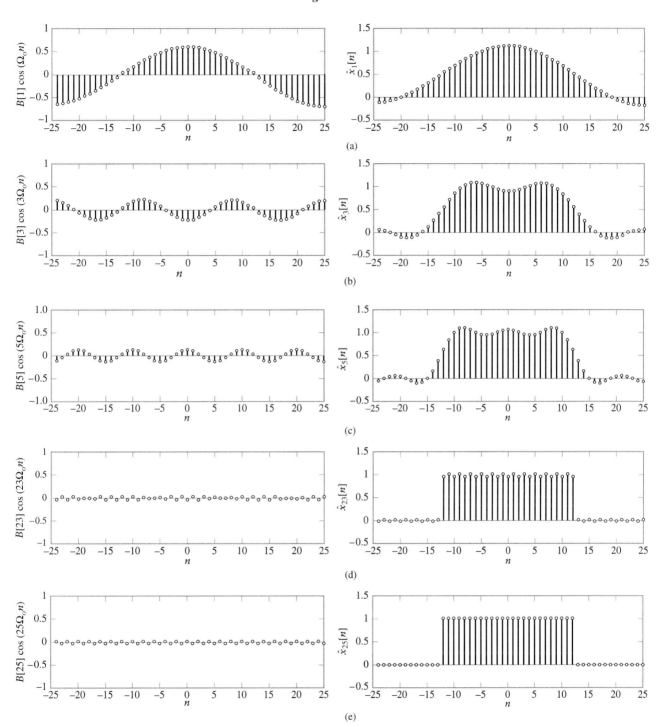

FIGURE 3.14 Individual terms in the DTFS expansion of a square wave (left panel) and the corresponding partial-sum approximations $\hat{x}_J[n]$ (right panel). The $J = 0$ term is $\hat{x}_0[n] = 1/2$ and is not shown. (a) $J = 1$. (b) $J = 3$. (c) $J = 5$. (d) $J = 23$. (e) $J = 25$.

The use of the DTFS as a numerical signal analysis tool is illustrated in the next example.

EXAMPLE 3.8 NUMERICAL ANALYSIS OF THE ECG   Evaluate the DTFS representations of the two electrocardiogram (ECG) waveforms depicted in Figs. 3.15(a) and (b). Figure 3.15(a) depicts a normal ECG, while Fig. 3.15(b) depicts the ECG of a heart experiencing ventricular tachycardia. The discrete-time signals are drawn as continuous functions, due to the difficulty of depicting all 2000 values in each case. Ventricular tachycardia is a serious cardiac rhythm disturbance (i.e., an arrhythmia) that can result in death. It is characterized by a rapid,

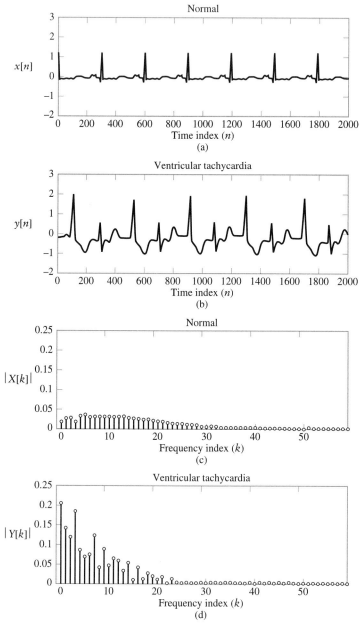

FIGURE 3.15   Electrocardiograms for two different heartbeats and the first 60 coefficients of their magnitude spectra. (a) Normal heartbeat. (b) Ventricular tachycardia. (c) Magnitude spectrum for the normal heartbeat. (d) Magnitude spectrum for ventricular tachycardia.

regular heart rate of approximately 150 beats per minute. Ventricular complexes are wide (about 160 ms in duration) compared with normal complexes (less than 110 ms) and have an abnormal shape. Both signals appear nearly periodic, with only slight variations in the amplitude and length of each period. The DTFS of one period of each ECG may be computed numerically. The period of the normal ECG is $N = 305$, while the period of the ECG showing ventricular tachycardia is $N = 421$. One period of each waveform is available. Evaluate the DTFS coefficients of each, and plot their magnitude spectrum.

Solution: The magnitude spectrum of the first 60 DTFS coefficients is depicted in Figs. 3.15(c) and (d). The higher indexed coefficients are very small and thus are not shown.

The time waveforms differ, as do the DTFS coefficients. The normal ECG is dominated by a sharp spike or impulsive feature. Recall that the DTFS coefficients of an impulse train have constant magnitude, as shown in Example 3.4. The DTFS coefficients of the normal ECG are approximately constant, exhibiting a gradual decrease in amplitude as the frequency increases. They also have a fairly small magnitude, since there is relatively little power in the impulsive signal. In contrast, the ventricular tachycardia ECG contains smoother features in addition to sharp spikes, and thus the DTFS coefficients have a greater dynamic range, with the low-frequency coefficients containing a large proportion of the total power. Also, because the ventricular tachycardia ECG has greater power than the normal ECG, the DTFS coefficients have a larger amplitude. ∎

3.5 Continuous-Time Periodic Signals: The Fourier Series

Continuous-time periodic signals are represented by the Fourier series (FS). We may write the FS of a signal $x(t)$ with fundamental period T and fundamental frequency $\omega_o = 2\pi/T$ as

$$x(t) = \sum_{k=-\infty}^{\infty} X[k]e^{jk\omega_o t}, \qquad (3.19)$$

where

$$X[k] = \frac{1}{T}\int_0^T x(t)e^{-jk\omega_o t}\, dt \qquad (3.20)$$

are the FS coefficients of the signal $x(t)$. We say that $x(t)$ and $X[k]$ are an FS *pair* and denote this relationship as

$$x(t) \xleftrightarrow{\ FS;\,\omega_o\ } X[k].$$

From the FS coefficients $X[k]$, we may determine $x(t)$ by using Eq. (3.19), and from $x(t)$, we may determine $X[k]$ by using Eq. (3.20). We shall see later that in some problems it is advantageous to represent the signal in the time domain as $x(t)$, while in others the FS coefficients $X[k]$ offer a more convenient description. The FS coefficients are known as a *frequency-domain representation* of $x(t)$ because each FS coefficient is associated with a complex sinusoid of a different frequency. As in the DTFS, the variable k determines the frequency of the complex sinusoid associated with $X[k]$ in Eq. (3.19).

The FS representation is most often used in electrical engineering to analyze the effect of systems on periodic signals.

The infinite series in Eq. (3.19) is not guaranteed to converge for all possible signals. In this regard, suppose we define

$$\hat{x}(t) = \sum_{k=-\infty}^{\infty} X[k]e^{jk\omega_o t}$$

and choose the coefficients $X[k]$ according to Eq. (3.20). Under what conditions does $\hat{x}(t)$ actually converge to $x(t)$? A detailed analysis of this question is beyond the scope of this text; however, we can state several results. First, if $x(t)$ is square integrable—that is, if

$$\frac{1}{T} \int_0^T |x(t)|^2 \, dt < \infty,$$

then the MSE between $x(t)$ and $\hat{x}(t)$ is zero, or, mathematically,

$$\frac{1}{T} \int_0^T |x(t) - \hat{x}(t)|^2 \, dt = 0.$$

This is a useful result that applies to a broad class of signals encountered in engineering practice. Note that, in contrast to the discrete-time case, an MSE of zero does not imply that $x(t)$ and $\hat{x}(t)$ are equal pointwise, or $x(t) = \hat{x}(t)$ at all values of t; it simply implies that there is zero power in their difference.

Pointwise convergence of $\hat{x}(t)$ to $x(t)$ is guaranteed at all values of t except those corresponding to discontinuities if the Dirichlet conditions are satisfied:

- ▸ $x(t)$ is bounded.
- ▸ $x(t)$ has a finite number of maxima and minima in one period.
- ▸ $x(t)$ has a finite number of discontinuities in one period.

If a signal $x(t)$ satisfies the Dirichlet conditions and is not continuous, then $\hat{x}(t)$ converges to the midpoint of the left and right limits of $x(t)$ at each discontinuity.

The next three examples illustrate how the FS representation is determined.

EXAMPLE 3.9 DIRECT CALCULATION OF FS COEFFICIENTS   Determine the FS coefficients for the signal $x(t)$ depicted in Fig. 3.16.

Solution: The period of $x(t)$ is $T = 2$, so $\omega_o = 2\pi/2 = \pi$. On the interval $0 \leq t \leq 2$, one period of $x(t)$ is expressed as $x(t) = e^{-2t}$, so Eq. (3.20) yields

$$X[k] = \frac{1}{2} \int_0^2 e^{-2t} e^{-jk\pi t} \, dt$$

$$= \frac{1}{2} \int_0^2 e^{-(2+jk\pi)t} \, dt.$$

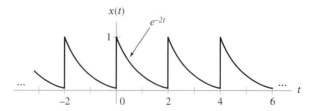

FIGURE 3.16   Time-domain signal for Example 3.9.

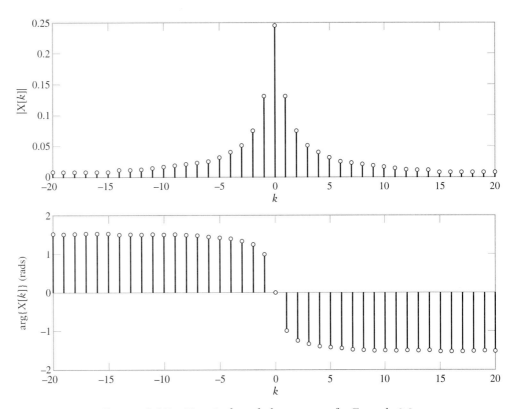

FIGURE 3.17 Magnitude and phase spectra for Example 3.9.

We evaluate the integral to obtain

$$X[k] = \frac{-1}{2(2 + jk\pi)} e^{-(2+jk\pi)t} \Big|_{0}^{2}$$

$$= \frac{1}{4 + jk2\pi}(1 - e^{-4}e^{-jk2\pi})$$

$$= \frac{1 - e^{-4}}{4 + jk2\pi},$$

since $e^{-jk2\pi} = 1$. Figure 3.17 depicts the magnitude spectrum $|X[k]|$ and the phase spectrum $\arg\{X[k]\}$. ∎

As with the DTFS, the magnitude of $X[k]$ is known as the *magnitude spectrum* of $x(t)$, while the phase of $X[k]$ is known as the *phase spectrum* of $x(t)$. Also, since $x(t)$ is periodic, the interval of integration in Eq. (3.20) may be chosen as any interval one period in length. Choosing the appropriate interval of integration often simplifies the problem, as illustrated in the next example.

EXAMPLE 3.10 FS COEFFICIENTS FOR AN IMPULSE TRAIN   Determine the FS coefficients for the signal defined by

$$x(t) = \sum_{l=-\infty}^{\infty} \delta(t - 4l).$$

Solution: The fundamental period is $T = 4$, and each period of this signal contains an impulse. The signal $x(t)$ has even symmetry, so it is easier to evaluate Eq. (3.20) by integrating over a period that is symmetric about the origin, $-2 < t \le 2$, to obtain

$$X[k] = \frac{1}{4} \int_{-2}^{2} \delta(t) e^{-jk(\pi/2)t} \, dt$$

$$= \frac{1}{4}.$$

In this case, the magnitude spectrum is constant and the phase spectrum is zero. Note that we cannot evaluate the infinite sum in Eq. (3.19) in this case and that $x(t)$ does not satisfy the Dirichlet conditions. However, the FS expansion of an impulse train is useful in spite of convergence difficulties.   ■

As with the DTFS, whenever $x(t)$ is expressed in terms of sinusoids, it is easier to obtain $X[k]$ by inspection. The method of inspection is based on expanding all real sinusoids in terms of complex sinusoids and then comparing each term in the resulting expansion to the corresponding terms of Eq. (3.19).

EXAMPLE 3.11 CALCULATION OF FS COEFFICIENTS BY INSPECTION   Determine the FS representation of the signal

$$x(t) = 3\cos(\pi t/2 + \pi/4),$$

using the method of inspection.

Solution: The fundamental period of $x(t)$ is $T = 4$. Hence, $\omega_o = 2\pi/4 = \pi/2$, and Eq. (3.19) is written as

$$x(t) = \sum_{k=-\infty}^{\infty} X[k] e^{jk\pi t/2}. \tag{3.21}$$

Using Euler's formula to expand the cosine yields

$$x(t) = 3\frac{e^{j(\pi t/2 + \pi/4)} + e^{-j(\pi t/2 + \pi/4)}}{2}$$

$$= \frac{3}{2} e^{j\pi/4} e^{j\pi t/2} + \frac{3}{2} e^{-j\pi/4} e^{-j\pi t/2}.$$

Equating each term in this expression to the terms in Eq. (3.21) gives the FS coefficients:

$$X[k] = \begin{cases} \frac{3}{2} e^{-j\pi/4}, & k = -1 \\ \frac{3}{2} e^{j\pi/4}, & k = 1 \\ 0, & \text{otherwise} \end{cases} \tag{3.22}$$

The magnitude and phase spectra are depicted in Fig. 3.18. Note that all of the power in this signal is concentrated at two frequencies: $\omega = \pi/2$ and $\omega = -\pi/2$.   ■

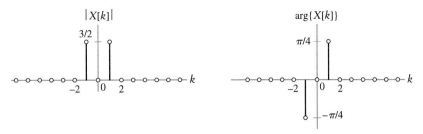

FIGURE 3.18   Magnitude and phase spectra for Example 3.11.

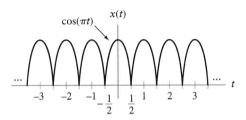

FIGURE 3.19   Full-wave rectified cosine for Problem 3.8.

▶ **Problem 3.7**   Determine the FS representation of

$$x(t) = 2\sin(2\pi t - 3) + \sin(6\pi t).$$

Answer:

$$x(t) \xleftrightarrow{\ FS;\,2\pi\ } X[k] = \begin{cases} j/2, & k = -3 \\ je^{j3}, & k = -1 \\ -je^{-j3}, & k = 1 \\ -j/2, & k = 3 \\ 0, & \text{otherwise} \end{cases} \qquad \blacktriangleleft$$

S & S Solutions

▶ **Problem 3.8**   Find the FS coefficients of the full-wave rectified cosine depicted in Fig. 3.19.

Answer:

$$X[k] = \frac{\sin(\pi(1 - 2k)/2)}{\pi(1 - 2k)} + \frac{\sin(\pi(1 + 2k)/2)}{\pi(1 + 2k)} \qquad \blacktriangleleft$$

The time-domain signal represented by a set of FS coefficients is obtained by evaluating Eq. (3.19), as illustrated in the next example.

EXAMPLE 3.12 INVERSE FS   Find the time-domain signal $x(t)$ corresponding to the FS coefficients

$$X[k] = (1/2)^{|k|}e^{jk\pi/20}.$$

Assume that the fundamental period is $T = 2$.

Solution: Substituting the values given for $X[k]$ and $\omega_o = 2\pi/T = \pi$ into Eq. (3.19) yields

$$x(t) = \sum_{k=0}^{\infty} (1/2)^k e^{jk\pi/20} e^{jk\pi t} + \sum_{k=-1}^{-\infty} (1/2)^{-k} e^{jk\pi/20} e^{jk\pi t}$$

$$= \sum_{k=0}^{\infty} (1/2)^k e^{jk\pi/20} e^{jk\pi t} + \sum_{l=1}^{\infty} (1/2)^l e^{-jl\pi/20} e^{-jl\pi t}.$$

The second geometric series is evaluated by summing from $l = 0$ to $l = \infty$ and subtracting the $l = 0$ term. The result of summing both infinite geometric series is

$$x(t) = \frac{1}{1 - (1/2)e^{j(\pi t + \pi/20)}} + \frac{1}{1 - (1/2)e^{-j(\pi t + \pi/20)}} - 1.$$

Putting the fractions over a common denominator results in

$$x(t) = \frac{3}{5 - 4\cos(\pi t + \pi/20)}$$

■

▶ **Problem 3.9**   Determine the time-domain signal represented by the following FS coefficients:

(a)

$$X[k] = -j\delta[k - 2] + j\delta[k + 2] + 2\delta[k - 3] + 2\delta[k + 3], \quad \omega_o = \pi$$

(b) $X[k]$ given in Fig. 3.20 with $\omega_o = \pi/2$

Answers:

(a)
$$x(t) = 2\sin(2\pi t) + 4\cos(3\pi t)$$

(b)
$$x(t) = \frac{\sin(9\pi(t - 1)/4)}{\sin(\pi(t - 1)/4)}$$

◀

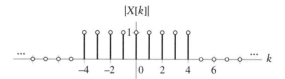

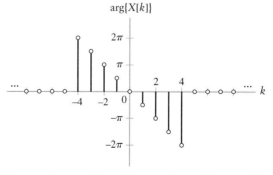

FIGURE 3.20   FS coefficients for Problem 3.9(b).

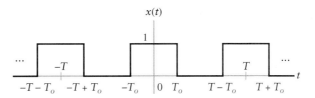

FIGURE 3.21 Square wave for Example 3.13.

EXAMPLE 3.13 FS FOR A SQUARE WAVE Determine the FS representation of the square wave depicted in Fig. 3.21.

Solution: The period is T, so $\omega_o = 2\pi/T$. Because the signal $x(t)$ has even symmetry, it is simpler to evaluate Eq. (3.20) by integrating over the range $-T/2 \leq t \leq T/2$. We obtain

$$
\begin{aligned}
X[k] &= \frac{1}{T} \int_{-T/2}^{T/2} x(t) e^{-jk\omega_o t} \, dt \\
&= \frac{1}{T} \int_{-T_o}^{T_o} e^{-jk\omega_o t} \, dt \\
&= \frac{-1}{Tjk\omega_o} e^{-jk\omega_o t} \Big|_{-T_o}^{T_o}, \quad k \neq 0 \\
&= \frac{2}{Tk\omega_o} \left(\frac{e^{jk\omega_o T_o} - e^{-jk\omega_o T_o}}{2j} \right), \quad k \neq 0 \\
&= \frac{2 \sin(k\omega_o T_o)}{Tk\omega_o}, \quad k \neq 0.
\end{aligned}
$$

For $k = 0$, we have

$$
X[0] = \frac{1}{T} \int_{-T_o}^{T_o} dt = \frac{2T_o}{T}.
$$

By means of L'Hôpital's rule, it is straightforward to show that

$$
\lim_{k \to 0} \frac{2 \sin(k\omega_o T_o)}{Tk\omega_o} = \frac{2T_o}{T},
$$

and thus we write

$$
X[k] = \frac{2 \sin(k\omega_o T_o)}{Tk\omega_o},
$$

with the understanding that $X[0]$ is obtained as a limit. In this problem, $X[k]$ is real valued. Using $\omega_o = 2\pi/T$ gives $X[k]$ as a function of the ratio T_o/T:

$$
X[k] = \frac{2 \sin(k2\pi T_o/T)}{k2\pi}. \tag{3.23}
$$

Figures 3.22(a)–(c) depict $X[k]$, $-50 \leq k \leq 50$, for $T_o/T = 1/4$, $T_o/T = 1/16$, and $T_o/T = 1/64$, respectively. Note that as T_o/T decreases, the energy within each period of the square-wave signal in Fig. 3.21 is concentrated over a narrower time interval, while the energy in the FS representations shown in Fig. 3.22 is distributed over a broader frequency interval. For example, the first zero crossing in $X[k]$ occurs at $k = 2$ for $T_o/T = 1/4$, $k = 8$ for $T_o/T = 1/16$, and $k = 32$ for $T_o/T = 1/64$. We shall explore the inverse relationship between the time- and frequency-domain extents of signals more fully in the sections that follow. ∎

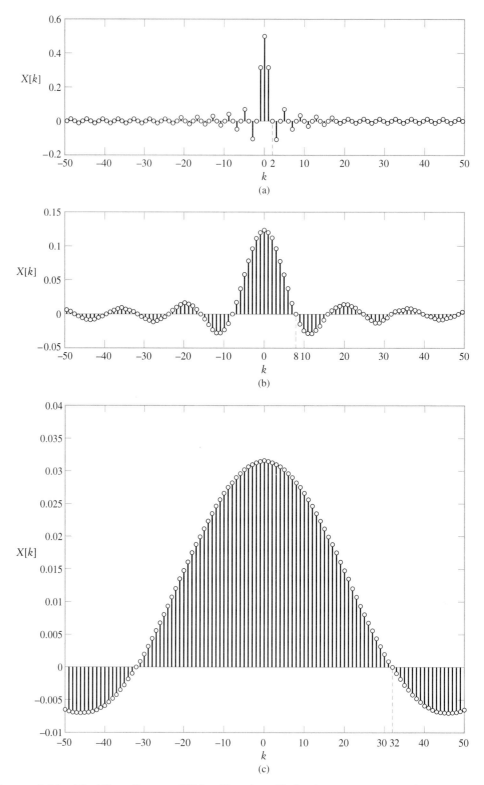

FIGURE 3.22 The FS coefficients, $X[k]$, $-50 \leq k \leq 50$, for three square waves. (See Fig. 3.21.) (a) $T_o/T = 1/4$. (b) $T_o/T = 1/16$. (c) $T_o/T = 1/64$.

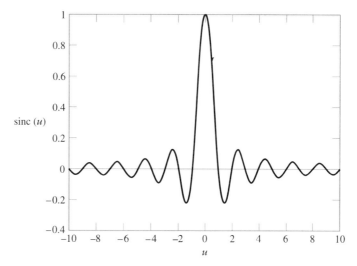

FIGURE 3.23   Sinc function $\text{sinc}(u) = \sin(\pi u)/(\pi u)$.

The functional form $\sin(\pi u)/(\pi u)$ occurs sufficiently often in Fourier analysis that we give it a special name:

$$\text{sinc}(u) = \frac{\sin(\pi u)}{\pi u}. \tag{3.24}$$

A graph of $\text{sinc}(u)$ is depicted in Fig. 3.23. The maximum of this function is unity at $u = 0$, the zero crossings occur at integers values of u, and the amplitude dies off as $1/u$. The portion of this function between the zero crossings at $u = \pm 1$ is known as the *mainlobe* of the sinc function. The smaller ripples outside the mainlobe are termed *sidelobes*. In sinc function notation, the FS coefficients in Eq. (3.23) are expressed as

$$X[k] = \frac{2T_o}{T} \text{sinc}\left(k \frac{2T_o}{T} \right).$$

▶ **Problem 3.10**   Find the FS representation of the sawtooth wave depicted in Fig. 3.24. (*Hint:* Use integration by parts.)

Answer: Integrate t from $-\frac{1}{2}$ to 1 in Eq. (3.20) to obtain

$$x(t) \xleftrightarrow{\ FS;\frac{4\pi}{3}\ } X[k] = \begin{cases} \dfrac{1}{4}, & k = 0 \\[2ex] \dfrac{-2}{3jk\omega_o}\left(e^{-jk\omega_o} + \dfrac{1}{2}e^{jk\omega_o/2} \right) + \dfrac{2}{3k^2\omega_o^2}(e^{-jk\omega_o} - e^{jk\omega_o/2}), & \text{otherwise} \end{cases}$$ ◀

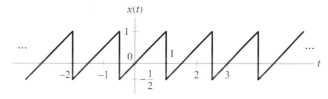

FIGURE 3.24   Periodic signal for Problem 3.10.

The form of the FS described by Eqs. (3.19) and (3.20) is termed the *exponential FS*. The *trigonometric FS* is often useful for real-valued signals and is expressed as

$$x(t) = B[0] + \sum_{k=1}^{\infty} B[k] \cos(k\omega_o t) + A[k] \sin(k\omega_o t), \qquad (3.25)$$

where the coefficients may be obtained from $x(t)$, using

and

$$B[0] = \frac{1}{T} \int_0^T x(t)\, dt$$

$$B[k] = \frac{2}{T} \int_0^T x(t) \cos(k\omega_o t)\, dt \qquad (3.26)$$

$$A[k] = \frac{2}{T} \int_0^T x(t) \sin(k\omega_o t)\, dt.$$

We see that $B[0] = X[0]$ represents the time-averaged value of the signal. Using Euler's formula to expand the cosine and sine functions in Eq. (3.26) and comparing the result with Eq. (3.20) shows that for $k \neq 0$,

$$B[k] = X[k] + X[-k]$$

and   (3.27)

$$A[k] = j(X[k] - X[-k]).$$

The relationships between the trigonometric, exponential, and polar forms of the FS are further studied in Problem 3.86.

The trigonometric FS coefficients of the square wave studied in Example 3.13 are obtained by substituting Eq. (3.23) into Eq. (3.27), yielding

$$B[0] = 2T_o/T,$$

$$B[k] = \frac{2 \sin(k 2\pi T_o/T)}{k\pi}, \quad k \neq 0, \qquad (3.28)$$

and

$$A[k] = 0.$$

The sine coefficients $A[k]$ are zero because $x(t)$ is an even function. Thus, the square wave may be expressed as a sum of harmonically related cosines:

$$x(t) = \sum_{k=0}^{\infty} B[k] \cos(k\omega_o t). \qquad (3.29)$$

This expression offers insight into the manner in which each FS component contributes to the representation of the signal, as is illustrated in the next example.

EXAMPLE 3.14 SQUARE-WAVE PARTIAL-SUM APPROXIMATION Let the partial-sum approximation to the FS in Eq. (3.29), be given by

$$\hat{x}_J(t) = \sum_{k=0}^{J} B[k]\cos(k\omega_o t).$$

This approximation involves the exponential FS coefficients with indices $-J \le k \le J$. Consider a square wave with $T = 1$ and $T_o/T = 1/4$. Depict one period of the Jth term in this sum, and find $\hat{x}_J(t)$ for $J = 1, 3, 7, 29,$ and 99.

Solution: In this case, we have

$$B[k] = \begin{cases} 1/2, & k = 0 \\ (2/(k\pi))(-1)^{(k-1)/2}, & k \text{ odd} , \\ 0, & k \text{ even} \end{cases}$$

so the even-indexed coefficients are zero. The individual terms and partial-sum approximations are depicted in Fig. 3.25, see page 226. The behavior of the partial-sum approximation in the vicinity of the square-wave discontinuities at $t = \pm T_o = \pm 1/4$ is of particular interest. We note that each partial-sum approximation passes through the average value $(1/2)$ of the discontinuity, as stated in our discussion of convergence. On each side of the discontinuity, the approximation exhibits ripple. As J increases, the maximum height of the ripples does not appear to change. In fact, it can be shown that, for any finite J, the maximum ripple is approximately 9% of the discontinuity. This ripple near discontinuities in partial-sum FS approximations is termed the *Gibbs phenomenon*, in honor of the mathematical physicist Josiah Gibbs for his explanation of it in 1899. The square wave satisfies the Dirichlet conditions, so we know that the FS approximation ultimately converges to the square wave for all values of t, except at the discontinuities. However, for finite J, the ripple is always present. As J increases, the ripple in the partial-sum approximations becomes more and more concentrated near the discontinuities. Hence, for any given J, the accuracy of the partial-sum approximation is best at times distant from discontinuities and worst near the discontinuities. ∎

The next example exploits linearity and the FS representation of the square wave in order to determine the output of an LTI system.

S & S
Solutions

EXAMPLE 3.15 RC CIRCUIT: CALCULATING THE OUTPUT BY MEANS OF FS Let us find the FS representation for the output $y(t)$ of the RC circuit depicted in Fig. 3.2 in response to the square-wave input depicted in Fig. 3.21, assuming that $T_o/T = 1/4$, $T = 1$ s, and $RC = 0.1$ s.

Solution: If the input to an LTI system is expressed as a weighted sum of sinusoids, then the output is also a weighted sum of sinusoids. As shown in Section 3.2, the kth weight in the output sum is given by the product of the kth weight in the input sum and the system frequency response evaluated at the kth sinusoid's frequency. Hence, if

$$x(t) = \sum_{k=-\infty}^{\infty} X[k]e^{jk\omega_o t},$$

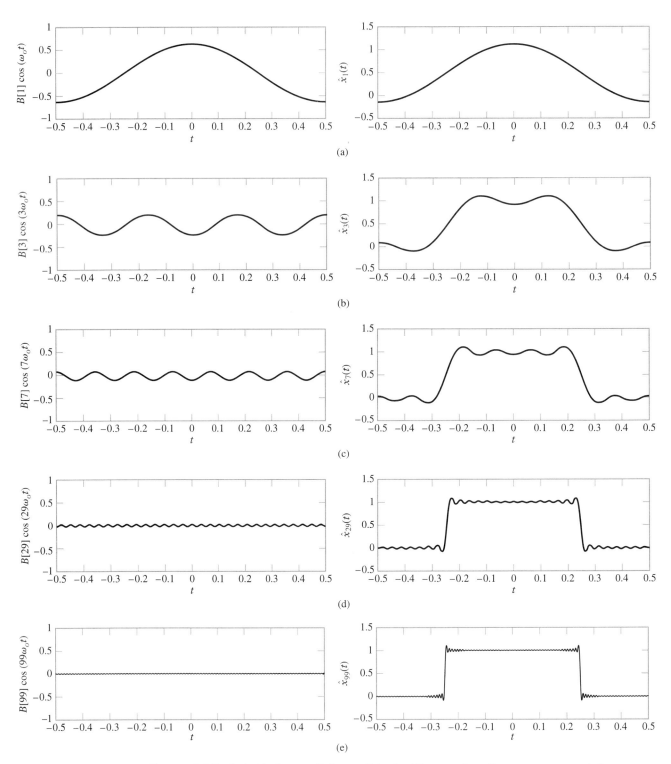

FIGURE 3.25 Individual terms (left panel) in the FS expansion of a square wave and the corresponding partial-sum approximations $\hat{x}_J(t)$ (right panel). The square wave has period $T = 1$ and $T_o/T = 1/4$. The $J = 0$ term is $\hat{x}_0(t) = 1/2$ and is not shown. (a) $J = 1$. (b) $J = 3$. (c) $J = 7$. (d) $J = 29$. (e) $J = 99$.

then the output is

$$y(t) = \sum_{k=-\infty}^{\infty} H(jk\omega_o)X[k]e^{jk\omega_o t},$$

where $H(j\omega)$ is the frequency response of the system. Thus,

$$y(t) \xleftrightarrow{\quad FS;\omega_o \quad} Y[k] = H(jk\omega_o)X[k].$$

In Example 3.1, the frequency response of the RC circuit was determined to be

$$H(j\omega) = \frac{1/RC}{j\omega + 1/RC}.$$

The FS coefficients of the square wave were given in Eq. (3.23). Substituting for $H(jk\omega_o)$ with $RC = 0.1$ s and $\omega_o = 2\pi$, and using $T_o/T = 1/4$, gives

$$Y[k] = \frac{10}{j2\pi k + 10} \frac{\sin(k\pi/2)}{k\pi}.$$

The $Y[k]$ go to zero in proportion to $1/k^2$ as k increases, so a reasonably accurate representation of $y(t)$ may be determined with the use of a modest number of terms in the FS. We plot the magnitude and phase spectra of $X[k]$ and $Y[k]$, and we determine $y(t)$ using the approximation

$$y(t) \approx \sum_{k=-100}^{100} Y[k]e^{jk\omega_o t}. \tag{3.30}$$

The magnitude and phase spectra are depicted in Figs. 3.26(a) and (b), respectively, for the range $-25 \le k \le 25$. The magnitude spectrum is very small outside this range and thus is not shown. Comparing $Y[k]$ with $X[k]$ as depicted in Fig. 3.22(a), we see that the circuit attenuates the amplitude of $X[k]$ when $|k| \ge 1$. The degree of attenuation increases as the frequency $k\omega_o$ increases. The circuit also introduces a frequency-dependent phase shift. One period of the waveform $y(t)$ is shown in Fig. 3.26(c). This result is consistent with our intuition from circuit analysis. When the input signal $x(t)$ switches from zero to unity, the charge in the capacitor increases, and the capacitor voltage $y(t)$ exhibits an exponential rise. When the input switches from unity to zero, the capacitor discharges, and the capacitor voltage exhibits an exponential decay. ∎

EXAMPLE 3.16 DC-TO-AC CONVERSION   A simple scheme for converting direct current (dc) to alternating current (ac) is based on applying a periodic switch to a dc power source and filtering out or removing the higher order harmonics in the switched signal. The switch in Fig. 3.27 changes position every $1/120$ second. We consider two cases: (a) The switch is either open or closed; (b) the switch reverses polarity. Figures 3.28(a) and (b) depict the output waveforms for these two cases. Define the conversion efficiency as the ratio of the power in the 60-Hz component of the output waveform $x(t)$ to the available dc power at the input. Evaluate the conversion efficiency for each case.

Solution: The results in Eq. (3.28) indicate that the trigonometric form of the FS for the square wave $x(t)$ shown in Fig. 3.28(a) with $T = 1/60$ second and $\omega_o = 2\pi/T = 120\pi$ rads/s is described by

$$B[0] = \frac{A}{2},$$

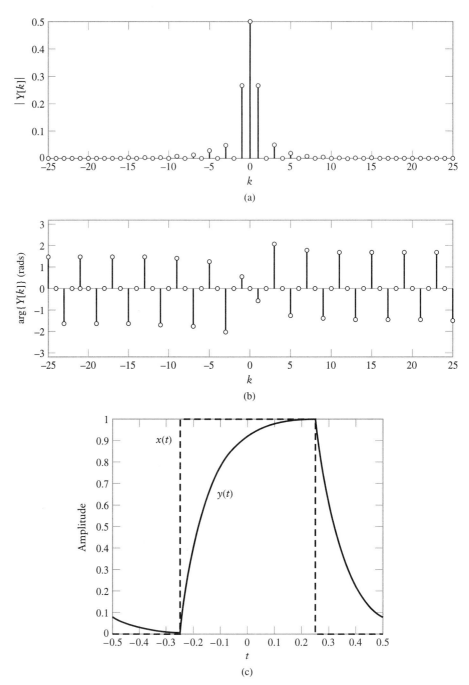

FIGURE 3.26   The FS coefficients $Y[k]$, $-25 \le k \le 25$, for the RC circuit output in response to a square-wave input. (a) Magnitude spectrum. (b) Phase spectrum. (c) One period of the input signal $x(t)$ (dashed line) and output signal $y(t)$ (solid line). The output signal $y(t)$ is computed from the partial-sum approximation given in Eq. (3.30).

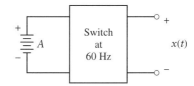

FIGURE 3.27 Switching power supply for DC-to-AC conversion.

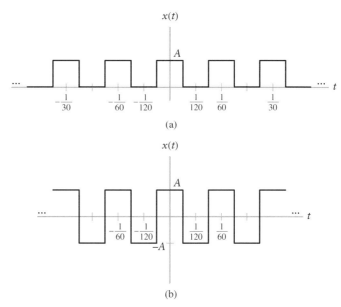

FIGURE 3.28 Switching power supply output waveforms with fundamental frequency $\omega_o = 2\pi/T = 120\pi$. (a) On–off switch. (b) Inverting switch.

and

$$B[k] = \frac{2A \sin(k\pi/2)}{k\pi}, \quad k \neq 0,$$

$$A[k] = 0.$$

The harmonic at 60 Hz in the trigonometric FS representation of $x(t)$ has amplitude given by $B[1]$ and contains power $B[1]^2/2$. The dc input has power A^2, so the conversion efficiency is

$$C_{\text{eff}} = \frac{(B[1])^2/2}{A^2}$$

$$= 2/\pi^2$$

$$\approx 0.20.$$

For the signal $x(t)$ shown in Fig. 3.28(b), the FS coefficients may also be determined with the use of the results in Eq. (3.28) by noting that $x(t)$ is a square wave of amplitude $2A$, but zero average value. Hence, the constant term $B[0]$ is zero, and the trigonometric FS coefficients are given by

$$B[0] = 0,$$

$$B[k] = \frac{4A \sin(k\pi/2)}{k\pi}, \quad k \neq 0,$$

and
$$A[k] = 0.$$

The conversion efficiency for the inverting switch is thus given by

$$C_{\text{eff}} = \frac{(B[1])^2/2}{A^2}$$
$$= 8/\pi^2$$
$$\approx 0.81.$$

The inverting switch offers a factor-of-four improvement in power conversion efficiency. ■

3.6 *Discrete-Time Nonperiodic Signals: The Discrete-Time Fourier Transform*

The DTFT is used to represent a discrete-time nonperiodic signal as a superposition of complex sinusoids. In Section 3.3, we reasoned that the DTFT would involve a continuum of frequencies on the interval $-\pi < \Omega \leq \pi$, where Ω has units of radians. Thus, the DTFT representation of a time-domain signal involves an integral over frequency, namely,

$$x[n] = \frac{1}{2\pi} \int_{-\pi}^{\pi} X(e^{j\Omega}) e^{j\Omega n} \, d\Omega, \tag{3.31}$$

where

$$X(e^{j\Omega}) = \sum_{n=-\infty}^{\infty} x[n] e^{-j\Omega n} \tag{3.32}$$

is the DTFT of the signal $x[n]$. We say that $X(e^{j\Omega})$ and $x[n]$ are a *DTFT pair* and write

$$x[n] \xleftrightarrow{\;\;DTFT\;\;} X(e^{j\Omega}).$$

The transform $X(e^{j\Omega})$ describes the signal $x[n]$ as a function of a sinusoidal frequency Ω and is termed the *frequency-domain representation* of $x[n]$. Equation (3.31) is usually termed the *inverse DTFT*, since it maps the frequency-domain representation back into the time domain.

The DTFT is used primarily to analyze the action of discrete-time systems on discrete-time signals.

The infinite sum in Eq. (3.32) converges if $x[n]$ has finite duration and is finite valued. If $x[n]$ is of infinite duration, then the sum converges only for certain classes of signals. If

$$\sum_{n=-\infty}^{\infty} |x[n]| < \infty$$

(i.e., if $x[n]$ is absolutely summable), then the sum in Eq. (3.32) converges uniformly to a continuous function of Ω. If $x[n]$ is not absolutely summable, but does satisfy

$$\sum_{n=-\infty}^{\infty} |x[n]|^2 < \infty$$

(i.e., if $x[n]$ has finite energy), then it can be shown that the sum in Eq. (3.32) converges in a mean-square error sense, but does not converge pointwise.

Many physical signals encountered in engineering practice satisfy these conditions. However, several common nonperiodic signals, such as the unit step $u[n]$, do not. In some of these cases, we can define a transform pair that behaves like the DTFT by including impulses in the transform. This enables us to use the DTFT as a problem-solving tool, even though, strictly speaking, it does not converge. One example of such usage is given later in the section; others are presented in Chapter 4.

We now consider several examples illustrating the determination of the DTFT for common signals.

EXAMPLE 3.17 DTFT OF AN EXPONENTIAL SEQUENCE   Find the DTFT of the sequence $x[n] = \alpha^n u[n]$.

Solution: Using Eq. (3.32), we have

$$X(e^{j\Omega}) = \sum_{n=-\infty}^{\infty} \alpha^n u[n] e^{-j\Omega n}$$

$$= \sum_{n=0}^{\infty} \alpha^n e^{-j\Omega n}.$$

This sum diverges for $|\alpha| \geq 1$. For $|\alpha| < 1$, we have the convergent geometric series

$$X(e^{j\Omega}) = \sum_{n=0}^{\infty} (\alpha e^{-j\Omega})^n$$

$$= \frac{1}{1 - \alpha e^{-j\Omega}}, \qquad |\alpha| < 1. \tag{3.33}$$

If α is real valued, we may expand the denominator of Eq. (3.33) using Euler's formula to obtain

$$X(e^{j\Omega}) = \frac{1}{1 - \alpha \cos \Omega + j\alpha \sin \Omega}.$$

From this form, we see that the magnitude and phase spectra are given by

$$|X(e^{j\Omega})| = \frac{1}{((1 - \alpha \cos \Omega)^2 + \alpha^2 \sin^2 \Omega)^{1/2}}$$

$$= \frac{1}{(\alpha^2 + 1 - 2\alpha \cos \Omega)^{1/2}}$$

and

$$\arg\{X(e^{j\Omega})\} = -\arctan\left(\frac{\alpha \sin \Omega}{1 - \alpha \cos \Omega}\right),$$

respectively. The magnitude and phase are depicted graphically in Fig. 3.29 for $\alpha = 0.5$ and $\alpha = 0.9$. The magnitude is even and the phase is odd. Note that both are 2π periodic. ∎

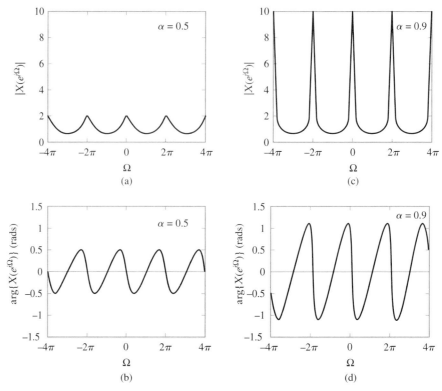

Figure 3.29   The DTFT of an exponential signal $x[n] = (\alpha)^n u[n]$. (a) Magnitude spectrum for $\alpha = 0.5$. (b) Phase spectrum for $\alpha = 0.5$. (c) Magnitude spectrum for $\alpha = 0.9$. (d) Phase spectrum for $\alpha = 0.9$.

As with the other Fourier representations, the *magnitude spectrum* of a signal is the magnitude of $X(e^{j\Omega})$ plotted as a function of Ω. The *phase spectrum* is the phase of $X(e^{j\Omega})$ plotted as a function of Ω.

▶ **Problem 3.11**   Find the DTFT of $x[n] = 2(3)^n u[-n]$.

Answer:

$$X(e^{j\Omega}) = \frac{2}{1 - e^{j\Omega}/3} \qquad ◀$$

Example 3.18 DTFT of a Rectangular Pulse   Let

$$x[n] = \begin{cases} 1, & |n| \leq M \\ 0, & |n| > M \end{cases},$$

as depicted in Fig. 3.30(a). Find the DTFT of $x[n]$.

Solution: We substitute for $x[n]$ in Eq. (3.32) to obtain

$$X(e^{j\Omega}) = \sum_{n=-M}^{M} 1 e^{-j\Omega n}.$$

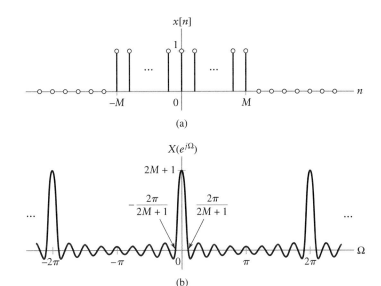

FIGURE 3.30 Example 3.18. (a) Rectangular pulse in the time domain. (b) DTFT in the frequency domain.

Now we perform the change of variable $m = n + M$, obtaining

$$X(e^{j\Omega}) = \sum_{m=0}^{2M} e^{-j\Omega(m-M)}$$

$$= e^{j\Omega M} \sum_{m=0}^{2M} e^{-j\Omega m}$$

$$= \begin{cases} e^{j\Omega M} \dfrac{1 - e^{-j\Omega(2M+1)}}{1 - e^{-j\Omega}}, & \Omega \neq 0, \pm 2\pi, \pm 4\pi, \ldots \\ 2M + 1, & \Omega = 0, \pm 2\pi, \pm 4\pi, \ldots \end{cases}.$$

The expression for $X(e^{j\Omega})$ when $\Omega \neq 0, \pm 2\pi, \pm 4\pi, \ldots$, may be simplified by symmetrizing the powers of the exponential in the numerator and denominator as follows:

$$X(e^{j\Omega}) = e^{j\Omega M} \frac{e^{-j\Omega(2M+1)/2}\left(e^{j\Omega(2M+1)/2} - e^{-j\Omega(2M+1)/2}\right)}{e^{-j\Omega/2}\left(e^{j\Omega/2} - e^{-j\Omega/2}\right)}$$

$$= \frac{e^{j\Omega(2M+1)/2} - e^{-j\Omega(2M+1)/2}}{e^{j\Omega/2} - e^{-j\Omega/2}}.$$

We may now write $X(e^{j\Omega})$ as a ratio of sine functions by dividing the numerator and denominator by $2j$ to obtain

$$X(e^{j\Omega}) = \frac{\sin(\Omega(2M + 1)/2)}{\sin(\Omega/2)}.$$

L'Hôpital's rule gives

$$\lim_{\Omega \to 0, \pm 2\pi, \pm 4\pi, \ldots,} \frac{\sin(\Omega(2M + 1)/2)}{\sin(\Omega/2)} = 2M + 1;$$

hence, rather than write $X(e^{j\Omega})$ as two forms dependent on the value of Ω, we simply write

$$X(e^{j\Omega}) = \frac{\sin(\Omega(2M + 1)/2)}{\sin(\Omega/2)},$$

with the understanding that $X(e^{j\Omega})$ for $\Omega = 0, \pm 2\pi, \pm 4\pi, \ldots$, is obtained as a limit. In this example, $X(e^{j\Omega})$ is purely real. A graph of $X(e^{j\Omega})$ as a function of Ω is given in Fig. 3.30(b). We see that as M increases, the time extent of $x[n]$ increases, while the energy in $X(e^{j\Omega})$ becomes more concentrated near $\Omega = 0$.   ■

EXAMPLE 3.19 INVERSE DTFT OF A RECTANGULAR SPECTRUM   Find the inverse DTFT of

$$X(e^{j\Omega}) = \begin{cases} 1, & |\Omega| < W \\ 0, & W < |\Omega| < \pi \end{cases},$$

which is depicted in Fig. 3.31(a).

Solution: First, note that $X(e^{j\Omega})$ is specified only for $-\pi < \Omega \leq \pi$. This is all that is needed, however, since $X(e^{j\Omega})$ is always 2π-periodic and the inverse DTFT depends solely on the values in the interval $-\pi < \Omega \leq \pi$. Substituting for $X(e^{j\Omega})$ in Eq. (3.31) gives

$$x[n] = \frac{1}{2\pi} \int_{-W}^{W} e^{j\Omega n} \, d\Omega$$

$$= \frac{1}{2\pi n j} e^{j\Omega n} \Big|_{-W}^{W}, \quad n \neq 0$$

$$= \frac{1}{\pi n} \sin(Wn), \quad n \neq 0.$$

For $n = 0$, the integrand is unity and we have $x[0] = W/\pi$. Using L'Hôpital's rule, we easily show that

$$\lim_{n \to 0} \frac{1}{\pi n} \sin(Wn) = \frac{W}{\pi},$$

and thus we usually write

$$x[n] = \frac{1}{\pi n} \sin(Wn)$$

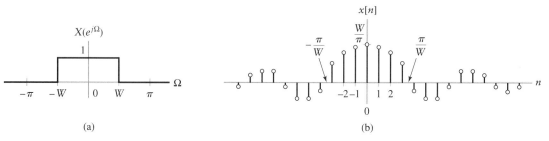

FIGURE 3.31   Example 3.19. (a) One period of rectangular pulse in the frequency domain. (b) Inverse DTFT in the time domain.

as the inverse DTFT of $X(e^{j\Omega})$, with the understanding that the value at $n = 0$ is obtained as the limit. We may also write

$$x[n] = \frac{W}{\pi} \operatorname{sinc}(Wn/\pi),$$

using the sinc function notation defined in Eq. (3.24). A graph depicting $x[n]$ versus time n is given in Fig. 3.31(b). ∎

EXAMPLE 3.20 DTFT OF THE UNIT IMPULSE   Find the DTFT of $x[n] = \delta[n]$.

Solution: For $x[n] = \delta[n]$, we have

$$X(e^{j\Omega}) = \sum_{n=-\infty}^{\infty} \delta[n]e^{-j\Omega n}$$
$$= 1.$$

Hence

$$\delta[n] \xleftrightarrow{\ DTFT\ } 1.$$

This DTFT pair is depicted in Fig. 3.32. ∎

EXAMPLE 3.21 INVERSE DTFT OF A UNIT IMPULSE SPECTRUM   Find the inverse DTFT of $X(e^{j\Omega}) = \delta(\Omega)$, $-\pi < \Omega \leq \pi$.

Solution: By definition, from Eq. (3.31),

$$x[n] = \frac{1}{2\pi} \int_{-\pi}^{\pi} \delta(\Omega)e^{j\Omega n}\, d\Omega.$$

We use the sifting property of the impulse function to obtain $x[n] = 1/(2\pi)$ and thus write

$$\frac{1}{2\pi} \xleftrightarrow{\ DTFT\ } \delta(\Omega), \quad -\pi < \Omega \leq \pi.$$

In this example, we have again defined only one period of $X(e^{j\Omega})$. Alternatively, we can define $X(e^{j\Omega})$ over all Ω by writing it as an infinite sum of delta functions shifted by integer multiples of 2π:

$$X(e^{j\Omega}) = \sum_{k=-\infty}^{\infty} \delta(\Omega - k2\pi).$$

Both definitions are common. This DTFT pair is depicted in Fig. 3.33. ∎

FIGURE 3.32   Example 3.20. (a) Unit impulse in the time domain. (b) DTFT of unit impulse in the frequency domain.

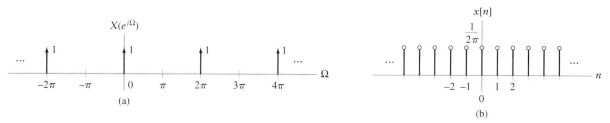

(a)

(b)

FIGURE 3.33   Example 3.21. (a) Unit impulse in the frequency domain. (b) Inverse DTFT in the time domain.

This last example presents an interesting dilemma: The DTFT of $x[n] = 1/(2\pi)$ does not converge, since it is not a square summable signal, yet $x[n]$ is a valid inverse DTFT! This is a direct consequence of allowing impulses in $X(e^{j\Omega})$. We shall treat $x[n]$ and $X(e^{j\Omega})$ as a DTFT pair despite this apparent quandry, because they do satisfy all the properties of a DTFT pair. Indeed, we can greatly expand the class of signals that can be represented by the DTFT if we allow impulses in the transform. Strictly speaking, the DTFTs of these signals do not exist, since the sum in Eq. (3.32) does not converge. However, as in this example, we can identify transform pairs by using the inverse transform of Eq. (3.31) and thus utilize the DTFT as a problem-solving tool. Additional examples illustrating the use of impulses in the DTFT are presented in Chapter 4.

S & S
Solutions

▶ **Problem 3.12**   Find the DTFT of the following time-domain signals:

(a)

$$x[n] = \begin{cases} 2^n, & 0 \le n \le 9 \\ 0, & \text{otherwise} \end{cases}$$

(b)

$$x[n] = a^{|n|}, \qquad |a| < 1$$

(c)

$$x[n] = \delta[6 - 2n] + \delta[6 + 2n]$$

(d) $x[n]$ as depicted in Fig. 3.34.

Answers:

(a)

$$X(e^{j\Omega}) = \frac{1 - 2^{10}e^{-j10\Omega}}{1 - 2e^{-j\Omega}}$$

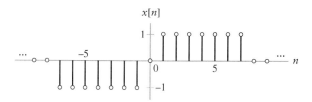

FIGURE 3.34   Signal $x[n]$ for Problem 3.12.

(b)

$$X(e^{j\Omega}) = \frac{1 - a^2}{1 - 2a\cos\Omega + a^2}$$

(c)

$$X(e^{j\Omega}) = 2\cos(3\Omega)$$

(d)

$$X(e^{j\Omega}) = -2j\sin(7\Omega/2)\frac{\sin(4\Omega)}{\sin(\Omega/2)}$$ ◀

▶ **Problem 3.13**   Find the inverse DTFT of the following frequency-domain signals:

(a)

$$X(e^{j\Omega}) = 2\cos(2\Omega)$$

(b)

$$X(e^{j\Omega}) = \begin{cases} e^{-j4\Omega}, & \pi/2 < |\Omega| \le \pi \\ 0, & \text{otherwise} \end{cases}, \quad \text{on } -\pi < \Omega \le \pi$$

(c) $X(e^{j\Omega})$ as depicted in Fig. 3.35.

Answers:

(a)

$$x[n] = \begin{cases} 1, & n = \pm 2 \\ 0, & \text{otherwise} \end{cases}$$

(b)

$$x[n] = \delta[n - 4] - \frac{\sin(\pi(n - 4)/2)}{\pi(n - 4)}$$

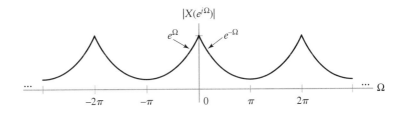

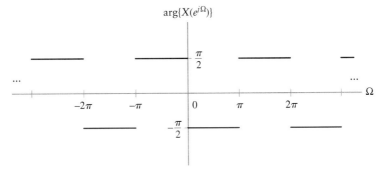

FIGURE 3.35   Frequency-domain signal for Problem 3.13(c).

(c)

$$x[n] = \frac{n(1 - e^{-\pi}(-1)^n)}{2\pi(n^2 + 1)}$$

◀

EXAMPLE 3.22   MOVING-AVERAGE SYSTEMS: FREQUENCY RESPONSE   Consider two different moving-average systems described by the input–output equations

$$y_1[n] = \frac{1}{2}(x[n] + x[n-1])$$

and

$$y_2[n] = \frac{1}{2}(x[n] - x[n-1]).$$

The first system averages successive inputs, while the second forms the difference. The impulse responses are

$$h_1[n] = \frac{1}{2}\delta[n] + \frac{1}{2}\delta[n-1]$$

and

$$h_2[n] = \frac{1}{2}\delta[n] - \frac{1}{2}\delta[n-1].$$

Find the frequency response of each system and plot the magnitude responses.

Solution: The frequency response is the DTFT of the impulse response, so we substitute $h_1[n]$ into Eq. (3.32) to obtain

$$H_1(e^{j\Omega}) = \frac{1}{2} + \frac{1}{2}e^{-j\Omega},$$

which may be rewritten as

$$H_1(e^{j\Omega}) = e^{-j\frac{\Omega}{2}}\frac{e^{j\frac{\Omega}{2}} + e^{-j\frac{\Omega}{2}}}{2}$$

$$= e^{-j\frac{\Omega}{2}}\cos\left(\frac{\Omega}{2}\right).$$

Hence, the magnitude response is expressed as

$$|H_1(e^{j\Omega})| = \left|\cos\left(\frac{\Omega}{2}\right)\right|,$$

and the phase response is expressed as

$$\arg\{H_1(e^{j\Omega})\} = -\frac{\Omega}{2}.$$

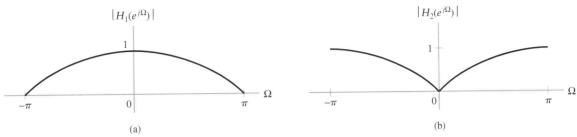

FIGURE 3.36 The magnitude responses of two simple discrete-time systems. (a) A system that averages successive inputs tends to attenuate high frequencies. (b) A system that forms the difference of successive inputs tends to attenuate low frequencies.

Similarly, the frequency response of the second system is given by

$$H_2(e^{j\Omega}) = \frac{1}{2} - \frac{1}{2}e^{-j\Omega}$$

$$= je^{-j\frac{\Omega}{2}}\frac{e^{j\frac{\Omega}{2}} - e^{-j\frac{\Omega}{2}}}{2j}$$

$$= je^{-j\frac{\Omega}{2}}\sin\left(\frac{\Omega}{2}\right).$$

In this case, the magnitude response is expressed as

$$\left|H_2(e^{j\Omega})\right| = \left|\sin\left(\frac{\Omega}{2}\right)\right|,$$

and the phase response is expressed as

$$\arg\{H_2(e^{j\Omega})\} = \begin{cases} \pi/2 - \dfrac{\Omega}{2}, & \Omega > 0 \\[2mm] -\dfrac{\Omega}{2} - \pi/2 & \Omega < 0 \end{cases}.$$

Figures 3.36(a) and (b) depict the magnitude responses of the two systems on the interval $-\pi < \Omega \leq \pi$. Since the system corresponding to $h_1[n]$ averages successive inputs, we expect it to pass low-frequency signals while attenuating high frequencies. This characteristic is reflected in the magnitude response. In contrast, the differencing operation implemented by $h_2[n]$ has the effect of attenuating low frequencies and passing high frequencies, as indicated by its magnitude response. ■

S & S
Solutions

EXAMPLE 3.23 MULTIPATH COMMUNICATION CHANNEL: FREQUENCY RESPONSE The input–output equation introduced in Section 1.10 describing a discrete-time model of a two-path propagation channel is

$$y[n] = x[n] + ax[n-1].$$

In Example 2.12, we identified the impulse response of this system as $h[n] = \delta[n] + a\delta[n-1]$ and determined that the impulse response of the inverse system was $h^{\text{inv}}[n] = (-a)^n u[n]$. The inverse system is stable, provided that $|a| < 1$. Compare the magnitude responses of both systems for $a = 0.5e^{j\pi/3}$ and $a = 0.9e^{j2\pi/3}$.

Solution: Recall that the frequency response of a system is given by the DTFT of the impulse response. The frequency response of the system modeling two-path propagation may be obtained from Eq. (3.32) as

$$H(e^{j\Omega}) = 1 + ae^{-j\Omega}.$$

Using $a = |a|e^{j\arg\{a\}}$, we rewrite the frequency response as

$$H(e^{j\Omega}) = 1 + |a|e^{-j(\Omega - \arg\{a\})}.$$

Now we apply Euler's formula to obtain

$$H(e^{j\Omega}) = 1 + |a|\cos(\Omega - \arg\{a\}) - j|a|\sin(\Omega - \arg\{a\}).$$

Hence, the magnitude response is given by

$$
\begin{aligned}
|H(e^{j\Omega})| &= ((1 + |a|\cos(\Omega - \arg\{a\}))^2 + |a|^2\sin^2(\Omega - \arg\{a\}))^{1/2} \\
&= (1 + |a|^2 + 2|a|\cos(\Omega - \arg\{a\}))^{1/2},
\end{aligned}
\tag{3.34}
$$

where we have used the identity $\cos^2\theta + \sin^2\theta = 1$. The frequency response of the inverse system may be obtained by replacing α with $-a$ in Eq. (3.33). The result is

$$H^{inv}(e^{j\Omega}) = \frac{1}{1 + ae^{-j\Omega}}, \qquad |a| < 1.$$

Expressing a in polar form yields

$$
\begin{aligned}
H^{inv}(e^{j\Omega}) &= \frac{1}{1 + |a|e^{-j(\Omega - \arg\{a\})}} \\
&= \frac{1}{1 + |a|\cos(\Omega - \arg\{a\}) - j|a|\sin(\Omega - \arg\{a\})}.
\end{aligned}
$$

Note that the frequency response of the inverse system is the inverse of the frequency response of the original system. This fact implies that the magnitude response of the inverse system is the inverse of the original system's magnitude response given in Eq. (3.34). Thus, we have

$$|H^{inv}(e^{j\Omega})| = \frac{1}{(1 + |a|^2 + 2|a|\cos(\Omega - \arg\{a\}))^{1/2}}.$$

Figure 3.37 depicts the magnitude response of $H(e^{j\Omega})$ for both $a = 0.5e^{j\pi/3}$ and $a = 0.9e^{j2\pi/3}$ on the interval $-\pi < \Omega \le \pi$. An examination of Eq. (3.34) indicates that the magnitude response attains a maximum of $1 + |a|$ when $\Omega = \arg\{a\}$ and a minimum of $1 - |a|$ when $\Omega = \arg\{a\} - \pi$. These conclusions are supported by Figs. 3.37(a) and (b). Hence, as $|a|$ approaches unity, complex sinusoids with frequencies close to $\arg\{a\} - \pi$ will be greatly attenuated by multipath propagation. The magnitude responses of the corresponding inverse systems are depicted in Fig. 3.38. Frequencies receiving significant attenuation are significantly amplified by the inverse system, so that, after passing through both systems, their amplitude is unchanged. Large amplification is problematic in practice, because any noise introduced by the channel or receiver will also be amplified. Note that if $|a| = 1$, then the multipath model applies zero gain to any sinusoid with frequency $\Omega = \arg\{a\} - \pi$. It is impossible to recover an input sinusoid whose amplitude is zero, and thus the multipath system cannot be inverted when $|a| = 1$. ∎

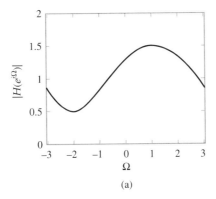 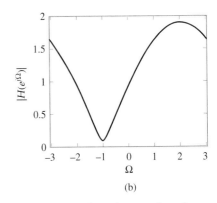

FIGURE 3.37 Magnitude response of the system in Example 3.23 describing multipath propagation. (a) Indirect path coefficient $a = 0.5e^{j\pi/3}$. (b) Indirect path coefficient $a = 0.9e^{j2\pi/3}$.

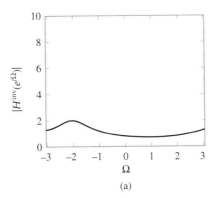 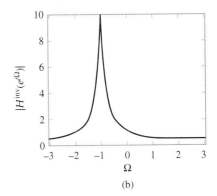

FIGURE 3.38 Magnitude response of the inverse system for multipath propagation in Example 3.23. (a) Indirect path coefficient $a = 0.5e^{j\pi/3}$. (b) Indirect path coefficient $a = 0.9e^{j2\pi/3}$.

3.7 Continuous-Time Nonperiodic Signals: The Fourier Transform

The Fourier transform (FT) is used to represent a continuous-time nonperiodic signal as a superposition of complex sinusoids. Recall from Section 3.3 that the continuous nonperiodic nature of a time signal implies that the superposition of complex sinusoids used in the Fourier representation of the signal involves a continuum of frequencies ranging from $-\infty$ to ∞. Thus, the FT representation of a continuous-time signal involves an integral over the entire frequency interval; that is,

$$x(t) = \frac{1}{2\pi} \int_{-\infty}^{\infty} X(j\omega)e^{j\omega t}\, d\omega, \qquad (3.35)$$

where

$$X(j\omega) = \int_{-\infty}^{\infty} x(t)e^{-j\omega t}\, dt \qquad (3.36)$$

is the FT of the signal $x(t)$. Note that in Eq. (3.35) we have expressed $x(t)$ as a weighted superposition of sinusoids having frequencies ranging from $-\infty$ to ∞. The superposition is an integral, and the weight on each sinusoid is $(1/(2\pi))X(j\omega)$. We say that $x(t)$ and $X(j\omega)$ are an *FT pair* and write

$$x(t) \xleftrightarrow{\quad FT \quad} X(j\omega).$$

The transform $X(j\omega)$ describes the signal $x(t)$ as a function of frequency ω and is termed the *frequency-domain representation* of $x(t)$. Equation (3.35) is termed the *inverse FT*, since it maps the frequency-domain representation $X(j\omega)$ back into the time domain.

The FT is used to analyze the characteristics of continuous-time systems and the interaction between continuous-time signals and systems. The FT is also used to analyze interactions between discrete- and continuous-time signals, such as occur in sampling. These topics are studied at length in Chapter 4.

The integrals in Eqs. (3.35) and (3.36) may not converge for all functions $x(t)$ and $X(j\omega)$. An analysis of convergence is beyond the scope of this book, so we simply state several convergence conditions on the time-domain signal $x(t)$. If we define

$$\hat{x}(t) = \frac{1}{2\pi} \int_{-\infty}^{\infty} X(j\omega)e^{j\omega t}\, d\omega,$$

where $X(j\omega)$ is expressed in terms of $x(t)$ by Eq. (3.36), it can be shown that the squared error between $x(t)$ and $\hat{x}(t)$, namely, the error energy, given by

$$\int_{-\infty}^{\infty} |x(t) - \hat{x}(t)|^2\, dt,$$

is zero if $x(t)$ is square integrable—that is, if

$$\int_{-\infty}^{\infty} |x(t)|^2\, dt < \infty.$$

Zero squared error does not imply pointwise convergence [i.e., $x(t) = \hat{x}(t)$ at all values of t]; it does, however, imply that there is zero energy in the difference of the terms.

Pointwise convergence is guaranteed at all values of t except those corresponding to discontinuities if $x(t)$ satisfies the Dirichlet conditions for nonperiodic signals:

▸ $x(t)$ is absolutely integrable:

$$\int_{-\infty}^{\infty} |x(t)|\, dt < \infty.$$

▸ $x(t)$ has a finite number of maxima, minima, and discontinuities in any finite interval.
▸ The size of each discontinuity is finite.

Almost all physical signals encountered in engineering practice satisfy the second and third conditions, but many idealized signals, such as the unit step, are neither absolutely nor square integrable. In some of these cases, we define a transform pair that satisfies FT properties through the use of impulses. In this way, we may still use the FT as a problem-solving tool, even though, in a strict sense, the FT does not converge for such signals.

The next five examples illustrate the determination of the FT and inverse FT for several common signals.

EXAMPLE 3.24 FT OF A REAL DECAYING EXPONENTIAL   Find the FT of $x(t) = e^{-at}u(t)$, shown in Fig. 3.39(a).

Solution: The FT does not converge for $a \leq 0$, since $x(t)$ is not absolutely integrable; that is,

$$\int_0^\infty e^{-at}\, dt = \infty, \qquad a \leq 0.$$

For $a > 0$, we have

$$\begin{aligned}
X(j\omega) &= \int_{-\infty}^{\infty} e^{-at} u(t) e^{-j\omega t}\, dt \\
&= \int_0^\infty e^{-(a+j\omega)t}\, dt \\
&= -\frac{1}{a + j\omega} e^{-(a+j\omega)t}\bigg|_0^\infty \\
&= \frac{1}{a + j\omega}.
\end{aligned}$$

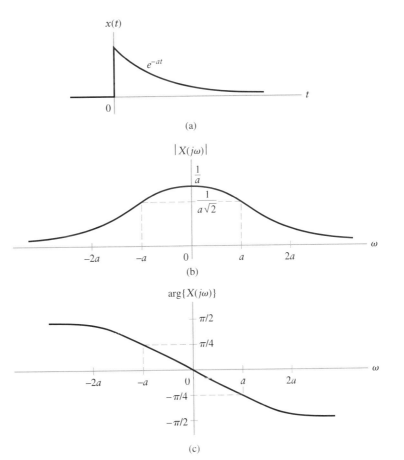

(a)

(b)

(c)

FIGURE 3.39   Example 3.24. (a) Real time-domain exponential signal. (b) Magnitude spectrum. (c) Phase spectrum.

Converting to polar form, we find that the magnitude and phase of $X(j\omega)$ are respectively given by

$$|X(j\omega)| = \frac{1}{(a^2 + \omega^2)^{\frac{1}{2}}}$$

and

$$\arg\{X(j\omega)\} = -\arctan(\omega/a),$$

as depicted in Figs. 3.39(b) and (c), respectively.   ■

As before, the magnitude of $X(j\omega)$ plotted against ω is termed the *magnitude spectrum* of the signal $x(t)$, and the phase of $X(j\omega)$ plotted as a function of ω is termed the *phase spectrum* of $x(t)$.

EXAMPLE 3.25 FT OF A RECTANGULAR PULSE   Consider the rectangular pulse depicted in Fig. 3.40(a) and defined as

$$x(t) = \begin{cases} 1, & -T_o < t < T_o \\ 0, & |t| > T_o \end{cases}.$$

Find the FT of $x(t)$.

Solution: The rectangular pulse $x(t)$ is absolutely integrable, provided that $T_o < \infty$. We thus have

$$
\begin{aligned}
X(j\omega) &= \int_{-\infty}^{\infty} x(t)e^{-j\omega t}\, dt \\
&= \int_{-T_o}^{T_o} e^{-j\omega t}\, dt \\
&= -\frac{1}{j\omega} e^{-j\omega t}\Big|_{-T_o}^{T_o} \\
&= \frac{2}{\omega}\sin(\omega T_o), \qquad \omega \neq 0.
\end{aligned}
$$

For $\omega = 0$, the integral simplifies to $2T_o$. L'Hôpital's rule straightforwardly shows that

$$\lim_{\omega \to 0} \frac{2}{\omega}\sin(\omega T_o) = 2T_o.$$

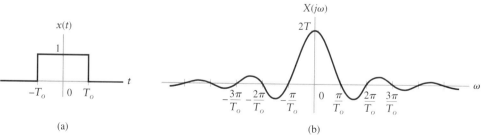

(a)   (b)

FIGURE 3.40   Example 3.25. (a) Rectangular pulse in the time domain. (b) FT in the frequency domain.

Thus, we usually write

$$X(j\omega) = \frac{2}{\omega}\sin(\omega T_o),$$

with the understanding that the value at $\omega = 0$ is obtained by evaluating a limit. In this case, $X(j\omega)$ is real. $X(j\omega)$ is depicted in Fig. 3.40(b). The magnitude spectrum is

$$|X(j\omega)| = 2\left|\frac{\sin(\omega T_o)}{\omega}\right|,$$

and the phase spectrum is

$$\arg\{X(j\omega)\} = \begin{cases} 0, & \sin(\omega T_o)/\omega > 0 \\ \pi, & \sin(\omega T_o)/\omega < 0 \end{cases}.$$

Using sinc function notation, we may write $X(j\omega)$ as

$$X(j\omega) = 2T_o\,\text{sinc}(\omega T_o/\pi). \qquad\blacksquare$$

The previous example illustrates a very important property of the Fourier transform. Consider the effect of changing T_o. As T_o increases, the nonzero time extent of $x(t)$ increases, while $X(j\omega)$ becomes more concentrated about the frequency origin. Conversely, as T_o decreases, the nonzero duration of $x(t)$ decreases, while $X(j\omega)$ becomes less concentrated about the frequency origin. In a certain sense, the duration of $x(t)$ is inversely related to the width or "bandwidth" of $X(j\omega)$. As a general principle, we shall see that signals which are concentrated in one domain are spread out in the other domain.

S & S
Solutions

▶ **Problem 3.14** Find the FT of the following signals:

(a) $x(t) = e^{2t}u(-t)$
(b) $x(t) = e^{-|t|}$
(c) $x(t) = e^{-2t}u(t-1)$
(d) $x(t)$ as shown in Fig. 3.41(a). (*Hint:* Use integration by parts.)
(e) $x(t)$ as shown in Fig. 3.41(b).

Answers:

(a) $X(j\omega) = -1/(j\omega - 2)$
(b) $X(j\omega) = 2/(1 + \omega^2)$
(c) $e^{-(j\omega+2)}/(j\omega + 2)$
(d) $X(j\omega) = j(2/\omega)\cos\omega - j(2/\omega^2)\sin\omega$
(e) $X(j\omega) = 2j(1 - \cos(2\omega))/\omega$

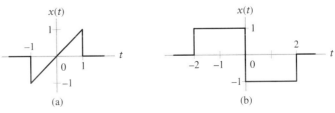

FIGURE 3.41 Time-domain signals for Problem 3.14. (a) Part (d). (b) Part (e).

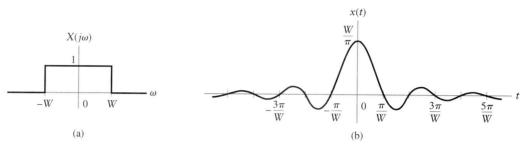

FIGURE 3.42   Example 3.26. (a) Rectangular spectrum in the frequency domain. (b) Inverse FT in the time domain.

EXAMPLE 3.26 INVERSE FT OF A RECTANGULAR SPECTRUM   Find the inverse FT of the rectangular spectrum depicted in Fig. 3.42(a) and given by

$$X(j\omega) = \begin{cases} 1, & -W < \omega < W \\ 0, & |\omega| > W \end{cases}.$$

Solution: Using Eq. (3.35) for the inverse FT gives

$$x(t) = \frac{1}{2\pi} \int_{-W}^{W} e^{j\omega t}\, d\omega$$

$$= \frac{1}{2j\pi t} e^{j\omega t} \bigg|_{-W}^{W}$$

$$= \frac{1}{\pi t} \sin(Wt), \qquad t \neq 0.$$

When $t = 0$, the integral simplifies to W/π. Since

$$\lim_{t \to 0} \frac{1}{\pi t} \sin(Wt) = W/\pi,$$

we usually write

$$x(t) = \frac{1}{\pi t} \sin(Wt),$$

or

$$x(t) = \frac{W}{\pi} \operatorname{sinc}\left(\frac{Wt}{\pi}\right),$$

with the understanding that the value at $t = 0$ is obtained as a limit. Figure 3.42(b) depicts $x(t)$.   ∎

Note again the inverse relationship between the concentration of the signal about the origin in the time domain and its concentration in the frequency domain: As W increases, the frequency-domain representation becomes less concentrated about $\omega = 0$, while the time-domain representation becomes more concentrated about $t = 0$. Another interesting observation can be made by considering the previous example and the one before it. In Example 3.25, a rectangular time-domain pulse is transformed to a sinc function in frequency. In Example 3.26, a sinc function in time is transformed to a rectangular pulse in frequency. This "duality" is a consequence of the similarity between the forward transform in Eq. (3.36) and inverse transform in Eq. (3.35) and is examined further in Section 3.18. The next two examples also exhibit the property of duality.

EXAMPLE 3.27 FT OF THE UNIT IMPULSE Find the FT of $x(t) = \delta(t)$.

Solution: This $x(t)$ does not satisfy the Dirichlet conditions, since the discontinuity at the origin is infinite. We attempt to proceed in spite of this potential problem, using Eq. (3.36) to write

$$X(j\omega) = \int_{-\infty}^{\infty} \delta(t)e^{-j\omega t}\, dt$$
$$= 1.$$

The evaluation to unity follows from the sifting property of the impulse function. Hence,

$$\delta(t) \xrightarrow{\;\;FT\;\;} 1,$$

and the impulse contains unity contributions from complex sinusoids of all frequencies, from $\omega = -\infty$ to $\omega = \infty$. ∎

EXAMPLE 3.28 INVERSE FT OF AN IMPULSE SPECTRUM Find the inverse FT of $X(j\omega) = 2\pi\delta(\omega)$.

Solution: Here again, we may expect convergence irregularities, since $X(j\omega)$ has an infinite discontinuity at the origin. Nevertheless, we may proceed by using Eq. (3.35) to write

$$x(t) = \frac{1}{2\pi} \int_{-\infty}^{\infty} 2\pi\delta(\omega)e^{j\omega t}\, d\omega$$
$$= 1.$$

Hence, we identify

$$1 \xleftarrow{\;\;FT\;\;} 2\pi\delta(\omega)$$

as an FT pair. This implies that the frequency content of a dc signal is concentrated entirely at $\omega = 0$, which is intuitively satisfying. ∎

Note the similarity between the preceding two examples and the DTFT Examples 3.20 and 3.21. In both discrete and continuous time, an impulse in time transforms to a constant frequency spectrum, while an impulse frequency spectrum has a constant as its inverse transform in the time domain. While the FT cannot be guaranteed to converge in those examples, the transform pairs do satisfy the properties of an FT pair and are thus useful for analysis. In both cases, the transform pairs are consequences of the properties of the impulse function. By permitting the use of impulses, we greatly expand the class of signals that are representable by the FT and thus enhance the power of the FT as a problem-solving tool. In Chapter 4, we shall use impulses to obtain FT representations of both periodic and discrete-time signals.

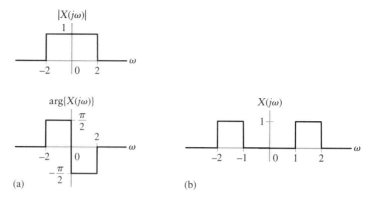

FIGURE 3.43   Frequency-domain signals for Problem 3.15. (a) Part (d). (b) Part (e).

S & S
Solutions

▶ **Problem 3.15**   Find the inverse FT of the following spectra:

(a)

$$X(j\omega) = \begin{cases} 2\cos\omega, & |\omega| < \pi \\ 0, & |\omega| > \pi \end{cases}$$

(b) $X(j\omega) = 3\delta(\omega - 4)$

(c) $X(j\omega) = \pi e^{-|\omega|}$

(d) $X(j\omega)$ as depicted in Fig. 3.43(a).

(e) $X(j\omega)$ as depicted in Fig. 3.43(b).

Answers:

(a)

$$x(t) = \frac{\sin(\pi(t + 1))}{\pi(t + 1)} + \frac{\sin(\pi(t - 1))}{\pi(t - 1)}$$

(b) $x(t) = (3/2\pi)e^{j4t}$

(c) $x(t) = 1/(1 + t^2)$

(d) $x(t) = (1 - \cos(2t))/(\pi t)$

(e) $x(t) = (\sin(2t) - \sin t)/(\pi t)$ ◀

EXAMPLE 3.29 CHARACTERISTICS OF DIGITAL COMMUNICATION SIGNALS   In a simple digital communication system, one signal or "symbol" is transmitted for each "1" in the binary representation of the message, while a different signal or symbol is transmitted for each "0." One common scheme, *binary phase-shift keying* (BPSK), assumes that the signal representing "0" is the negative of the signal representing "1." Figure 3.44 depicts two candidate signals for this approach: a rectangular pulse defined as

$$x_r(t) = \begin{cases} A_r, & |t| < T_o/2 \\ 0, & |t| > T_o/2 \end{cases}$$

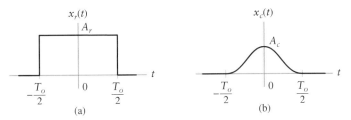

FIGURE 3.44 Pulse shapes used in BPSK communications. (a) Rectangular pulse. (b) Raised cosine pulse.

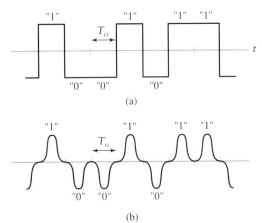

FIGURE 3.45 BPSK signals constructed by using (a) rectangular pulse shapes and (b) raised-cosine pulse shapes.

and a raised-cosine pulse defined as

$$x_c(t) = \begin{cases} (A_c/2)(1 + \cos(2\pi t/T_o)), & |t| < T_o/2 \\ 0, & |t| > T_o/2 \end{cases}.$$

The transmitted BPSK signals for communicating a sequence of bits using each pulse shape are illustrated in Figure 3.45. Note that each pulse is T_o seconds long, so this scheme has a transmission rate of $1/T_o$ bits per second. Each user's signal is transmitted within an assigned frequency band, as described in Chapter 5. In order to prevent interference with users of other frequency bands, governmental agencies place limits on the energy of a signal that any user transmits into adjacent frequency bands. Suppose the frequency band assigned to each user is 20 kHz wide. Then, to prevent interference with adjacent channels, we assume that the peak value of the magnitude spectrum of the transmitted signal outside the 20-kHz band is required to be −30 dB below the peak in-band magnitude spectrum. Choose the constants A_r and A_c so that both BPSK signals have unit power. Use the FT to determine the maximum number of bits per second that can be transmitted when the rectangular and raised-cosine pulse shapes are utilized.

Solution: Although the BPSK signals are not periodic, their magnitude squared is T_o periodic. Thus, their respective powers are calculated as

$$P_r = \frac{1}{T_o} \int_{-T_o/2}^{T_o/2} A_r^2 \, dt$$

$$= A_r^2$$

and

$$P_c = \frac{1}{T_o} \int_{-T_o/2}^{T_o/2} (A_c^2/4)(1 + \cos(2\pi t/T_o))^2 \, dt$$

$$= \frac{A_c^2}{4T_o} \int_{-T_o/2}^{T_o/2} [1 + 2\cos(2\pi t/T_o) + 1/2 + 1/2\cos(4\pi t/T_o)] \, dt$$

$$= \frac{3A_c^2}{8}.$$

Hence, unity transmission power is obtained by choosing $A_r = 1$ and $A_c = \sqrt{8/3}$.

Using the result of Example 3.25, we find that the FT of the rectangular pulse $x_r(t)$ is given by

$$X_r(j\omega) = 2\frac{\sin(\omega T_o/2)}{\omega}.$$

In this example, it is convenient to express frequency in units of hertz rather than radians per second. The conversion is explicitly indicated by substituting $\omega = 2\pi f$ and replacing $X_r(j\omega)$ with $X_r'(jf)$, yielding

$$X_r'(jf) = \frac{\sin(\pi f T_o)}{\pi f}.$$

The normalized magnitude spectrum of this signal in dB is given by $20\log_{10}\{|X_r'(jf)|/T_o\}$ and is shown in Fig. 3.46. The normalization by T_o removed the dependence of the magnitude on T_o. We find that the 10th sidelobe is the first one whose peak does not exceed -30 dB.

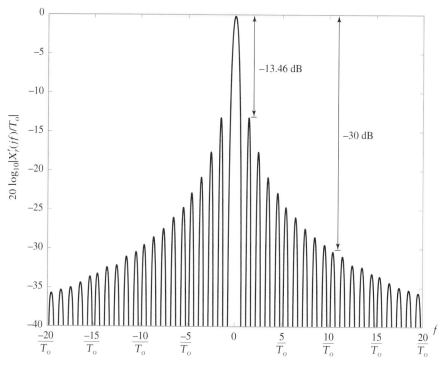

FIGURE 3.46   Spectrum of rectangular pulse in dB, normalized by T_o.

This implies that we must choose T_o so that the 10th zero crossing is at 10 kHz in order to satisfy the constraint that the peak value of the magnitude spectrum of the transmitted signal outside the 20-kHz band allotted to this user be less than -30 dB. The kth zero crossing occurs when $f = k/T_o$, so we require that $10,000 = 10/T_o$, or $T_o = 10^{-3}$ s, to satisfy the specifications, using the rectangular pulse. This implies a data transmission rate of 1000 bits per second.

The FT of the raised-cosine pulse $x_c(t)$ is given by

$$X_c(j\omega) = \frac{1}{2}\sqrt{\frac{8}{3}} \int_{-T_o/2}^{T_o/2} (1 + \cos(2\pi t/T_o))e^{-j\omega t}\,dt.$$

Using Euler's formula to expand the cosine gives

$$X_c(j\omega) = \sqrt{\frac{2}{3}} \int_{-T_o/2}^{T_o/2} e^{-j\omega t}\,dt + \frac{1}{2}\sqrt{\frac{2}{3}} \int_{-T_o/2}^{T_o/2} e^{-j(\omega - 2\pi/T_o)t}\,dt + \frac{1}{2}\sqrt{\frac{2}{3}} \int_{-T_o/2}^{T_o/2} e^{-j(\omega + 2\pi/T_o)t}\,dt.$$

Each of the three integrals is of the form

$$\int_{-T_o/2}^{T_o/2} e^{-j\gamma t}\,dt,$$

which may be evaluated by using the steps described in Example 3.25. The result is

$$2\,\frac{\sin(\gamma T_o/2)}{\gamma}.$$

Substituting the appropriate value of γ for each integral gives

$$X_c(j\omega) = 2\sqrt{\frac{2}{3}}\frac{\sin(\omega T_o/2)}{\omega} + \sqrt{\frac{2}{3}}\frac{\sin((\omega - 2\pi/T_o)T_o/2)}{\omega - 2\pi/T_o} + \sqrt{\frac{2}{3}}\frac{\sin((\omega + 2\pi/T_o)T_o/2)}{\omega + 2\pi/T_o},$$

which, using frequency f in hertz, may be expressed as

$$X_c'(jf) = \sqrt{\frac{2}{3}}\frac{\sin(\pi f T_o)}{\pi f} + 0.5\sqrt{\frac{2}{3}}\frac{\sin(\pi(f - 1/T_o)T_o)}{\pi(f - 1/T_o)} + 0.5\sqrt{\frac{2}{3}}\frac{\sin(\pi(f + 1/T_o)T_o)}{\pi(f + 1/T_o)}.$$

The first term in this expression corresponds to the spectrum of the rectangular pulse. The second and third terms have the exact same shape, but are shifted in frequency by $\pm 1/T_o$. Each of these three terms is depicted on the same graph in Fig. 3.47 for $T_o = 1$. Note that the second and third terms share zero crossings with the first term and have the opposite sign in the sidelobe region of the first term. Thus, the sum of these three terms has lower sidelobes than that of the spectrum of the rectangular pulse. The normalized magnitude spectrum in dB, $20\log\{|X_c'(jf)|/T_o\}$, is shown in Fig. 3.48. Here again, the normalization by T_o removed the dependence of the magnitude on T_o. In this case, the peak of the first sidelobe is below -30 dB, so we may satisfy the adjacent channel interference specifications by choosing the mainlobe to be 20 kHz wide, which implies that $10,000 = 2/T_o$, or $T_o = 2 \times 10^{-4}$ s. The corresponding data transmission rate is 5000 bits per second. The use of the raised-cosine pulse shape increases the data transmission rate by a factor of five relative to the rectangular pulse shape in this application. ∎

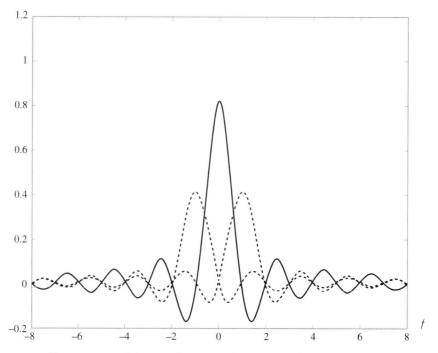

FIGURE 3.47   The spectrum of the raised-cosine pulse consists of a sum of three frequency-shifted and weighted sinc functions.

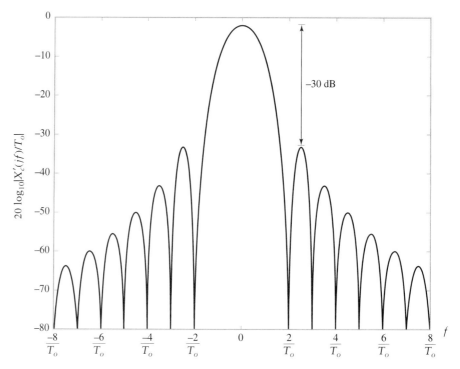

FIGURE 3.48   Spectrum of the raised-cosine pulse in dB, normalized by T_o.

3.8 *Properties of Fourier Representations*

The four Fourier representations discussed in this chapter are summarized in Table 3.2. This table provides a convenient reference for both the definition of each transform and the identification of the class of signals to which each applies. All four representations are based on complex sinusoids; consequently, they share a set of properties that follow from the characteristics of such sinusoids. The remainder of the chapter examines the properties of the four Fourier representations. In many cases, we derive a property of one representation and simply state it for the others. The reader is asked to prove some of these properties in the problem section of this chapter. A comprehensive table of all properties is given in Appendix C.

The borders of Table 3.2 summarize the periodicity properties of the four representations by denoting time-domain characteristics on the top and left sides, with the corresponding frequency-domain characteristics on the bottom and right sides. For example, the FS is continuous and periodic in time, but discrete and nonperiodic in the frequency index k.

Continuous- or discrete-time periodic signals have a series representation in which the signal is represented as a weighted sum of complex sinusoids having the same period as the signal. A discrete set of frequencies is involved in the series; hence, the frequency-domain representation involves a discrete set of weights or coefficients. In contrast, for nonperiodic signals, both continuous- and discrete-time Fourier transform representations involve weighted integrals of complex sinusoids over a continuum of frequencies. Accordingly, the frequency-domain representation for nonperiodic signals is a continuous function

TABLE 3.2   *The Four Fourier Representations.*

Time Domain	Periodic (t, n)	Non periodic (t, n)	
Continuous (t)	Fourier Series $x(t) = \sum_{k=-\infty}^{\infty} X[k]e^{jk\omega_o t}$ $X[k] = \frac{1}{T}\int_0^T x(t)e^{-jk\omega_o t}\,dt$ $x(t)$ has period T $\omega_o = \frac{2\pi}{T}$	Fourier Transform $x(t) = \frac{1}{2\pi}\int_{-\infty}^{\infty} X(j\omega)e^{j\omega t}\,d\omega$ $X(j\omega) = \int_{-\infty}^{\infty} x(t)e^{-j\omega t}\,dt$	Nonperiodic (k, ω)
Discrete (n)	Discrete-Time Fourier Series $x[n] = \sum_{k=0}^{N-1} X[k]e^{jk\Omega_o n}$ $X[k] = \frac{1}{N}\sum_{n=0}^{N-1} x[n]e^{-jk\Omega_o n}$ $x[n]$ and $X[k]$ have period N $\Omega_o = \frac{2\pi}{N}$	Discrete-Time Fourier Transform $x[n] = \frac{1}{2\pi}\int_{-\pi}^{\pi} X(e^{j\Omega})e^{j\Omega n}\,d\Omega$ $X(e^{j\Omega}) = \sum_{n=-\infty}^{\infty} x[n]e^{-j\Omega n}$ $X(e^{j\Omega})$ has period 2π	Periodic (k, Ω)
	Discrete (k)	Continuous (ω, Ω)	Frequency Domain

> **TABLE 3.3** *Periodicity Properties of Fourier Representations.*
>
> | *Time-Domain Property* | *Frequency-Domain Property* |
> |---|---|
> | continuous | nonperiodic |
> | discrete | periodic |
> | periodic | discrete |
> | nonperiodic | continuous |

of frequency. Signals that are periodic in time have discrete frequency-domain representations, while nonperiodic time signals have continuous frequency-domain representations. This correspondence is indicated on the top and bottom of Table 3.2.

   We also observe that the Fourier representations of discrete-time signals, either the DTFS or the DTFT, are periodic functions of frequency. This is because the discrete-time complex sinusoids used to represent discrete-time signals are 2π-periodic functions of frequency. That is, discrete-time sinusoids whose frequencies differ by integer multiples of 2π are identical. In contrast, Fourier representations of continuous-time signals involve superpositions of continuous-time sinusoids. Continuous-time sinusoids with distinct frequencies are always distinct; thus, the frequency-domain representations of continuous-time signals are nonperiodic. Summarizing, discrete-time signals have periodic frequency-domain representations, while continuous-time signals have nonperiodic frequency-domain representations. This correspondence is indicated on the left and right sides of Table 3.2.

   In general, representations that are continuous in one domain are nonperiodic in the other domain. Conversely, representations that are discrete in one domain are periodic in the other domain, as indicated in Table 3.3.

3.9 *Linearity and Symmetry Properties*

It is a straightforward excercise to show that all four Fourier representations involve linear operations. Specifically, they satisfy the linearity property:

$$
\begin{array}{lcl}
z(t) = ax(t) + by(t) & \xleftarrow{\;\;FT\;\;} & Z(j\omega) = aX(j\omega) + bY(j\omega) \\
z(t) = ax(t) + by(t) & \xleftarrow{\;\;FS;\,\omega_o\;\;} & Z[k] = aX[k] + bY[k] \\
z[n] = ax[n] + by[n] & \xleftarrow{\;\;DTFT\;\;} & Z(e^{j\Omega}) = aX(e^{j\Omega}) + bY(e^{j\Omega}) \\
z[n] = ax[n] + by[n] & \xleftarrow{\;\;DTFS;\,\Omega_o\;\;} & Z[k] = aX[k] + bY[k]
\end{array}
$$

In these relationships, we assume that the uppercase symbols denote the Fourier representation of the corresponding lowercase symbols. Furthermore, in the cases of the FS and DTFS, the signals being summed are assumed to have the same fundamental period. The linearity property is used to find Fourier representations of signals that are constructed as sums of signals whose representations are already known, as illustrated in the next example.

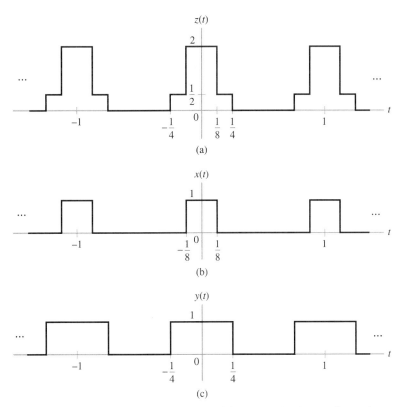

FIGURE 3.49 Representation of the periodic signal $z(t)$ as a weighted sum of periodic square waves: $z(t) = (3/2)x(t) + (1/2)y(t)$. (a) $z(t)$. (b) $x(t)$. (c) $y(t)$.

EXAMPLE 3.30 LINEARITY IN THE FS   Suppose $z(t)$ is the periodic signal depicted in Fig. 3.49(a). Use the linearity property and the results of Example 3.13 to determine the FS coefficients $Z[k]$.

Solution: Write $z(t)$ as a sum of signals; that is,

$$z(t) = \frac{3}{2}x(t) + \frac{1}{2}y(t),$$

where $x(t)$ and $y(t)$ are depicted in Figs. 3.49(b) and (c), respectively. From Example 3.13, we have

$$x(t) \xleftrightarrow{\;FS; 2\pi\;} X[k] = (1/(k\pi))\sin(k\pi/4)$$
$$y(t) \xleftrightarrow{\;FS; 2\pi\;} Y[k] = (1/(k\pi))\sin(k\pi/2)$$

The linearity property implies that

$$z(t) \xleftrightarrow{\;FS; 2\pi\;} Z[k] = \frac{3}{2k\pi}\sin(k\pi/4) + \frac{1}{2k\pi}\sin(k\pi/2)$$

■

▶ **Problem 3.16** Use the linearity property and Tables C.1–4 in Appendix C to determine the Fourier representations of the following signals:

(a) $x(t) = 2e^{-t}u(t) - 3e^{-2t}u(t)$

(b) $x[n] = 4(1/2)^n u[n] - \frac{1}{\pi n}\sin(\pi n/4)$

(c) $x(t) = 2\cos(\pi t) + 3\sin(3\pi t)$

Answers:

(a) $X(j\omega) = 2/(j\omega + 1) - 3/(j\omega + 2)$

(b)

$$X(e^{j\Omega}) = \begin{cases} \dfrac{3 + (1/2)e^{-j\Omega}}{1 - (1/2)e^{-j\Omega}} & |\Omega| \le \pi/4 \\[2mm] \dfrac{4}{1 - (1/2)e^{-j\Omega}} & \pi/4 < |\Omega| \le \pi \end{cases}$$

(c) $\omega_o = \pi, X[k] = \delta[k - 1] + \delta[k + 1] + 3/(2j)\delta[k - 3] - 3/(2j)\delta[k + 3]$ ◀

■ **3.9.1 SYMMETRY PROPERTIES: REAL AND IMAGINARY SIGNALS**

We use the FT to develop the symmetry properties. Results for the other three Fourier representations may be obtained in an analogous manner and are simply stated. First, consider

$$X^*(j\omega) = \left[\int_{-\infty}^{\infty} x(t)e^{-j\omega t}\,dt\right]^*$$
$$= \int_{-\infty}^{\infty} x^*(t)e^{j\omega t}\,dt. \tag{3.37}$$

Now, suppose $x(t)$ is real. Then $x(t) = x^*(t)$. Substitute $x(t)$ for $x^*(t)$ in Eq. (3.37) to obtain

$$X^*(j\omega) = \int_{-\infty}^{\infty} x(t)e^{-j(-\omega)t}\,dt,$$

which implies that

$$\boxed{X^*(j\omega) = X(-j\omega).} \tag{3.38}$$

Thus, $X(j\omega)$ is complex-conjugate symmetric, or $X^*(j\omega) = X(-j\omega)$. Taking the real and imaginary parts of this expression gives $\text{Re}\{X(j\omega)\} = \text{Re}\{X(-j\omega)\}$ and $\text{Im}\{X(j\omega)\} = -\text{Im}\{X(-j\omega)\}$. In words, if $x(t)$ is real valued, then the real part of the transform is an even function of frequency, while the imaginary part is an odd function of frequency. This also implies that the magnitude spectrum is an even function while the phase spectrum is an odd function. The symmetry conditions in all four Fourier representations of real-valued signals are indicated in Table 3.4. In each case, the real part of the Fourier representation has even symmetry and the imaginary part has odd symmetry. Hence, the magnitude spectrum has even symmetry and the phase spectrum has odd symmetry. Note that the conjugate symmetry property for the DTFS may also be written as $X^*[k] = X[N - k]$, because the DTFS coefficients are N periodic, and thus $X[-k] = X[N - k]$.

TABLE 3.4   *Symmetry Properties for Fourier Representation of Real- and Imaginary-Valued Time Signals.*

Representation	Real-Valued Time Signals	Imaginary-Valued Time Signals
FT	$X^*(j\omega) = X(-j\omega)$	$X^*(j\omega) = -X(-j\omega)$
FS	$X^*[k] = X[-k]$	$X^*[k] = -X[-k]$
DTFT	$X^*(e^{j\Omega}) = X(e^{-j\Omega})$	$X^*(e^{j\Omega}) = -X(e^{-j\Omega})$
DTFS	$X^*[k] = X[-k]$	$X^*[k] = -X[-k]$

The complex-conjugate symmetry in the FT leads to a simple characterization of the output of an LTI system with a real-valued impulse response when the input is a real-valued sinusoid. Let the input signal be

$$x(t) = A\cos(\omega t - \phi)$$

and the real-valued impulse response be denoted by $h(t)$. Then the frequency response $H(j\omega)$ is the FT of $h(t)$ and is thus conjugate symmetric. Using Euler's formula to expand $x(t)$ gives

$$x(t) = (A/2)e^{j(\omega t - \phi)} + (A/2)e^{-j(\omega t - \phi)}.$$

Appling Eq. (3.2) and linearity, we may write

$$y(t) = |H(j\omega)|(A/2)e^{j(\omega t - \phi + \arg\{H(j\omega)\})} + |H(-j\omega)|(A/2)e^{-j(\omega t - \phi - \arg\{H(-j\omega)\})}.$$

Exploiting the symmetry conditions $|H(j\omega)| = |H(-j\omega)|$ and $\arg\{H(j\omega)\} = -\arg\{H(-j\omega)\}$ and simplifying yields

$$y(t) = |H(j\omega)|A\cos(\omega t - \phi + \arg\{H(j\omega)\}).$$

Thus, the system modifies the amplitude of the input sinusoid by $|H(j\omega)|$ and the phase by $\arg\{H(j\omega)\}$. This modification, shown in Fig. 3.50, indicates that the frequency response of a system with a real-valued impulse response is easily measured using a sinusoidal oscillator to generate the system input and an oscilloscope to measure the amplitude and phase change between the input and output sinusoids for different oscillator frequencies.

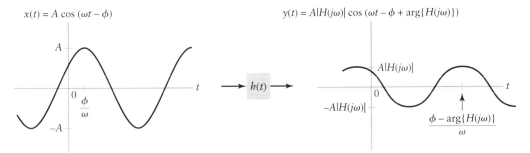

FIGURE 3.50   A sinusoidal input to an LTI system results in a sinusoidal output of the same frequency, with the amplitude and phase modified by the system's frequency response.

Similarly, if $x[n] = A\cos(\Omega n - \phi)$ is the input to a discrete-time LTI system with a real-valued impulse response $h[n]$, then

$$y[n] = |H(e^{j\Omega})|A\cos(\Omega n - \phi + \arg\{H(e^{j\Omega})\})$$

is the output signal. Once again, the system modifies the amplitude of the input sinusoid by $|H(e^{j\Omega})|$ and the phase by $\arg\{H(e^{j\Omega})\}$.

Now suppose that $x(t)$ is purely imaginary, so that $x^*(t) = -x(t)$. Substituting $x^*(t) = -x(t)$ into Eq. (3.37) results in

$$X^*(j\omega) = -\int_{-\infty}^{\infty} x(t)e^{-j(-\omega)t}\,dt.$$

That is,

$$X^*(j\omega) = -X(-j\omega). \tag{3.39}$$

Examining the real and imaginary parts of this relationship gives $\mathrm{Re}\{X(j\omega)\} = -\mathrm{Re}\{X(-j\omega)\}$ and $\mathrm{Im}\{X(j\omega)\} = \mathrm{Im}\{X(-j\omega)\}$. That is, if $x(t)$ is purely imaginary, then the real part of the FT has odd symmetry and the imaginary part has even symmetry. The corresponding symmetry relationships for all four Fourier representations are given in Table 3.4. In each case, the real part has odd symmetry and the imaginary part has even symmetry.

■ 3.9.2 SYMMETRY PROPERTIES: EVEN AND ODD SIGNALS

Suppose that $x(t)$ is real valued and has even symmetry. Then $x^*(t) = x(t)$ and $x(-t) = x(t)$, from which we deduce that $x^*(t) = x(-t)$. Substituting $x^*(t) = x(-t)$ into Eq. (3.37), we may write

$$X^*(j\omega) = \int_{-\infty}^{\infty} x(-t)e^{-j\omega(-t)}\,dt.$$

Now we perform the change of variable $\tau = -t$ to obtain

$$X^*(j\omega) = \int_{-\infty}^{\infty} x(\tau)e^{-j\omega\tau}\,d\tau$$

$$= X(j\omega).$$

The only way that the condition $X^*(j\omega) = X(j\omega)$ holds is for the imaginary part of $X(j\omega)$ to be zero. Hence, if $x(t)$ is real and even, then $X(j\omega)$ is real. Similarly, we may show that if $x(t)$ is real and odd, then $X^*(j\omega) = -X(j\omega)$ and $X(j\omega)$ is imaginary.

Identical symmetry relationships hold for all four Fourier representations. If the time signal is real and even, then the frequency-domain representation of it is also real. If the time signal is real and odd, then the frequency-domain representation is imaginary. Note that since we have assumed real-valued time-domain signals in deriving these symmetry properties, we may combine the results of this subsection with those of the previous subsection. That is, real and even time-domain signals have real and even frequency-domain representations, and real and odd time-domain signals have imaginary and odd frequency-domain representations.

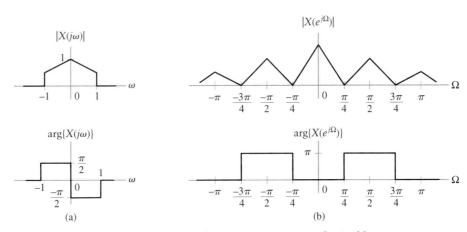

FIGURE 3.51 Frequency-domain representations for Problem 3.17.

▶ **Problem 3.17** Determine whether the time-domain signals corresponding to the following frequency-domain representations are real or complex valued and even or odd:

(a) $X(j\omega)$ as depicted in Fig. 3.51(a)

(b) $X(e^{j\Omega})$ as depicted in Fig. 3.51(b)

(c) FS: $X[k] = (1/2)^k u[k] + j2^k u[-k]$

(d) $X(j\omega) = \omega^{-2} + j\omega^{-3}$

(e) $X(e^{j\Omega}) = j\Omega^2 \cos(2\Omega)$

Answers:

(a) $x(t)$ is real and odd

(b) $x[n]$ is real and even

(c) $x(t)$ is complex valued

(d) $x(t)$ is real valued

(e) $x[n]$ is pure imaginary and even ◀

3.10 *Convolution Property*

Perhaps the most important property of Fourier representations is the convolution property. In this section, we show that convolution of signals in the time domain transforms to multiplication of their respective Fourier representations in the frequency domain. With the convolution property, we may analyze the input–output behavior of a linear system in the frequency domain by multiplying transforms instead of convolving time signals. This can significantly simplify system analysis and offers considerable insight into system behavior. The convolution property is a consequence of complex sinusoids being eigenfunctions of LTI systems. We begin by examining the convolution property as applied to nonperiodic signals.

■ 3.10.1 CONVOLUTION OF NONPERIODIC SIGNALS

Consider the convolution of two nonperiodic continuous-time signals $x(t)$ and $h(t)$. We define

$$y(t) = h(t) * x(t)$$
$$= \int_{-\infty}^{\infty} h(\tau)x(t - \tau)\,d\tau.$$

Now we express $x(t - \tau)$ in terms of its FT:

$$x(t - \tau) = \frac{1}{2\pi}\int_{-\infty}^{\infty} X(j\omega)e^{j\omega(t-\tau)}\,d\omega.$$

Substituting this expression into the convolution integral yields

$$y(t) = \int_{-\infty}^{\infty} h(\tau)\left[\frac{1}{2\pi}\int_{-\infty}^{\infty} X(j\omega)e^{j\omega t}e^{-j\omega\tau}\,d\omega\right]d\tau$$
$$= \frac{1}{2\pi}\int_{-\infty}^{\infty}\left[\int_{-\infty}^{\infty} h(\tau)e^{-j\omega\tau}\,d\tau\right]X(j\omega)e^{j\omega t}\,d\omega.$$

We recognize the inner integral over τ as the FT of $h(\tau)$, or $H(j\omega)$. Hence, $y(t)$ may be rewritten as

$$y(t) = \frac{1}{2\pi}\int_{-\infty}^{\infty} H(j\omega)X(j\omega)e^{j\omega t}\,d\omega,$$

and we identify $H(j\omega)X(j\omega)$ as the FT of $y(t)$. We conclude that convolution of $h(t)$ and $x(t)$ in the time domain corresponds to multiplication of their Fourier transforms, $H(j\omega)$ and $X(j\omega)$, in the frequency domain; that is,

$$y(t) = h(t) * x(t) \xleftrightarrow{\ FT\ } Y(j\omega) = X(j\omega)H(j\omega). \qquad (3.40)$$

The next two examples illustrate applications of this important property.

EXAMPLE 3.31 SOLVING A CONVOLUTION PROBLEM IN THE FREQUENCY DOMAIN   Let $x(t) = (1/(\pi t))\sin(\pi t)$ be the input to a system with impulse response $h(t) = (1/(\pi t))\sin(2\pi t)$. Find the output $y(t) = x(t) * h(t)$.

Solution: This problem is extremely difficult to solve in the time domain. However, it is simple to solve in the frequency domain if we use the convolution property. From Example 3.26, we have

$$x(t) \xleftrightarrow{\ FT\ } X(j\omega) = \begin{cases} 1, & |\omega| < \pi \\ 0, & |\omega| > \pi \end{cases}$$

and

$$h(t) \xleftrightarrow{\ FT\ } H(j\omega) = \begin{cases} 1, & |\omega| < 2\pi \\ 0, & |\omega| > 2\pi \end{cases}.$$

Since $y(t) = x(t) * h(t) \xleftrightarrow{\ FT\ } Y(j\omega) = X(j\omega)H(j\omega)$, it follows that

$$Y(j\omega) = \begin{cases} 1, & |\omega| < \pi \\ 0, & |\omega| > \pi \end{cases},$$

and we conclude that $y(t) = (1/(\pi t))\sin(\pi t)$.   ■

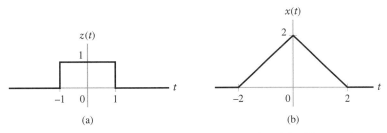

FIGURE 3.52 Signals for Example 3.32. (a) Rectangular pulse $z(t)$. (b) Convolution of $z(t)$ with itself gives $x(t)$.

EXAMPLE 3.32 FINDING INVERSE FT'S BY MEANS OF THE CONVOLUTION PROPERTY
Use the convolution property to find $x(t)$, where

$$x(t) \xleftrightarrow{\;FT\;} X(j\omega) = \frac{4}{\omega^2} \sin^2(\omega).$$

Solution: We may write $X(j\omega)$ as the product $Z(j\omega)Z(j\omega)$, where

$$Z(j\omega) = \frac{2}{\omega} \sin(\omega).$$

The convolution property states that $z(t) * z(t) \xleftrightarrow{\;FT\;} Z(j\omega)Z(j\omega)$, so $x(t) = z(t) * z(t)$. Using the result of Example 3.25, we have

$$z(t) = \begin{cases} 1, & |t| < 1 \\ 0, & |t| > 1 \end{cases} \xleftrightarrow{\;FT\;} Z(j\omega),$$

as depicted in Fig. 3.52(a). Performing the convolution of $z(t)$ with itself gives the triangular waveform depicted in Fig. 3.52(b) as the solution for $x(t)$. ∎

A similar property holds for convolution of discrete-time nonperiodic signals: If $x[n] \xleftrightarrow{\;DTFT\;} X(e^{j\Omega})$ and $h[n] \xleftrightarrow{\;DTFT\;} H(e^{j\Omega})$, then

$$\boxed{y[n] = x[n] * h[n] \xleftrightarrow{\;DTFT\;} Y(e^{j\Omega}) = X(e^{j\Omega})H(e^{j\Omega}).} \tag{3.41}$$

The proof of this result closely parallels that of the continuous-time case and is left as an exercise for the reader.

S&S
Solutions

▶ **Problem 3.18** Use the convolution property to find the FT of the system output, either $Y(j\omega)$ or $Y(e^{j\Omega})$, for the following inputs and system impulse responses:

(a) $x(t) = 3e^{-t}u(t)$ and $h(t) = 2e^{-2t}u(t)$
(b) $x[n] = (1/2)^n u[n]$ and $h[n] = (1/(\pi n)) \sin(\pi n/2)$

Answers:

(a)

$$Y(j\omega) = \left(\frac{2}{j\omega + 2} \right) \left(\frac{3}{j\omega + 1} \right)$$

(b)

$$Y(e^{j\Omega}) = \begin{cases} 1/(1 - (1/2)e^{-j\Omega}), & |\Omega| \leq \pi/2 \\ 0, & \pi/2 < |\Omega| \leq \pi \end{cases}$$ ◀

▶ **Problem 3.19**   Use the convolution property to find the time-domain signals corresponding to the following frequency-domain representations:

(a)

$$X(j\omega) = (1/(j\omega + 2))((2/\omega)\sin \omega)$$

(b)

$$X(e^{j\Omega}) = \left(\frac{1}{1 - (1/2)e^{-j\Omega}} \right)\left(\frac{1}{1 + (1/2)e^{-j\Omega}} \right)$$

Answers:

(a)

$$x(t) = \begin{cases} 0, & t < -1 \\ (1 - e^{-2(t+1)})/2, & -1 \leq t < 1 \\ (e^{-2(t-1)} - e^{-2(t+1)})/2, & 1 \leq t \end{cases}$$

(b)

$$x[n] = \begin{cases} 0, & n < 0, n = 1, 3, 5, \ldots \\ (1/2)^n, & n = 0, 2, 4, \ldots \end{cases}$$ ◀

▶ **Problem 3.20**   Find the outputs of the following systems with the stated impulse response and input:

(a) $h[n] = (1/(\pi n))\sin(\pi n/4)$ and $x[n] = (1/(\pi n))\sin(\pi n/8)$

(b) $h(t) = (1/(\pi t))\sin(\pi t)$ and $x(t) = (3/(\pi t))\sin(2\pi t)$

Answers:

(a)

$$y[n] = (1/(\pi n))\sin(\pi n/8)$$

(b)

$$y(t) = (3/(\pi t))\sin(\pi t)$$ ◀

■ **3.10.2 FILTERING**

The multiplication that occurs in the frequency-domain representation gives rise to the notion of *filtering*. A system performs filtering on the input signal by presenting a different response to components of the input that are at different frequencies. Typically, the term "filtering" implies that some frequency components of the input are eliminated while others are passed by the system unchanged. We may describe systems in terms of the type of filtering that they perform on the input signal. A *low-pass filter* attenuates high-frequency

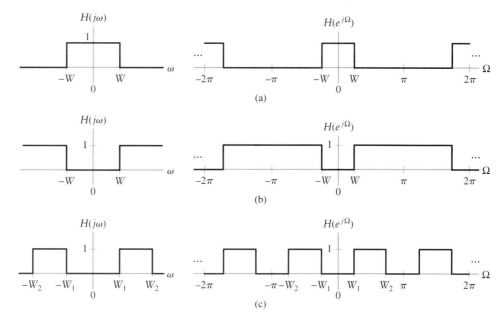

FIGURE 3.53   Frequency response of ideal continuous- (left panel) and discrete-time (right panel) filters. (a) Low-pass characteristic. (b) High-pass characteristic. (c) Band-pass characteristic.

components of the input and passes the lower frequency components. In contrast, a *high-pass filter* attenuates low frequencies and passes the high frequencies. A *band-pass filter* passes signals within a certain frequency band and attenuates signals outside that band. Figures 3.53(a)–(c) illustrate ideal low-pass, high-pass, and band-pass filters, respectively, corresponding to both continuous- and discrete-time systems. Note that the characterization of the discrete-time filter is based on its behavior in the frequency range $-\pi < \Omega \leq \pi$ because its frequency response is 2π-periodic. Hence a high-pass discrete-time filter passes frequencies near π and attenuates frequencies near zero.

The *passband* of a filter is the band of frequencies that are passed by the system, while the *stopband* refers to the range of frequencies that are attenuated by the system. It is impossible to build a practical system that has the discontinuous frequency response characteristics of the ideal systems depicted in Fig. 3.53. Realistic filters always have a gradual transition from the passband to the stopband. The range of frequencies over which this occurs is known as the *transition band*. Furthermore, realistic filters do not have zero gain over the entire stopband, but instead have a very small gain relative to that of the passband. In general, filters with sharp transitions from passband to stopband are more difficult to implement. (A detailed treatment of filters is deferred to Chapter 8.)

The magnitude response of a filter is commonly described in units of decibels, or dB, defined as

$$20 \log|H(j\omega)| \quad \text{or} \quad 20 \log|H(e^{j\Omega})|.$$

The magnitude response in the stopband is normally much smaller than that in the passband, and the details of the stopband response are difficult to visualize on a linear scale. By using decibels, we display the magnitude response on a logarithmic scale and are able to examine the details of the response in both the passband and the stopband. Note that unity gain corresponds to zero dB. Hence, the magnitude response in the filter passband is normally close to zero dB. The edge of the passband is usually defined by the frequencies for which the response is -3 dB, corresponding to a magnitude response of $(1/\sqrt{2})$. Since the energy spectrum of the filter output is given by

$$|Y(j\omega)|^2 = |H(j\omega)|^2 |X(j\omega)|^2,$$

the -3-dB point corresponds to frequencies at which the filter passes only half of the input power. The -3-dB points are usually termed the *cutoff frequencies* of the filter. The majority of filtering applications involve real-valued impulse responses, which implies magnitude responses with even symmetry. In this case, the passband, stopband, and cutoff frequencies are defined by using positive frequencies, as illustrated in the next example.

S & S
Solutions

▶ **EXAMPLE 3.33 RC CIRCUIT: FILTERING**   The RC circuit depicted in Fig. 3.54 may be used with two different outputs: the voltage across the resistor, $y_R(t)$, or the voltage across the capacitor, $y_C(t)$. The impulse response for the case where $y_C(t)$ is the output is given by (see Example 1.21)

$$h_C(t) = \frac{1}{RC} e^{-t/(RC)} u(t).$$

Since $y_R(t) = x(t) - y_C(t)$, the impulse response for the case where $y_R(t)$ is the output is given by

$$h_R(t) = \delta(t) - \frac{1}{RC} e^{-t/(RC)} u(t).$$

Plot the magnitude responses of both systems on a linear scale and in dB, and characterize the filtering properties of the systems.

Solution: The frequency response corresponding to $h_C(t)$ is

$$H_C(j\omega) = \frac{1}{j\omega RC + 1},$$

while that corresponding to $h_R(t)$ is

$$H_R(j\omega) = \frac{j\omega RC}{j\omega RC + 1}.$$

Figures 3.55(a) and (b) depict the magnitude responses $|H_C(j\omega)|$ and $|H_R(j\omega)|$, respectively. Figures 3.55(c) and (d) illustrate the magnitude responses in dB. The system corresponding to output $y_C(t)$ has unit gain at low frequencies and tends to attenuate high frequencies. Hence, it has a low-pass filtering characteristic. We see that the cutoff frequency is $\omega_c = 1/(RC)$, since the magnitude response is -3 dB at ω_c. Therefore, the filter passband is from 0 to $1/(RC)$. The system corresponding to output $y_R(t)$ has zero gain at low frequencies and unit gain at high frequencies and thus has a high-pass filtering characteristic. The cutoff frequency is $\omega_c = 1/(RC)$, and thus the filter passband is $|\omega| > \omega_c$. ■

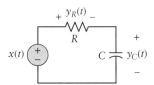

FIGURE 3.54 RC circuit with input $x(t)$ and outputs $y_C(t)$ and $y_R(t)$.

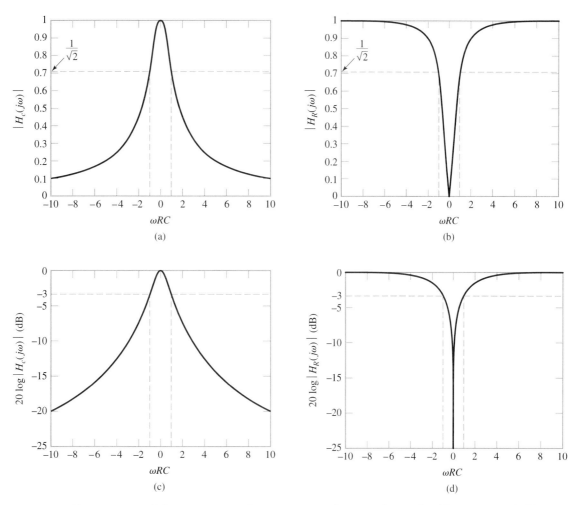

FIGURE 3.55 *RC* circuit magnitude responses as a function of normalized frequency ωRC. (a) Magnitude response of the system corresponding to $y_C(t)$, linear scale. (b) Magnitude response of the system corresponding to $y_R(t)$, linear scale. (c) Magnitude response of the system corresponding to $y_C(t)$, dB scale. (d) Magnitude response of the system corresponding to $y_R(t)$, dB scale, shown on the range from 0 dB to -25 dB.

The convolution property implies that the frequency response of a system may be expressed as the ratio of the FT or DTFT of the output to that of the input. Specifically, we may write, for a continuous-time system,

$$H(j\omega) = \frac{Y(j\omega)}{X(j\omega)} \qquad\qquad (3.42)$$

and, for a discrete-time system,

$$H(e^{j\Omega}) = \frac{Y(e^{j\Omega})}{X(e^{j\Omega})}. \qquad\qquad (3.43)$$

Both $H(j\omega)$ and $H(e^{j\Omega})$ are of the indeterminate form $\frac{0}{0}$ at frequencies where $X(j\omega)$ or $X(e^{j\Omega})$ are zero. Hence, if the input spectrum is nonzero at all frequencies, the frequency response of a system may be determined from knowledge of the input and output spectra.

EXAMPLE 3.34 IDENTIFYING A SYSTEM, GIVEN ITS INPUT AND OUTPUT   The output of an LTI system in response to an input $x(t) = e^{-2t}u(t)$ is $y(t) = e^{-t}u(t)$. Find the frequency response and the impulse response of this system.

Solution: We take the FT of $x(t)$ and $y(t)$, obtaining

$$X(j\omega) = \frac{1}{j\omega + 2}$$

and

$$Y(j\omega) = \frac{1}{j\omega + 1}.$$

Now we use the definition

$$H(j\omega) = \frac{Y(j\omega)}{X(j\omega)}$$

to obtain the system frequency response

$$H(j\omega) = \frac{j\omega + 2}{j\omega + 1}.$$

This equation may be rewritten as

$$H(j\omega) = \left(\frac{j\omega + 1}{j\omega + 1}\right) + \frac{1}{j\omega + 1}$$

$$= 1 + \frac{1}{j\omega + 1}.$$

We take the inverse FT of each term to obtain the impulse response of the system:

$$h(t) = \delta(t) + e^{-t}u(t). \qquad ■$$

Note that Eqs. (3.42) and (3.43) also imply that we can recover the input of the system from the output as

$$X(j\omega) = H^{\text{inv}}(j\omega)Y(j\omega)$$

and

$$X(e^{j\Omega}) = H^{\text{inv}}(e^{j\Omega})Y(e^{j\Omega}),$$

where $H^{\text{inv}}(j\omega) = 1/H(j\omega)$ and $H^{\text{inv}}(e^{j\Omega}) = 1/H(e^{j\Omega})$ are the frequency responses of the respective inverse systems. An inverse system is also known as an *equalizer*, and the process of recovering the input from the output is known as *equalization*. In practice, causality restrictions often make it difficult or impossible to build an exact inverse system, so an approximate inverse is used. For example, a communication channel may introduce a time delay in addition to distorting the signal's magnitude and phase spectra. In order to compensate for the time delay, an exact equalizer would have to introduce a time advance, which implies that the equalizer is noncausal and cannot in fact be implemented. However, we may choose to build an approximate equalizer, one that compensates for all the distortion except for the time delay. Approximate equalizers are also frequently used when noise is present in $Y(j\omega)$ or $Y(e^{j\Omega})$, to prevent excessive amplification of the noise. (An introduction to equalizer design is given in Chapter 8.)

S & S
Solutions

EXAMPLE 3.35 MULTIPATH COMMUNICATION CHANNEL: EQUALIZATION   Consider again the problem addressed in Example 2.13. In this problem, a distorted received signal $y[n]$ is expressed in terms of a transmitted signal $x[n]$ as

$$y[n] = x[n] + ax[n-1], \qquad |a| < 1.$$

Use the convolution property to find the impulse response of an inverse system that will recover $x[n]$ from $y[n]$.

Solution: In Example 2.13, we expressed the received signal as the convolution of the input with the system impulse response, or $y[n] = x[n] * h[n]$, where the impulse response is given by

$$h[n] = \begin{cases} 1, & n = 0 \\ a, & n = 1 \\ 0, & \text{otherwise} \end{cases}.$$

The impulse response of an inverse system, $h^{inv}[n]$, must satisfy the equation

$$h^{inv}[n] * h[n] = \delta[n].$$

Taking the DTFT of both sides of this equation and using the convolution property gives

$$H^{inv}(e^{j\Omega})H(e^{j\Omega}) = 1,$$

which implies that the frequency response of the inverse system is given by

$$H^{inv}(e^{j\Omega}) = \frac{1}{H(e^{j\Omega})}.$$

Substituting $h[n]$ into the definition of the DTFT yields

$$h[n] \xleftarrow{\;DTFT\;} H(e^{j\Omega}) = 1 + ae^{-j\Omega}.$$

Hence,

$$H^{inv}(e^{j\Omega}) = \frac{1}{1 + ae^{-j\Omega}}.$$

Taking the inverse DTFT of $H^{inv}(e^{j\Omega})$ gives the impulse response of the inverse system:

$$h^{inv}[n] = (-a)^n u[n]. \qquad \blacksquare$$

▶ **Problem 3.21**   Find the input of the following systems, given the impulse response and system output:

(a) $h(t) = e^{-4t}u(t)$ and $y(t) = e^{-3t}u(t) - e^{-4t}u(t)$
(b) $h[n] = (1/2)^n u[n]$ and $y[n] = 4(1/2)^n u[n] - 2(1/4)^n u[n]$

Answers:

(a) $x(t) = e^{-3t}u(t)$
(b) $x[n] = 2(1/4)^n u[n]$ ◀

■ 3.10.3 CONVOLUTION OF PERIODIC SIGNALS

This subsection addresses the convolution of two signals that are periodic functions of time. The convolution of periodic signals does not occur naturally in the context of evaluating the input–output relationships of systems, since any system with a periodic impulse response is unstable. However, the convolution of periodic signals often occurs in the context of signal analysis and manipulation.

We define the periodic convolution of two continuous-time signals $x(t)$ and $z(t)$, each having period T, as

$$y(t) = x(t) \circledast z(t)$$
$$= \int_0^T x(\tau)z(t - \tau)\, d\tau,$$

where the symbol $\circledast$ denotes that integration is performed over a single period of the signals involved. The result $y(t)$ is also periodic with period T; hence, the FS is the appropriate representation for all three signals: $x(t), z(t),$ and $y(t)$.

Substituting the FS representation of $z(t)$ into the convolution integral leads to the property

$$y(t) = x(t) \circledast z(t) \xleftrightarrow{\quad FS;\frac{2\pi}{T}\quad} Y[k] = TX[k]Z[k]. \tag{3.44}$$

Again, we see that convolution in time transforms to multiplication of the frequency-domain representations.

EXAMPLE 3.36 CONVOLUTION OF TWO PERIODIC SIGNALS   Evaluate the periodic convolution of the sinusoidal signal

$$z(t) = 2\cos(2\pi t) + \sin(4\pi t)$$

with the periodic square wave $x(t)$ depicted in Fig. 3.56.

Solution: Both $x(t)$ and $z(t)$ have fundamental period $T = 1$. Let $y(t) = x(t) \circledast z(t)$. The convolution property indicates that $y(t) \xleftrightarrow{\;FS;2\pi\;} Y[k] = X[k]Z[k]$. The FS representation of $z(t)$ has coefficients

$$Z[k] = \begin{cases} 1, & k = \pm 1 \\ 1/(2j), & k = 2 \\ -1/(2j), & k = -2 \\ 0, & \text{otherwise} \end{cases}.$$

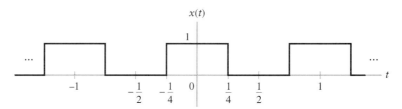

FIGURE 3.56   Square wave for Example 3.36.

The FS coefficients for $x(t)$ may be obtained from Example 3.13 as

$$X[k] = \frac{2\sin(k\pi/2)}{k2\pi}.$$

Hence, the FS coefficients for $y(t)$ are

$$Y[k] = X[k]Z[k] = \begin{cases} 1/\pi, & k = \pm1 \\ 0, & \text{otherwise} \end{cases},$$

which implies that

$$y(t) = (2/\pi)\cos(2\pi t). \qquad\blacksquare$$

The convolution property explains the origin of the Gibbs phenomenon observed in Example 3.14. A partial-sum approximation to the FS representation for $x(t)$ may be obtained by using the FS coefficients

$$\hat{X}_J[k] = X[k]W[k],$$

where

$$W[k] = \begin{cases} 1, & -J \le k \le J \\ 0, & \text{otherwise} \end{cases}$$

and $T = 1$. In the time domain, $\hat{x}_J(t)$ is the periodic convolution of $x(t)$ and

$$w(t) = \frac{\sin(\pi t(2J+1))}{\sin(\pi t)}. \qquad (3.45)$$

One period of the signal $w(t)$ is depicted in Fig. 3.57 for $J = 10$. The periodic convolution of $x(t)$ and $w(t)$ is the area under time-shifted versions of $w(t)$ on $|t| < \frac{1}{2}$. Time-shifting the sidelobes of $w(t)$ into and out of the interval $|t| < \frac{1}{2}$ introduces ripples into the partial-sum approximation $\hat{x}_J(t)$. As J increases, the mainlobe and sidelobe widths in $w(t)$ decrease, but the size of the sidelobes does not change. This is why the ripples in $\hat{x}_J(t)$ become more concentrated near the discontinuity of $x(t)$, but retain the same magnitude.

The discrete-time convolution of two N-periodic sequences $x[n]$ and $z[n]$ is defined as

$$y[n] = x[n] \circledast z[n]$$
$$= \sum_{k=0}^{N-1} x[k]z[n-k].$$

This is the periodic convolution of $x[n]$ and $z[n]$. The signal $y[n]$ is N periodic, so the DTFS is the appropriate representation for all three signals—$x[n]$, $z[n]$, and $y[n]$. Substitution of the DTFS representation for $z[n]$ results in the property

$$y[n] = x[n] \circledast z[n] \xleftarrow{\quad DTFS;\frac{2\pi}{N}\quad} Y[k] = NX[k]Z[k]. \qquad (3.46)$$

Thus, convolution of time signals transforms to multiplication of DTFS coefficients.

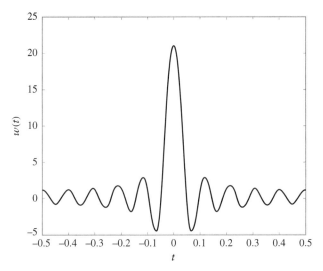

FIGURE 3.57   The signal $w(t)$ defined in Eq. (3.45) when $J = 10$.

TABLE 3.5   *Convolution Properties.*

$$x(t) * z(t) \xleftrightarrow{\ FT\ } X(j\omega)Z(j\omega)$$

$$x(t) \circledast z(t) \xleftrightarrow{\ FS;\omega_o\ } TX[k]Z[k]$$

$$x[n] * z[n] \xleftrightarrow{\ DTFT\ } X(e^{j\Omega})Z(e^{j\Omega})$$

$$x[n] \circledast z[n] \xleftrightarrow{\ DTFS;\Omega_o\ } NX[k]Z[k]$$

The convolution properties of all four Fourier representations are summarized in Table 3.5.

We have not yet considered several important cases of convolution that arise when classes of signals are mixed—for example, the convolution of a periodic and nonperiodic signal such as occurs when applying a periodic signal to a linear system. The properties derived here can be applied to these cases if we use a Fourier transform representation for periodic signals. This representation is developed in Chapter 4.

∎ 3.11 *Differentiation and Integration Properties*

Differentiation and integration are operations that apply to continuous functions. Hence, we may consider the effect of differentiation and integration with respect to time for a continuous-time signal or with respect to frequency in the FT and DTFT, since these representations are continuous functions of frequency. We derive integration and differentiation properties for several of these cases.

■ 3.11.1 DIFFERENTIATION IN TIME

Consider the effect of differentiating a nonperiodic signal $x(t)$. First, recall that $x(t)$ and its FT, $X(j\omega)$, are related by

$$x(t) = \frac{1}{2\pi} \int_{-\infty}^{\infty} X(j\omega)e^{j\omega t}\, d\omega.$$

Differentiating both sides of this equation with respect to t yields

$$\frac{d}{dt}x(t) = \frac{1}{2\pi} \int_{-\infty}^{\infty} X(j\omega)j\omega e^{j\omega t}\, d\omega,$$

from which it follows that

$$\boxed{\frac{d}{dt}x(t) \xleftrightarrow{\;\;FT\;\;} j\omega X(j\omega).}$$

That is, differentiating a signal in the time domain corresponds to multiplying its FT by $j\omega$ in the frequency domain. Thus, differentiation accentuates the high-frequency components of the signal. Note that differentiation destroys any dc component of $x(t)$, and consequently, the FT of the differentiated signal at $\omega = 0$ is zero.

EXAMPLE 3.37 VERIFYING THE DIFFERENTIATION PROPERTY   The differentiation property implies that

$$\frac{d}{dt}\left(e^{-at}u(t)\right) \xleftrightarrow{\;\;FT\;\;} \frac{j\omega}{a + j\omega}.$$

Verify this result by differentiating and taking the FT of the result.

Solution: Using the product rule for differentiation, we have

$$\frac{d}{dt}\left(e^{-at}u(t)\right) = -ae^{-at}u(t) + e^{-at}\delta(t)$$

$$= -ae^{-at}u(t) + \delta(t).$$

Taking the FT of each term and using linearity, we may write

$$\frac{d}{dt}\left(e^{-at}u(t)\right) \quad \xleftarrow{\;\;FT\;\;} \quad \frac{-a}{a + j\omega} + 1$$

$$\xleftarrow{\;\;FT\;\;} \quad \frac{j\omega}{a + j\omega}.$$ ■

▶ **Problem 3.22**   Use the differentiation property to find the FT of the following signals:

(a) $$x(t) = \frac{d}{dt}e^{-2|t|}$$

(b) $x(t) = \dfrac{d}{dt}\left(2te^{-2t}u(t)\right)$

Answers:

(a) $X(j\omega) = (4j\omega)/(4 + \omega^2)$
(b) $X(j\omega) = (2j\omega)/(2 + j\omega)^2$ ◀

▶ **Problem 3.23**   Use the differentiation property to find $x(t)$ if

$$X(j\omega) = \begin{cases} j\omega, & |\omega| < 1 \\ 0, & |\omega| > 1 \end{cases}.$$

Answer: $x(t) = (1/\pi t)\cos t - (1/\pi t^2)\sin t$ ◀

The differentiation property may be used to find the frequency response of a continuous-time system described by the differential equation

$$\sum_{k=0}^{N} a_k \frac{d^k}{dt^k} y(t) = \sum_{k=0}^{M} b_k \frac{d^k}{dt^k} x(t).$$

First, we take the FT of both sides of this equation and repeatedly apply the differentiation property to obtain

$$\sum_{k=0}^{N} a_k (j\omega)^k Y(j\omega) = \sum_{k=0}^{M} b_k (j\omega)^k X(j\omega).$$

Now we rearrange this equation as the ratio of the FT of the output to the FT of the input:

$$\frac{Y(j\omega)}{X(j\omega)} = \frac{\sum_{k=0}^{M} b_k (j\omega)^k}{\sum_{k=0}^{N} a_k (j\omega)^k}$$

Equation (3.42) implies that the frequency response of the system is

$$H(j\omega) = \frac{\sum_{k=0}^{M} b_k (j\omega)^k}{\sum_{k=0}^{N} a_k (j\omega)^k}. \tag{3.47}$$

Thus, the frequency response of a system described by a linear constant-coefficient differential equation is a ratio of two polynomials in $j\omega$. Note that we can reverse this process and determine a differential-equation description of the system from the frequency response, provided that the frequency response is expressed as a ratio of polynomials in $j\omega$.

By definition, the frequency response is the amplitude and phase change that the system imparts to a complex sinusoid. The sinusoid is assumed to exist for all time; it does not have a starting or ending time. This implies that the frequency response is the system's steady-state response to a sinusoid. In contrast to differential- and difference-equation descriptions of a system, the frequency-response description cannot represent initial conditions; it can only describe a system that is in a steady-state condition.

S & S
Solutions

EXAMPLE 3.38 MEMS ACCELEROMETER: FREQUENCY RESPONSE AND RESONANCE   The MEMS accelerometer introduced in Section 1.10 is described by the differential equation

$$\frac{d^2}{dt^2} y(t) + \frac{\omega_n}{Q} \frac{d}{dt} y(t) + \omega_n^2 y(t) = x(t).$$

Find the frequency response of this system and plot the magnitude response in dB for $\omega_n = 10,000$ rads/s for (a) $Q = 2/5$, (b) $Q = 1$, and (c) $Q = 200$.

Solution: Applying Eq. (3.47) gives

$$H(j\omega) = \frac{1}{(j\omega)^2 + \dfrac{\omega_n}{Q}(j\omega) + \omega_n^2}.$$

Figure 3.58 depicts the magnitude response in dB for the specified values of ω_n and Q. The magnitude response for case (a), $Q = 2/5$, decreases as ω increases, while that for case (b), $Q = 1$, is approximately constant for $\omega < \omega_n$ and then decays with increasing frequency for $\omega > \omega_n$. Note that in case (c), when Q is large, there is a sharp peak in the

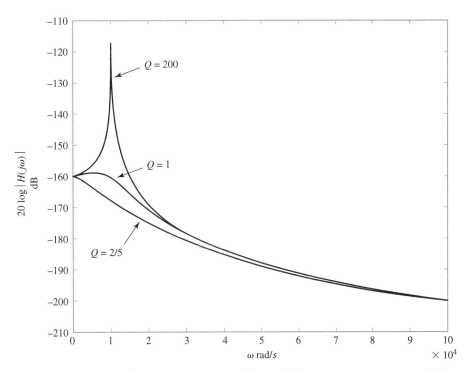

FIGURE 3.58   Magnitude of frequency response in dB for MEMS accelerometer for $\omega_n = 10{,}000$ rad/s, and (a) $Q = 2/5$, (b) $Q = 1$, and (c) $Q = 200$.

frequency response at $\omega_n = 10{,}000$ rad/s, which indicates a *resonant condition*. That is, the system exhibits a strong tendency toward an oscilliatory motion of ω_n. The accelerometer application favors $Q = 1$, so that all accelerations in the frequency range of interest, $\omega < \omega_n$, are characterized by approximately equal gain. If $Q < 1$, then the system bandwidth decreases, so the accelerometer response to input components near ω_n is reduced. If $Q \gg 1$, then the device acts as a very narrowband bandpass filter and the accelerometer response is dominated by a sinusoid of frequency ω_n. Note that while the resonant effect associated with $Q \gg 1$ is undesirable in the accelerometer application, other applications of this basic MEMS structure, such as narrowband filtering, utilize large values of Q. Very large values of Q can be obtained by packaging the basic structure in a vacuum.

A resonant condition occurs when the damping factor D representing frictional forces is small, since Q varies inversely with D, as shown in Eq. (1.107). In this case, the dominant forces are due to the spring and the inertia of the mass, both of which store energy. Note that the potential energy associated with the spring is maximal when the mass is at maximum displacement and zero when the mass passes through the equilibrium position. Conversely, the kinetic energy is maximal when the mass passes through the equilibrium position, since that is when the velocity reaches a maximum, and is zero when the mass is at maximum displacement. The mechanical energy in the system is constant, so, at resonance, kinetic energy is being exchanged with potential energy as the mass oscillates in a sinusoidal fashion. The frequency of motion for which the maximum kinetic energy is equal to the maximum potential energy determines the resonant frequency ω_n. Analogous resonant behavior occurs in a series RLC circuit, although here the resistor is the loss mechanism, so resonance occurs when the resistance is small. The capacitor and inductor are energy storage devices and an oscillatory current–voltage behavior results from energy exchange between the capacitor and the inductor.   ∎

▶ **Problem 3.24**   Write the differential equation relating the input $x(t)$ to the output $y_C(t)$ for the RC circuit depicted in Fig. 3.54, and identify the frequency response.

Answer: See Example 3.33   ◀

If $x(t)$ is a periodic signal, then we have the FS representation

$$x(t) = \sum_{k=-\infty}^{\infty} X[k]e^{jk\omega_o t}.$$

Differentiating both sides of this equation gives

$$\frac{d}{dt}x(t) = \sum_{k=-\infty}^{\infty} X[k]jk\omega_o e^{jk\omega_o t},$$

and thus we conclude that

$$\boxed{\frac{d}{dt}x(t) \xleftrightarrow{\;FS;\,\omega_o\;} jk\omega_o X[k].}$$

Once again, differentiation forces the time-averaged value of the differentiated signal to be zero; hence, the FS coefficient for $k = 0$ is zero.

EXAMPLE 3.39   Use the differentiation property to find the FS representation of the triangular wave depicted in Fig. 3.59(a).

Solution: Define a waveform $z(t) = \frac{d}{dt}y(t)$. Figure 3.59(b) illustrates $z(t)$. The FS coefficients for a periodic square wave were derived in Example 3.13. The signal $z(t)$ corresponds to the square wave $x(t)$ of that example with $T_o/T = 1/4$, provided that we first scale the amplitude of $x(t)$ by a factor of four and then subtract a constant term of two units. That is, $z(t) = 4x(t) - 2$. Thus, $Z[k] = 4X[k] - 2\delta[k]$, and we may write

$$z(t) \xleftrightarrow{\;FS;\,\omega_o\;} Z[k] = \begin{cases} 0, & k = 0 \\ \dfrac{4\sin\left(\frac{k\pi}{2}\right)}{k\pi}, & k \neq 0 \end{cases}.$$

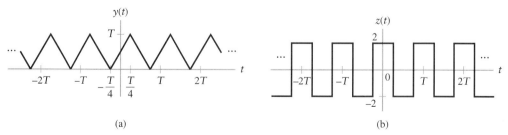

(a)   (b)

FIGURE 3.59   Signals for Example 3.39. (a) Triangular wave $y(t)$. (b) The derivative of $y(t)$ is the square wave $z(t)$.

The differentiation property implies that $Z[k] = jk\omega_o Y[k]$. Hence, we may determine $Y[k]$ from $Z[k]$ as $Y[k] = \frac{1}{jk\omega_o}Z[k]$, except for $k = 0$. The quantity $Y[0]$ is the average value of $y(t)$ and is determined by inspection of Fig. 3.59(a) to be $T/2$. Therefore,

$$y(t) \xleftrightarrow{\;FS;\,\omega_o\;} Y[k] = \begin{cases} T/2, & k = 0 \\ \dfrac{2T\sin\left(\frac{k\pi}{2}\right)}{jk2\pi^2}, & k \neq 0 \end{cases}.$$

∎

■ 3.11.2 DIFFERENTIATION IN FREQUENCY

Consider next the effect of differentiating the frequency-domain representation of a signal. Beginning with the FT

$$X(j\omega) = \int_{-\infty}^{\infty} x(t)e^{-j\omega t}\,dt,$$

we differentiate both sides of this equation with respect to ω and obtain

$$\frac{d}{d\omega}X(j\omega) = \int_{-\infty}^{\infty} -jtx(t)e^{-j\omega t}\,dt,$$

from which it follows that

$$\boxed{-jtx(t) \xleftrightarrow{\;FT\;} \frac{d}{d\omega}X(j\omega).}$$

Thus, differentiation of an FT in the frequency domain corresponds to multiplication of the signal by $-jt$ in the time domain.

EXAMPLE 3.40 FT OF A GAUSSIAN PULSE   Use the differentiation-in-time and differentiation-in-frequency properties to determine the FT of the *Gaussian pulse*, defined by $g(t) = (1/\sqrt{2\pi})e^{-t^2/2}$ and depicted in Fig. 3.60.

Solution: We note that the derivative of $g(t)$ with respect to time is given by

$$\frac{d}{dt}g(t) = (-t/\sqrt{2\pi})e^{-t^2/2} \tag{3.48}$$

$$= -tg(t).$$

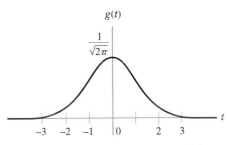

FIGURE 3.60   Gaussian pulse $g(t)$.

The differentiation-in-time property says that

$$\frac{d}{dt}g(t) \xleftrightarrow{FT} j\omega G(j\omega),$$

and thus Eq. (3.48) implies that

$$-tg(t) \xleftrightarrow{FT} j\omega G(j\omega). \qquad (3.49)$$

The differentiation-in-frequency property, namely,

$$-jtg(t) \xleftrightarrow{FT} \frac{d}{d\omega}G(j\omega)$$

indicates that

$$-tg(t) \xleftrightarrow{FT} \frac{1}{j}\frac{d}{d\omega}G(j\omega). \qquad (3.50)$$

Since the left-hand sides of Eqs. (3.49) and (3.50) are equal, the right-hand sides must also be equal; thus,

$$\frac{d}{d\omega}G(j\omega) = -\omega G(j\omega).$$

This is a differential-equation description of $G(j\omega)$ that has the same mathematical form as the differential-equation description of $g(t)$ given in Eq. (3.48). Therefore, the functional form of $G(j\omega)$ is the same as that of $g(t)$, and we have

$$G(j\omega) = ce^{-\omega^2/2}.$$

The constant c is determined by noting that (see Appendix A-4)

$$G(j0) = \int_{-\infty}^{\infty}(1/\sqrt{2\pi})e^{-t^2/2}\,dt$$
$$= 1.$$

Thus, $c = 1$, and we conclude that the FT of a Gaussian pulse is also a Gaussian pulse, so that

$$(1/\sqrt{2\pi})e^{-t^2/2} \xleftrightarrow{FT} e^{-\omega^2/2}. \qquad ▪$$

▶ **Problem 3.25**   Use the frequency-differentiation property to find the FT of

$$x(t) = te^{-at}u(t),$$

given that $e^{-at}u(t) \xleftrightarrow{FT} 1/(j\omega + a)$.

Answer:

$$X(j\omega) = \frac{1}{(a + j\omega)^2} \qquad ◀$$

▶ **Problem 3.26** Use the time-differentiation and convolution properties to find the FT of

$$y(t) = \frac{d}{dt}\{te^{-3t}u(t) * e^{-2t}u(t)\}.$$

Answer:

$$Y(j\omega) = \frac{j\omega}{(3 + j\omega)^2(j\omega + 2)} \qquad \blacktriangleleft$$

The operation of differentiation does not apply to discrete-valued quantities, and thus a frequency-differentiation property for the FS or DTFS does not exist. However, a frequency-differentiation property does exist for the DTFT. By definition,

$$X(e^{j\Omega}) = \sum_{n=-\infty}^{\infty} x[n]e^{-j\Omega n}.$$

Differentiation of both sides of this expression with respect to frequency leads to the property

$$\boxed{-jnx[n] \xleftrightarrow{DTFT} \frac{d}{d\Omega}X(e^{j\Omega}).}$$

▶ **Problem 3.27** Use the frequency-differentiation property to find the DTFT of $x[n] = (n + 1)\alpha^n u[n]$.

Answer:

$$X(e^{j\Omega}) = 1/(1 - \alpha e^{-j\Omega})^2 \qquad \blacktriangleleft$$

▶ **Problem 3.28** Determine $x[n]$, given the DTFT

$$X(e^{j\Omega}) = j\frac{d}{d\Omega}\left(\frac{\sin(11\Omega/2)}{\sin(\Omega/2)}\right).$$

Answer:

$$x[n] = \begin{cases} n, & |n| \leq 5 \\ 0, & \text{otherwise} \end{cases} \qquad \blacktriangleleft$$

■ **3.11.3 INTEGRATION**

The operation of integration applies only to continuous dependent variables. Hence, we may integrate with respect to time in both the FT and FS and with respect to frequency in the FT and DTFT. We limit our consideration here to integrating nonperiodic signals with respect to time. We define

$$y(t) = \int_{-\infty}^{t} x(\tau)\, d\tau;$$

that is, the value of y at time t is the integral of x over all time prior to t. Note that

$$\frac{d}{dt}y(t) = x(t), \qquad (3.51)$$

so the differentiation property would suggest that

$$Y(j\omega) = \frac{1}{j\omega}X(j\omega). \qquad (3.52)$$

This relationship is indeterminate at $\omega = 0$, a consequence of the differentiation operation in Eq. (3.51) destroying any dc component of $y(t)$ and implying that $X(j0)$ must be zero. Thus, Eq. (3.52) applies only to signals with a zero time-averaged value; that is, $X(j0) = 0$.

In general, we desire to apply the integration property to signals that do not have a zero time-averaged value. However, if the average value of $x(t)$ is not zero, then it is possible that $y(t)$ is not square integrable, and consequently, the FT of $y(t)$ may not converge. We can get around this problem by including impulses in the transform. We know that Eq. (3.52) holds for all ω, except possibly $\omega = 0$. The value at $\omega = 0$ is modified by adding a term $c\delta(\omega)$, where the constant c depends on the average value of $x(t)$. The correct result is obtained by setting $c = \pi X(j0)$. This gives the integration property:

$$\int_{-\infty}^{t} x(\tau)\, d\tau \xleftarrow{\;\; FT \;\;} \frac{1}{j\omega} X(j\omega) + \pi X(j0)\delta(\omega). \qquad (3.53)$$

Note that it is understood that the first term on the right-hand side is zero at $\omega = 0$. Integration may be viewed as an averaging operation, and thus it tends to smooth signals in time, a property that corresponds to deemphasizing the high-frequency components of the signal, as indicated in Eq. (3.53).

We can demonstrate this property by deriving the FT of the unit step, which may be expressed as the integral of the unit impulse:

$$u(t) = \int_{-\infty}^{t} \delta(\tau)\, d\tau.$$

Since $\delta(t) \xleftarrow{\;\; FT \;\;} 1$, Eq. (3.53) suggests that

$$u(t) \xleftarrow{\;\; FT \;\;} U(j\omega) = \frac{1}{j\omega} + \pi\delta(\omega).$$

Let us check this result by independently deriving $U(j\omega)$. First, we express the unit step as the sum of two functions:

$$u(t) = \frac{1}{2} + \frac{1}{2}\mathrm{sgn}(t). \qquad (3.54)$$

Here, the *signum* function is defined as

$$\mathrm{sgn}(t) = \begin{cases} -1, & t < 0 \\ 0, & t = 0 \\ 1, & t > 0 \end{cases}.$$

This representation is illustrated in Fig. 3.61. Using the results of Example 3.28, we have $\frac{1}{2} \xleftarrow{\;\; FT \;\;} \pi\delta(\omega)$. The transform of $\mathrm{sgn}(t)$ may be derived using the differentiation property because it has a zero time-averaged value. Let $\mathrm{sgn}(t) \xleftarrow{\;\; FT \;\;} S(j\omega)$. Then

$$\frac{d}{dt}\mathrm{sgn}(t) = 2\delta(t).$$

Hence,

$$j\omega S(j\omega) = 2.$$

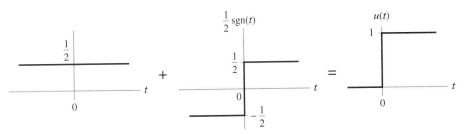

FIGURE 3.61   Representation of a step function as the sum of a constant and a signum function.

We know that $S(j0) = 0$ because $\text{sgn}(t)$ is an odd function and thus has zero average value. This knowledge removes the indeterminacy at $\omega = 0$ associated with the differentiation property, and we conclude that

$$S(j\omega) = \begin{cases} \dfrac{2}{j\omega}, & \omega \neq 0 \\ 0, & \omega = 0 \end{cases}.$$

It is common to write this relationship as $S(j\omega) = 2/(j\omega)$, with the understanding that $S(j0) = 0$. Now we apply the linearity property to Eq. (3.54) and obtain the FT of $u(t)$:

$$u(t) \xleftrightarrow{\ FT\ } \frac{1}{j\omega} + \pi\delta(\omega).$$

This agrees exactly with the transform of the unit step obtained by using the integration property.

▶ **Problem 3.29**   Use the integration property to find the inverse FT of

$$X(j\omega) = \frac{1}{j\omega(j\omega + 1)} + \pi\delta(\omega).$$

Answer:

$$x(t) = (1 - e^{-t})u(t) \qquad\qquad ◀$$

Table 3.6 summarizes the differentiation and integration properties of Fourier representations.

TABLE 3.6   *Commonly Used Differentiation and Integration Properties.*

$$\frac{d}{dt}x(t) \xleftrightarrow{\ FT\ } j\omega X(j\omega)$$

$$\frac{d}{dt}x(t) \xleftrightarrow{\ FS;\,\omega_o\ } jk\omega_o X[k]$$

$$-jtx(t) \xleftrightarrow{\ FT\ } \frac{d}{d\omega}X(j\omega)$$

$$-jnx[n] \xleftrightarrow{\ DTFT\ } \frac{d}{d\Omega}X(e^{j\Omega})$$

$$\int_{-\infty}^{t} x(\tau)\,d\tau \xleftrightarrow{\ FT\ } \frac{1}{j\omega}X(j\omega) + \pi X(j0)\delta(\omega)$$

3.12 Time- and Frequency-Shift Properties

In this section, we consider the effect of time and frequency shifts on the Fourier representation. As before, we derive the result for the FT and state the results for the other three representations.

■ 3.12.1 TIME-SHIFT PROPERTY

Let $z(t) = x(t - t_o)$ be a time-shifted version of $x(t)$. The goal is to relate the FT of $z(t)$ to the FT of $x(t)$. We have

$$Z(j\omega) = \int_{-\infty}^{\infty} z(t)e^{-j\omega t}\, dt$$

$$= \int_{-\infty}^{\infty} x(t - t_o)e^{-j\omega t}\, dt.$$

Next, we effect the change of variable $\tau = t - t_o$, obtaining

$$Z(j\omega) = \int_{-\infty}^{\infty} x(\tau)e^{-j\omega(\tau + t_o)}\, d\tau$$

$$= e^{-j\omega t_o}\int_{-\infty}^{\infty} x(\tau)e^{-j\omega\tau}\, d\tau$$

$$= e^{-j\omega t_o}X(j\omega).$$

Thus, the result of time-shifting the signal $x(t)$ by t_o is to multiply the FT $X(j\omega)$ by $e^{-j\omega t_o}$. Note that $|Z(j\omega)| = |X(j\omega)|$ and $\arg\{Z(j\omega)\} = \arg\{X(j\omega)\} - \omega t_o$. Hence, a shift in time leaves the magnitude spectrum unchanged and introduces a phase shift that is a linear function of frequency. The slope of this linear phase shift is equal to the time delay. A similar property holds for the other three Fourier representations, as indicated in Table 3.7. These properties are a direct consequence of the time-shift properties of the complex sinusoids used in Fourier representations. Time-shifting a complex sinusoid results in a complex sinusoid of the same frequency and magnitude, with the phase shifted by the product of the time shift and the sinusoid's frequency.

TABLE 3.7   Time-Shift Properties of Fourier Representations.

$$x(t - t_o) \xleftrightarrow{\text{FT}} e^{-j\omega t_o}X(j\omega)$$

$$x(t - t_o) \xleftrightarrow{\text{FS};\omega_o} e^{-jk\omega_o t_o}X[k]$$

$$x[n - n_o] \xleftrightarrow{\text{DTFT}} e^{-j\Omega n_o}X(e^{j\Omega})$$

$$x[n - n_o] \xleftrightarrow{\text{DTFS};\Omega_o} e^{-jk\Omega_o n_o}X[k]$$

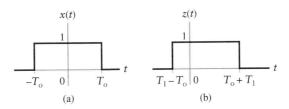

FIGURE 3.62   Application of the time-shift property for Example 3.41.

EXAMPLE 3.41   FINDING AN FT USING THE TIME-SHIFT PROPERTY   Use the FT of the rectangular pulse $x(t)$ depicted in Fig. 3.62 (a) to determine the FT of the time-shifted rectangular pulse $z(t)$ depicted in Fig. 3.62(b).

Solution: First, we note that $z(t) = x(t - T_1)$, so the time-shift property implies that $Z(j\omega) = e^{-j\omega T_1} X(j\omega)$. In Example 3.25, we obtained

$$X(j\omega) = \frac{2}{\omega} \sin(\omega T_o).$$

Thus, we have

$$Z(j\omega) = e^{-j\omega T_1} \frac{2}{\omega} \sin(\omega T_o). \qquad \blacksquare$$

▶ **Problem 3.30**   Use the DTFS of the periodic square-wave depicted in Fig. 3.63(a), as derived in Example 3.6, to determine the DTFS of the periodic square-wave depicted in Fig. 3.63(b).

Answer:

$$Z[k] = e^{-jk6\pi/7} \frac{1}{7} \frac{\sin(k5\pi/7)}{\sin(k\pi/7)} \qquad \blacktriangleleft$$

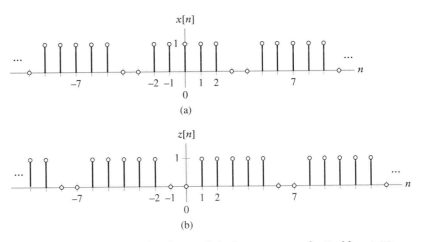

FIGURE 3.63   Original and time-shifted square waves for Problem 3.30.

▶ **Problem 3.31**   Find the Fourier representation of the following time-domain signals:

(a) $x(t) = e^{-2t}u(t - 3)$

(b) $y[n] = \sin(\pi(n + 2)/3)/(\pi(n + 2))$

Answers:

(a)

$$X(j\omega) = e^{-6}e^{-j3\omega}/(j\omega + 2)$$

(b)

$$Y(e^{j\Omega}) = \begin{cases} e^{j2\Omega}, & |\Omega| \leq \pi/3 \\ 0, & \pi/3 < |\Omega| \leq \pi \end{cases}$$

◀

S & S
Solutions

▶ **Problem 3.32**   Find the time-domain signals corresponding to the following Fourier representations:

(a) $X(j\omega) = e^{j4\omega}/(2 + j\omega)^2$

(b) $Y[k] - e^{-jk4\pi/5}/10$, DTFS with $\Omega_o = 2\pi/10$

Answers:

(a)

$$x(t) = (t + 4)e^{-2(t+4)}u(t + 4)$$

(b)

$$y[n] = \sum_{p=-\infty}^{\infty} \delta[n - 4 - 10p]$$

◀

The time-shift property may be used to find the frequency response of a system described by a difference equation. To see this, consider the difference equation

$$\sum_{k=0}^{N} a_k y[n - k] = \sum_{k=0}^{M} b_k x[n - k].$$

First, we take the DTFT of both sides of this equation, using the time-shift property

$$z[n - k] \xleftarrow{\quad DTFT \quad} e^{-jk\Omega}Z(e^{j\Omega})$$

to obtain

$$\sum_{k=0}^{N} a_k(e^{-j\Omega})^k Y(e^{j\Omega}) = \sum_{k=0}^{M} b_k(e^{-j\Omega})^k X(e^{j\Omega}).$$

Next, we rewrite this equation as the ratio

$$\frac{Y(e^{j\Omega})}{X(e^{j\Omega})} = \frac{\sum_{k=0}^{M} b_k(e^{-j\Omega})^k}{\sum_{k=0}^{N} a_k(e^{-j\Omega})^k}.$$

Identifying this ratio with Eq. (3.43), we have

$$H(e^{j\Omega}) = \frac{\sum_{k=0}^{M} b_k(e^{-j\Omega})^k}{\sum_{k=0}^{N} a_k(e^{-j\Omega})^k}. \tag{3.55}$$

The frequency response of a discrete-time system is a ratio of polynomials in $e^{-j\Omega}$. Given a frequency response of the form described in Eq. (3.55), we may reverse our derivation to determine a difference-equation description of the system if so desired.

▶ **Problem 3.33** Find the difference equation corresponding to the system with frequency response

$$H(e^{j\Omega}) = \frac{1 + 2e^{-j2\Omega}}{3 + 2e^{-j\Omega} - 3e^{-j3\Omega}}.$$

Answer:

$$3y[n] + 2y[n-1] - 3y[n-3] = x[n] + 2x[n-2]$$ ◀

■ 3.12.2 FREQUENCY-SHIFT PROPERTY

In the previous subsection, we considered the effect of a time shift on the frequency-domain representation. In the current subsection, we consider the effect of a frequency shift on the time-domain signal. Suppose $x(t) \overset{FT}{\longleftrightarrow} X(j\omega)$. The problem is to express the inverse FT of $Z(j\omega) = X(j(\omega - \gamma))$ in terms of $x(t)$. Let $z(t) \overset{FT}{\longleftrightarrow} Z(j\omega)$. By the definition of the inverse FT, we have

$$z(t) = \frac{1}{2\pi} \int_{-\infty}^{\infty} Z(j\omega)e^{j\omega t}\, d\omega$$

$$= \frac{1}{2\pi} \int_{-\infty}^{\infty} X(j(\omega - \gamma))e^{j\omega t}\, d\omega.$$

We effect the substitution of variables $\eta = \omega - \gamma$, obtaining

$$z(t) = \frac{1}{2\pi} \int_{-\infty}^{\infty} X(j\eta)e^{j(\eta + \gamma)t}\, d\eta$$

$$= e^{j\gamma t}\frac{1}{2\pi} \int_{-\infty}^{\infty} X(j\eta)e^{j\eta t}\, d\eta$$

$$= e^{j\gamma t}x(t).$$

Hence, a frequency shift corresponds to multiplication in the time domain by a complex sinusoid whose frequency is equal to the shift.

This property is a consequence of the frequency-shift properties of the complex sinusoid. A shift in the frequency of a complex sinusoid is equivalent to a multiplication of the original complex sinusoid by another complex sinusoid whose frequency is equal to the shift. Since all the Fourier representations are based on complex sinusoids, they all share this property, as summarized in Table 3.8. Note that the frequency shift must be integer valued in both Fourier series cases. This leads to multiplication by a complex sinusoid whose frequency is an integer multiple of the fundamental frequency. The other observation is that the frequency-shift property is the "dual" of the time-shift property. We may summarize both properties by stating that a shift in one domain, either frequency or time, leads to a multiplication by a complex sinusoid in the other domain.

TABLE 3.8 *Frequency-Shift Properties of Fourier Representations.*

$$e^{j\gamma t}x(t) \xleftarrow{\quad FT \quad} X(j(\omega - \gamma))$$

$$e^{jk_o\omega_o t}x(t) \xleftarrow{\quad FS;\omega_o \quad} X[k - k_o]$$

$$e^{j\Gamma n}x[n] \xleftarrow{\quad DTFT \quad} X(e^{j(\Omega-\Gamma)})$$

$$e^{jk_o\Omega_o n}x[n] \xleftarrow{\quad DTFS;\Omega_o \quad} X[k - k_o]$$

EXAMPLE 3.42 FINDING AN FT BY USING THE FREQUENCY-SHIFT PROPERTY   Use the frequency-shift property to determine the FT of the complex sinusoidal pulse

$$z(t) = \begin{cases} e^{j10t}, & |t| < \pi \\ 0, & |t| > \pi \end{cases}.$$

Solution: We may express $z(t)$ as the product of a complex sinusoid e^{j10t} and a rectangular pulse

$$x(t) = \begin{cases} 1, & |t| < \pi \\ 0, & |t| > \pi \end{cases}.$$

Using the results of Example 3.25, we write

$$x(t) \xleftarrow{\quad FT \quad} X(j\omega) = \frac{2}{\omega}\sin(\omega\pi),$$

and employing the frequency-shift property

$$e^{j10t}x(t) \xleftarrow{\quad FT \quad} X(j(\omega - 10)),$$

we obtain

$$z(t) \xleftarrow{\quad FT \quad} \frac{2}{\omega - 10}\sin((\omega - 10)\pi). \qquad ■$$

▶ **Problem 3.34**   Use the frequency-shift property to find the time-domain signals corresponding to the following Fourier representations:

(a)

$$Z(e^{j\Omega}) = \frac{1}{1 - \alpha e^{-j(\Omega + \pi/4)}}, \qquad |\alpha| < 1$$

(b)

$$X(j\omega) = \frac{1}{2 + j(\omega - 3)} + \frac{1}{2 + j(\omega + 3)}$$

Answers:

(a)

$$z[n] = e^{-j\pi/4n}\alpha^n u[n]$$

(b)

$$x(t) = 2\cos(3t)e^{-2t}u(t) \qquad ◀$$

EXAMPLE 3.43 USING MULTIPLE PROPERTIES TO FIND AN FT   Find the FT of the signal

$$x(t) = \frac{d}{dt}\{(e^{-3t}u(t)) * (e^{-t}u(t-2))\}.$$

Solution: We identify three properties required to solve this problem: differentiation in time, convolution, and time shifting. These must be applied in the order corresponding to mathematical precedence rules in order to obtain the correct result. Let $w(t) = e^{-3t}u(t)$ and $v(t) = e^{-t}u(t-2)$. Then we may write

$$x(t) = \frac{d}{dt}\{w(t) * v(t)\}.$$

Hence, applying the convolution and differentiation properties from Tables 3.5 and 3.6, we obtain

$$X(j\omega) = j\omega\{W(j\omega)V(j\omega)\}.$$

The transform pair

$$e^{-at}u(t) \xleftrightarrow{\ FT\ } \frac{1}{a+j\omega}$$

implies that

$$W(j\omega) = \frac{1}{3+j\omega}.$$

We use the same transform pair and the time-shift property to find $V(j\omega)$ by first writing

$$v(t) = e^{-2}e^{-(t-2)}u(t-2).$$

Thus,

$$V(j\omega) = e^{-2}\frac{e^{-j2\omega}}{1+j\omega}$$

and

$$X(j\omega) = e^{-2}\frac{j\omega e^{-j2\omega}}{(1+j\omega)(3+j\omega)}.\qquad\blacksquare$$

▶ **Problem 3.35**   Find the Fourier representations of the following time-domain signals:

(a)
$$x[n] = ne^{j\pi/8n}\alpha^{n-3}u[n-3]$$

(b)
$$x(t) = (t-2)\frac{d}{dt}\left[e^{-j5t}e^{-2|t-3|}\right]$$

Answers:

(a)
$$X(e^{j\Omega}) = j\frac{d}{d\Omega}\left\{\frac{e^{-j3(\Omega-\pi/8)}}{1-\alpha e^{-j(\Omega-\pi/8)}}\right\}$$

(b)
$$X(j\omega) = \frac{-8j\omega e^{-j3(\omega+5)}}{4+(\omega+5)^2} + j\frac{d}{d\omega}\left[\frac{4j\omega e^{-j3(\omega+5)}}{4+(\omega+5)^2}\right] \qquad ◀$$

3.13 Finding Inverse Fourier Transforms by Using Partial-Fraction Expansions

In Section 3.11, we showed that the frequency response of a system described by a linear constant-coefficient differential equation is given by a ratio of two polynomials in $j\omega$. Similarly, in Section 3.12, we showed that the frequency response of a system described by a linear constant-coefficient difference equation is given by a ratio of two polynomials in $e^{j\Omega}$. FT's and DTFT's of this form occur frequently in the analysis of system and signal interaction, because of the importance of linear constant-coefficient differential and difference equations. In order to find inverse transforms for ratios of polynomials, we use partial-fraction expansions.

■ 3.13.1 INVERSE FOURIER TRANSFORM

Suppose $X(j\omega)$ is expressed as a ratio of polynomials in $j\omega$:

$$X(j\omega) = \frac{b_M(j\omega)^M + \cdots + b_1(j\omega) + b_0}{(j\omega)^N + a_{N-1}(j\omega)^{N-1} + \cdots + a_1(j\omega) + a_0} = \frac{B(j\omega)}{A(j\omega)}.$$

Then we may determine the inverse FT of such ratios by using a partial-fraction expansion. The partial-fraction expansion expresses $X(j\omega)$ as a sum of terms for which the inverse FT is known. Since the FT is linear, the inverse FT of $X(j\omega)$ is the sum of the inverse FT's of each term in the expansion.

We assume that $M < N$. If $M \geq N$, then we may use long division to express $X(j\omega)$ in the form

$$X(j\omega) = \sum_{k=0}^{M-N} f_k(j\omega)^k + \frac{\widetilde{B}(j\omega)}{A(j\omega)}.$$

The numerator polynomial $\widetilde{B}(j\omega)$ now has order one less than that of the denominator, and the partial-fraction expansion is applied to determine the inverse Fourier transform of $\widetilde{B}(j\omega)/A(j\omega)$. The inverse Fourier transform of the terms in the sum are obtained from the pair $\delta(t) \xleftrightarrow{\ FT\ } 1$ and the differentiation property.

Let the roots of the denominator polynomial $A(j\omega)$ be $d_k, k = 1, 2, \ldots, N$. These roots are found by replacing $j\omega$ with a generic variable v and determining the roots of the polynomial

$$v^N + a_{N-1}v^{N-1} + \cdots + a_1 v + a_0 = 0.$$

We may then write

$$X(j\omega) = \frac{\sum_{k=0}^{M} b_k(j\omega)^k}{\prod_{k=1}^{N}(j\omega - d_k)}.$$

Assuming that all the roots $d_k, k = 1, 2, \ldots, N$, are distinct, we may write

$$X(j\omega) = \sum_{k=1}^{N} \frac{C_k}{j\omega - d_k},$$

where the coefficients $C_k, k = 1, 2, \ldots, N$, are determined either by solving a system of linear equations or by the method of residues. These methods and the expansion for repeated roots are reviewed in Appendix B. In Example 3.24, we derived the FT pair

$$e^{dt}u(t) \xleftrightarrow{\ FT\ } \frac{1}{j\omega - d} \qquad \text{for } d < 0.$$

The reader may verify that this pair is valid even if d is complex, provided that $\text{Re}\{d\} < 0$. Assuming that the real part of each d_k, $k = 1, 2, \ldots, N$, is negative, we use linearity to write

$$x(t) = \sum_{k=1}^{N} C_k e^{d_k t} u(t) \xrightarrow{\quad FT \quad} X(j\omega) = \sum_{k=1}^{N} \frac{C_k}{j\omega - d_k}.$$

The next example illustrates this technique.

S & S
Solutions

EXAMPLE 3.44 MEMS ACCELEROMETER: IMPULSE RESPONSE   Find the impulse response for the MEMS accelerometer introduced in Section 1.10, assuming that $\omega_n = 10,000$, and (a) $Q = 2/5$, (b) $Q = 1$, and (c) $Q = 200$.

Solution: The frequency response of this system was determined from the differential equation in Example 3.38. In case (a), substituting $\omega_n = 10,000$ and $Q = 2/5$, we have

$$H(j\omega) = \frac{1}{(j\omega)^2 + 25,000(j\omega) + (10,000)^2}.$$

The impulse response is obtained by finding the inverse FT of $H(j\omega)$. This is accomplished by first finding the partial-fraction expansion of $H(j\omega)$. The roots of the denominator polynomial are $d_1 = -20,000$ and $d_2 = -5,000$. Hence, we write $H(j\omega)$ as a sum:

$$\frac{1}{(j\omega)^2 + 25,000(j\omega) + (10,000)^2} = \frac{C_1}{j\omega + 20,000} + \frac{C_2}{j\omega + 5,000}.$$

We can solve for C_1 and C_2 by using the method of residues described in Appendix B. We obtain

$$C_1 = (j\omega + 20,000) \frac{1}{(j\omega)^2 + 25,000(j\omega) + (10,000)^2} \Big|_{j\omega = -20,000}$$

$$= \frac{1}{j\omega + 5,000} \Big|_{j\omega = -20,000}$$

$$= -1/15,000$$

and

$$C_2 = (j\omega + 5,000) \frac{1}{(j\omega)^2 + 25,000(j\omega) + (10,000)^2} \Big|_{j\omega = -5,000}$$

$$= \frac{1}{j\omega + 20,000} \Big|_{j\omega = -5,000}$$

$$= 1/15,000.$$

Thus, the partial-fraction expansion of $H(j\omega)$ is

$$H(j\omega) = \frac{-1/15,000}{j\omega + 20,000} + \frac{1/15,000}{j\omega + 5,000}.$$

Taking the inverse FT of each term yields the impulse response:

$$h(t) = (1/15,000)(e^{-5,000t} - e^{-20,000t})u(t).$$

Next, in case (b) $Q = 1$ and we have

$$H(j\omega) = \frac{1}{(j\omega)^2 + 10,000(j\omega) + (10,000)^2}.$$

In this case, the roots of the denominator polynomial are $d_1 = -5000 + j5000\sqrt{3}$ and $d_2 = -5000 - j5000\sqrt{3}$, and the partial-fraction expansion is given by

$$H(j\omega) = \frac{-j/(10,000\sqrt{3})}{j\omega + 5000 - j5000\sqrt{3}} + \frac{j/(10,000\sqrt{3})}{j\omega + 5000 + j5000\sqrt{3}}$$

Again, taking the inverse FT of each term yields the impulse response

$$h(t) = -j/(10,000\sqrt{3})(e^{-5000t}e^{j5000\sqrt{3}t} - e^{-5000t}e^{-j5000\sqrt{3}t})$$
$$= 1/(5000\sqrt{3})e^{-5000t}\sin(5000\sqrt{3}t)u(t).$$

Now, for case (c) $Q = 200$ and we have

$$H(j\omega) = \frac{1}{(j\omega)^2 + 50(j\omega) + (10,000)^2}.$$

In this case, the roots of the denominator polynomial are $d_1 = -25 + j10,000$ and $d_2 = -25 - j10,000$. Performing the partial-fraction expansion, taking the inverse FT, and simplifying gives the impulse response:

$$h(t) = 1/(10,000)e^{-25t}\sin(10,000t)u(t).$$

The first 2 ms of the impulse responses for $Q = 2/5$, $Q = 1$, and $Q = 200$ are shown in Fig. 3.64(a)–(c). Comparing the impulse responses with the corresponding magnitude responses in Fig. 3.58, we see that the increased bandwidth associated with $Q = 1$ results in more of the energy in the impulse response being concentrated near $t = 0$. Increased bandwidth corresponds to faster response. Note that for both $Q = 2/5$ and $Q = 1$, the impulse response is approximately zero for $t > 1$ ms, which indicates that these accelerometers have a submillisecond response time. The *resonant* nature of the case where $Q = 200$ is exemplified in the impulse response of Fig. 3.64(c) as a sinusoidal oscillation of $\omega_n = 10,000$ rads/s. An impulse input causes the system to resonate at ω_n, as suggested by the magnitude response. This resonant behavior is undesirable in the accelerometer, since it precludes the system from reacting to sudden changes in applied acceleration.   ■

▶ **Problem 3.36**   Use partial-fraction expansions to determine the time-domain signals corresponding to the following FT's:

(a)
$$X(j\omega) = \frac{-j\omega}{(j\omega)^2 + 3j\omega + 2}$$

(b)
$$X(j\omega) = \frac{5j\omega + 12}{(j\omega)^2 + 5j\omega + 6}$$

(c)
$$X(j\omega) = \frac{2(j\omega)^2 + 5j\omega - 9}{(j\omega + 4)(-\omega^2 + 4j\omega + 3)}$$

Answers:

(a) $x(t) = e^{-t}u(t) - 2e^{-2t}u(t)$
(b) $x(t) = 3e^{-3t}u(t) + 2e^{-2t}u(t)$
(c) $x(t) = e^{-4t}u(t) - 2e^{-t}u(t) + 3e^{-3t}u(t)$ ◀

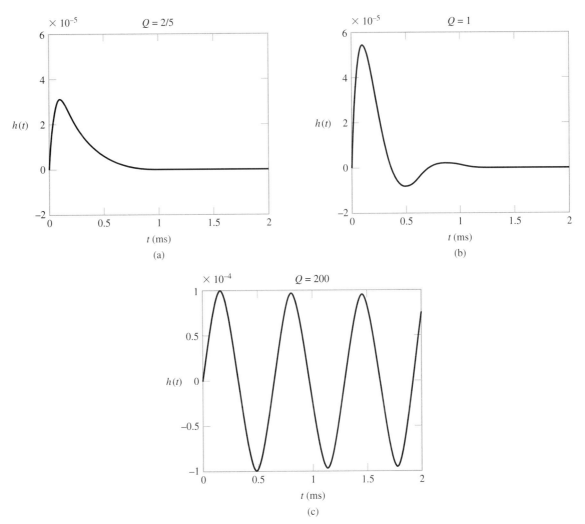

FIGURE 3.64   Impulse response of MEMS accelerometer. (a) $Q = 2/5$. (b) $Q = 1$. (c) $Q = 200$.

S & S
Solutions

▶ **Problem 3.37**   Use the frequency response to find the output $y_C(t)$ of the RC circuit depicted in Fig. 3.54 if $RC = 1$ s and the input is $x(t) = 3e^{-2t}u(t)$.

Answer: $y(t) = 3e^{-t}u(t) - 3e^{-2t}u(t)$ ◀

▆ 3.13.2 INVERSE DISCRETE-TIME FOURIER TRANSFORM

Suppose $X(e^{j\Omega})$ is given by a ratio of polynomials in $e^{j\Omega}$; that is,

$$X(e^{j\Omega}) = \frac{\beta_M e^{-j\Omega M} + \cdots + \beta_1 e^{-j\Omega} + \beta_0}{\alpha_N e^{-j\Omega N} + \alpha_{N-1} e^{-j\Omega(N-1)} + \cdots + \alpha_1 e^{-j\Omega} + 1}.$$

Note that the constant term in the denominator polynomial has been normalized to unity. As in the continuous-time case, using a partial-fraction expansion, we rewrite $X(e^{j\Omega})$ as a sum of terms whose inverse DTFT is known. We factor the denominator polynomial as

$$\alpha_N e^{-j\Omega N} + \alpha_{N-1} e^{-j\Omega(N-1)} + \cdots + \alpha_1 e^{-j\Omega} + 1 = \prod_{k=1}^{N}(1 - d_k e^{-j\Omega}).$$

Partial-fraction expansions based on this factorization are reviewed in Appendix B. In this case we replace $e^{j\Omega}$ with the generic variable v and find the d_k from the roots of the polynomial

$$v^N + \alpha_1 v^{N-1} + \alpha_2 v^{N-2} + \cdots + \alpha_{N-1} v + \alpha_N = 0.$$

Assuming that $M < N$ and all the d_k are distinct, we may express $X(e^{j\Omega})$ as

$$X(e^{j\Omega}) = \sum_{k=1}^{N} \frac{C_k}{1 - d_k e^{-j\Omega}}.$$

Expansions for repeated roots are treated in Appendix B. Since

$$(d_k)^n u[n] \xleftarrow{\quad DTFT \quad} \frac{1}{1 - d_k e^{-j\Omega}},$$

the linearity property implies that

$$x[n] = \sum_{k=1}^{N} C_k (d_k)^n u[n].$$

EXAMPLE 3.45 INVERSION BY PARTIAL-FRACTION EXPANSION   Find the inverse DTFT of

$$X(e^{j\Omega}) = \frac{-\frac{5}{6} e^{-j\Omega} + 5}{1 + \frac{1}{6} e^{-j\Omega} - \frac{1}{6} e^{-j2\Omega}}.$$

Solution: The roots of the polynomial

$$v^2 + \frac{1}{6} v - \frac{1}{6} = 0$$

are $d_1 = -1/2$ and $d_2 = 1/3$. We seek coefficients C_1 and C_2 such that

$$\frac{-\frac{5}{6} e^{-j\Omega} + 5}{1 + \frac{1}{6} e^{-j\Omega} - \frac{1}{6} e^{-j2\Omega}} = \frac{C_1}{1 + \frac{1}{2} e^{-j\Omega}} + \frac{C_2}{1 - \frac{1}{3} e^{-j\Omega}}.$$

Using the method of residues described in Appendix B, we obtain

$$C_1 = \left(1 + \frac{1}{2} e^{-j\Omega}\right) \frac{-\frac{5}{6} e^{-j\Omega} + 5}{1 + \frac{1}{6} e^{-j\Omega} - \frac{1}{6} e^{-j2\Omega}} \Bigg|_{e^{-j\Omega} = -2}$$

$$= \frac{-\frac{5}{6} e^{-j\Omega} + 5}{1 - \frac{1}{3} e^{-j\Omega}} \Bigg|_{e^{-j\Omega} = -2}$$

$$= 4$$

and

$$C_2 = \left(1 - \frac{1}{3} e^{-j\Omega}\right) \frac{-\frac{5}{6} e^{-j\Omega} + 5}{1 + \frac{1}{6} e^{-j\Omega} - \frac{1}{6} e^{-j2\Omega}} \Bigg|_{e^{-j\Omega} = 3}$$

$$= \frac{-\frac{5}{6} e^{-j\Omega} + 5}{1 + \frac{1}{2} e^{-j\Omega}} \Bigg|_{e^{-j\Omega} = 3}$$

$$= 1.$$

Hence,

$$x[n] = 4(-1/2)^n u[n] + (1/3)^n u[n].$$
■

▶ **Problem 3.38**   Find the frequency and impulse responses of the discrete-time systems described by the following difference equations:

(a)

$$y[n-2] + 5y[n-1] + 6y[n] = 8x[n-1] + 18x[n]$$

(b)

$$y[n-2] - 9y[n-1] + 20y[n] = 100x[n] - 23x[n-1]$$

Answers:

(a)

$$H(e^{j\Omega}) = \frac{8e^{-j\Omega} + 18}{(e^{-j\Omega})^2 + 5e^{-j\Omega} + 6}$$

$$h[n] = 2(-1/3)^n u[n] + (-1/2)^n u[n]$$

(b)

$$H(e^{j\Omega}) = \frac{100 - 23e^{-j\Omega}}{20 - 9e^{-j\Omega} + e^{-j2\Omega}}$$

$$h[n] = 2(1/4)^n u[n] + 3(1/5)^n u[n]$$ ◀

3.14 *Multiplication Property*

The multiplication property defines the Fourier representation of a product of time-domain signals. We begin by considering the product of nonperiodic continuous-time signals.

If $x(t)$ and $z(t)$ are nonperiodic signals, then we wish to express the FT of the product $y(t) = x(t)z(t)$ in terms of the FT of $x(t)$ and $z(t)$. We represent $x(t)$ and $z(t)$ in terms of their respective FT's as

$$x(t) = \frac{1}{2\pi} \int_{-\infty}^{\infty} X(j\nu)e^{j\nu t}\, d\nu$$

and

$$z(t) = \frac{1}{2\pi} \int_{-\infty}^{\infty} Z(j\eta)e^{j\eta t}\, d\eta.$$

The product term, $y(t)$, may thus be written in the form

$$y(t) = \frac{1}{(2\pi)^2} \int_{-\infty}^{\infty} \int_{-\infty}^{\infty} X(j\nu)Z(j\eta)e^{j(\eta+\nu)t}\, d\eta\, d\nu.$$

Now we effect the change of variable $\eta = \omega - \nu$ to obtain

$$y(t) = \frac{1}{2\pi} \int_{-\infty}^{\infty} \left[\frac{1}{2\pi} \int_{-\infty}^{\infty} X(j\nu)Z(j(\omega - \nu))\, d\nu \right] e^{j\omega t}\, d\omega.$$

The inner integral over ν represents the convolution of $Z(j\omega)$ and $X(j\omega)$, while the outer integral over ω is of the form of the Fourier representation for $y(t)$. Hence, we identify this convolution, scaled by $1/(2\pi)$, as $Y(j\omega)$; that is,

$$y(t) = x(t)z(t) \xleftrightarrow{\ FT\ } Y(j\omega) = \frac{1}{2\pi}X(j\omega) * Z(j\omega), \qquad (3.56)$$

where

$$X(j\omega) * Z(j\omega) = \int_{-\infty}^{\infty} X(j\nu)Z(j(\omega - \nu))\,d\nu.$$

Multiplication of two signals in the time domain corresponds to convolution of their FT's in the frequency domain and multiplication by the factor $1/(2\pi)$.

Similarly, if $x[n]$ and $z[n]$ are discrete-time nonperiodic signals, then the DTFT of the product $y[n] = x[n]z[n]$ is given by the convolution of the their DTFT's and multiplication by $1/(2\pi)$; that is,

$$y[n] = x[n]z[n] \xleftrightarrow{\ DTFT\ } Y(e^{j\Omega}) = \frac{1}{2\pi}X(e^{j\Omega}) \circledast Z(e^{j\Omega}), \qquad (3.57)$$

where, as before, the symbol $\circledast$ denotes periodic convolution. Here, $X(e^{j\Omega})$ and $Z(e^{j\Omega})$ are 2π-periodic, so we evaluate the convolution over a 2π interval:

$$X(e^{j\Omega}) \circledast Z(e^{j\Omega}) = \int_{-\pi}^{\pi} X(e^{j\theta})Z(e^{j(\Omega-\theta)})\,d\theta.$$

The multiplication property enables us to study the effects of truncating a time-domain signal on its frequency-domain representation. The process of truncating a signal is also known as *windowing*, since it corresponds to viewing the signal through a window. The portion of the signal that is not visible through the window is truncated or assumed to be zero. The windowing operation is represented mathematically by multiplying the signal, say, $x(t)$, by a window function $w(t)$ that is zero outside the time range of interest. Denoting the windowed signal by $y(t)$, we have $y(t) = x(t)w(t)$. This operation is illustrated in Fig. 3.65(a) for a window function that truncates $x(t)$ to the time interval $-T_o < t < T_o$. The FT of $y(t)$ is related to the FTs of $x(t)$ and $w(t)$ through the multiplication property:

$$y(t) \xleftrightarrow{\ FT\ } Y(j\omega) = \frac{1}{2\pi}X(j\omega) * W(j\omega).$$

If $w(t)$ is the rectangular window depicted in Fig. 3.65(a), then, from Example 3.25, we have

$$W(j\omega) = \frac{2}{\omega}\sin(\omega T_o).$$

Figure 3.65(b) illustrates the frequency-domain effect of windowing with a rectangular time-domain window. Note that $X(j\omega)$ is arbitrarily chosen and is not the actual FT of the time-domain signal $x(t)$ depicted in Fig. 3.65(a). The general effect of the window is to smooth details in $X(j\omega)$ and introduce oscillations near discontinuities in $X(j\omega)$, as illustrated in Fig. 3.65(b). The smoothing is a consequence of the $2\pi/T_o$ width of the mainlobe of $W(j\omega)$, while the oscillations near discontinuities are due to the oscillations in the side-lobes of $W(j\omega)$. The next example illustrates the effect of windowing the impulse response of an ideal discrete-time system.

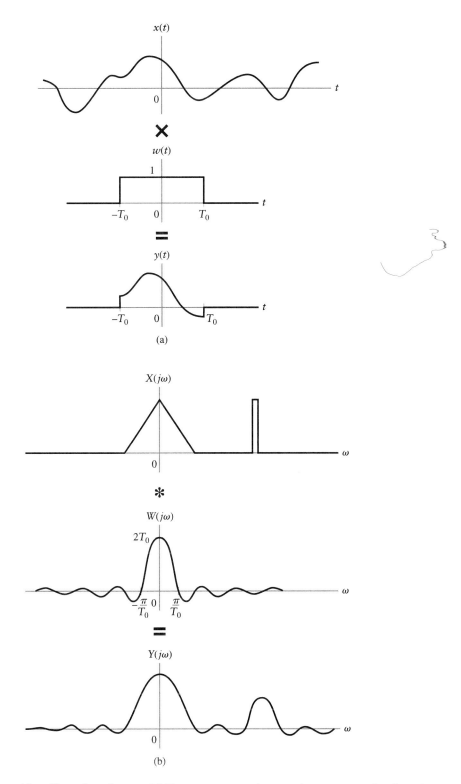

FIGURE 3.65   The effect of windowing. (a) Truncating a signal in time by using a window function $w(t)$. (b) Convolution of the signal and window FTs resulting from truncation in time.

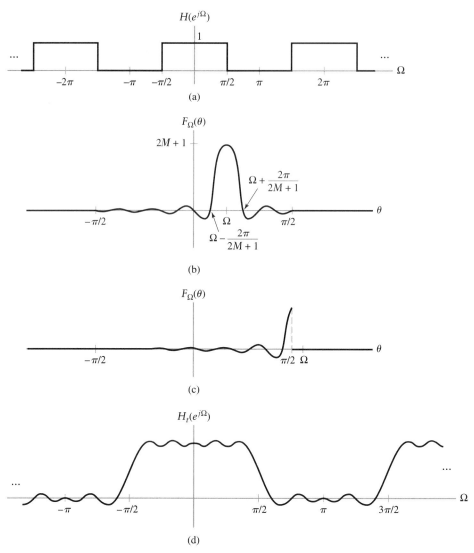

FIGURE 3.66   The effect of truncating the impulse response of a discrete-time system. (a) Frequency response of ideal system. (b) $F_\Omega(\theta)$ for Ω near zero. (c) $F_\Omega(\theta)$ for Ω slightly greater than $\pi/2$. (d) Frequency response of system with truncated impulse response.

EXAMPLE 3.46   TRUNCATING THE IMPULSE RESPONSE   The frequency response $H(e^{j\Omega})$ of an ideal discrete-time system is depicted in Fig. 3.66(a). Describe the frequency response of a system whose impulse response is the ideal system impulse response truncated to the interval $-M \le n \le M$.

Solution: The ideal impulse response is the inverse DTFT of $H(e^{j\Omega})$. Using the result of Example 3.19, we write

$$h[n] = \frac{1}{\pi n}\sin\left(\frac{\pi n}{2}\right).$$

This response is infinite in extent. Let $h_t[n]$ be the truncated impulse response:

$$h_t[n] = \begin{cases} h[n], & |n| \leq M \\ 0, & \text{otherwise} \end{cases}.$$

We may express $h_t[n]$ as the product of $h[n]$ and a window function $w[n]$, where

$$w[n] = \begin{cases} 1, & |n| \leq M \\ 0, & \text{otherwise} \end{cases}.$$

Let $h_t[n] \xleftrightarrow{\ DTFT\ } H_t(e^{j\Omega})$, and use the multiplication property given by Eq. (3.57) to obtain

$$H_t(e^{j\Omega}) = \frac{1}{2\pi} \int_{-\pi}^{\pi} H(e^{j\theta}) W(e^{j(\Omega-\theta)}) \, d\theta.$$

Since

$$H(e^{j\theta}) = \begin{cases} 1, & |\theta| < \pi/2 \\ 0, & \pi/2 < |\theta| < \pi \end{cases},$$

and because, on the basis of Example 3.18, we have

$$W(e^{j(\Omega-\theta)}) = \frac{\sin((\Omega - \theta)(2M + 1)/2)}{\sin((\Omega - \theta)/2)},$$

it follows that

$$H_t(e^{j\Omega}) = \frac{1}{2\pi} \int_{-\pi/2}^{\pi/2} F_\Omega(\theta) \, d\theta,$$

where we have defined

$$F_\Omega(\theta) = H(e^{j\theta}) W(e^{j(\Omega-\theta)}) = \begin{cases} W(e^{j(\Omega-\theta)}), & |\theta| < \pi/2 \\ 0, & |\theta| > \pi/2 \end{cases}.$$

Figure 3.66(b) depicts $F_\Omega(\theta)$. $H_t(e^{j\Omega})$ is the area under $F_\Omega(\theta)$ between $\theta = -\pi/2$ and $\theta = \pi/2$. To visualize the behavior of $H_t(e^{j\Omega})$, consider the area under $F_\Omega(\theta)$ as Ω increases, starting from $\Omega = 0$. As Ω increases, the small oscillations in $F_\Omega(\theta)$ move through the boundary at $\theta = \pi/2$. When a positive oscillation moves through the boundary at $\theta = \pi/2$, the net area under $F_\Omega(\theta)$ decreases. When a negative oscillation moves through the boundary at $\theta = \pi/2$, the net area increases. Oscillations also move through the boundary at $\theta = -\pi/2$. However, these are smaller than those on the right because they are further away from Ω and thus have much less of an effect. The effect of the oscillations in $F_\Omega(\theta)$ moving through the boundary at $\theta = \pi/2$ is to introduce oscillations in $H_t(e^{j\Omega})$. These oscillations increase in size as Ω increases. As Ω approaches $\pi/2$, the area under $F_\Omega(\theta)$ decreases rapidly because the main lobe moves through $\theta = \pi/2$. Figure 3.66(c) depicts $F_\Omega(\theta)$ for Ω slightly larger than $\pi/2$. As Ω continues to increase, the oscillations to the left of the main lobe move through the boundary at $\theta = \pi/2$, causing additional oscillations in the area under $F_\Omega(\theta)$. However, now the net area ocillates about zero because the main lobe of $F_\Omega(\theta)$ is no longer included in the integral.

Thus, $H_t(e^{j\Omega})$ takes on the form depicted in Fig. 3.66(d). We see, then, that truncation of the ideal impulse response introduces ripples into the frequency response and widens the transitions at $\Omega = \pm\pi/2$. These effects decrease as M increases, since the main lobe of $W(e^{j\Omega})$ then becomes narrower and the oscillations decay more quickly. ∎

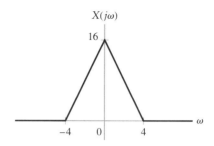

FIGURE 3.67   Solution to Problem 3.39.

▶ **Problem 3.39**   Use the multiplication property to find the FT of

$$x(t) = \frac{4}{\pi^2 t^2} \sin^2(2t).$$

Answer: See Fig. 3.67   ◄

The multiplication property for periodic signals is analogous to that for nonperiodic signals. Multiplication of periodic time-domain signals thus corresponds to convolution of the Fourier representations. Specifically, in continuous time, we have

$$\boxed{y(t) = x(t)z(t) \xleftarrow{\;FT;2\pi/T\;} Y[k] = X[k] * Z[k],}$$ (3.58)

where

$$X[k] * Z[k] = \sum_{m=-\infty}^{\infty} X[m]Z[k-m]$$

is the nonperiodic convolution of the FS coefficients. Note that this property applies, provided that $x(t)$ and $z(t)$ have a common period. If the fundamental period of $x(t)$ is different from that of $y(t)$, then the FS coefficients $X[k]$ and $Y[k]$ must be determined by using the fundamental period of the product of the signals—that is, the least common multiple of each signal's fundamental period.

S & S
Solutions

EXAMPLE 3.47 RADAR RANGE MEASUREMENT: SPECTRUM OF RF PULSE TRAIN   The RF pulse train used to measure range and introduced in Section 1.10 may be defined as the product of a square wave $p(t)$ and a sine wave $s(t)$, as shown in Fig. 3.68. Assume that $s(t) = \sin(1000\pi t/T)$. Find the FS coefficients of $x(t)$.

Solution: Since $x(t) = p(t)s(t)$, the multiplication property given in Eq. (3.58) implies that $X[k] = P[k] * S[k]$. In order to apply this result, the FS expansions for both $p(t)$ and $s(t)$ must use the same fundamental frequency. For $p(t)$, the fundamental frequency is $\omega_o = 2\pi/T$. We may thus write $s(t) = \sin(500\omega_o t)$; that is, the frequency of $s(t)$ is the 500th harmonic of the fundamental frequency for $p(t)$. Using ω_o as the fundamental frequency for $s(t)$ gives the FS coefficients

$$S[k] = \begin{cases} 1/(2j), & k = 500 \\ -1/(2j), & k = -500. \\ 0, & \text{otherwise} \end{cases}$$

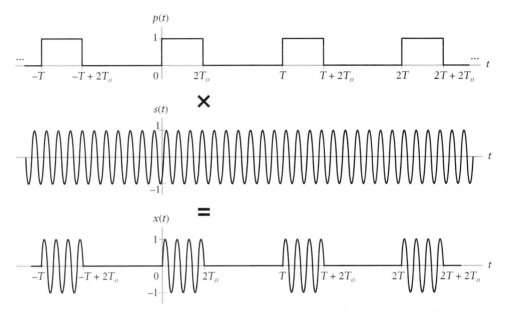

FIGURE 3.68 The RF pulse is expressed as the product of a periodic square wave and a sine wave.

We may also write $S[k] = 1/(2j)\delta[k - 500] - 1/(2j)\delta[k + 500]$. The FS coefficients of $p(t)$ are obtained by using the result of Example 3.13 and the time-shift property and are given by

$$P[k] = e^{-jkT_o\omega_o}\frac{\sin(k\omega_o T_o)}{k\pi}.$$

Convolution of a signal with a shifted impulse simply shifts the signal to the position of the impulse. We use this result to evaluate the convolution $X[k] = P[k] * S[k]$ and obtain

$$X[k] = \frac{1}{2j}e^{-j(k-500)T_o\omega_o}\frac{\sin((k - 500)\omega_o T_o)}{(k - 500)\pi} - \frac{1}{2j}e^{-j(k+500)T_o\omega_o}\frac{\sin((k + 500)\omega_o T_o)}{(k + 500)\pi}.$$

Figure 3.69 depicts the magnitude spectrum for $0 \le k \le 1000$. The power in the RF pulse is concentrated about the harmonic associated with the sinusoid $s(t)$. ∎

The multiplication property for discrete-time periodic signals is

$$y[n] = x[n]z[n] \xleftrightarrow{DTFS;2\pi/N} Y[k] = X[k] \circledast Z[k], \tag{3.59}$$

where

$$X[k] \circledast Z[k] = \sum_{m=0}^{N-1} X[m]Z[k - m]$$

is the periodic convolution of DTFS coefficients. Again, all three time-domain signals have a common fundamental period N.

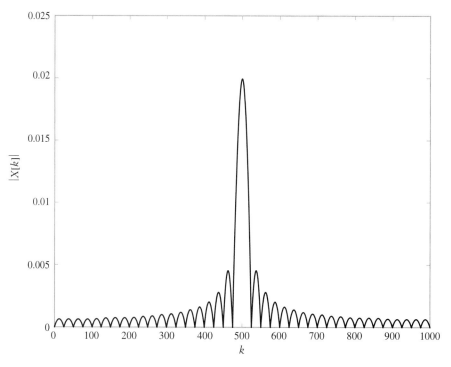

FIGURE 3.69   FS magnitude spectrum of RF pulse train for $0 \leq k \leq 1000$. The result is depicted as a continuous curve, due to the difficulty of displaying 1000 stems.

TABLE 3.9   *Multiplication Properties of Fourier Representations.*

$$x(t)z(t) \xleftrightarrow{\ FT\ } \frac{1}{2\pi} X(j\omega) * Z(j\omega)$$

$$x(t)z(t) \xleftrightarrow{\ FS;\omega_o\ } X[k] * Z[k]$$

$$x[n]z[n] \xleftrightarrow{\ DTFT\ } \frac{1}{2\pi} X(e^{j\Omega}) \circledast Z(e^{j\Omega})$$

$$x[n]z[n] \xleftrightarrow{\ DTFS;\Omega_o\ } X[k] \circledast Z[k]$$

The multiplication properties are summarized for all four Fourier representations in Table 3.9.

▶ **Problem 3.40**   Find the time-domain signals corresponding to the following Fourier representations:

(a)

$$X(e^{j\Omega}) = \left(\frac{e^{-j3\Omega}}{1 + \frac{1}{2}e^{-j\Omega}} \right) \circledast \left(\frac{\sin(21\Omega/2)}{\sin(\Omega/2)} \right)$$

(b)

$$X(j\omega) = \frac{2\sin(\omega - 2)}{\omega - 2} * \frac{e^{-j2\omega}\sin(2\omega)}{\omega}$$

Answers:

(a)

$$x[n] = 2\pi(-1/2)^{n-3}(u[n-3] - u[n-11])$$

(b)

$$x(t) = \pi e^{j2t}(u(t) - u(t-1)) \qquad \blacktriangleleft$$

3.15 *Scaling Properties*

Consider the effect of scaling the time variable on the frequency-domain representation of a signal. Beginning with the FT, let $z(t) = x(at)$, where a is a constant. By definition, we have

$$Z(j\omega) = \int_{-\infty}^{\infty} z(t)e^{-j\omega t}\, dt$$

$$= \int_{-\infty}^{\infty} x(at)e^{-j\omega t}\, dt.$$

We effect the change of variable $\tau = at$ to obtain

$$Z(j\omega) = \begin{cases} (1/a)\int_{-\infty}^{\infty} x(\tau)e^{-j(\omega/a)\tau}\, d\tau, & a > 0 \\ (1/a)\int_{\infty}^{-\infty} x(\tau)e^{-j(\omega/a)\tau}\, d\tau, & a < 0 \end{cases}.$$

These two integrals may be combined into the single integral

$$Z(j\omega) = (1/|a|)\int_{-\infty}^{\infty} x(\tau)e^{-j(\omega/a)\tau}\, d\tau,$$

from which we conclude that

$$\boxed{z(t) = x(at) \xleftrightarrow{\;\;FT\;\;} (1/|a|)X(j\omega/a).} \qquad (3.60)$$

Hence, scaling the signal in time introduces the inverse scaling in the frequency-domain representation and an amplitude change, as illustrated in Fig. 3.70.

This effect may be experienced by playing a recorded sound at a speed different from that at which it was recorded. If we play the sound back at a higher speed, corresponding to $a > 1$, we compress the time signal. The inverse scaling in the frequency domain expands the Fourier representation over a broader frequency band and explains the increase in the perceived pitch of the sound. Conversely, playing the sound back at a slower speed corresponds to expanding the time signal, since $a < 1$. The inverse scaling in the frequency domain compresses the Fourier representation and explains the decrease in the perceived pitch of the sound.

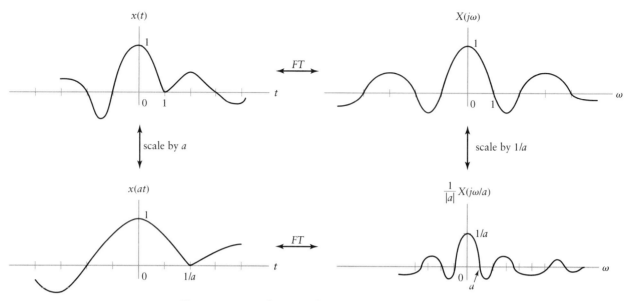

FIGURE 3.70   The FT scaling property. The figure assumes that $0 < a < 1$.

EXAMPLE 3.48 SCALING A RECTANGULAR PULSE   Let the rectangular pulse

$$x(t) = \begin{cases} 1, & |t| < 1 \\ 0, & |t| > 1 \end{cases}.$$

Use the FT of $x(t)$ and the scaling property to find the FT of the scaled rectangular pulse

$$y(t) = \begin{cases} 1, & |t| < 2 \\ 0, & |t| > 2 \end{cases}.$$

Solution: Substituting $T_o = 1$ into the result of Example 3.25 gives

$$X(j\omega) = \frac{2}{\omega}\sin(\omega).$$

Note that $y(t) = x(t/2)$. Hence, application of the scaling property of Eq. (3.60) with $a = 1/2$ gives

$$Y(j\omega) = 2X(j2\omega)$$
$$= 2\left(\frac{2}{2\omega}\right)\sin(2\omega)$$
$$= \frac{2}{\omega}\sin(2\omega).$$

This answer can also be obtained by substituting $T_o = 2$ into the result of Example 3.25. Figure 3.71 illustrates the scaling between time and frequency that occurs in this example.   ■

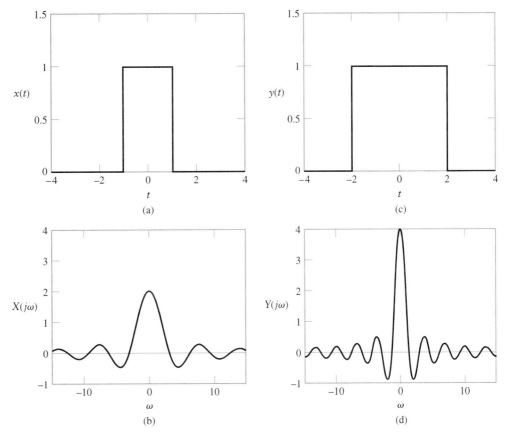

FIGURE 3.71   Application of the FT scaling property in Example 3.48. (a) Original time signal.
(b) Original FT. (c) Scaled time signal $y(t) = x(t/2)$. (d) Scaled FT $Y(j\omega) = 2X(j2\omega)$.

EXAMPLE 3.49 USING MULTIPLE PROPERTIES TO FIND AN INVERSE FT   Find $x(t)$ if

$$X(j\omega) = j\frac{d}{d\omega}\left\{\frac{e^{j2\omega}}{1 + j(\omega/3)}\right\}.$$

Solution: We identify three different properties that may be of use in finding $x(t)$: differentiation in frequency, time shifting, and scaling. These must be applied according to
their mathematical precedence given in $X(j\omega)$ to obtain the correct result. We use the
transform pair

$$s(t) = e^{-t}u(t) \xleftarrow{\ \ FT\ \ } S(j\omega) = \frac{1}{1 + j\omega}$$

to express $X(j\omega)$ as

$$X(j\omega) = j\frac{d}{d\omega}\{e^{j2\omega}S(j\omega/3)\}.$$

Applying the innermost property first, we scale, then time shift, and, lastly, differentiate. If we define $Y(j\omega) = S(j\omega/3)$, then application of the scaling property given in Eq. (3.60) yields

$$y(t) = 3s(3t)$$
$$= 3e^{-3t}u(3t)$$
$$= 3e^{-3t}u(t).$$

Now we define $W(j\omega) = e^{j2\omega}Y(j\omega)$ and apply the time-shift property from Table 3.7 to obtain

$$w(t) = y(t + 2)$$
$$= 3e^{-3(t+2)}u(t + 2).$$

Finally, since $X(j\omega) = j\frac{d}{d\omega}W(j\omega)$, the differentiation property given in Table 3.6 yields

$$x(t) = tw(t)$$
$$= 3te^{-3(t+2)}u(t + 2).$$ ■

If $x(t)$ is a periodic signal, then $z(t) = x(at)$ is also periodic, and the FS is the appropriate Fourier representation. For convenience, we assume that a is positive. In this case, scaling changes the fundamental period of the signal: If $x(t)$ has fundamental period T, then $z(t)$ has fundamental period T/a. Hence, if the fundamental frequency of $x(t)$ is ω_o, then the fundamental frequency of $z(t)$ is $a\omega_o$. From Eq. (3.20), the FS coefficients for $z(t)$ are given by

$$Z[k] = \frac{a}{T}\int_0^{T/a} z(t)e^{-jka\omega_o t}\,dt.$$

Substituting $x(at)$ for $z(t)$ and effecting the change of variable as in the FT case, we obtain

$$z(t) = x(at) \xleftrightarrow{\ FS; a\omega_o\ } Z[k] = X[k], \qquad a > 0. \tag{3.61}$$

That is, the FS coefficients of $x(t)$ and $x(at)$ are identical; the scaling operation simply changes the harmonic spacing from ω_o to $a\omega_o$.

▶ **Problem 3.41**   A signal has FT $x(t) \xleftrightarrow{\ FT\ } X(j\omega) = e^{-j\omega}|\omega|e^{-2|\omega|}$. Without determining $x(t)$, use the scaling property to find the FT representation of $y(t) = x(-2t)$.

Answer:

$$Y(j\omega) = (1/2)e^{j\omega/2}|\omega/2|e^{-|\omega|}$$ ◀

S & S
Solutions

▶ **Problem 3.42**   A periodic signal has FS $x(t) \xleftrightarrow{\ FS; \pi\ } X[k] = e^{-jk\pi/2}|k|e^{-2|k|}$. Without determining $x(t)$, use the scaling property to find the FS representation of $y(t) = x(3t)$.

Answer:

$$y(t) \xleftrightarrow{\ FS; \pi/3\ } Y[k] = e^{-jk\pi/2}|k|e^{-2|k|}$$ ◀

The scaling operation has a slightly different character in discrete time than in continuous time. First of all, $z[n] = x[pn]$ is defined only for integer values of p. Second, if $|p| > 1$, then the scaling operation discards information, since it retains only every pth value of $x[n]$. This loss of information prevents us from expressing the DTFT or DTFS of $z[n]$ in terms of the DTFT or DTFS of $x[n]$ in a manner similar to the way we did for the continuous-time results derived earlier. The scaling of discrete-time signals is further addressed in Problem 3.80.

3.16 *Parseval Relationships*

The Parseval relationships state that the energy or power in the time-domain representation of a signal is equal to the energy or power in the frequency-domain representation. Hence, energy and power are conserved in the Fourier representation. We derive this result for the FT and simply state it for the other three cases.

The energy in a continuous-time nonperiodic signal is

$$W_x = \int_{-\infty}^{\infty} |x(t)|^2\, dt,$$

where it is assumed that $x(t)$ may be complex valued in general. Note that $|x(t)|^2 = x(t)x^*(t)$. Taking the conjugate of both sides of Eq. (3.35), we may express $x^*(t)$ in terms of its FT $X(j\omega)$ as

$$x^*(t) = \frac{1}{2\pi} \int_{-\infty}^{\infty} X^*(j\omega)e^{-j\omega t}\, d\omega.$$

Substituting this formula into the expression for W_x, we obtain

$$W_x = \int_{-\infty}^{\infty} x(t)\left[\frac{1}{2\pi} \int_{-\infty}^{\infty} X^*(j\omega)e^{-j\omega t}\, d\omega\right] dt.$$

Now we interchange the order of integration:

$$W_x = \frac{1}{2\pi} \int_{-\infty}^{\infty} X^*(j\omega)\left\{ \int_{-\infty}^{\infty} x(t)e^{-j\omega t}\, dt\right\} d\omega.$$

Observing that the integral inside the braces is the FT of $x(t)$, we obtain

$$W_x = \frac{1}{2\pi} \int_{-\infty}^{\infty} X^*(j\omega)X(j\omega)\, d\omega$$

and so conclude that

$$\boxed{\int_{-\infty}^{\infty} |x(t)|^2\, dt = \frac{1}{2\pi} \int_{-\infty}^{\infty} |X(j\omega)|^2\, d\omega.}\tag{3.62}$$

Hence, the energy in the time-domain representation of the signal is equal to the energy in the frequency-domain representation, normalized by 2π. The quantity $|X(j\omega)|^2$ plotted against ω is termed the *energy spectrum* of the signal.

Analogous results hold for the other three Fourier representations, as summarized in Table 3.10. The energy or power in the time-domain representation is equal to the energy or power in the frequency-domain representation. Energy is used for nonperiodic time-domain signals, while power applies to periodic time-domain signals. Recall that power is defined as the integral or sum of the magnitude squared over one period, normalized by the length of the period. The power or energy spectrum of a signal is defined as the square of the magnitude spectrum. These relationships indicate how the power or energy in the signal is distributed as a function of frequency.

TABLE 3.10   *Parseval Relationships for the Four Fourier Representations.*

Representation	Parseval Relation
FT	$\int_{-\infty}^{\infty} |x(t)|^2\, dt = \frac{1}{2\pi}\int_{-\infty}^{\infty} |X(j\omega)|^2\, d\omega$
FS	$\frac{1}{T}\int_0^T |x(t)|^2\, dt = \sum_{k=-\infty}^{\infty} |X[k]|^2$
DTFT	$\sum_{n=-\infty}^{\infty} |x[n]|^2 = \frac{1}{2\pi}\int_{-\pi}^{\pi} |X(e^{j\Omega})|^2\, d\Omega$
DTFS	$\frac{1}{N}\sum_{n=0}^{N-1} |x[n]|^2 = \sum_{k=0}^{N-1} |X[k]|^2$

EXAMPLE 3.50 CALCULATING THE ENERGY IN A SIGNAL   Let

$$x[n] = \frac{\sin(Wn)}{\pi n}.$$

Use Parseval's theorem to evaluate

$$\chi = \sum_{n=-\infty}^{\infty} |x[n]|^2$$

$$= \sum_{n=-\infty}^{\infty} \frac{\sin^2(Wn)}{\pi^2 n^2}.$$

Solution: Using the DTFT Parseval relationship in Table 3.10, we have

$$\chi = \frac{1}{2\pi}\int_{-\pi}^{\pi} |X(e^{j\Omega})|^2\, d\Omega.$$

Since

$$x[n] \xleftarrow{\ DTFT\ } X(e^{j\Omega}) = \begin{cases} 1, & |\Omega| \le W \\ 0, & W < |\Omega| \le \pi \end{cases},$$

it follows that

$$\chi = \frac{1}{2\pi}\int_{-W}^{W} 1\, d\Omega,$$

$$= W/\pi.$$

Note that a direct calculation of χ using the time-domain signal $x[n]$ is very difficult. ▪

▶ **Problem 3.43**   Use Parseval's theorem to evaluate the following quantities:

(a)

$$\chi_1 = \int_{-\infty}^{\infty} \frac{2}{|j\omega + 2|^2}\, d\omega$$

(b)

$$\chi_2 = \sum_{k=0}^{29} \frac{\sin^2(11\pi k/30)}{\sin^2(\pi k/30)}$$

Answers:

   (a) $\chi_1 = \pi$
   (b) $\chi_2 = 330$ ◀

3.17 *Time–Bandwidth Product*

Earlier, we observed an inverse relationship between the time and frequency extent of a signal. From Example 3.25, recall that

$$x(t) = \begin{cases} 1, & |t| \le T_o \\ 0, & |t| > T_o \end{cases} \xleftrightarrow{\text{FT}} X(j\omega) = 2\sin(\omega T_o)/\omega.$$

As depicted in Fig. 3.72, the signal $x(t)$ has time extent $2T_o$. The FT of $x(t)$, $X(j\omega)$, is actually of infinite extent in frequency, but has the majority of its energy contained in the interval associated with the mainlobe of the sinc function, $|\omega| < \pi/T_o$. As T_o decreases, the signal's time extent decreases, while the frequency extent increases. In fact, the product of the time extent T_o and mainlobe width $2\pi/T_o$ is a constant.

The general nature of the inverse relationship between time and frequency extent is demonstrated by the scaling property: Compressing a signal in time leads to expansion in the frequency domain and vice versa. This inverse relationship may be formally stated in terms of the signal's time–bandwidth product.

The *bandwidth* of a signal is the extent of the signal's significant frequency content. It is difficult to define bandwidth, especially for signals having infinite frequency extent, because the meaning of the term "significant" is not mathematically precise. In spite of this difficulty, several definitions for "bandwidth" are in common use. One such definition applies to real-valued signals that have a frequency-domain representation characterized by a mainlobe bounded by nulls. If the signal is *low pass* (i.e., if the mainlobe is centered on the origin), then the bandwidth is defined as the frequency corresponding to the first null, which is one-half the width of the mainlobe. With this definition, the signal depicted in Fig. 3.72 has bandwidth π/T_o. If the signal is *band pass*, meaning that the mainlobe is centered on ω_c, then the bandwidth is equal to the distance between nulls, which is equal to the width of the mainlobe. Another commonly used definition of the bandwidth is based on the frequency at which the magnitude spectrum is $1/\sqrt{2}$ times its peak value. At this frequency, the energy spectrum has a value of one-half its peak value. Note that similar difficulty is encountered in precisely defining the time extent or duration of a signal.

The preceding definitions of the bandwidth and duration are not well suited for analytic evaluation. We may analytically describe the inverse relationship between the

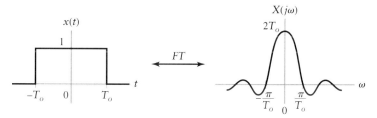

FIGURE 3.72   Rectangular pulse illustrating the inverse relationship between the time and frequency extent of a signal.

duration and bandwidth of arbitrary signals by defining root-mean-square measures of effective duration and bandwidth. We formally define the effective duration of a signal $x(t)$ as

$$T_d = \left[\frac{\int_{-\infty}^{\infty} t^2 |x(t)|^2 \, dt}{\int_{-\infty}^{\infty} |x(t)|^2 \, dt} \right]^{1/2} \tag{3.63}$$

and the bandwidth as

$$B_w = \left[\frac{\int_{-\infty}^{\infty} \omega^2 |X(j\omega)|^2 \, d\omega}{\int_{-\infty}^{\infty} |X(j\omega)|^2 \, d\omega} \right]^{1/2}. \tag{3.64}$$

These definitions assume that $x(t)$ is centered about the origin and is low pass. The interpretation of T_d as an effective duration follows from an examination of Eq. (3.63). The integral in the numerator is the second moment of the signal about the origin. The integrand weights the square of the value of $x(t)$ at each instant of time by the square of the distance of $x(t)$ from $t = 0$. Hence, if $x(t)$ is large for large values of t, the duration will be larger than if $x(t)$ is large for small values of t. This integral is normalized by the total energy in $x(t)$. A similar interpretation applies to B_w. Note that while the root-mean-square definitions offer certain analytic tractability, they are not easily measured from a given signal and its magnitude spectrum.

It can be shown that the time–bandwidth product for any signal is lower bounded according to the relationship

$$T_d B_w \geq 1/2. \tag{3.65}$$

This bound indicates that we cannot simultaneously decrease the duration and bandwidth of a signal. Gaussian pulses are the only signals that satisfy this relationship with equality. Equation (3.65) is also known as the *uncertainty principle* after its application in modern physics, which states that the exact position and exact momentum of an electron cannot be determined simultaneously. This result generalizes to alternative definitions of bandwidth and duration: The product of bandwidth and duration is always lower bounded by a constant, with the value of this constant dependent on the definitions of bandwidth and duration.

EXAMPLE 3.51 BOUNDING THE BANDWIDTH OF A RECTANGULAR PULSE   Let

$$x(t) = \begin{cases} 1, & |t| \leq T_o \\ 0, & |t| > T_o \end{cases}.$$

Use the uncertainty principle to place a lower bound on the effective bandwidth of $x(t)$.

Solution: First use Eq. (3.63) to calculate T_d for $x(t)$:

$$
\begin{aligned}
T_d &= \left[\frac{\int_{-T_o}^{T_o} t^2 \, dt}{\int_{-T_o}^{T_o} dt} \right]^{1/2} \\
&= \left[\left(1/(2T_o) \right)(1/3) t^3 \Big|_{-T_o}^{T_o} \right]^{1/2} \\
&= T_o / \sqrt{3}.
\end{aligned}
$$

The uncertainty principle given by Eq. (3.65) states that $B_w \geq 1/(2T_d)$, so we conclude that

$$B_w \geq \sqrt{3}/(2T_o).$$ ■

Bounds on the time–bandwidth product analogous to Eq. (3.65) can be derived for the other Fourier representations.

3.18 *Duality*

Throughout this chapter, we have observed a consistent symmetry between the time- and frequency-domain representations of signals. For example, a rectangular pulse in either time or frequency corresponds to a sinc function in either frequency or time, as illustrated in Fig. 3.73. An impulse in time transforms to a constant in frequency, while a constant in time transforms to an impulse in frequency. We have also observed symmetries in Fourier representation properties: Convolution in one domain corresponds to modulation in the other domain, differentiation in one domain corresponds to multiplication by the independent variable in the other domain, and so on. These symmetries are a consequence of the symmetry in the definitions of time- and frequency-domain representations. If we are careful, we may interchange time and frequency. This interchangeability property is termed *duality*.

■ 3.18.1 THE DUALITY PROPERTY OF THE FT

Begin with the FT, and recall Eqs. (3.35) and (3.36), respectively:

$$x(t) = \frac{1}{2\pi} \int_{-\infty}^{\infty} X(j\omega)e^{j\omega t}\, d\omega$$

and

$$X(j\omega) = \int_{-\infty}^{\infty} x(t)e^{-j\omega t}\, dt.$$

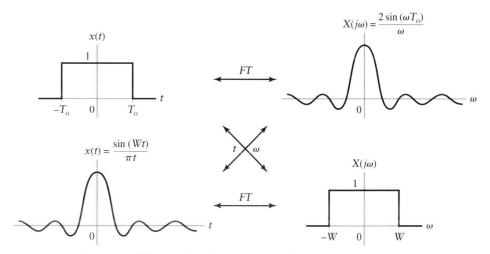

FIGURE 3.73   Duality of rectangular pulses and sinc functions.

The difference between the expression for $x(t)$ and that for $X(j\omega)$ is the factor 2π and the sign change in the complex sinusoid. Both can be expressed in terms of the general equation

$$y(\nu) = \frac{1}{2\pi} \int_{-\infty}^{\infty} z(\eta)e^{j\nu\eta}\, d\nu. \tag{3.66}$$

If we choose $\nu = t$ and $\eta = \omega$, then Eq. (3.66) implies that

$$y(t) = \frac{1}{2\pi} \int_{-\infty}^{\infty} z(\omega)e^{j\omega t}\, d\omega.$$

Therefore, we conclude that

$$y(t) \xleftrightarrow{\;FT\;} z(\omega). \tag{3.67}$$

Conversely, if we interchange the roles of time and frequency by setting $\nu = -\omega$ and $\eta = t$, then Eq. (3.66) implies that

$$y(-\omega) = \frac{1}{2\pi} \int_{-\infty}^{\infty} z(t)e^{-j\omega t}\, dt,$$

and we have

$$z(t) \xleftrightarrow{\;FT\;} 2\pi y(-\omega). \tag{3.68}$$

The relationships of Eqs. (3.67) and (3.68) imply a certain symmetry between the roles of time and frequency. Specifically, if we are given an FT pair

$$f(t) \xleftrightarrow{\;FT\;} F(j\omega), \tag{3.69}$$

then we may interchange the roles of time and frequency to obtain the new FT pair

$$\boxed{F(jt) \xleftrightarrow{\;FT\;} 2\pi f(-\omega).} \tag{3.70}$$

The notation $F(jt)$ means that $F(j\omega)$ in Eq. (3.69) is evaluated with the frequency ω replaced by time t, while $f(-\omega)$ means that $f(t)$ is evaluated with time t replaced by the reflected frequency $-\omega$. The duality relationship described by Eqs. (3.69) and (3.70) is illustrated in Fig. 3.74.

EXAMPLE 3.52 APPLYING DUALITY Find the FT of

$$x(t) = \frac{1}{1 + jt}.$$

Solution: First, recognize that

$$f(t) = e^{-t}u(t) \xleftrightarrow{\;FT\;} F(j\omega) = \frac{1}{1 + j\omega}.$$

Replacing ω by t, we obtain

$$F(jt) = \frac{1}{1 + jt}.$$

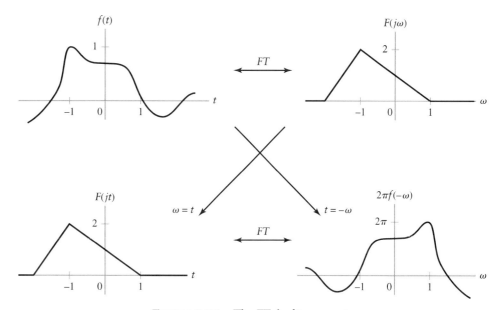

FIGURE 3.74   The FT duality property.

Hence, we have expressed $x(t)$ as $F(jt)$. The duality property given by Eqs. (3.69) and (3.70) states that

$$F(jt) \xleftrightarrow{FT} 2\pi f(-\omega),$$

which implies that

$$X(j\omega) = 2\pi f(-\omega)$$
$$= 2\pi e^{\omega} u(-\omega). \qquad \blacksquare$$

▶ **Problem 3.44**   Use duality to evaluate the inverse FT of the step function in frequency, $X(j\omega) = u(\omega)$.

Answer:

$$x(t) = \frac{-1}{2\pi jt} + \frac{\delta(-t)}{2} \qquad \blacktriangleleft$$

S & S
Solutions

▶ **Problem 3.45**   Use duality to evaluate the FT of

$$x(t) = \frac{1}{1 + t^2}.$$

Answer:

$$X(j\omega) = \pi e^{-|\omega|} \qquad \blacktriangleleft$$

■ **3.18.2 THE DUALITY PROPERTY OF THE DTFS**

The FT stays entirely within its signal class, mapping a continuous-time nonperiodic function into a continuous-frequency nonperiodic function. The DTFS also stays entirely within its signal class, since discrete periodic functions are mapped into discrete periodic functions. The DTFS possesses a duality property analogous to the FT. Recall that

$$x[n] = \sum_{k=0}^{N-1} X[k]e^{jk\Omega_o n}$$

and

$$X[k] = \frac{1}{N}\sum_{n=0}^{N-1} x[n]e^{-jk\Omega_o n}.$$

Here, the difference between the form of the forward transform and that of the inverse transform is the factor N and the change in sign of the complex sinusoidal frequencies. The DTFS duality property is stated as follows: If

$$x[n] \xleftrightarrow{\ DTFS;\, 2\pi/N\ } X[k], \tag{3.71}$$

then

$$X[n] \xleftrightarrow{\ DTFS;\, 2\pi/N\ } \frac{1}{N}x[-k], \tag{3.72}$$

where n is the time index and k is the frequency index. The notation $X[n]$ indicates that $X[k]$ in Eq. (3.71) is evaluated as a function of the time index n, while the notation $x[-k]$ indicates that $x[n]$ in Eq. (3.71) is evaluated as a function of the frequency index $-k$.

■ **3.18.3 THE DUALITY PROPERTY OF THE DTFT AND FS**

The DTFT and FS do not stay within their signal class, so the duality relationship in this case is between the FS and DTFT, as we now show. Recall that the FS maps a continuous periodic function into a discrete nonperiodic function, while the DTFT maps a discrete nonperiodic function into a continuous periodic function. Compare the FS expansion of a periodic continous time signal $z(t)$, given by

$$z(t) = \sum_{k=-\infty}^{\infty} Z[k]e^{jk\omega_o t}$$

and the DTFT of an nonperiodic discrete-time signal $x[n]$, given by

$$X(e^{j\Omega}) = \sum_{n=-\infty}^{\infty} x[n]e^{-j\Omega n}.$$

In order to identify a duality relationship between $z(t)$ and $X(e^{j\Omega})$, we require $z(t)$ to have the same period as $X(e^{j\Omega})$; that is, we require that $T = 2\pi$. With this assumption, $\omega_o = 1$, and we see that Ω in the DTFT corresponds to t in the FS, while n in the DTFT corresponds to $-k$ in the FS. Similarly, the expression for the FS coefficients $Z[k]$ parallels the expression for the DTFT representation of $x[n]$, as shown by

$$Z[k] = \frac{1}{2\pi}\int_{-\pi}^{\pi} z(t)e^{-jkt}\, dt$$

TABLE 3.11	***Duality Properties of Fourier Representations.***

FT	$f(t) \xleftrightarrow{\;FT\;} F(j\omega)$	$F(jt) \xleftrightarrow{\;FT\;} 2\pi f(-\omega)$
DTFS	$x[n] \xleftrightarrow{\;DTFS;2\pi/N\;} X[k]$	$X[n] \xleftrightarrow{\;DTFS;2\pi/N\;} (1/N)x[-k]$
FS–DTFT	$x[n] \xleftrightarrow{\;DTFT\;} X(e^{j\Omega})$	$X(e^{it}) \xleftrightarrow{\;FS;1\;} x[-k]$

and

$$x[n] = \frac{1}{2\pi} \int_{-\pi}^{\pi} X(e^{j\Omega})e^{jn\Omega}\,d\Omega.$$

The roles of Ω and n in the DTFT again correspond to those of t and $-k$ in the FS. We may now state the duality property between the FS and the DTFT: If

$$x[n] \xleftrightarrow{\;DTFT\;} X(e^{j\Omega}), \tag{3.73}$$

then

$$X(e^{it}) \xleftrightarrow{\;FS;1\;} x[-k]. \tag{3.74}$$

The notation $X(e^{it})$ indicates that $X(e^{j\Omega})$ is evaluated as a function of the time index t, while the notation $x[-k]$ indicates that $x[n]$ is evaluated as a function of the frequency index $-k$.
    The duality properties of Fourier representations are summarized in Table 3.11.

EXAMPLE 3.53 FS–DTFT DUALITY   Use the duality property and the results of Example 3.39 to determine the inverse DTFT of the triangular spectrum $X(e^{j\Omega})$ depicted in Fig. 3.75(a).

Solution: Define a time function $z(t) = X(e^{it})$. The duality property of Eq. (3.74) implies that if $z(t) \xleftrightarrow{\;FS;1\;} Z[k]$, then $x[n] = Z[-n]$. Hence, we seek the FS coefficients $Z[k]$ associated with $z(t)$. Now $z(t)$ is a time-shifted version of the triangular wave $y(t)$

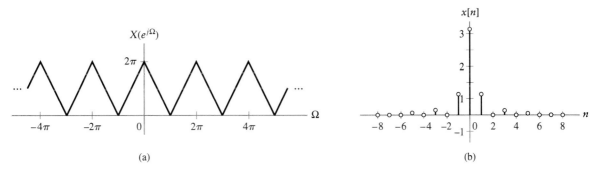

(a)   (b)

FIGURE 3.75   Example 3.53. (a) Triangular spectrum. (b) Inverse DTFT.

considered in Example 3.39, assuming that $T = 2\pi$. Specifically, $z(t) = y(t + \pi/2)$. Using the time-shift property, we have

$$Z[k] = e^{jk\pi/2}Y[k]$$

$$= \begin{cases} \pi, & k = 0 \\ \dfrac{4j^{k-1}\sin(k\pi/2)}{\pi k^2}, & k \neq 0 \end{cases}.$$

Consequently, from $x[n] = Z[-n]$, we obtain

$$x[n] = \begin{cases} \pi, & n = 0 \\ \dfrac{-4(-j)^{n+1}\sin(n\pi/2)}{\pi n^2}, & n \neq 0 \end{cases}.$$

Figure 3.75(b) depicts $x[n]$.   ■

3.19 *Exploring Concepts with MATLAB*

■ 3.19.1 FREQUENCY RESPONSE OF LTI SYSTEMS FROM IMPULSE RESPONSE

The frequency response of a system is a continuous function of frequency. Numerically, however, we can evaluate the frequency response only at discrete values of frequency. Thus, a large number of values are normally used to capture the details in the system's frequency response. Recall that the impulse and frequency response of a continuous-time system are related through the FT, while the DTFT relates the impulse and frequency response of discrete-time systems. Hence, determining the frequency response directly from a description of the impulse response requires approximating either the DTFT or the FT with the DTFS, a topic that is discussed in Sections 4.8 and 4.9, respectively.

We may identify the frequency response of a discrete-time LTI system by measuring the amplitude and phase change of the infinite-duration complex sinusoidal input signal $x[n] = e^{j\Omega n}$. The frequency response of a discrete-time LTI system with finite-duration impulse response may be determined with the use of a finite-duration input sinusoid that is sufficiently long to drive the system to a steady state. To demonstrate this idea, suppose $h[n] = 0$ for $n < k_h$ and $n > l_h$, and let the system input be the finite-duration sinusoid $v[n] = e^{j\Omega n}(u[n] - u[n - l_v])$. Then we may write the system output as

$$\begin{aligned} y[n] &= h[n] * v[n] \\ &= \sum_{k=k_h}^{l_h} h[k]e^{j\Omega(n-k)}, \quad l_h \leq n < k_h + l_v \\ &= h[n] * e^{j\Omega n}, \quad l_h \leq n < k_h + l_v \\ &= H(e^{j\Omega})e^{j\Omega n}, \quad l_h \leq n < k_h + l_v. \end{aligned}$$

Hence, the system output in response to a finite-duration sinusoidal input corresponds to the output in response to an infinite-duration sinusoidal input on the interval $l_b \leq n < k_b + l_v$. The magnitude and phase response of the system may be determined from $y[n]$, $l_b \leq n < k_b + l_v$, by noting that

$$y[n] = |H(e^{j\Omega})|e^{j(\Omega n + \arg\{H(e^{j\Omega})\})}, \qquad l_b \leq n < k_b + l_v.$$

We take the magnitude and phase of $y[n]$ to obtain

$$|y[n]| = |H(e^{j\Omega})|, \qquad l_b \leq n < k_b + l_v$$

and

$$\arg\{y[n]\} - \Omega n = \arg\{H(e^{j\Omega})\}, \qquad l_b \leq n < k_b + l_v.$$

We may use this approach to evaluate the frequency response of one of the systems given in Example 3.22. Consider the system with impulse response

$$h_2[n] = \frac{1}{2}\delta[n] - \frac{1}{2}\delta[n-1].$$

Let us determine the frequency response and 50 values of the steady-state output of this system for input frequencies $\Omega = \frac{\pi}{4}$ and $\frac{3\pi}{4}$.

Here, $k_b = 0$ and $l_b = 1$, so, to obtain 50 values of the sinusoidal steady-state response, we require that $l_v \geq 51$. The output signals are obtained by MATLAB commands:

```
>> Omega1 = pi/4;   Omega2 = 3*pi/4;
>> v1 = exp(j*Omega1*[0:50]);
>> v2 = exp(j*Omega2*[0:50]);
>> h = [0.5, -0.5];
>> y1 = conv(v1, h);   y2 = conv(v2, h);
```

Figures 3.76(a) and (b) depict the real and imaginary components of y1, respectively, and may be obtained with the following commands:

```
>> subplot(2, 1, 1)
>> stem([0:51], real(y1))
>> xlabel('Time'); ylabel('Amplitude');
>> title('Real(y1)')
>> subplot(2, 1, 2)
>> stem([0:51], imag(y1))
>> xlabel('Time'); ylabel('Amplitude');
   title('Imag(y1)')
```

The steady-state outputs are represented by the values at time indices 1 through 50.

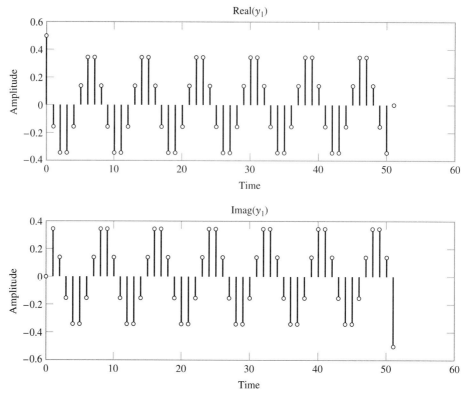

FIGURE 3.76   Sinusoidal steady-state response computed with the use of MATLAB. The values at times 1 through 50 represent the sinusoidal steady-state response.

We may now obtain the magnitude and phase responses from any element of the vectors y1 and y2, except for the first one or the last one. We use the fifth element and the following MATLAB commands:

```
>> H1mag = abs(y1(5))
H1mag =
    0.3287
>> H2mag = abs(y2(5))
H2mag =
    0.9239
>> H1phs = angle(y1(5)) - Omega1*5
H1phs =
    -5.8905
>> H2phs = angle(y2(5)) - Omega2*5
H2phs =
    -14.5299
```

The phase response is measured in radians. Note that the angle command always returns a value between $-\pi$ and π radians. Hence, measuring the phase with the command angle(y1(n)) - Omega1*n may result in answers that differ by integer multiples of 2π when different values of n are used.

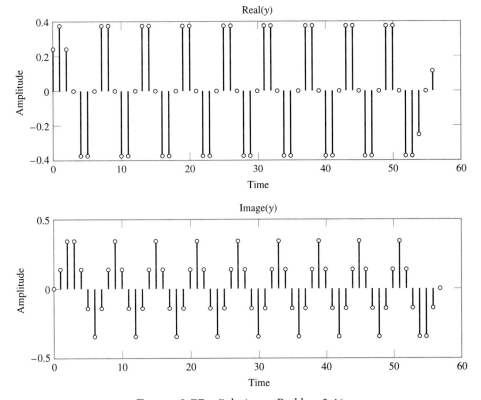

FIGURE 3.77   Solution to Problem 3.46.

▶ **Problem 3.46**   Evaluate the frequency response at the frequency $\Omega = \frac{\pi}{3}$ and 50 values of the steady-state output in response to a complex sinusoidal input of frequency $\Omega = \frac{\pi}{3}$ for the moving-average system with impulse response

$$h[n] = \begin{cases} \dfrac{1}{4} & 0 \le n \le 3 \\ 0 & \text{otherwise} \end{cases}.$$

Answer: See Fig. 3.77   ◀

■ 3.19.2 THE DTFS

The DTFS is the only Fourier representation that is discrete valued in both time and frequency and hence is suited for direct MATLAB implementation. While Eqs. (3.10) and (3.11) are easily implemented as M-files, the built in MATLAB commands **fft** and **ifft** may also be used to evaluate the DTFS. Given a length–N vector **x** representing one period of an N periodic signal $x[n]$, the command

```
>> X = fft(x)/N
```

produces a length-N vector **X** containing the DTFS coefficients $X[k]$. MATLAB assumes that the summations in Eqs. (3.10) and (3.11) run from 0 to $N - 1$, so the first elements of **x** and **X** correspond to $x[0]$ and $X[0]$, respectively, while the last elements correspond to $x[N - 1]$ and $X[N - 1]$. Note that division by N is necessary because **fft** evaluates the sum in Eq. (3.11) without dividing by N. Similarly, given DTFS coefficients in a vector **X**, the command

```
>> x = ifft(X)*N
```

produces a vector **x** that represents one period of the time-domain waveform. Note that **ifft** must be multiplied by N to evaluate Eq. (3.10). Both **fft** and **ifft** are computed by means of a numerically efficient or fast algorithm termed the *fast Fourier transform*. The development of this algorithm is discussed in Section 4.10.

Consider using MATLAB to solve Problem 3.3(a) for the DTFS coefficients. The signal is

$$x[n] = 1 + \sin\left(n\pi/12 + \frac{3\pi}{8}\right)$$

This signal has period 24, so we define one period and evaluate the DTFS coefficients using the following commands:

```
>> x = ones(1,24) + sin([0:23]*pi/12 + 3*pi/8);
>> X = fft(x)/24
X =
Columns 1 through 4
1.0000   0.4619 - 0.1913i   0.0000 + 0.0000i
     -0.0000 + 0.0000i
Columns 5 through 8
0.0000 + 0.0000i   -0.0000 - 0.0000i   0.0000 - 0.0000i
     -0.0000 - 0.0000i
Columns 9 through 12
-0.0000 - 0.0000i   -0.0000 - 0.0000i   -0.0000 - 0.0000i
     0.0000 - 0.0000i
Columns 13 through 16
0.0000 + 0.0000i   0.0000 + 0.0000i   -0.0000 + 0.0000i
     0.0000 - 0.0000i
Columns 17 through 20
-0.0000 - 0.0000i   -0.0000 - 0.0000i   0.0000 + 0.0000i
     -0.0000 + 0.0000i
Columns 21 through 24
-0.0000 + 0.0000i   -0.0000 - 0.0000i   0.0000 - 0.0000i
     0.4619 + 0.1913i
```

(Note that MATLAB uses **i** to denote the square root of -1.) We conclude that

$$X[k] = \begin{cases} 1, & k = 0 \\ 0.4619 - j0.1913, & k = 1 \\ 0.4619 + j0.1913, & k = 23 \\ 0, & \text{otherwise on } 0 \le k \le 23 \end{cases},$$

which corresponds to the answer to Problem 3.3(a) expressed in rectangular form. Note that since $X[k]$ has period 24, using indices $-11 \le k \le 12$, we may also write the answer by specifying one period as

$$X[k] = \begin{cases} 1, & k = 0 \\ 0.4619 - j0.1913, & k = 1 \\ 0.4619 + j0.1913, & k = -1 \\ 0, & \text{otherwise on } -11 \le k \le 12 \end{cases}.$$

Using ifft, we may reconstruct the time-domain signal and evaluate the first four values of the reconstructed signal with the commands

```
>> xrecon = ifft(X)*24;
>> xrecon(1:4);

ans =
1.9239 - 0.0000i   1.9914 + 0.0000i   1.9914 + 0.0000i
1.9239 - 0.0000i
```

Note that the reconstructed signal has an imaginary component (albeit a very small one), even though the original signal was purely real. The imaginary component is an artifact of numerical rounding errors in the computations performed by fft and ifft and may be ignored.

▶ **Problem 3.47**   Repeat Problem 3.2, using MATLAB.

The partial-sum approximation used in Example 3.7 is easily evaluated in MATLAB:

```
>> k = 1:24:
>> n = -24:25:
>> B(1) = 25/50;    % coeff for k = 0
>> B(2:25) = 2*sin(k*pi*25/50)./(50*sin(k*pi/50));
>> B(26) = sin(25*pi*25/50)/(50*sin(25*pi/50));
    % coeff for k = N/2
>> xJhat(1,:) = B(1)*cos(n*0*pi/25);
    % term in sum for k = 0
    % accumulate partial sums
>> for k = 2:26
xJhat(k,:) = xJhat(k-1,:) + B(k)*cos(n*(k-1)*pi/25);
end
```

This set of commands produces a matrix xJhat whose $(J + 1)$st row corresponds to $\hat{x}_J[n]$. ◀

■ 3.19.3 THE FS

The partial-sum approximation to the trigonometric FS in Example 3.14 is evaluated analogously to that of the DTFS, but with one important additional consideration: The signal $\hat{x}_J(t)$ and the cosines in the partial-sum approximation are continuous functions of time. Since MATLAB represents these functions as vectors consisting of discrete points, we must use sufficiently closely spaced samples to capture the details in $\hat{x}_J(t)$. This is assured by sampling the functions closely enough so that the highest-frequency term in the sum, $\cos(J_{max}\omega_o t)$, is well approximated by the sampled signal, $\cos(J_{max}\omega_o n T_s)$. With MATLAB's plot command, the sampled cosine provides a visually pleasing approximation to the continuous cosine if there are on the order of 20 samples per period. Using 20 samples per period, we obtain $T_s = T/(20 J_{max})$. Note that the total number of samples in one period is then $20 J_{max}$. Assuming $J_{max} = 99$ and $T = 1$, we may compute the partial sums, given $B[k]$, by using the following commands:

```
>> t = [-(10*Jmax-1):10*Jmax]*(1/(20*99));
>> xJhat(1,:) = B(1)*cos(t*0*2*pi/T);
>> for k = 2:100
xJhat(k,:) = xJhat(k-1,:) + B(k)*cos(t*(k-1)*2*pi/T);
end
```

Since the rows of **xJhat** represent samples of a continuous-valued function, we display them by means of **plot** instead of **stem**. For example, the partial sum for $J = 5$ is displayed with the command **plot(t, xJhat(6,:))**.

■ 3.19.4 FREQUENCY RESPONSE OF LTI SYSTEMS DESCRIBED BY DIFFERENTIAL OR DIFFERENCE EQUATIONS

The MATLAB Signal Processing and Control System Toolboxes contain the commands **freqs** and **freqz**, which evaluate the frequency response for systems described by differential and difference equations, respectively. The command **h = freqs(b,a,w)** returns the values of the continuous-time system frequency response given by Eq. (3.47) at the frequencies specified in the vector **w**. Here, we assume that vectors **b** = $[b_M, b_{M-1}, \ldots, b_0]$ and **a** = $[a_N, a_{N-1}, \ldots, a_0]$ represent the coefficients of the differential equation. The frequency response of the MEMS accelerometer depicted in Fig. 3.58 for $Q = 1$ is obtained via the commands

```
>> w = 0:100:100000
>> b = 1;
>> a = [1 10000 10000*10000];
>> H = freqs(b,a,w);
>> plot(w,20*log10(abs(H)))
```

The syntax for **freqz** is different from that for **freqs** in a subtle way. The command **h = freqz(b,a,w)** evaluates the discrete-time system frequency response given by Eq. (3.55) at the frequencies specified in the vector **w**. In the discrete-time case, the entries of **w** must lie between 0 and 2π, and the vectors **b** = $[b_0, b_1, \ldots, b_M]$ and **a** = $[a_0, a_1, \ldots, a_N]$ contain the difference-equation coefficients in the reverse order of that required by **freqs**.

■ 3.19.5 TIME–BANDWIDTH PRODUCT

The **fft** command may be used to evaluate the DTFS and explore the time–bandwidth product property for discrete-time periodic signals. Since the DTFS applies to signals that are periodic in both time and frequency, we define both duration and bandwidth on the basis of the extent of the signal within one period. For example, consider the period-N square wave studied in Example 3.6. One period of the time-domain signal is defined as

$$x[n] = \begin{cases} 1, & |n| \leq M \\ 0, & M < n < N - M \end{cases},$$

and the DTFS coefficients are given by

$$X[k] = \frac{1}{N} \frac{\sin\left(k\frac{\pi}{N}(2M + 1)\right)}{\sin\left(k\frac{\pi}{N}\right)}.$$

If we define the duration T_d as the nonzero portion of one period of $x[n]$, then $T_d = 2M + 1$. If we further define the bandwidth B_w as the "frequency" of the first null of $X[k]$, then we have $B_w \approx N/(2M + 1)$, and we see that the time–bandwidth product for the square wave, $T_d B_w \approx N$, is independent of M.

The following set of MATLAB commands may be used to verify this result:

```
>> x = [ones(1,M+1), zeros(1,N-2M-1), ones(1,M)];
>> X = fft(x)/N;
>> k = [0:N-1];   % frequency index
>> stem(k, real(fftshift(X)))
```

Here, we define one period of an even square wave on the interval $0 \le n \le N - 1$, find the DTFS coefficients by means of the `fft` command, and display them by using `stem`. The `real` command is used to suppress any small imaginary components resulting from numerical rounding. The `fftshift` command reorders the elements of the vector X to generate the DTFS coefficients centered on $k = 0$. We then determine the effective bandwidth by counting the number of DTFS coefficients before the first zero crossing. One of the computer experiments at the end of the chapter evaluates the time–bandwidth product in this fashion.

The formal definitions of effective duration and bandwidth given in Eqs. (3.63) and (3.64), respectively, may be generalized to discrete-time periodic signals by replacing the integrals with sums over one period. We get

$$T_d = \left[\frac{\sum_{n=-(N-1)/2}^{(N-1)/2} n^2 |x[n]|^2}{\sum_{n=-(N-1)/2}^{(N-1)/2} |x[n]|^2} \right]^{\frac{1}{2}} \tag{3.75}$$

and

$$B_w = \left[\frac{\sum_{k=-(N-1)/2}^{(N-1)/2} k^2 |X[k]|^2}{\sum_{k=-(N-1)/2}^{(N-1)/2} |X[k]|^2} \right]^{\frac{1}{2}}. \tag{3.76}$$

Here, we assume that N is odd and the majority of the energy in $x[n]$ and $X[k]$ within one period is centered around the origin.

The following MATLAB function evaluates the product $T_d B_w$ on the basis of Eqs. (3.75) and (3.76):

```
function TBP = TdBw(x)
% Compute the Time-Bandwidth product using the DTFS
% One period must be less than 1025 points
% N=1025;
M = (N - max(size(x)))/2;
xc = [zeros(1,M),x,zeros(1,M)];
    % center pulse within a period
n = [-(N-1)/2:(N-1)/2];
n2 = n.*n;
Td = sqrt((xc.*xc)*n2'/(xc*xc'));
X = fftshift(fft(xc)/N);   % evaluate DTFS and center
Bw = sqrt(real((X.*conj(X))*n2'/(X*X')));
TBP = Td*Bw;
```

This function assumes that the length of the input signal x is odd and centers x within a 1025-point period before computing `Td` and `Bw`. Note that `.*` is used to perform the element-by-element product. Placed between a row vector and a column vector, the `*` operation computes the inner product. The apostrophe ' indicates the complex-conjugate transpose. Hence, the command `X*X'` performs the inner product of X and the complex conjugate of X—that is, the sum of the magnitude squared of each element of X.

We may use the function `TdBw` to evaluate the time–bandwidth product for two rectangular, raised cosine, and Gaussian pulse trains as follows:

```
>> x = ones(1,101);    % 101 point rectangular pulse
>> TdBw(x)
ans =
788.0303
>> x = ones(1,301);    % 301 point rectangular pulse
>> TdBw(x)
ans =
1.3604e+03
>> x = 0.5*ones(1,101) + cos(2*pi*[-50:50]/101);
    % 101 point raised cosine
>> TdBw(x)
ans =
277.7327
>> x = 0.5*ones(1,301) + cos(2*pi*[-150:150]/301);
    % 301 point raised cosine
>> TdBw(x)
ans =
443.0992
>> n = [-500:500];
>> x = exp(-0.001*(n.*n));    % narrow Gaussian pulse
>> TdBw(x)
ans =
81.5669
>> x = exp(-0.0001*(n.*n));    % broad Gaussian pulse
>> TdBw(x)
ans =
81.5669
```

Note that the Gaussian pulse trains have the smallest time–bandwidth product. Furthermore, the time–bandwidth product is identical for both the narrow and broad Gaussian pulse trains. These observations offer evidence that the time–bandwidth product for periodic discrete-time signals is lower bounded by that of a Gaussian pulse train. Such a result would not be too surprising, given that the Gaussian pulses attain the lower bound for continuous-time non-periodic signals. (This issue is revisited as a computer experiment in Chapter 4.)

▌3.20 *Summary*

In this chapter, we developed techniques for representing signals as weighted superpositions of complex sinusoids. The weights are a function of the complex sinusoidal frequencies and provide a frequency-domain description of the signal. There are four distinct representations applicable to four different signal classes:

▶ The DTFS applies to discrete-time N-periodic signals and represents the signal as a weighted sum of N discrete-time complex sinusoids whose frequencies are integer multiples of the fundamental frequency of the signal. This frequency-domain representation is a discrete and N-periodic function of frequency. The DTFS is the only Fourier representation that can be computed numerically.

▶ The FS applies to continuous-time periodic signals and represents the signal as a weighted sum of an infinite number of continuous-time complex sinusoids whose frequencies are integer multiples of the signal's fundamental frequency. Here, the frequency-domain representation is a discrete and nonperiodic function of frequency.

▶ The DTFT represents nonperiodic discrete-time signals as a weighted integral of discrete-time complex sinusoids whose frequencies vary continuously over an interval of 2π. This frequency-domain representation is a continuous and 2π-periodic function of frequency.

▶ The FT represents nonperiodic continuous-time signals as a weighted integral of continuous-time complex sinusoids whose frequencies vary continuously from $-\infty$ to ∞. Here, the frequency-domain representation is a continuous and nonperiodic function of frequency.

Fourier representation properties, a consequence of the properties of complex sinusoids, relate the effect of an action on a signal in the time domain to a corresponding change in the frequency-domain representation. Since all four representations employ complex sinusoids, all four share similar properties. The properties afford an insight into the nature of both time- and frequency-domain signal representations, as well as providing a powerful set of tools for manipulating signals in both the time and frequency domain. Often, it is much simpler to use the properties to determine a time- or frequency-domain signal representation than it is to use the defining equation.

The frequency domain offers an alternative perspective of signals and the systems they interact with. Certain characteristics of signals are more easily identified in the frequency domain than in the time domain and vice versa. Also, some systems' problems are more easily solved in the frequency domain than in the time domain and vice versa. For example, convolution of time-domain signals corresponds to multiplication of the respective frequency-domain representations. Depending on the problem of interest, one or the other of these two operations is relatively easy to accomplish, and that dictates which approach is adopted to solve the problem. Both the time- and frequency-domain representations have their own advantages and disadvantages. Where one may excel, the other may be cumbersome. Determining which domain is the most advantageous for solving a particular problem is an important skill to develop and can be accomplished only through experience. We continue our journey in the next chapter by studying Fourier analysis for problems involving a mixture of different classes of signals.

FURTHER READING

1. Joseph Fourier studied the flow of heat in the early 19th century. Understanding heat flow was a problem of both practical and scientific significance at that time and required solving a partial-differential equation called the heat equation. Fourier developed a technique for solving partial-differential equations that was based on the assumption that the solution was a weighted sum of harmonically related sinusoids with unknown coefficients, which we now term the Fourier series. Fourier's initial work on heat conduction was submitted as a paper to the Academy of Sciences of Paris in 1807 and rejected after review by Lagrange, Laplace, and Legendre. Fourier persisted in developing his ideas in spite of being criticized for a lack of rigor by his contemporaries. Eventually, in 1822, he published a book containing much of his work, *Theorie analytique de la chaleur*, which is now regarded as one of the classics of mathematics.

2. The DTFS differs from the DFT in the signal-processing literature by a factor of N. For example, the MATLAB command `fft` computes the DFT—hence the need to divide by N

when using **fft** to compute DTFS coefficients. We have adopted the DTFS in this text because the terminology involved is more descriptive and less likely to lead to confusion with the DTFT. The reader should be aware that he or she will likely encounter DFT terminology in other texts and references.

3. A general treatment of Fourier analysis is presented in

   ▶ Kammler, D. W., *A First Course in Fourier Analysis* (Prentice-Hall, 2000)

   ▶ Bracewell, R. N., *The Fourier Transform and Its Applications*, 2nd ed. (McGraw-Hill, 1978)

   ▶ Papoulis, A. *The Fourier Integral and Its Applications* (McGraw-Hill, 1962)

   The text by Kammler provides a mathematical treatment of the FT, FS, DTFT, and DTFS. The texts by Bracewell and Papoulis are application oriented and focus on the FT.

4. The role of the FS and FT in solving partial-differential equations such as the heat equation, wave equation, and potential equation is described in

   ▶ Powers, D. L., *Boundary Value Problems* 2nd ed.(Academic Press, 1979)

5. The uncertainty principle, Eq. (3.65), is proved in Bracewell, *op. cit.*

ADDITIONAL PROBLEMS

3.48 Use the defining equation for the DTFS coefficients to evaluate the DTFS representation of the following signals:

(a) $x[n] = \cos\left(\frac{6\pi}{17}n + \frac{\pi}{3}\right)$

(b) $x[n] = 2\sin\left(\frac{14\pi}{19}n\right) + \cos\left(\frac{10\pi}{19}n\right) + 1$

(c) $x[n] = \sum_{m=-\infty}^{\infty}(-1)^m(\delta[n-2m] + \delta[n+3m])$

(d) $x[n]$ as depicted in Figure P3.48(a).

(e) $x[n]$ as depicted in Figure P3.48(b).

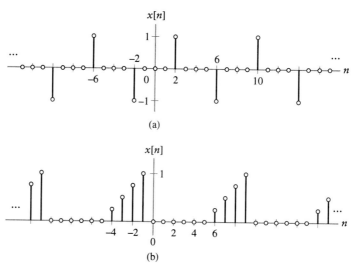

(a)

(b)

FIGURE P3.48

3.49 Use the definition of the DTFS to determine the time-domain signals represented by the following DTFS coefficients:

(a) $X[k] = \cos\left(\frac{8\pi}{21}k\right)$

(b) $X[k] = \cos\left(\frac{10\pi}{19}k\right) + j2\sin\left(\frac{4\pi}{19}k\right)$

(c) $X[k] = \sum_{m=-\infty}^{\infty}(-1)^m(\delta[k-2m] - 2\delta[k+3m])$

(d) $X[k]$ as depicted in Figure P3.49(a).

(e) $X[k]$ as depicted in Figure P3.49(b).

(f) $X[k]$ as depicted in Figure P3.49(c).

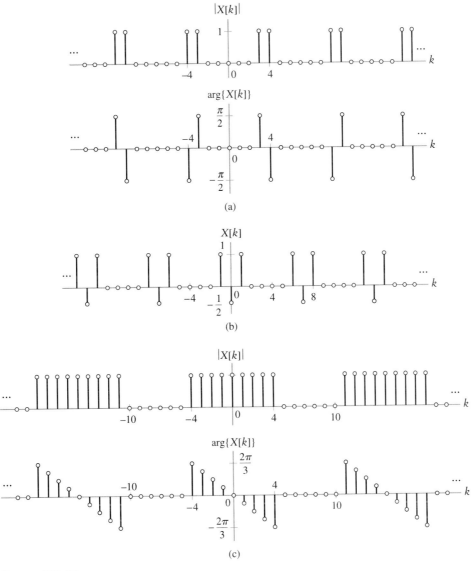

FIGURE P3.49

3.50 Use the defining equation for the FS coefficients to evaluate the FS representation of the following signals:

(a) $x(t) = \sin(3\pi t) + \cos(4\pi t)$

(b) $x(t) = \sum_{m=-\infty}^{\infty} \delta(t - m/3) + \delta(t - 2m/3)$

(c) $x(t) = \sum_{m=-\infty}^{\infty} e^{j\frac{2\pi}{7}m} \delta(t - 2m)$

(d) $x(t)$ as depicted in Figure P3.50(a).

(e) $x(t)$ as depicted in Figure P3.50(b).

(f) $x(t)$ as depicted in Figure P3.50(c).

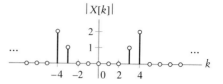

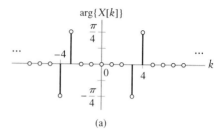

(a)

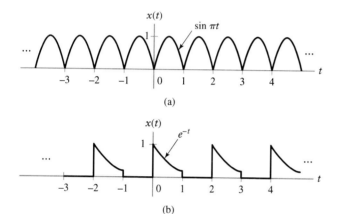

(a)

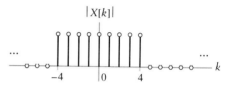

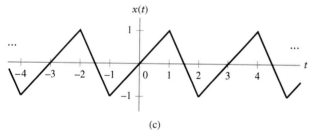

(b)

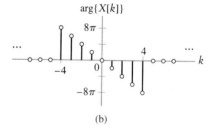

(b)

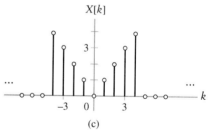

(c)

(c)

FIGURE P3.50

FIGURE P3.51

3.51 Use the definition of the FS to determine the time-domain signals represented by the following FS coefficients:

(a) $X[k] = j\delta[k - 1] - j\delta[k + 1] + \delta[k - 3] + \delta[k + 3]$, $\omega_o = 2\pi$

(b) $X[k] = j\delta[k - 1] - j\delta[k + 1] + \delta[k - 3] + \delta[k + 3]$, $\omega_o = 4\pi$

(c) $X[k] = \left(\frac{-1}{3}\right)^{|k|}$, $\omega_o = 1$

(d) $X[k]$ as depicted in Figure P.3.51(a), $\omega_o = \pi$.

(e) $X[k]$ as depicted in Figure P.3.51(b), $\omega_o = 2\pi$.

(f) $X[k]$ as depicted in Figure P.3.51(c), $\omega_o = \pi$.

3.52 Use the defining equation for the DTFT to evaluate the frequency-domain representations of the following signals:

(a) $x[n] = \left(\frac{3}{4}\right)^n u[n - 4]$

(b) $x[n] = a^{|n|}$ $|a| < 1$

(c) $x[n] = \begin{cases} \frac{1}{2} + \frac{1}{2}\cos\left(\frac{\pi}{N}n\right), & |n| \le N \\ 0, & \text{otherwise} \end{cases}$

(d) $x[n] = 2\delta[4 - 2n]$

(e) $x[n]$ as depicted in Figure P3.52(a).

(f) $x[n]$ as depicted in Figure P3.52(b).

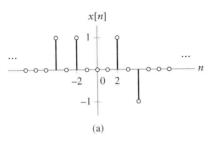

FIGURE P3.52

3.53 Use the equation describing the DTFT representation to determine the time-domain signals corresponding to the following DTFTs:

(a) $X(e^{j\Omega}) = \cos(2\Omega) + j\sin(2\Omega)$

(b) $X(e^{j\Omega}) = \sin(\Omega) + \cos\left(\frac{\Omega}{2}\right)$.

(c) $|X(e^{j\Omega})| = \begin{cases} 1, & \pi/4 < |\Omega| < 3\pi/4, \\ 0, & \text{otherwise} \end{cases}$

   $\arg\{X(e^{j\Omega})\} = -4\Omega$

(d) $X(e^{j\Omega})$ as depicted in Figure P3.53(a).

(e) $X(e^{j\Omega})$ as depicted in Figure P3.53(b).

(f) $X(e^{j\Omega})$ as depicted in Figure P3.53(c).

3.54 Use the defining equation for the FT to evaluate the frequency-domain representations of the following signals:

(a) $x(t) = e^{-2t}u(t - 3)$

(b) $x(t) = e^{-4|t|}$

(c) $x(t) = te^{-t}u(t)$

(d) $x(t) = \sum_{m=0}^{\infty} a^m\delta(t - m), |a| < 1$

(e) $x(t)$ as depicted in Figure P3.54(a).

(f) $x(t)$ as depicted in Figure P3.54(b).

3.55 Use the equation describing the FT representation to determine the time-domain signals corresponding to the following FTs:

(a) $X(j\omega) = \begin{cases} \cos(2\omega), & |\omega| < \frac{\pi}{4} \\ 0, & \text{otherwise} \end{cases}$

(b) $X(j\omega) = e^{-2\omega}u(\omega)$

(c) $X(j\omega) = e^{-2|\omega|}$

(d) $X(j\omega)$ as depicted in Figure P3.55(a).

(e) $X(j\omega)$ as depicted in Figure P3.55(b).

(f) $X(j\omega)$ as depicted in Figure P3.55(c).

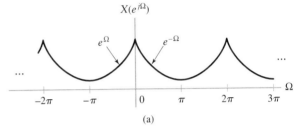

(a)

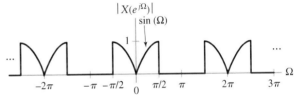

(b)

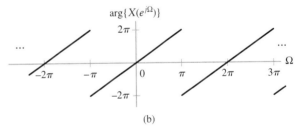

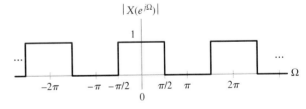

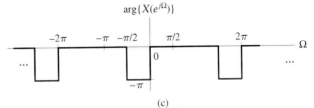

(c)

FIGURE P3.53

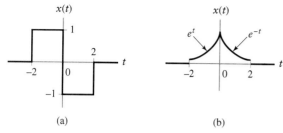

FIGURE P3.54

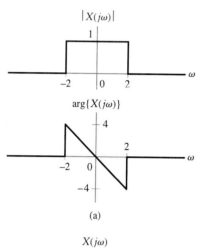

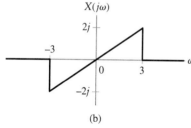

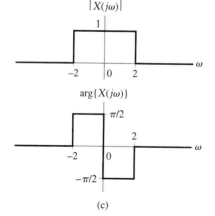

FIGURE P3.55

3.56 Determine the appropriate Fourier representations of the following time-domain signals, using the defining equations:

(a) $x(t) = e^{-t}\cos(2\pi t)u(t)$

(b) $x[n] = \begin{cases} \cos(\frac{\pi}{10}n) + j\sin(\frac{\pi}{10}n), & |n| < 10 \\ 0, & \text{otherwise} \end{cases}$

(c) $x[n]$ as depicted in Figure P3.56(a).

(d) $x(t) = e^{1+t}u(-t+2)$

(e) $x(t) = |\sin(2\pi t)|$

(f) $x[n]$ as depicted in Figure P3.56(b).

(g) $x(t)$ as depicted in Figure P3.56(c).

3.57 Determine the time-domain signal corresponding to each of the following frequency-domain representations:

(a) $X[k] = \begin{cases} e^{-jk\pi/2}, & |k| < 10 \\ 0, & \text{otherwise} \end{cases}$

Fundamental period of time domain signal is $T = 1$.

(b) $X[k]$ as depicted in Figure P3.57(a).

(c) $X(j\omega) = \begin{cases} \cos(\frac{\omega}{4}) + j\sin(\frac{\omega}{4}), & |\omega| < \pi \\ 0, & \text{otherwise} \end{cases}$

(d) $X(j\omega)$ as depicted in Figure P3.57(b).

(e) $X(e^{j\Omega})$ as depicted in Figure P3.57(c).

(f) $X[k]$ as depicted in Figure P3.57(d).

(g) $X(e^{j\Omega}) = |\sin(\Omega)|$

3.58 Use the tables of transforms and properties to find the FT's of the following signals:

(a) $x(t) = \sin(2\pi t)e^{-t}u(t)$

(b) $x(t) = te^{-3|t-1|}$

(c) $x(t) = \left[\frac{2\sin(3\pi t)}{\pi t}\right]\left[\frac{\sin(2\pi t)}{\pi t}\right]$

(d) $x(t) = \frac{d}{dt}(te^{-2t}\sin(t)u(t))$

(e) $x(t) = \int_{-\infty}^{t}\frac{\sin(2\pi\tau)}{\pi\tau}d\tau$

(f) $x(t) = e^{-t+2}u(t-2)$

(g) $x(t) = \left(\frac{\sin(t)}{\pi t}\right) * \frac{d}{dt}\left[\left(\frac{\sin(2t)}{\pi t}\right)\right]$

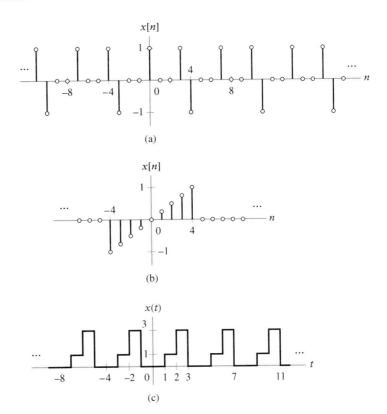

FIGURE P3.56

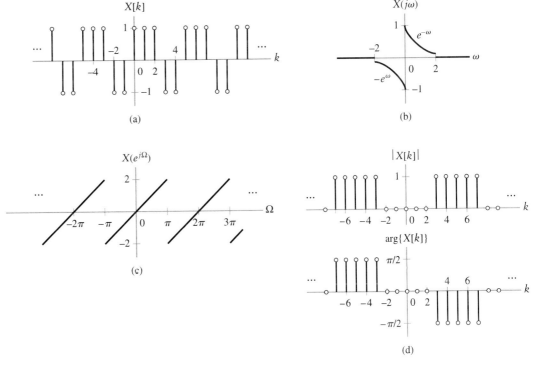

FIGURE P3.57

3.59 Use the tables of transforms and properties to find the inverse FTs of the following signals:

(a) $X(j\omega) = \dfrac{j\omega}{(1 + j\omega)^2}$

(b) $X(j\omega) = \dfrac{4\sin(2\omega - 4)}{2\omega - 4} - \dfrac{4\sin(2\omega + 4)}{2\omega + 4}$

(c) $X(j\omega) = \dfrac{1}{j\omega(j\omega + 2)} - \pi\delta(\omega)$

(d) $X(j\omega) = \dfrac{d}{d\omega}\left[4\sin(4\omega)\dfrac{\sin(2\omega)}{\omega}\right]$

(e) $X(j\omega) = \dfrac{2\sin(\omega)}{\omega(j\omega + 2)}$

(f) $X(j\omega) = \dfrac{4\sin^2(\omega)}{\omega^2}$

3.60 Use the tables of transforms and properties to find the DTFTs of the following signals:

(a) $x[n] = \left(\frac{1}{3}\right)^n u[n + 2]$

(b) $x[n] = (n - 2)(u[n + 4] - u[n - 5])$

(c) $x[n] = \cos\left(\frac{\pi}{4}n\right)\left(\frac{1}{2}\right)^n u[n - 2]$

(d) $x[n] = \left[\dfrac{\sin\left(\frac{\pi}{4}n\right)}{\pi n}\right] * \left[\dfrac{\sin\left(\frac{\pi}{4}(n - 8)\right)}{\pi(n - 8)}\right]$

(e) $x[n] = \left[\dfrac{\sin\left(\frac{\pi}{2}n\right)}{\pi n}\right]^2 * \dfrac{\sin\left(\frac{\pi}{2}n\right)}{\pi n}$

3.61 Use the tables of transforms and properties to find the inverse DTFTs of the following signals:

(a) $X(e^{j\Omega}) = j\sin(4\Omega) - 2$

(b) $X(e^{j\Omega}) = \left[e^{-j2\Omega}\dfrac{\sin\left(\frac{15}{2}\Omega\right)}{\sin\left(\frac{\Omega}{2}\right)}\right] \circledast \left[\dfrac{\sin\left(\frac{7}{2}\Omega\right)}{\sin\left(\frac{\Omega}{2}\right)}\right]$

(c) $X(e^{j\Omega}) = \cos(4\Omega)\left[\dfrac{\sin\left(\frac{3}{2}\Omega\right)}{\sin\left(\frac{\Omega}{2}\right)}\right]$

(d) $X(e^{j\Omega}) = \begin{cases} e^{-j4\Omega} & \frac{\pi}{4} < |\Omega| < \frac{3\pi}{4} \\ 0 & \text{otherwise} \end{cases}$, for $|\Omega| < \pi$

(e) $X(e^{j\Omega}) = e^{-j\left(4\Omega + \frac{\pi}{2}\right)}\dfrac{d}{d\Omega}\left[\dfrac{2}{1 + \frac{1}{4}e^{-j\left(\Omega - \frac{\pi}{4}\right)}} + \dfrac{2}{1 + \frac{1}{4}e^{-j\left(\Omega + \frac{\pi}{4}\right)}}\right]$

3.62 Use the FT pair

$$x(t) = \begin{cases} 1 & |t| < 1 \\ 0 & \text{otherwise} \end{cases} \xleftrightarrow{\ FT\ } X(j\omega) = \dfrac{2\sin(\omega)}{\omega}$$

and the FT properties to evaluate the frequency-domain representations of the signals depicted in Figures P3.62(a)–(g).

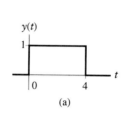

(a)

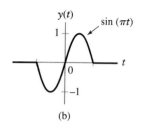

(b)

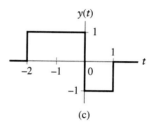

(c)

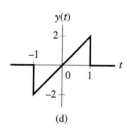

(d)

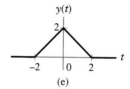

(e)

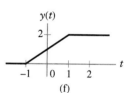

(f)

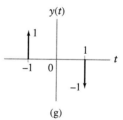

(g)

FIGURE P3.62

3.63 You are given $x[n] = n\left(\frac{3}{4}\right)^{|n|} \xleftrightarrow{\ DTFT\ } X(e^{j\Omega})$. Without evaluating $X(e^{j\Omega})$, find $y[n]$ if

(a) $Y(e^{j\Omega}) = e^{-j4\Omega}X(e^{j\Omega})$

(b) $Y(e^{j\Omega}) = \text{Re}\{X(e^{j\Omega})\}$

(c) $Y(e^{j\Omega}) = \dfrac{d}{d\Omega}X(e^{j\Omega})$

(d) $Y(e^{j\Omega}) = X(e^{j\Omega}) \circledast X(e^{j(\Omega - \pi/2)})$

(e) $Y(e^{j\Omega}) = \dfrac{d}{d\Omega}X(e^{j2\Omega})$

(f) $Y(e^{j\Omega}) = X(e^{j\Omega}) + X(e^{-j\Omega})$

(g) $Y(e^{j\Omega}) = \dfrac{d}{d\Omega}\left\{e^{-j4\Omega}\left[X\left(e^{j\left(\Omega+\frac{\pi}{4}\right)}\right) + X\left(e^{j\left(\Omega-\frac{\pi}{4}\right)}\right)\right]\right\}$

3.64 A periodic signal has the FS representation
$x(t) \xleftrightarrow{\;FS;\,\pi\;} X[k] = -k2^{-|k|}$. Without determining $x(t)$, find the FS representation ($Y[k]$ and ω_o) if

(a) $y(t) = x(3t)$ (b) $y(t) = \dfrac{d}{dt}x(t)$

(c) $x(t) = x(t-1)$ (d) $y(t) = \text{Re}\{x(t)\}$

(e) $y(t) = \cos(4\pi t)x(t)$

(f) $y(t) = x(t) \circledast x(t-1)$

3.65 Given
$$x[n] = \frac{\sin\left(\frac{11\pi}{20}n\right)}{\sin\left(\frac{\pi}{20}n\right)} \xleftrightarrow{\;DTFS;\,\frac{\pi}{10}\;} X[k],$$

evaluate the time signal $y[n]$ with the following DTFS coefficients, using only DTFS properties:

(a) $Y[k] = X[k-5] + X[k+5]$

(b) $Y[k] = \cos\left(k\frac{\pi}{5}\right)X[k]$

(c) $Y[k] = X[k] \circledast X[k]$

(d) $Y[k] = \text{Re}\{X[k]\}$

3.66 Sketch the frequency response of the systems described by the following impulse responses:

(a) $h(t) = \delta(t) - 2e^{-2t}u(t)$

(b) $h(t) = 4e^{-2t}\cos(50t)u(t)$

(c) $h[n] = \frac{1}{8}\left(\frac{7}{8}\right)^n u[n]$

(d) $h[n] = \begin{cases} (-1)^n & |n| \le 10 \\ 0 & \text{otherwise} \end{cases}$

Characterize each system as low pass, band pass, or high pass.

3.67 Find the frequency response and the impulse response of the systems having the output $y(t)$ for the input $x(t)$:

(a) $x(t) = e^{-t}u(t)$, $y(t) = e^{-2t}u(t) + e^{-3t}u(t)$

(b) $x(t) = e^{-3t}u(t)$, $y(t) = e^{-3(t-2)}u(t-2)$

(c) $x(t) = e^{-2t}u(t)$, $y(t) = 2te^{-2t}u(t)$

(d) $x[n] = \left(\frac{1}{2}\right)^n u[n]$, $y[n] = \frac{1}{4}\left(\frac{1}{2}\right)^n u[n] + \left(\frac{1}{4}\right)^n u[n]$

(e) $x[n] = \left(\frac{1}{4}\right)^n u[n]$,
$$y[n] = \left(\frac{1}{4}\right)^n u[n] - \left(\frac{1}{4}\right)^{n-1}u[n-1]$$

3.68 Determine the frequency response and the impulse response for the systems described by the following differential and difference equations:

(a) $\dfrac{d}{dt}y(t) + 3y(t) = x(t)$

(b) $\dfrac{d^2}{dt^2}y(t) + 5\dfrac{d}{dt}y(t) + 6y(t) = -\dfrac{d}{dt}x(t)$

(c) $y[n] - \frac{1}{4}y[n-1] - \frac{1}{8}y[n-2] = 3x[n]$
$$- \tfrac{3}{4}x[n-1]$$

(d) $y[n] + \frac{1}{2}y[n-1] = x[n] - 2x[n-1]$

3.69 Determine the differential- or difference-equation descriptions for the systems with the following impulse responses:

(a) $h[t] = \dfrac{1}{a}e^{-\frac{t}{a}}u(t)$

(b) $h(t) = 2e^{-2t}u(t) - 2te^{-2t}u(t)$

(c) $h[n] = \alpha^n u[n]$, $|\alpha| < 1$

(d) $h[n] = \delta[n] + 2\left(\frac{1}{2}\right)^n u[n] + \left(\frac{-1}{2}\right)^n u[n]$

3.70 Determine the differential- or difference-equation descriptions for the systems with the following frequency responses:

(a) $H(j\omega) = \dfrac{2 + 3j\omega - 3(j\omega)^2}{1 + 2j\omega}$

(b) $H(j\omega) = \dfrac{1 - j\omega}{-\omega^2 - 4}$

(c) $H(j\omega) = \dfrac{1 + j\omega}{(j\omega + 2)(j\omega + 1)}$

(d) $H(e^{j\Omega}) = \dfrac{1 + e^{-j\Omega}}{e^{-j2\Omega} + 3}$

(e) $H(e^{j\Omega}) = 1 + \dfrac{e^{-j\Omega}}{\left(1 - \frac{1}{2}e^{-j\Omega}\right)\left(1 + \frac{1}{4}e^{-j\Omega}\right)}$

3.71 Consider the RL circuit depicted in Fig. P3.71.

(a) Let the output be the voltage across the inductor, $y_L(t)$. Write a differential-equation description for this system and find the frequency response. Characterize the system as a filter.

(b) Determine and plot the voltage across the inductor, using circuit analysis techniques, if the input is the square wave depicted in Fig. 3.21 with $T = 1$ and $T_o = 1/4$.

(c) Let the output be the voltage across the resistor, $y_R(t)$. Write a differential-equation description for this system and find the frequency response. Characterize the system as a filter.

(d) Determine and plot the voltage across the resistor, using circuit analysis techniques, if the input is the square wave depicted in Fig. 3.21 with $T = 1$ and $T_o = 1/4$.

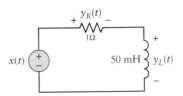

FIGURE P3.71

3.72 Consider the RLC circuit depicted in Fig. P3.72 with input $x(t)$ and output $y(t)$.

S&S Solutions

(a) Write a differential-equation description for this system and find the frequency response. Characterize the system as a filter.

(b) Determine and plot the output if the input is the square wave depicted in Fig. 3.21 with $T = 2\pi \times 10^{-3}$ and $T_o = (\pi/2) \times 10^{-3}$, assuming that $L = 10$ mH.

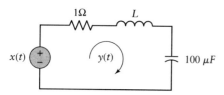

FIGURE P3.72

3.73 Use partial-fraction expansions to determine the inverse FT for the following signals:

S&S Solutions

(a) $X(j\omega) = \dfrac{6j\omega + 16}{(j\omega)^2 + 5j\omega + 6}$

(b) $X(j\omega) = \dfrac{j\omega - 2}{-\omega^2 + 5j\omega + 4}$

(c) $X(j\omega) = \dfrac{j\omega}{(j\omega)^2 + 6j\omega + 8}$

(d) $X(j\omega) = \dfrac{-(j\omega)^2 - 4j\omega - 6}{((j\omega)^2 + 3j\omega + 2)(j\omega + 4)}$

(e) $X(j\omega) = \dfrac{2(j\omega)^2 + 12j\omega + 14}{(j\omega)^2 + 6j\omega + 5}$

(f) $X(j\omega) = \dfrac{j\omega + 3}{(j\omega + 1)^2}$

3.74 Use partial-fraction expansions to determine the inverse DTFT for the following signals:

S&S Solutions

(a) $X(e^{j\Omega}) = \dfrac{2e^{-j\Omega}}{-\frac{1}{4}e^{-j2\Omega} + 1}$

(b) $X(e^{j\Omega}) = \dfrac{2 + \frac{1}{4}e^{-j\Omega}}{-\frac{1}{8}e^{-j2\Omega} + \frac{1}{4}e^{-j\Omega} + 1}$

(c) $X(e^{j\Omega}) = \dfrac{12}{-e^{-j2\Omega} + e^{-j\Omega} + 6}$

(d) $X(e^{j\Omega}) = \dfrac{6 - 2e^{-j\Omega} + \frac{1}{2}e^{-j2\Omega}}{\left(-\frac{1}{4}e^{-j2\Omega} + 1\right)\left(1 - \frac{1}{4}e^{-j\Omega}\right)}$

(e) $X(e^{j\Omega}) = \dfrac{6 - \frac{2}{3}e^{-j\Omega} - \frac{1}{6}e^{-j2\Omega}}{-\frac{1}{6}e^{-j2\Omega} + \frac{1}{6}e^{-j\Omega} + 1}$

3.75 Evaluate the following quantities:

(a) $\displaystyle\int_{-\pi}^{\pi} \dfrac{4}{\left|1 - \frac{1}{3}e^{-j\Omega}\right|^2}\,d\Omega$

(b) $\displaystyle\sum_{k=-\infty}^{\infty} \dfrac{\sin^2(k\pi/8)}{k^2}$

(c) $\displaystyle\int_{-\infty}^{\infty} \dfrac{8}{(\omega^2 + 4)^2}\,d\omega$

(d) $\displaystyle\int_{-\infty}^{\infty} \dfrac{\sin^2(\pi t)}{\pi t^2}\,dt$

3.76 Use the duality property to evaluate

S&S Solutions

(a) $x(t) \xleftrightarrow{\ FT\ } e^{-2\omega}u(\omega)$

(b) $\dfrac{1}{(2 + jt)^2} \xleftrightarrow{\ FT\ } X(j\omega)$

(c) $\dfrac{\sin\left(\frac{11\pi}{20}n\right)}{\sin\left(\frac{\pi}{20}n\right)} \xleftrightarrow{\ DTFS; \frac{\pi}{10}\ } X[k]$

3.77 For the FT $X(j\omega)$ shown in Figure P3.77, evaluate the following quantities without explicitly computing $x(t)$:

(a) $\int_{-\infty}^{\infty} x(t)\,dt$

(b) $\int_{-\infty}^{\infty} |x(t)|^2\,dt$

(c) $\int_{-\infty}^{\infty} x(t)e^{j3t}\,dt$

(d) $\arg\{x(t)\}$

(e) $x(0)$

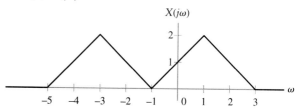

FIGURE P3.77

3.78 Let $x[n] \xleftrightarrow{\ DTFT\ } X(e^{j\Omega})$, where $x[n]$ is depicted in Figure P3.78. Evaluate the following without explicitly computing $X(e^{j\Omega})$:

(a) $X(e^{j0})$

(b) $\arg\{X(e^{j\Omega})\}$

(c) $\int_{-\pi}^{\pi} |X(e^{j\Omega})|^2\,d\Omega$

(d) $\int_{-\pi}^{\pi} X(e^{j\Omega})e^{j3\Omega}\,d\Omega$

(e) $y[n] \xleftrightarrow{\ DTFT\ } \mathrm{Re}\{e^{j2\Omega}X(e^{j\Omega})\}$

3.79 Prove the following properties:

(a) The FS symmetry properties for
 (i) Real-valued time signals.
 (ii) Real and even time signals.

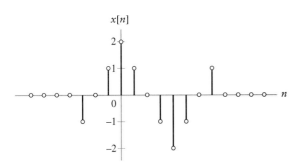

FIGURE P3.78

(b) The DTFT time-shift property.
(c) The DTFS frequency-shift property.
(d) Linearity for the FT.
(e) The DTFT convolution property.
(f) The DTFT modulation property.
(g) The DTFS convolution property.
(h) The FS modulation property.
(i) The Parseval relationship for the FS.

3.80 Define a signal that is zero, except at integer multiples of the scaling parameter p. That is, let

$$x_z[n] = 0, \text{ unless } n/p \text{ is integer.}$$

Figure P3.80(a) illustrates such a signal for $p = 3$.

(a) Show that the DTFT of $z[n] = x_z[pn]$ is given by $Z(e^{j\Omega}) = X_z(e^{j\Omega/p})$.
(b) In Fig. P3.80(b), use the DTFT of the signal $w[n]$ and the scaling property to determine the DTFT of the signal $f[n]$.
(c) Assume that $x_z[n]$ is periodic with fundamental period N, so that $z[n] = x_z[pn]$ has fundamental period N/p, a positive integer. Show that the DTFS of $z[n]$ satisfies $Z[k] = pX_z[k]$.

3.81 In this problem we show that Gaussian pulses achieve the lower bound in the time–bandwidth product. (*Hint:* Use the definite integrals in Appendix A.4.)

(a) Let $x(t) = e^{-(t^2/2)}$. Find the effective duration T_d and the bandwidth B_w, and evaluate the time–bandwidth product.

(b) Let $x(t) = e^{-t^2/2a^2}$. Find the effective duration T_d and the bandwidth B_w, and evaluate the time–bandwidth product. What happens to T_d, B_w, and $T_d B_w$ as a increases?

3.82 Let

$$x(t) = \begin{cases} 1, & |t| < T_o \\ 0, & \text{otherwise} \end{cases}.$$

Use the uncertainty principle to bound the effective bandwidth of $x(t) * x(t)$.

3.83 Use the uncertainty principle to bound the effective bandwidth of $x(t) = e^{-|t|}$.

3.84 Show that the time–bandwidth product $T_d B_w$ of a signal $x(t)$ is invariant to scaling. That is, use the definitions of T_d and B_w to show that $x(t)$ and $x(at)$ have the same time–bandwidth product.

3.85 A key property of the complex sinusoids used in the DTFS and FS expansions is orthogonality, according to which the inner product of two harmonically related sinuoids is zero. The inner product is defined as the sum or integral of the product of one signal and the conjugate of the other over a fundamental period.

(a) Show that discrete-time complex sinusoids are orthogonal; that is, prove that

$$\frac{1}{N} \sum_{n=0}^{N-1} e^{jk\frac{2\pi}{N}n} e^{-jl\frac{2\pi}{N}n} = \begin{cases} 1, & k = l \\ 0, & k \neq l \end{cases}$$

where we assume that $|k - l| < N$.

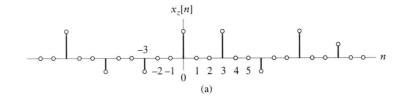

(a)

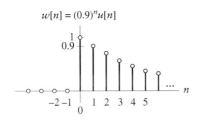

$w[n] = (0.9)^n u[n]$

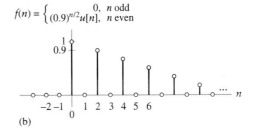

$f(n) = \begin{cases} 0, & n \text{ odd} \\ (0.9)^{n/2} u[n], & n \text{ even} \end{cases}$

(b)

FIGURE P3.80

(b) Show that harmonically related continuous-time complex sinusoids are orthogonal; that is, prove that

$$\frac{1}{T}\int_0^T e^{jk\frac{2\pi}{T}t}e^{-jl\frac{2\pi}{T}t}\,dt = \begin{cases} 1, & k = l \\ 0, & k \neq l \end{cases}$$

(c) Show that of harmonically related sines and cosines are orthogonal; that is, prove that

$$\frac{1}{T}\int_0^T \sin\left(k\frac{2\pi}{T}t\right)\sin\left(l\frac{2\pi}{T}t\right)\,dt = \begin{cases} 1/2, k = l \\ 0, \ k \neq l \end{cases}$$

$$\frac{1}{T}\int_0^T \cos\left(k\frac{2\pi}{T}t\right)\cos\left(l\frac{2\pi}{T}t\right)\,dt = \begin{cases} 1/2, k = l \\ 0, \ k \neq l \end{cases}$$

and

$$\frac{1}{T}\int_0^T \cos\left(k\frac{2\pi}{T}t\right)\sin\left(l\frac{2\pi}{T}t\right)\,dt = 0.$$

3.86 The form of the FS representation presented in this chapter, namely

$$x(t) = \sum_{k=-\infty}^{\infty} X[k]e^{jk\omega_o t}$$

is termed the exponential FS. In this problem, we explore several alternative, yet equivalent, ways of expressing the FS representation for real-valued periodic signals.

(a) Trigonometric form.

(i) Show that the FS for a real-valued signal $x(t)$ can be written as

$$x(t) = B[0] + \sum_{k=1}^{\infty} B[k]\cos(k\omega_o t)$$
$$+ A[k]\sin(k\omega_o t),$$

where $B[k]$ and $A[k]$ are real-valued coefficients.

(ii) Express $X[k]$ in terms of $B[k]$ and $A[k]$.

(iii) Use the orthogonality of harmonically related sines and cosines (See Problem 3.85) to show that

$$B[0] = \frac{1}{T}\int_0^T x(t)\,dt,$$

$$B[k] = \frac{2}{T}\int_0^T x(t)\cos k\omega_o t\,dt,$$

and

$$A[k] = \frac{2}{T}\int_0^T x(t)\sin k\omega_o t\,dt.$$

(iv) Show that $A[k] = 0$ if $x(t)$ is even and $B[k] = 0$ if $x(t)$ is odd.

(b) Polar form.

(i) Show that the FS for a real-valued signal $x(t)$ can be written as

$$x(t) = C[0] + \sum_{k=1}^{\infty} C[k]\cos(k\omega_o t + \theta[k]),$$

where $C[k]$ is the (positive) magnitude and $\theta[k]$ is the phase of the kth harmonic.

(ii) Express $C[k]$ and $\theta[k]$ as a function of $X[k]$.

(iii) Express $C[k]$ and $\theta[k]$ as a function of $B[k]$ and $A[k]$ from (a).

3.87 In this problem, we derive the frequency response of continuous- and discrete-time LTI systems described by state-variable representations.

(a) Define $\mathbf{q}(j\omega)$ as the FT of each element of the state vector in the state-variable representation for a continuous-time LTI system. That is,

$$\mathbf{q}(j\omega) = \begin{bmatrix} Q_1(j\omega) \\ Q_2(j\omega) \\ \vdots \\ Q_N(j\omega) \end{bmatrix},$$

where the ith entry in $\mathbf{q}(j\omega)$ is the FT of the ith state variable, $q_i(t) \xleftarrow{\ FT\ } Q_i(j\omega)$. Take the FT of the state equation $\frac{d}{dt}\mathbf{q}(t) = \mathbf{Aq}(t) + \mathbf{b}x(t)$, and use the differentiation property to express $\mathbf{q}(j\omega)$ as a function of ω, $\mathbf{A}$, $\mathbf{b}$, and $X(j\omega)$. Next, take the FT of the output equation $y(t) = \mathbf{cq}(t) + \mathbf{D}x(t)$, and substitute for $\mathbf{q}(j\omega)$ to show that

$$H(j\omega) = \mathbf{c}(j\omega\mathbf{I} - \mathbf{A})^{-1}\mathbf{b} + \mathbf{D}.$$

(b) Use the time-shift property to express the frequency response of a discrete-time LTI system in terms of the state-variable representation as

$$H(e^{j\Omega}) = \mathbf{c}(e^{j\Omega}\mathbf{I} - \mathbf{A})^{-1}\mathbf{b} + \mathbf{D}.$$

3.88 Use the result of Problem 3.87 to determine the frequency response, impulse response, and differential-equation descriptions for the continuous-time LTI systems described by the following state-variable matrices:

(a) $\mathbf{A} = \begin{bmatrix} -2 & 0 \\ 0 & -1 \end{bmatrix}$, $\mathbf{b} = \begin{bmatrix} 0 \\ 2 \end{bmatrix}$,

$\mathbf{c} = \begin{bmatrix} 1 & 1 \end{bmatrix}$, $\mathbf{D} = [0]$

(b) $\mathbf{A} = \begin{bmatrix} 1 & 2 \\ -3 & -4 \end{bmatrix}$, $\mathbf{b} = \begin{bmatrix} 1 \\ 2 \end{bmatrix}$,

$\mathbf{c} = \begin{bmatrix} 0 & 1 \end{bmatrix}$, $\mathbf{D} = [0]$

3.89 Use the result of Problem 3.87 to determine the frequency response, impulse response, and difference-equation descriptions for the discrete-time systems described by the following state variable matrices:

(a) $\mathbf{A} = \begin{bmatrix} -\frac{1}{2} & 1 \\ 0 & \frac{1}{4} \end{bmatrix}$, $\mathbf{b} = \begin{bmatrix} 0 \\ 1 \end{bmatrix}$,

$\mathbf{c} = \begin{bmatrix} 1 & 0 \end{bmatrix}$, $\mathbf{D} = \begin{bmatrix} 1 \end{bmatrix}$

(b) $\mathbf{A} = \begin{bmatrix} \frac{1}{4} & \frac{3}{4} \\ \frac{1}{4} & -\frac{1}{4} \end{bmatrix}$, $\mathbf{b} = \begin{bmatrix} 1 \\ 1 \end{bmatrix}$,

$\mathbf{c} = \begin{bmatrix} 0 & 1 \end{bmatrix}$, $\mathbf{D} = \begin{bmatrix} 0 \end{bmatrix}$

3.90 A continuous-time system is described by the state-variable matrices

$$\mathbf{A} = \begin{bmatrix} -1 & 0 \\ 0 & -3 \end{bmatrix}, \quad \mathbf{b} = \begin{bmatrix} 0 \\ 2 \end{bmatrix},$$

$$\mathbf{c} = \begin{bmatrix} 0 & 1 \end{bmatrix}, \quad \text{and} \quad \mathbf{D} = \begin{bmatrix} 0 \end{bmatrix}.$$

Transform the state vector associated with this system by using the matrix

$$\mathbf{T} = \begin{bmatrix} 1 & -1 \\ 1 & 1 \end{bmatrix}$$

to find a new state-variable description for the system. Show that the frequency response of the original and transformed systems are equal.

ADVANCED PROBLEMS

3.91 A signal with fundamental period T is said to possess half-wave symmetry if it satisfies the relationship $x(t) = -x\left(t - \frac{T}{2}\right)$. That is, half of one period of the signal is the negative of the other half. Show that the FS coefficients associated with even harmonics, $X[2k]$, are zero for all signals with half-wave symmetry.

3.92 The FS of piecewise-constant signals may be determined by using the differentiation and time-shift properties from the FS of the impulse train as follows: Differentiate the time-domain signal to obtain a sum of time-shifted impulse trains. Note that differentiation introduces an impulse at each discontinuity in the time-domain signal. Next, use the time-shift property and the FS of the impulse train to find the FS of the differentiated signal. Finally, use the differentiation property to obtain the FS coefficient of the original signal from the differentiated signal.

(a) Can this method be used to determine the FS coefficient for $k = 0$? How can you find it?

(b) Use the method to find the FS coefficients for the piecewise-constant waveforms in Fig. P3.92.

3.93 The method for finding the FS coefficients that was described in the previous problem may be extended to signals that are piecewise linear by differentiating twice to obtain a sum of impulse trains and doublet trains. The time-shift property and FS of the impulse train and doublet train are then used to find the FS for the twice-differentiated signal, and the differentiation property is used to obtain the FS coefficients of the original signal from the twice-differentiated signal.

(a) Find the FS coefficients for the doublet train

$$d(t) = \sum_{l=-\infty}^{\infty} \delta^{(1)}(t - lT),$$

where $\delta^{(1)}(t)$ denotes the doublet.

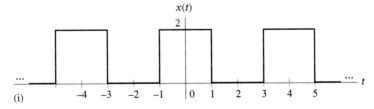

(i)

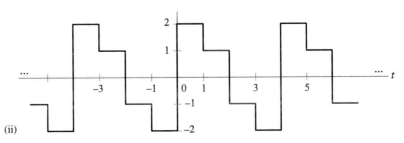

(ii)

FIGURE P3.92

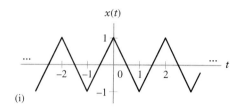

(i)

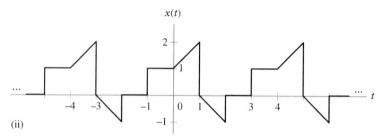

(ii)

FIGURE P3.93

(b) Use this method to find the FS coefficients of the waveforms in Fig. P3.93.

3.94 The FT relates the electromagnetic field at a distant point to the electric-field distribution at the antenna. This problem derives that result for a one-dimensional antenna with a monochromatic (single-frequency) excitation of ω_o. Let the electric field at a point z within the antenna aperture have amplitude $a(z)$ and phase $\phi(z)$, so that the electric field as a function of z and t is $x(z, t) = a(z) \cos(\omega_o t + \phi(z))$. Define the complex amplitude of the field as $w(z) = a(z)e^{j\phi(z)}$, so that

$$x(z, t) = \text{Re}\{w(z)e^{j\omega_o t}\}.$$

Huygen's principle states that the electric field at a distant point is the superposition of the effects of each differential component of the electric field at the aperture. Suppose the point of interest is at a distance r. It takes time $t_o = r/c$ for the differential component between z and $z + dz$ to propagate a distance r, where c is the propagation velocity. Thus, the contribution to the field at r from this differential component of the field at the aperture is given by

$$y(z, t)\, dz = x(z, t - t_o)\, dz$$
$$= \text{Re}\{w(z)e^{-j\omega_o t_o}\, dz\, e^{j\omega_o t}\}.$$

Since the wavelength $\lambda = 2\pi c/\omega_o$, we have $\omega_o t_o = 2\pi r/\lambda$, and the complex amplitude associated with this differential component is $w(z)e^{-j2\pi r/\lambda}$.

(a) Consider a point P at an angle θ with respect to the axis normal to the aperture and at a distance R from $z = 0$, as shown in Fig. P3.94. If R is much greater than the maximum extent of the

aperture along the z-axis, then we may approximate r as $r = R + zs$, where $s = \sin\theta$. Use this approximation to determine the contribution of the differential component between z and $z + dz$ to P.

(b) Integrate all differential components of the electric field at the antenna aperture to show that the field at P is given by

$$Y(s, R) = \text{Re}\{G(s)e^{-j2\pi R/\lambda}e^{j\omega_o t}\},$$

where

$$G(s) = \int_{-\infty}^{\infty} w(z)e^{-j2\pi zs/\lambda}\, dz$$

represents the complex amplitude of the field as a function of $\sin\theta$. A comparison with Eq. (3.36) indicates that $G(s)$ is the FT of $w(z)$ evaluated at $2\pi s/\lambda$.

(c) Use the FT relationship developed in (b) to determine the far-field pattern, $|G(s)|$, for the following aperture distributions $w(z)$:

(i) $w(z) = \begin{cases} 1, & |z| < 5 \\ 0, & \text{otherwise} \end{cases}$

(ii) $w(z) = \begin{cases} e^{j\pi z/4}, & |z| < 5 \\ 0, & \text{otherwise} \end{cases}$

(iii)

$$w(z) = \begin{cases} 1/2 + (1/2)\cos(\pi z/5), & |z| < 5 \\ 0, & \text{otherwise} \end{cases}$$

(iv) $w(z) = e^{-z^2}$.

Assume that $\lambda = 1$, and sketch $|G(s)|$ for $-\pi/2 < \theta < \pi/2$.

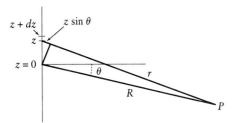

FIGURE P3.94

3.95 Figure P3.95 depicts a system known as a beam-former. The output of the beamformer is the weighted sum of signals measured at each antenna in the array. We assume that the antenna measures the complex amplitude of propagating plane waves of a single frequency ω_o and that the antennas are equally spaced by a distance d along a vertical line. A plane wave $p(t) = e^{j\omega_o t}$ is shown arriving at the array from the direction θ. If the top antenna measures $p(t)$, then the second antenna measures $p(t - \tau(\theta))$ where $\tau(\theta) = (d \sin \theta)/c$ is the time delay required for the plane wave front to propagate from the top to the second antenna and c is the speed of light. Since the antennas are equally spaced, the kth antenna measures the signal $p(t - k\tau(\theta))$, and the output of the beamformer is

$$y(t) = \sum_{k=0}^{N-1} w_k p(t - k\tau(\theta))$$
$$= e^{j\omega_o t} \sum_{k=0}^{N-1} w_k e^{-j\omega_o k\tau(\theta)}$$
$$= e^{j\omega_o t} \sum_{n=0}^{N-1} w_k e^{-j(\omega_o k d \sin \theta)/c}.$$

We may interpret this as a complex sinusoidal input from the direction θ resulting in a complex sinusoidal output of the same frequency. The beamformer introduces a magnitude and phase change given by the complex number

$$b(\theta) = \sum_{k=0}^{N-1} w_k e^{-j(\omega_o k d \sin \theta)/c}.$$

The gain of the beamformer, $|b(\theta)|$, is termed the beam pattern. Note that the gain is a function of the direction of arrival, and thus the beamformer offers the potential for discriminating between signals arriving from different directions. For convenience, we assume that the operating frequency and spacing are chosen so that $\omega_o d/c = \pi$. We also assume that θ is in the range $-\pi/2 < \theta < \pi/2$.

(a) Compare the expression for the beam pattern with the frequency response of a discrete-time system having only N nonzero impulse response coefficients. That is, assume that $h[k] = 0$ for $k < 0$ and $k \geq N$.

(b) Evaluate and plot the beam pattern for $N = 2$ with $w_0 = w_1 = 0.5$, and $w_1 = 0.5$, $w_2 = -0.5$.

(c) Evaluate and plot the beam pattern for $N = 4$ with $w_k = 0.25, k = 0, 1, 2, 3$.

(d) Compare the beam patterns obtained for $N = 8$ with $w_k = 1/8$, $k = 0, 1, \ldots, 7$, and $w_k = 1/8 e^{jk\pi/2}, k = 0, 1, \ldots, 7$.

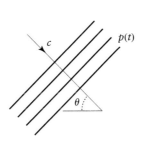

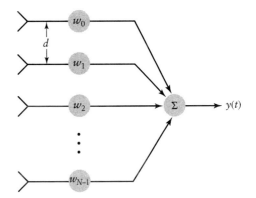

FIGURE P3.95

3.96 In Problem 2.81, we determined that the solution of the equations of motion for a vibrating string have the form $y_k(l, t) = \phi_k(l)f_k(t)$, $0 \le l \le a$, where

$$f_k(t) = a_k \cos(\omega_k ct) + b_k \sin(\omega_k ct),$$
$$\phi_k(l) = \sin(\omega_k l)$$

and $\omega_k = k\pi/a$. Since $y_k(l, t)$ is a solution for any a_k and b_k, the most general solution has the form

$$y(l, t) = \sum_{k=1}^{\infty} \sin(\omega_k l)(a_k \cos(\omega_k ct) + b_k \sin(\omega_k ct)).$$

We can use the initial conditions $y(l, 0) = x(l)$ to find the a_k and $\frac{\partial}{\partial t} y(l, t)\big|_{t=0}$ to find the b_k.

(a) Express the solution for a_k in terms of the FS coefficients of $x(l)$. (*Hint:* Consider Eqs. (3.25) and (3.26), with l replacing t.)

(b) Express the solution for b_k in terms of the FS coefficients of $g(l)$.

(c) Find $y(l, t)$, assuming that $g(l) = 0$ and $x(l)$ is as shown in Fig. P3.96.

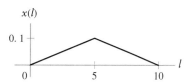

$x(l)$

0.1

0   5   10 l

FIGURE P3.96

3.97 In this problem, we explore a matrix representation for the DTFS. The DTFS expresses the N time-domain values of an N-periodic signal $x[n]$ as a function of N frequency domain values $X[k]$. Define vectors

$$\mathbf{x} = \begin{bmatrix} x[0] \\ x[1] \\ \vdots \\ x[N-1] \end{bmatrix} \quad \text{and} \quad \mathbf{X} = \begin{bmatrix} X[0] \\ X[1] \\ \vdots \\ X[N-1] \end{bmatrix}.$$

(a) Show that the DTFS representation

$$x[n] = \sum_{k=0}^{N-1} X[k]e^{jk\Omega_o n}, \quad n = 0, 1, \ldots, N-1$$

can be written in matrix vector form as $\mathbf{x} = \mathbf{VX}$, where $\mathbf{V}$ is an N-by-N matrix. Find the elements of $\mathbf{V}$.

(b) Show that the expression for the DTFS coefficients,

$$X[k] = \frac{1}{N} \sum_{n=0}^{N-1} x[n]e^{-jk\Omega_o n}, \quad k = 0, 1, \ldots, N-1,$$

can be written in matrix vector form as $\mathbf{X} = \mathbf{Wx}$, where $\mathbf{W}$ is an N-by-N matrix. Find the elements of $\mathbf{W}$.

(c) The expression $\mathbf{x} = \mathbf{VX}$ implies that $\mathbf{X} = \mathbf{V}^{-1}\mathbf{x}$ provided that $\mathbf{V}$ is a nonsingular matrix. Comparing this equation with the results of (b), we conclude that $\mathbf{W} = \mathbf{V}^{-1}$. Show that this is true by establishing that $\mathbf{WV} = \mathbf{I}$. (*Hint:* Use the definitions of $\mathbf{V}$ and $\mathbf{W}$ determined in (a) and (b) to obtain an expression for the element in the lth row and mth column of $\mathbf{WV}$, and use the result of Problem 3.85.

3.98 We may find the FS coefficients by forming the inner product of the series expansion and the conjugate of the basis functions. Let

$$x(t) = \sum_{k=-\infty}^{\infty} X[k]e^{jk\omega_o t}.$$

Using the result of Problem 3.85, derive the expression for $X[k]$ by multiplying both sides of this equation by $e^{-jk\omega_o t}$ and integrating over one period.

3.99 In this problem, we find the FS coefficients $X[k]$ by minimizing the mean square error (MSE) between the signal $x(t)$ and its FS approximation. Define the J-term FS

$$\hat{x}_J(t) = \sum_{k=-J}^{J} A[k]e^{jk\omega_o t}$$

and the J-term MSE as the average squared difference over one period:

$$\text{MSE}_J = \frac{1}{T} \int_0^T |x(t) - \hat{x}_J(t)|^2 \, dt.$$

(a) Substitute the series representation of $\hat{x}_J(t)$, and expand the magnitude squared, using the identity $|a + b|^2 = (a + b)(a^* + b^*)$, to obtain

$$\text{MSE}_J = \frac{1}{T} \int_0^T |x(t)|^2 \, dt$$
$$- \sum_{k=-J}^{J} A^*[k]\left(\frac{1}{T} \int_0^T x(t)e^{-jk\omega_o t} \, dt\right)$$
$$- \sum_{k=-J}^{J} A[k]\left(\frac{1}{T} \int_0^T x^*(t)e^{jk\omega_o t} \, dt\right)$$
$$+ \sum_{m=-J}^{J} \sum_{k=-J}^{J} A^*[k]A[m]\left(\frac{1}{T} \int_0^T e^{-jk\omega_o t}e^{jm\omega_o t} \, dt\right).$$

(b) Define

$$X[k] = \frac{1}{T} \int_0^T x(t)e^{-jk\omega_o t} \, dt$$

and use the orthogonality of $e^{jk\omega_o t}$ and $e^{jm\omega_o t}$ (see Problem 3.85) to show that

$$MSE_J = \frac{1}{T}\int_0^T |x(t)|^2\, dt - \sum_{k=-J}^{J} A^*[k]X[k]$$
$$- \sum_{k=-J}^{J} A[k]X^*[k] + \sum_{k=-J}^{J} |A[k]|^2.$$

(c) Use the technique of completing the square to show that

$$MSE_J = \frac{1}{T}\int_0^T |x(t)|^2\, dt$$
$$- \sum_{k=-J}^{J} |A[k] - X[k]|^2 - \sum_{k=-J}^{J} |X[k]|^2.$$

(d) Find the value of $A[k]$ that minimizes MSE_J.

(e) Express the minimum MSE_J as a function of $x(t)$ and $X[k]$. What happens to MSE_J as J increases?

3.100 *Generalized Fourier Series.* The concept of the Fourier Series may be generalized to sums of signals other than complex sinusoids. That is, we may approximate a signal $x(t)$ on an interval $[t_1, t_2]$ as a weighted sum of N functions $\phi_0(t)$, $\phi_2(t), \ldots, \phi_{N-1}(t)$:

$$x(t) \approx \sum_{k=0}^{N-1} c_k \phi_k(t)$$

We assume that these N functions are mutually orthogonal on $[t_1, t_2]$; that is,

$$\int_{t_1}^{t_2} \phi_k(t)\phi_l^*(t)\, dt = \begin{cases} 0, & k \neq l, \\ f_k, & k = l. \end{cases}$$

In this approximation, the mean squared error is

$$MSE = \frac{1}{t_2 - t_1}\int_{t_1}^{t_2} \left| x(t) - \sum_{k=1}^{N} c_k \phi_k(t) \right|^2 dt$$

(a) Show that the MSE is minimized by choosing $c_k = \frac{1}{f_k}\int_{t_2}^{t_1} x(t)\phi_k^*(t)\, dt$. (*Hint:* Generalize steps outlined in Problem 3.99 (a)–(d) to this problem.)

(b) Show that the MSE is zero if

$$\int_{t_1}^{t_2} |x(t)|^2\, dt = \sum_{k=0}^{N-1} f_k |c_k|^2.$$

If this relationship holds for all $x(t)$ in a given class of functions, then the basis functions $\phi_0(t)$, $\phi_2(t), \ldots, \phi_{N-1}(t)$ are said to be "complete" for that class.

(c) The Walsh functions are one set of orthogonal functions that are used for representing a signal on $[0, 1]$. Determine the c_k and MSE obtained by approximating the following signals with the first six Walsh functions depicted in

Fig. P3.100:

(i) $x(t) = \begin{cases} 2, & \frac{1}{2} < t < \frac{3}{4} \\ 0, & 0 < t < \frac{1}{2}, \frac{3}{4} < t < 1 \end{cases}$

(ii) $x(t) = \sin(2\pi t)$

Sketch the signal and the Walsh function approximation.

(d) The Legendre polynomials are another set of orthogonal functions on the interval $[-1, 1]$. These functions are obtained from the difference equation

$$\phi_k(t) = \frac{2k-1}{k}t\phi_{k-1}(t) - \frac{k-1}{k}\phi_{k-2}(t),$$

using the initial functions $\phi_0(t) = 1$ and $\phi_1(t) = t$. Determine the c_k and MSE obtained by approximating the following signals with the first six Legendre polynomials:

(i) $x(t) = \begin{cases} 2, & 0 < t < \frac{1}{2} \\ 0, & -1 < t < 0, \frac{1}{2} < t < 1 \end{cases}$

(ii) $x(t) = \sin(\pi t)$

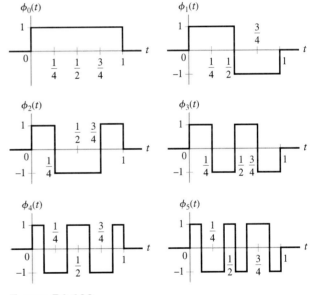

FIGURE P3.100

3.101 We may derive the FT from the FS by describing a nonperiodic signal as the limiting form of a periodic signal whose period T approaches infinity. In order to take this approach, we assume that the FS of the periodic version of the signal exists, that the nonperiodic signal is zero for $|t| > \frac{T}{2}$, and that the limit as T approaches infinity is taken in a symmetric manner. Define the finite-duration nonperiodic signal $x(t)$ as one period of the T-periodic signal $\widetilde{x}(t)$; that is,

$$
x(t) = \begin{cases} \widetilde{x}(t), & -\frac{T}{2} < t < \frac{T}{2} \\ 0, & |t| > \frac{T}{2} \end{cases}
$$

(a) Graph an example of $x(t)$ and $\widetilde{x}(t)$ to demonstrate that, as T increases, the periodic replicates of $x(t)$ in $\widetilde{x}(t)$ are moved farther and farther away from the origin. Eventually, as T approaches infinity, these replicates are removed to infinity. Thus, we write

$$
x(t) = \lim_{T \to \infty} \widetilde{x}(t).
$$

(b) The FS representation for the periodic signal $\widetilde{x}(t)$ is

$$
\widetilde{x}(t) = \sum_{k=-\infty}^{\infty} X[k] e^{jk\omega_o t},
$$

where

$$
X[k] = \frac{1}{T} \int_{-\frac{T}{2}}^{\frac{T}{2}} \widetilde{x}(t) e^{-jk\omega_o t}\, dt.
$$

Show that $X[k] = \frac{1}{T} X(jk\omega_o)$, where

$$
X(j\omega) = \int_{-\infty}^{\infty} x(t) e^{j\omega t}\, dt.
$$

(c) Substitute the preceding definition of $X[k]$ into the expression for $\widetilde{x}(t)$ in (b), and show that

$$
\widetilde{x}(t) = \frac{1}{2\pi} \sum_{k=-\infty}^{\infty} X(jk\omega_o) e^{jk\omega_o t} \omega_o.
$$

(d) Use the limiting expression for $x(t)$ in (a), and define $\omega \approx k\omega_o$ to express the limiting form of the sum in (c) as the integral

$$
x(t) = \frac{1}{2\pi} \int_{-\infty}^{\infty} X(j\omega) e^{j\omega t}\, d\omega.
$$

COMPUTER EXPERIMENTS

3.102 Use MATLAB to repeat Example 3.7 for $N = 50$ and (a) $M = 12$, (b) $M = 5$, and (c) $M = 20$.

3.103 Use MATLAB's `fft` command to repeat Problem 3.48.

3.104 Use MATLAB's `ifft` command to repeat Problem 3.49.

3.105 Use MATLAB's `fft` command to repeat Example 3.8.

3.106 Use MATLAB to repeat Example 3.14. Evaluate the peak overshoot for $J = 29, 59$, and 99.

3.107 Let $x(t)$ be the triangular wave depicted in Fig. P3.107.
(a) Find the FS coefficients $X[k]$.
(b) Show that the FS representation for $x(t)$ can be expressed in the form
$$
x(t) = \sum_{k=0}^{\infty} B[k] \cos(k\omega_o t).
$$
(c) Define the J-term partial-sum approximation to $x(t)$ as
$$
\hat{x}_J(t) = \sum_{k=0}^{J} B[k] \cos(k\omega_o t).
$$

Use MATLAB to evaluate and plot one period of the Jth term in this sum and $\hat{x}_J(t)$ for $J = 1, 3, 7, 29$, and 99.

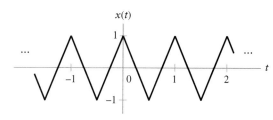

FIGURE P3.107

3.108 Repeat Problem 3.107 for the impulse train given by
$$
x(t) = \sum_{n=-\infty}^{\infty} \delta(t - n).
$$

3.109 Use MATLAB to repeat Example 3.15, with the following values for the time constant:
(a) $RC = 0.01$ s.
(b) $RC = 0.1$ s.
(c) $RC = 1$ s.

3.110 This experiment builds on Problem 3.71.
(a) Graph the magnitude response of the circuit depicted in Fig. P3.71, assuming that the voltage across the inductor is the output. Use logarithmically spaced frequencies from 0.1 rad/s to 1000 rad/s. You can generate N logarithmically spaced values between 10^{d1} and 10^{d2} by using the MATLAB command `logspace(d1,d2,N)`.

(b) Determine and plot the voltage across the inductor, using at least 99 harmonics in a truncated FS expansion for the output, if the input is the square wave depicted in Fig. 3.21 with $T = 1$ and $T_o = 1/4$.

(c) Graph the magnitude response of the circuit depicted in Fig. P3.71, assuming that the voltage across the resistor is the output. Use logarithmically spaced frequencies from 0.1 rad/s to 1000 rad/s.

(d) Determine and plot the voltage across the resistor, using at least 99 harmonics in a truncated FS expansion for the output if the input is the square wave depicted in Fig. 3.21 with $T = 1$ and $T_o = 1/4$.

3.111 This experiment builds on Problem 3.72.

(a) Graph the magnitude response of the circuit depicted in Fig. P3.72. Use 501 logarithmically spaced frequencies from 1 rad/s to 10^5 rad/s. You can generate N logarithmically spaced values between 10^{d1} and 10^{d2} by using the MATLAB command `logspace(d1,d2,N)`.
  (i) Assume that $L = 10$ mH.
  (ii) Assume that $L = 4$ mH.

(b) Determine and plot the output, using at least 99 harmonics in a truncated FS expansion, if the input is the square wave depicted in Fig. 3.21 with $T = 2\pi \times 10^{-3}$ and $T_o = (\pi/2) \times 10^{-3}$.
  (i) Assume that $L = 10$ mH.
  (ii) Assume that $L = 4$ mH.

3.112 Evaluate the frequency response of the truncated filter in Example 3.46. You can do this in MATLAB by writing an m-file to evaluate

$$H_t(e^{j\Omega}) = \sum_{n=-M}^{M} h[n]e^{-j\Omega n}$$

for a large number (> 1000) of values of Ω in the interval $-\pi < \Omega \leq \pi$. Plot the frequency response magnitude in dB ($20 \log_{10}|H_t(e^{j\Omega})|$) for the following values of M:

(a) $M = 4$

(b) $M = 10$

(c) $M = 25$

(d) $M = 50$

Discuss the effect of increasing M on the accuracy with which $H_t(e^{j\Omega})$ approximates $H(e^{j\Omega})$.

3.113 Use the MATLAB command `freqs` or `freqz` to plot the magnitude response of the following systems:

(a) $H(j\omega) = \dfrac{8}{(j\omega)^3 + 4(j\omega)^2 + 8j\omega + 8}$

(b) $H(j\omega) = \dfrac{(j\omega)^3}{(j\omega)^3 + 2(j\omega)^2 + 2j\omega + 1}$

(c) $H(e^{j\Omega}) = \dfrac{1 + 3e^{-j\Omega} + 3e^{-j2\Omega} + e^{-j3\Omega}}{6 + 2e^{-j2\Omega}}$

(d) $H(e^{j\Omega}) =$

$$\frac{0.02426(1 - e^{-j\Omega})^4}{(1 + 1.10416e^{-j\Omega} + 0.4019e^{-j2\Omega})(1 + 0.56616e^{-j\Omega} + 0.7657e^{-j2\Omega})}$$

Determine whether the system has a low-pass, high-pass, or band-pass characteristic.

3.114 Use MATLAB to verify that the time–bandwidth product for a discrete-time square wave is approximately independent of the number of nonzero values in each period when the duration is defined as the number of nonzero values in the square wave and the bandwidth is defined as the mainlobe width. Define one period of the square wave as

$$x[n] = \begin{cases} 1, & 0 \leq n < M \\ 0, & M \leq n \leq 999 \end{cases}.$$

Evaluate the bandwidth by first using `fft` and `abs` to obtain the magnitude spectrum and then counting the number of DTFS coefficients in the mainlobe for $M = 10, 20, 40, 50, 100$, and 200.

3.115 Use the MATLAB function `TdBw` introduced in Section 3.19 to evaluate and plot the time–bandwidth product as a function of duration for the following classes of signals:

(a) *Rectangular pulse trains.* Let the pulse in a single period be of length M, and vary M from 51 to 701 in steps of 50.

(b) *Raised-cosine pulse trains.* Let the pulse in a single period be of length M, and vary M from 51 to 701 in steps of 50.

(c) *Gaussian pulse trains.* Let $x[n] = e^{-an^2}$, $-500 \leq n \leq 500$, represent the Gaussian pulse in a single period. Vary the pulse duration by letting a take the following values: 0.00005, 0.0001, 0.0002, 0.0005, 0.001, 0.002, and 0.005.

3.116 Use MATLAB to evaluate and plot the solution to Problem 3.96 on the interval $0 \leq l \leq 1$ at $t = 0, 0.25, 0.5, 0.75, 1, 1.25, 1.5, 1.75$, assuming that $c = 1$. Use at least 99 harmonics in the sum.

3.117 The frequency response of an either a continuous- or discrete-time system described in state-variable form (see Problem 3.87) may be computed with the use of the MATLAB command `freqresp`. The syntax is `h = freqresp(sys,w)`, where `sys` is the object containing the state-variable description (see Section 2.14) and `w` is a vector containing the frequencies at which to evaluate the frequency response. Note that `freqresp` applies in general to multiple-input, multiple-output systems, so the output `h` is a multidimensional array. For the class of single-input, single-output systems considered in this text and N frequency points in `w`, `h` is of dimension 1 by 1 by N. The command `squeeze(h)` converts `h`

to an N-dimensional vector that may be displayed with the `plot` command. Thus, we may obtain the frequency response by using the following commands:

```
>> h = freqresp(sys,w);
>> hmag = abs(squeeze(h));
>> plot(w,hmag)
>> title('System Magnitude
   Response')
>> xlabel('Frequency
(rads/sec)'); ylabel('Magnitude')
```

Use MATLAB to plot the magnitude and phase response for the systems with state-variable descriptions given in Problems 3.88 and 3.89.

4 | Applications of Fourier Representations to Mixed Signal Classes

4.1 Introduction

In the previous chapter, we developed the Fourier representations of four distinct classes of signals: the discrete-time Fourier series (DTFS) for periodic discrete-time signals, the Fourier series (FS) for periodic continuous-time signals, the discrete-time Fourier transform (DTFT) for nonperiodic discrete-time signals, and the Fourier transform (FT) for nonperiodic continuous-time signals. We now focus on applications of Fourier representations to situations in which the classes of signals are mixed. In particular, we consider mixing of the following classes of signals:

- ▶ periodic and nonperiodic signals
- ▶ continuous- and discrete-time signals

Such mixing occurs most commonly when one uses Fourier methods to (1) analyze the interaction between signals and systems or (2) numerically evaluate properties of signals or the behavior of a system. For example, if we apply a periodic signal to a stable LTI system, the convolution representation of the system output involves a mixing of nonperiodic (impulse response) and periodic (input) signals. As another example, a system that samples continuous-time signals involves both continuous- and discrete-time signals.

In order to use Fourier methods to analyze such interactions, we must build bridges between the Fourier representations of different classes of signals. We establish these relationships in this chapter. The FT and DTFT are most commonly used for analysis applications. Hence, we develop FT and DTFT representations of continuous- and discrete-time periodic signals, respectively. We may then use the FT to analyze continuous-time applications that involve a mixture of periodic and nonperiodic signals. Similarly, the DTFT may be used to analyze mixtures of discrete-time periodic and nonperiodic signals. We develop an FT representation for discrete-time signals to analyze problems involving mixtures of continuous- and discrete-time signals. The DTFS is the primary representation used for computational applications, so we also examine the manner in which the DTFS represents the FT, FS, and DTFT. The first and major portion of this chapter is devoted to the presentation of analysis applications; computational applications are discussed briefly at the end of the chapter.

We begin the chapter by deriving FT and DTFT representations of periodic signals and then revisit convolution and modulation, considering applications in which periodic and nonperiodic signals interact. Next, we develop the FT representation of discrete-time signals and analyze the process of sampling signals and reconstructing continuous-time signals from samples. These issues are of fundamental importance whenever a computer is used to manipulate continuous-time signals, in particular in communication systems (Chapter 5) and for the purpose of filtering (Chapter 8) and control (Chapter 9). Our analysis reveals the limitations associated with the discrete-time processing of continuous-time signals and suggests a practical system that minimizes these limitations.

Recall that the DTFS is the only Fourier representation that can be evaluated numerically on a computer. Consequently, the DTFS finds extensive use in numerical algorithms for processing signals. We conclude the chapter by examining two common uses of the DTFS: numerical approximation of the FT and the efficient implementation of discrete-time convolution. In both of these applications, a clear understanding of the relationship between the Fourier representations of different classes of signals is essential to a correct interpretation of the results.

A thorough understanding of the relationships between the four Fourier representations is a critical step in using Fourier methods to solve problems involving signals and systems.

4.2 Fourier Transform Representations of Periodic Signals

Recall that the FS and DTFS have been derived as the Fourier representations of periodic signals. Strictly speaking, neither the FT nor the DTFT converges for periodic signals. However, by incorporating impulses into the FT and DTFT in the appropriate manner, we may develop FT and DTFT representations of such signals. These representations satisfy the properties expected of the FT and DTFT; hence, we may use them and the properties of the FT or DTFT to analyze problems involving mixtures of periodic and nonperiodic signals. This development establishes the relationship between Fourier series representations and Fourier transform representations. We begin the discussion with the continuous-time case.

▪ 4.2.1 RELATING THE FT TO THE FS

The FS representation of a periodic signal $x(t)$ is

$$x(t) = \sum_{k=-\infty}^{\infty} X[k]e^{jk\omega_o t}, \qquad (4.1)$$

where ω_o is the fundamental frequency of the signal. Recall from Section 3.7 that $1 \xleftarrow{\ FT\ } 2\pi\delta(\omega)$. Using this result and the frequency-shift property from Section 3.12, we obtain the inverse FT of a frequency-shifted impulse $\delta(\omega - k\omega_o)$ as a complex sinusoid with frequency $k\omega_o$:

$$e^{jk\omega_o t} \xleftarrow{\ FT\ } 2\pi\delta(\omega - k\omega_o). \qquad (4.2)$$

We substitute the FT pair given in Eq. (4.2) into the FS representation (4.1) and use the linearity property of the FT to obtain

$$x(t) = \sum_{k=-\infty}^{\infty} X[k]e^{jk\omega_o t} \xleftarrow{\ FT\ } X(j\omega) = 2\pi \sum_{k=-\infty}^{\infty} X[k]\delta(\omega - k\omega_o). \qquad (4.3)$$

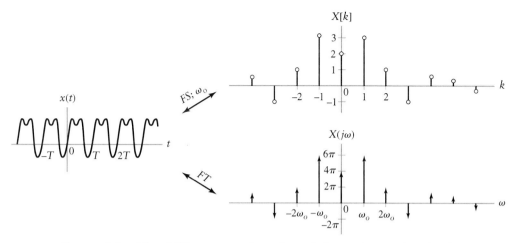

FIGURE 4.1   FS and FT representations of a periodic continuous-time signal.

Thus, the FT of a periodic signal is a series of impulses spaced by the fundamental frequency ω_o. The kth impulse has strength $2\pi X[k]$, where $X[k]$ is the kth FS coefficient. Figure 4.1 illustrates this relationship. Note that the shape of $X(j\omega)$ is identical to that of $X[k]$.

Equation (4.3) also indicates how to convert between FT and FS representations of periodic signals. The FT is obtained from the FS by placing impulses at integer multiples of ω_o and weighting them by 2π times the corresponding FS coefficient. Given an FT consisting of impulses that are uniformly spaced in ω, we obtain the corresponding FS coefficients by dividing the impulse strengths by 2π. The fundamental frequency corresponds to the spacing between impulses.

EXAMPLE 4.1 FT OF A COSINE   Find the FT representation of $x(t) = \cos(\omega_o t)$.

Solution: The FS representation of $x(t)$ is

$$\cos(\omega_o t) \xleftrightarrow{\ FS;\ \omega_o\ } X[k] = \begin{cases} \frac{1}{2}, & k = \pm 1 \\ 0, & k \neq \pm 1 \end{cases}.$$

Substituting these coefficients into Eq. (4.3) gives

$$\cos(\omega_o t) \xleftrightarrow{\ FT\ } X(j\omega) = \pi\delta(\omega - \omega_o) + \pi\delta(\omega + \omega_o).$$

This pair is depicted graphically in Fig. 4.2.   ∎

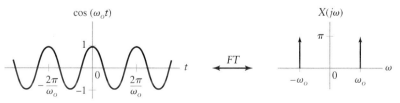

FIGURE 4.2   FT of a cosine.

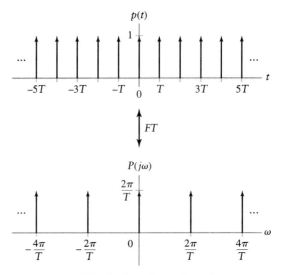

FIGURE 4.3   An impulse train and its FT.

EXAMPLE 4.2 FT OF A UNIT IMPULSE TRAIN   Find the FT of the impulse train

$$p(t) = \sum_{n=-\infty}^{\infty} \delta(t - nT).$$

Solution: We note that $p(t)$ is periodic with fundamental period T, so $\omega_o = 2\pi/T$, and the FS coefficients are given by

$$\begin{aligned} P[k] &= \frac{1}{T} \int_{-T/2}^{T/2} \delta(t) e^{-jk\omega_o t}\, dt \\ &= 1/T. \end{aligned}$$

We substitute these values into Eq. (4.3) to obtain

$$P(j\omega) = \frac{2\pi}{T} \sum_{k=-\infty}^{\infty} \delta(\omega - k\omega_o).$$

Hence, the FT of $p(t)$ is also an impulse train; that is, an impulse train is its own FT. The spacing between the impulses in the frequency domain is inversely related to the spacing between the impulses in the time domain, and the strengths of the impulses differ by a factor of $2\pi/T$. This FT pair is depicted in Fig. 4.3.   ■

▶ **Problem 4.1**   Find the FT representation of the following periodic signals:

(a) $x(t) = \sin(\omega_o t)$

(b) The periodic square wave depicted in Fig. 4.4

(c) $x(t) = |\sin(\pi t)|$

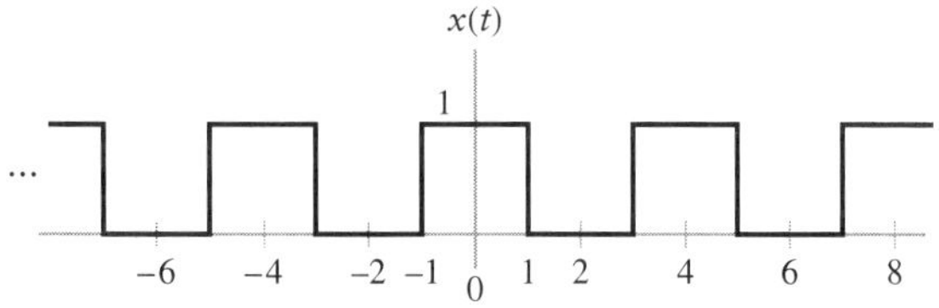

$x(t)$

FIGURE 4.4 Square wave for Problem 4.1.

Answers:

(a)
$$X(j\omega) = (\pi/j)\delta(\omega - \omega_o) - (\pi/j)\delta(\omega + \omega_o)$$

(b)
$$X(j\omega) = \sum_{k=-\infty}^{\infty} \frac{2\sin(k\pi/2)}{k}\delta(\omega - k\pi/2)$$

(c)
$$X(j\omega) = \sum_{k=-\infty}^{\infty} 4/(1 - 4k^2)\delta(\omega - k2\pi)$$ ◀

▶ **Problem 4.2**   Find the time-domain signal $x(t)$ corresponding to the following FT representations:

(a) $X(j\omega) = 4\pi\delta(\omega - 3\pi) + 2j\pi\delta(\omega - 5\pi) + 4\pi\delta(\omega + 3\pi) - 2j\pi\delta(\omega + 5\pi)$

(b)
$$X(j\omega) = \sum_{k=0}^{6} \frac{\pi}{1 + |k|}\{\delta(\omega - k\pi/2) + \delta(\omega + k\pi/2)\}$$

Answers:

(a)
$$x(t) = 4\cos(3\pi t) - 2\sin(5\pi t)$$

(b)
$$x(t) = \sum_{k=0}^{6} \frac{1}{1 + |k|}\cos(k\pi t/2)$$ ◀

■ **4.2.2 RELATING THE DTFT TO THE DTFS**

The method for deriving the DTFT of a discrete-time periodic signal parallels that given in the previous subsection. The DTFS expression for an N-periodic signal $x[n]$ is

$$x[n] = \sum_{k=0}^{N-1} X[k]e^{jk\Omega_o n}.$$ (4.4)

As in the FS case, the key observation is that the inverse DTFT of a frequency-shifted impulse is a discrete-time complex sinusoid. The DTFT is a 2π-periodic function of frequency, so we may express a frequency-shifted impulse either by expressing one period, such as

$$e^{jk\Omega_o n} \xleftarrow{\quad DTFT \quad} \delta(\Omega - k\Omega_o), \quad -\pi < \Omega \le \pi, \quad -\pi < k\Omega_o \le \pi,$$

or by using an infinite series of shifted impulses separated by an interval of 2π to obtain the 2π-periodic function

$$e^{jk\Omega_o n} \xleftarrow{\quad DTFT \quad} \sum_{m=-\infty}^{\infty} \delta(\Omega - k\Omega_o - m2\pi),$$ (4.5)

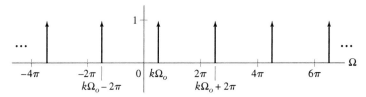

FIGURE 4.5   Infinite series of frequency-shifted impulses that is 2π periodic in frequency Ω.

which is depicted in Fig. 4.5. The inverse DTFT of Eq. (4.5) is evaluated by means of the sifting property of the impulse function. We have

$$\frac{1}{2\pi}e^{jk\Omega_o n} \xleftarrow{\;DTFT\;} \sum_{m=-\infty}^{\infty} \delta(\Omega - k\Omega_o - m2\pi). \qquad (4.6)$$

Hence, we identify the complex sinusoid and the frequency-shifted impulse as a DTFT pair. This relationship is a direct consequence of the properties of impulse functions.

Next, we use linearity and substitute Eq. (4.6) into Eq. (4.4) to obtain the DTFT of the periodic signal $x[n]$:

$$x[n] = \sum_{k=0}^{N-1} X[k]e^{jk\Omega_o n} \xleftarrow{\;DTFT\;} X(e^{j\Omega}) = 2\pi \sum_{k=0}^{N-1} X[k] \sum_{m=-\infty}^{\infty} \delta(\Omega - k\Omega_o - m2\pi). \quad (4.7)$$

Since $X[k]$ is N periodic and $N\Omega_o = 2\pi$, we may combine the two sums on the right-hand side of Eq. (4.7) and rewrite the DTFT of $x[n]$ as

$$\boxed{x[n] = \sum_{k=0}^{N-1} X[k]e^{jk\Omega_o n} \xleftarrow{\;DTFT\;} X(e^{j\Omega}) = 2\pi \sum_{k=-\infty}^{\infty} X[k]\delta(\Omega - k\Omega_o).} \qquad (4.8)$$

Thus, the DTFT representation of a periodic signal is a series of impulses spaced by the fundamental frequency Ω_o. The kth impulse has strength $2\pi X[k]$, where $X[k]$ is the kth DTFS coefficient for $x[n]$. Figure 4.6 depicts both the DTFS and the DTFT representations of a periodic discrete-time signal. Here again, we see that the DTFS $X[k]$ and the corresponding DTFT $X(e^{j\Omega})$ have similar shape.

Equation (4.8) establishes the relationship between the DTFS and DTFT. Given the DTFS coefficients and the fundamental frequency Ω_o, we obtain the DTFT representation

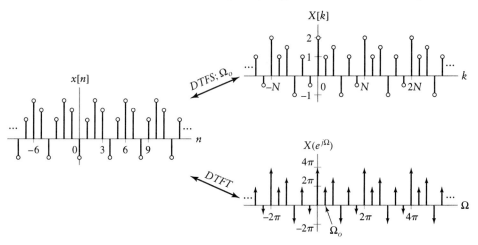

FIGURE 4.6   DTFS and DTFT representations of a periodic discrete-time signal.

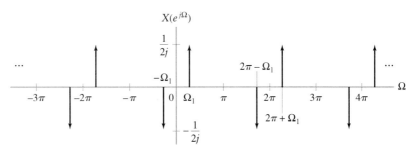

FIGURE 4.7 DTFT of periodic signal for Example 4.3.

by placing impulses at integer multiples of Ω_o and weighting them by 2π times the corresponding DTFS coefficients. We reverse this process to obtain the DTFS coefficients from the DTFT representation. If the DTFT consists of impulses that are uniformly spaced in Ω, then we obtain DTFS coefficients by dividing the impulse strengths by 2π. The fundamental frequency of $x[n]$ is the spacing between the impulses.

EXAMPLE 4.3 DTFT OF A PERIODIC SIGNAL Determine the inverse DTFT of the frequency-domain representation depicted in Fig. 4.7, where $\Omega_1 = \pi/N$.

Solution: We express one period of $X(e^{j\Omega})$ as

$$X(e^{j\Omega}) = \frac{1}{2j}\delta(\Omega - \Omega_1) - \frac{1}{2j}\delta(\Omega + \Omega_1), \quad -\pi < \Omega \le \pi,$$

from which we infer that

$$X[k] = \begin{cases} 1/(4\pi j), & k = 1 \\ -1/(4\pi j), & k = -1 \\ 0, & \text{otherwise on } -1 \le k \le N - 2 \end{cases}.$$

Then we take the inverse DTFT to obtain

$$x[n] = \frac{1}{2\pi}\left[\frac{1}{2j}\left(e^{j\Omega_1 n} - e^{-j\Omega_1 n}\right)\right]$$

$$= \frac{1}{2\pi}\sin(\Omega_1 n).$$

S & S
Solutions

▶ **Problem 4.3** Find the DTFT representations of the following periodic signals:

(a) $x[n] = \cos(7\pi n/16)$
(b) $x[n] = 2\cos(3\pi n/8 + \pi/3) + 4\sin(\pi n/2)$
(c) $x[n] = \sum_{k=-\infty}^{\infty}\delta[n - 10k]$

Answers: DTFTs on $\pi < \Omega \le \pi$

(a) $X(e^{j\Omega}) = \pi\delta(\Omega - 7\pi/16) + \pi\delta(\Omega + 7\pi/16)$
(b) $X(e^{j\Omega}) = -(4\pi/j)\delta(\Omega + \pi/2) + 2\pi e^{-j\pi/3}\delta(\Omega + 3\pi/8) + 2\pi e^{j\pi/3}\delta(\Omega - 3\pi/8) + (4\pi/j)\delta(\Omega - \pi/2)$
(c)

$$X(e^{j\Omega}) = \frac{2\pi}{10}\sum_{k=-4}^{5}\delta(\Omega - k\pi/5)$$

◀

4.3 Convolution and Multiplication with Mixtures of Periodic and Nonperiodic Signals

In this section, we use the FT and DTFT representations of periodic signals to analyze problems involving mixtures of periodic and nonperiodic signals. It is common to have mixing of periodic and nonperiodic signals in convolution and multiplication problems. For example, if a periodic signal is applied to a stable filter, the output is expressed as the convolution of the periodic input signal and the nonperiodic impulse response. The tool we use to analyze problems involving mixtures of periodic and nonperiodic continuous-time signals is the FT; the DTFT applies to mixtures of periodic and nonperiodic discrete-time signals. This analysis is possible, since we now have FT and DTFT representations of both periodic and nonperiodic signals. We begin by examining the convolution of periodic and nonperiodic signals and then focus on multiplication applications.

■ 4.3.1 CONVOLUTION OF PERIODIC AND NONPERIODIC SIGNALS

In Section 3.10, we established the fact that convolution in the time domain corresponds to multiplication in the frequency domain. That is,

$$y(t) = x(t) * h(t) \overset{FT}{\longleftrightarrow} Y(j\omega) = X(j\omega)H(j\omega).$$

This property may be applied to problems in which one of the time-domain signals—say, $x(t)$—is periodic by using its FT representation. Recall from Eq. (4.3) that the FT of a periodic signal $x(t)$ is

$$x(t) \overset{FT}{\longleftrightarrow} X(j\omega) = 2\pi \sum_{k=-\infty}^{\infty} X[k]\delta(\omega - k\omega_o)$$

where $X[k]$ are the FS coefficients. We substitute this representation into the convolution property to obtain

$$y(t) = x(t) * h(t) \overset{FT}{\longleftrightarrow} Y(j\omega) = 2\pi \sum_{k=-\infty}^{\infty} X[k]\delta(\omega - k\omega_o)H(j\omega). \qquad (4.9)$$

Now we use the sifting property of the impulse to write

$$\boxed{y(t) = x(t) * h(t) \overset{FT}{\longleftrightarrow} Y(j\omega) = 2\pi \sum_{k=-\infty}^{\infty} H(jk\omega_o)X[k]\delta(\omega - k\omega_o).} \qquad (4.10)$$

Figure 4.8 illustrates the multiplication of $X(j\omega)$ and $H(j\omega)$ that occurs in Eq. (4.10). The strength of the kth impulse in $X(j\omega)$ is adjusted by the value of $H(j\omega)$ evaluated at the frequency at which it is located, or $H(jk\omega_o)$, to yield an impulse in $Y(j\omega)$ at $\omega = k\omega_o$. The form of $Y(j\omega)$ corresponds to a periodic signal. Hence, $y(t)$ is periodic with the same period as $x(t)$. The most common application of this property is in determining the output of a filter with impulse response $h(t)$ and periodic input $x(t)$.

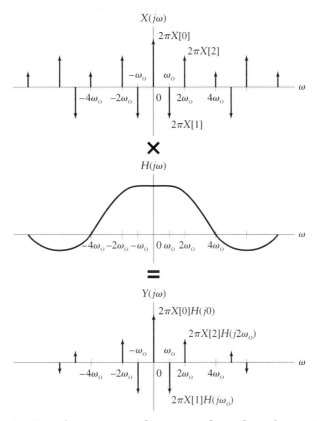

FIGURE 4.8   Convolution property for mixture of periodic and nonperiodic signals.

EXAMPLE 4.4 PERIODIC INPUT TO AN LTI SYSTEM   Let the input signal applied to an LTI system with impulse response $h(t) = (1/(\pi t)) \sin(\pi t)$ be the periodic square wave depicted in Fig. 4.4. Use the convolution property to find the output of this system.

Solution: The frequency response of the LTI system is obtained by taking the FT of the impulse response $h(t)$, as given by

$$h(t) \xleftarrow{\quad FT \quad} H(j\omega) = \begin{cases} 1, & |\omega| \le \pi \\ 0, & |\omega| > \pi \end{cases}.$$

Using Eq. (4.3), we may write the FT of the periodic square wave:

$$X(j\omega) = \sum_{k=-\infty}^{\infty} \frac{2\sin(k\pi/2)}{k} \delta\left(\omega - k\frac{\pi}{2}\right).$$

The FT of the system output is $Y(j\omega) = H(j\omega)X(j\omega)$. This product is depicted in Fig. 4.9, where

$$Y(j\omega) = 2\delta\left(\omega + \frac{\pi}{2}\right) + \pi\delta(\omega) + 2\delta\left(\omega - \frac{\pi}{2}\right),$$

which follows from the fact that $H(j\omega)$ acts as a low-pass filter, passing the harmonics at $-\pi/2, 0$, and $\pi/2$, while suppressing all others. Taking the inverse FT of $Y(j\omega)$ gives the output:

$$y(t) = (1/2) + (2/\pi)\cos(t\pi/2). \qquad \blacksquare$$

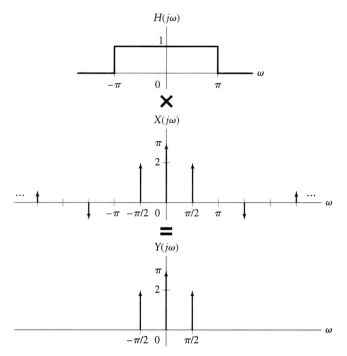

FIGURE 4.9 Application of convolution property in Example 4.4.

▶ **Problem 4.4** An LTI system has impulse response $h(t) = 2\cos(4\pi t)\sin(\pi t)/(\pi t)$. Use the FT to determine the output if the input is

(a) $x(t) = 1 + \cos(\pi t) + \sin(4\pi t)$

(b) $x(t) = \sum_{m=-\infty}^{\infty} \delta(t - m)$

(c) $x(t)$ as depicted in Fig. 4.10

Answers:

(a) $y(t) = \sin(4\pi t)$

(b) $y(t) = 2\cos(4\pi t)$

(c) $y(t) = 0$ ◀

An analogous result is obtained in the discrete-time case. The convolution property is

$$y[n] = x[n] * h[n] \xleftarrow{\text{DTFT}} Y(e^{j\Omega}) = X(e^{j\Omega})H(e^{j\Omega}).$$

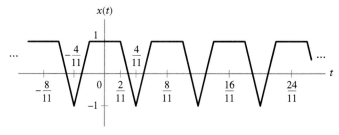

FIGURE 4.10 Signal $x(t)$ for Problem 4.4.

We may use this property when $x[n]$ is periodic with fundamental frequency Ω_o by replacing $X(e^{j\Omega})$ with the DTFT representation of periodic signals given in Eq. (4.8) to obtain

$$y[n] = x[n] * h[n] \xleftrightarrow{\quad DTFT \quad} Y(e^{j\Omega}) = 2\pi \sum_{k=-\infty}^{\infty} H(e^{jk\Omega_o})X[k]\delta(\Omega - k\Omega_o). \qquad (4.11)$$

The form of $Y(e^{j\Omega})$ indicates that $y[n]$ is also periodic with the same period as $x[n]$. This property finds application in evaluating the input–output behavior of discrete-time LTI systems.

▶ **Problem 4.5**   Let the input to a discrete-time LTI system be

$$x[n] = 3 + \cos(\pi n + \pi/3).$$

Determine the output of this system for the following impulse responses:

(a) $h[n] = \left(\frac{1}{2}\right)^n u[n]$

(b) $h[n] = \sin(\pi n/4)/(\pi n)$

(c) $h[n] = (-1)^n \sin(\pi n/4)/(\pi n)$

Answers:

(a) $y[n] = 6 + (2/3)\cos(\pi n + \pi/3)$

(b) $y[n] = 3$

(c) $y[n] = \cos(\pi n + \pi/3)$ ◀

■ **4.3.2 MULTIPLICATION OF PERIODIC AND NONPERIODIC SIGNALS**

Consider again the multiplication property of the FT, repeated here as

$$y(t) = g(t)x(t) \xleftrightarrow{\quad FT \quad} Y(j\omega) = \frac{1}{2\pi} G(j\omega) * X(j\omega).$$

If $x(t)$ is periodic, we may apply the multiplication property by employing the FT representation. Using Eq. (4.3) for $X(j\omega)$ gives

$$y(t) = g(t)x(t) \xleftrightarrow{\quad FT \quad} Y(j\omega) = G(j\omega) * \sum_{k=-\infty}^{\infty} X[k]\delta(\omega - k\omega_o).$$

The sifting property of the impulse function implies that the convolution of any function with a shifted impulse results in a shifted version of the original function. Hence, we have

$$y(t) = g(t)x(t) \xleftrightarrow{\quad FT \quad} Y(j\omega) = \sum_{k=-\infty}^{\infty} X[k]G(j(\omega - k\omega_o)). \qquad (4.12)$$

Multiplication of $g(t)$ with the periodic function $x(t)$ gives an FT consisting of a weighted sum of shifted versions of $G(j\omega)$. This result is illustrated in Fig. 4.11. As expected, the form of $Y(j\omega)$ corresponds to the FT of a nonperiodic signal, since the product of periodic and nonperiodic signals is nonperiodic.

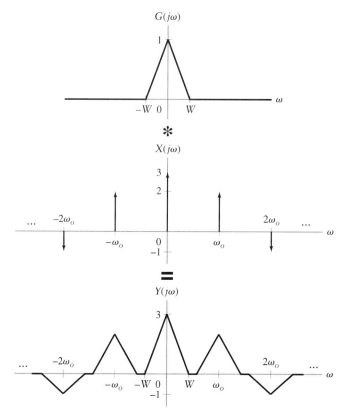

FIGURE 4.11   Multiplication of periodic and nonperiodic time-domain signals corresponds to convolution of the corresponding FT representations.

EXAMPLE 4.5 MULTIPLICATION WITH A SQUARE WAVE   Consider a system with output $y(t) = g(t)x(t)$. Let $x(t)$ be the square wave depicted in Fig. 4.4. **(a)** Find $Y(j\omega)$ in terms of $G(j\omega)$. **(b)** Sketch $Y(j\omega)$ if $g(t) = \cos(t/2)$.

Solution: The square wave has the FS representation

$$x(t) \xleftrightarrow{\ FS; \pi/2\ } X[k] = \frac{\sin(k\pi/2)}{\pi k}.$$

(a) Substituting this result into Eq. (4.12) gives

$$Y(j\omega) = \sum_{k=-\infty}^{\infty} \frac{\sin(k\pi/2)}{\pi k} G(j(\omega - k\pi/2)).$$

(b) Here, we have

$$G(j\omega) = \pi\delta(\omega - 1/2) + \pi\delta(\omega + 1/2)$$

and thus $Y(j\omega)$ may be expressed as

$$Y(j\omega) = \sum_{k=-\infty}^{\infty} \frac{\sin(k\pi/2)}{k} [\delta(\omega - 1/2 - k\pi/2) + \delta(\omega + 1/2 - k\pi/2)].$$

Figure 4.12 depicts the terms constituting $Y(j\omega)$ in the sum near $k = 0$.   ■

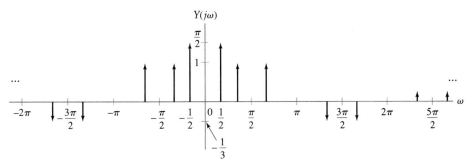

FIGURE 4.12 Solution for Example 4.5(b).

EXAMPLE 4.6 AM RADIO The multiplication property forms the basis for understanding the principles behind a form of amplitude modulation (AM) radio. (A more detailed discussion of AM systems is given in Chapter 5.) A simplified transmitter and receiver are depicted in Fig. 4.13(a). The effect of propagation and channel noise are ignored in this system: The signal at the receiver antenna, $r(t)$, is assumed equal to the transmitted signal. The passband of the low-pass filter in the receiver is equal to the message bandwidth, $-W < \omega < W$. Analyze this system in the frequency domain.

Solution: Assume that the spectrum of the message is as depicted in Fig. 4.13(b). The transmitted signal is expressed as

$$r(t) = m(t)\cos(\omega_c t) \xleftrightarrow{\;FT\;} R(j\omega) = (1/2)M(j(\omega - \omega_c)) + (1/2)M(j(\omega + \omega_c)),$$

where we have used Eq. (4.12) to obtain $R(j\omega)$. Figure 4.14(a) depicts $R(j\omega)$. Note that multiplication by the cosine centers the frequency content of the message on the carrier frequency ω_c.

In the receiver, $r(t)$ is multiplied by the identical cosine used in the transmitter to obtain

$$q(t) = r(t)\cos(\omega_c t) \xleftrightarrow{\;FT\;} Q(j\omega) = (1/2)R(j(\omega - \omega_c)) + (1/2)R(j(\omega + \omega_c)).$$

Expressing $R(j\omega)$ in terms of $M(j\omega)$, we have

$$Q(j\omega) = (1/4)M(j(\omega - 2\omega_c)) + (1/2)M(j\omega) + (1/4)M(j(\omega + 2\omega_c)),$$

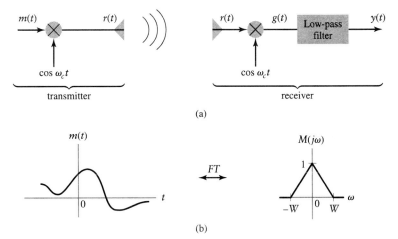

(a)

(b)

FIGURE 4.13 (a) Simplified AM radio transmitter and receiver. (b) Spectrum of message signal with normalized amplitude.

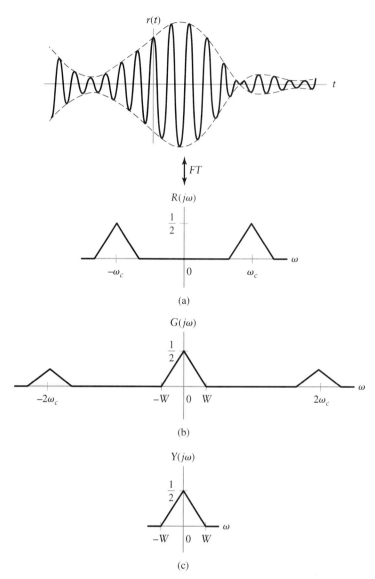

FIGURE 4.14   Signals in the AM transmitter and receiver. (a) Transmitted signal $r(t)$ and spectrum $R(j\omega)$. (b) Spectrum of $g(t)$ in the receiver. (c) Spectrum of receiver output $y(t)$.

as shown in Fig. 4.14(b). Multiplication by the cosine in the receiver centers a portion of the message back on the origin and a portion of the message at twice the carrier frequency. The original message is recovered by low-pass filtering to remove the message replicates centered at twice the carrier frequency. The result of such filtering is an amplitude-scaled version of the original message, as depicted in Fig. 4.14(c).

As explained in Section 1.3.1, the motivation for shifting the message frequency band so that it is centered on a carrier includes the following: (1) Multiple messages may be transmitted simultaneously without interfering with one another, and (2) the physical size of a practical antenna is inversely proportional to the carrier frequency, so at higher carrier frequencies, smaller antennas are required. ■

▶ **Problem 4.6**   Use the multiplication property to determine the frequency response of a system with impulse response

$$h(t) = \frac{\sin(\pi t)}{\pi t} \cos(3\pi t).$$

Compare your answer with that obtained by using the frequency-shift property.

Answer:

$$H(j\omega) = \begin{cases} 1/2, & 2\pi \le |\omega| \le 4\pi \\ 0, & \text{otherwise} \end{cases}$$

◀

The discrete-time multiplication property may be restated as

$$y[n] = x[n]z[n] \xleftarrow{\ \ DTFT\ \ } Y(e^{j\Omega}) = \frac{1}{2\pi} X(e^{j\Omega}) \circledast Z(e^{j\Omega}). \tag{4.13}$$

If $x[n]$ is periodic, then this property is still applicable, provided that we use the DTFT representation of $x[n]$, given in Eq. (4.8), namely,

$$X(e^{j\Omega}) = 2\pi \sum_{k=-\infty}^{\infty} X[k]\delta(\Omega - k\Omega_o),$$

where $X[k]$ are the DTFS coefficients. We substitute $X(e^{j\Omega})$ into the definition of periodic convolution to obtain

$$Y(e^{j\Omega}) = \int_{-\pi}^{\pi} \sum_{k=-\infty}^{\infty} X[k]\delta(\theta - k\Omega_o)Z(e^{j(\Omega-\theta)}) \, d\theta.$$

In any 2π interval of θ, there are exactly N impulses of the form $\delta(\theta - k\Omega_o)$. This is because $\Omega_o = 2\pi/N$. Hence, we may reduce the infinite sum to any N consecutive values of k. Interchanging the order of summation and integration gives

$$Y(e^{j\Omega}) = \sum_{k=0}^{N-1} X[k] \int_{-\pi}^{\pi} \delta(\theta - k\Omega_o)Z(e^{j(\Omega-\theta)}) \, d\theta.$$

Now we apply the sifting property of the impulse function to evaluate the integral and obtain

$$y[n] = x[n]z[n] \xleftarrow{\ \ DTFT\ \ } Y(e^{j\Omega}) = \sum_{k=0}^{N-1} X[k]Z(e^{j(\Omega-k\Omega_o)}). \tag{4.14}$$

Multiplication of $z[n]$ with the periodic sequence $x[n]$ results in a DTFT consisting of a weighted sum of shifted versions of $Z(e^{j\Omega})$. Note that $y[n]$ is nonperiodic, since the product of a periodic signal and a nonperiodic signal is nonperiodic. Consequently, the form of $Y(e^{j\Omega})$ corresponds to a nonperiodic signal.

EXAMPLE 4.7 APPLICATION: WINDOWING DATA It is common in data-processing applications to have access only to a portion of a data record. In this example, we use the multiplication property to analyze the effect of truncating a signal on the DTFT. Consider the signal

$$x[n] = \cos\left(\frac{7\pi}{16}n\right) + \cos\left(\frac{9\pi}{16}n\right).$$

Evaluate the effect of computing the DTFT, using only the $2M + 1$ values $x[n]$, $|n| \leq M$.

Solution: The DTFT of $x[n]$ is obtained from the FS coefficients of $x[n]$ and Eq. (4.8) as

$$X(e^{j\Omega}) = \pi\delta\left(\Omega + \frac{9\pi}{16}\right) + \pi\delta\left(\Omega + \frac{7\pi}{16}\right) + \pi\delta\left(\Omega - \frac{7\pi}{16}\right) + \pi\delta\left(\Omega - \frac{9\pi}{16}\right),$$
$$-\pi < \Omega \leq \pi$$

which consists of impulses at $\pm 7\pi/16$ and $\pm 9\pi/16$. Now define a signal $y[n] = x[n]w[n]$, where

$$w[n] = \begin{cases} 1, & |n| \leq M \\ 0, & |n| > M \end{cases}.$$

Multiplication of $x[n]$ by $w[n]$ is termed *windowing*, since it simulates viewing $x[n]$ through a window. The window $w[n]$ selects the $2M + 1$ values of $x[n]$ centered on $n = 0$. We compare the DTFTs of $y[n] = x[n]w[n]$ and $x[n]$ to establish the effect of windowing. The discrete-time multiplication property Eq. (4.13) implies that

$$Y(e^{j\Omega}) = \frac{1}{2}\{W(e^{j(\Omega+9\pi/16)}) + W(e^{j(\Omega+7\pi/16)}) + W(e^{j(\Omega-7\pi/16)}) + W(e^{j(\Omega-9\pi/16)})\},$$

where the DTFT of the window $w[n]$ is given by

$$W(e^{j\Omega}) = \frac{\sin(\Omega(2M + 1)/2)}{\sin(\Omega/2)}.$$

We see that windowing introduces replicas of $W(e^{j\Omega})$ centered at the frequencies $7\pi/16$ and $9\pi/16$, instead of the impulses that are present in $X(e^{j\Omega})$. We may view this state of affairs as a smearing or broadening of the original impulses: The energy in $Y(e^{j\Omega})$ is now smeared over a band centered on the frequencies of the cosines. The extent of the smearing depends on the width of the mainlobe of $W(e^{j\Omega})$, which is given by $4\pi/(2M + 1)$. (See Figure 3.30.)

Figure 4.15(a)–(c) depicts $Y(e^{j\Omega})$ for several decreasing values of M. If M is large enough so that the width of the mainlobe of $W(e^{j\Omega})$ is small relative to the separation between the frequencies $7\pi/16$ and $9\pi/16$, then $Y(e^{j\Omega})$ is a fairly good approximation to $X(e^{j\Omega})$. This case is depicted in Fig. 4.15(a), using $M = 80$. However, as M decreases and the mainlobe width becomes about the same as the separation between frequencies $7\pi/16$ and $9\pi/16$, the peaks associated with each shifted version of $W(e^{j\Omega})$ begin to overlap and merge into a single peak. This merging is illustrated in Fig. 4.15(b) and (c) by using the values $M = 12$ and $M = 8$, respectively. ▪

The problem of identifying sinusoidal signals of different frequencies in data is very important and occurs often in signal analysis. The preceding example illustrates that our ability to distinguish distinct sinusoids is limited by the length of the data record. If the number of available data points is small relative to the frequency separation, the DTFT is unable to distinguish the presence of two distinct sinusoids. In practice, we are always restricted

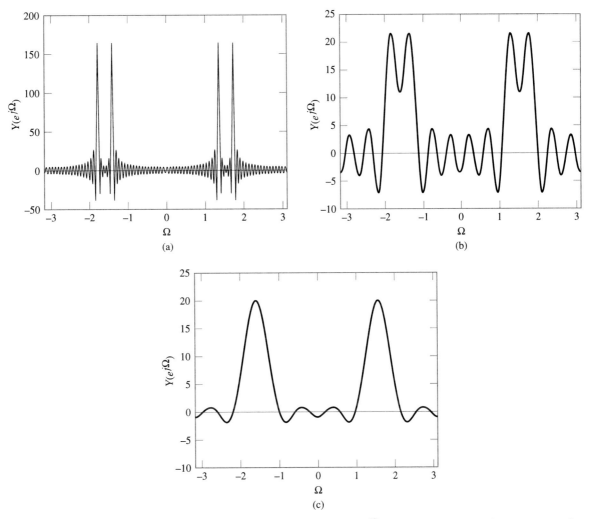

FIGURE 4.15   Effect of windowing a data record. $Y(e^{j\Omega})$ for different values of M, assuming that $\Omega_1 = 7\pi/16$ and $\Omega_2 = 9\pi/16$. (a) $M = 80$, (b) $M = 12$, (c) $M = 8$.

to finite-length data records in any signal analysis application. Thus, it is important to recognize the effects of windowing and take the proper precautions. These issues are discussed in greater detail in Section 4.9.

▶ **Problem 4.7**   Consider the LTI system depicted in Fig. 4.16. Determine an expression for $Y(e^{j\Omega})$, the DTFT of the output, $y[n]$, and sketch $Y(e^{j\Omega})$, assuming that $X(e^{j\Omega})$ is as depicted in Fig. 4.16 and (a) $z[n] = (-1)^n$, (b) $z[n] = 2\cos(\pi n/2)$.

Answers:

(a)
$$Y(e^{j\Omega}) = X(e^{j\Omega}) + X(e^{j(\Omega - \pi)})$$

(b)
$$Y(e^{j\Omega}) = X(e^{j\Omega}) + X(e^{j(\Omega - \pi/2)}) + X(e^{j(\Omega + \pi/2)})$$

as shown in Figs. 4.17(a) and (b), respectively.   ◀

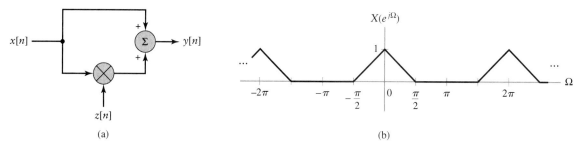

FIGURE 4.16 Problem 4.7 (a) System. (b) Input spectrum.

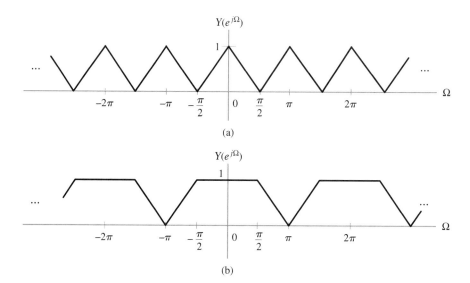

FIGURE 4.17 Solutions to Problem 4.7.

4.4 Fourier Transform
Representation of Discrete-Time Signals

In this section, we derive an FT representation of discrete-time signals by incorporating impulses into the description of the signal in the appropriate manner. This representation satisfies all the properties of the FT and thus converts the FT into a powerful tool for analyzing problems involving mixtures of discrete- and continuous-time signals. Our derivation also indicates the relationship between the FT and DTFT. Combining the results of this section with the Fourier transform representations of periodic signals derived in Section 4.2 enables the FT to be used as an analysis tool for any of the four classes of signals.

We begin the discussion by establishing a correspondence between the continuous-time frequency ω and the discrete-time frequency Ω. Let us define the complex sinusoids $x(t) = e^{j\omega t}$ and $g[n] = e^{j\Omega n}$. A connection between the frequencies of these sinusoids is established by requiring $g[n]$ to correspond to $x(t)$. Suppose we force $g[n]$ to be equal to the samples of $x(t)$ taken at intervals of T_s; that is, $g[n] = x(nT_s)$. This implies that

$$e^{j\Omega n} = e^{j\omega T_s n},$$

from which we conclude that $\Omega = \omega T_s$. In words, the dimensionless discrete-time frequency Ω corresponds to the continuous-time frequency ω, multiplied by the sampling interval T_s.

■ 4.4.1 RELATING THE FT TO THE DTFT

Now consider the DTFT of an arbitrary discrete-time signal $x[n]$. We have

$$X(e^{j\Omega}) = \sum_{n=-\infty}^{\infty} x[n]e^{-j\Omega n}. \tag{4.15}$$

We seek an FT pair $x_\delta(t) \xleftrightarrow{\;FT\;} X_\delta(j\omega)$ that corresponds to the DTFT pair $x[n] \xleftrightarrow{\;DTFT\;} X(e^{j\Omega})$. Substituting $\Omega = \omega T_s$ into Eq. (4.15), we obtain the following function of continuous-time frequency ω:

$$
\begin{aligned}
X_\delta(j\omega) &= X(e^{j\Omega})\big|_{\Omega=\omega T_s}, \\
&= \sum_{n=-\infty}^{\infty} x[n]e^{-j\omega T_s n}.
\end{aligned}
\tag{4.16}
$$

Taking the inverse FT of $X_\delta(j\omega)$, using linearity and the FT pair

$$\delta(t - nT_s) \xleftrightarrow{\;FT\;} e^{-j\omega T_s n},$$

yields the continuous-time signal description

$$x_\delta(t) = \sum_{n=-\infty}^{\infty} x[n]\delta(t - nT_s). \tag{4.17}$$

Hence,

$$\boxed{x_\delta(t) = \sum_{n=-\infty}^{\infty} x[n]\delta(t - nT_s) \xleftrightarrow{\;FT\;} X_\delta(j\omega) = \sum_{n=-\infty}^{\infty} x[n]e^{-j\omega T_s n},} \tag{4.18}$$

where $x_\delta(t)$ is a continuous-time signal that corresponds to $x[n]$, while the Fourier transform $X_\delta(j\omega)$ corresponds to the discrete-time Fourier transform $X(e^{j\Omega})$. We refer to Eq. (4.17) as the *continuous-time representation of $x[n]$*. This representation has an associated sampling interval T_s that determines the relationship between continuous- and discrete-time frequency: $\Omega = \omega T_s$. Figure 4.18 illustrates the relationships between the

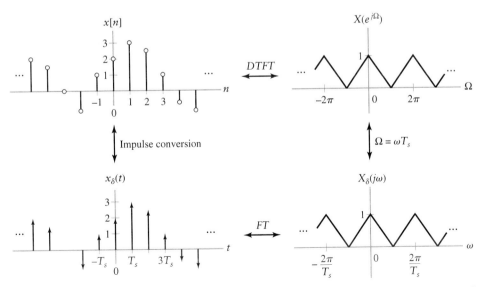

FIGURE 4.18   Relationship between FT and DTFT representations of a discrete-time signal.

signals $x[n]$ and $x_\delta(t)$ and the corresponding Fourier representations $X(e^{j\Omega})$ and $X_\delta(j\omega)$. The DTFT $X(e^{j\Omega})$ is 2π periodic in Ω, while the FT $X_\delta(j\omega)$ is $2\pi/T_s$ periodic in ω. The discrete-time signal has values $x[n]$, while the corresponding continuous-time signal consists of a series of impulses separated by T_s, with the nth impulse having strength $x[n]$.

EXAMPLE 4.8 FT FROM THE DTFT   Determine the FT pair associated with the DTFT pair

$$x[n] = a^n u[n] \xleftrightarrow{\text{DTFT}} X(e^{j\Omega}) = \frac{1}{1 - ae^{-j\Omega}}.$$

This pair is derived in Example 3.17 assuming that $|a| < 1$ so the DTFT converges.

Solution: We substitute for $x[n]$ in Eq. (4.17) to define the continuous-time signal

$$x_\delta(t) = \sum_{n=0}^{\infty} a^n \delta(t - nT_s).$$

Using $\Omega = \omega T_s$ gives

$$x_\delta(t) \xleftrightarrow{\text{FT}} X_\delta(j\omega) = \frac{1}{1 - ae^{-j\omega T_s}}. \qquad\blacksquare$$

Note the many parallels between the continuous-time representation of a discrete-time signal given in Eq. (4.17) and the FT representation of a periodic signal given in Eq. (4.3). The FT representation set forth in Eq.(4.3) is obtained from the FS coefficients by introducing impulses at integer multiples of the fundamental frequency ω_o, with the strength of the kth impulse determined by the kth FS coefficient. The FS representation $X[k]$ is discrete valued, while the corresponding FT representation $X(j\omega)$ is continuous in frequency. In Eq. (4.18), $x[n]$ is discrete valued, while $x_\delta(t)$ is continuous. The parameter T_s determines the separation between impulses in $x_\delta(t)$, just as ω_o does in $X(j\omega)$. These parallels between $x_\delta(t)$ and $X(j\omega)$ are a direct consequence of the FS–DTFT duality property discussed in Section 3.18. Duality states that the roles of time and frequency are interchangeable. Here, $x_\delta(t)$ is a continuous-time signal whose FT, given by Eq. (4.18), is a $(2\pi/T_s)$-periodic function of frequency, while $X(j\omega)$ is a continuous-frequency signal whose inverse FT is a $(2\pi/\omega_o)$-periodic function of time.

▶ **Problem 4.8**   Sketch the FT representation $X_\delta(j\omega)$ of the discrete-time signal

$$x[n] = \frac{\sin(3\pi n/8)}{\pi n},$$

assuming that (a) $T_s = 1/2$, (b) $T_s = 3/2$.

Answer: See Fig. 4.19   ◀

■ 4.4.2 RELATING THE FT TO THE DTFS

In Section 4.2, we derived the FT representation of a periodic continuous-time signal. In the current section, we have shown how to represent a discrete-time nonperiodic signal with the FT. The remaining case, the representation of a discrete-time periodic signal with the FT, is obtained by combining the DTFT representation of a discrete-time periodic signal with the results of the previous subsection. Once this is accomplished, we may use the FT to represent any of the four classes of signals.

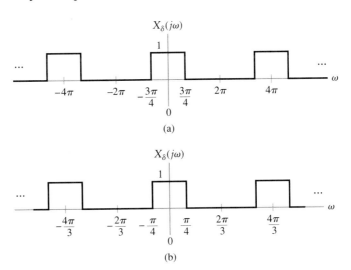

FIGURE 4.19   Solution to Problem 4.8.

Recall that the DTFT representation of an N-periodic signal $x[n]$ is given in Eq. (4.8) as

$$X(e^{j\Omega}) = 2\pi \sum_{k=-\infty}^{\infty} X[k]\delta(\Omega - k\Omega_o),$$

where $X[k]$ are the DTFS coefficients. Substituting $\Omega = \omega T_s$ into this equation yields the FT representation

$$X_\delta(j\omega) = X(e^{j\omega T_s})$$

$$= 2\pi \sum_{k=-\infty}^{\infty} X[k]\delta(\omega T_s - k\Omega_o)$$

$$= 2\pi \sum_{k=-\infty}^{\infty} X[k]\delta(T_s(\omega - k\Omega_o/T_s)).$$

Now we use the scaling property of the impulse, $\delta(a\nu) = (1/a)\delta(\nu)$, to rewrite $X_\delta(j\omega)$ as

$$X_\delta(j\omega) = \frac{2\pi}{T_s} \sum_{k=-\infty}^{\infty} X[k]\delta(\omega - k\Omega_o/T_s). \tag{4.19}$$

Recall that $X[k]$ is an N-periodic function, which implies that $X_\delta(j\omega)$ is periodic with period $N\Omega_o/T_s = 2\pi/T_s$. The signal $x_\delta(t)$ corresponding to this FT is most easily obtained by substituting the periodic signal $x[n]$ into Eq. (4.17); that is,

$$x_\delta(t) = \sum_{n=-\infty}^{\infty} x[n]\delta(t - nT_s). \tag{4.20}$$

Note that the N-periodic nature of $x[n]$ implies that $x_\delta(t)$ is also periodic with fundamental period NT_s. Hence, both $x_\delta(t)$ and $X_\delta(j\omega)$ are N-periodic impulse trains, as depicted in Fig. 4.20.

▶ **Problem 4.9**   Determine the FT pair associated with the discrete-time periodic signal

$$x[n] = \cos\left(\frac{2\pi}{N}n\right).$$

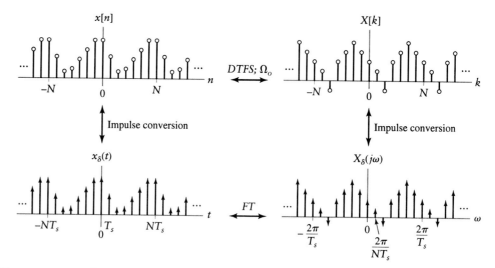

FIGURE 4.20 Relationship between FT and DTFS representations of a discrete-time periodic signal.

Answer:

$$x_\delta(t) = \sum_{n=-\infty}^{\infty} \cos\left(\frac{2\pi}{N}n\right)\delta(t - nT_s) \xleftrightarrow{\;FT\;}$$

$$X_\delta(j\omega) = \frac{\pi}{T_s} \sum_{m=-\infty}^{\infty} \delta\left(\omega + \frac{2\pi}{NT_s} - \frac{m2\pi}{T_s}\right) + \delta\left(\omega - \frac{2\pi}{NT_s} - \frac{m2\pi}{T_s}\right) \quad \blacktriangleleft$$

4.5 *Sampling*

In this section, we use the FT representation of discrete-time signals to analyze the effects of uniformly sampling a signal. The sampling operation generates a discrete-time signal from a continuous-time signal. Sampling of continuous-time signals is often performed in order to manipulate the signal on a computer or microprocessor. Such manipulations are common in communication, control, and signal-processing systems. We shall show how the DTFT of the sampled signal is related to the FT of the continuous-time signal. Sampling is also frequently performed on discrete-time signals to change the effective data rate, an operation termed *subsampling*. In this case, the sampling process discards certain values of the signal. We examine the impact of subsampling by comparing the DTFT of the sampled signal with the DTFT of the original signal.

■ 4.5.1 SAMPLING CONTINUOUS-TIME SIGNALS

Let $x(t)$ be a continuous-time signal. We define a discrete-time signal $x[n]$ that is equal to the "samples" of $x(t)$ at integer multiples of a sampling interval T_s; that is, $x[n] = x(nT_s)$. The impact of sampling is evaluated by relating the DTFT of $x[n]$ to the FT of $x(t)$. Our tool for exploring this relationship is the FT representation of discrete-time signals.

We begin with the continuous-time representation of the discrete-time signal $x[n]$ given in Eq. (4.17):

$$x_\delta(t) = \sum_{n=-\infty}^{\infty} x[n]\delta(t - nT_s).$$

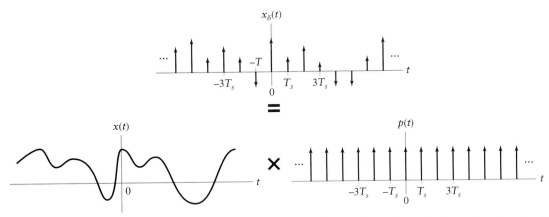

FIGURE 4.21   Mathematical representation of sampling as the product of a given time signal and an impulse train.

Now we use $x(nT_s)$ for $x[n]$ to obtain

$$x_\delta(t) = \sum_{n=-\infty}^{\infty} x(nT_s)\delta(t - nT_s).$$

Since $x(t)\delta(t - nT_s) = x(nT_s)\delta(t - nT_s)$, we may rewrite $x_\delta(t)$ as a product of time functions:

$$x_\delta(t) = x(t)p(t). \tag{4.21}$$

Here,

$$p(t) = \sum_{n=-\infty}^{\infty} \delta(t - nT_s). \tag{4.22}$$

Hence, Eq. (4.21) implies that we may mathematically represent the sampled signal as the product of the original continuous-time signal and an impulse train, as depicted in Fig. 4.21. This representation is commonly termed *impulse sampling* and is a mathematical tool used only to analyze sampling.

The effect of sampling is determined by relating the FT of $x_\delta(t)$ to the FT of $x(t)$. Since multiplication in the time domain corresponds to convolution in the frequency domain, we have

$$X_\delta(j\omega) = \frac{1}{2\pi}X(j\omega) * P(j\omega).$$

Substituting the value for $P(j\omega)$ determined in Example 4.2 into this relationship, we obtain

$$X_\delta(j\omega) = \frac{1}{2\pi}X(j\omega) * \frac{2\pi}{T_s}\sum_{k=-\infty}^{\infty} \delta(\omega - k\omega_s),$$

where $\omega_s = 2\pi/T_s$ is the sampling frequency. Now we convolve $X(j\omega)$ with each of the frequency-shifted impulses to obtain

$$X_\delta(j\omega) = \frac{1}{T_s}\sum_{k=-\infty}^{\infty} X(j(\omega - k\omega_s)). \tag{4.23}$$

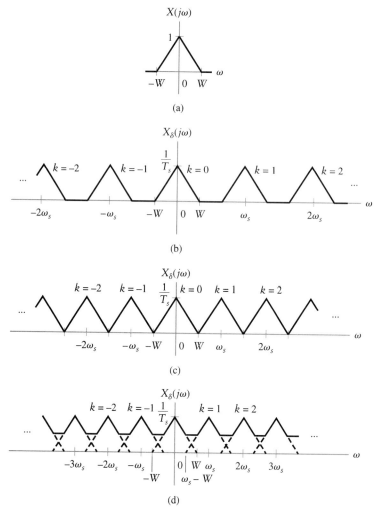

FIGURE 4.22 The FT of a sampled signal for different sampling frequencies. (a) Spectrum of continuous-time signal. (b) Spectrum of sampled signal when $\omega_s = 3W$. (c) Spectrum of sampled signal when $\omega_s = 2W$. (d) Spectrum of sampled signal when $\omega_s = 1.5W$.

Thus, the FT of the sampled signal is given by an infinite sum of shifted versions of the original signal's FT. The shifted versions are offset by integer multiples of ω_s. The shifted versions of $X(j\omega)$ may overlap with each other if ω_s is not large enough compared with the frequency extent, or bandwidth, of $X(j\omega)$. This effect is demonstrated in Fig. 4.22 by depicting Eq. (4.23) for several different values of $T_s = 2\pi/\omega_s$. The frequency content of the signal $x(t)$ is assumed to lie within the frequency band $-W < \omega < W$ for purposes of illustration. In Fig. 4.22(b)–(d), we depict the cases $\omega_s = 3W$, $\omega_s = 2W$, and $\omega_s = 3W/2$, respectively. The shifted replicates of $X(j\omega)$ associated with the kth term in Eq. (4.23) are labeled. Note that as T_s increases and ω_s decreases, the shifted replicates of $X(j\omega)$ move closer together, finally overlapping one another when $\omega_s < 2W$.

Overlap in the shifted replicas of the original spectrum is termed *aliasing*, which refers to the phenomenon of a high-frequency continuous-time component taking on the identity of a low-frequency discrete-time component. Aliasing distorts the spectrum of the sampled signal. The effect is illustrated in Fig. 4.22(d). Overlap between the replicas of $X(j\omega)$

centered at $\omega = 0$ (the $k = 0$ term in Eq. (4.23)) and at $\omega = \omega_s$ (the $k = 1$ term in Eq. (4.23)) occurs for frequencies between $\omega_s - W$ and W. These replicas add, and thus the basic shape of the spectrum changes from portions of a triangle to a constant. The spectrum of the sampled signal no longer has a one-to-one correspondence with that of the original continuous-time signal. This means that we cannot use the spectrum of the sampled signal to analyze the continuous-time signal, and we cannot uniquely reconstruct the original continuous-time signal from its samples. The reconstruction problem is addressed in the next section. As Fig. 4.22 illustrates, aliasing is prevented by choosing the sampling interval T_s so that $\omega_s > 2W$, where W is the highest nonzero frequency component in the signal. This implies that the sampling interval must satisfy the condition $T_s < \pi/W$ for reconstruction of the original signal to be feasible.

The DTFT of the sampled signal is obtained from $X_\delta(j\omega)$ by using the relationship $\Omega = \omega T_s$; that is,

$$x[n] \xleftrightarrow{\;\;DTFT\;\;} X(e^{j\Omega}) = X_\delta(j\omega)\big|_{\omega=\frac{\Omega}{T_s}}.$$

This scaling of the independent variable implies that $\omega = \omega_s$ corresponds to $\Omega = 2\pi$. Figure 4.23(a)–(c) depicts the DTFTs of the sampled signals corresponding to the FTs in Fig. 4.22(b)–(d). Note that the shape is the same in each case, the only difference being a scaling of the frequency axis. The FTs have period ω_s, while the DTFTs have period 2π.

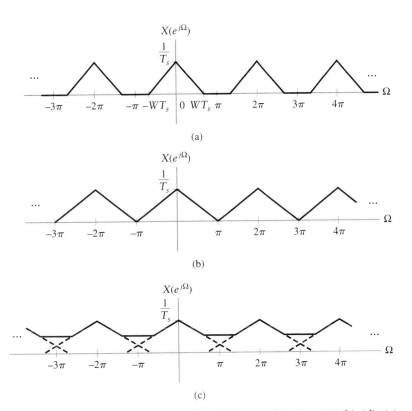

FIGURE 4.23   The DTFTs corresponding to the FTs depicted in Fig. 4.22(b)–(d). (a) $\omega_s = 3W$. (b) $\omega_s = 2W$. (c) $\omega_s = 1.5W$.

EXAMPLE 4.9 SAMPLING A SINUSOID   Consider the effect of sampling the sinusoidal signal

$$x(t) = \cos(\pi t).$$

Determine the FT of the sampled signal for the following sampling intervals: (i) $T_s = 1/4$, (ii) $T_s = 1$, and (iii) $T_s = 3/2$.

Solution: Use Eq. (4.23) for each value of T_s. In particular, note from Example 4.1 that

$$x(t) \xleftrightarrow{\ FT\ } X(j\omega) = \pi\delta(\omega + \pi) + \pi\delta(\omega - \pi).$$

Substitution of $X(j\omega)$ into Eq. (4.23) gives

$$X_\delta(j\omega) = \frac{\pi}{T_s} \sum_{k=-\infty}^{\infty} \delta(\omega + \pi - k\omega_s) + \delta(\omega - \pi - k\omega_s).$$

Hence, $X_\delta(j\omega)$ consists of pairs of impulses separated by 2π, centered on integer multiples of the sampling frequency ω_s. This frequency differs in each of the three cases. Using $\omega_s = 2\pi/T_s$ gives (i) $\omega_s = 8\pi$, (ii) $\omega_s = 2\pi$, and (iii) $\omega_s = 4\pi/3$, respectively. The impulse-sampled representations for the continuous-time signals and their FTs are depicted in Fig. 4.24.

In case (i), in which $T_s = 1/4$, the impulses are clearly paired about multiples of 8π, as depicted in Fig. 4.24(b). As T_s increases and ω_s decreases, pairs of impulses associated with different values of k become closer together. In the second case, in which $T_s = 1$, impulses associated with adjacent indices k are superimposed on one another, as illustrated in Fig. 4.24(c). This corresponds to a sampling interval of one-half period, as shown on the left-hand side of the figure. There is an ambiguity here, since we cannot uniquely determine the original signal from either $x_\delta(t)$ or $X_\delta(j\omega)$. For example, both the original signal $x(t) = \cos(\pi t)$ and $x_1(t) = e^{j\pi t}$ result in the same sequence $x[n] = (-1)^n$ for $T_s = 1$. In the last case, in which $T_s = 3/2$, shown in Fig. 4.24(d), the pairs of impulses associated with each index k are interspersed, and we have another ambiguity. Both the original signal $x(t) = \cos(\pi t)$, shown as the solid line on the left-hand side of the figure, and the signal $x_2(t) = \cos(\pi t/3)$, shown as the dashed line on the left-hand side of the figure, are consistent with the sampled signal $x_\delta(t)$ and spectrum $X_\delta(j\omega)$. Consequently, sampling has caused the original sinusoid with frequency π to alias or appear as a new sinusoid of frequency $\pi/3$. ■

EXAMPLE 4.10 ALIASING IN MOVIES   Film-based movies are produced by recording 30 still frames of a scene every second. Hence, the sampling interval for the video portion of a movie is $T_s = 1/30$ second. Consider taking a movie of a wheel with $r = 1/4$ m radius shown in Fig. 4.25(a). The wheel rotates counterclockwise at a rate of ω radians per second and thus moves across the scene from right to left at a linear velocity of $v = \omega r = \omega/4$ meters per second. Show that the sampling involved in making the movie can cause the wheel to appear as though it is rotating backwards or not at all.

Solution: Suppose the center of the wheel corresponds to the origin of a complex plane. This origin is translated from right to left as the wheel moves across the scene, since it is fixed to the center of the wheel. At a given time t, the angular position of the mark on the wheel forms an angle of ωt with respect to one of the coordinate axes, so the position of the radial mark $x(t)$ relative to the origin is described by the complex sinusoid $x(t) = e^{j\omega t}$. The position of the mark relative to the origin in the movie is described by the sampled version of this sinusoid, $x[n] = e^{j\omega n T_s}$.

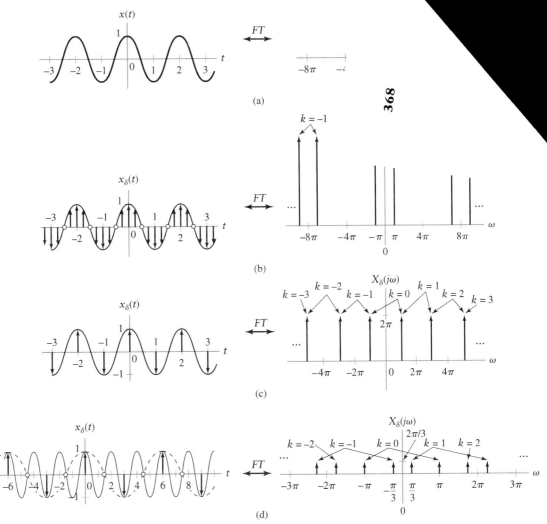

FIGURE 4.24   The effect of sampling a sinusoid at different rates (Example 4.9). (a) Original signal and FT. (b) Original signal, impulse-sampled representation and FT for $T_s = 1/4$. (c) Original signal, impulse-sampled representation and FT for $T_s = 1$. (d) Original signal, impulse-sampled representation and FT for $T_s = 3/2$. A cosine of frequency $\pi/3$ is shown as the dashed line.

If the wheel rotates an angle less than π radians between frames, then the apparent rotation of the wheel is visually consistent with its left-to-right motion, as shown in Fig. 4.25(b). This implies that $\omega T_s < \pi$, or $\omega < 30\pi$ radians per second, which is one-half the movie's sampling frequency. If the rotational rate of the wheel satisfies this condition, then no aliasing occurs. If the wheel rotates between π and 2π radians between frames, then the wheel appears to be rotating clockwise, as shown in Fig. 4.25(c), and the rotation of the wheel appears to be inconsistent with its linear motion. This occurs when $\pi < \omega T_s < 2\pi$ or $30\pi < \omega < 60\pi$ radians per second and for linear velocities of $23.56 < v < 47.12$ meters per second. If there is exactly one revolution between frames, then the wheel does not appear to be rotating at all, as shown in Fig. 4.25(d). This occurs when $\omega = 60\pi$ radians per second and $v = 47.12$ meters per second, or approximately 170 kilometers per hour. ∎

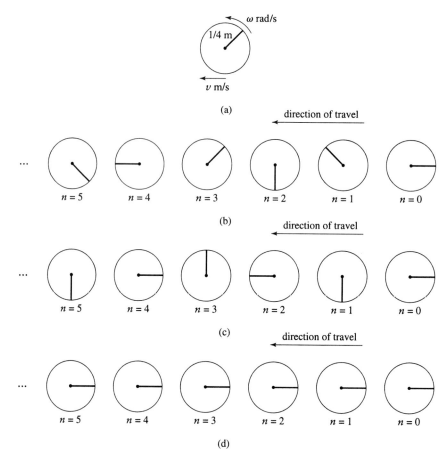

FIGURE 4.25 Aliasing in a movie. (a) Wheel rotating at ω radians per second and moving from right to left at v meters per second. (b) Sequence of movie frames, assuming that the wheel rotates less than one-half turn between frames. (c) Sequence of movie frames, assuming that the wheel rotates three-fourths of a turn between frames. (d) Sequence of movie frames, assuming that the wheel rotates one turn between frames.

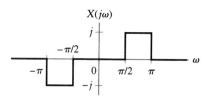

FIGURE 4.26 Spectrum of original signal for Problem 4.10.

▶ **Problem 4.10** Draw the FT of a sampled version of the continuous-time signal having the FT depicted in Fig. 4.26 for (a) $T_s = 1/2$ and (b) $T_s = 2$.

Answer: See Figs. 4.27(a) and (b) ◀

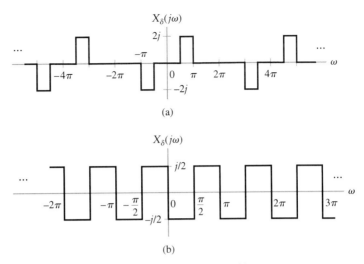

$X_\delta(j\omega)$

(a)

$X_\delta(j\omega)$

(b)

FIGURE 4.27   Solution to Problem 4.10.

S & S
Solutions

EXAMPLE 4.11   MULTIPATH COMMUNICATION CHANNEL: DISCRETE-TIME MODEL   The two-path communication channel introduced in Section 1.10 is described by the equation

$$y(t) = x(t) + \alpha x(t - T_{\text{diff}}). \qquad (4.24)$$

A discrete-time model for this channel was also introduced in that section. Let the channel input $x(t)$ and output $y(t)$ be sampled at $t = nT_s$, and consider the discrete-time multipath-channel model

$$y[n] = x[n] + ax[n - 1]. \qquad (4.25)$$

Let the input signal $x(t)$ have bandwidth π/T_s. Evaluate the approximation error of the discrete-time model.

Solution: Take the FT of both sides of Eq. (4.24) to obtain the frequency response of the two-path channel as

$$H(j\omega) = 1 + \alpha e^{-j\omega T_{\text{diff}}}.$$

Similarly, take the DTFT of Eq. (4.25) to obtain the frequency response of the discrete-time channel model as

$$H(e^{j\Omega}) = 1 + ae^{-j\Omega}.$$

Now use $\Omega = \omega T_s$ to express the FT of the discrete-time channel model as

$$H_\delta(j\omega) = 1 + ae^{-j\omega T_s}.$$

When comparing $H(j\omega)$ and $H_\delta(j\omega)$, we consider only frequencies within the bandwidth of the input signal $x(t)$, $-\pi/T_s \leq \omega \leq \pi/T_s$, since this is the only portion of the channel frequency response that affects the input signal. We compute the mean-squared error between $H(j\omega)$ and $H_\delta(j\omega)$ over this frequency band:

$$\text{MSE} = \frac{T_s}{2\pi} \int_{-\pi/T_s}^{\pi/T_s} |H(j\omega) - H_\delta(j\omega)|^2 \, d\omega$$

$$= \frac{T_s}{2\pi} \int_{-\pi/T_s}^{\pi/T_s} |\alpha e^{-j\omega T_{\text{diff}}} - ae^{-j\omega T_s}|^2 \, d\omega \qquad (4.26)$$

$$= |\alpha|^2 + |a|^2 - \alpha a^* \gamma - \alpha^* a \gamma^*.$$

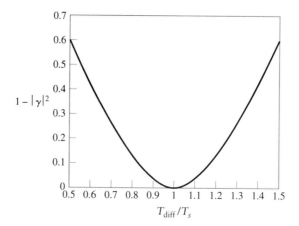

FIGURE 4.28   Factor $1 - |\gamma|^2$ determining the mean-squared error of the discrete-time model for the two-path communication channel.

In this equation,

$$\gamma = \frac{T_s}{2\pi} \int_{-\pi/T_s}^{\pi/T_s} e^{-j\omega(T_{\text{diff}} - T_s)} \, d\omega$$

$$= \text{sinc}\left(\frac{T_{\text{diff}} - T_s}{T_s}\right).$$

Here, γ characterizes the effect of a difference between the true path delay, T_{diff}, and the discrete-time model path delay, T_s. In order to see the relationship between α and a, it is convenient to rewrite the MSE as a perfect square in a:

$$\text{MSE} = |a - \alpha\gamma|^2 + |\alpha|^2(1 - |\gamma|^2).$$

The equivalence between this expression and Eq. (4.26) is easily verified by expanding the square and canceling terms. In this form, we see that the mean-squared error is minimized by choosing the discrete-time model path gain as $a = \alpha\gamma$. The resulting minimum mean-squared error is

$$\text{MSE}_{\text{min}} = |\alpha|^2(1 - |\gamma|^2).$$

Thus, the quality of the discrete-time-channel model depends only on the relationship between T_{diff} and T_s, as determined by the quantity $1 - |\gamma|^2$. Figure 4.28 depicts $1 - |\gamma|^2$ as a function of T_{diff}/T_s. Note that this factor is less than 0.1 for $0.83 \leq T_{\text{diff}}/T_s \leq 1.17$, which indicates that the discrete-time model is reasonably accurate, provided that $T_s \approx T_{\text{diff}}$.   ■

■ 4.5.2 SUBSAMPLING: SAMPLING DISCRETE-TIME SIGNALS

The FT is also very helpful in analyzing the effect of *subsampling* a discrete-time signal. Let $y[n] = x[qn]$ be a subsampled version of $x[n]$. We require q to be a positive integer for this operation to be meaningful. Our goal is to relate the DTFT of $y[n]$ to the DTFT of $x[n]$. We accomplish this by using the FT to represent $x[n]$ as a sampled version of a continuous-time signal $x(t)$. We then express $y[n]$ as a sampled version of the same underlying continuous-time signal $x(t)$ obtained using a sampling interval q times that associated with $x[n]$.

The result relates the DTFT of $y[n]$ to the DTFT of $x[n]$ as

$$Y(e^{j\Omega}) = \frac{1}{q}\sum_{m=0}^{q-1} X(e^{j(\Omega-m2\pi)/q}). \qquad (4.27)$$

The reader is asked to derive this result in Problem 4.42.

Equation (4.27) indicates that $Y(e^{j\Omega})$ is obtained by summing versions of the scaled DTFT $X_q(e^{j\Omega}) = X(e^{j\Omega/q})$ that are shifted by integer multiples of 2π. We may write this result explicitly as

$$Y(e^{j\Omega}) = \frac{1}{q}\sum_{m=0}^{q-1} X_q(e^{j(\Omega-m2\pi)}).$$

Figure 4.29 illustrates the relationship between $Y(e^{j\Omega})$ and $X(e^{j\Omega})$ described in Eq. (4.27). Figure 4.29(a) depicts $X(e^{j\Omega})$. Figures 4.29(b)–(d) show the individual terms in the sum of Eq. (4.27) corresponding to $m = 0, m = 1$, and $m = q - 1$. In Fig. 4.29(e), we depict $Y(e^{j\Omega})$, assuming that $W < \pi/q$, while Fig. 4.29(f) shows $Y(e^{j\Omega})$, assuming that $W > \pi/q$. In this last case, there is overlap between the scaled and shifted versions of $X(e^{j\Omega})$ involved in Eq. (4.27), and aliasing occurs. We conclude that aliasing can be prevented if W, the highest frequency component of $X(e^{j\Omega})$, is less than π/q.

▶ **Problem 4.11**   Depict the DTFT of the subsampled signal $y[n] = x[qn]$ for $q = 2$ and $q = 5$, assuming that

$$x[n] = 2\cos\left(\frac{\pi}{3}n\right).$$

Answer: See Figs. 4.30(a) and (b)   ◀

4.6 *Reconstruction of Continuous-Time Signals from Samples*

The problem of reconstructing a continuous-time signal from samples involves a mixture of continuous- and discrete-time signals. As illustrated in the block diagram of Fig. 4.31, a device that performs this operation has a discrete-time input signal and a continuous-time output signal. The FT is well suited for analyzing this problem, since it may be used to represent both continuous- and discrete-time signals. In the current section, we first consider the conditions that must be met in order to uniquely reconstruct a continuous-time signal from its samples. Assuming that these conditions are satisfied, we establish a method for perfect reconstruction. Unfortunately, the perfect-reconstruction approach cannot be implemented in any practical system. Hence, the section concludes with an analysis of practical reconstruction techniques and their limitations.

■ 4.6.1 SAMPLING THEOREM

Our discussion of sampling in the previous section indicated that the samples of a signal do not always uniquely determine the corresponding continuous-time signal. For example, if we sample a sinusoid at intervals of a period, then the sampled signal appears to be a constant, and we cannot determine whether the original signal was a constant or the sinusoid. Figure 4.32 illustrates this problem by depicting two different continuous-time signals having the same set of samples. We have

$$x[k] = x_1(nT_s) = x_2(nT_s).$$

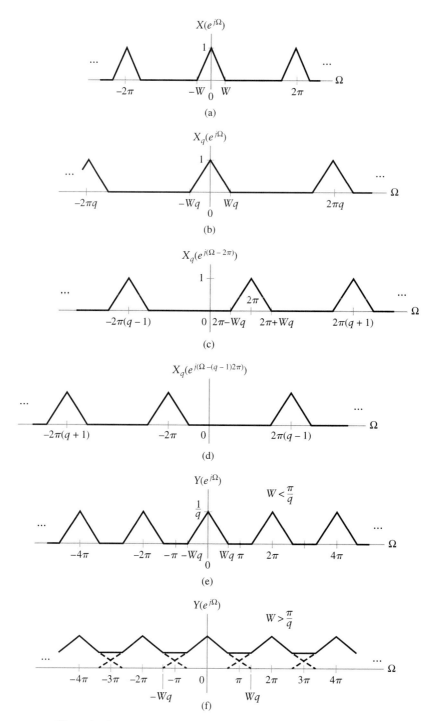

FIGURE 4.29 Effect of subsampling on the DTFT. (a) Original signal spectrum. (b) $m = 0$ term, $X_q(e^{j\Omega})$, in Eq. (4.27). (c) $m = 1$ term in Eq. (4.27). (d) $m = q - 1$ term in Eq. (4.27). (e) $Y(e^{j\Omega})$, assuming that $W < \pi/q$. (f) $Y(e^{j\Omega})$, assuming that $W > \pi/q$.

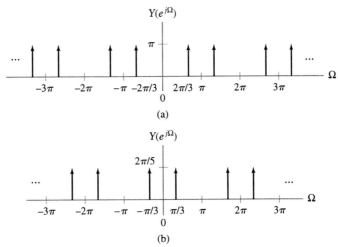

FIGURE 4.30 Solution to Problem 4.11.

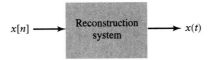

FIGURE 4.31 Block diagram illustrating conversion of a discrete-time signal to a continuous-time signal.

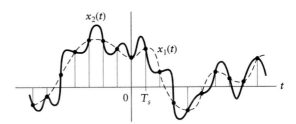

FIGURE 4.32 Two continuous-time signals $x_1(t)$ (dashed line) and $x_2(t)$ (solid line) that have the same set of samples.

Note that the samples do not tell us anything about the behavior of the signal in between the times it is sampled. In order to determine how the signal behaves in between those times, we must specify additional constraints on the continuous-time signal. One such set of constraints that is very useful in practice involves requiring the signal to make smooth transitions from one sample to another. The smoothness, or rate at which the time-domain signal changes, is directly related to the maximum frequency that is present in the signal. Hence, constraining smoothness in the time domain corresponds to limiting the bandwidth of the signal.

Because there is a one-to-one correspondence between the time-domain and frequency-domain representations of a signal, we may also consider the problem of reconstructing the continuous-time signal in the frequency domain. To reconstruct a continuous-time signal uniquely from its samples, there must be a unique correspondence between the FTs of the continuous-time signal and the sampled signal. These FTs are uniquely related if the sampling process does not introduce aliasing. As we discovered in the previous section, aliasing distorts the spectrum of the original signal and destroys the one-to-one relationship between the FT's of the continuous-time signal and the sampled signal. This suggests that the condition for a unique correspondence between the continuous-time signal and its samples is equivalent to a condition for the prevention of aliasing, a requirement that is formally stated in the following theorem:

> *Sampling Theorem*   Let $x(t) \xleftrightarrow{\;FT\;} X(j\omega)$ represent a band-limited signal, so that $X(j\omega) = 0$ for $|\omega| > \omega_m$. If $\omega_s > 2\omega_m$, where $\omega_s = 2\pi/T_s$ is the sampling frequency, then $x(t)$ is uniquely determined by its samples $x(nT_s)$, $n = 0, \pm1 \pm2, \ldots$

The minimum sampling frequency, $2\omega_m$, is termed the *Nyquist sampling rate* or *Nyquist rate*. The actual sampling frequency, ω_s, is commonly referred to as the *Nyquist frequency* when discussing the FT of either the continuous-time or sampled signal. We note that in many problems it is more convenient to evaluate the sampling theorem with frequency expressed in hertz. If $f_m = \omega_m/(2\pi)$ is the highest frequency present in the signal and f_s denotes the sampling frequency, both expressed in hertz, then the sampling theorem states that $f_s > 2f_m$, where $f_s = 1/T_s$. Alternatively, we must have $T_s < 1/(2f_m)$ to satisfy the conditions of the theorem.

EXAMPLE 4.12 SELECTING THE SAMPLING INTERVAL   Suppose $x(t) = \sin(10\pi t)/(\pi t)$. Determine the condition on the sampling interval T_s so that $x(t)$ is uniquely represented by the discrete-time sequence $x[n] = x(nT_s)$.

Solution: In order to apply the sampling theorem, we must first determine the maximum frequency ω_m present in $x(t)$. Taking the FT (see Example 3.26), we have

$$X(j\omega) = \begin{cases} 1, & |\omega| \le 10\pi \\ 0, & |\omega| > 10\pi \end{cases},$$

as depicted in Fig. 4.33. Also, $\omega_m = 10\pi$. Hence, we require that

$$2\pi/T_s > 20\pi,$$

or

$$T_s < (1/10). \qquad\blacksquare$$

▶ **Problem 4.12**   Determine the conditions on the sampling interval T_s so that each $x(t)$ is uniquely represented by the discrete-time sequence $x[n] = x(nT_s)$.

(a) $x(t) = \cos(\pi t) + 3\sin(2\pi t) + \sin(4\pi t)$

(b) $x(t) = \cos(2\pi t)\dfrac{\sin(\pi t)}{\pi t} + 3\sin(6\pi t)\dfrac{\sin(2\pi t)}{\pi t}$

(c) The signal $x(t)$ with FT given in Fig. 4.34.

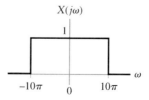

FIGURE 4.33   FT of continuous-time signal for Example 4.12.

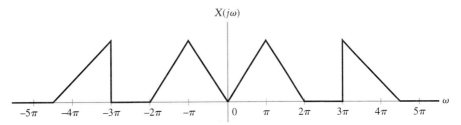

$X(j\omega)$

FIGURE 4.34   FT of $x(t)$ for Problem 4.12(c).

Answers:

  (a) $T_s < 1/4$

  (b) $T_s < 1/8$

  (c) $T_s < 2/9$ ◀

We are often interested only in the lower frequency components of a signal and wish to sample the signal at a rate ω_s less than twice the highest frequency that is actually present in the signal. A reduced sampling rate can be used if the signal is passed through a continuous-time low-pass filter prior to sampling. Ideally, this filter passes frequency components below $\omega_s/2$ without distortion and suppresses any frequency components above $\omega_s/2$. Such a filter prevents aliasing and is thus termed an *antialiasing filter*. A practical antialiasing filter changes from passband to stopband gradually. To compensate for the filter's transition band, the passband is usually chosen to include the maximum signal frequency that is of interest, and the sampling frequency ω_s is chosen so that $\omega_s/2$ is in the stopband of the antialiasing filter. (This issue is discussed further in Section 4.7.) Even if the signal of interest is band limited to less than $\omega_s/2$, an antialiasing filter is normally used to avoid aliasing associated with the presence of measurement or electronic noise.

■ **4.6.2 IDEAL RECONSTRUCTION**

The sampling theorem indicates how fast we must sample a signal so that the samples uniquely represent the continuous-time signal. Now we consider the problem of reconstructing the continuous-time signal from the samples. This problem is most easily solved in the frequency domain with the use of the FT. Recall that if $x(t) \xleftrightarrow{FT} X(j\omega)$, then the FT representation of the sampled signal is given by Eq. (4.23), or

$$X_\delta(j\omega) = \frac{1}{T_s} \sum_{k=-\infty}^{\infty} X(j\omega - jk\omega_s).$$

Figures 4.35(a) and (b) depict $X(j\omega)$ and $X_\delta(j\omega)$, respectively, assuming that the conditions of the sampling theorem are satisfied.

The goal of reconstruction is to apply some operation to $X_\delta(j\omega)$ that converts it back to $X(j\omega)$. Any such operation must eliminate the replicas, or *images*, of $X(j\omega)$ that are centered at $k\omega_s$. This is accomplished by multiplying $X_\delta(j\omega)$ by

$$H_r(j\omega) = \begin{cases} T_s, & |\omega| \leq \omega_s/2 \\ 0, & |\omega| > \omega_s/2 \end{cases}, \tag{4.28}$$

as depicted in Fig. 4.35(c). We then have

$$X(j\omega) = X_\delta(j\omega)H_r(j\omega). \tag{4.29}$$

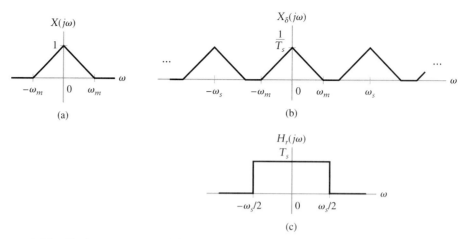

FIGURE 4.35   Ideal reconstruction. (a) Spectrum of original signal. (b) Spectrum of sampled signal. (c) Frequency response of reconstruction filter.

Note that multiplication by $H_r(j\omega)$ will not recover $X(j\omega)$ from $X_\delta(j\omega)$ if the conditions of the sampling theorem are not met and aliasing occurs.

Multiplication in the frequency domain transforms to convolution in the time domain, so Eq. (4.29) implies that

$$x(t) = x_\delta(t) * h_r(t),$$

where $h_r(t) \xleftrightarrow{\ FT\ } H_r(j\omega)$. Substituting Eq. (4.17) for $x_\delta(t)$ in this relation gives

$$x(t) = h_r(t) * \sum_{n=-\infty}^{\infty} x[n]\delta(t - nT_s),$$

$$= \sum_{n=-\infty}^{\infty} x[n]h_r(t - nT_s).$$

Now we use

$$h_r(t) = \frac{T_s \sin\left(\dfrac{\omega_s}{2}t\right)}{\pi t}$$

on the basis of the result of Example 3.26 to obtain

$$x(t) = \sum_{n=-\infty}^{\infty} x[n]\,\mathrm{sinc}(\omega_s(t - nT_s)/(2\pi)). \tag{4.30}$$

In the time domain, we reconstruct $x(t)$ as a weighted sum of sinc functions shifted by the sampling interval. The weights correspond to the values of the discrete-time sequence. This reconstruction operation is illustrated in Fig. 4.36. The value of $x(t)$ at $t = nT_s$ is given by $x[n]$ because all of the shifted sinc functions are zero at nT_s, except the nth one, and its value is unity. The value of $x(t)$ in between integer multiples of T_s is determined by all of the values of the sequence $x[n]$.

The operation described in Eq. (4.30) is commonly referred to as *ideal band-limited interpolation*, since it indicates how to interpolate in between the samples of a bandlimited

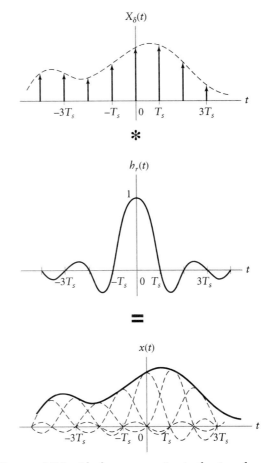

FIGURE 4.36   Ideal reconstruction in the time domain.

signal. In practice, Eq. (4.30) cannot be implemented, for two reasons: First of all, it represents a noncausal system, because the output, $x(t)$, depends on past and future values of the input, $x[n]$; second, the influence of each sample extends over an infinite amount of time, because $h_r(t)$ has infinite duration.

▪ 4.6.3 A PRACTICAL RECONSTRUCTION: THE ZERO-ORDER HOLD

In practice, a continuous-time signal is often reconstructed by means of a device known as a *zero-order hold*, which simply maintains or holds the value $x[n]$ for T_s seconds, as depicted in Fig. 4.37. This causes sharp transitions in $x_o(t)$ at integer multiples of T_s and produces

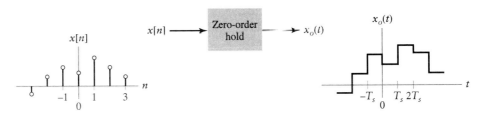

FIGURE 4.37   Reconstruction via a zero-order hold.

Figure 4.38   Rectangular pulse used to analyze zero-order hold reconstruction.

a stair-step approximation to the continuous-time signal. Once again, the FT offers a means for analyzing the quality of this approximation.

The zero-order hold is represented mathematically as a weighted sum of rectangular pulses shifted by integer multiples of the sampling interval. Let

$$h_o(t) = \begin{cases} 1, & 0 < t < T_s \\ 0, & t < 0, t > T_s \end{cases},$$

as depicted in Fig. 4.38. The output of the zero-order hold is expressed in terms of $h_o(t)$ as

$$x_o(t) = \sum_{n=-\infty}^{\infty} x[n] h_o(t - nT_s). \tag{4.31}$$

We recognize Eq. (4.31) as the convolution of the impulse-sampled signal $x_\delta(t)$ with $h_o(t)$:

$$x_o(t) = h_o(t) * \sum_{n=-\infty}^{\infty} x[n]\delta(t - nT_s)$$
$$= h_o(t) * x_\delta(t).$$

Now we take the FT of $x_o(t)$, using the convolution–multiplication property of the FT to obtain

$$X_o(j\omega) = H_o(j\omega)X_\delta(j\omega),$$

from which, on the basis of the result of Example 3.25 and the FT time-shift property, we obtain

$$h_o(t) \xleftarrow{\quad FT \quad} H_o(j\omega) = 2e^{-j\omega T_s/2}\frac{\sin(\omega T_s/2)}{\omega}.$$

Figure 4.39 depicts the effect of the zero-order hold in the frequency domain, assuming that T_s is chosen to satisfy the sampling theorem. Comparing $X_o(j\omega)$ with $X(j\omega)$, we see that the zero-order hold introduces three forms of modification:

1. A linear phase shift corresponding to a time delay of $T_s/2$ seconds.
2. A distortion of the portion of $X_\delta(j\omega)$ between $-\omega_m$ and ω_m. [The distortion is produced by the curvature of the mainlobe of $H_o(j\omega)$.]
3. Distorted and attenuated versions of the images of $X(j\omega)$, centered at nonzero multiples of ω_s.

By holding each value $x[n]$ for T_s seconds, we introduce a time shift of $T_s/2$ seconds into $x_o(t)$. This is the source of modification 1. Modifications 2 and 3 are associated with the stair-step approximation. Note that the sharp transitions in $x_o(t)$ suggest the presence of high-frequency components and are consistent with modification 3. Both modifications 1 and 2 are reduced by increasing ω_s or, equivalently, decreasing T_s.

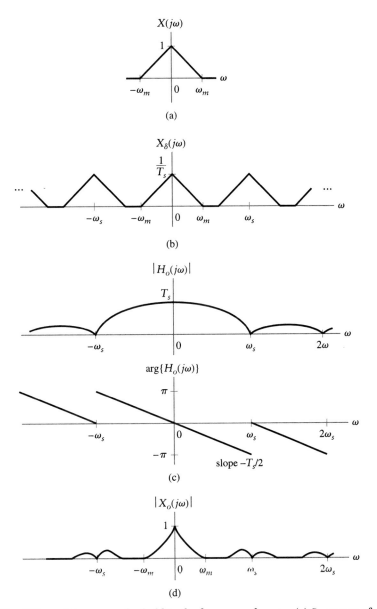

FIGURE 4.39 Effect of the zero-order hold in the frequency domain. (a) Spectrum of original continuous-time signal. (b) FT of sampled signal. (c) Magnitude and phase of $H_o(j\omega)$. (d) Magnitude spectrum of signal reconstructed using zero-order hold.

In some applications, the modifications associated with the zero-order hold may be acceptable. In others, further processing of $x_o(t)$ may be desirable to reduce the distortion associated with modifications 2 and 3. In most situations, a delay of $T_s/2$ seconds is of no real consequence. Modifications 2 and 3 may be eliminated by passing $x_o(t)$ through a continuous-time compensation filter with frequency response

$$H_c(j\omega) = \begin{cases} \dfrac{\omega T_s}{2\sin(\omega T_s/2)}, & |\omega| < \omega_m \\[2ex] 0, & |\omega| > \omega_s - \omega_m \end{cases}.$$

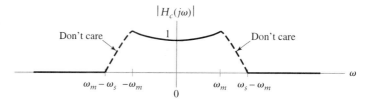

FIGURE 4.40   Frequency response of a compensation filter used to eliminate some of the distortion introduced by the zero-order hold.

The magnitude of this frequency response is depicted in Fig. 4.40. On $|\omega| < \omega_m$, the compensation filter reverses the distortion introduced by the mainlobe curvature of $H_o(j\omega)$. For $|\omega| > \omega_s - \omega_m$, $H_c(j\omega)$ removes the energy in $X_o(j\omega)$ centered at nonzero multiples of ω_s. The value of $H_c(j\omega)$ does not matter on the frequency band $\omega_m < |\omega| < \omega_s - \omega_m$, since $X_o(j\omega)$ is zero there. $H_c(j\omega)$ is often termed an *anti-imaging filter*, because it eliminates the distorted images of $X(j\omega)$ that are present at nonzero multiples of ω_s. A block diagram representing the compensated zero-order hold reconstruction process is depicted in Fig. 4.41. The anti-imaging filter smooths out the step discontinuities in $x_o(t)$.

Several practical issues arise in designing and building an anti-imaging filter. We cannot obtain a causal anti-imaging filter that has zero phase; hence a practical filter will introduce some phase distortion. In many cases, a linear phase in the passband, $|\omega| < \omega_m$, is acceptable, since linear-phase distortion corresponds to an additional time delay. The difficulty of approximating $|H_c(j\omega)|$ depends on the separation between ω_m and $\omega_s - \omega_m$. First of all, if this distance, $\omega_s - 2\omega_m$, is large, then the mainlobe curvature of $H_o(j\omega)$ is very small, and a good approximation is obtained simply by setting $|H_c(j\omega)| = 1$. Second, the region $\omega_m < \omega < \omega_s - \omega_m$ is used to make the transition from passband to stopband. If $\omega_s - 2\omega_m$ is large, then the transition band of the filter is large. Filters with large transition bands are much easier to design and build than those with small transition bands. Hence, the requirements on an anti-imaging filter are greatly reduced by choosing T_s sufficiently small so that $\omega_s \gg 2\omega_m$. (A more detailed discussion of filter design is given in Chapter 8.)

In practical reconstruction schemes, it is common to increase the effective sampling rate of the discrete-time signal prior to the zero-order hold. This technique, known as *oversampling*, is done to relax the requirements on the anti-imaging filter, as illustrated in the next example. Although doing so increases the complexity of the discrete-time hardware, it usually produces a decrease in overall system cost for a given level of reconstruction quality.

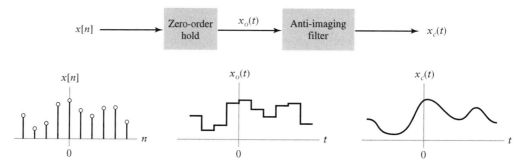

FIGURE 4.41   Block diagram of a practical reconstruction system.

EXAMPLE 4.13 OVERSAMPLING IN CD PLAYERS   In this example, we explore the benefits of oversampling in reconstructing a continuous-time audio signal using an audio compact disc player. Assume that the maximum signal frequency is $f_m = 20$ kHz. Consider two cases: (a) reconstruction using the standard digital audio rate of $1/T_{s1} = 44.1$ kHz, and (b) reconstruction using eight-times oversampling, for an effective sampling rate of $1/T_{s2} = 352.8$ kHz. In each case, determine the constraints on the magnitude response of an anti-imaging filter so that the overall magnitude response of the zero-order hold reconstruction system is between 0.99 and 1.01 in the signal passband and the images of the original signal's spectrum centered at multiples of the sampling frequency [the $k = \pm1$, $\pm2, \ldots$ terms in Eq. (4.23)] are attenuated by a factor of 10^{-3} or more.

Solution: In this example, it is convenient to express frequency in units of hertz rather than radians per second. This is explicitly indicated by replacing ω with f and by representing the frequency responses $H_o(j\omega)$ and $H_c(j\omega)$ as $H'_o(jf)$ and $H'_c(jf)$, respectively. The overall magnitude response of the zero-order hold followed by an anti-imaging filter $H'_c(jf)$ is $|H'_o(jf)||H'_c(jf)|$. Our goal is to find the acceptable range of $|H'_c(jf)|$ so that the product $|H'_o(jf)||H'_c(jf)|$ satisfies the constraints on the response. Figures 4.42(a) and (b) depict $|H'_o(jf)|$, assuming sampling rates of 44.1 kHz and 352.8 kHz, respectively. The dashed lines in each figure denote the signal passband and its images. At the lower sampling rate [Fig. 4.42(a)], we see that the signal and its images occupy the majority of the spectrum; they are separated by 4.1 kHz. In the eight-times oversampling case [Fig. 4.42(b)], the signal and its images occupy a very small portion of the much wider spectrum; they are separated by 312.8 kHz.

The passband constraint is $0.99 < |H'_o(jf)||H'_c(jf)| < 1.01$, which implies that

$$\frac{0.99}{|H'_o(jf)|} < |H'_c(jf)| < \frac{1.01}{|H'_o(jf)|}, \quad -20 \text{ kHz} < f < 20 \text{ kHz}.$$

Figure 4.42(c) depicts these constraints for both cases. Here, we have multiplied $|H'_c(jf)|$ by the sampling interval T_{s1} or T_{s2}, so that both cases are displayed with the same vertical scale. Note that case (a) requires substantial curvature in $|H'_c(jf)|$ to eliminate the passband distortion introduced by the mainlobe of $H'_o(jf)$. At the edge of the passband, the bounds are as follows:

Case (a):

$$1.4257 < T_{s1}|H'_c(jf_m)| < 1.4545, \quad f_m = 20 \text{ kHz}$$

Case (b):

$$0.9953 < T_{s2}|H'_c(jf_m)| < 1.0154, \quad f_m = 20 \text{ kHz}$$

The image-rejection constraint implies that $|H'_o(jf)||H'_c(jf)| < 10^{-3}$ for all frequencies at which images are present. This condition is simplified somewhat by considering only the frequency at which $|H'_o(jf)|$ is largest. The maximum value of $|H'_o(jf)|$ in the image frequency bands occurs at the smallest frequency in the first image: 24.1 kHz in case (a) and 332.8 kHz in case (b). The value of $|H'_o(jf)|/T_{s1}$ and $|H'_o(jf)|/T_{s2}$ at these frequencies is 0.5763 and 0.0598, respectively, which implies that the bounds are

$$T_{s1}|H'_c(jf)| < 0.0017, \quad f > 24.1 \text{ kHz},$$

and

$$T_{s2}|H'_c(jf)| < 0.0167, \quad f > 332.8 \text{ kHz},$$

for cases (a) and (b), respectively. Hence, the anti-imaging filter for case (a) must show a transition from a value of $1.4257/T_{s1}$ to $0.0017/T_{s1}$ over an interval of 4.1 kHz. In contrast,

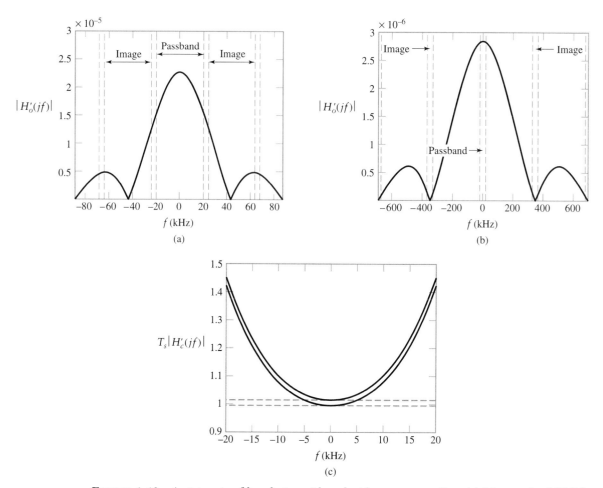

FIGURE 4.42   Anti-imaging filter design with and without oversampling. (a) Magnitude of $H_o'(jf)$ for 44.1-kHz sampling rate. Dashed lines denote signal passband and images. (b) Magnitude of $H_o'(jf)$ for eight-times oversampling (352.8-kHz sampling rate). Dashed lines denote signal passband and images. (c) Normalized constraints on passband response of anti-imaging filter. Solid lines assume a 44.1-kHz sampling rate; dashed lines assume eight-times oversampling. The normalized filter response must lie between each pair of lines.

with eight-times oversampling the filter must show a transition from $0.9953/T_{s2}$ to $0.0167/T_{s2}$ over a frequency interval of 312.8 kHz. Thus, oversampling not only increases the transition width by a factor of almost 80, but also relaxes the stopband attenuation constraint by a factor of more than 10. ■

4.7 Discrete-Time Processing of Continuous-Time Signals

In this section, we use Fourier methods to discuss and analyze a typical system for the discrete-time processing of continuous-time signals. There are several advantages to processing a continuous-time signal with a discrete-time system. These advantages result from the power and flexibility of discrete-time computing devices. First, a broad class of signal manipulations are more easily performed by using the arithmetic operations of a computer than through the use of analog components. Second, implementing a system in a computer only involves writing a set of instructions or program for the computer to execute. Third, the discrete-time system is easily changed by modifying the computer program. Often, the system can be modified in real time to optimize some criterion associated with the processed signal. Yet another advantage of discrete-time processing is the direct dependence of the dynamic range and signal-to-noise ratio on the number of bits used to represent the discrete-time signal. These advantages have led to a proliferation of computing devices designed specifically for discrete-time signal processing.

A minimal system for the discrete-time processing of continuous-time signals must contain a sampling device, as well as a computing device for implementing the discrete-time system. In addition, if the processed signal is to be converted back to continuous time, then reconstruction is necessary. More sophisticated systems may also utilize oversampling, decimation, and interpolation. *Decimation* and *interpolation* are methods for changing the effective sampling rate of a discrete-time signal. Decimation reduces the effective sampling rate, while interpolation increases the effective sampling rate. Judicious use of these methods can reduce the cost of the overall system. We begin with an analysis of a basic system for processing continuous-time signals. We conclude by revisiting oversampling and examining the role of interpolation and decimation in systems that process continuous-time signals.

■ 4.7.1 A BASIC DISCRETE-TIME SIGNAL-PROCESSING SYSTEM

A typical system for processing continuous-time signals in discrete time is illustrated in Fig. 4.43(a). A continuous-time signal is first passed through a low-pass anti-aliasing filter and then sampled at intervals of T_s to convert it to a discrete-time signal. The sampled signal is then processed by a discrete-time system to impart some desired effect to the signal. For example, the discrete-time system may represent a filter designed to have a specific frequency response, such as an equalizer. After processing, the signal is converted back to continuous-time format. A zero-order-hold device converts the discrete-time signal back to continuous time, and an anti-imaging filter removes the distortion introduced by the zero-order hold.

This combination of operations may be reduced to an equivalent continuous-time filter by using the FT as an analysis tool. The idea is to find a continuous-time system $g(t) \xleftrightarrow{\;FT\;} G(j\omega)$ such that $Y(j\omega) = G(j\omega)X(j\omega)$, as depicted in Fig. 4.43(b). Hence, $G(j\omega)$ has the same effect on the input as the system in Fig. 4.43(a). We assume for this

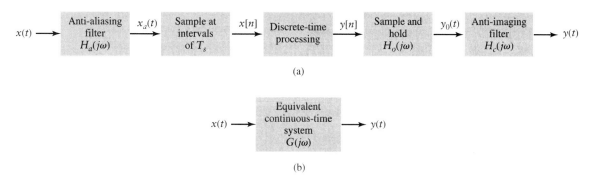

(a)

(b)

FIGURE 4.43   Block diagram for discrete-time processing of continuous-time signals. (a) A basic system. (b) Equivalent continuous-time system.

analysis that the discrete-time processing operation is represented by a discrete-time system with frequency response $H(e^{j\Omega})$. Recall that $\Omega = \omega T_s$, where T_s is the sampling interval, so the discrete-time system has a continuous-time frequency response $H(e^{j\omega T_s})$. Also, the frequency response associated with the zero-order-hold device is

$$H_o(j\omega) = 2e^{-j\omega T_s/2}\frac{\sin(\omega T_s/2)}{\omega}.$$

The first operation applied to $x(t)$ is the continuous-time anti-aliasing filter, whose output has FT given by

$$X_a(j\omega) = H_a(j\omega)X(j\omega).$$

Equation (4.23) indicates that, after sampling, the FT representation for $x[n]$ is

$$
\begin{aligned}
X_\delta(j\omega) &= \frac{1}{T_s}\sum_{k=-\infty}^{\infty} X_a(j(\omega - k\omega_s)) \\
&= \frac{1}{T_s}\sum_{k=-\infty}^{\infty} H_a(j(\omega - k\omega_s))X(j(\omega - k\omega_s)),
\end{aligned}
\tag{4.32}
$$

where $\omega_s = 2\pi/T_s$ is the sampling frequency. The discrete-time system modifies $X_\delta(j\omega)$ by $H(e^{j\omega T_s})$, producing

$$Y_\delta(j\omega) = \frac{1}{T_s}H(e^{j\omega T_s})\sum_{k=-\infty}^{\infty} H_a(j(\omega - k\omega_s))X(j(\omega - k\omega_s)).$$

The reconstruction process modifies $Y_\delta(j\omega)$ by the product $H_o(j\omega)H_c(j\omega)$; thus, we may write

$$Y(j\omega) = \frac{1}{T_s}H_o(j\omega)H_c(j\omega)H(e^{j\omega T_s})\sum_{k=-\infty}^{\infty} H_a(j(\omega - k\omega_s))X(j(\omega - k\omega_s)).$$

Assuming that aliasing does not occur, the anti-imaging filter $H_c(j\omega)$ eliminates frequency components above $\omega_s/2$, hence eliminating all the terms in the infinite sum except for the $k = 0$ term. We therefore have

$$Y(j\omega) = \frac{1}{T_s}H_o(j\omega)H_c(j\omega)H(e^{j\omega T_s})H_a(j\omega)X(j\omega).$$

This expression indicates that the overall system is equivalent to a continuous-time LTI system having the frequency response

$$\boxed{G(j\omega) = \frac{1}{T_s}H_o(j\omega)H_c(j\omega)H(e^{j\omega T_s})H_a(j\omega).}
\tag{4.33}$$

If the anti-aliasing and anti-imaging filters are chosen to compensate for the effects of sampling and reconstruction, as discussed in the previous sections, then $(1/T_s)H_o(j\omega)H_c(j\omega)H_a(j\omega) \approx 1$ on the frequency band of interest, and we see that $G(j\omega) \approx H(e^{j\omega T_s})$. That is, we may implement a continuous-time system in discrete time by choosing sampling parameters appropriately and designing a corresponding discrete-time system. Note that this correspondence to a continuous-time LTI system assumes the absence of aliasing.

▪ 4.7.2 Oversampling

In Section 4.6, we noted that increasing the effective sampling rate associated with a discrete-time signal prior to the use of a zero-order hold for converting the discrete-time signal back to continuous time relaxes the requirements on the anti-imaging filter. Similarly, the requirements on the anti-aliasing filter are relaxed if the sampling rate is chosen to be significantly greater than the Nyquist rate. This allows a wide transition band in the anti-aliasing filter.

An anti-aliasing filter prevents aliasing by limiting the bandwidth of the signal prior to sampling. While the signal of interest may have maximum frequency W, the continuous-time signal will, in general, have energy at higher frequencies due to the presence of noise and other nonessential characteristics. Such a situation is illustrated in Fig. 4.44(a). The shaded area of the spectrum represents energy at frequencies above the maximum frequency of the signal; we shall refer to this component as noise. The anti-aliasing filter is chosen to prevent such noise from moving back down into the band of interest and producing aliases there. The magnitude response of a practical anti-aliasing filter cannot go from unit gain to zero at frequency W, but instead goes from passband to stopband over a range of frequencies, as depicted in Fig. 4.44(b). Here, the stopband of the filter is W_s, and $W_t = W_s - W$ denotes the width of the transition band. The spectrum of the filtered signal $X_a(j\omega)$ now has maximum frequency W_s, as shown in Fig. 4.44(c). This signal is sampled at a rate ω_s, resulting in the spectrum $X_\delta(j\omega)$ illustrated in Fig. 4.44(d). Note that we have drawn $X_\delta(j\omega)$ assuming that ω_s is large enough to prevent aliasing. As ω_s decreases, replicas of the original signal's spectrum begin to overlap and aliasing occurs.

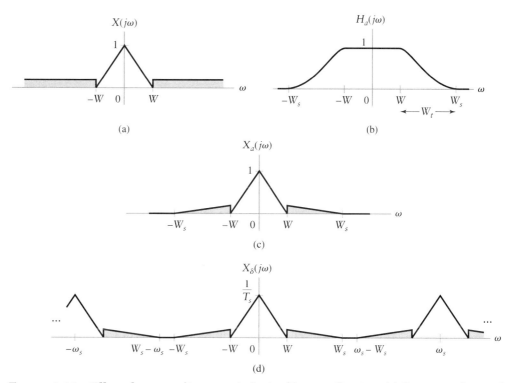

Figure 4.44   Effect of oversampling on anti-aliasing filter specifications. (a) Spectrum of original signal. (b) Anti-aliasing filter frequency response magnitude. (c) Spectrum of signal at the anti-aliasing filter output. (d) Spectrum of the anti-aliasing filter output after sampling. The graph depicts the case of $\omega_s > 2W_s$.

In order to prevent the noise from aliasing with itself, we require that $\omega_s - W_s > W_s$ or $\omega_s > 2W_s$, as predicted by the sampling theorem. However, because of the subsequent discrete-time processing, we often do not care whether the noise aliases with itself, but rather wish to prevent the noise from aliasing back into the signal band $-W < \omega < W$. This implies, however, that we must have

$$\omega_s - W_s > W.$$

Using $W_s = W_t + W$ in the preceding inequality and rearranging terms to obtain the relationship between the transition band of the anti-aliasing filter and the sampling frequency, we have

$$W_t < \omega_s - 2W.$$

Hence, the transition band of the anti-aliasing filter must be less than the sampling frequency minus twice the frequency of the highest frequency component of interest in the signal. Filters with small transition bands are difficult to design and expensive. By oversampling, or choosing $\omega_s \gg 2W$, we can greatly relax the requirements on the anti-aliasing filter transition band and, consequently, reduce its complexity and cost.

In both sampling and reconstruction, the difficulties of implementing practical analog filters suggests using the highest possible sampling rate. However, if the data set in question is processed with a discrete-time system, as depicted in Fig. 4.43(a), then high sampling rates lead to increased discrete-time system cost, because the discrete-time system must perform its computations at a faster rate. This conflict over the sampling rate is mitigated if we can somehow change the sampling rate such that a high rate is used for sampling and reconstruction and a lower rate is used for discrete-time processing. Decimation and interpolation, discussed next, offer such a capability.

■ 4.7.3 DECIMATION

Consider the DTFTs obtained by sampling an identical continuous-time signal at different intervals T_{s1} and T_{s2}. Let the sampled signals be denoted as $x_1[n]$ and $x_2[n]$. We assume that $T_{s1} = qT_{s2}$, where q is integer, and that aliasing does not occur at either sampling rate. Figure 4.45 depicts the FT of a representative continuous-time signal and the DTFTs $X_1(e^{j\Omega})$ and $X_2(e^{j\Omega})$ associated with the sampling intervals T_{s1} and T_{s2}. Decimation corresponds to changing $X_2(e^{j\Omega})$ to $X_1(e^{j\Omega})$. One way to do this is to convert the discrete-time sequence back to a continuous-time signal and then resample. Such an approach is subject to distortion introduced in the reconstruction operation. We can avoid the distortion by using methods that operate directly on the discrete-time signals to change the sampling rate.

Subsampling is the key to reducing the sampling rate. If the sampling interval is T_{s2} and we wish to increase it to $T_{s1} = qT_{s2}$, we may do so by selecting every qth sample of the sequence $x_2[n]$; that is, we set $g[n] = x_2[qn]$. Equation (4.27) indicates that the relationship between $G(e^{j\Omega})$ and $X_2(e^{j\Omega})$ is

$$G(e^{j\Omega}) = \frac{1}{q}\sum_{m=0}^{q-1} X_2\big(e^{j((\Omega-m2\pi)/q)}\big).$$

That is, $G(e^{j\Omega})$ is a sum of shifted versions of $X_2(e^{j\Omega/q})$. The scaling spreads out $X_2(e^{j\Omega})$ by the factor q. Shifting these scaled versions of $X_2(e^{j\Omega})$ gives $G(e^{j\Omega})$, as depicted in Fig. 4.46. Identifying $T_{s1} = qT_{s2}$, we see that $G(e^{j\Omega})$ corresponds to $X_1(e^{j\Omega})$ in Fig. 4.45(b). Hence, subsampling by q changes the effective sampling rate by q.

The preceding analysis assumes that the maximum frequency component of $X_2(e^{j\Omega})$ satisfies $WT_{s2} < \pi/q$, so that aliasing does not occur as a consequence of subsampling.

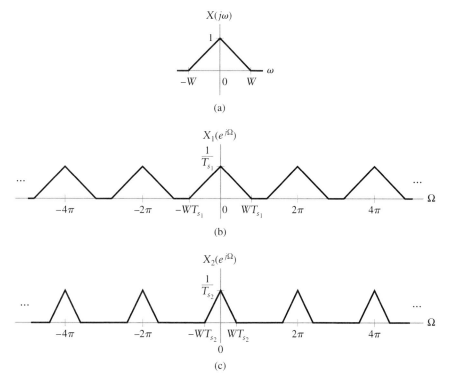

FIGURE 4.45 Effect of changing the sampling rate. (a) Underlying continuous-time signal FT. (b) DTFT of sampled data at sampling interval T_{s1}. (c) DTFT of sampled data at sampling interval T_{s2}.

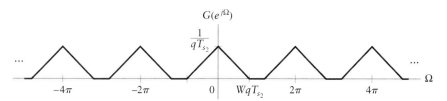

FIGURE 4.46 The spectrum that results from subsampling the DTFT $X_2(e^{j\Omega})$ depicted in Fig. 4.45 (c) by a factor of q.

This assumption is rarely satisfied in practice: Even if the signal of interest is band limited in such a manner, there will often be noise or other components present at higher frequencies. For example, if oversampling is used to obtain $x_2[n]$, then noise that passed through the transition band of the anti-aliasing filter will be present at frequencies above π/q. If we subsample $x_2[n]$ directly, then this noise will alias into frequencies $|\Omega| < WT_{s1}$ and distort the signal of interest. This aliasing problem is prevented by applying a low-pass discrete-time filter to $x_2[n]$ prior to subsampling.

Figure 4.47(a) depicts a decimation system that includes a low-pass discrete-time filter. The input signal $x[n]$ with DTFT shown in Fig. 4.47(b) corresponds to the oversampled signal, whose FT is depicted in Fig. 4.47(d). The shaded regions indicate noise energy.

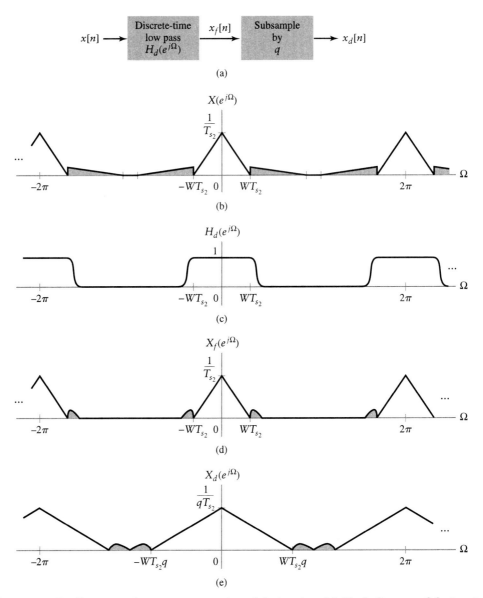

FIGURE 4.47   Frequency-domain interpretation of decimation. (a) Block diagram of decimation system. (b) Spectrum of oversampled input signal. Noise is depicted as the shaded portions of the spectrum. (c) Filter frequency response. (d) Spectrum of filter output. (e) Spectrum after subsampling.

The low-pass filter characterized in Fig. 4.47(c) removes most of the noise in producing the output signal depicted in Fig. 4.47(d). After subsampling, the noise does not alias into the signal band, as illustrated in Fig. 4.47(e). Note that this procedure is effective only if the discrete-time filter has a rapid transition from passband to stopband. Fortunately, a discrete-time filter with a narrow transition band is much easier to design and implement than a comparable continuous-time filter.

Decimation is also known as *downsampling*. It is often denoted by a downwards arrow followed by the decimation factor, as illustrated in the block diagram of Fig. 4.48.

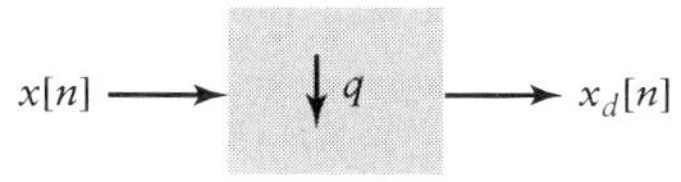

$$x[n] \longrightarrow \boxed{\downarrow q} \longrightarrow x_d[n]$$

FIGURE 4.48 Symbol for decimation by a factor of q.

4.7.4 INTERPOLATION

Interpolation increases the sampling rate and requires that we somehow produce values between the samples of the signal. In the frequency domain, we seek to convert $X_1(e^{j\Omega})$ of Fig. 4.45(b) into $X_2(e^{j\Omega})$ of Fig. 4.45(c). We shall assume that we are increasing the sampling rate by an integer factor; that is, $T_{s1} = qT_{s2}$.

The DTFT scaling property derived in Problem 3.80 is the key to developing an interpolation procedure. Let $x_1[n]$ be the sequence to be interpolated by the factor q. Define a new sequence

$$x_z[n] = \begin{cases} x_1[n/q], & n/q \text{ integer} \\ 0, & \text{otherwise} \end{cases}. \tag{4.34}$$

With this definition, we have $x_1[n] = x_z[qn]$, and the DTFT scaling property implies that

$$X_z(e^{j\Omega}) = X_1(e^{jq\Omega}).$$

That is, $X_z(e^{j\Omega})$ is a scaled version of $X_1(e^{j\Omega})$, as illustrated in Figs. 4.49(a) and (b). Identifying $T_{s2} = T_{s1}/q$, we find that $X_z(e^{j\Omega})$ corresponds to $X_2(e^{j\Omega})$ in Fig. 4.45(c), except for the spectrum replicas centered at $\pm\frac{2\pi}{q}, \pm\frac{4\pi}{q}, \ldots \pm\frac{(q-1)2\pi}{q}$. These can be removed by passing the signal $x_z[n]$ through a low-pass filter whose frequency response is depicted in Fig. 4.49(c). The passband of this filter is defined by $|\Omega| < WT_{s2}$ and the transition band must lie in the region $WT_{s2} < |\Omega| < \frac{2\pi}{q} - WT_{s2}$. The passband gain is chosen to be q so that the interpolated signal has the correct amplitude. Figure 4.49(d) illustrates the spectrum of the filter output, $X_i(e^{j\Omega})$.

Hence, interpolation by the factor q is accomplished by inserting $q - 1$ zeros in between each sample of $x_1[n]$ and then low-pass filtering. A block diagram illustrating this procedure is depicted in Fig. 4.50(a). Interpolation is also known as *upsampling* and is often denoted by an upwards arrow followed by the interpolation factor, as depicted in the block diagram of Fig. 4.50(b). The time-domain interpretation of the interpolation procedure just presented is developed in Problem 4.52.

Figure 4.51 depicts a block diagram for a discrete-time signal-processing system that uses decimation and interpolation.

4.8 *Fourier Series Representations of Finite-Duration Nonperiodic Signals*

The DTFS and FS are the Fourier representations of periodic signals. In this section, we explore their use for representing finite-duration *non*periodic signals. The primary motivation for doing this has to do with the numerical computation of Fourier representations. Recall that the DTFS is the only Fourier representation that can be evaluated numerically. As a result, we often apply the DTFS to signals that are not periodic. It is important to understand the implications of applying a periodic representation to nonperiodic signals.

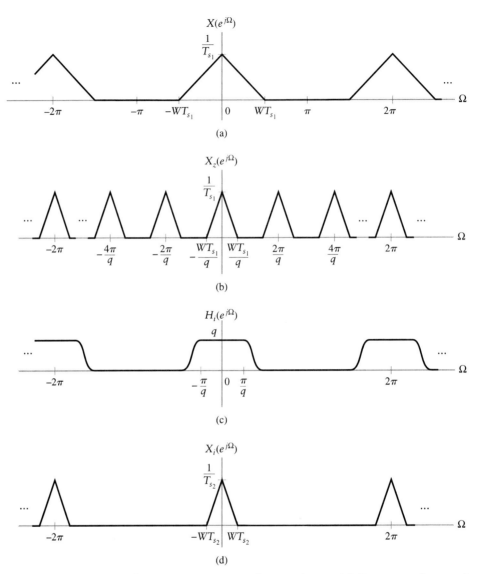

FIGURE 4.49   Frequency-domain interpretation of interpolation. (a) Spectrum of original sequence. (b) Spectrum after inserting $q - 1$ zeros in between every value of the original sequence. (c) Frequency response of a filter for removing undesired replicates located at $\pm 2\pi/q, \pm 4\pi/q, \ldots,$ $\pm (q - 1)2\pi/q$.   (d) Spectrum of interpolated sequence.

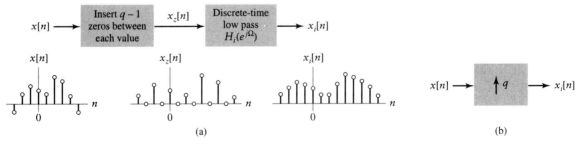

FIGURE 4.50   (a) Block diagram of an interpolation system. (b) Symbol denoting interpolation by a factor of q.

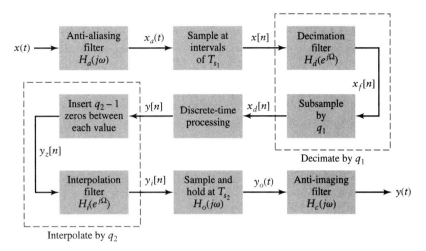

FIGURE 4.51 Block diagram of a system for discrete-time processing of continuous-time signals including decimation and interpolation.

A secondary benefit is an increase in our understanding of the relationship between the Fourier transform and corresponding Fourier series representations. We begin the discussion with the discrete-time case.

■ 4.8.1 RELATING THE DTFS TO THE DTFT

Let $x[n]$ be a finite-duration signal of length M; that is,

$$x[n] = 0, \quad n < 0 \quad \text{or} \quad n \geq M.$$

The DTFT of this signal is

$$X(e^{j\Omega}) = \sum_{n=0}^{M-1} x[n]e^{-j\Omega n}.$$

Now suppose we introduce a periodic discrete-time signal $\widetilde{x}[n]$ with period $N \geq M$ such that one period of $\widetilde{x}[n]$ is given by $x[n]$, as shown in the top half of Fig. 4.52. The DTFS coefficients of $\widetilde{x}[n]$ are given by

$$\widetilde{X}[k] = \frac{1}{N} \sum_{n=0}^{N-1} x[n]e^{-jk\Omega_o n}, \tag{4.35}$$

where $\Omega_o = 2\pi/N$. Since $x[n] = 0$ for $n \geq M$, we have

$$\widetilde{X}[k] = \frac{1}{N} \sum_{n=0}^{M-1} x[n]e^{-jk\Omega_o n}.$$

A comparison of $\widetilde{X}[k]$ and $X(e^{j\Omega})$ reveals that

$$\widetilde{X}[k] = \frac{1}{N}X(e^{j\Omega})\bigg|_{\Omega=k\Omega_o}. \tag{4.36}$$

The DTFS coefficients of $\widetilde{x}[n]$ are samples of the DTFT of $x[n]$, divided by N and evaluated at intervals of $2\pi/N$.

Although $x[n]$ is not periodic, we define DTFS coefficients using $x[n]$, $n = 0$, $1, \ldots, N - 1$ according to

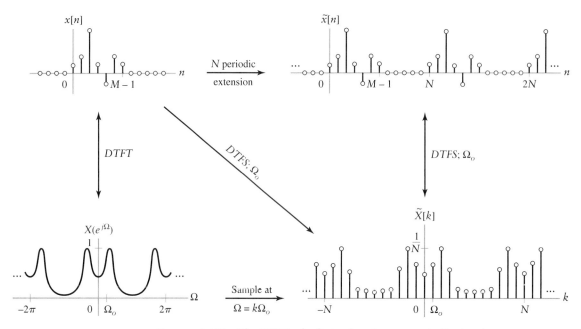

FIGURE 4.52 The DTFS of a finite-duration nonperiodic signal.

$$X[k] = \frac{1}{N}\sum_{n=0}^{N-1} x[n]e^{-jk\Omega_o n}.$$

With this definition, we see that $X[k] = \tilde{X}[k]$ given in Eq. (4.35) and thus write the DTFS of the finite-duration signal $x[n]$, using Eq. (4.36), as $X[k] = (1/N)X(e^{jk\Omega_o})$.

The latter equation implies that the DTFS coefficients of $x[n]$ correspond to the DTFS coefficients of a periodically extended signal $\tilde{x}[n]$. In other words, *the effect of sampling the DTFT of a finite-duration nonperiodic signal is to periodically extend the signal in the time domain.* That is,

$$\tilde{x}[n] = \sum_{m=-\infty}^{\infty} x[n + mN] \xleftrightarrow{\;DTFS;\Omega_o\;} \tilde{X}[k] = \frac{1}{N}X(e^{jk\Omega_o}). \qquad (4.37)$$

Figure 4.52 illustrates these relationships in both the time and frequency domains. They are the dual to sampling in frequency. Recall that sampling a signal in time generates shifted replicas of the spectrum of the original signal in the frequency domain. Sampling a signal in frequency generates shifted replicas of the original time signal in the time-domain representation. In order to prevent overlap, or aliasing, of these shifted replicas in time, we require the frequency sampling interval Ω_o to be less than or equal to $2\pi/M$. In essence, this result corresponds to the sampling theorem applied in the frequency domain.

EXAMPLE 4.14 SAMPLING THE DTFT OF A COSINE PULSE   Consider the signal

$$x[n] = \begin{cases} \cos\left(\dfrac{3\pi}{8}n\right), & 0 \le n \le 31 \\ 0, & \text{otherwise} \end{cases}.$$

Derive both the DTFT, $X(e^{j\Omega})$, and the DTFS, $X[k]$, of $x[n]$, assuming a period $N > 31$. Evaluate and plot $|X(e^{j\Omega})|$ and $N|X[k]|$ for $N = 32, 60$, and 120.

Solution: First we evaluate the DTFT. Write $x[n] = g[n]w[n]$, where $g[n] = \cos(3\pi n/8)$ and

$$w[n] = \begin{cases} 1, & 0 \leq n \leq 31 \\ 0, & \text{otherwise} \end{cases}$$

is the window function. We have

$$G(e^{j\Omega}) = \pi\delta\left(\Omega + \frac{3\pi}{8}\right) + \pi\delta\left(\Omega - \frac{3\pi}{8}\right), \quad -\pi < \Omega \leq \pi,$$

as one 2π period of $G(e^{j\Omega})$, and we take the DTFT of $w[n]$ to obtain

$$W(e^{j\Omega}) = e^{-j31\Omega/2}\frac{\sin(16\Omega)}{\sin(\Omega/2)}.$$

The multiplication property implies that $X(e^{j\Omega}) = (1/(2\pi))G(e^{j\Omega})\circledast W(e^{j\Omega})$; for the problem at hand, this property yields

$$X(e^{j\Omega}) = \frac{e^{-j31(\Omega+3\pi/8)/2}}{2}\frac{\sin(16(\Omega + 3\pi/8))}{\sin((\Omega + 3\pi/8)/2)} + \frac{e^{-j\frac{31}{2}(\Omega-3\pi/8)}}{2}\frac{\sin(16(\Omega - 3\pi/8))}{\sin((\Omega - 3\pi/8)/2)}.$$

Now let $\Omega_o = 2\pi/N$, so that the N DTFS coefficients are given by

$$X[k] = \frac{1}{N}\sum_{n=0}^{31}\cos(3\pi/8n)e^{-jk\Omega_o n}$$

$$= \frac{1}{2N}\sum_{n=0}^{31}e^{-j(k\Omega_o+3\pi/8)n} + \frac{1}{2N}\sum_{n=0}^{31}e^{-j(k\Omega_o-3\pi/8)n}.$$

Summing each geometric series produces

$$X[k] = \frac{1}{2N}\frac{1 - e^{-j(k\Omega_o+3\pi/8)32}}{1 - e^{-j(k\Omega_o+3\pi/8)}} + \frac{1}{2N}\frac{1 - e^{-j(k\Omega_o-3\pi/8)32}}{1 - e^{-j(k\Omega_o-3\pi/8)}},$$

which we rewrite as

$$X[k] = \left(\frac{e^{-j(k\Omega_o+3\pi/8)16}}{2Ne^{-j\frac{1}{2}(k\Omega_o+3\pi/8)}}\right)\frac{e^{j(k\Omega_o+3\pi/8)16} - e^{-j(k\Omega_o+3\pi/8)16}}{e^{j(k\Omega_o+3\pi/8)/2} - e^{-j(k\Omega_o+3\pi/8)/2}}$$

$$+ \left(\frac{e^{-j(k\Omega_o-3\pi/8)16}}{2Ne^{-j(k\Omega_o-3\pi/8)/2}}\right)\frac{e^{j(k\Omega_o-3\pi/8)16} - e^{-j(k\Omega_o-3\pi/8)16}}{e^{j(k\Omega_o-3\pi/8)/2} - e^{-j(k\Omega_o-3\pi/8)/2}}$$

$$= \left(\frac{e^{-j31(k\Omega_o+3\pi/8)/2}}{2N}\right)\frac{\sin(16(k\Omega_o + 3\pi/8))}{\sin((k\Omega_o + 3\pi/8)/2)}$$

$$+ \left(\frac{e^{-j31(k\Omega_o-3\pi/8)/2}}{2N}\right)\frac{\sin(16(k\Omega_o - 3\pi/8))}{\sin((k\Omega_o - 3\pi/8)/2)}.$$

A comparison of $X[k]$ and $X(e^{j\Omega})$ indicates that Eq. (4.36) holds for this example. Hence, the DTFS of the finite-duration cosine pulse is given by samples of the DTFT.

Figures 4.53(a)–(c) depict $|X(e^{j\Omega})|$ and $N|X[k]|$ for $N = 32, 60$, and 120. As N increases, $X[k]$ samples $X(e^{j\Omega})$ more densely, and the shape of the DTFS coefficients resembles that of the underlying DTFT more closely. ∎

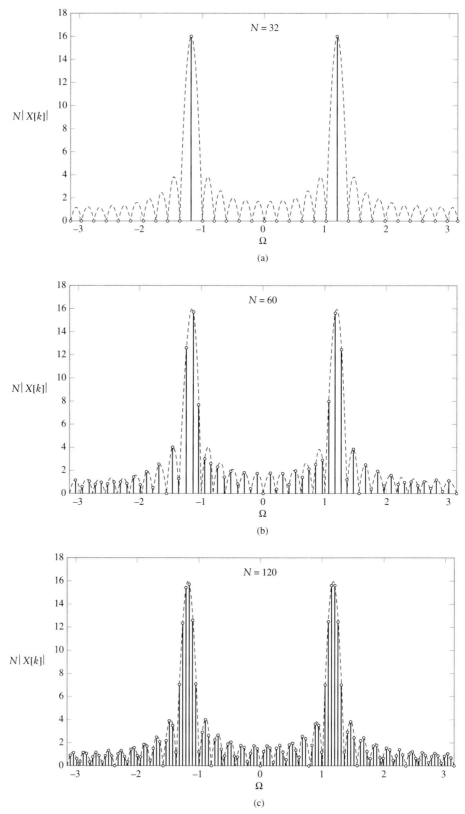

Figure 4.53 The DTFT and length-N DTFS of a 32-point cosine. The dashed line denotes $|X(e^{j\Omega})|$, while the stems represent $N|X[k]|$. (a) $N = 32$, (b) $N = 60$, (c) $N = 120$.

In many applications, only M values of a signal $x[n]$ are available, and we have no knowledge of the signal's behavior outside the set of M values. The DTFS provides samples of the DTFT of the length-M sequence. The practice of choosing $N > M$ when evaluating the DTFS is known as *zero padding*, since it can be viewed as augmenting or padding the M available values of $x[n]$ with $N - M$ zeros. We emphasize that zero padding does not overcome any of the limitations associated with knowing only M values of $x[n]$; it simply samples the underlying length M DTFT more densely, as illustrated in the previous example.

▶ **Problem 4.13**   Use the DTFT of the finite-duration nonperiodic signal

$$x[n] = \begin{cases} 1, & 0 \le n \le 31 \\ 0, & \text{otherwise} \end{cases}$$

to find the DTFS coefficients of the period-N signal

$$\widetilde{x}[n] = \begin{cases} 1, & 0 \le n \le 31 \\ 0, & 32 \le n \le N \end{cases}$$

for (a) $N = 40$ and (b) $N = 64$.

Answers:

$$\widetilde{x}[n] \xleftrightarrow{\quad DTFS; 2\pi/N \quad} \widetilde{X}[k]$$

(a)

$$\widetilde{X}[k] = e^{-jk31\pi/40} \frac{\sin(k32\pi/40)}{40\sin(k\pi/40)}$$

(b)

$$\widetilde{X}[k] = e^{-jk31\pi/64} \frac{\sin(k32\pi/64)}{64\sin(k\pi/64)}$$ ◀

■ **4.8.2 RELATING THE FS TO THE FT**

The relationship between the FS coefficients and the FT of a finite-duration nonperiodic continuous-time signal is analogous to that of the discrete-time case discussed in the previous subsection. Let $x(t)$ have duration T_o, so that

$$x(t) = 0, \quad t < 0 \quad \text{or} \quad t \ge T_o.$$

Construct a periodic signal

$$\widetilde{x}(t) = \sum_{m=-\infty}^{\infty} x(t + mT)$$

with $T \ge T_o$ by periodically extending $x(t)$. The FS coefficients of $\widetilde{x}(t)$ are

$$\widetilde{X}[k] = \frac{1}{T} \int_0^T \widetilde{x}(t) e^{-jk\omega_o t}\, dt$$

$$= \frac{1}{T} \int_0^{T_o} x(t) e^{-jk\omega_o t}\, dt,$$

where we have used the relationship $\widetilde{x}(t) = x(t)$ for $0 \le t \le T_o$ and $\widetilde{x}(t) = 0$ for $T_o < t < T$. The FT of $x(t)$ is defined by

$$X(j\omega) = \int_{-\infty}^{\infty} x(t)e^{-j\omega t}\,dt$$

$$= \int_{0}^{T_o} x(t)e^{-j\omega t}\,dt.$$

In the second line, we used the finite duration of $x(t)$ to change the limits on the integral. Hence, comparing $\widetilde{X}[k]$ with $X(j\omega)$, we conclude that

$$\widetilde{X}[k] = \frac{1}{T}X(j\omega)\Big|_{\omega=k\omega_o}.$$

The FS coefficients are samples of the FT, normalized by T.

4.9 The Discrete-Time Fourier Series Approximation to the Fourier Transform

The DTFS involves a finite number of discrete-valued coefficients in both the frequency and time domains. All the other Fourier representations are continuous in either the time or frequency domain or both. Hence, the DTFS is the only Fourier representation that can be evaluated on a computer, and it is widely applied as a computational tool for manipulating signals. In this section, we consider using the DTFS to approximate the FT of a continuous-time signal.

The FT applies to continuous-time nonperiodic signals. The DTFS coefficients are computed by using N values of a discrete-time signal. In order to use the DTFS to approximate the FT, we must sample the continuous-time signal and retain at most N samples. We assume that the sampling interval is T_s and that $M < N$ samples of the continuous-time signal are retained. Figure 4.54 depicts this sequence of steps. The problem at hand is to determine how well the DTFS coefficients $Y[k]$ approximate $X(j\omega)$, the FT of $x(t)$. Both the sampling and windowing operations are potential sources of error in the approximation.

The error introduced by sampling is due to aliasing. Let $x_\delta(t) \xleftrightarrow{\ FT\ } X_\delta(j\omega)$. Equation (4.23) indicates that

$$X_\delta(j\omega) = \frac{1}{T_s}\sum_{k=-\infty}^{\infty} X(j(\omega - k\omega_s)), \tag{4.38}$$

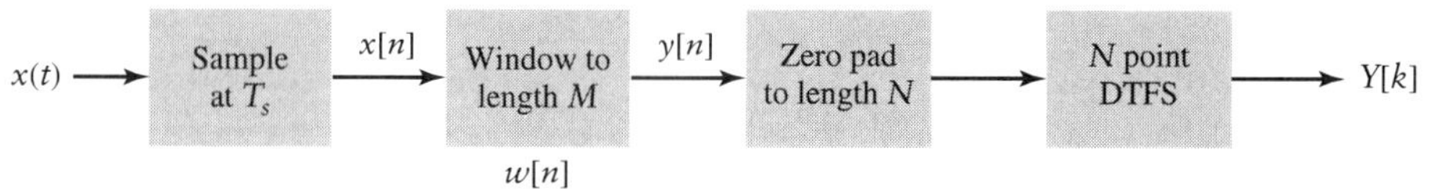

$x(t) \longrightarrow$ [Sample at T_s] $\xrightarrow{x[n]}$ [Window to length M] $\xrightarrow{y[n]}$ [Zero pad to length N] $\longrightarrow$ [N point DTFS] $\longrightarrow Y[k]$

$w[n]$

FIGURE 4.54   Block diagram depicting the sequence of operations involved in approximating the FT with the DTFS.

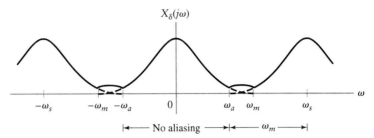

FIGURE 4.55   Effect of aliasing.

where $\omega_s = 2\pi/T_s$. Suppose we wish to approximate $X(j\omega)$ on the interval $-\omega_a < \omega < \omega_a$, and suppose further that $x(t)$ is band limited with maximum frequency $\omega_m \geq \omega_a$. Aliasing in the band $-\omega_a < \omega < \omega_a$ is prevented by choosing T_s such that $\omega_s > \omega_m + \omega_a$, as illustrated in Fig. 4.55. That is, we require that

$$\boxed{T_s < \frac{2\pi}{\omega_m + \omega_a}.} \tag{4.39}$$

The windowing operation of length M corresponds to the periodic convolution

$$Y(e^{j\Omega}) = \frac{1}{2\pi} X(e^{j\Omega}) \circledast W(e^{j\Omega}),$$

where $x[n] \xleftrightarrow{\;DTFT\;} X(e^{j\Omega})$ and $W(e^{j\Omega})$ is the window frequency response. We may rewrite this periodic convolution in terms of the continuous-time frequency ω by performing the change of variable $\Omega = \omega T_s$ in the convolution integral. We then have

$$Y_\delta(j\omega) = \frac{1}{\omega_s} X_\delta(j\omega) \circledast W_\delta(j\omega), \tag{4.40}$$

where $X_\delta(j\omega)$ is given in Eq. (4.38), $y_\delta(t) \xleftrightarrow{\;FT\;} Y_\delta(j\omega)$, and $w_\delta(t) \xleftrightarrow{\;FT\;} W_\delta(j\omega)$. Both $X_\delta(j\omega)$ and $W_\delta(j\omega)$ have the same period ω_s; hence, the periodic convolution is performed over an interval of that length. Since

$$w[n] = \begin{cases} 1, & 0 \leq n \leq M - 1, \\ 0, & \text{otherwise} \end{cases},$$

we have

$$W_\delta(j\omega) = e^{-j\omega T_s(M-1)/2} \frac{\sin(M\omega T_s/2)}{\sin\left(\dfrac{\omega T_s}{2}\right)}. \tag{4.41}$$

A plot of $|W_\delta(j\omega)|$ is given in Fig. 4.56. The effect of the convolution in Eq. (4.40) is to smear, or smooth, the spectrum of $X_\delta(j\omega)$. This smearing limits our ability to resolve details in the spectrum. The degree of smearing depends on the mainlobe width of $W_\delta(j\omega)$. It

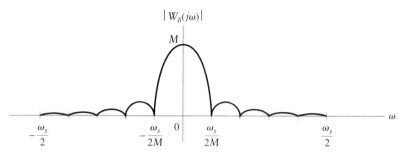

FIGURE 4.56   Magnitude response of M-point window.

is difficult to quantify precisely the loss in resolution resulting from windowing. Since we cannot resolve details in the spectrum that are closer than a mainlobe width apart, we define the *resolution* as the mainlobe width ω_s/M. Hence, to achieve a specified resolution ω_r, we require that

$$M \geq \frac{\omega_s}{\omega_r}.$$

(4.42)

Using $\omega_s = 2\pi/T_s$, we may rewrite this inequality explicitly as

$$MT_s \geq \frac{2\pi}{\omega_r}.$$

Recognizing that MT_s is the total time over which we sample $x(t)$, we see that that time interval must exceed $2\pi/\omega_r$.

The DTFS $y[n] \xleftrightarrow{\;DTFS;\,2\pi/N\;} Y[k]$ samples the DTFT $Y(e^{j\Omega})$ at intervals of $2\pi/N$. That is, $Y[k] = (1/N)Y(e^{jk2\pi/N})$. In terms of the continuous-time frequency ω, the samples are spaced at intervals of $2\pi/(NT_s) = \omega_s/N$, so

$$Y[k] = \frac{1}{N}Y_\delta(jk\omega_s/N).$$

(4.43)

If the desired sampling interval is at least $\Delta\omega$, then we require that

$$N \geq \frac{\omega_s}{\Delta\omega}.$$

(4.44)

Consequently, if aliasing docs not occur and M is chosen large enough to prevent loss of resolution due to windowing, then the DTFS approximation is related to the spectrum of the original signal according to

$$Y[k] \approx \frac{1}{NT_s}X(jk\omega_s/N).$$

The next example illustrates the use of the guidelines given in Eqs. (4.39), (4.42), and (4.44) to approximate the FT with the DTFS.

EXAMPLE 4.15 DTFS APPROXIMATION OF THE FT FOR DAMPED SINUSOIDS Use the DTFS to approximate the FT of the signal

$$x(t) = e^{-t/10}u(t)(\cos(10t) + \cos(12t)).$$

Assume that the frequency band of interest is $-20 < \omega < 20$ and the desired sampling interval is $\Delta\omega = \pi/20$ rads/s. Compare the DTFS approximation with the underlying FT for resolutions of (a) $\omega_r = 2\pi$ rad/s, (b) $\omega_r = 2\pi/5$ rad/s, and (c) $\omega_r = 2\pi/25$ rad/s.

Solution: In order to evaluate the quality of the DTFS approximation, we first determine the FT of $x(t)$. Let $f(t) = e^{-t/10}u(t)$ and $g(t) = (\cos(10t) + \cos(12t))$, so that $x(t) = f(t)g(t)$. Use

$$F(j\omega) = \frac{1}{j\omega + \frac{1}{10}}$$

and

$$G(j\omega) = \pi\delta(\omega + 10) + \pi\delta(\omega - 10) + \pi\delta(\omega + 12) + \pi\delta(\omega - 12),$$

together with the multiplication property, to obtain

$$X(j\omega) = \frac{1}{2}\left(\frac{1}{j(\omega + 10) + \frac{1}{10}} + \frac{1}{j(\omega - 10) + \frac{1}{10}} + \frac{1}{j(\omega + 12) + \frac{1}{10}} + \frac{1}{j(\omega - 12) + \frac{1}{10}}\right).$$

Now put the first two terms and last two terms of $X(j\omega)$ over common denominators:

$$X(j\omega) = \frac{\frac{1}{10} + j\omega}{\left(\frac{1}{10} + j\omega\right)^2 + 10^2} + \frac{\frac{1}{10} + j\omega}{\left(\frac{1}{10} + j\omega\right)^2 + 12^2}. \tag{4.45}$$

The maximum frequency of interest is given as 20, so $\omega_a = 20$ rad/s. In order to use Eq. (4.39) to find the sampling interval, we must also determine ω_m, the highest frequency present in $x(t)$. While $X(j\omega)$ in Eq. (4.45) is not strictly band limited, for $\omega \gg 12$ the magnitude spectrum $|X(j\omega)|$ decreases as $1/\omega$. We shall assume that $X(j\omega)$ is effectively band limited to $\omega_m = 500$, since $|X(j500)|$ is more than a factor of 10 less than $|X(j20)|$, the highest frequency of interest and the nearest frequency at which aliasing occurs. This will not prevent aliasing in $-20 < \omega < 20$, but will ensure that the effect of aliasing in this region is small for all practical purposes. We require that

$$T_s < 2\pi/520$$

$$= 0.0121 \text{ s.}$$

To satisfy this requirement, we choose $T_s = 0.01$ s.

Given the sampling interval T_s, we determine the number of samples, M, using Eq. (4.42):

$$M \geq \frac{200\pi}{\omega_r}.$$

Hence, for (a), $\omega_r = 2\pi$ rad/s, we choose $M = 100$; for (b), $\omega_r = 2\pi/5$ rad/s, we choose $M = 500$; and for (c), $\omega_r = 2\pi/25$ rad/s, we choose $M = 2500$.

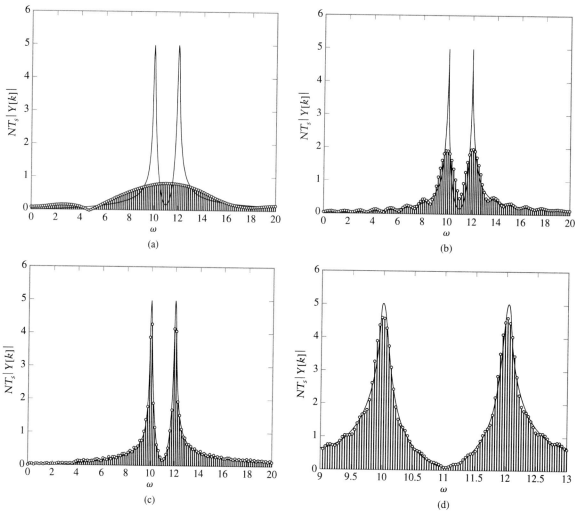

FIGURE 4.57 The DTFS approximation to the FT of $x(t) = e^{-1/10}u(t)(\cos(10t) + \cos(12t))$. The solid line is the FT $|X(j\omega)|$, and the stems denote the DTFS approximation $NT_s|Y[k]|$. Both $|X(j\omega)|$ and $NT_s|Y[k]|$ have even symmetry, so only $0 < \omega < 20$ is displayed. (a) $M = 100, N = 4000$. (b) $M = 500, N = 4000$. (c) $M = 2500, N = 4000$. (d) $M = 2500, N = 16,000$ for $9 < \omega < 13$.

Finally, the length of the DTFS, N, must satisfy Eq. (4.44):

$$N \geq \frac{200\pi}{\Delta\omega}$$

Substitution of $\Delta\omega = \pi/20$ into this relation gives $N \geq 4000$, so we choose $N = 4000$.

We compute the DTFS coefficients $Y[k]$ using these values of $T_s, M,$ and N. Figure 4.57 compares the FT with the DTFS approximation. The solid line in each plot is $|X(j\omega)|$, and the stems represent the DTFS approximation, $NT_s|Y[k]|$. Both $|X(j\omega)|$ and $|Y[k]|$ have even symmetry because $x(t)$ is real, so we only need to depict the interval $0 < \omega < 20$. Figure 4.57(a) depicts $M = 100$, (b) depicts $M = 500$, and (c) depicts $M = 2500$. As M increases and the resolution ω_r decreases, the quality of the approximation improves. In the case of $M = 100$, the resolution ($2\pi \approx 6$) is larger than the separation between the two peaks, and we cannot distinguish the presence of separate peaks. The only portions of the spectrum that are reasonably well approximated are the smooth sections away from the peaks. When $M = 500$, the resolution ($2\pi/5 \approx 1.25$) is less than the separation between the peaks, and distinct peaks are evident, although each is still blurred. As we move away from the peaks, the quality of the approximation improves. In case (c), the resolution ($2\pi/25 \approx 0.25$) is much less than the peak separation, and a much better approximation is obtained over the entire frequency range.

It appears that the values at each peak are still not represented accurately in case (c). This could be due to the resolution limit imposed by M or because we have not sampled the DTFT at small enough intervals. In Fig. 4.57(d), we increase N to 16,000 while keeping $M = 2500$. The region of the spectrum near the peaks, $9 < \omega < 13$, is depicted. Increasing N by a factor of 4 reduces the frequency sampling interval by that same factor. We see that there is still some error in representing each peak value, although less than suggested by Fig. 4.57(c).   ■

▶ **Problem 4.14**   Given a sampling interval $T_s = 2\pi \times 10^{-3}$ s, number of samples $M = 1000$, and zero padding to $N = 2000$, if the signal $X(j\omega)$ is bandlimited to $\omega_m = 600$ rad/s, find (a) the frequency band ω_a on which the DTFS provides an accurate approximation to the FT, (b) the resolution ω_r, and (c) frequency-domain sampling interval $\Delta\omega$.

Answers: (a) $\omega_a = 400$ rad/s, (b) $\omega_r = 1$ rad/s, (c) $\Delta\omega = 0.5$ rad/s   ◀

The quality of the DTFS approximation to the FT improves as T_s decreases, MT_s increases, and N increases. However, practical considerations such as memory limitations and hardware costs generally limit the range over which we can choose these parameters and force compromises. For example, if memory is limited, then we can increase MT_s to obtain better resolution only if we increase T_s and reduce the range of frequencies over which the approximation is valid.

Recall that the FT of periodic signals contains continuous-valued impulse functions whose area is proportional to the value of the corresponding FS coefficients. The nature of the DTFS approximation to the FT of a periodic signal differs slightly from that of the nonperiodic case because the DTFS coefficients are discrete and thus are not well suited to approximating continuous-valued impulses. In this case, the DTFS coefficients are proportional to the area under the impulses in the FT.

To illustrate, consider using the DTFS to approximate the FT of a complex sinusoid $x(t) = ae^{j\omega_o t}$ with amplitude a and frequency ω_o. We have

$$x(t) \xleftrightarrow{\quad FT \quad} X(j\omega) = 2\pi a\delta(\omega - \omega_o).$$

Substitution of $X(j\omega)$ into Eq. (4.38) yields

$$X_\delta(j\omega) = \frac{2\pi}{T_s} a \sum_{k=-\infty}^{\infty} \delta(\omega - \omega_o - k\omega_s).$$

Recognizing that $\omega_s = 2\pi/T_s$ and substituting for $X_\delta(j\omega)$ in Eq. (4.40) gives the FT of the sampled and windowed complex sinusoid as

$$Y_\delta(j\omega) = a \sum_{k=-\infty}^{\infty} W_\delta(j(\omega - \omega_o - k\omega_s)),$$

where $W_\delta(j\omega)$ is given by Eq. (4.41). Using the fact that $W_\delta(j\omega)$ has period ω_s, we may simplify this expression to obtain

$$Y_\delta(j\omega) = a W_\delta(j(\omega - \omega_o)). \tag{4.46}$$

Application of Eq. (4.43) indicates that the DTFS coefficients associated with the sampled and windowed complex sinusoid are given by

$$Y[k] = \frac{a}{N} W_\delta\left(j\left(k\frac{\omega_s}{N} - \omega_o\right)\right). \tag{4.47}$$

Hence, the DTFS approximation to the FT of a complex sinusoid consists of samples of the FT of the window frequency response centered on ω_o, with amplitude proportional to a.

If we choose $N = M$ (no zero padding) and if the frequency of the complex sinusoid satisfies $\omega_o = m\omega_s/M$, then the DTFS samples $W_\delta(j(\omega - \omega_o))$ at the peak of its mainlobe and at its zero crossings. Consequently, we have

$$Y[k] = \begin{cases} a, & k = m \\ 0, & \text{otherwise for } 0 \le k \le M - 1 \end{cases}.$$

In this special case, the continuous-valued impulse with area $2\pi a$ in the FT is approximated by a discrete-valued impulse of amplitude a.

An arbitrary periodic signal is represented by the FS as a weighted sum of harmonically related complex sinusoids, so, in general, the DTFS approximation to the FT consists of samples of a weighted sum of shifted window frequency responses. The next example illustrates this effect.

EXAMPLE 4.16 DTFS APPROXIMATION OF SINUSOIDS   Use the DTFS to approximate the FT of the periodic signal

$$x(t) = \cos(2\pi(0.4)t) + \frac{1}{2}\cos(2\pi(0.45)t).$$

Assume that the frequency band of interest is $-10\pi < \omega < 10\pi$ and the desired sampling interval is $\Delta\omega = 20\pi/M$. Evaluate the DTFS approximation for resolutions of (a) $\omega_r = \pi/2$ rad/s and (b) $\omega_r = \pi/100$ rad/s.

Solution: First note that the FT of $x(t)$ is given by

$$X(j\omega) = \pi\delta(\omega + 0.8\pi) + \pi\delta(\omega - 0.8\pi) + \frac{\pi}{2}\delta(\omega + 0.9\pi) + \frac{\pi}{2}\delta(\omega - 0.9\pi).$$

The maximum frequency of interest is $\omega_a = 10\pi$ rad/s, and this is much larger than the highest frequency in $X(j\omega)$, so aliasing is not a concern and we choose $\omega_s = 2\omega_a$. This gives $T_s = 0.1$ s. The number of samples, M, is determined by substituting ω_s into Eq. (4.42):

$$M \ge \frac{20\pi}{\omega_r}.$$

To obtain the resolution specified in case (a), we require that $M \ge 40$ samples, while in case (b) we need $M \ge 2000$ samples. We shall choose $M = 40$ for case (a) and $M = 2000$ for case (b). We substitute $\Delta\omega = 20\pi/M$ into Eq. (4.44) with equality to obtain $N = M$, and thus no zero padding is required.

The signal is a weighted sum of complex sinusoids, so the underlying FT is a weighted sum of shifted window frequency responses and is given by

$$Y_\delta(j\omega) = \frac{1}{2}W_\delta(j(\omega + 0.8\pi)) + \frac{1}{2}W_\delta(j(\omega - 0.8\pi)) + \frac{1}{4}W_\delta(j(\omega + 0.9\pi))$$

$$+ \frac{1}{4}W_\delta(j(\omega - 0.9\pi)).$$

In case (a),

$$W_\delta(j\omega) = e^{-j\omega 39/20}\frac{\sin(2\omega)}{\sin(\omega/20)}.$$

In case (b),

$$W_\delta(j\omega) = e^{-j\omega 1999/20}\frac{\sin(100\omega)}{\sin(\omega/20)}.$$

The DTFS coefficients $Y[k]$ are obtained by sampling $Y_\delta(j\omega)$ at intervals of $\Delta\omega$. The stems in Fig. 4.58(a) depict $|Y[k]|$ for $M = 40$, while the solid line depicts $(1/M)|Y_\delta(j\omega)|$ for positive frequencies. We have chosen to label the axis in units of Hz rather than rad/s for convenience. In this case, the minimum resolution of $\omega_r = \pi/2$ rad/s, or 0.25 Hz, is five times greater than the separation between the two sinusoidal components. Hence, we cannot identify the presence of two sinusoids in either $|Y[k]|$ or $(1/M)|Y_\delta(j\omega)|$.

Figure 4.58(b) illustrates $|Y[k]|$ for $M = 2000$. We zoom in on the frequency band containing the sinusoids in Fig. 4.58(c), depicting $|Y[k]|$ with the stems and $(1/M)|Y_\delta(j\omega)|$ with the solid line. In this case, the minimum resolution is a factor of 10 times smaller than the separation between the two sinusoidal components, and we clearly see the presence of two sinusoids. The interval for which the DTFS samples $Y_\delta(j\omega)$ is $2\pi/200$ rad/s, or 0.005 Hz. The frequency of each sinusoid is an integer multiple of the sampling interval, so $Y[k]$ samples $Y_\delta(j\omega)$ once at the peak of each mainlobe, with the remainder of samples occuring at the zero crossings. Thus, the amplitude of each component is correctly reflected in $|Y[k]|$.

Figure 4.58(d) depicts $|Y[k]|$ and $(1/M)|Y_\delta(j\omega)|$, assuming that $M = 2010$. This results in slightly better resolution than $M = 2000$. However, now the frequency of each sinusoid is not an integer multiple of the interval at which the DTFS samples $Y_\delta(j\omega)$. Consequently, $Y_\delta(j\omega)$ is not sampled at the peak of each mainlobe and the zero crossings. While the resolution is sufficient to reveal the presence of two components, we can no longer determine the amplitude of each component directly from $|Y[k]|$.

In practice, it is unusual for the frequencies of the sinusoids to be known and thus impossible to choose M so that $Y_\delta(j\omega)$ is sampled at the mainlobe peak and zero crossings. In many applications we seek to determine both the frequency and amplitude of one or more sinusoids in a data record. In this case, the sinusoid amplitude and frequency may be determined by zero padding so that $Y[k]$ samples $Y_\delta(j\omega)$ sufficiently densely to capture the peak amplitude and location of the mainlobe. It is not unusual to choose $N \geq 10M$ so that the mainlobe is represented by 10 or more samples of $Y[k]$.   ∎

▶ **Problem 4.15**   Let $x(t) = a\cos(2.4\pi t)$, and assume that the maximum frequency of interest is $\omega_a = 5\pi$ and that there is no zero padding. Find the largest sampling interval T_s and minimum number of samples M so that the coefficient of peak magnitude in the DTFS approximation may be used to determine a. Determine which DTFS coefficient has the largest magnitude.

Answer: $T_s = 0.2$ s, $M = 25$, and $X[6]$ has largest magnitude. Note that $X[k]$ is periodic with period 25, and by symmetry, $X[-6] - X[6]$ ◀

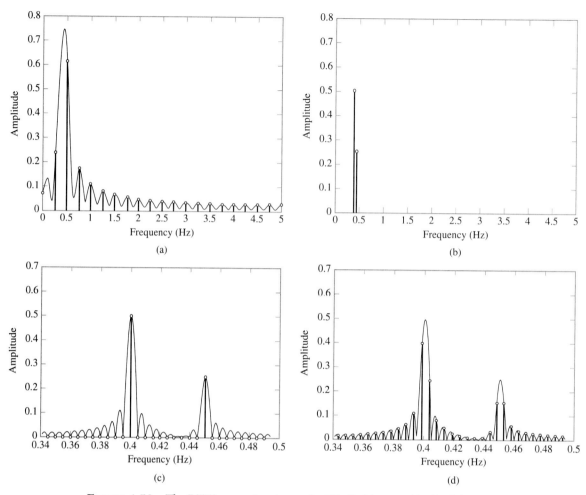

FIGURE 4.58   The DTFS approximation to the FT of $x(t) = \cos(2\pi(0.4)t) + \cos(2\pi(0.45)t)$. The stems denote $|Y[k]|$, while the solid lines denote $(1/M)|Y_\delta(j\omega)|$. The frequency axis is displayed in units of Hz for convenience, and only positive frequencies are illustrated. (a) $M = 40$. (b) $M = 2000$. Only the stems with nonzero amplitude are depicted. (c) Behavior in the vicinity of the sinusoidal frequencies for $M = 2000$. (d) Behavior in the vicinity of the sinusoidal frequencies for $M = 2010$.

4.10 *Efficient Algorithms for Evaluating the DTFS*

The role of the DTFS as a computational tool is greatly enhanced by the availability of efficient algorithms for evaluating the forward and inverse DTFS. Such algorithms are collectively termed *fast Fourier transform (FFT) algorithms*. These algorithms exploit the "divide and conquer" principle by splitting the DTFS into a series of lower order DTFSs and using the symmetry and periodicity properties of the complex sinusoid $e^{jk2\pi n}$. Less computation is required to evaluate and combine the lower order DTFS than to evaluate the original DTFS, hence the designation "fast". We shall demonstrate the computational savings that accrue from the splitting process.

Recall that the DTFS pair may be evaluated with the use of the expressions

$$X[k] = \frac{1}{N} \sum_{n=0}^{N-1} x[n] e^{-jk\Omega_o n}$$

and

$$x[n] = \sum_{k=0}^{N-1} X[k] e^{jk\Omega_o n}. \tag{4.48}$$

These expressions are virtually identical, differing only in the normalization by N and the sign of the complex exponential. Hence, the same basic algorithm can be used to compute either relationship; only minor changes are required. We shall consider evaluating Eq. (4.48).

Evaluating Eq. (4.48) directly for a single value of n requires N complex multiplications and $N - 1$ complex additions. Thus, the computation of $x[n]$, $0 \leq n \leq N - 1$, requires N^2 complex multiplications and $N^2 - N$ complex additions. In order to demonstrate how this number of operations can be reduced, we assume that N is even. We split $X[k]$, $0 \leq k \leq N - 1$, into even- and odd-indexed signals, respectively, shown by

$$X_e[k] = X[2k], \quad 0 \leq k \leq N' - 1$$

and

$$X_o[k] = X[2k + 1], \quad 0 \leq k \leq N' - 1,$$

where $N' = N/2$ and

$$x_e[n] \xleftrightarrow{\;DTFS;\,\Omega_o'\;} X_e[k], \quad x_o[n] \xleftrightarrow{\;DTFS;\,\Omega_o'\;} X_o[k],$$

with $\Omega_o' = 2\pi/N'$. Now we express Eq. (4.48) as a combination of the N' DTFS coefficients $X_e[k]$ and $X_o[k]$:

$$x[n] = \sum_{k=0}^{N-1} X[k] e^{jk\Omega_o n}$$

$$= \sum_{k\,\text{even}} X[k] e^{jk\Omega_o n} + \sum_{k\,\text{odd}} X[k] e^{jk\Omega_o n}.$$

We write the even and odd indices as $2m$ and $2m + 1$, respectively, to obtain

$$x[n] = \sum_{m=0}^{N'-1} X[2m] e^{jm2\Omega_o n} + \sum_{m=0}^{N'-1} X[2m + 1] e^{j(m2\Omega_o n + \Omega_o n)}.$$

Substituting the definitions of $X_e[k]$, $X_o[k]$, and $\Omega_o' = 2\Omega_o$ into the previous equation yields

$$x[n] = \sum_{m=0}^{N'-1} X_e[m] e^{jm\Omega_o' n} + e^{j\Omega_o n} \sum_{m=0}^{N'-1} X_o[m] e^{jm\Omega_o' n}$$

$$= x_e[n] + e^{j\Omega_o n} x_o[n], \quad 0 \leq n \leq N - 1.$$

This indicates that $x[n]$ is a weighted combination of $x_e[n]$ and $x_o[n]$.

We may further simplify our result by exploiting the periodicity properties of $x_e[n]$ and $x_o[n]$. Using $x_e[n + N'] = x_e[n]$, $x_o[n + N'] = x_o[n]$, and $e^{j(n+N')\Omega_o} = -e^{jn\Omega_o}$, we obtain

$$x[n] = x_e[n] + e^{jn\Omega_o} x_o[n], \quad 0 \leq n \leq N' - 1 \tag{4.49}$$

as the first N' values of $x[n]$ and

$$x[n + N'] = x_e[n] - e^{jn\Omega_o} x_o[n], \quad 0 \leq n \leq N' - 1 \tag{4.50}$$

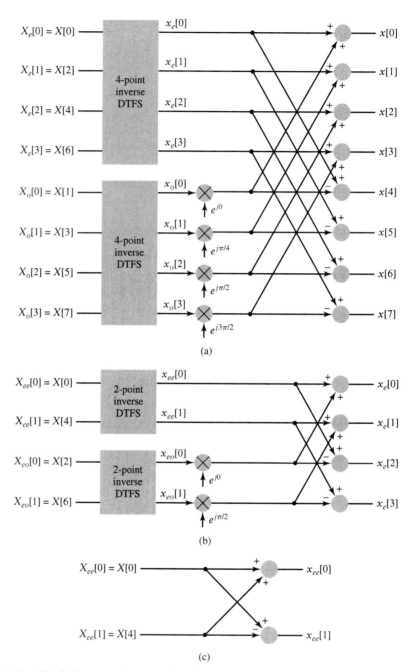

FIGURE 4.59 Block diagrams depicting the decomposition of an inverse DTFS as a combination of lower order inverse DTFSs for $N - 8$. (a) Eight-point inverse DTFS represented in terms of two four-point inverse DTFSs. (b) Four-point inverse DTFS represented in terms of two-point inverse DTFSs. (c) Two-point inverse DTFS.

as the second N' values of $x[n]$. Figure 4.59(a) depicts the computation described in Eqs. (4.49) and (4.50) graphically for $N = 8$. We see that we need only multiply by $e^{jn\Omega_o}$ once in computing both equations. The remaining operations are addition and subtraction.

    Let us consider the computation required to evaluate Eqs. (4.49) and (4.50). The evaluation of each of $x_e[n]$ and $x_o[n]$ requires $(N')^2$ complex multiplications, for a total of $N^2/2$ such multiplications. An additional N' multiplications are required to compute $e^{-jn\Omega_o}x_o[n]$. Thus, the total number of complex multiplications is $N^2/2 + N/2$. For large N, this is approximately $N^2/2$, about one-half the number of multiplications required to evaluate $x[n]$ directly. Further reductions in computational requirements are obtained if we split $X_e[k]$ and $X_o[k]$ again, this time into even- and odd-indexed sequences. For example, Fig. 4.59(b) illustrates how to split the four-point inverse DTFS used to calculate $x_e[n]$ into two two-point inverse DTFS's for $N = 8$. The greatest savings is when N is a power of 2. In that case, we can continue subdividing until the size of each inverse DTFS is 2. The two-point inverse DTFS requires no multiplications, as illustrated in Fig. 4.59(c).

Figure 4.60 shows the FFT computation for $N = 8$. The repeated partitioning into even- and odd-indexed sequences permutes the order of the DTFS coefficients at the input. This permutation is termed *bit reversal*, since the location of $X[k]$ may be determined by reversing the bits in a binary representation of the index k. For example, $X[6]$ has index $k = 6$. Representing $k = 6$ in binary form gives $k = 110_2$. Now reversing the bits gives $k' = 011_2$, or $k' = 3$, so $X[6]$ appears in the fourth position. The basic two-input, two-output structure depicted in Fig. 4.59(c) that is duplicated in each stage of the FFT (see Fig. 4.60) is termed a *butterfly* because of its appearance.

FFT algorithms for N a power of 2 require on the order of $N\log_2(N)$ complex multiplications. This can represent an extremely large savings in computation relative to N^2 when N is large. For example, if $N = 8192$, or 2^{13}, the direct approach requires approximately 630 times as many arithmetic operations as the FFT algorithm.

A word of caution is in order here. Many software packages contain routines that implement FFT algorithms. Unfortunately, the location of the $1/N$ factor is not standardized. Some routines place the $1/N$ in the expression for the DTFS coefficients $X[k]$, as we have done here, while others place the $1/N$ in the expression for the time signal $x[n]$. Yet another convention is to place $1/\sqrt{N}$ in each of the expressions for $X[k]$ and $x[n]$. The only effect of these alternative conventions is to multiply the DTFS coefficients $X[k]$ by either N or $\sqrt{N}$.

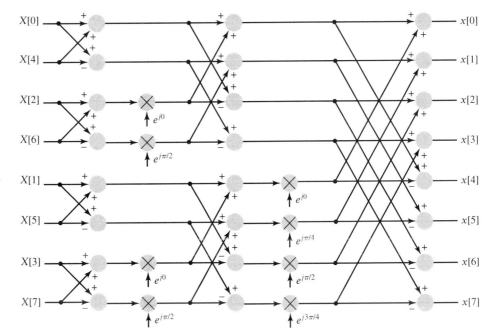

FIGURE 4.60 Diagram of the FFT algorithm for computing $x[n]$ from $X[k]$ for $N = 8$.

4.11 *Exploring Concepts with MATLAB*

■ 4.11.1 DECIMATION AND INTERPOLATION

Recall that decimation reduces the effective sampling rate of a discrete-time signal, while interpolation increases the effective sampling rate. Decimation is accomplished by subsampling a low-pass filtered version of the signal, while interpolation is performed by inserting zeros in between samples and then applying a low-pass filter. MATLAB's Signal Processing Toolbox contains several routines for performing decimation and interpolation. All of them automatically design and apply the low-pass filter required for both operations. The command `y = decimate(x,r)` decimates the signal represented by `x` by a positive integer factor `r` to produce the vector `y`, which is a factor of `r` shorter than `x`. Similarly, `y = interp(x,r)` interpolates `x` by a positive integer factor `r`, producing a vector `y` that is `r` times as long as `x`. The command `y = resample(x,p,q)` resamples the signal in vector `x` at `p/q` times the original sampling rate, where `p` and `q` are positive integers. This is conceptually equivalent to first interpolating by a factor `p` and then decimating by a factor `q`. The vector `y` is `p/q` times the length of `x`. The values of the resampled sequence may be inaccurate near the beginning and end of `y` if `x` contains large deviations from zero at its beginning and end.

Suppose the discrete-time signal

$$x[n] = e^{-\frac{n}{15}} \sin\left(\frac{2\pi}{13}n + \frac{\pi}{8}\right), \quad 0 \le n \le 59$$

results from sampling a continuous-time signal at a rate of 45 kHz and that we wish to find the discrete-time signal resulting from sampling the underlying continuous-time signal at 30 kHz. This corresponds to changing the sampling rate by the factor $\frac{30}{45} = \frac{2}{3}$. The `resample` command is used to effect this change as follows:

```
>> x = exp(-[0:59]/15).*sin([0:59]*2*pi/13 + pi/8);
>> y = resample(x,2,3);
>> subplot(2,1,1)
>> stem([0:59],x);
>> title('Signal Sampled at 45kHz'); xlabel('Time');
>> ylabel('Amplitude')
>> subplot(2,1,2)
>> stem([0:39],y);
>> title('Signal Sampled at 30kHz'); xlabel('Time');
   ylabel('Amplitude')
```

The original and resampled signals resulting from these commands are depicted in Figure 4.61.

■ 4.11.2 RELATING THE DTFS TO THE DTFT

Equation (4.36) states that the DTFS coefficients of a finite-duration signal correspond to samples of the DTFT, divided by the number of DTFS coefficients, N. As discussed in Section 3.19, the MATLAB routine `fft` calculates N times the DTFS coefficients. Hence, `fft` directly evaluates samples of the DTFT of a finite-duration signal. The zero-padding process involves appending zeros to the finite-duration signal before computing the DTFS and results in a denser sampling of the underlying DTFT. Zero padding is easily accom-

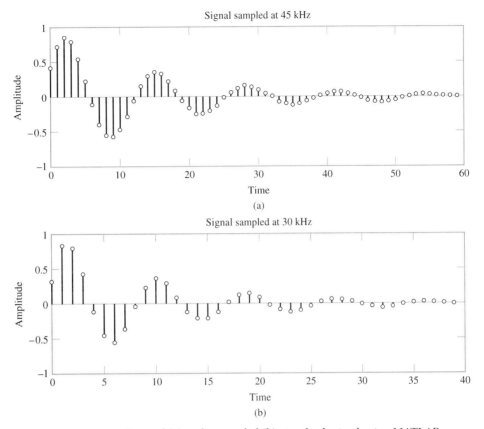

FIGURE 4.61   Original (a) and resampled (b) signals obtained using MATLAB.

plished with **fft** by adding an argument that specifies the number of coefficients to compute. If **x** is a length-*M* vector representing a finite-duration time signal and **n** is greater than *M*, then the command **X = fft(x,n)** evaluates *n* samples of the DTFT of **x** by first padding **x** with trailing zeros to length **n**. If **n** is less than *M*, then **fft(x,n)** first truncates **x** to length **n**.

The frequency values corresponding to the samples in **X** are represented by a vector **n** points long with the first element zero and the remaining entries spaced at intervals of $2\pi/n$. For example, the command **w=[0:(n-1)]*2*pi/n** generates the appropriate vector of frequencies. Note that this describes the DTFT for $0 \le \Omega < 2\pi$. It is sometimes more convenient to view the DTFT over a period centered on zero—that is $-\pi < \Omega \le \pi$. The MATLAB command **Y=fftshift(X)** swaps the left and right halves of **X** in order to put the zero-frequency value in the center. The vector of frequency values corresponding to the values in **Y** may be generated by using **w = [-n/2:(n/2-1)]*2*pi/n**.

Suppose we revisit Example 4.14, using MATLAB to evaluate $|X(e^{j\Omega})|$ at intervals in frequency of (a) $\frac{2\pi}{32}$, (b) $\frac{2\pi}{60}$, and (c) $\frac{2\pi}{120}$. Recall that

$$x[n] = \begin{cases} \cos\left(\frac{3\pi}{8}n\right), & 0 \le n \le 31 \\ 0, & \text{otherwise} \end{cases}.$$

For case (a) we use a 32-point DTFS computed from the 32 nonzero values of the signal. In cases (b) and (c), we zero pad to length 60 and 120, respectively, to sample the DTFT at the specified intervals. We evaluate and display the results on $-\pi < \Omega \le \pi$, using the following commands:

```
>> n = [0:31];
>> x = cos(3*pi*n/8);
>> X32 = abs(fftshift(fft(x)));        %magnitude for 32
   point DTFS
>> X60 = abs(fftshift(fft(x,60)));      %magnitude for 60
   point DTFS
>> X120 = abs(fftshift(fft(x,120)));     %magnitude for
   120 point DTFS
>> w32 = [-16:15]*2*pi/32;     w60=[-30:29]*2*pi/60;
>> w120 = [-60:59]*2*pi/120;
>> stem(w32,X32);       % stem plot for Fig. 4.53 (a)
>> stem(w60,X60);       % stem plot for Fig. 4.53 (b)
>> stem(w120,X120);     % stem plot for Fig. 4.53 (c)
```

The results are depicted as the stem plots in Figs. 4.53 (a)–(c).

■ 4.11.3 COMPUTATIONAL APPLICATIONS OF THE DTFS

As previously noted, MATLAB's `fft` command may be used to evaluate the DTFS and thus is used for approximating the FT. In particular, the `fft` is used to generate the DTFS approximations in Examples 4.15 and 4.16. To repeat Example 4.16, we use the following commands:

```
>> ta = 0:0.1:3.9;    % time samples for case (a)
>> tb = 0:0.1:199.9;   % time samples for case (b)
>> xa = cos(0.8*pi*ta) + 0.5*cos(0.9*pi*ta);
>> xb = cos(0.8*pi*tb) + 0.5*cos(0.9*pi*tb);
>> Ya = abs(fft(xa)/40);    Yb = abs(fft(xb)/2000);
>> Ydela = abs(fft(xa,8192)/40);    % evaluate 1/M
   Y_delta(j omega) for case (a)
>> Ydelb = abs(fft(xa,16000)/2000);    % evaluate 1/M
   Y_delta(j omega) for case (b)
>> fa = [0:19]*5/20;    fb = [0:999]*5/1000;
>> fdela = [0:4095]*5/4096;    fdelb = [0:7999]*5/8000;
>> plot(fdela,Ydela(1:4192))    % Fig. 4.58a
>> hold on
>> stem(fa,Ya(1:20))
>> xlabel('Frequency (Hz)');    ylabel('Amplitude')
>> hold off
>> plot(fdelb(560:800),Ydelb(560:800))    %Fig. 4.58c
>> hold on
>> stem(fb(71:100),Yb(71:100))
>> xlabel('Frequency (Hz)');    ylabel('Amplitude')
```

Note that here we evaluated $\frac{1}{M} Y_\delta(j\omega)$ by using `fft`, and zero padding with a large number of zeros relative to the length of $x[n]$. Recall that zero padding decreases the spacing between the samples of the DTFT that are obtained by the DTFS. Hence, by padding with a large number of zeros, we capture sufficient detail such that `plot` provides a smooth approximation to the underlying DTFT. If `plot` is used to display the DTFS coefficients without zero padding, then a much coarser approximation to the underlying DTFT is obtained. Figure 4.62 depicts the DTFS coefficients for case (b) of Example 4.16, using both plot and stem. The coefficients are obtained via the following commands:

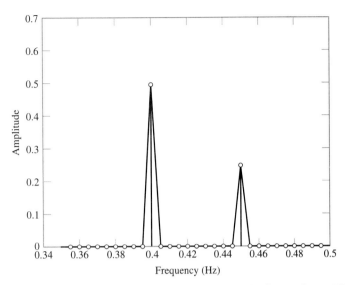

FIGURE 4.62   The use of the MATLAB command `plot` for displaying the DTFS coefficients in case (b) of Example 4.16.

```
>> plot(fb(71:100),Yb(71:100))
>> hold on
>> stem(fb(71:100),Yb(71:100))
```

Here, the `plot` command produces triangles centered on the frequencies associated with the sinusoids. The triangles are a consequence of `plot` drawing straight lines in between the values in `Yb`.

The `fft` command is implemented using the numerically efficient, or fast Fourier transform, algorithm based on the divide-and-conquer principle discussed in Section 4.10.

4.12 *Summary*

The mixing of classes of signals is frequently encountered in the course of applying Fourier representations. We have established relationships between different Fourier representations in this chapter in order to address situations in which there is a mixing of different classes of signals:

► periodic and nonperiodic signals

► continuous- and discrete-time signals

Periodic and nonperiodic signals often interact in the context of the interaction between signals and LTI systems (e.g., filtering) and in performing other basic manipulations of signals (e.g., multiplying two signals). Mixtures of continuous- and discrete-time classes of signals are encountered in sampling continuous-time signals or in reconstructing continuous-time signals from samples. Use of the DTFS to numerically approximate the FT also involves a mixing of signal classes. Since each class has its own Fourier representation, such situations cannot be addressed without extending our set of Fourier representation tools.

The FT is the most versatile representation for analysis, since all four signal classes have FT representations, a situation made possible by permitting the use of impulses in the time and frequency domains. The FT is most often used to analyze continuous-time LTI systems and systems that sample continuous-time signals or reconstruct continuous-time signals from samples. The primary use of the DTFT is to analyze discrete-time systems. We have developed a DTFT representation of discrete-time periodic signals to facilitate this role. The DTFS is used to approximate both the FT and the DTFT for computational purposes. We have established various relationships between the DTFS and the FT, as well as between the DTFS and DTFT, in order to correctly interpret the results of numerical computations.

The existence of FFT algorithms or otherwise computationally efficient algorithms for evaluating the DTFS greatly expands the range of problems in which Fourier analysis may be used. These algorithms are based on dividing the DTFS into a nested set of lower order DTFS computations and are available in almost all commercial software packages for processing data.

Fourier methods provide a powerful set of analytic and numerical tools for solving problems involving signals and systems and for studying communication systems, as we will see in the next chapter. They also have extensive application in the context of filtering, the topic of Chapter 8.

FURTHER READING

1. The topics of sampling, reconstruction, discrete-time signal-processing systems, computational applications of the DTFS, and fast algorithms for the DTFS are discussed in greater detail in the following texts:

   ▶ Proakis, J. G., and D. G. Manolakis, *Digital Signal Processing: Principles, Algorithms and Applications*, 3rd ed. (Prentice Hall, 1995)

   ▶ Oppenheim, A. V., R. W. Schafer, and J. R. Buck, *Discrete-Time Signal Processing*, 2nd ed. (Prentice Hall, 1999)

   ▶ Jackson, L. B., *Digital Filters and Signal Processing*, 3rd ed. (Kluwer, 1996)

   ▶ Roberts, R. A. and C. T. Mullis, *Digital Signal Processing* (Addison-Wesley, 1987)

2. In the literature discussing numerical-computation applications, the discrete Fourier transform, or DFT, terminology is usually used in place of the DTFS terminology adopted in this text. The DFT coefficients are N times the DTFS coefficients. We have chosen to retain the DTFS terminology for consistency and to avoid confusion with the DTFT.

3. The modern discovery of the FFT algorithm for evaluating the DTFS is attributed to J. W. Cooley and J. W. Tukey for their 1965 publication "An algorithm for the machine calculation of complex Fourier series," *Mat. Comput.*, vol. 19, pp. 297–301. This paper greatly accelerated the development of a field called digital signal processing, which was in its infancy in the mid-1960s. The availability of a fast algorithm for computing the DTFS opened up a tremendous number of new applications for digital signal processing and resulted in explosive growth of the new field. Indeed, the majority of this chapter and a substantial portion of Chapter 8 concern the field of digital signal processing. One very important application of the FFT is a computationally efficient implementation of linear convolution for filtering signals. Two basic approaches, "overlap and add" and "overlap and save," implement convolution via the multiplication of DTFS coefficients computed from segments of the input signal. The basis of the "overlap and save" algorithm is explored in Problem 4.54.

   Carl Friedrich Gauss, the eminent German mathematician, has been credited with developing an equivalent efficient algorithm for computing DTFS coefficients as early as 1805, predating Joseph Fourier's work on harmonic analysis. Additional reading on the history of the FFT and its impact on digital signal processing is found in the following two articles:

▶ Heideman, M. T., D. H. Johnson, and C. S. Burrus, "Gauss and the history of the fast Fourier transform," *IEEE ASSP Magazine*, vol. 1, no. 4, pp. 14–21, October 1984.

▶ J. W. Cooley, "How the FFT gained acceptance," *IEEE Signal Processing Magazine*, vol. 9, no. 1, pp. 10–13, January 1992.

The following book is devoted to a detailed treatment of the FFT algorithm:

▶ E. O. Brigham, *The Fast Fourier Transform and Its Applications* (Prentice Hall, 1988)

ADDITIONAL PROBLEMS

4.16 Find the FT representations of the following periodic signals:

(a) $x(t) = 2\cos(\pi t) + \sin(2\pi t)$

(b) $x(t) = \sum_{k=0}^{4} \frac{(-1)^k}{k+1}\cos((2k+1)\pi t)$

(c) $x(t)$ as depicted in Fig. P4.16(a).

(d) $x(t)$ as depicted in Fig. P4.16(b).

Sketch the magnitude and phase spectra.

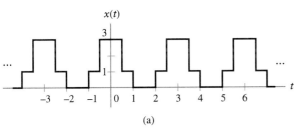

(a)

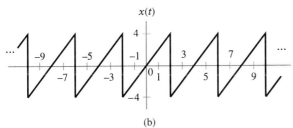

(b)

FIGURE P4.16

4.17 Find the DTFT representations of the following periodic signals:

(a) $x[n] = \cos\left(\frac{\pi}{8}n\right) + \sin\left(\frac{\pi}{5}n\right)$

(b) $x[n] = 1 + \sum_{m=-\infty}^{\infty} \cos\left(\frac{\pi}{4}m\right)\delta[n-m]$

(c) $x[n]$ as depicted in Fig. P4.17(a).

(d) $x[n]$ as depicted in Fig. P4.17(b).

(e) $x[n]$ as depicted in Fig. P4.17(c).

Sketch the magnitude and phase spectra.

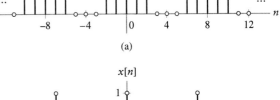

(a)

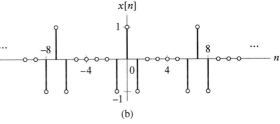

(b)

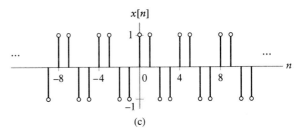

(c)

FIGURE P4.17

4.18 An LTI system has the impulse response

$$h(t) = 2\frac{\sin(2\pi t)}{\pi t}\cos(7\pi t).$$

Use the FT to determine the system output if the input is

(a) $x(t) = \cos(2\pi t) + \sin(6\pi t)$

(b) $x(t) = \sum_{m=-\infty}^{\infty} (-1)^m \delta(t-m)$

(c) $x(t)$ as depicted in Fig. P4.18(a).

(d) $x(t)$ as depicted in Fig. P4.18(b).

(e) $x(t)$ as depicted in Fig. P4.18(c).

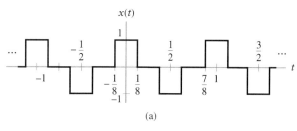

(a)

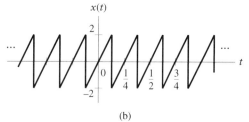

(b)

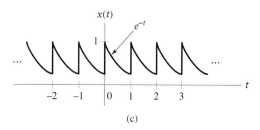

(c)

FIGURE P4.18

4.19 We may design a dc power supply by cascading a full-wave rectifier and an RC circuit as depicted in Fig. P4.19. The full-wave rectifier output is given by

$$z(t) = |x(t)|.$$

Let $H(j\omega)$

$$H(j\omega) = \frac{Y(j\omega)}{Z(j\omega)} = \frac{1}{j\omega RC + 1}$$

be the frequency response of the RC circuit. Suppose the input is $x(t) = \cos(120\pi t)$.

(a) Find the FT representation of $z(t)$.

(b) Find the FT representation of $y(t)$.

(c) Find the range for the time constant RC such that the first harmonic of the ripple in $y(t)$ is less than 1% of the average value.

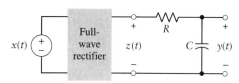

FIGURE P4.19

4.20 Consider the system depicted in Fig. P4.20(a). The FT of the input signal is depicted in Fig. P4.20(b). Let $z(t) \xleftrightarrow{\text{FT}} Z(j\omega)$ and $y(t) \xleftrightarrow{\text{FT}} Y(j\omega)$. Sketch $Z(j\omega)$ and $Y(j\omega)$ for the following cases:

(a) $w(t) = \cos(5\pi t)$ and $h(t) = \frac{\sin(6\pi t)}{\pi t}$

(b) $w(t) = \cos(5\pi t)$ and $h(t) = \frac{\sin(5\pi t)}{\pi t}$

(c) $w(t)$ depicted in Fig. P4.20(c) and $h(t) = \frac{\sin(2\pi t)}{\pi t} \cos(5\pi t)$

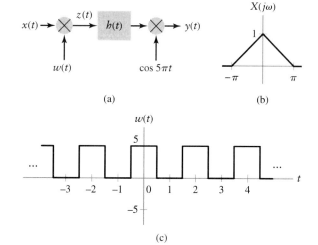

(a)   (b)

(c)

FIGURE P4.20

4.21 Consider the system depicted in Fig. P4.21 The impulse response is given by

$$h(t) = \frac{\sin(11\pi t)}{\pi t},$$

and we have

$$x(t) = \sum_{k=1}^{\infty} \frac{1}{k^2} \cos(k5\pi t)$$

and

$$g(t) = \sum_{k=1}^{10} \cos(k8\pi t)$$

Use the FT to determine $y(t)$.

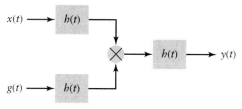

FIGURE P4.21

4.22 The input to a discrete-time system is given by

$$x[n] = \cos\left(\frac{\pi}{8}n\right) + \sin\left(\frac{3\pi}{4}n\right).$$

Use the DTFT to find the output of the system, $y[n]$, if the impulse response is given by

(a) $h[n] = \dfrac{\sin\left(\frac{\pi}{4}n\right)}{\pi n}$

(b) $h[n] = (-1)^n \dfrac{\sin\left(\frac{\pi}{2}n\right)}{\pi n}$

(c) $h[n] = \cos\left(\frac{\pi}{2}n\right)\dfrac{\sin\left(\frac{\pi}{5}n\right)}{\pi n}$

4.23 Consider the discrete-time system depicted in Fig. P4.23. Let $h[n] = \dfrac{\sin\left(\frac{\pi}{2}n\right)}{\pi n}$. Use the DTFT to determine the output $y[n]$ for the following cases:

(a) $x[n] = \dfrac{\sin\left(\frac{\pi}{4}n\right)}{\pi n}, \quad w[n] = (-1)^n$

(b) $x[n] = \delta[n] - \dfrac{\sin\left(\frac{\pi}{4}n\right)}{\pi n}, \quad w[n] = (-1)^n$

(c) $x[n] = \dfrac{\sin\left(\frac{\pi}{2}n\right)}{\pi n}, \quad w[n] = \cos\left(\frac{\pi}{2}n\right)$

(d) $x[n] = 1 + \sin\left(\frac{\pi}{16}n\right) + 2\cos\left(\frac{3\pi}{4}n\right),$

$w[n] = \cos\left(\frac{3\pi}{8}n\right)$

Also, sketch $G(e^{j\Omega})$, the DTFT of $g[n]$.

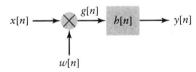

FIGURE P4.23

4.24 Determine and sketch the FT representation, $X_\delta(j\omega)$, for the following discrete-time signals with the sampling interval T_s as given:

(a) $x[n] = \dfrac{\sin\left(\frac{\pi}{3}n\right)}{\pi n}, \quad T_s = 2$

(b) $x[n] = \dfrac{\sin\left(\frac{\pi}{3}n\right)}{\pi n}, \quad T_s = \frac{1}{4}$

(c) $x[n] = \cos\left(\frac{\pi}{2}n\right)\dfrac{\sin\left(\frac{\pi}{4}n\right)}{\pi n}, \quad T_s = 2$

(d) $x[n]$ depicted in Fig. P4.17(a) with $T_s = 4$.

(e) $x[n] = \sum_{p=-\infty}^{\infty} \delta[n - 4p], \quad T_s = \frac{1}{8}$

4.25 Consider sampling the signal $x(t) = \frac{1}{\pi t}\sin(2\pi t)$.

(a) Sketch the FT of the sampled signal for the following sampling intervals:

(i) $T_s = \frac{1}{8}$

(ii) $T_s = \frac{1}{3}$

(iii) $T_s = \frac{1}{2}$

(iv) $T_s = \frac{2}{3}$

(b) Let $x[n] = x(nT_s)$. Sketch the DTFT of $x[n]$, $X(e^{j\Omega})$, for each of the sampling intervals given in (a).

4.26 The continuous-time signal $x(t)$ with FT as depicted in Fig. P4.26 is sampled.

(a) Sketch the FT of the sampled signal for the following sampling intervals:

(i) $T_s = \frac{1}{14}$

(ii) $T_s = \frac{1}{7}$

(iii) $T_s = \frac{1}{5}$

In each case, identify whether aliasing occurs.

(b) Let $x[n] = x(nT_s)$. Sketch the DTFT of $x[n]$, $X(e^{j\Omega})$, for each of the sampling intervals given in (a).

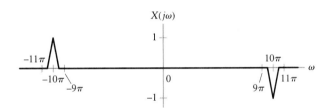

FIGURE P4.26

4.27 Consider subsampling the signal $x[n] = \dfrac{\sin\left(\frac{\pi}{6}n\right)}{\pi n}$ so that $y[n] = x[qn]$. Sketch $Y(e^{j\Omega})$ for the following choices of q:

(a) $q = 2$

(b) $q = 4$

(c) $q = 8$

4.28 The discrete-time signal $x[n]$ with DTFT depicted in Fig. P4.28 is subsampled to obtain $y[n] = x[qn]$. Sketch $Y(e^{j\Omega})$ for the following choices of q:

(a) $q = 3$

(b) $q = 4$

(c) $q = 8$

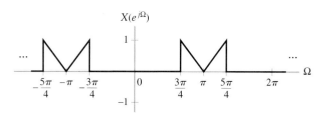

FIGURE P4.28

4.29 For each of the following signals, sampled with sampling interval T_s, determine the bounds on T_s, which guarantee that there will be no aliasing:

(a) $x(t) = \frac{1}{t}\sin 3\pi t + \cos(2\pi t)$

(b) $x(t) = \cos(12\pi t)\dfrac{\sin(\pi t)}{2t}$

(c) $x(t) = e^{-6t}u(t) * \dfrac{\sin(Wt)}{\pi t}$

(d) $x(t) = w(t)z(t)$, where the FT's $W(j\omega)$ and $Z(j\omega)$ are depicted in Fig. P4.29.

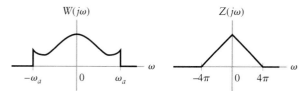

FIGURE P4.29

4.30 Consider the system depicted in Fig. P4.30. Let $|X(j\omega)| = 0$ for $|\omega| > \omega_m$. Find the largest value of T such that $x(t)$ can be reconstructed from $y(t)$. Determine a system that will perform the reconstruction for this maximum value of T.

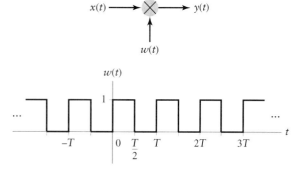

FIGURE P4.30

4.31 Let $|X(j\omega)| = 0$ for $|\omega| > \omega_m$. Form the signal $y(t) = x(t)[\cos(3\pi t) + \sin(10\pi t)]$. Determine the maximum value of ω_m for which $x(t)$ can be reconstructed from $y(t)$, and specify a system that will perform the reconstruction.

4.32 A reconstruction system consists of a zero-order hold followed by a continuous-time anti-imaging filter with frequency response $H_c(j\omega)$. The original signal $x(t)$ is band limited to ω_m (i.e., $X(j\omega) = 0$ for $|\omega| > \omega_m$) and is sampled with a sampling interval of T_s. Determine the constraints on the magnitude response of the anti-imaging filter so that the overall magnitude response of this reconstruction system is between 0.99 and 1.01 in the signal passband and less than 10^{-4} on all bands containing the images of the signal spectrum for the following values:

(a) $\omega_m = 10\pi$, $\quad T_s = 0.1$

(b) $\omega_m = 10\pi$, $\quad T_s = 0.05$

(c) $\omega_m = 10\pi$, $\quad T_s = 0.02$

(d) $\omega_m = 2\pi$, $\quad T_s = 0.05$

4.33 The zero-order hold produces a stair-step approximation to the sampled signal $x(t)$ from samples $x[n] = x(nT_s)$. A device termed a *first-order hold* linearly interpolates between the samples $x[n]$ and thus produces a smoother approximation to $x(t)$. The output of the first-order hold may be described as

$$x_1(t) = \sum_{n=-\infty}^{\infty} x[n]h_1(t - nT_s),$$

where $h_1(t)$ is the triangular pulse shown in Fig. P4.33(a). The relationship between $x[n]$ and $x_1(t)$ is depicted in Fig. P4.33(b).

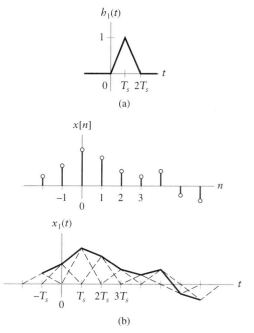

FIGURE P4.33

(a) Identify the distortions introduced by the first-order hold, and compare them with those introduced by the zero-order hold. [*Hint:* $h_1(t) = h_o(t) * h_o(t)$.]

(b) Consider a reconstruction system consisting of a first-order hold followed by an anti-imaging filter with frequency response $H_c(j\omega)$. Find $H_c(j\omega)$ so that perfect reconstruction is obtained.

(c) Determine the constraints on $|H_c(j\omega)|$ so that the overall magnitude response of this reconstruction system is between 0.99 and 1.01 in the signal passband and is less than 10^{-4} on all bands containing the images of the signal spectrum for the following values:

(i) $T_s = 0.05$

(ii) $T_s = 0.02$

Assume that $x(t)$ is band limited to 12π; that is, $X(j\omega) = 0$ for $|\omega| > 12\pi$.

4.34 Determine the maximum factor q by which a discrete-time signal $x[n]$ with DTFT $X(e^{j\Omega})$ depicted in Fig. P4.34 can be decimated without aliasing. Sketch the DTFT of the sequence that results when $x[n]$ is decimated by the factor q.

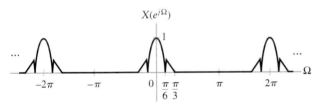

$X(e^{j\Omega})$

FIGURE P4.34

4.35 A discrete-time system for processing continuous-time signals is shown in Fig. P4.35. Sketch the magnitude of the frequency response of an equivalent continuous-time system for the following cases:

(a) $\Omega_1 = \frac{\pi}{4}$, $W_c = 20\pi$

(b) $\Omega_1 = \frac{3\pi}{4}$, $W_c = 20\pi$

(c) $\Omega_1 = \frac{\pi}{4}$, $W_c = 2\pi$

4.36 Let $X(e^{j\Omega}) = \dfrac{\sin\left(\frac{11\Omega}{2}\right)}{\sin\left(\frac{\Omega}{2}\right)}$ and define $\widetilde{X}[k] = X(e^{jk\Omega_o})$.

Find and sketch $\widetilde{x}[n]$, where $\widetilde{x}[n] \xleftrightarrow{\ DTFS;\ \Omega_o\ } \widetilde{X}[k]$

for the following values of Ω_o:

(a) $\Omega_o = \frac{2\pi}{15}$

(b) $\Omega_o = \frac{\pi}{10}$

(c) $\Omega_o = \frac{\pi}{3}$

4.37 Let $X(j\omega) = \frac{\sin(2\omega)}{\omega}$ and define $\widetilde{X}[k] = X(jk\omega_o)$.

Find and sketch $\widetilde{x}(t)$, where $\widetilde{x}(t) \xleftrightarrow{\ FS;\ \omega_o\ } \widetilde{X}[k]$

for the following values of ω_o:

(a) $\omega_o = \frac{\pi}{8}$

(b) $\omega_o = \frac{\pi}{4}$

(c) $\omega_o = \frac{\pi}{2}$

4.38 A signal $x(t)$ is sampled at intervals of $T_s = 0.01$ s. One hundred samples are collected, and a 200-point DTFS is taken in an attempt to approximate $X(j\omega)$. Assume that $|X(j\omega)| \approx 0$ for $|\omega| > 120\pi$ rad/s. Determine the frequency range $-\omega_a < \omega < \omega_a$ over which the DTFS offers a reasonable approximation to $X(j\omega)$, the effective resolution ω_r of this approximation and the frequency interval $\Delta\omega$ between each DTFS coefficient.

4.39 A signal $x(t)$ is sampled at intervals of $T_s = 0.1$ s. Assume that $|X(j\omega)| \approx 0$ for $|\omega| > 12\pi$ rad/s. Determine the frequency range $-\omega_a < \omega < \omega_a$ over which the DTFS offers a reasonable approximation to $X(j\omega)$, the minimum number of samples required to obtain an effective resolution $\omega_r = 0.01\pi$ rad/s, and the length of the DTFS required so that the frequency interval between DTFS coefficients is $\Delta\omega = 0.001\pi$ rad/s.

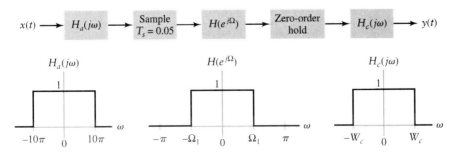

FIGURE P4.35

4.40 Let $x(t) = a\sin(\omega_o t)$ be sampled at intervals of $T_s = 0.1$ s. Assume that 100 samples of $x(t)$, $x[n] = x(nT_s)$, $n = 0, 1, \ldots 99$, are available. We use the DTFS of $x[n]$ to approximate the FT of $x(t)$ and wish to determine a from the DTFS coefficient of largest magnitude. The samples $x[n]$ are zero padded to length N before taking the DTFS. Determine the minimum value of N for the following values of ω_o:

(a) $\omega_o = 3.2\pi$

(b) $\omega_o = 3.1\pi$

(c) $\omega_o = 3.15\pi$

Determine which DTFS coefficient has the largest magnitude in each case.

ADVANCED PROBLEMS

4.41 A continuous-time signal lies in the frequency band $|\omega| < 5\pi$. The signal is contaminated by a large sinusoidal signal of frequency 120π. The contaminated signal is sampled at a sampling rate of $\omega_s = 13\pi$.

(a) After sampling, at what frequency does the sinusoidal intefering signal appear?

(b) The contaminated signal is passed through an anti-aliasing filter consisting of the RC circuit depicted in Fig. P4.41. Find the value of the time constant RC required so that the contaminating sinusoid is attenuated by a factor of 1000 prior to sampling.

(c) Sketch the magnitude response in dB that the anti-aliasing filter presents to the signal of interest for the value of RC identified in (b).

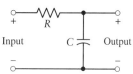

FIGURE P4.41

4.42 This problem derives the frequency-domain relationship for subsampling given in Eq. (4.27). Use Eq. (4.17) to represent $x[n]$ as the impulse-sampled continuous-time signal with sampling interval T_s, and thus write

$$x_\delta(t) = \sum_{n=-\infty}^{\infty} x[n]\delta(t - nT_s).$$

Suppose $x[n]$ are the samples of a continuous-time signal $x(t)$ obtained at integer multiples of T_s. That is, $x[n] = x(nT_s)$. Let $x(t) \overset{FT}{\longleftrightarrow} X(j\omega)$. Define the subsampled signal $y[n] = x[qn]$ so that $y[n] = x(nqT_s)$ is also expressed as samples of $x(t)$.

(a) Apply Eq. (4.23) to express $X_\delta(j\omega)$ as a function of $X(j\omega)$. Show that

$$Y_\delta(j\omega) = \frac{1}{qT_s} \sum_{k=-\infty}^{\infty} X\left(j\left(\omega - \frac{k}{q}\omega_s\right)\right).$$

(b) The goal is to express $Y_\delta(j\omega)$ as a function of $X_\delta(j\omega)$ so that $Y(e^{j\Omega})$ can be expressed in terms

of $X(e^{j\Omega})$. To that end, write k/q in $Y_\delta(j\omega)$ as the proper fraction

$$\frac{k}{q} = l + \frac{m}{q},$$

where l is the integer portion of k/q and m is the remainder. Show that we may thus rewrite $Y_\delta(j\omega)$ as

$$Y_\delta(j\omega) = \frac{1}{q}\sum_{m=0}^{q-1}\left\{\frac{1}{T_s}\sum_{l=-\infty}^{\infty} X\left(j\left(\omega - l\omega_s - \frac{m}{q}\omega_s\right)\right)\right\}.$$

Next, show that

$$Y_\delta(j\omega) = \frac{1}{q}\sum_{m=0}^{q-1} X_\delta\left(j\left(\omega - \frac{m}{q}\omega_s\right)\right).$$

(c) Now we convert from the FT representation back to the DTFT in order to express $Y(e^{j\Omega})$ as a function of $X(e^{j\Omega})$. The sampling interval associated with $Y_\delta(j\omega)$ is qT_s. Using the relationship $\Omega = \omega q T_s$ in

$$Y(e^{j\Omega}) = Y_\delta(j\omega)\big|_{\omega = \frac{\Omega}{qT_s}},$$

show that

$$Y(e^{j\Omega}) = \frac{1}{q}\sum_{m=0}^{q-1} X_\delta\left(\frac{j}{T_s}\left(\frac{\Omega}{q} - \frac{m}{q}2\pi\right)\right).$$

(d) Last, use $X(e^{j\Omega}) = X_\delta(j\frac{\Omega}{T_s})$ to obtain

$$Y(e^{j\Omega}) = \frac{1}{q}\sum_{m=0}^{q-1} X\left(e^{j\frac{1}{q}(\Omega - m2\pi)}\right).$$

4.43 A band-limited signal $x(t)$ satisfies $|X(j\omega)| = 0$ for $|\omega| < \omega_1$ and $|\omega| > \omega_2$. Assume that $\omega_1 > \omega_2 - \omega_1$. In this case, we can sample $x(t)$ at a rate less than that indicated by the sampling interval and still perform perfect reconstruction by using a band-pass reconstruction filter $H_r(j\omega)$. Let $x[n] = x(nT_s)$. Determine the maximum sampling interval T_s such that $x(t)$ can be perfectly reconstructed from $x[n]$. Sketch the frequency response of the reconstruction filter required for this case.

4.44 Suppose a periodic signal $x(t)$ has FS coefficients

$$X[k] = \begin{cases} \left(\frac{3}{4}\right)^k, & |k| \le 4 \\ 0, & \text{otherwise} \end{cases}.$$

The period of this signal is $T = 1$.

(a) Determine the minimum sampling interval for the signal that will prevent aliasing.

(b) The constraints of the sampling theorem can be relaxed somewhat in the case of periodic signals if we allow the reconstructed signal to be a time-scaled version of the original. Suppose we choose a sampling interval $T_s = \frac{20}{19}$ and use a reconstruction filter

$$H_r(j\omega) = \begin{cases} 1, & |\omega| < \pi \\ 0, & \text{otherwise} \end{cases}.$$

Show that the reconstructed signal is a time-scaled version of $x(t)$, and identify the scaling factor.

(c) Find the constraints on the sampling interval T_s so that the use of $H_r(j\omega)$ in (b) results in the reconstruction filter being a time-scaled version of $x(t)$, and determine the relationship between the scaling factor and T_s.

4.45 In this problem, we reconstruct a signal $x(t)$ from its samples $x[n] = x(nT_s)$, using pulses of width less than T_s followed by an anti-imaging filter with frequency response $H_c(j\omega)$. Specifically, we apply

$$x_p(t) = \sum_{n=-\infty}^{\infty} x[n]h_p(t - nT_s)$$

to the anti-imaging filter, where $h_p(t)$ is a pulse of width T_o, as depicted in Fig. P4.45 (a). An example of $x_p(t)$ is depicted in Fig. P4.45 (b). Determine the constraints on $|H_c(j\omega)|$ so that the overall magnitude response of this reconstruction system is between 0.99 and 1.01 in the signal passband and less than 10^{-4} on the band containing the images of the signal spectrum for the following values, with $x(t)$ band limited to 10π—that is, $X(j\omega) = 0$ for $\omega > 10\pi$:

(a) $T_s = 0.08$, $\quad T_o = 0.04$

(b) $T_s = 0.08$, $\quad T_o = 0.02$

(c) $T_s = 0.04$, $\quad T_o = 0.02$

(d) $T_s = 0.04$, $\quad T_o = 0.01$

4.46 A nonideal sampling operation obtains $x[n]$ from $x(t)$ as

$$x[n] = \int_{(n-1)T_s}^{nT_s} x(t)\, dt.$$

(a) Show that this equation can be written as ideal sampling of a filtered signal $y(t) = x(t) * h(t)$ [i.e., $x[n] = y(nT_s)$], and find $h(t)$.

(b) Express the FT of $x[n]$ in terms of $X(j\omega)$, $H(j\omega)$, and T_s.

(c) Assume that $x(t)$ is band limited to the frequency range $|\omega| < 3\pi/(4T_s)$. Determine the frequency response of a discrete-time system that will correct the distortion in $x[n]$ introduced by nonideal sampling.

4.47 The system depicted in Fig. P4.47(a) converts a continuous-time signal $x(t)$ to a discrete-time signal $y[n]$. We have

$$H(e^{j\Omega}) = \begin{cases} 1, & |\Omega| < \frac{\pi}{4} \\ 0, & \text{otherwise} \end{cases}.$$

Find the sampling frequency $\omega_s = 2\pi/T_s$ and the constraints on the anti-aliasing filter frequency response $H_a(j\omega)$ so that an input signal with FT $X(j\omega)$ shown in Fig. P4.47(b) results in the output signal with DTFT $Y(e^{j\Omega})$.

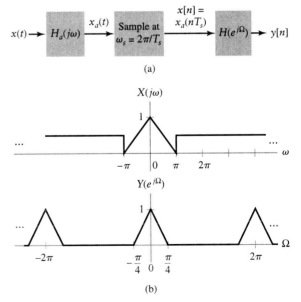

(a)

(b)

FIGURE P4.47

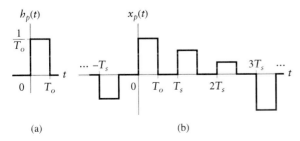

(a)

(b)

FIGURE P4.45

4.48 The discrete-time signal $x[n]$ with DTFT $X(e^{j\Omega})$ shown in Fig. P4.48(a) is decimated by first passing $x[n]$ through the filter with frequency response $H(e^{j\Omega})$ shown in Fig. P4.48(b) and then subsampling by the factor q. For the following values of q and W, determine the minimum value of Ω_p and maximum value of Ω_s such that the subsampling operation does not change the shape of the portion of $X(e^{j\Omega})$ on $|\Omega| < W$:

(a) $q = 2, W = \frac{\pi}{3}$
(b) $q = 2, W = \frac{\pi}{4}$
(c) $q = 3, W = \frac{\pi}{4}$

In each case, sketch the DTFT of the subsampled signal.

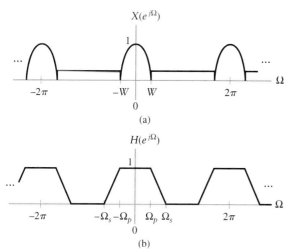

FIGURE P4.48

4.49 A signal $x[n]$ is interpolated by the factor q by first inserting $q - 1$ zeros between each sample and then passing the zero-stuffed sequence through a filter with frequency response $H(e^{j\Omega})$ depicted in Fig. P4.48(b). The DTFT of $x[n]$ is depicted in Fig. P4.49. Determine the minimum value of Ω_p and maximum value of Ω_s such that ideal interpolation is obtained for the following cases:

(a) $q = 2, W = \frac{\pi}{2}$
(b) $q = 2, W = \frac{3\pi}{4}$
(c) $q = 3, W = \frac{3\pi}{4}$

In each case, sketch the DTFT of the interpolated signal.

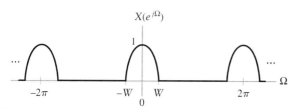

FIGURE P4.49

4.50 Consider interpolating a signal $x[n]$ by repeating each value q times, as depicted in Fig. P4.50. That is, we define $x_o[n] = x\left[\text{floor}\left(\frac{n}{q}\right)\right]$, where floor$(z)$ is the greatest integer less than or equal to z. Let $x_z[n]$ be derived from $x[n]$ by inserting $q - 1$ zeros between each value of $x[n]$; that is,

$$x_z[n] = \begin{cases} x\left[\frac{n}{q}\right], & \frac{n}{q} \text{ integer} \\ 0, & \text{otherwise} \end{cases}.$$

We may now write $x_o[n] = x_z[n] * h_o[n]$, where

$$h_o[n] = \begin{cases} 1, & 0 \leq n \leq q - 1 \\ 0, & \text{otherwise} \end{cases}.$$

Note that this is the discrete-time analog of the zero-order hold. The interpolation process is completed by passing $x_o[n]$ through a filter with frequency response $H(e^{j\Omega})$.

(a) Express $X_o(e^{j\Omega})$ in terms of $X(e^{j\Omega})$ and $H_o(e^{j\Omega})$. Sketch $\left|X_o(e^{j\Omega})\right|$ if $x[n] = \dfrac{\sin\left(\frac{3\pi}{4}n\right)}{\pi n}$.

(b) Assume that $X(e^{j\Omega})$ is as shown in Fig. P4.49. Specify the constraints on $H(e^{j\Omega})$ so that ideal interpolation is obtained for the following cases:

(i) $q = 2, W = \frac{3\pi}{4}$
(ii) $q = 4, W = \frac{3\pi}{4}$

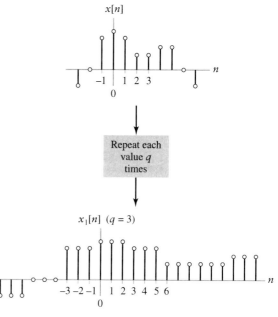

FIGURE P4.50

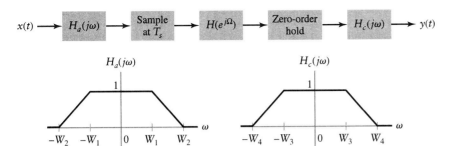

FIGURE P4.51

4.51 The system shown in Fig. P4.51 is used to implement a band-pass filter. The frequency response of discrete-time filter is

$$H(e^{j\Omega}) = \begin{cases} 1, & \Omega_a \le |\Omega| \le \Omega_b \\ 0, & \text{otherwise} \end{cases}$$

on $-\pi < \Omega \le \pi$. Find the sampling interval T_s, Ω_a, Ω_b, W_1, W_2, W_3, and W_4 so that the equivalent continuous-time frequency response $G(j\omega)$ satisfies

$$0.9 < |G(j\omega)| < 1.1, \quad \text{for} \quad 100\pi < \omega < 200\pi$$
$$G(j\omega) = 0 \quad \text{elsewhere}$$

In solving this problem, choose W_1 and W_3 as small as possible, and choose T_s, W_2, and W_4 as large as possible.

4.52 A time-domain interpretation of the interpolation procedure described in Fig. 4.50(a) is derived in this problem. Let $h_i[n] \overset{FT}{\longleftrightarrow} H_i(e^{j\Omega})$ be an ideal low-pass filter with transition band of zero width. That is,

$$H_i(e^{j\Omega}) = \begin{cases} q, & |\Omega| < \frac{\pi}{q} \\ 0, & \frac{\pi}{q} < |\Omega| < \pi \end{cases}.$$

(a) Substitute for $h_i[n]$ in the convolution sum:

$$x_i[n] = \sum_{k=-\infty}^{\infty} x_z[k] * h_i[n-k].$$

(b) The zero-insertion procedure implies that $x_z[k] = 0$, unless $k = qm$, where m is integer. Rewrite $x_i[n]$, using only the nonzero terms in the sum, as a sum over m, and substitute $x[m] = x_z[qm]$ to obtain the following expression for ideal discrete-time interpolation:

$$x_i[n] = \sum_{m=-\infty}^{\infty} x[m] \frac{q \sin\left(\frac{\pi}{q}(n-qm)\right)}{\pi(n-qm)}.$$

4.53 The continuous-time representation of a periodic discrete-time signal $x[n] \overset{DTFS; \frac{2\pi}{N}}{\longleftrightarrow} X[k]$ is periodic and thus has an FS representation. This FS representation is a function of the DTFS coefficients $X[k]$, as we show in this problem. The result establishes the relationship between the FS and DTFS representations. Let $x[n]$ have period N and let $x_\delta(t) = \sum_{n=-\infty}^{\infty} x[n]\delta(t - nT_s)$.

(a) Show that $x_\delta(t)$ is periodic and find the period T.

(b) Begin with the definition of the FS coefficients:

$$X_\delta[k] = \frac{1}{T} \int_0^T x_\delta(t) e^{-jk\omega_o t} \, dt.$$

Substitute for T, ω_o, and one period of $x_\delta(t)$ to show that

$$X_\delta[k] = \frac{1}{T_s} X[k].$$

4.54 The fast Fourier transform (FFT) algorithm for evaluating the DTFS may be used to develop a computationally efficient algorithm for determining the output of a discrete-time system with a finite-length impulse reponse. Instead of directly computing the convolution sum, the DTFS is used to compute the output by performing multiplication in the frequency domain. This requires that we develop a correspondence between the periodic convolution implemented by the DTFS and the linear convolution associated with the system output. That is the goal of this problem. Let $h[n]$ be an impulse response of length M so that $h[n] = 0$ for $n < 0$, $n \ge M$. The system output $y[n]$ is related to the input via the convolution sum

$$y[n] = \sum_{k=0}^{M-1} h[k] x[n-k].$$

4.54 (*Continued*)

(a) Consider the N-point periodic convolution of $h[n]$ with N consecutive values of the input sequence $x[n]$, and assume that $N > M$. Let $\tilde{x}[n]$ and $\tilde{h}[n]$ be N-periodic versions of $x[n]$ and $h[n]$, respectively:

$$\tilde{x}[n] = x[n], \quad \text{for } 0 \le n \le N - 1;$$
$$\tilde{x}[n + mN] = \tilde{x}[n], \quad \text{for all integer } m, 0 \le n \le N - 1;$$

$$\tilde{h}[n] = h[n], \quad \text{for } 0 \le n \le N - 1;$$
$$\tilde{h}[n + mN] = \tilde{h}[n], \quad \text{for all integer } m, 0 \le n \le N - 1.$$

The periodic convolution between $\tilde{h}[n]$ and $\tilde{x}[n]$ is

$$\tilde{y}[n] = \sum_{k=0}^{N-1} \tilde{h}[k]\tilde{x}[n - k].$$

Use the relationship between $h[n], x[n]$ and $\tilde{h}[n], \tilde{x}[n]$ to prove that $\tilde{y}[n] = y[n]$, $M - 1 \le n \le N - 1$. That is, the periodic convolution is equal to the linear convolution at $L = N - M + 1$ values of n.

(b) Show that we may obtain values of $y[n]$ other than those on the interval $M - 1 \le n \le N - 1$ by shifting $x[n]$ prior to defining $\tilde{x}[n]$. That is, show that if

$$\tilde{x}_p[n] = x[n + pL], \quad 0 \le n \le N - 1,$$
$$\tilde{x}_p[n + mN] = \tilde{x}_p[n],$$
$$\text{for all integer } m, 0 \le n \le N - 1,$$

and

$$\tilde{y}_p[n] = \tilde{h}[n] \circledast \tilde{x}_p[n],$$

then

$$\tilde{y}_p[n] = y[n + pL], \quad M - 1 \le n \le N - 1.$$

This implies that the last L values in one period of $\tilde{y}_p[n]$ correspond to $y[n]$ for $M - 1 + pL \le n \le N - 1 + pL$. Each time we increment p, the N-point periodic convolution gives us L new values of the linear convolution. This result is the basis for the so-called *overlap-and-save method* for evaluating a linear convolution with the DTFS.

COMPUTER EXPERIMENTS

4.55 Repeat Example 4.7, using zero padding and the MATLAB commands `fft` and `fftshift` to sample and plot $Y(e^{j\Omega})$ at 512 points on $-\pi < \Omega \le \pi$ for each case.

4.56 The rectangular window is defined as

$$w_r[n] = \begin{cases} 1, & 0 \le n \le M \\ 0, & \text{otherwise} \end{cases}.$$

We may truncate the duration of a signal to the interval $0 \le n \le M$ by multiplying the signal with $w[n]$. In the frequency domain, we convolve the DTFT of the signal with

$$W_r(e^{j\Omega}) = e^{-j\frac{M}{2}\Omega} \frac{\sin\left(\frac{\Omega(M + 1)}{2}\right)}{\sin\left(\frac{\Omega}{2}\right)}.$$

The effect of this convolution is to smear detail and introduce ripples in the vicinity of discontinuities. The smearing is proportional to the mainlobe width, while the ripple is proportional to the size of the sidelobes. A variety of alternative windows are used in practice to reduce sidelobe height in return for increased mainlobe width. In this problem, we evaluate the effect of windowing time-domain signals on their DTFT. The role of windowing in filter design is explored in Chapter 8.

The Hanning window is defined as

$$w_h[n] = \begin{cases} 0.5 - 0.5 \cos\left(\frac{2\pi n}{M}\right), & 0 \le n \le M \\ 0, & \text{otherwise} \end{cases}.$$

(a) Assume that $M = 50$ and use the MATLAB command `fft` to evaluate the magnitude spectrum of the rectangular window in dB at intervals of $\frac{\pi}{50}$, $\frac{\pi}{100}$, and $\frac{\pi}{200}$.

(b) Assume that $M = 50$ and use the MATLAB command `fft` to evaluate the magnitude spectrum of the Hanning window in dB at intervals of $\frac{\pi}{50}$, $\frac{\pi}{100}$, and $\frac{\pi}{200}$.

(c) Use the results from (a) and (b) to evaluate the mainlobe width and peak sidelobe height in dB for each window.

(d) Let $y_r[n] = x[n]w_r[n]$ and $y_h[n] = x[n]w_h[n]$, where $x[n] = \cos\left(\frac{26\pi}{100}n\right) + \cos\left(\frac{29\pi}{100}n\right)$ and $M = 50$. Use the the MATLAB command `fft` to evaluate $|Y_r(e^{j\Omega})|$ in dB and $|Y_h(e^{j\Omega})|$ in dB at intervals of $\frac{\pi}{200}$. Does the choice of window affect whether you can identify the presence of two sinusoids? Why?

(e) Let $y_r[n] = x[n]w_r[n]$ and $y_h[n] = x[n]w_h[n]$, where $x[n] = \cos\left(\frac{26\pi}{100}n\right) + 0.02 \cos\left(\frac{51\pi}{100}n\right)$ and $M = 50$. Use the the MATLAB command `fft` to evaluate $|Y_r(e^{j\Omega})|$ in dB and $|Y_b(e^{j\Omega})|$ in dB at intervals of $\frac{\pi}{200}$. Does the choice of window affect whether you can identify the presence of two sinusoids? Why?

4.57 Let the discrete-time signal

$$x[n] = \begin{cases} e^{-\frac{(0.1n)^2}{2}}, & |n| \leq 50 \\ 0, & \text{otherwise} \end{cases}.$$

Use the MATLAB commands `fft` and `fft-shift` to numerically evaluate and plot the DTFT of $x[n]$ and the following subsampled signals at 500 values of Ω on the interval $-\pi < \Omega \leq \pi$:

(a) $y[n] = x[2n]$

(b) $z[n] = x[4n]$

4.58 Repeat Problem 4.57, assuming that

$$x[n] = \begin{cases} \cos\left(\frac{\pi}{2}n\right)e^{-\frac{(0.1n)^2}{2}}, & |n| \leq 50 \\ 0, & \text{otherwise} \end{cases}.$$

4.59 A signal is defined as

$$x(t) = \cos\left(\frac{3\pi}{2}t\right)e^{-\frac{t^2}{2}}.$$

(a) Evaluate the FT $X(j\omega)$, and show that $|X(j\omega)| \approx 0$ for $|\omega| > 3\pi$.

In parts (b)–(d), we compare $X(j\omega)$ with the FT of the sampled signal, $x[n] = x(nT_s)$, for several sampling intervals. Let $x[n] \xleftrightarrow{\;FT\;} X_\delta(j\omega)$ be the FT of the sampled version of $x(t)$. Use MATLAB to numerically determine $X_\delta(j\omega)$ by evaluating

$$X_\delta(j\omega) = \sum_{n=-25}^{25} x[n]e^{-j\omega T_s}$$

at 500 values of ω on the interval $-3\pi < \omega < 3\pi$. In each case, compare $X(j\omega)$ and $X_\delta(j\omega)$ and explain any differences.

(b) $T_s = \frac{1}{3}$

(c) $T_s = \frac{2}{5}$

(d) $T_s = \frac{1}{2}$

4.60 Use the MATLAB command `fft` to repeat Example 4.14.

4.61 Use the MATLAB command `fft` to repeat Example 4.15.

4.62 Use the MATLAB command `fft` to repeat Example 4.16. Also, depict the DTFS approximation and the underlying DTFT for $M = 2001$ and $M = 2005$.

4.63 Consider the sum of sinusoids,

$$x(t) = \cos(2\pi t) + 2\cos(2\pi(0.8)t)$$
$$+ \frac{1}{2}\cos(2\pi(1.1)t).$$

Assume that the frequency band of interest is

$$-5\pi < \omega < 5\pi.$$

(a) Determine the sampling interval T_s so that the DTFS approximation to the FT of $x(t)$ spans the desired frequency band.

(b) Determine the minimum number of samples M_o so that the DTFS approximation consists of discrete-valued impulses located at the frequency corresponding to each sinusoid.

(c) Use MATLAB to plot $\frac{1}{M}|Y_\delta(j\omega)|$ and $|Y[k]|$ for the value of T_s chosen in part (a) and for $M = M_o$.

(d) Repeat part (c), using $M = M_o + 5$ and $M = M_o + 8$.

4.64 We desire to use the DTFS to approximate the FT of a continuous-time signal $x(t)$ on the band $-\omega_a < \omega < \omega_a$ with resolution ω_r and a maximum sampling interval in frequency of $\Delta\omega$. Find the sampling interval T_s, the number of samples M, and the DTFS length N. You may assume that the signal is effectively band limited to a frequency ω_m for which $|X(j\omega_a)| \geq 10|X(j\omega)|$, $\omega > \omega_m$. Plot the FT and the DTFS approximation for each of the following cases, using the MATLAB command `fft`:

(a) $x(t) = \begin{cases} 1, & |t| < 1 \\ 0, & \text{otherwise} \end{cases}$, $\omega_a = \frac{3\pi}{2}$, $\omega_r = \frac{3\pi}{4}$, and $\Delta\omega = \frac{\pi}{8}$

(b) $x(t) = \frac{1}{2\pi}e^{-\frac{t^2}{2}}$, $\omega_a = 3$, $\omega_r = \frac{1}{2}$, and $\Delta\omega = \frac{1}{8}$

(c) $x(t) = \cos(20\pi t) + \cos(21\pi t)$, $\omega_a = 40\pi$, $\omega_r = \frac{\pi}{3}$, and $\Delta\omega = \frac{\pi}{10}$

(d) Repeat case (c) using $\omega_r = \frac{\pi}{10}$.

[*Hint:* Be sure to sample the pulses in (a) and (b) symmetrically about $t = 0$.]

4.65 The overlap-and-save method for linear filtering is discussed in Problem 4.54. Write a MATLAB m-file that implements the method, using `fft` to evaluate the convolution $y[n] = h[n] * x[n]$ on $0 \leq n < L$ for the following signals:

(a) $h[n] = \frac{1}{5}(u[n] - u[n-5])$,
$x[n] = \cos\left(\frac{\pi}{6}n\right)$, $L = 30$

(b) $h[n] = \frac{1}{5}(u[n] - u[n-5])$,
$x[n] = \left(\frac{1}{2}\right)^n u[n]$, $L = 20$

4.66 Plot the ratio of the number of multiplications in the direct method for computing the DTFS coefficients to that of the FFT approach when $N = 2^p$ for $p = 2, 3, 4, \ldots, 16$.

4.67 In this experiment, we investigate the evaluation of the time–bandwidth product with the DTFS. Let $x(t) \xleftrightarrow{\text{FT}} X(j\omega)$.

(a) Use the Riemann sum approximation to an integral,

$$\int_a^b f(u)\,du \approx \sum_{m=m_a}^{m_b} f(m\Delta u)\,\Delta u,$$

to show that

$$T_d = \left[\frac{\int_{-\infty}^{\infty} t^2 |x(t)|^2\,dt}{\int_{-\infty}^{\infty} |x(t)|^2\,dt}\right]^{\frac{1}{2}}$$

$$\approx T_s \left[\frac{\sum_{n=-M}^{M} n^2 |x[n]|^2}{\sum_{n=-M}^{M} |x[n]|^2}\right]^{\frac{1}{2}},$$

provided that $x[n] = x(nT_s)$ represents the samples of $x(t)$ and $x(nT_s) \approx 0$ for $|n| > M$.

(b) Use the DTFS approximation to the FT and the Riemann sum approximation to an integral to show that

$$B_w = \left[\frac{\int_{-\infty}^{\infty} \omega^2 |X(j\omega)|^2\,d\omega}{\int_{-\infty}^{\infty} |X(j\omega)|^2\,d\omega}\right]^{\frac{1}{2}}$$

$$\approx \frac{\omega_s}{2M + 1}\left[\frac{\sum_{k=-M}^{M} |k|^2 |X[k]|^2}{\sum_{k=-M}^{M} |X[k]|^2}\right]^{\frac{1}{2}},$$

where $x[n] \xleftrightarrow{\text{DTFS}; \frac{2\pi}{2M+1}} X[k]$, $\omega_s = 2\pi/T_s$ is the sampling frequency, and $X\left(jk\dfrac{\omega_s}{2M+1}\right) \approx 0$ for $|k| > M$.

(c) Use the result from (a) and (b) and Eq. (3.65) to show that the time–bandwidth product computed by using the DTFS approximation satisfies the relationship

$$\left[\frac{\sum_{n=-M}^{M} n^2 |x[n]|^2}{\sum_{n=-M}^{M} |x[n]|^2}\right]^{\frac{1}{2}}\left[\frac{\sum_{k=-M}^{M} |k|^2 |X[k]|^2}{\sum_{k=-M}^{M} |X[k]|^2}\right]^{\frac{1}{2}}$$

$$\geq \frac{2M + 1}{4\pi}.$$

(d) Repeat Computer Experiment 3.115 to demonstrate that the bound in (c) is satisfied and that Gaussian pulses satisfy the bound with equality.

5 Application to Communication Systems

5.1 Introduction

The purpose of a communication system is to transport a signal representing a message (generated by a source of information) over a channel and deliver a reliable estimate of that signal to a user. For example, the message signal may be a speech signal, and the channel may be a cellular telephone channel or a satellite channel. As mentioned in Chapter 1, modulation is basic to the operation of a communication system. Modulation provides the means for (1) shifting the range of frequencies contained in the message signal into another frequency range suitable for transmission over the channel and (2) performing a corresponding shift back to the original frequency range after reception of the signal. Formally, *modulation* is defined as *the process by which some characteristic of a carrier wave is varied in accordance with the message signal*. The message signal is referred to as the *modulating wave*, and the result of the modulation process is referred to as the *modulated wave*. In the receiver, *demodulation* is used to recover the message signal from the modulated wave. Demodulation is the inverse of the modulation process.

In this chapter, we present an introductory treatment of modulation from a system-theoretic viewpoint, building on Fourier analysis as discussed in the previous two chapters. We begin the discussion with a description of the basic types of modulation, followed by the practical benefits derived from their use. This sets the stage for a discussion of so-called *amplitude modulation*, which is widely used in practice for analog communications by virtue of its simplicity. One common application of amplitude modulation is in radio broadcasting. We then discuss some important variants of amplitude modulation. The counterpart of amplitude modulation that is used in digital communications is known as *pulse-amplitude modulation*, which is discussed in the latter part of the chapter. In reality, pulse-amplitude modulation is another manifestation of the sampling process that we studied in Chapter 4.

5.2 Types of Modulation

The specific type of modulation employed in a communication system is determined by the form of carrier wave used to perform the modulation. The two most commonly used forms of carrier are a *sinusoidal wave* and a *periodic pulse train*. Correspondingly, we may identify two main classes of modulation: *continuous-wave (CW) modulation* and *pulse modulation*.

1. *Continuous-wave (CW) modulation.*

Consider the sinusoidal carrier wave

$$c(t) = A_c \cos(\phi(t)), \tag{5.1}$$

which is uniquely defined by the carrier amplitude A_c and angle $\phi(t)$. Depending on which of these parameters is chosen for modulation, we may identify two subclasses of CW modulation:

- ▶ *Amplitude modulation*, in which the carrier amplitude is varied with the message signal, and
- ▶ *Angle modulation*, in which the angle of the carrier is varied with the message signal.

Figure 5.1 gives examples of amplitude-modulated and angle-modulated signals for the case of sinusoidal modulation.

Amplitude modulation itself can be implemented in several different forms. For a given message signal, the frequency content of the modulated wave depends on the form of amplitude modulation used. Specifically, we have the following four types:

- ▶ Full amplitude modulation (double sideband-transmitted carrier)
- ▶ Double sideband-suppressed carrier modulation
- ▶ Single sideband modulation
- ▶ Vestigial sideband modulation

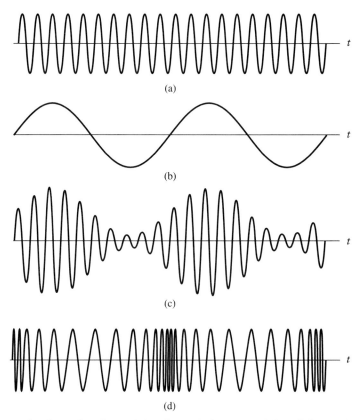

FIGURE 5.1 Amplitude- and angle-modulated signals for sinusoidal modulation. (a) Carrier wave. (b) Sinusoidal modulating signal. (c) Amplitude-modulated signal. (d) Angle-modulated signal.

The last three types of amplitude modulation are examples of *linear modulation*, in the sense that if the amplitude of the message signal is scaled by a certain factor, then the amplitude of the modulated wave is scaled by exactly the same factor. In this strict sense, full amplitude modulation fails to meet the definition of linear modulation with respect to the message signal for reasons that will become apparent later. Nevertheless, the departure from linearity in the case of full amplitude modulation is of a rather mild sort, such that many of the mathematical procedures applicable to the analysis of linear modulation may be retained. Most importantly from our present perspective, all four different forms of amplitude modulation mentioned here lend themselves to mathematical analysis using the tools presented in this book. Subsequent sections of the chapter develop the details of this analysis.

In contrast, angle modulation is a *nonlinear* modulation process. To describe it in a formal manner, we need to introduce the notion of *instantaneous radian frequency*, denoted by $\omega_i(t)$ and defined as the derivative of the angle $\phi(t)$ with respect to time t:

$$\omega_i(t) = \frac{d\phi(t)}{dt}. \tag{5.2}$$

Equivalently, we may write

$$\phi(t) = \int_0^t \omega_i(\tau)\,d\tau, \tag{5.3}$$

where it is assumed that the initial value

$$\phi(0) = \int_{-\infty}^0 \omega_i(\tau)\,d\tau$$

is zero.

Equation (5.2) includes the usual definition of radian frequency as a special case. Consider the ordinary form of a sinusoidal wave, written as

$$c(t) = A_c \cos(\omega_c t + \theta),$$

where A_c is the amplitude, ω_c is the radian frequency, and θ is the phase. For this simple case, the angle

$$\phi(t) = \omega_c t + \theta,$$

in which case the use of Eq. (5.2) yields the expected result,

$$\omega_i(t) = \omega_c \qquad \text{for all } t.$$

Returning to the general definition of Eq. (5.2), we find that when the instantaneous radian frequency $\omega_i(t)$ is varied in accordance with a message signal denoted by $m(t)$, we may write

$$\omega_i(t) = \omega_c + k_f m(t), \tag{5.4}$$

where k_f is the frequency sensitivity factor of the modulator. Hence, substituting Eq. (5.4) into Eq. (5.3), we get

$$\phi(t) = \omega_c t + k_f \int_0^t m(\tau)\,d\tau.$$

The resulting form of angle modulation is known as *frequency modulation* (FM) and is written as

$$s_{FM}(t) = A_c \cos\left(\omega_c t + k_f \int_0^t m(\tau)\, d\tau \right), \qquad (5.5)$$

where the carrier amplitude is maintained constant.

When the angle $\phi(t)$ is varied in accordance with the message signal $m(t)$, we may write

$$\phi(t) = \omega_c t + k_p m(t),$$

where k_p is the phase sensitivity factor of the modulator. This time we have a different form of angle modulation known as *phase modulation* (PM), defined by

$$s_{PM}(t) = A_c \cos(\omega_c t + k_p m(t)), \qquad (5.6)$$

where the carrier amplitude is again maintained constant.

Although Eqs. (5.5) and (5.6), for FM and PM signals, respectively, look different, they are in fact intimately related to each other. For the present, it suffices to say that both of them are nonlinear functions of the message signal $m(t)$, which makes their mathematical analysis more difficult than that of amplitude modulation. Since the primary emphasis in this book is on linear analysis of signals and systems, we will devote much of the discussion in this chapter to amplitude modulation and its variants.

2. *Pulse modulation.*
Consider next a carrier wave

$$c(t) = \sum_{n=-\infty}^{\infty} p(t - nT)$$

that consists of a periodic train of narrow pulses, where T is the period and $p(t)$ denotes a pulse of relatively short duration (compared with the period T) and centered on the origin. When some characteristic parameter of $p(t)$ is varied in accordance with the message signal, we have *pulse modulation*. Figure 5.2 gives an example of pulse amplitude modulation for the case of a sinusoidal modulating wave.

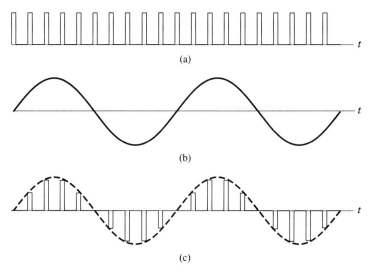

(a)

(b)

(c)

FIGURE 5.2   Pulse-amplitude modulation. (a) Train of rectangular pulses as the carrier wave. (b) Sinusoidal modulating signal. (c) Pulse-amplitude modulated signal.

Depending on how pulse modulation is actually accomplished, we may distinguish the following two subclasses:

▶ *Analog pulse modulation*, in which a characteristic parameter such as the amplitude, duration, or position of a pulse is varied continuously with the message signal. We thus speak of pulse-amplitude modulation, pulse-duration modulation, and pulse-position modulation as different realizations of analog pulse modulation. This type of pulse modulation may be viewed as the counterpart of CW modulation.

▶ *Digital pulse modulation*, in which the modulated signal is represented in coded form. This representation can be accomplished in a number of different ways. The standard method involves two operations. First, the amplitude of each modulated pulse is approximated by the nearest member of a set of discrete levels that occupies a compatible range of values. This operation is called *quantization*, and the device for performing it is called a *quantizer*. Second, the quantizer output is *coded* (e.g., in binary form). This particular form of digital pulse modulation is known as *pulse-code modulation* (PCM). Quantization is a nonlinear process that results in a loss of information, but the loss is under the designer's control in that it can be made as small as desired simply by using a large enough number of discrete (quantization) levels. In any event, PCM has no CW counterpart. As with angle modulation, a complete discussion of PCM is beyond the scope of this book. Insofar as pulse modulation is concerned, the primary emphasis in the current chapter is on pulse-amplitude modulation, which is a linear process.

5.3 *Benefits of Modulation*

The use of modulation is not confined exclusively to communication systems. Rather, modulation in one form or another is used in signal processing, radiotelemetry, radar, sonar, control systems, and general-purpose instruments such as spectrum analyzers and frequency synthesizers. However, it is in the study of communication systems that we find modulation playing a dominant role.

In the context of communication systems, we may identify four practical benefits that result from the use of modulation:

1. *Modulation is used to shift the spectral content of a message signal so that it lies inside the operating frequency band of a communication channel.*
Consider, for example, telephonic communication over a cellular radio channel. In such an application, the frequency components of a speech signal from about 300 to 3100 Hz are considered adequate for the purpose of audio communication. In North America, the band of frequencies assigned to cellular radio systems is 800–900 MHz. The subband 824–849 MHz is used to receive signals from mobile units, and the subband 869–894 MHz is used for transmitting signals to mobile units. For this form of telephonic communication to be feasible, we clearly need to do two things: shift the essential spectral content of a speech signal so that it lies inside the prescribed subband for transmission, and shift it back to its original frequency band on reception. The first of these two operations is one of modulation, and the second is one of demodulation.

As another example, consider the transmission of high-speed digital data over an optical fiber. When we speak of high-speed digital data, we mean a combination of digitized audio, digitized video, and computer data, whose overall rate is on the order of megabits per second and higher. Optical fibers have unique characteristics that make them highly attractive as a transmission mechanism, offering the following advantages:

- An enormous potential bandwidth, resulting from the use of optical carrier frequencies around 2×10^{14} Hz

- Low transmission losses, on the order of 0.2 dB/km and less

- Immunity to electromagnetic interference

- Small size and weight, characterized by a diameter no greater than that of a human hair

- Ruggedness and flexibility, exemplified by very high tensile strengths and the possibility of being bent or twisted without damage

Information-bearing signals are modulated onto a photonic energy source, either a light-emitting diode (LED) or a laser diode. A simple form of modulation involves switching back and forth between two different values of light intensity.

2. *Modulation provides a mechanism for putting the information content of a message signal into a form that may be less vulnerable to noise or interference.*
In a communication system, the received signal is ordinarily corrupted by noise generated at the front end of the receiver or by interference picked up in the course of transmission. Some specific forms of modulation, such as frequency modulation and pulse-code modulation, have the inherent ability to trade off increased transmission bandwidth for improved system performance in the presence of noise. We are careful here to say that this important property is not shared by all modulation techniques. In particular, those modulation techniques which vary the amplitude of a CW or pulsed carrier provide absolutely no protection against noise or interference in the received signal.

3. *Modulation permits the use of multiplexing.*
A communication channel (e.g., a telephone channel, mobile radio channel, or satellite communications channel) represents a major capital investment and must therefore be deployed in a cost-effective manner. *Multiplexing* is a signal-processing operation that makes this possible. In particular, it permits the simultaneous transmission of information-bearing signals from a number of independent sources over the channel and on to their respective destinations. It can take the form of frequency-division multiplexing for use with CW modulation techniques or time-division multiplexing for use with digital pulse modulation techniques.

4. *Modulation makes it possible for the physical size of the transmitting or receiving antenna to assume a practical value.*
In this context, we first note from electromagnetic theory that the physical aperture of an antenna is directly comparable to the wavelength of the radiated or incident electromagnetic signal. Alternatively, since wavelength and frequency are inversely related, we may say that the aperture of the antenna is inversely proportional to the operating frequency. Modulation elevates the spectral content of the modulating signal (e.g., a voice signal) by an amount equal to the carrier frequency. Hence, the larger the carrier frequency, the smaller will be the physical aperture of the transmitting antenna, as well as that of the receiving antenna, in radio-based forms of communication.

In this chapter, we will discuss the frequency-shifting and multiplexing aspects of modulation. However, a study of the issues relating to noise in modulation systems is beyond the scope of the text.

5.4 *Full Amplitude Modulation*

Consider a sinusoidal carrier wave

$$c(t) = A_c \cos(\omega_c t). \tag{5.7}$$

For convenience of presentation, we have assumed that the phase of the carrier wave is zero in Eq. (5.7). We are justified in making this assumption, as the primary emphasis here is on variations imposed on the carrier amplitude. Let $m(t)$ denote a message signal of interest. *Amplitude modulation (AM) is defined as a process in which the amplitude of the carrier is varied in proportion to a message signal $m(t)$*, according to the formula

$$s(t) = A_c[1 + k_a m(t)]\cos(\omega_c t), \tag{5.8}$$

where k_a is a constant called the *amplitude sensitivity* factor of the modulator. The modulated wave $s(t)$ so defined is said to be a "full" AM wave, for reasons explained later in the section. (See Subsection 5.4.5.) Note that the radian frequency ω_c of the carrier is maintained constant.

■ 5.4.1 PERCENTAGE OF MODULATION

The amplitude of the time function multiplying $\cos(\omega_c t)$ in Eq. (5.8) is called the *envelope* of the AM wave $s(t)$. Using $a(t)$ to denote this envelope, we may write

$$a(t) = A_c|1 + k_a m(t)|. \tag{5.9}$$

Two cases arise, depending on the magnitude of $k_a m(t)$, compared with unity:

1. *Undermodulation*, which is governed by the condition
$$|k_a m(t)| \leq 1 \qquad \text{for all } t.$$

   Under this condition, the term $1 + k_a m(t)$ is always nonnegative. We may therefore simplify the expression for the envelope of the AM wave by writing
$$a(t) = A_c[1 + k_a m(t)] \qquad \text{for all } t. \tag{5.10}$$

2. *Overmodulation*, which is governed by the weaker condition
$$|k_a m(t)| > 1 \qquad \text{for some } t.$$

   Under this second condition, we must use Eq. (5.9) in evaluating the envelope of the AM wave.

The maximum absolute value of $k_a m(t)$, multiplied by 100, is referred to as the *percentage modulation*. Accordingly, case 1 corresponds to a percentage modulation less than or equal to 100%, whereas case 2 corresponds to a percentage modulation in excess of 100%.

■ 5.4.2 GENERATION OF AM WAVE

Various schemes have been devised for the generation of an AM wave. Here we consider a simple circuit that follows from the defining equation (5.8). First, we rewrite this equation in the equivalent form

$$s(t) = k_a[m(t) + B]A_c\cos(\omega_c t). \tag{5.11}$$

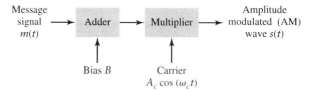

FIGURE 5.3 System involving an adder and multiplier for generating an AM wave.

The constant B, equal to $1/k_a$, represents a *bias* that is added to the message signal $m(t)$ before modulation. Equation (5.11) suggests the scheme described in the block diagram of Fig. 5.3 for generating an AM wave. Basically, it consists of two functional blocks:

▸ An adder, which adds the bias B to the incoming message signal $m(t)$.

▸ A multiplier, which multiplies the adder output $(m(t) + B)$ by the carrier wave $A_c \cos(\omega_c t)$, producing the AM wave $s(t)$.

The percentage modulation is controlled simply by adjusting the bias B.

▶ **Problem 5.1**   Assuming that $M_{\max}$ is the maximum absolute value of the message signal, what condition must the bias B satisfy to avoid overmodulation?

Answer: $B \geq M_{\max}$ ◀

■ 5.4.3 POSSIBLE WAVEFORMS OF AM WAVE

The waveforms of Fig. 5.4 illustrate the amplitude modulation process. Part (a) of the figure depicts the waveform of a message signal $m(t)$. Part (b) depicts an AM wave produced by this message signal for a value of k_a for which the percentage modulation is 66.7% (i.e., a case of undermodulation).

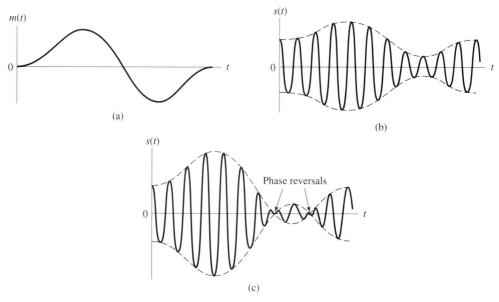

FIGURE 5.4 Amplitude modulation for a varying percentage of modulation. (a) Message signal $m(t)$. (b) AM wave for $|k_a m(t)| < 1$ for all t, where k_a is the amplitude sensitivity of the modulator. This case represents undermodulation. (c) AM wave for $|k_a m(t)| > 1$ some of the time. This second case represents overmodulation.

By contrast, the AM wave shown in Fig. 5.4(c) corresponds to a value of k_a for which the percentage modulation is 166.7% (i.e., a case of overmodulation). Comparing the waveforms of these two AM waves with that of the message signal, we draw an important conclusion:

> The envelope of the AM wave has a waveform that bears a one-to-one correspondence with that of the message signal if and only if the percentage modulation is less than or equal to 100%.

This correspondence is destroyed if the percentage modulation is permitted to exceed 100%, in which case the modulated wave is said to suffer from *envelope distortion*.

▶ **Problem 5.2**   For 100% modulation, is it possible for the envelope $a(t)$ to become zero for some time t? Justify your answer.

Answer: If $k_a m(t) = -1$ for some time t, then $a(t) = 0$ ◀

■ 5.4.4 DOES FULL-AMPLITUDE MODULATION SATISFY THE LINEARITY PROPERTY?

Earlier, we defined linear modulation to be that form of modulation in which, if the amplitude of the message signal (i.e., modulating wave) is scaled by a certain factor, then the amplitude of the modulated wave is scaled by exactly the same factor. This definition of linear modulation is consistent with the notion of linearity of a system that was introduced in Section 1.8. Amplitude modulation, as defined in Eq. (5.8), fails the linearity test in a strict sense. To demonstrate this, suppose the message signal $m(t)$ consists of the sum of two components, $m_1(t)$ and $m_2(t)$. Let $s_1(t)$ and $s_2(t)$ denote the AM waves produced by these two components acting separately. With the operator H denoting the amplitude modulation process, we may then write

$$H\{m_1(t) + m_2(t)\} = A_c[1 + k_a(m_1(t) + m_2(t))]\cos(\omega_c t)$$
$$\neq s_1(t) + s_2(t),$$

where

$$s_1(t) = A_c[1 + k_a m_1(t)]\cos(\omega_c t)$$

and

$$s_2(t) = A_c[1 + k_a m_2(t)]\cos(\omega_c t).$$

The presence of the carrier wave $A_c \cos(\omega_c t)$ in the AM wave causes the principle of superposition to be violated.

However, as pointed out earlier, the failure of amplitude modulation to meet the criterion for linearity is of a rather mild sort. From the definition given in Eq. (5.8), we see that the AM signal $s(t)$ is, in fact, a linear combination of the carrier component $A_c \cos(\omega_c t)$ and the modulated component $A_c \cos(\omega_c t)m(t)$. Accordingly, amplitude modulation does permit the use of Fourier analysis without difficulty.

■ 5.4.5 FREQUENCY-DOMAIN DESCRIPTION OF AMPLITUDE MODULATION

Equation (5.8) defines the full AM wave $s(t)$ as a function of time. To develop the frequency description of this AM wave, we take the Fourier transform of both sides of Eq. (5.8). Let $S(j\omega)$ denote the Fourier transform of $s(t)$ and $M(j\omega)$ denote the Fourier

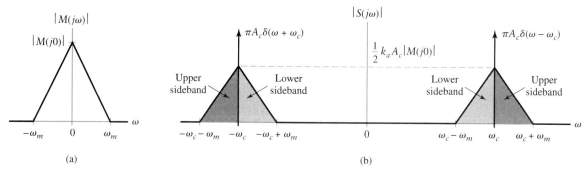

FIGURE 5.5   Spectral content of AM wave. (a) Magnitude spectrum of message signal. (b) Magnitude spectrum of the AM wave, showing the compositions of the carrier and the upper and lower sidebands.

transform of $m(t)$; we refer to $M(j\omega)$ as the *message spectrum*. Recall the following results from Chapter 4:

1. The Fourier transform of $A_c \cos(\omega_c t)$ is (see Example 4.1)

$$\pi A_c[\delta(\omega - \omega_c) + \delta(\omega + \omega_c)].$$

2. The Fourier transform of $m(t) \cos(\omega_c t)$ is (see Example 4.6)

$$\frac{1}{2}[M(j\omega - j\omega_c) + M(j\omega + j\omega_c)].$$

Using these results and invoking the linearity and scaling properties of the Fourier transform, we may express the Fourier transform of the AM wave of Eq. (5.8) as follows:

$$\begin{aligned} S(j\omega) = {} & \pi A_c[\delta(\omega - \omega_c) + \delta(\omega + \omega_c)] \\ & + \tfrac{1}{2}k_a A_c[M(j(\omega - \omega_c)) + M(j(\omega + \omega_c))] \end{aligned} \tag{5.12}$$

Let the message signal $m(t)$ be band limited to the interval $-\omega_m \leq \omega \leq \omega_m$, as in Fig. 5.5(a). We refer to the highest frequency component ω_m of $m(t)$ as the *message bandwidth*, which is measured in rad/s. The shape of the spectrum shown in the figure is intended for the purpose of illustration only. We find from Eq. (5.12) that the spectrum $S(j\omega)$ of the AM wave is as shown in Fig. 5.5(b) for the case when $\omega_c > \omega_m$. This spectrum consists of two impulse functions weighted by the factor πA_c and occurring at $\pm\omega_c$, and two versions of the message spectrum shifted in frequency by $\pm\omega_c$ and scaled in amplitude by $\frac{1}{2}k_a A_c$. The spectrum of Fig. 5.5(b) may be described as follows:

1. For positive frequencies, the portion of the spectrum of the modulated wave lying above the carrier frequency ω_c is called the *upper sideband*, whereas the symmetric portion below ω_c is called the *lower sideband*. For negative frequencies, the image of the upper sideband is represented by the portion of the spectrum below $-\omega_c$ and the image of the lower sideband by the portion above $-\omega_c$. The condition $\omega_c > \omega_m$ is a necessary condition for the sidebands not to overlap.

2. For positive frequencies, the highest frequency component of the AM wave is $\omega_c + \omega_m$, and the lowest frequency component is $\omega_c - \omega_m$. The difference between these two frequencies defines the *transmission bandwidth* ω_T for an AM wave, which is exactly twice the message bandwidth ω_m; that is,

$$\omega_T = 2\omega_m. \tag{5.13}$$

The spectrum of the AM wave as depicted in Fig. 5.5(b) is *full*, in that the carrier, the upper sideband, and the lower sideband are all completely represented. Accordingly, we refer to this form of modulation as "full amplitude modulation."

The upper sideband of the AM wave represents the positive frequency components of the message spectrum $M(j\omega)$, shifted upward in frequency by the carrier frequency ω_c. The lower sideband of the AM wave represents the negative frequency components of the message spectrum $M(j\omega)$, also shifted upward in frequency by ω_c. Herein lies the importance of admitting the use of negative frequencies in the Fourier analysis of signals. In particular, the use of amplitude modulation reveals the negative frequency components of $M(j\omega)$ completely, provided that $\omega_c > \omega_m$.

EXAMPLE 5.1 FULL AMPLITUDE MODULATION FOR SINUSOIDAL MODULATING SIGNAL
Consider a modulating wave $m(t)$ that consists of a single tone or frequency component; that is,

$$m(t) = A_0 \cos(\omega_0 t),$$

where A_0 is the amplitude of the modulating wave and ω_0 is its radian frequency. (See Fig. 5.6(a).) The sinusoidal carrier wave $c(t)$ has amplitude A_c and radian frequency ω_c (See Fig. 5.6(b).) Evaluate the time-domain and frequency-domain characteristics of the AM wave.

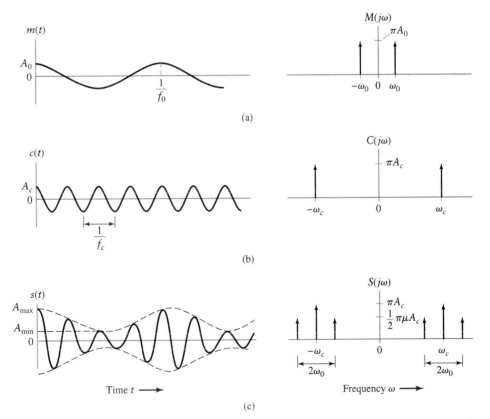

FIGURE 5.6 Time-domain (on the left) and frequency-domain (on the right) characteristics of AM produced by a sinusoidal modulating wave. (a) Modulating wave. (b) Carrier wave. (c) AM wave.

1off1

Solution: The AM wave is described by

$$s(t) = A_c[1 + \mu\cos(\omega_0 t)]\cos(\omega_c t), \qquad (5.14)$$

where

$$\mu = k_a A_0.$$

The dimensionless constant μ for a sinusoidal modulating wave is called the *modulation factor* and equals the percentage modulation when it is expressed numerically as a percentage. To avoid envelope distortion due to overmodulation, the modulation factor μ must be kept below unity. Figure 5.6(c) is a sketch of $s(t)$ for μ less than unity.

Let $A_{\max}$ and $A_{\min}$ denote the maximum and minimum values, respectively, of the envelope of the modulated wave. Then, from Eq. (5.14), we get

$$\frac{A_{\max}}{A_{\min}} = \frac{A_c(1 + \mu)}{A_c(1 - \mu)}.$$

Solving for μ yields

$$\mu = \frac{A_{\max} - A_{\min}}{A_{\max} + A_{\min}}.$$

Expressing the product of the two cosines in Eq. (5.14) as the sum of two sinusoidal waves, one having frequency $\omega_c + \omega_0$ and the other having frequency $\omega_c - \omega_0$, we get

$$s(t) = A_c\cos(\omega_c t) + \tfrac{1}{2}\mu A_c\cos[(\omega_c + \omega_0)t]$$
$$+ \tfrac{1}{2}\mu A_c\cos[(\omega_c - \omega_0)t].$$

In light of the Fourier transform pairs derived in Example 4.1, the Fourier transform of $s(t)$ is therefore

$$S(j\omega) = \pi A_c[\delta(\omega - \omega_c) + \delta(\omega + \omega_c)]$$
$$+ \tfrac{1}{2}\pi\mu A_c[\delta(\omega - \omega_c - \omega_0) + \delta(\omega + \omega_c + \omega_0)]$$
$$+ \tfrac{1}{2}\pi\mu A_c[\delta(\omega - \omega_c + \omega_0) + \delta(\omega + \omega_c - \omega_0)].$$

Thus, in ideal terms, the spectrum of a full AM wave, for the special case of sinusoidal modulation, consists of impulse functions at $\pm\omega_c$, $\omega_c \pm \omega_0$, and $-\omega_c \pm \omega_0$, as depicted in Fig. 5.6(c). ■

EXAMPLE 5.2 AVERAGE POWER OF SINUSOIDALLY MODULATED SIGNAL   Continuing with Example 5.1, investigate the effect of varying the modulation factor μ on the power content of the AM wave.

Solution: In practice, the AM wave $s(t)$ is a voltage or current signal. In either case, the average power delivered to a 1-ohm load resistor by $s(t)$ is composed of three components, whose values are derived from Eq. (1.15) as follows:

$$\text{Carrier power} = \tfrac{1}{2}A_c^2;$$
$$\text{Upper side-frequency power} = \tfrac{1}{8}\mu^2 A_c^2;$$
$$\text{Lower side-frequency power} = \tfrac{1}{8}\mu^2 A_c^2.$$

The ratio of the total sideband power to the total power in the modulated wave is therefore equal to $\mu^2/(2 + \mu^2)$, which depends only on the modulation factor μ. If $\mu = 1$ (i.e., if 100% modulation is used), the total power in the two side frequencies of the resulting AM wave is only one-third of the total power in the modulated wave.

Figure 5.7 shows the percentage of total power in both side frequencies and in the carrier, plotted against the percentage modulation. ■

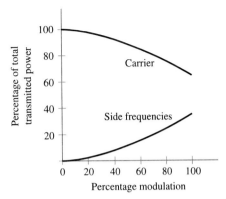

FIGURE 5.7   Variations of carrier power and side-frequency power with percentage modulation of AM wave for the case of sinusoidal modulation.

▶ **Problem 5.3**   For a particular case of sinusoidal modulation, the percentage modulation is 20%. Calculate the average power in (a) the carrier and (b) each of the two side frequencies.

Answers:

(a) $\frac{1}{2} A_c^2$

(b) $\frac{1}{200} A_c^2$ ◀

▶ **Problem 5.4**   Refer back to the transmitted radar signal described in Section 1.10. The signal may be viewed as a form of full AM modulation. Justify this statement and identify the modulating signal.

Answer: The modulating signal consists of a rectangular wave of period T and pulse duration T_0. The transmitted signal is simply the product of this modulating signal and the carrier wave; however, the modulating signal has a dc component—hence the presence of the carrier in the transmitted radar signal.   ◀

■ **5.4.6 SPECTRAL OVERLAP**

As mentioned previously, the spectral description of full amplitude modulation depicted in Fig. 5.5 presupposes that the carrier frequency ω_c is greater than the highest frequency ω_m of the message signal $m(t)$. But what if this condition is *not* satisfied? The answer is that the modulated signal $s(t)$ undergoes *frequency distortion* due to *spectral overlap*. This phenomenon is depicted in Fig. 5.8, where, for the purpose of illustration, it is assumed that the Fourier transform $S(j\omega)$ is real valued. The spectral overlap is produced by two movements:

▶ Movement of the lower sideband into the negative frequency range.

▶ Movement of the image of the lower sideband into the positive frequency range.

Although the upper sideband extending from ω_c to $\omega_c + \omega_m$ and its mirror image for negative frequencies remain intact, the movement of the lower sideband and it image interferers with the recovery of the original message signal. We therefore conclude that the condition

$$\boxed{\omega_c \geq \omega_m}$$

is a necessary condition for avoiding the spectral overlap problem.

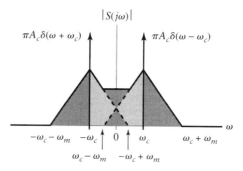

FIGURE 5.8   Spectral overlap phenomenon in amplitude modulation. The phenomenon arises when the carrier frequency ω_c is less than the highest frequency component ω_m of the modulating signal.

■ 5.4.7 DEMODULATION OF AM WAVE

The so-called *envelope detector* provides a simple, yet effective, device for the demodulation of a narrowband AM signal for which the percentage modulation is less than 100%. By "narrowband," we mean that the carrier frequency is large compared with the message bandwidth; this condition makes visualization of the envelope of the modulated signal a relatively easy task. Ideally, an envelope detector produces an output signal that follows the envelope of the input signal waveform exactly—hence the name. Some version of this circuit is used in almost all commercial AM radio receivers.

Figure 5.9(a) shows the circuit diagram of an envelope detector that consists of a diode and a resistor-capacitor filter. The operation of this envelope detector is as follows: On the positive half-cycle of the input signal, the diode is forward biased and the capacitor C charges up rapidly to the peak value of the input signal. When the input signal falls below this value, the diode becomes reverse biased and the capacitor C discharges slowly through the load resistor R_l. The discharging process continues until the next positive half-

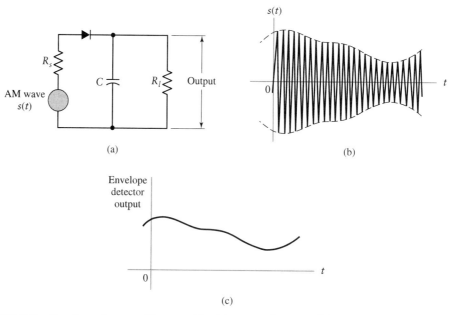

FIGURE 5.9   Envelope detector, illustrated by (a) circuit diagram, (b) AM wave input, and (c) envelope detector output, assuming ideal conditions.

cycle. When the input signal becomes greater than the voltage across the capacitor, the diode conducts again and the process is repeated. We make the following assumptions:

- ▶ The diode is ideal, presenting zero impedance to current flow in the forward-biased region and infinite impedance in the reverse-biased region.
- ▶ The AM signal applied to the envelope is supplied by a voltage source of internal resistance R_s.
- ▶ The load resistance R_l is large compared with the source resistance R_s. During the charging process, the time constant is effectively equal to $R_s C$. This time constant must be short compared with the carrier period $2\pi/\omega_c$; that is,

$$R_s C \ll \frac{2\pi}{\omega_c}. \tag{5.15}$$

Accordingly, the capacitor C charges rapidly and thereby follows the applied voltage up to the positive peak when the diode is conducting. In contrast, when the diode is reverse biased, the discharging time constant is equal to $R_l C$. This second time constant must be long enough to ensure that the capacitor discharges slowly through the load resistor R_l between positive peaks of the carrier wave, but not so long that the capacitor voltage will not discharge at the maximum rate of change of the modulating wave; that is,

$$\frac{2\pi}{\omega_c} \ll R_l C \ll \frac{2\pi}{\omega_m}, \tag{5.16}$$

where ω_m is the message bandwidth. The result is that the capacitor voltage or detector output is very nearly the same as the envelope of the AM wave, as we can see from Figs. 5.9(b) and (c). The detector output usually has a small ripple (not shown in Fig. 5.9(c)) at the carrier frequency; this ripple is easily removed by low-pass filtering.

▶ **Problem 5.5**   An envelope detector has a source resistance $R_s = 75\,\Omega$ and a load resistance $R_l = 10\,\text{k}\Omega$. Suppose $\omega_c = 2\pi \times 10^5$ rad/s and $\omega_m = 2\pi \times 10^3$ rad/s. Suggest a suitable value for the capacitor C.

Answer: $C = 0.01\,\mu\text{F}$ ◀

▶ **Problem 5.6**

(a) Following the three assumptions made on the diode, source resistance, and load resistance, write the equations that describe the charging and discharging process involving the capacitor C. Hence, justify the two design equations (5.15) and (5.16).

(b) How are these design equations modified if the forward resistance r_f and backward resistance r_b of the diode are significant enough to be taken into account?

Answers:

(a) Charging process (normalized with respect to the pertinent amplitude of $s(t)$):
$$1 - e^{-t/R_s C}.$$
Discharging process (normalized with respect to the pertinent amplitude of $s(t)$):
$$e^{-t/R_l C}.$$

(b) $(R_s + r_f)C \ll \dfrac{2\pi}{\omega_c}$

$$\frac{2\pi}{\omega_c} \ll \left(\frac{R_l r_b}{R_l + r_b}\right)C \ll \frac{2\pi}{\omega_m}, \text{ assuming that } r_b \gg R_s \qquad ◀$$

5.5 Double Sideband-Suppressed Carrier Modulation

In full AM, the carrier wave $c(t)$ is completely independent of the message signal $m(t)$, which means that the transmission of the carrier wave represents a waste of power. This points to a shortcoming of amplitude modulation, namely, that only a fraction of the total transmitted power is affected by $m(t)$, an effect that was well demonstrated in Example 5.2. To overcome this shortcoming, we may suppress the carrier component from the modulated wave, resulting in *double sideband-suppressed carrier* (DSB-SC) *modulation*. By suppressing the carrier, we obtain a modulated signal that is proportional to the product of the carrier wave and the message signal. Thus, to describe a DSB-SC modulated signal as a function of time, we simply write

$$\boxed{\begin{aligned} s(t) &= c(t)m(t) \\ &= A_c\cos(\omega_c t)m(t). \end{aligned}} \tag{5.17}$$

This modulated signal undergoes a phase reversal whenever the message signal $m(t)$ crosses zero, as illustrated in Fig. 5.10; part (a) of the figure depicts the waveform of a message signal, and part (b) depicts the corresponding DSB-SC modulated signal. Accordingly, unlike amplitude modulation, the envelope of a DSB-SC modulated signal is entirely different from that of the message signal.

▶ **Problem 5.7** In what manner is the envelope of the DSB-SC modulated signal shown in Fig. 5.10(b) different from that of the full AM signal of Fig. 5.4(b).

Answer: In Fig. 5.4(b), the envelope is a scaled version of the modulating wave. On the other hand, the envelope in Fig. 5.10(b) is a rectified version of the modulating wave. ◀

▪ 5.5.1 FREQUENCY-DOMAIN DESCRIPTION

The suppression of the carrier from the modulated signal of Eq. (5.17) may be well appreciated by examining its spectrum. Specifically, the Fourier transform of $s(t)$ was determined previously in Chapter 4. (See Example 4.6). We may thus write

$$S(j\omega) = \frac{1}{2}A_c[M(j(\omega - \omega_c)) + M(j(\omega + \omega_c))], \tag{5.18}$$

where, as before, $S(j\omega)$ is the Fourier transform of the modulated signal $s(t)$ and $M(j\omega)$ is the Fourier transform of the message signal $m(t)$. When the message signal $m(t)$ is limited to the interval $-\omega_m \leq \omega \leq \omega_m$, as in Fig. 5.11(a), we find that the spectrum $S(j\omega)$ is as

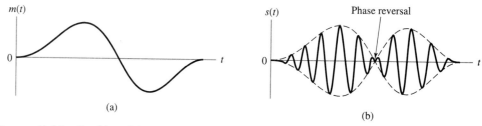

(a)

(b)

FIGURE 5.10 Double sideband-suppressed carrier modulation. (a) Message signal. (b) DSB-SC modulated wave, resulting from multiplication of the message signal by the sinusoidal carrier wave.

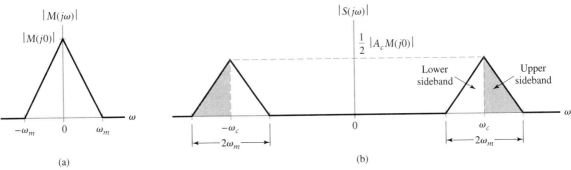

FIGURE 5.11 Spectral content of DSB-SC modulated wave. (a) Magnitude spectrum of message signal. (b) Magnitude spectrum of DSB-SC modulated wave, consisting of upper and lower sidebands only.

illustrated in part (b) of the figure. Except for a change in scale factor, the modulation process simply translates the spectrum of the message signal by $\pm\omega_c$. Of course, the transmission bandwidth required by DSB-SC modulation is the same as that for full amplitude modulation, namely, $2\omega_m$. However, comparing the spectrum shown in Fig. 5.11(b) for DSB-SC modulation with that of Fig. 5.5(b) for full AM, we see clearly that the carrier is suppressed in the DSB-SC case, whereas it is present in the full AM case, as exemplified by the existence of the pair of impulse functions at $\pm\omega_c$.

The generation of a DSB-SC modulated wave consists simply of the product of the message signal $m(t)$ and the carrier wave $A_c \cos(\omega_c t)$, as indicated in Eq. (5.17). A device for achieving this requirement is called a *product modulator*, which is another term for a straightforward multiplier. Figure 5.12(a) shows the block diagram representation of a product modulator.

■ 5.5.2 COHERENT DETECTION

The message signal $m(t)$ may be recovered from a DSB-SC modulated signal $s(t)$ by first multiplying $s(t)$ with a locally generated sinusoidal wave and then applying a low-pass filter to the product, as depicted in Fig. 5.12(b). It is assumed that the source of this locally generated sinusoidal wave, called a *local oscillator*, is exactly coherent or synchronized, in both frequency and phase, with the carrier wave $c(t)$ used in the product modulator of the transmitter to generate $s(t)$. This method of demodulation is known as *coherent detection* or *synchronous demodulation*.

It is instructive to derive coherent detection as a special case of the more general demodulation process, using a local oscillator whose output has the same frequency, but arbitrary phase difference ϕ, measured with respect to the carrier wave $c(t)$. Thus, denoting

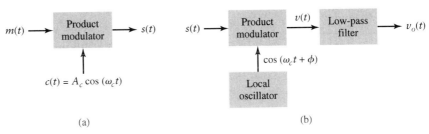

FIGURE 5.12 (a) Product modulator for generating the DSB-SC modulated wave. (b) Coherent detector for demodulation of the DSB-SC modulated wave.

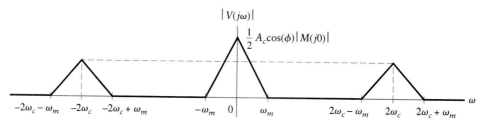

FIGURE 5.13   Magnitude spectrum of the product modulator output $v(t)$ in the coherent detector of Fig. 5.12(b).

the local oscillator output in the receiver by $\cos(\omega_c t + \phi)$, assumed to be of unit amplitude for convenience, and using Eq. (5.17) for the DSB-SC modulated signal $s(t)$, we find that the product modulator output in Fig. 5.12(b) is given by

$$
\begin{aligned}
v(t) &= \cos(\omega_c t + \phi)s(t) \\
&= A_c \cos(\omega_c t)\cos(\omega_c t + \phi)m(t) \\
&= \tfrac{1}{2} A_c \cos(\phi)m(t) + \tfrac{1}{2} A_c \cos(2\omega_c t + \phi)m(t).
\end{aligned}
\tag{5.19}
$$

The first term on the right-hand side of Eq. (5.19), namely, $\tfrac{1}{2} A_c \cos(\phi)m(t)$, represents a scaled version of the original message signal $m(t)$. The second term, $\tfrac{1}{2} A_c \cos(2\omega_c t + \phi)m(t)$, represents a new DSB-SC modulated signal with carrier frequency $2\omega_c$. Figure 5.13 shows the magnitude spectrum of $v(t)$. The clear separation between the spectra of the two components of $v(t)$ indicated in Fig. 5.13 hinges on the assumption that the original carrier frequency ω_c satisfies the condition

$$
2\omega_c - \omega_m > \omega_m,
$$

or, equivalently,

$$
\omega_c > \omega_m,
\tag{5.20}
$$

where ω_m is the message bandwidth; Eq. (5.20) is a restatement of the condition derived in Subsection 5.4.6 for avoiding spectral overlap. Provided that this condition is satisfied, we may use a low-pass filter to suppress the unwanted second term of $v(t)$. To accomplish this, the passband of the low-pass filter must extend over the entire message spectrum and no more. More precisely, its specifications must satisfy two requirements:

1. The cutoff frequency must be ω_m.
2. The transition band must be $\omega_m \leq \omega \leq 2\omega_c - \omega_m$.

Thus, the overall output in Fig. 5.12(b) is given by

$$
v_o(t) = \tfrac{1}{2} A_c \cos(\phi)m(t).
\tag{5.21}
$$

The demodulated signal $v_o(t)$ is proportional to $m(t)$ when the phase error ϕ is a constant. The amplitude of this demodulated signal is maximum when $\phi = 0$ and has a minimum of zero when $\phi = \pm\pi/2$. The zero demodulated signal, which occurs for $\phi = \pm\pi/2$, represents the *quadrature null effect* of the coherent detector. The phase error ϕ in the local oscillator causes the detector output to be attenuated by a factor equal to $\cos\phi$. As long as the phase error ϕ is constant, the detector output provides an undistorted version of the original message signal $m(t)$. In practice, however, we usually find that the phase error ϕ varies randomly with time, owing to random variations in the communication channel. The result is that, at the detector output, the multiplying factor $\cos\phi$ also varies randomly with time, which is obviously undesirable. Therefore, circuitry must be provided in the receiver to maintain the local oscillator in perfect synchronism, in both frequency and phase, with the carrier wave used to generate the DSB-SC modulated wave in the transmitter. The resulting increase in complexity of the receiver is the price that must be paid for suppressing the carrier wave to save transmitter power. Subsection 5.5.3 describes one such receiver.

▶ **Problem 5.8**   For the coherent detector of Fig. 5.12(b) to operate properly, Eq. (5.20) must be satisfied. What would happen if it is not?

Answer: The lower and upper sidebands would overlap, in which case the coherent detector would fail to operate properly   ◀

EXAMPLE 5.3 SINUSOIDAL DSB-SC MODULATION   Consider again the sinusoidal modulating signal

$$m(t) = A_0 \cos(\omega_0 t)$$

with amplitude A_0 and frequency ω_0; see Fig. 5.14(a). The carrier wave is

$$c(t) = A_c \cos(\omega_c t)$$

with amplitude A_c and frequency ω_c; see Fig. 5.14(b). Investigate the time-domain and frequency-domain characteristics of the corresponding DSB-SC modulated wave.

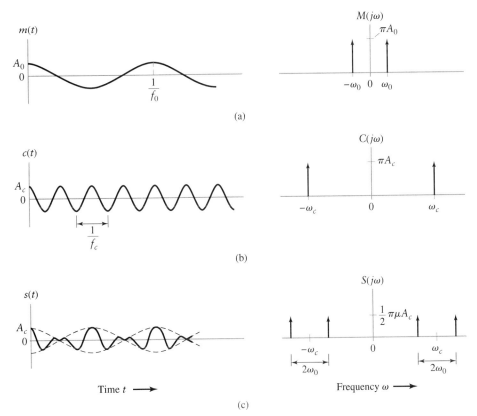

FIGURE 5.14   Time-domain (on the left) and frequency-domain (on the right) characteristics of DSB-SC modulation produced by a sinusoidal modulating wave. (a) Modulating wave. (b) Carrier wave. (c) DSB-SC modulated wave.

Solution: The modulated DSB-SC signal is defined by

$$s(t) = A_c A_0 \cos(\omega_c t) \cos(\omega_0 t)$$
$$= \tfrac{1}{2} A_c A_0 \cos[(\omega_c + \omega_0)t] + \tfrac{1}{2} A_c A_0 \cos[(\omega_c - \omega_0)t].$$

The Fourier transform of $s(t)$ is given by

$$S(j\omega) = \tfrac{1}{2}\pi A_c A_0 [\delta(\omega - \omega_c - \omega_0) + \delta(\omega + \omega_c + \omega_0)$$
$$+ \delta(\omega - \omega_c + \omega_0) + \delta(\omega + \omega_c - \omega_0)],$$

which consists of four weighted impulse functions at the frequencies $\omega_c + \omega_0$, $-\omega_c - \omega_0$, $\omega_c - \omega_0$, and $-\omega_c + \omega_0$, as illustrated in the right-hand side of Fig. 5.14(c). This Fourier transform differs from that depicted in the right-hand side of Fig. 5.6(c) for the corresponding example of full AM in one important respect: The impulse functions at $\pm\omega_c$ due to the carrier are removed.

Application of the sinusoidally modulated DSB-SC signal to the product modulator of Fig. 5.12(b) yields the output (assuming $\phi = 0$)

$$v(t) = \tfrac{1}{2} A_c A_0 \cos(\omega_c t)\{\cos[(\omega_c + \omega_0)t] + \cos[(\omega_c - \omega_0)t]\}$$
$$= \tfrac{1}{4} A_c A_0 \{\cos[(2\omega_c + \omega_0)t] + \cos(\omega_0 t)$$
$$+ \cos[(2\omega_c - \omega_0)t] + \cos(\omega_0 t)\}.$$

The first two sinusoidal terms of $v(t)$ are produced by the upper side frequency, and the last two sinusoidal terms are produced by the lower side frequency. With $\omega_c > \omega_0$, the first and third sinusoidal terms, of frequencies $2\omega_c + \omega_0$ and $2\omega_c - \omega_0$, respectively, are removed by the low-pass filter of Fig. 5.12(b), which leaves the second and fourth sinusoidal terms, of frequency ω_0, as the only output of the filter. The coherent detector output thus reproduces the original modulating wave. Note, however, that this output appears as two equal terms, one derived from the upper side frequency and the other from the lower side frequency. We therefore conclude that, for the transmission of the sinusoidal signal $A_0 \cos(\omega_0 t)$, only one side frequency is necessary. (The issue will be discussed further in Section 5.7.) ■

S & S Solutions

► **Problem 5.9** For the sinusoidal modulation considered in Example 5.3, what is the average power in the lower or upper side frequency, expressed as a percentage of the total power in the DSB-SC modulated wave?

Answer: 50% ◄

■ 5.5.3 COSTAS RECEIVER

One method of obtaining a practical synchronous receiver system suitable for demodulating DSB-SC waves is to use the *Costas receiver* shown in Fig. 5.15. This receiver consists of two coherent detectors supplied with the same input signal, namely, the incoming DSB-SC wave $A_c \cos(2\pi f_c t)m(t)$, but with individual local oscillator signals that are in phase quadrature with respect to each other. The frequency of the local oscillator is adjusted to be the same as the carrier frequency f_c, which is assumed to be known a priori. The detector in the upper path is referred to as the *in-phase coherent detector*, or *I-channel*, and that in the lower path is referred to as the *quadrature-phase coherent detector*, or *Q-channel*. These two detectors are coupled together to form a *negative-feedback* system designed in such a way as to maintain the local oscillator synchronous with the carrier wave.

To understand the operation of this receiver, suppose that the local oscillator signal is of the same phase as the carrier wave $A_c \cos(2\pi f_c t)$ used to generate the incoming DSB-SC wave. Under these conditions, we find that the *I*-channel output contains the desired de-

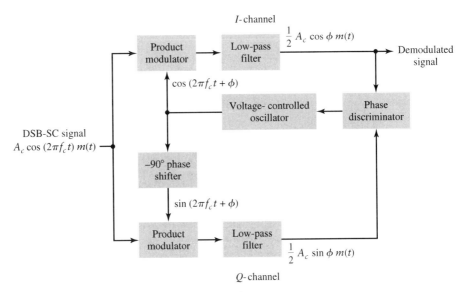

FIGURE 5.15   Costas receiver.

modulated signal $m(t)$, whereas the Q-channel output is zero due to the quadrature null effect. Suppose next that the local oscillator phase drifts from its proper value by a small angle of ϕ radians. Then the I-channel output will remain essentially unchanged, but at the Q-channel output there will now be some signal that is proportional to $\sin \phi \simeq \phi$ for small ϕ. This Q-channel output will have the same polarity as the I-channel output for one direction of local oscillator phase drift and will have opposite polarity for the opposite direction of local oscillator phase drift. Thus, by combining the I- and Q-channel outputs in a *phase discriminator* (which consists of a multiplier followed by a low-pass filter), as shown in Fig. 5.15, a dc control signal is obtained that automatically corrects for local phase errors in the *voltage-controlled oscillator*.

It is apparent that phase control in the Costas receiver ceases when the modulating signal $m(t)$ is zero and that phase lock has to be reestablished when the modulating signal is nonzero. This is not a serious problem when one is receiving voice transmission, because the lock-up process normally occurs so rapidly that no distortion is perceptible.

5.6 *Quadrature-Carrier Multiplexing*

A *quadrature-carrier multiplexing*, or *quadrature-amplitude modulation* (QAM), system enables two DSB-SC modulated waves (resulting from the application of two *independent* message signals) to occupy the same transmission bandwidth, and yet it allows for their separation at the receiver output. It is therefore a *bandwidth-conservation scheme*.

Figure 5.16 is a block diagram of the quadrature-carrier multiplexing system. The transmitter of the system, shown in part (a) of the figure, involves the use of two separate product modulators that are supplied with two carrier waves of the same frequency, but differing in phase by $-90°$. The multiplexed signal $s(t)$ consists of the sum of these two product modulator outputs; that is,

$$s(t) = A_c m_1(t) \cos(\omega_c t) + A_c m_2(t) \sin(\omega_c t), \qquad (5.22)$$

where $m_1(t)$ and $m_2(t)$ denote the two different message signals applied to the product modulators. Since each term in Eq. (5.22) has a transmission bandwidth of $2\omega_m$ and is centered on ω_c, we see that the multiplexed signal occupies a transmission bandwidth of $2\omega_m$,

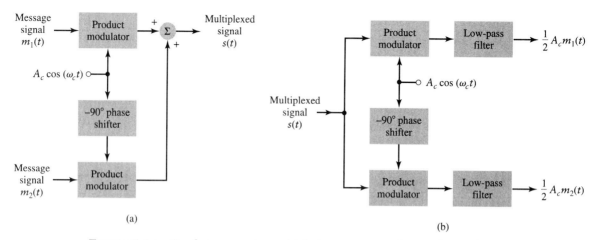

(a)

(b)

Figure 5.16   Quadrature-carrier multiplexing system, exploiting the quadrature null effect of DSB-SC modulation. (a) Transmitter. (b) Receiver, assuming perfect synchronization with the transmitter.

centered on the carrier frequency ω_c, where ω_m is the common message bandwidth of $m_1(t)$ and $m_2(t)$.

The receiver of the system is shown in Fig. 5.16(b). The multiplexed signal is applied simultaneously to two separate coherent detectors that are supplied with two local carriers of the same frequency, but differing in phase by $-90°$. The output of the top detector is $\frac{1}{2}A_c m_1(t)$, whereas the output of the bottom detector is $\frac{1}{2}A_c m_2(t)$.

For the quadrature-carrier multiplexing system to operate satisfactorily, it is important to maintain the correct phase and frequency relationships between the local oscillators used in the transmitter and receiver parts of the system, which may be achieved by using the Costas receiver. This increase in system complexity is the price that must be paid for the practical benefit gained from bandwidth conservation.

▶ **Problem 5.10**   Verify that the outputs of the receiver in Fig. 5.16 in response to $s(t)$ of Eq. (5.22) are as indicated in the figure, assuming perfect synchronism.   ◀

5.7 Other Variants of Amplitude Modulation

The full AM and DSB-SC forms of modulation are wasteful of bandwidth, because they both require a transmission bandwidth equal to twice the message bandwidth. In either case, one-half the transmission bandwidth is occupied by the upper sideband of the modulated wave, whereas the other half is occupied by the lower sideband. Indeed, the upper and lower sidebands are uniquely related to each other by virtue of their symmetry about the carrier frequency, as illustrated in Figs. 5.5 and 5.11; note that this symmetry only holds for real-valued signals. That is, given the amplitude and phase spectra of either sideband, we can uniquely determine the other. This means that insofar as the transmission of information is concerned, only one sideband is necessary, and if both the carrier and the other sideband are suppressed at the transmitter, no information is lost. In this way, the channel needs to provide only the same bandwidth as the message signal, a conclusion that is intuitively satisfying. When only one sideband is transmitted, the modulation is referred to as *single sideband* (SSB) *modulation*.

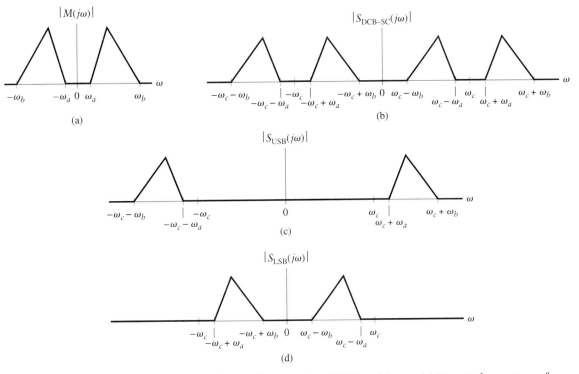

FIGURE 5.17   Frequency-domain characteristics of SSB modulation. (a) Magnitude spectrum of message signal, with energy gap from $-\omega_a$ to ω_a. (b) Magnitude spectrum of DSB-SC signal. (c) Magnitude spectrum of SSB modulated wave, containing upper sideband only. (d) Magnitude spectrum of SSB modulated wave, containing lower sideband only.

■ 5.7.1 FREQUENCY-DOMAIN DESCRIPTION OF SSB MODULATION

The precise frequency-domain description of an SSB modulated wave depends on which sideband is transmitted. To investigate this issue, consider a message signal $m(t)$ with a spectrum $M(j\omega)$ limited to the band $\omega_a \leq |\omega| \leq \omega_b$, as in Fig. 5.17(a). The spectrum of the DSB-SC modulated wave, obtained by multiplying $m(t)$ by the carrier wave $A_c \cos(\omega_c t)$, is as shown in Fig. 5.17(b). The upper sideband is represented in duplicate by the frequencies above ω_c and those below $-\omega_c$; when only the upper sideband is transmitted, the resulting SSB modulated wave has the spectrum shown in Fig. 5.17(c). Likewise, the lower sideband is represented in duplicate by the frequencies below ω_c (for positive frequencies) and those above $-\omega_c$ (for negative frequencies); when only the lower sideband is transmitted, the spectrum of the corresponding SSB modulated wave is as shown in Fig. 5.17(d). Thus, the essential function of SSB modulation is to *translate* the spectrum of the modulating wave, either with or without inversion, to a new location in the frequency domain. Moreover, the transmission bandwidth requirement of an SSB modulation system is one-half that of a full AM or DSB-SC modulation system. The benefit of using SSB modulation is therefore derived principally from the reduced bandwidth requirement and the elimination of the high-power carrier wave, two features that make SSB modulation the *optimum* form of linear CW modulation. The principal disadvantage of SSB modulation is the cost and complexity of implementing both the transmitter and the receiver. Here again, we have a trade-off between increased system complexity and improved system performance.

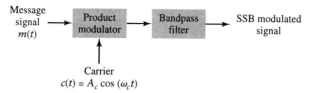

FIGURE 5.18   System consisting of product modulator and bandpass filter, for generating SSB modulated wave.

Using the frequency-domain descriptions in Fig. 5.17, we may readily deduce the *frequency-discrimination scheme* shown in Fig. 5.18 for producing SSB modulation. The scheme consists of a product modulator followed by a bandpass filter. The filter is designed to pass the sideband selected for transmission and suppress the remaining sideband. For a filter to be physically realizable, the transition band separating the passband from the stopband must have a non-zero width. In the context of the scheme shown in Fig. 5.18, this requirement demands that there be an adequate separation between the lower sideband and upper sideband of the DSB-SC modulated wave produced at the output of the product modulator. Such a requirement can be satisfied only if the message signal $m(t)$ applied to the product modulator has an *energy gap* in its spectrum, as indicated in Fig. 5.17(a). Fortunately, speech signals for telephonic communication do exhibit an energy gap extending from -300 to 300 Hz, a feature of speech signals that makes SSB modulation well suited for the transmission of speech signals. Indeed, analog telephony, which was dominant for a good part of the 20th century, relied on SSB modulation for its transmission needs.

S & S
Solutions

▶ **Problem 5.11**   An SSB modulated wave $s(t)$ is generated by means of a carrier of frequency ω_c and a sinusoidal modulating wave of frequency ω_0. The carrier amplitude is A_c, and that of the modulating wave is A_0. Define $s(t)$, assuming that (a) only the upper side frequency is transmitted and (b) only the lower side frequency is transmitted.

Answers:

(a) $s(t) = \frac{1}{2}A_cA_0\cos[(\omega_c + \omega_0)t]$

(b) $s(t) = \frac{1}{2}A_cA_0\cos[(\omega_c - \omega_0)t]$ ◀

▶ **Problem 5.12**   The spectrum of a speech signal lies inside the band $\omega_1 \leq |\omega| \leq \omega_2$. The carrier frequency is ω_c. Specify the passband, transition band, and stopband of the bandpass filter in Fig. 5.18 so as to transmit (a) the lower sideband and (b) the upper sideband. (You may refer to Subsection 3.10.2 for definitions of these bands.)

Answers:

(a)   Passband: $\omega_c - \omega_2 \leq |\omega| \leq \omega_c - \omega_1$
      Transition band: $\omega_c - \omega_1 \leq |\omega| \leq \omega_c + \omega_1$
      Stopband: $\omega_c + \omega_1 \leq |\omega| \leq \omega_c + \omega_2$

(b)   Passband: $\omega_c + \omega_1 \leq |\omega| \leq \omega_c + \omega_2$
      Transition band: $\omega_c - \omega_1 \leq |\omega| \leq \omega_c + \omega_1$
      Stopband: $\omega_c - \omega_2 \leq |\omega| \leq \omega_c - \omega_1$ ◀

■ 5.7.2 TIME-DOMAIN DESCRIPTION OF SSB MODULATION

The frequency-domain description of SSB modulation depicted in Fig. 5.17 and its generation using the frequency-discrimination scheme shown in Fig. 5.18 build on our knowledge of DSB-SC modulation in a straightforward fashion. However, unlike the situation with DSB-SC modulation, the time-domain description of SSB modulation is not as straightforward. To develop the time-domain description of SSB modulation, we need a mathematical tool known as the Hilbert transform. The device used to perform this transformation is known as the *Hilbert transformer*, the frequency response of which is characterized as follows:

▶ The magnitude response is unity for all frequencies, both positive and negative.

▶ The phase response is $-90°$ for positive frequencies and $+90°$ for negative frequencies.

The Hilbert transformer may therefore be viewed as a wideband $-90°$ phase shifter, wideband in the sense that its frequency response occupies a band of frequencies that, in theory, is infinite in extent. Further consideration of the time-domain description of SSB modulation is beyond the scope of this book. (See Note 4 of the section on Further Reading on page 475.)

■ 5.7.3 VESTIGIAL SIDEBAND MODULATION

Single sideband modulation is well suited for the transmission of speech because of the energy gap that exists in the spectrum of speech signals between zero and a few hundred hertz for positive frequencies. When the message signal contains significant components at extremely low frequencies (as in the case of television signals and wideband data), the upper and lower sidebands meet at the carrier frequency. This means that the use of SSB modulation is inappropriate for the transmission of such message signals, owing to the practical difficulty of building a filter to isolate one sideband completely. This difficulty suggests another scheme known as *vestigial sideband (VSB) modulation*, which is a compromise between SSB and DSB-SC forms of modulation. In VSB modulation, one sideband is passed almost completely, whereas just a trace, or *vestige*, of the other sideband is retained.

Figure 5.19 illustrates the spectrum of a *VSB modulated* wave $s(t)$ in relation to that of the message signal $m(t)$, assuming that the lower sideband is modified into the vestigial sideband. The transmitted vestige of the lower sideband compensates for the amount removed from the upper sideband. The transmission bandwidth required by the VSB modulated wave is therefore given by

$$\omega_T = \omega_m + \omega_v, \tag{5.23}$$

where ω_m is the message bandwidth and ω_v is the width of the vestigial sideband.

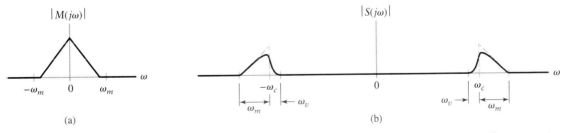

(a)   (b)

FIGURE 5.19 Spectral content of VSB modulated wave. (a) Magnitude spectrum of message signal. (b) Magnitude spectrum of VSB modulated wave containing a vestige of the lower sideband.

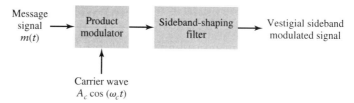

FIGURE 5.20   System consisting of product modulator and sideband shaping filter, for generating VSB modulated wave.

To generate a VSB modulated wave, we pass a DSB-SC modulated wave through a *sideband-shaping filter*, as in Fig. 5.20. Unlike the bandpass filter used for SSB modulation, the sideband-shaping filter does not have a "flat" magnitude response in its passband, because the upper and lower sidebands have to be shaped differently. The filter response is designed so that the original message spectrum $M(j\omega)$ [i.e., the Fourier transform of the message signal $m(t)$] is reproduced on demodulation as a result of the superposition of two spectra:

▶ The positive-frequency part of $S(j\omega)$ (i.e., the Fourier transform of the transmitted signal $s(t)$), shifted downward in frequency by ω_c.

▶ The negative-frequency part of $S(j\omega)$, shifted upward in frequency by ω_c.

The magnitudes of these two spectral contributions are illustrated in Figs. 5.21(a) and (b), respectively. In effect, a reflection of the vestige of the lower sideband makes up for the missing part of the upper sideband.

The design requirement described herein makes the implementation of the sideband-shaping filter a challenging task.

Vestigial sideband modulation has the virtue of conserving bandwidth almost as efficiently as does single sideband modulation, while retaining the excellent low-frequency characteristics of double sideband modulation. Thus, VSB modulation has become standard for the analog transmission of television and similar signals, in which good phase characteristics and the transmission of low-frequency components are important, but the bandwidth required for double sideband transmission is unavailable or uneconomical.

In the transmission of television signals in practice, a controlled amount of carrier is added to the VSB modulated signal. This is done to permit the use of an envelope detector for demodulation. The design of the receiver is thereby considerably simplified.

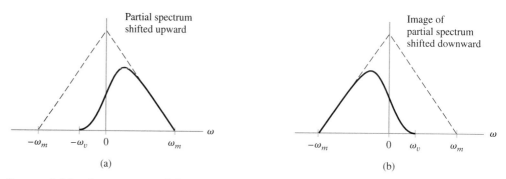

FIGURE 5.21   Superposition of the two spectra shown in parts (a) and (b) of the figure results in the original message spectrum (shown dashed) upon demodulation.

5.8 *Pulse-Amplitude Modulation*

Having familiarized ourselves with continuous-wave AM and its variants, we now turn our attention to *pulse-amplitude modulation* (PAM), a widely used form of pulse modulation. Whereas frequency shifting plays a basic role in the operation of AM systems, the basic operation in PAM systems is that of sampling.

■ 5.8.1 SAMPLING REVISITED

The sampling process, including a derivation of the sampling theorem and related issues of aliasing and reconstructing the message signal from its sampled version, is covered in detail in Sections 4.5 and 4.6. In this subsection, we tie the discussion of sampling for PAM to the material covered therein. To begin with, we may restate the sampling theorem in the context of PAM in two equivalent parts as follows:

1. *A band-limited signal of finite energy that has no radian frequency components higher than ω_m is uniquely determined by the values of the signal at instants of time separated by π/ω_m seconds.*

2. *A band-limited signal of finite energy that has no radian frequency components higher than ω_m may be completely recovered from a knowledge of its samples taken at the rate of ω_m/π per second.*

Part 1 of the sampling theorem is exploited in the transmitter of a PAM system, part 2 in the receiver of the system. The special value of the sampling rate ω_m/π is referred to as the *Nyquist rate*, in recognition of the pioneering work done by U.S. physicist Harry Nyquist (1889–1976) on data transmission.

Typically, the spectrum of a message signal is not strictly band limited, contrary to what is required by the sampling theorem. Rather, it approaches zero asymptotically as the frequency approaches infinity, which gives rise to aliasing and therefore distorts the signal. Recall that aliasing consists of a high-frequency component in the spectrum of the message signal apparently taking on the identity of a lower frequency in the spectrum of a sampled version of the message signal. To combat the effects of aliasing in practice, we use two corrective measures:

▶ Prior to sampling, a low-pass antialiasing filter is used to attenuate those high-frequency components of the signal which lie outside the band of interest.

▶ The filtered signal is sampled at a rate slightly higher than the Nyquist rate.

On this basis, the generation of a PAM signal as a sequence of flat-topped pulses whose amplitudes are determined by the corresponding samples of the filtered message signal follows the block diagram shown in Fig. 5.22.

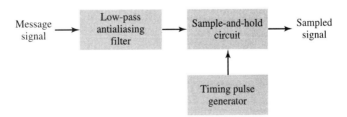

FIGURE 5.22   System consisting of antialiasing filter and sample-and-hold circuit, for converting a message signal into a flat-topped PAM signal.

EXAMPLE 5.4 TELEPHONIC COMMUNICATION   The highest frequency component of a speech signal needed for telephonic communication is about 3.1 kHz. Suggest a suitable value for the sampling rate.

Solution: The highest frequency component of 3.1 kHz corresponds to

$$\omega_m = 6.2\pi \times 10^3 \, \text{rad/s}.$$

Correspondingly, the Nyquist rate is

$$\frac{\omega_m}{\pi} = 6.2 \, \text{kHz}.$$

A suitable value for the sampling rate—one slightly higher than the Nyquist rate—may be 8 kHz, a rate that is the international standard for telephone speech signals.   ■

■ 5.8.2 MATHEMATICAL DESCRIPTION OF PAM

The carrier wave used in PAM consists of a sequence of short pulses of fixed duration in terms of which PAM is formally defined as follows: PAM is a *form of pulse modulation in which the amplitude of the pulsed carrier is varied in accordance with instantaneous sample values of the message signal*; the duration of the pulsed carrier is maintained constant throughout. Figure 5.23 illustrates the waveform of such a PAM signal. Note that the fundamental frequency of the carrier wave (i.e., the pulse repetition frequency) is the same as the sampling rate.

For a mathematical representation of the PAM signal $s(t)$ for a message signal $m(t)$, we may write

$$s(t) = \sum_{n=-\infty}^{\infty} m[n]h(t - nT_s), \tag{5.24}$$

where T_s is the sampling period, $m[n]$ is the value of the message signal $m(t)$ at time $t = nT_s$, and $h(t)$ is a rectangular pulse of unit amplitude and duration T_0, defined as follows (see Fig. 5.24(a)):

$$h(t) = \begin{cases} 1, & 0 < t < T_0 \\ 0, & \text{otherwise} \end{cases}. \tag{5.25}$$

In physical terms, Eq. (5.24) represents a *sample-and-hold operation* analogous to the zero-order-hold-based reconstruction described in Section 4.6. These two operations differ from each other in that the impulse response $h(t)$ in Eq. (5.25) is T_0 wide instead of T_s.

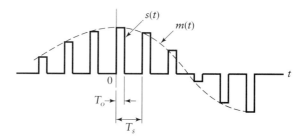

FIGURE 5.23   Waveform of flat-topped PAM signal with pulse duration T_o and sampling period T_s.

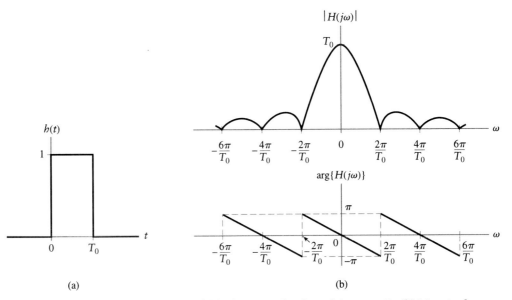

FIGURE 5.24 (a) Rectangular pulse $h(t)$ of unit amplitude and duration T_0. (b) Magnitude spectrum $|H(j\omega)|$ and phase spectrum $\arg\{H(j\omega)\}$ of pulse $h(t)$.

Bearing this difference in mind, we may follow the material presented in Section 4.6 to derive the spectrum of the PAM signal $s(t)$.

The impulse-sampled version of the message signal $m(t)$ is given by

$$m_\delta(t) = \sum_{n=-\infty}^{\infty} m[n]\delta(t - nT_s). \tag{5.26}$$

The PAM signal is expressed as

$$\boxed{\begin{aligned} s(t) &= \sum_{n=-\infty}^{\infty} m[n]h(t - nT_s) \\ &= m_\delta(t) * h(t). \end{aligned}} \tag{5.27}$$

Equation (5.27) states that $s(t)$ is mathematically equivalent to the convolution of $m_\delta(t)$—the impulse-sampled version of $m(t)$—and the pulse $h(t)$.

The convolution described herein is a time-domain operation. Recall from Chapter 3 that convolution in the time domain is Fourier transformed into multiplication in the frequency domain. Thus, taking the Fourier transform of both sides of Eq. (5.27), we get

$$S(j\omega) = M_\delta(j\omega)H(j\omega), \tag{5.28}$$

where $S(j\omega) \xleftrightarrow{FT} s(t)$, $M_\delta(j\omega) \xleftrightarrow{FT} m_\delta(t)$, and $H(j\omega) \xleftrightarrow{FT} h(t)$. Recall further from Eq. (4.23) that impulse sampling of the message signal $m(t)$ introduces periodicity into the spectrum as given by

$$M_\delta(j\omega) = \frac{1}{T_s} \sum_{k=-\infty}^{\infty} M(j(\omega - k\omega_s)), \tag{5.29}$$

where $1/T_s$ is the sampling rate and $\omega_s = 2\pi/T_s$ rad/s. Therefore, substitution of Eq. (5.29) into (5.28) yields

$$S(j\omega) = \frac{1}{T_s} \sum_{k=-\infty}^{\infty} M(j(\omega - k\omega_s))H(j\omega), \qquad (5.30)$$

where $M(j\omega) \overset{FT}{\longleftrightarrow} m(t)$.

Finally, suppose that $m(t)$ is strictly band limited and that the sampling rate $1/T_s$ is greater than the Nyquist rate. Then passing $s(t)$ through a reconstruction filter chosen as an ideal low-pass filter with cutoff frequency ω_m and gain T_s, we find that the spectrum of the resulting filter output is equal to $M(j\omega)H(j\omega)$. This result is equivalent to that which would be obtained by passing the original message signal $m(t)$ through a low-pass filter with frequency response $H(j\omega)$.

From Eq. (5.25) we find that

$$H(j\omega) = T_0 \operatorname{sinc}(\omega T_0/(2\pi))e^{-j\omega T_0/2}; \qquad (5.31)$$

the magnitude and phase components of $H(j\omega)$ are plotted in Fig. 5.24(b). Hence, in light of Eqs. (5.28) and (5.31), we see that by using PAM to represent a continuous-time message signal, we introduce amplitude distortion as well as a delay of $T_0/2$. Both of these effects are present as well in the sample-and-hold reconstruction scheme described in Section 4.6. A similar form of amplitude distortion is caused by the finite size of the scanning aperture in television and facsimile. Accordingly, the frequency distortion caused by the use of flat-topped samples in the generation of a PAM wave, illustrated in Fig. 5.23, is referred to as the *aperture effect*.

▶ **Problem 5.13**   What happens to the scaled frequency response $H(j\omega)/T_0$ of Eq. (5.31) as the pulse duration T_0 approaches zero?

Answer: $\displaystyle \lim_{T_0 \to 0} \frac{H(j\omega)}{T_0} = 1$ ◀

■ **5.8.3 DEMODULATION OF PAM SIGNAL**

Given a sequence of flat-topped samples $s(t)$, we may reconstruct the original message signal $m(t)$ by using the scheme shown in Fig. 5.25. The system consists of two components connected in cascade. The first component is a low-pass filter with a cutoff frequency that equals the highest frequency component ω_m of the message signal. The second component is an *equalizer* that corrects for the aperture effect due to flat-topped sampling in the sample-and-hold circuit. The equalizer has the effect of decreasing the in-band loss of the interpolation filter as the frequency increases in such a manner as to compensate for the aperture effect. Ideally, the amplitude response of the equalizer is given by

$$\frac{1}{|H(j\omega)|} = \frac{1}{T_0 |\operatorname{sinc}(\omega T_0/2)|} = \frac{1}{2T_0} \frac{\omega T_0}{|\sin(\omega T_0/2)|},$$

where $H(j\omega)$ is the frequency response defined in Eq. (5.31). The amount of equalization needed in practice is usually small.

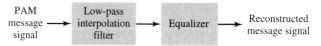

FIGURE 5.25   System consisting of low-pass interpolation filter and equalizer, for reconstructing a message signal from its flat-topped sampled version.

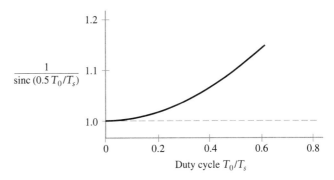

FIGURE 5.26   Normalized equalizer gain (to compensate for aperture effect) plotted against the duty cycle T_0/T_s.

EXAMPLE 5.5   EQUALIZATION FOR PAM TRANSMISSION   The duty cycle in a PAM signal, namely, T_0/T_s, is 10%. Evaluate the peak amplification required for equalization.

Solution: At $\omega_m = \pi/T_s$, which corresponds to the highest frequency component of the message signal for a sampling rate equal to the Nyquist rate, we find from Eq. (5.31) that the magnitude response of the equalizer at ω_m, normalized to that at zero frequency, is

$$\frac{1}{\text{sinc}(0.5T_0/T_s)} = \frac{(\pi/2)(T_0/T_s)}{\sin[(\pi/2)(T_0/T_s)]},$$

where the ratio T_0/T_s is equal to the duty cycle of the sampling pulses. In Fig. 5.26, this result is plotted as a function of the *duty cycle* T_0/T_s. Ideally, it should be equal to unity for all values of T_0/T_s. For a duty cycle of 10%, it is equal to 1.0041. It follows that, for duty cycles of less than 10%, the magnitude equalization required is less than 1.0041, in which case the aperture effect is usually considered to be negligible.   ∎

5.9 Multiplexing

In Section 5.3, we pointed out that modulation provides a method for multiplexing whereby message signals derived from independent sources are combined into a composite signal suitable for transmission over a common channel. In a telephone system, for example, multiplexing is used to transmit multiple conversations over a single long-distance line. The signals associated with different speakers are combined in such a way as to not interfere with each other during transmission and so that they can be separated at the receiving end of the system. Multiplexing can be accomplished by separating the different message signals either in frequency or in time or through the use of coding techniques. We thus have three basic types of multiplexing:

1. *Frequency-division multiplexing*, in which the signals are separated by allocating them to different frequency bands. This is illustrated in Fig. 5.27(a) for the case of six different message signals. Frequency-division multiplexing favors the use of CW modulation, where each message signal is able to use the channel on a continuous-time basis.

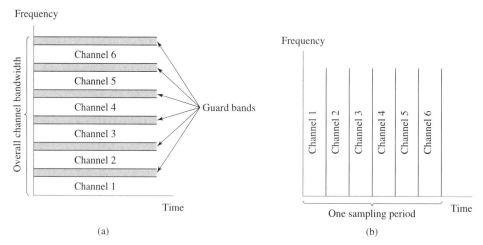

FIGURE 5.27   Two basic forms of multiplexing. (a) Frequency-division multiplexing (with guardbands). (b) Time-division multiplexing; no provision is shown here for synchronizing pulses.

2. *Time-division multiplexing*, wherein the signals are separated by allocating them to different time slots within a sampling interval. This second type of multiplexing is illustrated in Fig. 5.27(b) for the case of six different message signals. Time-division multiplexing favors the use of pulse modulation, whereby each message signal has access to the complete frequency response of the channel.

3. *Code-division multiplexing*, which relies on the assignment of different codes to the individual users of the channel.

The first two methods of multiplexing are described in the rest of the section; a discussion of code-division multiplexing is beyond the scope of the book. (See Note 5 of the section on Further Reading on page 476.)

■ 5.9.1 FREQUENCY-DIVISION MULTIPLEXING (FDM)

A block diagram of an FDM system is shown in Fig. 5.28. The incoming message signals are assumed to be of the low-pass variety, but their spectra do not necessarily have nonzero values all the way down to zero frequency. Following each input signal is a low-pass filter, which is designed to remove high-frequency components that do not contribute significantly to representing the signal, but that are capable of disturbing other message signals which share the common channel. These low-pass filters may be omitted only if the input signals are sufficiently band limited initially. The filtered signals are applied to modulators that shift the frequency ranges of the signals so as to occupy mutually exclusive frequency intervals. The carrier frequencies needed to perform these translations are obtained from a carrier supply. For the modulation, we may use any one of the methods described in previous sections of this chapter. However, the most widely used method of modulation in frequency-division multiplexing is single sideband modulation, which, in the case of voice signals, requires a bandwidth that is approximately equal to that of the original voice signal. In practice, each voice input is usually assigned a bandwidth of 4 kHz. The band-pass filters following the modulators are used to restrict the band of each modulated wave to its prescribed range. Next, the resulting band-pass filter outputs are summed to form the input to the common channel. At the receiving terminal, a bank of band-pass filters, with their inputs connected in parallel, is used to separate the message signals on a frequency-

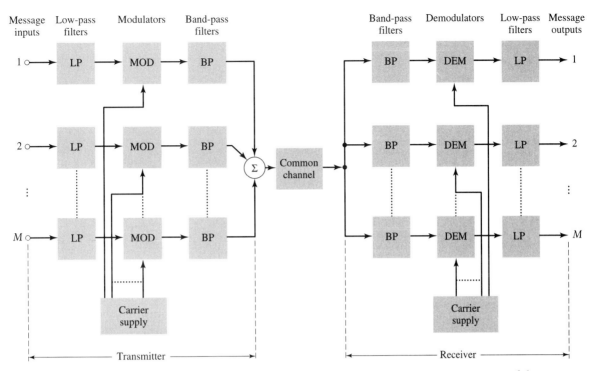

FIGURE 5.28   Block diagram of FDM system, showing the important constituents of the transmitter and receiver.

occupancy basis. Finally, the original message signals are recovered by individual demodulators. Note that the FDM system shown in Fig. 5.28 operates in only one direction. To provide for two-way transmission, as in telephony, for example, we have to duplicate the multiplexing facilities, with the components connected in reverse order and the signal waves proceeding from right to left.

EXAMPLE 5.6 SSB-FDM SYSTEM   An FDM system is used to multiplex 24 independent voice signals. SSB modulation is used for the transmission. Given that each voice signal is allotted a bandwidth of 4 kHz, calculate the overall transmission bandwidth of the channel.

Solution: With each voice signal allotted a bandwidth of 4 kHz, the use of SSB modulation requires a bandwidth of 4 kHz for its transmission. Accordingly, the overall transmission bandwidth provided by the channel is $24 \times 4 = 96$ kHz.   ∎

■ 5.9.2 TIME-DIVISION MULTIPLEXING (TDM)

Basic to the operation of a TDM system is the sampling theorem, which states that we can transmit all the information contained in a band-limited message signal by using samples of the signal taken uniformly at a rate that is usually slightly higher than the Nyquist rate. An important feature of the sampling process has to do with conservation of time. That is, the transmission of the message samples engages the transmission channel for only a fraction of the sampling interval on a periodic basis, equal to the width T_0 of a PAM modulating pulse. In this way, some of the time interval between adjacent samples is cleared for use by other independent message sources on a time-shared basis.

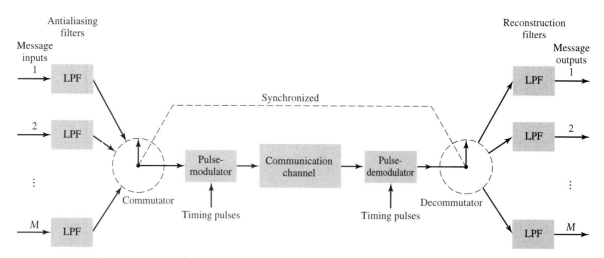

FIGURE 5.29   Block diagram of TDM system, showing the important constituents of the transmitter and receiver.

The concept of TDM is illustrated by the block diagram shown in Fig. 5.29. Each input message signal is first restricted in bandwidth by a low-pass filter to remove the frequencies that are nonessential to an adequate representation of the signal. The low-pass filter outputs are then applied to a *commutator* that is usually implemented by means of electronic switching circuitry. The function of the commutator is twofold: (1) to take a narrow sample of each of the M input message signals at a rate $1/T_s$ that is slightly higher than ω_c/π, where ω_c is the cutoff frequency of the input low-pass filter; and (2) to sequentially interleave these M samples inside a sampling interval T_s. Indeed, the latter function is the essence of the time-division multiplexing operation. Following commutation, the multiplexed signal is applied to a *pulse modulator* (e.g., a pulse-amplitude modulator), the purpose of which is to transform the multiplexed signal into a form suitable for transmission over the common channel. The use of time-division multiplexing introduces a *bandwidth expansion factor M*, because the scheme must squeeze M samples derived from M independent message sources into a time slot equal to one sampling interval. At the receiving end of the system, the signal is applied to a *pulse demodulator*, which performs the inverse operation of the pulse modulator. The narrow samples produced at the pulse demodulator output are distributed to the appropriate low-pass reconstruction filters by means of a *decommutator*, which operates in synchronism with the commutator in the transmitter.

Synchronization between the timing operations of the transmitter and receiver in a TDM system is essential for satisfactory performance of the system. In the case of a TDM system using PAM, synchronization may be achieved by inserting an extra pulse into each sampling interval on a regular basis. The combination of M PAM signals and a synchronization pulse contained in a single sampling period is referred to as a *frame*. In PAM, the feature of a message signal that is used for modulation is its amplitude. Accordingly, a simple way of identifying the synchronizing pulse train at the receiver is to make sure that its constant amplitude is large enough to stand above every one of the PAM signals. On this basis, the synchronizing pulse train is identified at the receiver by using a threshold device set at the appropriate level. Note that the use of time synchronization in the manner described here increases the bandwidth expansion factor to $M + 1$, where M is the number of message signals being multiplexed.

The TDM system is highly sensitive to dispersion in the common transmission channel—that is, to variations of amplitude with frequency or a nonlinear phase response. Accordingly, accurate equalization of both the amplitude and phase responses of the channel is necessary to ensure satisfactory operation of the system. Equalization of a communication channel is discussed in Chapter 8.

EXAMPLE 5.7   COMPARISON OF TDM WITH FDM   A TDM system is used to multiplex four independent voice signals using PAM. Each voice signal is sampled at the rate of 8 kHz. The system incorporates a synchronizing pulse train for its proper operation.

  (a) Determine the timing relationships between the synchronizing pulse train and the impulse trains used to sample the four voice signals.
  (b) Calculate the transmission bandwidth of the channel for the TDM system, and compare the result with a corresponding FDM system using SSB modulation.

Solution:

  (a) The sampling period is

$$T_s = \frac{1}{8 \times 10^3} \, \text{s} = 125 \, \mu\text{s}.$$

In this example, the number of voice signals is $M = 4$. Hence, dividing the sampling period of 125 μs among these voice signals and the synchronizing pulse train, we obtain the time slot allocated to each one of them:

$$T_0 = \frac{T_s}{M + 1}$$

$$= \frac{125}{5} = 25 \, \mu\text{s}.$$

Figure 5.30 shows the timing relationships between the synchronizing pulse train and the four impulse trains used to sample the different voice signals in a single frame. Each frame includes time slots of common duration $T_0 = 25 \, \mu$s, which are allocated to the pulse-modulated signals and synchronizing pulse.

  (b) As a consequence of the time–bandwidth product (see Section 3.17), there is an inverse relationship between the duration of a pulse and the bandwidth (i.e., cutoff

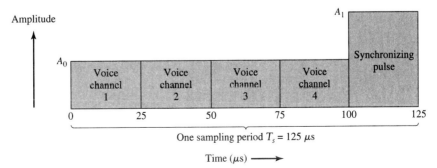

FIGURE 5.30   Composition of one frame of a multiplexed PAM signal incorporating four voice signals and a synchronizing pulse.

frequency) of the channel needed for its transmission. Accordingly, the overall transmission bandwidth of the channel is

$$f_T = \frac{\omega_T}{2\pi}$$

$$= \frac{1}{T_0}$$

$$= \frac{1}{25}\,\text{MHz} = 40\,\text{kHz}.$$

In contrast, the use of an FDM system based on SSB modulation requires a channel bandwidth equal to M times that of a single voice signal—that is, $4 \times 4 = 16$ kHz. Thus, the use of PAM-TDM requires a channel bandwidth that is $40/16 = 2.5$ times that of SSB-FDM.

In practice, pulse-code modulation is commonly used as the method of modulation for TDM; this results in a further increase in channel bandwidth, depending on the length of the code word used in the digital representation of each pulse in the PAM signal.   ■

5.10 *Phase and Group Delays*

Whenever a signal is transmitted through a dispersive (i.e., frequency-selective) system, such as a communication channel, some *delay* is introduced into the output signal in relation to the input signal. The delay is determined by the phase response of the system.

For convenience of presentation, let

$$\phi(\omega) = \arg\{H(j\omega)\} \tag{5.32}$$

denote the phase response of a dispersive communication channel, where $H(j\omega)$ is the frequency response of the channel. Suppose that a sinusoidal signal is transmitted through the channel at a frequency ω_c. The signal received at the channel output lags the transmitted signal by $\phi(\omega_c)$ radians. The time delay corresponding to this phase lag is simply equal to $-\phi(\omega_c)/\omega_c$, where the minus sign accounts for the lag. The time delay is called the *phase delay* of the channel, formally defined as

$$\tau_p = -\frac{\phi(\omega_c)}{\omega_c}. \tag{5.33}$$

It is important to realize, however, that the phase delay is *not* necessarily the true signal delay. This follows from the fact that a sinusoidal signal has infinite duration, with each cycle exactly like the preceding one. Such a signal does not convey information, except for the fact that it is there, so to speak. It would therefore be incorrect to deduce from the preceding reasoning that the phase delay is the true signal delay. In actuality, as we have seen from the material presented in this chapter, information can be transmitted through a channel by only applying some form of modulation to a carrier.

Suppose that we have a transmitted signal

$$s(t) = A\cos(\omega_c t)\cos(\omega_0 t) \tag{5.34}$$

consisting of a DSB-SC modulated wave with carrier frequency ω_c and sinusoidal modulation frequency ω_0. This signal corresponds to the one considered in Example 5.3. (For convenience of presentation, we have set $A = A_c A_0$.) Expressing the modulated signal $s(t)$ in terms of its upper and lower side frequencies, we may write

$$s(t) = \tfrac{1}{2} A \cos(\omega_1 t) + \tfrac{1}{2} A \cos(\omega_2 t),$$

where

$$\omega_1 = \omega_c + \omega_0 \tag{5.35}$$

and

$$\omega_2 = \omega_c - \omega_0. \tag{5.36}$$

Now let the signal $s(t)$ be transmitted through the channel with phase response $\phi(\omega)$. For illustrative purposes, we assume that the magnitude response of the channel is essentially constant (equal to unity) over the frequency range from ω_1 to ω_2. Accordingly, the signal received at the channel output is

$$y(t) = \tfrac{1}{2} A \cos(\omega_1 t + \phi(\omega_1)) + \tfrac{1}{2} A \cos(\omega_2 t + \phi(\omega_2)),$$

where $\phi(\omega_1)$ and $\phi(\omega_2)$ are the phase shifts produced by the channel at frequencies ω_1 and ω_2, respectively. Equivalently, we may express $y(t)$ as

$$y(t) = A \cos\left(\omega_c t + \frac{\phi(\omega_1) + \phi(\omega_2)}{2}\right) \cos\left(\omega_0 t + \frac{\phi(\omega_1) - \phi(\omega_2)}{2}\right), \tag{5.37}$$

where we have invoked the definitions of ω_1 and ω_2 given in Eqs. (5.35) and (5.36), respectively. Comparing the sinusoidal carrier and message components of the received signal $y(t)$ in Eq. (5.37) with those of the transmitted signal $s(t)$ in Eq. (5.34), we make the following two observations:

1. The carrier component at frequency ω_c in $y(t)$ lags its counterpart in $s(t)$ by $\tfrac{1}{2}(\phi(\omega_1) + \phi(\omega_2))$, which represents a time delay equal to

$$-\frac{\phi(\omega_1) + \phi(\omega_2)}{2\omega_c} = -\frac{\phi(\omega_1) + \phi(\omega_2)}{\omega_1 + \omega_2}. \tag{5.38}$$

2. The message component at frequency ω_0 in $y(t)$ lags its counterpart in $s(t)$ by $\tfrac{1}{2}(\phi(\omega_1) - \phi(\omega_2))$, which represents a time delay equal to

$$-\frac{\phi(\omega_1) - \phi(\omega_2)}{2\omega_0} = -\frac{\phi(\omega_1) - \phi(\omega_2)}{\omega_1 - \omega_2}. \tag{5.39}$$

Suppose that the modulation frequency ω_0 is small compared with the carrier frequency ω_c, which implies that the side frequencies ω_1 and ω_2 are close together, with ω_c between them. Such a modulated signal is said to be a *narrowband signal*. Then we may approximate the phase response $\phi(\omega)$ in the vicinity of $\omega = \omega_c$ by the two-term Taylor series expansion

$$\phi(\omega) = \phi(\omega_c) + \frac{d\phi(\omega)}{d\omega}\bigg|_{\omega = \omega_c} \times (\omega - \omega_c). \tag{5.40}$$

Using this expansion to evaluate $\phi(\omega_1)$ and $\phi(\omega_2)$ for substitution into Eq. (5.38), we see that the *carrier delay* is equal to $-\phi(\omega_c)/\omega_c$, which is identical to the formula given in Eq. (5.33) for the phase delay. Treating Eq. (5.39) in a similar way, we find that the time delay incurred by the message signal (i.e., the "envelope" of the modulated signal) is given by

$$\boxed{\tau_g = -\frac{d\phi(\omega)}{d\omega}\bigg|_{\omega = \omega_c}.} \tag{5.41}$$

The time delay τ_g is called the *envelope delay* or *group delay*. Thus, the group delay is defined as the negative of the derivative of the phase response $\phi(\omega)$ of the channel with respect to ω, evaluated at the carrier frequency ω_c.

In general, then, we find that when a modulated signal is transmitted through a communication channel, there are two different delays to be considered:

1. The carrier or phase delay τ_p, defined by Eq. (5.33)
2. The envelope or group delay τ_g, defined by Eq. (5.41)

The group delay is the true signal delay.

▶ **Problem 5.14**   What are the conditions for which the phase delay and group delay assume a common value?

Answer: The phase response $\phi(\omega)$ must be linear in ω, and $\phi(\omega_c) = 0$ ◀

EXAMPLE 5.8 PHASE AND GROUP DELAYS FOR BAND-PASS CHANNEL   The phase response of a band-pass communication channel is defined by

$$\phi(\omega) = -\tan^{-1}\left(\frac{\omega^2 - \omega_c^2}{\omega\omega_c}\right).$$

The signal $s(t)$ defined in Eq. (5.34) is transmitted through this channel with

$$\omega_c = 4.75 \text{ rad/s} \quad \text{and} \quad \omega_0 = 0.25 \text{ rad/s}.$$

Calculate (**a**) the phase delay and (**b**) the group delay.

Solution:

(a) At $\omega = \omega_c$, $\phi(\omega_c) = 0$. According to Eq. (5.33), the phase delay τ_p is zero.

(b) Differentiating $\phi(\omega)$ with respect to ω, we get

$$\frac{d\phi(\omega)}{d\omega} = -\frac{\omega_c(\omega^2 + \omega_c^2)}{\omega_c^2\omega^2 + (\omega^2 - \omega_c^2)^2}.$$

Using this result in Eq. (5.41), we find that the group delay is

$$\tau_g = \frac{2}{\omega_c} = \frac{2}{4.75} = 0.4211 \text{ s}.$$

To display the results obtained in parts (a) and (b) in graphical form, Fig. 5.31 shows a superposition of two waveforms obtained as follows:

1. One waveform, shown as a solid curve, was obtained by multiplying the transmitted signal $s(t)$ by the carrier wave $\cos(\omega_c t)$.

2. The second waveform, shown as a dotted curve, was obtained by multiplying the received signal $y(t)$ by the carrier wave $\cos(\omega_c t)$.

The figure clearly shows that the carrier (phase) delay τ_p is zero and the envelope of the received signal $y(t)$ is lagging behind that of the transmitted signal by τ_g seconds. For the presentation of waveforms in this figure, we purposely did not use a filter to suppress the high-frequency components resulting from the multiplications described under points 1 and 2, because of the desire to retain a contribution due to the carrier for display.

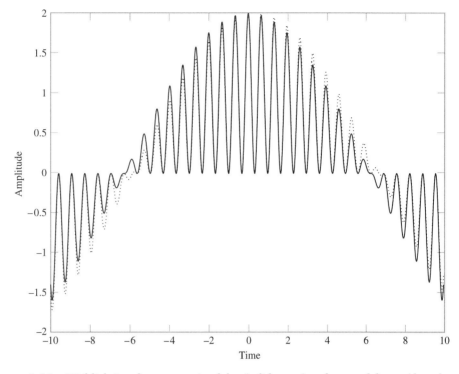

FIGURE 5.31   Highlighting the zero carrier delay (solid curve) and group delay τ_g (dotted curve), which are determined in accordance with Example 5.8.

Note also that the separation between the upper side frequency $\omega_1 = \omega_c + \omega_0$ = 5.00 rad/s and the lower side frequency $\omega_2 = \omega_c - \omega_0$ = 4.50 rad/s is about 10% of the carrier frequency ω_c = 4.75 rad/s, which justifies referring to the modulated signal in this example as a narrowband signal.   ■

■ 5.10.1 SOME PRACTICAL CONSIDERATIONS

Having established that group delay is the true signal delay when a modulated signal is transmitted through a communication channel, we now need to address the following question: What is the practical importance of group delay? To deal with this question, we first have to realize that the formula of Eq. (5.41) for determining group delay applies strictly to modulated signals that are narrowband; that is, the bandwidth of the message signal is small compared with the carrier frequency. It is only when this condition is satisfied that we would be justified in using the two-term approximation of Eq. (5.40) for the phase response $\phi(\omega)$, on the basis of which Eq. (5.41) was derived.

However, there are many practical situations in which this narrowband assumption is not satisfied because the message bandwidth is comparable to the carrier frequency. In situations of this kind, the group delay is formulated as a *frequency-dependent parameter*; that is,

$$\tau_g(\omega) = -\frac{d\phi(\omega)}{d\omega}, \tag{5.42}$$

which includes Eq. (5.41) as a special case. Now we begin to see the real importance of group delay: When a *wideband* modulated signal is transmitted through a dispersive

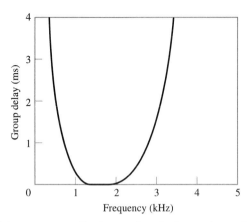

FIGURE 5.32   Group delay response of voice-grade telephone channel. (Adapted from Bellamy, J. C., *Digital Telephony*, Wiley, 1982.)

channel, the frequency components of the message signal are delayed by different amounts at the channel output. Consequently, the message signal undergoes a form of linear distortion known as *delay distortion*. To reconstruct a faithful version of the original message signal in the receiver, we have to use a *delay equalizer*. This equalizer has to be designed in such a way that when it is connected in cascade with the channel, the overall group delay is constant (i.e., the overall phase is linear with frequency).

As an illustrative example, consider the ubiquitous telephone channel, the useful frequency band of which extends from about 0.1 to 3.1 kHz. Over this band of frequencies, the magnitude response of the channel is considered to be essentially constant, so that there is little amplitude distortion. In contrast, the group delay of the channel is highly dependent on frequency, as shown in Fig. 5.32. Insofar as telephonic communication is concerned, the variation of group delay in the channel with frequency is of no real consequence, because our ears are relatively insensitive to delay distortion. The story is dramatically different, however, when wideband data are transmitted over a telephone channel. For example, for a data rate of 4 kilobits per second, the bit duration is about 25 μs. From the figure, we see that over the useful frequency band of the telephone channel, the group delay varies from zero to several milliseconds. Accordingly, delay distortion is extremely harmful to wideband data transmission over a telephone channel. In such an application, delay equalization is essential for satisfactory operation.

5.11 *Exploring Concepts with MATLAB*

Earlier, we discussed the idea of modulation for the transmission of a message signal over a band-pass channel. To illustrate this idea, we used a sinusoidal wave as the message (modulating) signal. In this regard, we used Examples 5.1 and 5.3 to illustrate the spectra of sinusoidally modulated waves based on full AM and DSB-SC modulation, assuming ideal conditions. In this section, we use MATLAB to expand on those examples by considering modulated waves of finite duration, which is how they always are in real-life situations. In particular, we build on the results presented in Example 4.16, in which we used the DTFS to approximate the Fourier transform of a finite-duration signal consisting of a pair of sinusoidal components.

■ 5.11.1 FULL AM

In the time-domain description of amplitude modulation, the mo
the carrier plus a product of the message signal (i.e., the modulat.
er. Thus, for the case of sinusoidal modulation considered in Exan

$$s(t) = A_c[1 + \mu \cos(\omega_0 t)] \cos(\omega_c t)$$

where μ is the modulation factor. The term $1 + \mu \cos(\omega_0 t)$ is a mc
modulating signal, and $A_c \cos(\omega_c t)$ is the carrier.

For the AM experiment described here, we have the following data:

Carrier amplitude,	$A_c = 1$;
Carrier frequency,	$\omega_c = 0.8\pi$ rad/s;
Modulation frequency,	$\omega_0 = 0.1\pi$ rad/s.

We wish to display and analyze 10 full cycles of the AM wave, corresponding to a total duration of 200 s. Choosing a sampling rate $1/T_s = 10$ Hz, we have a total of $N = 2000$
time samples. The frequency band of interest is $-10\pi \le \omega \le 10\pi$. Since the separation between the carrier and either side frequency is equal to the modulation frequency
$\omega_0 = 0.1\pi$ rad/s, we would like to have a frequency resolution $\omega_r = 0.01\pi$ rad/s. To
achieve this resolution, we require the following number of frequency samples (see
Eq. (4.42)):

$$M \ge \frac{\omega_s}{\omega_r} = \frac{20\pi}{0.01\pi} = 2000.$$

We therefore choose $M = 2000$. To approximate the Fourier transform of the AM wave
$s(t)$, we may use a 2000-point DTFS. The only variable in the AM experiment is the modulation factor μ, with respect to which we wish to investigate three different situations:

- $\mu = 0.5$, corresponding to undermodulation
- $\mu = 1.0$, for which the AM system is on the verge of overmodulation
- $\mu = 2.0$, corresponding to overmodulation

Putting all of these points together, we may now formulate the MATLAB commands
for generating the AM wave and analyzing its frequency content as follows:

```
>> Ac = 1;  % carrier amplitude
>> wc = 0.8*pi;  % carrier frequency
>> w0 = 0.1*pi;  % modulation frequency
>> mu = 0.5;  % modulation factor
>> t = 0:0.1:199.9;
>> s = Ac*(1 + mu*cos(w0*t)).*cos(wc*t);
>> plot(t,s)
>> Smag = abs(fftshift(fft(s,2000)))/2000;
   % Smag denotes the magnitude spectrum of the AM wave
>> w = 10*[-1000:999]*2*pi/2000;
>> plot(w,Smag)
>> axis ([-30  30  0  0.8])
```

The fourth command is written for $\mu = 0.5$. The computations are repeated for $\mu = 1, 2$.

We next describe the effect of varying the modulation factor μ on the time-domain
and frequency-domain characteristics of the AM wave:

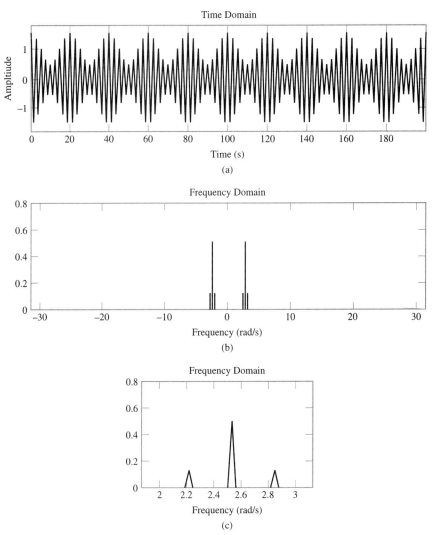

FIGURE 5.33 Amplitude modulation with 50% modulation. (a) AM wave, (b) magnitude spectrum of the AM wave, and (c) expanded spectrum around the carrier frequency.

1. $\mu = 0.5$.

Figure 5.33(a) shows 10 cycles of the full AM wave $s(t)$ corresponding to $\mu = 0.5$. The envelope of $s(t)$ is clearly seen to follow the sinusoidal modulating wave faithfully. This means that we can use an envelope detector for demodulation. Figure 5.33(b) shows the magnitude spectrum of $s(t)$. In Fig. 5.33(c), we have zoomed in on the fine structure of the spectrum of $s(t)$ around the carrier frequency. This figure clearly displays the exact relationships between the side frequencies and the carrier, in accordance with modulation theory. In particular, the lower side frequency, the carrier, and the upper side frequency are located at $\omega_c - \omega_0 = \pm 0.7\pi$ rad/s, $\omega_c = \pm 0.8\pi$ rad/s, and $\omega_c + \omega_0 = \pm 0.9\pi$ rad/s, respectively. Moreover, the amplitude of both sidebands is $(\mu/2) = 0.25$ times that of the carrier.

2. $\mu = 1.0$.

Figure 5.34(a) shows 10 cycles of the AM wave $s(t)$ with the same parameters as in Fig. 5.33(a), except for the fact that $\mu = 1.0$. This figure shows that the AM wave is now

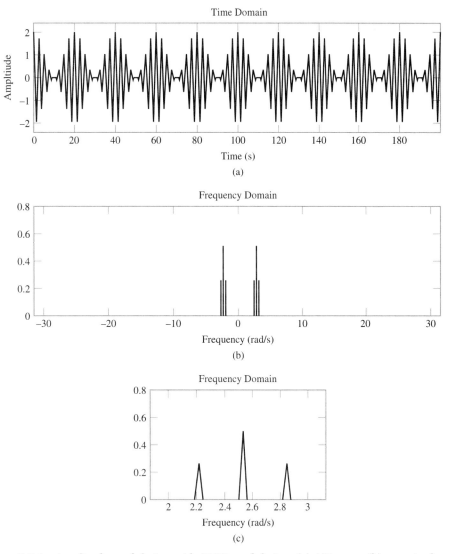

FIGURE 5.34 Amplitude modulation with 100% modulation. (a) AM wave, (b) magnitude spectrum of the AM wave, and (c) expanded spectrum around the carrier frequency.

on the verge of overmodulation. The magnitude spectrum of $s(t)$ is shown in Fig. 5.34(b), and its zoomed version (around the carrier frequency) is shown in Fig. 5.34(c). Here again, we see that the basic structure of the magnitude spectrum of the full AM wave is in perfect accord with the theory.

3. $\mu = 2.0.$
Figure 5.35(a) demonstrates the effect of overmodulation by using a modulation factor of $\mu = 2$. Here we see that there is no clear relationship between the envelope of the overmodulated wave $s(t)$ and the sinusoidal modulating wave. This implies that an envelope detector will not work, so we must use a coherent detector to perform the process of demodulation. Note, however, that the basic spectral content of the AM wave displayed in Figs. 5.35(b) and (c) follows exactly what the theory predicts.

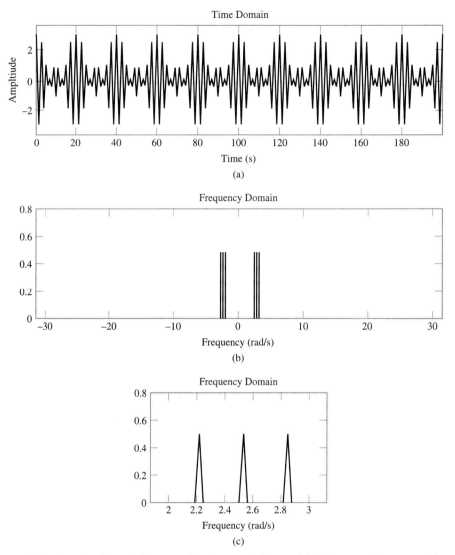

FIGURE 5.35   Amplitude modulation with 200% modulation. (a) AM wave, (b) magnitude spectrum of the AM wave, and (c) expanded spectrum around the carrier frequency.

■ 5.11.2 DSB-SC MODULATION

In a DSB-SC modulated wave, the carrier is suppressed and both sidebands are transmitted in full. This signal is produced simply by multiplying the modulating wave by the carrier wave. Thus, for the case of sinusoidal modulation, we have

$$s(t) = A_c A_0 \cos(\omega_c t) \cos(\omega_0 t).$$

The MATLAB commands for generating $s(t)$ and analyzing its frequency content are as follows:

```
>> Ac = 1;  % carrier amplitude
>> wc = 0.8*pi;  % carrier frequency in rad/s
>> A0 = 1;  % amplitude of modulating signal
>> w0 = 0.1*pi;  % frequency of modulating signal
>> t = 0:.1:199.9;
```

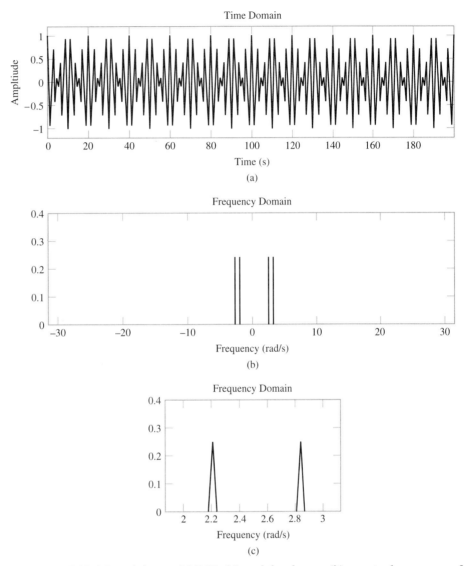

FIGURE 5.36 DSB-SC modulation. (a) DSB-SC modulated wave, (b) magnitude spectrum of the modulated wave, and (c) expanded spectrum around the carrier frequency.

```
>> s = Ac*A0*cos(wc*t).*cos(w0*t);
>> plot(t,s)
>> Smag = abs(fftshift(fft(s,2000)))/2000;
>> w = 10*[-1000:999]*2*pi/2000;
>> plot(w,Smag)
```

These commands were used to investigate the following different aspects of DSB-SC modulation:

1. Figure 5.36(a) shows 10 cycles of the DSB-SC modulated wave $s(t)$ produced by the sinusoidal modulating wave. As expected, the envelope of the modulated wave bears no clear relationship to the sinusoidal modulating wave. Accordingly, we must use coherent detection for demodulation, which is discussed further under point 2, next. Figure 5.36(b) shows the magnitude spectrum of $s(t)$. An expanded view of the spectrum

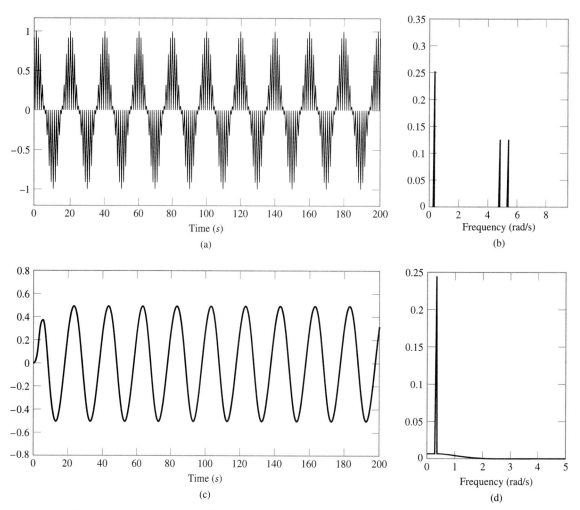

FIGURE 5.37   Coherent detection of DSB-SC modulated wave. (a) Waveform of signal produced at the output of product modulator; (b) magnitude spectrum of the signal in part (a); (c) waveform of low-pass filter output; (d) magnitude spectrum of signal in part (c).

around the carrier frequency is shown in Fig. 5.36(c). These two figures clearly show that the carrier is indeed suppressed and that the upper and lower side frequencies are located exactly where they should be, namely, at 0.9π and 0.7π rad/s, respectively.

2. To perform coherent detection, we multiply the DSB-SC modulated wave $s(t)$ by a replica of the carrier and then pass the result through a low-pass filter, as described in Section 5.5.2. Assuming perfect synchronism between the transmitter and receiver, we define the output of the product modulator in Fig. 5.12(b) with $\phi = 0$ as

$$v(t) = s(t)\cos(\omega_c t).$$

Correspondingly, the MATLAB command is

```
>> v = s.*cos(wc*t);
```

where **s** is itself as computed previously. Figure 5.37(a) shows the waveform of $v(t)$. Applying the **fft** command to **v** and taking the absolute value of the result, we obtain the magnitude spectrum of Fig. 5.37(b), which readily shows that $v(t)$ consists of the following components:

▶ A sinusoidal component with frequency 0.1π rad/s, representing the modulating wave.

▶ A new DSB-SC modulated wave with double carrier frequency of 1.6π rad/s; in actuality, the side frequencies of this modulated wave are located at 1.5π and 1.7π rad/s.

Accordingly, we may recover the sinusoidal modulating signal by passing $v(t)$ through a low-pass filter with the following requirements:

▶ The frequency of the modulating wave lies inside the passband of the filter.

▶ The upper and lower side frequencies of the new DSB-SC modulated wave lie inside the stopband of the filter.

The issue of how to design a filter with these requirements will be considered in detail in Chapter 8. For the present, it suffices to say that the preceding requirements can be met by using the following MATLAB commands:

```
>> [b,a] = butter(3,0.025);
>> output = filter(b,a,v);
```

The first command produces a special type of filter called a Butterworth filter. For the experiment considered here, the filter order is 3 and its *normalized cutoff frequency* of 0.025 is calculated as follows:

$$\frac{\text{Actual cutoff frequency of filter}}{\text{Half the sampling rate}} = \frac{0.25\pi \text{ rad/s}}{10\pi \text{ rad/s}}$$
$$= 0.025.$$

The second command computes the filter's output in response to the product modulator output $v(t)$. (We will revisit the design of this filter in Chapter 8.) Figure 5.37(c) displays the waveform of the low-pass filter output; this waveform represents a sinusoidal signal of frequency 0.05 Hz, an observation that is confirmed by using the `fft` command to approximate the spectrum of the filter output. The result of the computation is shown in Fig. 5.37(d).

3. In Fig. 5.38, we explore another aspect of DSB-SC modulation, namely, the effect of varying the modulation frequency. Figure 5.38(a) shows five cycles of a DSB-SC modulated wave that has the same carrier frequency as that in Fig. 5.36(a), but the modulation frequency has been reduced to 0.025 Hz (i.e., a radian frequency of 0.05π). Figure 5.38(b) shows the magnitude spectrum of this second DSB-SC modulated wave, and its zoomed-in version is shown in Fig. 5.38(c). Comparing this latter figure with Fig. 5.36(c), we see clearly that decreasing the modulation frequency has the effect of moving the upper and lower side frequencies closer together, which is consistent with modulation theory.

▶ **Problem 5.15**   A *radiofrequency (RF) pulse* is defined as the product of a rectangular pulse and a sinusoidal carrier wave. Using MATLAB, plot the waveform of this pulse for each of the following two cases:

(a) Pulse duration = 1 s
   Carrier frequency = 5 Hz

(b) Pulse duration = 1 s
   Carrier frequency = 25 Hz

Use a sampling frequency of 1 kHz.   ◀

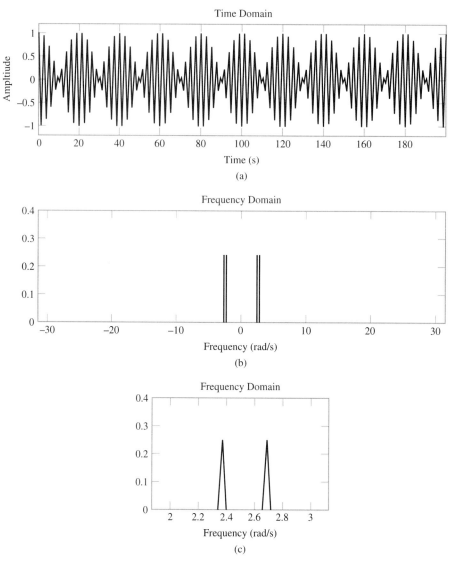

FIGURE 5.38   Effect of varying the modulation frequency, compared with that in Fig. 5.36. (a) Waveform of DSB-SC modulated wave with a modulation frequency one-half that used in Fig. 5.36; (b) magnitude spectrum of signal in part (a); (c) expanded spectrum around the carrier frequency.

▶ **Problem 5.16**   Using the `fft` command, plot the magnitude spectrum of the RF pulse for each of the two cases described in Problem 5.15. Hence, demonstrate the following:

(a) For case (a), corresponding to carrier frequency 5 Hz, the lower sidebands for positive and negative frequencies overlap each other. This effect is the spectral overlap discussed in Section 5.4.6.

(b) For case (b), corresponding to carrier frequency 25 Hz, the spectrum is essentially free from spectral overlap.

For a radar perspective of Problems 5.15 and 5.16, the reader may refer to the Theme Example on Radar described in Section 1.10.   ◀

◼ 5.11.3 PHASE AND GROUP DELAYS

In Example 5.8, we studied the phase and group delays for a band-pass channel with phase response

$$\phi(\omega) = -\tan^{-1}\left(\frac{\omega^2 - \omega_c^2}{\omega\omega_c}\right).$$

At $\omega = \omega_c$, the phase delay is $\tau_p = 0$ and the group delay is $\tau_g = 0.4211s$. One of the waveforms, displayed in Fig. 5.31, is

$$x_1(t) = s(t)\cos(\omega_c t)$$

where (see page 461)

$$s(t) = \frac{A}{2}[\cos(\omega_1 t) + \cos(\omega_2 t)]$$

and $\omega_1 = \omega_c + \omega_0$ and $\omega_2 = \omega_c - \omega_0$. The waveform shown in the figure as a solid curve is a plot of $x_1(t)$. The other waveform, also displayed in Fig. 5.31, is

$$x_2(t) = y(t)\cos(\omega_c t)$$

$$= \frac{A}{2}[\cos(\omega_1 t + \phi(\omega_1)) + \cos(\omega_2 t + \phi(\omega_2))]\cos(\omega_c t),$$

where the angles $\phi(\omega_1)$ and $\phi(\omega_2)$ are the values of the phase response $\phi(\omega)$ at $\omega = \omega_1$ and $\omega = \omega_2$, respectively. The waveform shown as a dotted curve in the figure is a plot of $x_2(t)$.

The generation of $x_1(t)$ and $x_2(t)$ in MATLAB is achieved with the following commands with $A/2$ set equal to unity:

```
>> wc = 4.75;
>> w0 = 0.25;
>> w1 = wc + w0;
>> w2 = wc - w0;
>> t = -10 : 0.001 = 10;
>> o1 = -atan((w1^2 - wc^2)/(w1*wc));
>> o2 = -atan((w2^2 - wc^2)/(w2*wc));
>> s = cos(w1*t) + cos(w2*t);
>> y = cos(w1*t + o1) + cos(w2*t + o2);
>> x1 = s.*cos(4.75*t);
>> x2 = y.*cos(4.75*t);
>> plot (t, x1, `b´)
>> hold on
>> plot (t, x2, `k´)
>> hold off
>> xlabel(`Time´)
>> ylabel(`Amplitude´)
```

Note that we have set $(A/2) = 1$ for convenience of presentation. The function `atan` in the first two commands returns the arctangent. Note also that the computation of both `x1` and `x2` involve element-by-element multiplications—hence the use of a period followed by an asterisk.

5.12 *Summary*

In this chapter, we presented a discussion of linear modulation techniques for the transmission of a message signal over a communication channel.

In particular, we described amplitude modulation (AM) and its variants, summarized as follows:

▶ In full AM, the spectrum consists of two sidebands (one termed the upper sideband, the other the lower sideband) and the carrier. The primary advantage of full AM is the simplicity of its implementation, which explains its popular use in radio broadcasting. The disadvantages include a wastage of transmission bandwidth and power.

▶ In double sideband-suppressed carrier (DSB-SC) modulation, the carrier is suppressed, saving transmission power. However, the transmission bandwidth for DSB-SC modulation is the same as that of full AM—that is, twice the message bandwidth.

▶ In single sideband (SSB) modulation, only one of the sidebands is transmitted. SSB modulation is therefore the optimum form of continuous-wave (CW) modulation, in that it requires the least amount of channel bandwidth and power for its transmission. The use of SSB modulation requires the presence of an energy gap at around zero frequency in the spectrum of the message signal.

▶ In vestigial sideband (VSB) modulation, a modified version of one sideband and appropriately designed vestige of the other sideband are transmitted. This form of AM is well suited for transmitting wideband signals whose spectra extend down to zero frequency. VSB modulation is the standard analog method for the transmission of television signals.

The other form of linear modulation that was discussed in the chapter is pulse-amplitude modulation (PAM), which represents the simplest form of pulse modulation. PAM may be viewed as a direct manifestation of the sampling process; accordingly, it is commonly used as a method of modulation in its own right. Moreover, it constitutes an operation that is basic to all the other forms of pulse modulation, including pulse-code modulation.

We then discussed the notion of multiplexing, which permits the sharing of a common communication channel among a number of independent users. In frequency-division multiplexing (FDM), the sharing is performed in the frequency domain; in time-division multiplexing (TDM), the sharing is performed in the time domain.

The other topic that was discussed in the chapter is that of phase (carrier) delay and group (envelope) delay, both of which are defined in terms of the phase response of a channel over which a modulated signal is transmitted. The group delay is the true signal delay; it becomes of paramount importance when a wideband modulated signal is transmitted over the channel.

One final comment is in order: In discussing the modulation systems presented in this chapter, we made use of two functional blocks:

▶ Filters for the suppression of spurious signals

▶ Equalizers for correcting signal distortion produced by physical transmission systems

The approach taken herein was from a system-theoretic viewpoint, and we did not concern ourselves with the design of these functional blocks. Design considerations of filters and equalizers are taken up in Chapter 8.

FURTHER READING

1. Communication technology has an extensive history that dates back to the invention of the telegraph (the predecessor to digital communication) by Samuel Morse in 1837. This was followed by the invention of the telephone by Alexander Graham Bell in 1875, in whose honor the decibel is named. Other notable contributors to the subject include Harry Nyquist, who published a classic paper on the theory of signal transmission in telegraphy in 1928, and Claude Shannon, who laid down the foundations of *information theory* in 1948. Information theory is a broad subject, encompassing the transmission, processing, and utilization of information.

   For a historical account of communication systems, see Chapter 1 of

   ▶ Haykin, S., *Communication Systems*, 4th ed. (Wiley, 2001)

2. The subbands detailed under point 1 of Section 5.3 on Benefits of Modulation apply to the first-generation (analog) and second-generation (digital) cellular radio systems. The evolution of third-generation systems began in the late 1980s. *Universal Mobile Telecommunications System* (UMTS) is the term introduced for the third-generation wireless mobile communication systems. The subbands for UMTS corresponding to those detailed in Section 5.3 are 1885–2025 MHz and 2110–2200 MHz, respectively. Third-generation systems are wideband, whereas both first-generation and second-generation systems are narrowband.

   For detailed treatment of wireless mobile communication systems, see the book Steel, R., and L. Hanzo, *Mobile Radio Communications*, 2nd ed., Wiley, 1999.

3. For a more complete treatment of modulation theory, see the following books:

   ▶ Carlson, A. B., *Communication Systems: An Introduction to Signals and Noise in Electrical Communications*, 3rd ed. (McGraw-Hill, 1986)

   ▶ Couch, L. W., III, *Digital and Analog Communication Systems*, 3rd ed. (Prentice Hall, 1990)

   ▶ Haykin, *op. cit.*

   ▶ Schwartz, M., *Information Transmission Modulation and Noise: A Unified Approach*, 3rd ed. (McGraw-Hill, 1980)

   ▶ Stremler, F. G., *Introduction to Communication Systems*, 3rd ed. (Addison-Wesley, 1990)

   ▶ Ziemer, R. E., and W. H. Tranter, *Principles of Communication Systems*, 3rd ed. (Houghton Mifflin, 1990)

   These books cover both continuous-wave modulation and pulse modulation techniques. The books listed here also include the study of how noise affects the performance of modulation systems.

4. The Hilbert transform of a signal $x(t)$ is defined by

$$\hat{x}(t) = \frac{1}{\pi} \int_{-\infty}^{\infty} \frac{x(\tau)}{t - \tau} \, d\tau.$$

   Equivalently, we may define the Hilbert transform $\hat{x}(t)$ as the convolution of $x(t)$ with $1/(\pi t)$. The Fourier transform of $1/(\pi t)$ is $-j$ times the signum function, denoted

$$\text{sgn}(\omega) = \begin{cases} +1, & \text{for } \omega > 0 \\ 0, & \text{for } \omega = 0. \\ -1, & \text{for } \omega < 0 \end{cases}$$

   (See subsection 3.11.3.) Passing $x(t)$ through a Hilbert transformer is therefore equivalent to the combination of the following two operations in the frequency domain:

   ▶ Keeping $|X(j\omega)|$ (i.e., the magnitude spectrum of $x(t)$) unchanged for all ω

> ▸ Shifting $\arg\{X(j\omega)\}$ (i.e., the phase spectrum of $x(t)$) by $+90°$ for negative frequencies and $-90°$ for positive frequencies

For a more complete discussion of the Hilbert transform and its use in the time-domain description of single sideband modulation, see Haykin, *op. cit.*

5. For a discussion of code-division multiplexing, see Haykin, *op. cit.*

6. For an advanced treatment of phase delay and group delay, see Haykin, *op. cit.*

ADDITIONAL PROBLEMS

5.17 Using the message signal

$$m(t) = \frac{1}{1 + t^2},$$

sketch the modulated waves for the following methods of modulation:

(a) Amplitude modulation with 50% modulation

(b) Double sideband-suppressed carrier modulation

5.18 The message signal $m(t)$ is applied to a full amplitude modulator. The carrier frequency is 100 kHz. Determine the frequency components generated by the modulator for the following message signals, where time t is measured in seconds:

(a) $m(t) = A_0 \cos(2\pi \times 10^3 t)$

(b) $m(t) = A_0 \cos(2\pi \times 10^3 t)$
$\qquad + A_1 \sin(4\pi \times 10^3 t)$

(c) $m(t) = A_0 \cos(2\pi \times 10^3 t)$
$\qquad \times \sin(4\pi \times 10^3 t)$

(d) $m(t) = A_0 \cos^2(2\pi \times 10^3 t)$

(e) $m(t) = \cos^2(2\pi \times 10^3 t) + \sin^2(4\pi \times 10^3 t)$

(f) $m(t) = A_0 \cos^3(2\pi \times 10^3 t)$

5.19 Repeat Problem 5.18 for $m(t)$ consisting of a square wave with a fundamental frequency equal to 500 Hz. The amplitude of the square wave takes one of two forms: (a) alternating between 0 and 1 and (b) alternating between -1 and $+1$.

5.20 Repeat Problem 5.18 for $m(t)$ consisting of the following:

(a) a voice signal whose spectral content extends from 300 Hz to 3100 Hz, and

(b) an audio signal whose spectral content extends from 50 Hz to 15 kHz.

5.21 The sinusoidal modulating signal

$$m(t) = A_0 \sin(\omega_0 t)$$

is applied to a full amplitude modulator. The carrier wave is $A_c \cos(\omega_c t)$. The maximum and minimum values of the envelope of the resulting modulated wave are

$$A_{\max} = 9.75 \text{ V}$$

and

$$A_{\min} = 0.25 \text{ V}.$$

Calculate the percentage of average power in (a) each of the two side frequencies and (b) the carrier.

5.22 The message signal $m(t)$ is applied to a double sideband-suppressed carrier modulator. The carrier frequency is 100 kHz. Determine the frequency components generated by the modulator for the following message signals, where time t is measured in seconds:

(a) $m(t) = A_0 \cos(2\pi \times 10^3 t)$

(b) $m(t) = A_0 \cos(2\pi \times 10^3 t)$
$\qquad + A_1 \sin(4\pi \times 10^3 t)$

(c) $m(t) = A_0 \cos(2\pi \times 10^3 t)$
$\qquad \times \sin(4\pi \times 10^3 t)$

(d) $m(t) = A_0 \cos^2(2\pi \times 10^3 t)$

(e) $m(t) = A_0 \cos^2(2\pi \times 10^3 t)$
$\qquad + A_1 \sin^2(4\pi \times 10^3 t)$

(f) $m(t) = A_0 \cos^3(2\pi \times 10^3 t)$

5.23 Repeat Problem 5.22 for $m(t)$ consisting of a square wave with a fundamental frequency equal to 500 Hz. The amplitude of the square wave takes one of two forms: (a) alternating between 0 and 1 and (b) alternating between -1 and $+1$.

5.24 Repeat Problem 5.22 for $m(t)$ consisting of the following:

(a) a voice signal whose spectral content extends from 300 Hz to 3100 Hz.

(b) an audio signal whose spectral content extends from 50 Hz to 15 kHz.

5.25 The message signal $m(t)$ is applied to a single-sideband modulator. The carrier frequency is 100 kHz. Determine the frequency components generated by the modulator for the following message signals, where time t is measured in seconds:

(a) $m(t) = A_0 \cos(2\pi \times 10^3 t)$

(b) $m(t) = A_0 \cos(2\pi \times 10^3 t)$
$\qquad + A_1 \sin(4\pi \times 10^3 t)$

(c) $m(t) = A_0 \cos(2\pi \times 10^3 t)$
$\qquad \times \sin(4\pi \times 10^3 t)$

(d) $m(t) = A_0 \cos^2(2\pi \times 10^3 t)$

(e) $m(t) = A_0 \cos^2(2\pi \times 10^3 t)$
$\qquad + A_1 \sin^2(4\pi \times 10^3 t)$

(f) $m(t) = A_0 \cos^3(2\pi \times 10^3 t)$

For your calculations, consider (i) transmission of the upper sideband and (ii) transmission of the lower sideband.

5.26 Repeat Problem 5.25 for $m(t)$ consisting of a square wave with a fundamental frequency of 500 Hz. The amplitude of the square wave takes one of two forms: (a) alternating between 0 and 1 and (b) alternating between -1 and $+1$.

5.27 Repeat Problem 5.26 for $m(t)$ consisting of the following:

(a) a voice signal whose spectral content extends from 300 Hz to 3100 Hz.

(b) an audio signal whose spectral content extends from 50 Hz to 15 kHz.

5.28 A full amplitude modulator has the following specifications:

Modulating signal:	sinusoidal
Modulation frequency:	4 kHz
Carrier frequency:	2 kHz

Determine the frequency content of the resulting modulated signal. Explain why this modulator will not function properly.

5.29 A double sideband-suppressed carrier modulator has the following specifications:

Modulating signal:	sinusoidal
Modulation frequency:	4 kHz
Carrier frequency:	2 kHz

(a) Determine the frequency content of the resulting modulated signal.

(b) To demonstrate that the modulator does not function properly, apply the modulated signal to a coherent detector that is supplied with a local oscillator of frequency 2 kHz. Show that the demodulated signal contains two different sinusoidal components, and determine their individual frequencies.

5.30 Consider a message signal $m(t)$ with the spectrum shown in Fig. P5.30. The message bandwidth $\omega_m = 2\pi \times 10^3$ rad/s. The signal is applied to a product modulator, together with a carrier wave $A_c \cos(\omega_c t)$, producing the DSB-SC modulated signal $s(t)$. The modulated signal is next applied to a coherent detector. Assuming perfect synchronism between the carrier waves in the modulator and detector, determine the spectrum of the detector output when (a) the carrier frequency $\omega_c = 2.5\pi \times 10^3$ rad/s and (b) the carrier frequency $\omega_c = 1.5\pi \times 10^3$ rad/s.

What is the lowest carrier frequency for which each component of the modulated signal $s(t)$ is uniquely determined by $m(t)$?

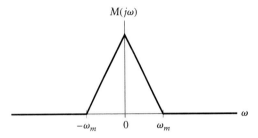

FIGURE P5.30

5.31 Figure P5.31 shows the circuit diagram of a balanced modulator. The input applied to the top AM modulator is $m(t)$, whereas that applied to the lower AM modulator is $-m(t)$; these two modulators have the same amplitude sensitivity. Show that the output $s(t)$ of the balanced modulator consists of a DSB-SC modulated signal.

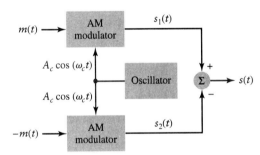

FIGURE P5.31

5.32 A pulse amplitude modulator has the following specifications:

Operation:	Sample and hold
Pulse duration =	$10\ \mu s$
Modulating signal:	sinusoidal
Modulation frequency =	1 kHz

Determine the side frequencies of the modulated signal.

5.33 Let a radio-frequency (RF) pulse be defined by

$$s(t) = \begin{cases} A_c \cos(\omega_c t), & -T/2 \le t \le T/2 \\ 0, & \text{otherwise} \end{cases}.$$

(a) Derive a formula for the spectrum of $s(t)$, assuming that $\omega_c T \gg 2\pi$.

(b) Sketch the magnitude spectrum of $s(t)$ for $\omega_c T = 20\pi$.

5.34 The transmitted signal $s(t)$ of a radar system consists of a periodic sequence of short RF pulses. The fundamental period of the sequence is T_0. Each RF pulse has duration T_1 and frequency ω_c. Typical values are

$$T = 1 \text{ ms,}$$
$$T_0 = 1 \text{ } \mu s,$$

and

$$\omega_c = 2\pi \times 10^9 \text{ rad/s.}$$

Using the results of Problem 5.33, sketch the magnitude spectrum of $s(t)$.

5.35 A DSB-SC modulated signal is demodulated by applying it to a coherent detector. Evaluate the effect of a frequency error $\Delta\omega$ in the local carrier frequency of the detector, measured with respect to the carrier frequency of the incoming DSB-SC signal.

5.36 Figure P5.36 shows the block diagram of a *frequency synthesizer*, which makes possible the generation of many frequencies, each with the same high accuracy as the *master oscillator*. With frequency 1 MHz, the master oscillator feeds two *spectrum generators*, one directly and the other through a *frequency divider*. Spectrum generator 1 produces a signal rich in the following harmonics: 1, 2, 3, 4, 5, 6, 7, 8, and 9 MHz. The frequency divider provides a 100-kHz output, in response to which spectrum generator 2 produces a second signal rich in the following harmonics: 100, 200, 300, 400, 500, 600, 700, 800, and 900 kHz. The harmonic selectors are designed to feed two signals into the *mixer*, one from spectrum generator 1 and the other from spectrum generator 2. (The mixer is another term for single-sideband modulator.) Find the range of possible frequency outputs and the resolution (i.e., the separation between adjacent frequency outputs) of the synthesizer.

5.37 Compare full AM with PAM, emphasizing their similarities and differences.

5.38 Specify the Nyquist rate for each of the following signals:

(a) $g(t) = \text{sinc}(200t)$
(b) $g(t) = \text{sinc}^2(200t)$
(c) $g(t) = \text{sinc}(200t) + \text{sinc}^2(200t)$

5.39 Twenty-four voice signals are sampled uniformly and are then time-division multiplexed, using PAM. The

PAM signal is reconstructed from flat-topped pulses with 1-μs duration. The multiplexing operation provides for synchronization by adding an extra pulse of sufficient amplitude and also 1-μs duration. The highest frequency component of each voice signal is 3.4 kHz.

(a) Assuming a sampling rate of 8 kHz, calculate the spacing between successive pulses of the multiplexed signal.

(b) Repeat your calculation, assuming the use of Nyquist rate sampling.

5.40 Twelve different message signals, each with a bandwidth of 10 kHz, are to be multiplexed and transmitted. Determine the minimum bandwidth required for each method if the multiplexing and modulation method used are

(a) FDM and SSB
(b) TDM and PAM

5.41 A PAM *telemetry* system involves the multiplexing of four input signals, $s_i(t)$, $i = 1, 2, 3, 4$. Two of the signals, $s_1(t)$ and $s_2(t)$, have bandwidths of 80 Hz each, whereas the remaining two signals, $s_3(t)$ and $s_4(t)$, have bandwidths of 1 kHz each. The signals $s_3(t)$ and $s_4(t)$ are each sampled at the rate of 2400 samples per second. This sampling rate is divided by 2^R (i.e., an integer power of 2) in order to derive the sampling rate for $s_1(t)$ and $s_2(t)$.

(a) Find the maximum permissible value of R.

(b) Using the value of R found in part (a), design a multiplexing system that first multiplexes $s_1(t)$ and $s_2(t)$ into a new sequence, $s_5(t)$, and then multiplexes $s_3(t)$, $s_4(t)$, and $s_5(t)$.

5.42 In Chapter 3, we presented the fundamentals of Fourier analysis. Chapter 4 examined the application of Fourier analysis to mixtures of signal classes. Write an essay beginning "Fourier analysis is an indispensable tool for the design of continuous-wave amplitude modulation and pulse-amplitude modulation systems." Your essay should emphasize two basic points:

(i) Spectral analysis of the modulated signal produced in the transmitter.

(ii) Recovery of the message signal in the receiver.

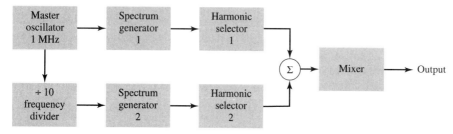

FIGURE P5.36

ADVANCED PROBLEMS

5.43 Suppose you are given a nonlinear device whose input–output relation is described by

$$i_o = a_1 v_i + a_2 v_i^2,$$

where a_1 and a_2 are constants, v_i is the input voltage, and i_o is the output current. Let

$$v_i(t) = A_c \cos(\omega_c t) + A_m \cos(\omega_m t),$$

where the first term represents a sinusoidal carrier and the second term represents a sinusoidal modulating signal.

(a) Determine the frequency content of $i_o(t)$.

(b) The output current $i_o(t)$ contains an AM signal produced by the two components of $v_i(t)$. Describe the specification of a filter that extracts this AM signal from $i_o(t)$.

5.44 In this problem, we discuss and compare two different methods of transmitting binary data over a band-pass channel. The two methods are *on–off keying* (OOK) and *binary phase-shift keying* (BPSK). In OOK, the binary symbols 0 and 1 are respectively represented by amplitude levels 0 volt and 1 volt. By contrast, in BPSK, the binary symbols are respectively represented by −1 volt and +1 volt. In the communications literature, these two representations of binary data are referred to as *unipolar* and *polar nonreturn-to-zero* sequences, respectively. In both cases, the sequence of binary symbols is multiplied by a sinusoidal carrier of fixed frequency and is transmitted over the channel.

(a) Consider the special case of binary data consisting of an infinite sequence of alternating symbols 0 and 1, each of duration T_0. Sketch the corresponding waveform of the resulting OOK signal and determine its magnitude spectrum.

(b) For the special binary sequence considered in part (a), plot the waveform of the BPSK signal and determine its magnitude spectrum.

(c) The alternating sequences of 0's and 1's considered in both parts (a) and (b) are square waves, hence permitting the application of the FS version of Parseval's theorem discussed in Section 3.16. (See Table 3.10.) Using this theorem, determine the average power of the transmitted OOK and BPSK signals.

(d) Building on the results derived in Part (c), how would you modify the sequence of binary symbols so that the average power of the OOK signal is the same as that of the BPSK signal?

(e) The OOK and BPSK signals may be viewed as digital versions of full AM and DSB-SC modulated waveforms. Justify the validity of this statement and discuss its practical implications.

(*Note:* The BPSK signal was considered briefly in Example 3.29, where the use of a square pulse for representing the symbol 0 or 1 was compared with the use of a raised-cosine pulse.)

5.45 Consider the quadrature-carrier multiplex system of Fig. 5.16. The multiplexed signal $s(t)$ produced at the transmitter input in Fig. 5.16(a) is applied to a communication channel with frequency response $H(j\omega)$. The output of this channel is in turn applied to the receiver input in Fig. 5.16(b). Prove that the condition

$$H(j\omega_c + j\omega) = H^*(j\omega_c - j\omega), \quad 0 < \omega < \omega_m$$

where ω_c is the carrier frequency and ω_m is the message bandwidth, is necessary for recovery of the message signal $m_1(t)$ and $m_2(t)$ at the receiver outputs. (*Hint:* Evaluate the spectra of the two receiver outputs.)

5.46 The spectrum of a voice signal $m(t)$ is zero outside the interval $\omega_a \leq |\omega| \leq \omega_b$. To ensure communication privacy, this signal is applied to a *scrambler* that consists of the following cascade of components: a product modulator, a high-pass filter, a second product modulator, and a low-pass filter. The carrier wave applied to the first product modulator has a frequency equal to ω_c, whereas that applied to the second product modulator has a frequency equal to $\omega_b + \omega_c$; both waves have unit amplitude. The high-pass and low-pass filters have the same cutoff frequency $\omega_c > \omega_b$.

(a) Derive an expression for the scrambler output $s(t)$, and sketch its spectrum.

(b) Show that the original voice signal $m(t)$ may be recovered from $s(t)$ by using an *unscrambler* that is identical to the scrambler.

5.47 A single sideband modulated wave $s(t)$ is applied to the coherent detector shown in Fig. P5.47. The cutoff frequency of the low-pass filter is set equal to the highest frequency component of the message signal. Using frequency-domain ideas, show that this detector produces an output that is a scaled version of the original message signal. You may assume that the carrier frequency ω_c satisfies the condition $\omega_c > \omega_m$.

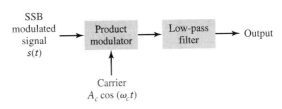

FIGURE P5.47

5.48 Consider a multiplex system in which four input signals $m_1(t)$, $m_2(t)$, $m_3(t)$, and $m_4(t)$ are respectively multiplied by the carrier waves

$$[\cos(\omega_a t) + \cos(\omega_b t)],$$
$$[\cos(\omega_a t + \alpha_1) + \cos(\omega_b t + \beta_1)],$$
$$[\cos(\omega_a t + \alpha_2) + \cos(\omega_b t + \beta_2)],$$

and

$$[\cos(\omega_a t + \alpha_3) + \cos(\omega_b t + \beta_3)]$$

and the resulting DSB-SC signals are summed and then transmitted over a common channel. In the receiver, demodulation is achieved by multiplying the sum of the DSB-SC signals by the four carrier waves separately and then using filtering to remove the unwanted components. Determine the conditions that the phase angles α_1, α_2, α_3 and β_1, β_2, β_3 must satisfy in order that the output of the kth demodulator be $m_k(t)$, where $k = 1, 2, 3, 4$.

5.49 In this problem, we study the idea of mixing utilized in a *superheterodyne receiver*. Specifically, consider the block diagram of the *mixer* shown in Fig. P5.49 that consists of a product modulator with a local oscillator of variable frequency, followed by a bandpass filter. The input signal is an AM wave of bandwidth 10 kHz and a carrier frequency that may

lie anywhere in the range 0.535–1.605 MHz; these parameters are typical of AM radio broadcasting. The signal is to be translated to a frequency band centered at a fixed *intermediate frequency* (IF) of 0.455 MHz. Find the range of tuning that must be provided in the local oscillator in order to meet this requirement.

5.50 In *natural sampling*, an analog signal $g(t)$ is multiplied by a periodic train of rectangular pulses, $c(t)$. The pulse repetition frequency of the train is ω_s, and the duration of each rectangular pulse is T_0 (with $\omega_s T_0 \gg 2\pi$). Find the spectrum of the signal $s(t)$ that results from the use of natural sampling; you may assume that time $t = 0$ corresponds to the midpoint of a rectangular pulse in $c(t)$.

5.51 In this problem, we explore the discrete-time version of DSB-SC modulation, which uses the sinusoidal carrier

$$c[n] = \cos(\Omega_c n), \qquad n = \pm 1, \pm 2, \ldots,$$

where the carrier frequency Ω_c is fixed and n denotes discrete time. Given a discrete-time message signal $m[n]$, with zero time average, the *discrete-time DSB-SC modulated signal* is defined by

$$s[n] = c[n]m[n].$$

(a) The spectrum of $m[n]$ is depicted in Fig. P5.51, where the highest message frequency Ω_m is less than the carrier frequency Ω_c. Plot the spectrum of the modulated signal $s[n]$.

(b) Following the treatment of continuous-time DSB-SC modulation presented in Section 5.5, describe a discrete-time coherent detector for the demodulation of $s[n]$.

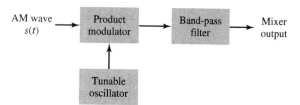

FIGURE P5.49

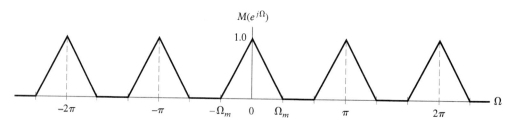

FIGURE P5.51

COMPUTER EXPERIMENTS

Note: The reader is expected to choose sampling rates for the computer experiments described next. A thorough understanding of the material presented in Chapter 4 is needed.

5.52 Use MATLAB to generate and display an AM wave with the following specifications:

Modulating wave	Sinusoidal
Modulation frequency	1 kHz
Carrier frequency	20 kHz
Percentage modulation	75%

Compute and display the magnitude spectrum of the AM wave.

5.53 (a) Generate a symmetric triangular wave $m(t)$ with a fundamental frequency of 1 Hz, alternating between -1 and $+1$.

(b) Use $m(t)$ to modulate a carrier of frequency $f_c = 25$ Hz, generating a full AM wave with 80% modulation. Compute the magnitude spectrum of the AM wave.

5.54 Continuing with Problem 5.53, investigate the effect of varying the carrier frequency f_c on the spectrum of the AM wave. Determine the minimum value of f_c necessary to ensure that there is no overlap between the lower and upper sidebands of the AM wave.

5.55 The triangular wave described in Problem 5.53(a) is used to perform DSB-SC modulation on a carrier of frequency $f_c = 25$ Hz.

(a) Generate and display the DSB-SC modulated wave so produced.

(b) Compute and display the spectrum of the modulated wave. Investigate the use of coherent detection for demodulation.

5.56 Use MATLAB to do the following:

(a) Generate a PAM wave, using a sinusoidal modulating signal of frequency $\omega_m = 0.5\pi$ rad/s, sampling period $T_s = 1$ s, and pulse duration $T_0 = 0.05$ s.

(b) Compute and display the magnitude spectrum of the PAM wave.

(c) Repeat the experiment for pulse duration $T_0 = 0.1, 0.2, 0.3, 0.4$, and 0.5 s.

Comment on the results of your experiment.

5.57 *Natural sampling* involves the multiplication of a message signal by a rectangular pulse train, as discussed in Problem 5.50. The fundamental period of the pulse train is T, and the pulse duration is T_0.

(a) Generate and display the modulated wave for a sinusoidal modulating wave, given the following specifications:

Modulation frequency	1 kHz
Pulse-repetition frequency	$(1/T_c) = 10$ kHz
Pulse duration	$T = 10\,\mu$s

(b) Compute and display the spectrum of the modulated wave. Hence, verify that the original modulating wave can be recovered without distortion by passing the modulated wave through a low-pass filter. Specify the requirements that this filter must satisfy.

6 Representing Signals by Using Continuous-Time Complex Exponentials: the Laplace Transform

6.1 Introduction

In Chapters 3 and 4, we developed representations of signals and LTI systems by using super-positions of complex sinusoids. We now consider a more general continuous-time signal and system representation based on complex exponential signals. The *Laplace transform* provides a broader characterization of continuous-time LTI systems and their interaction with signals than is possible with Fourier methods. For example, the Laplace transform can be used to analyze a large class of continuous-time problems involving signals that are not absolutely integrable, such as the impulse response of an unstable system. The FT does not exist for signals that are not absolutely integrable, so FT-based methods cannot be employed in this class of problems.

The Laplace transform possesses a distinct set of properties for analyzing signals and LTI systems. Many of these properties parallel those of the FT. For example, we shall see that continuous-time complex exponentials are eigenfunctions of LTI systems. As with complex sinusoids, one consequence of this property is that the convolution of time signals becomes multiplication of the associated Laplace transforms. Hence, the output of an LTI system is obtained by multiplying the Laplace transform of the input by the Laplace trans-form of the impulse response, which is defined as the transfer function of the system. The transfer function generalizes the frequency response characterization of an LTI system's input–output behavior and offers new insights into system characteristics.

The Laplace transform comes in two varieties: (1) unilateral, or one sided, and (2) bi-lateral, or two sided. The unilateral Laplace transform is a convenient tool for solving dif-ferential equations with initial conditions. The bilateral Laplace transform offers insight into the nature of system characteristics such as stability, causality, and frequency response. The primary role of the Laplace transform in engineering is the transient and stability analysis of causal LTI systems described by differential equations. We shall develop the Laplace transform with these roles in mind throughout this chapter.

6.2 The Laplace Transform

Let e^{st} be a complex exponential with *complex frequency* $s = \sigma + j\omega$. We may write

$$e^{st} = e^{\sigma t} \cos(\omega t) + j e^{\sigma t} \sin(\omega t). \tag{6.1}$$

The real part of e^{st} is an exponentially damped cosine, and the imaginary part is an expo-nentially damped sine, as depicted in Fig. 6.1. In this figure it is assumed that σ is negative.

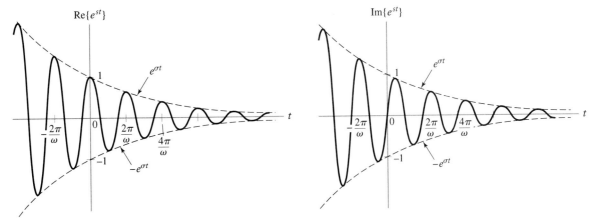

FIGURE 6.1 Real and imaginary parts of the complex exponential e^{st}, where $s = \sigma + j\omega$.

The real part of s is the exponential damping factor σ, and the imaginary part of s is the frequency of the cosine and sine factor, namely, ω.

■ 6.2.1 EIGENFUNCTION PROPERTY OF e^{st}

Consider applying an input of the form $x(t) = e^{st}$ to an LTI system with impulse response $h(t)$. The system output is given by

$$
\begin{aligned}
y(t) &= H\{x(t)\} \\
&= h(t) * x(t) \\
&= \int_{-\infty}^{\infty} h(\tau)x(t-\tau)d\tau.
\end{aligned}
$$

We use $x(t) = e^{st}$ to obtain

$$
\begin{aligned}
y(t) &= \int_{-\infty}^{\infty} h(\tau)e^{s(t-\tau)}d\tau \\
&= e^{st} \int_{-\infty}^{\infty} h(\tau)e^{-s\tau}d\tau.
\end{aligned}
$$

We define the *transfer function*

$$
\boxed{H(s) = \int_{-\infty}^{\infty} h(\tau)e^{-s\tau}d\tau} \tag{6.2}
$$

so that we may write

$$
y(t) = H\{e^{st}\} = H(s)e^{st}.
$$

The action of the system on an input e^{st} is multiplication by the transfer function $H(s)$. Recall that an eigenfunction is a signal that passes through the system without being modified except for multiplication by a scalar. Hence, we identify e^{st} as an eigenfunction of the LTI system and $H(s)$ as the corresponding eigenvalue.

Next, we express the complex-valued transfer function $H(s)$ in polar form as $H(s) = |H(s)|e^{j\phi(s)}$ where $|H(s)|$ and $\phi(s)$ are the magnitude and phase of $H(s)$, respectively. Now we rewrite the LTI system output as

$$
y(t) = |H(s)|e^{j\phi(s)}e^{st}.
$$

We use $s = \sigma + j\omega$ to obtain

$$
\begin{aligned}
y(t) &= |H(\sigma + j\omega)|e^{\sigma t}e^{j\omega t + \phi(\sigma + j\omega)} \\
&= |H(\sigma + j\omega)|e^{\sigma t}\cos(\omega t + \phi(\sigma + j\omega)) + j|H(\sigma + j\omega)|e^{\sigma t}\sin(\omega t + \phi(\sigma + j\omega)).
\end{aligned}
$$

Since the input $x(t)$ has the form given in Eq. (6.1), we see that the system changes the amplitude of the input by $|H(\sigma + j\omega)|$ and shifts the phase of the sinusoidal components by $\phi(\sigma + j\omega)$. The system does not change the damping factor σ or the sinusoidal frequency ω of the input.

■ 6.2.2 LAPLACE TRANSFORM REPRESENTATION

Given the simplicity of describing the action of the system on inputs of the form e^{st}, we now seek a representation of arbitrary signals as a weighted superposition of eigenfunctions e^{st}. Substituting $s = \sigma + j\omega$ into Eq. (6.2) and using t as the variable of integration, we obtain

$$
\begin{aligned}
H(\sigma + j\omega) &= \int_{-\infty}^{\infty} h(t)e^{-(\sigma + j\omega)t}\,dt \\
&= \int_{-\infty}^{\infty} [h(t)e^{-\sigma t}]e^{-j\omega t}\,dt.
\end{aligned}
$$

This indicates that $H(\sigma + j\omega)$ is the Fourier transform of $h(t)e^{-\sigma t}$. Hence, the inverse Fourier transform of $H(\sigma + j\omega)$ must be $h(t)e^{-\sigma t}$; that is,

$$
h(t)e^{-\sigma t} = \frac{1}{2\pi}\int_{-\infty}^{\infty} H(\sigma + j\omega)e^{j\omega t}\,d\omega.
$$

We may recover $h(t)$ by multiplying both sides of this equation by $e^{\sigma t}$:

$$
\begin{aligned}
h(t) &= e^{\sigma t}\frac{1}{2\pi}\int_{-\infty}^{\infty} H(\sigma + j\omega)e^{j\omega t}\,d\omega, \\
&= \frac{1}{2\pi}\int_{-\infty}^{\infty} H(\sigma + j\omega)e^{(\sigma + j\omega)t}\,d\omega.
\end{aligned}
\tag{6.3}
$$

Now, substituting $s = \sigma + j\omega$ and $d\omega = ds/j$ into Eq. (6.3) we get

$$
h(t) = \frac{1}{2\pi j}\int_{\sigma - j\infty}^{\sigma + j\infty} H(s)e^{st}\,ds.
\tag{6.4}
$$

The limits on the integral are also a result of the substitution $s = \sigma + j\omega$. Equation (6.2) indicates how to determine $H(s)$ from $h(t)$, while Eq. (6.4) expresses $h(t)$ as a function of $H(s)$. We say that $H(s)$ is the *Laplace transform* of $h(t)$ and that $h(t)$ is the *inverse Laplace transform* of $H(s)$.

We have obtained the Laplace transform of the impulse response of a system. This relationship holds for an arbitrary signal. The Laplace transform of $x(t)$ is

$$
\boxed{X(s) = \int_{-\infty}^{\infty} x(t)e^{-st}\,dt,}
\tag{6.5}
$$

and the inverse Laplace transform of $X(s)$ is

$$x(t) = \frac{1}{2\pi j} \int_{\sigma-j\infty}^{\sigma+j\infty} X(s)e^{st}ds. \qquad (6.6)$$

We express this relationship with the notation

$$x(t) \xleftrightarrow{\;\mathcal{L}\;} X(s).$$

Note that Eq. (6.6) represents the signal $x(t)$ as a weighted superposition of complex exponentials e^{st}. The weights are proportional to $X(s)$. In practice, we usually do not evaluate this integral directly, since it requires techniques of contour integration. Instead, we determine inverse Laplace transforms by exploiting the one-to-one relationship between $x(t)$ and $X(s)$.

■ 6.2.3 CONVERGENCE

Our development indicates that the Laplace transform is the Fourier transform of $x(t)e^{-\sigma t}$. Hence, a necessary condition for convergence of the Laplace transform is the absolute integrability of $x(t)e^{-\sigma t}$. That is, we must have

$$\int_{-\infty}^{\infty} |x(t)e^{-\sigma t}|dt < \infty.$$

The range of σ for which the Laplace transform converges is termed the *region of convergence* (ROC).

Note that the Laplace transform exists for signals that do not have a Fourier transform. By limiting ourselves to a certain range of σ, we may ensure that $x(t)e^{-\sigma t}$ is absolutely integrable, even though $x(t)$ is not absolutely integrable by itself. For example, the Fourier transform of $x(t) = e^t u(t)$ does not exist, since $x(t)$ is an increasing real exponential signal and is thus not absolutely integrable. However, if $\sigma > 1$, then $x(t)e^{-\sigma t} = e^{(1-\sigma)t}u(t)$ is absolutely integrable, and so the Laplace transform, which is the Fourier transform of $x(t)e^{-\sigma t}$, does exist. This scenario is illustrated in Fig. 6.2. The existence of the Laplace transform for signals that have no Fourier transform is a significant advantage gained from using the complex exponential representation.

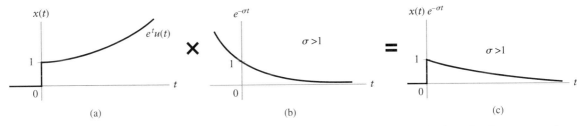

FIGURE 6.2   The Laplace transform applies to more general signals than the Fourier transform does. (a) Signal for which the Fourier transform does not exist. (b) Attenuating factor associated with Laplace transform. (c) The modified signal $x(t)e^{-\sigma t}$ is absolutely integrable for $\sigma > 1$.

■ 6.2.4 THE s-PLANE

It is convenient to represent the complex frequency s graphically in terms of a complex plane termed the *s-plane*, as depicted in Fig. 6.3. The horizontal axis represents the real part of s (i.e., the exponential damping factor σ), and the vertical axis represents the imaginary part of s (i.e., the sinusoidal frequency ω). Note that if $x(t)$ is absolutely integrable, then we may obtain the Fourier transform from the Laplace transform by setting $\sigma = 0$:

$$X(j\omega) = X(s)|_{\sigma=0}. \tag{6.7}$$

In the s-plane, $\sigma = 0$ corresponds to the imaginary axis. We thus say that the Fourier transform is given by the Laplace transform evaluated along the imaginary axis.

The $j\omega$-axis divides the s-plane in half. The region of the s-plane to the left of the $j\omega$-axis is termed the *left half of the s-plane*, while the region to the right of the $j\omega$-axis is termed the *right half of the s-plane*. The real part of s is negative in the left half of the s-plane and positive in the right half of the s-plane.

■ 6.2.5 POLES AND ZEROS

The most commonly encountered form of the Laplace transform in engineering is a ratio of two polynomials in s; that is,

$$X(s) = \frac{b_M s^M + b_{M-1} s^{M-1} + \cdots + b_0}{s^N + a_{N-1} s^{N-1} + \cdots + a_1 s + a_0}.$$

It is useful to factor $X(s)$ as a product of terms involving the roots of the denominator and numerator polynomials:

$$X(s) = \frac{b_M \prod_{k=1}^{M} (s - c_k)}{\prod_{k=1}^{N} (s - d_k)}.$$

The c_k are the roots of the numerator polynomial and are termed the *zeros* of $X(s)$. The d_k are the roots of the denominator polynomial and are termed the *poles* of $X(s)$. We denote the locations of zeros in the s-plane with the " ○ " symbol and the locations of poles with the "×" symbol, as illustrated in Fig. 6.3. The locations of poles and zeros in the s-plane uniquely specify $X(s)$, except for the constant gain factor b_M.

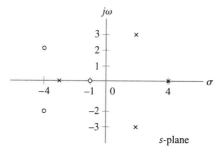

FIGURE 6.3   The s-plane. The horizontal axis is Re$\{s\}$ and the vertical axis is Im$\{s\}$. Zeros are depicted at $s = -1$ and $s = -4 \pm 2j$, and poles are depicted at $s = -3$, $s = 2 \pm 3j$, and $s = 4$.

EXAMPLE 6.1 LAPLACE TRANSFORM OF A CAUSAL EXPONENTIAL SIGNAL   Determine the Laplace transform of

$$x(t) = e^{at}u(t),$$

and depict the ROC and the locations of poles and zeros in the *s*-plane. Assume that *a* is real.

Solution: Substitute $x(t)$ into Eq. (6.5), obtaining

$$X(s) = \int_{-\infty}^{\infty} e^{at}u(t)e^{-st}dt$$

$$= \int_{0}^{\infty} e^{-(s-a)t}\,dt$$

$$= \frac{-1}{s-a}e^{-(s-a)t}\bigg|_{0}^{\infty}.$$

To evaluate $e^{-(s-a)t}$ at the limits, we use $s = \sigma + j\omega$ to write

$$X(s) = \frac{-1}{\sigma + j\omega - a}e^{-(\sigma-a)t}e^{-j\omega t}\bigg|_{0}^{\infty}.$$

Now, if $\sigma > a$, then $e^{-(\sigma-a)t}$ goes to zero as t approaches infinity, and

$$X(s) = \frac{-1}{\sigma + j\omega - a}(0 - 1), \quad \sigma > a,$$

$$= \frac{1}{s-a}, \quad \text{Re}(s) > a.$$

(6.8)

The Laplace transform $X(s)$ does not exist for $\sigma \le a$, since the integral does not converge. The ROC for this signal is thus $\sigma > a$, or equivalently, $\text{Re}(s) > a$. The ROC is depicted as the shaded region of the *s*-plane in Fig. 6.4. The pole is located at $s = a$.   ∎

The expression for the Laplace transform does not uniquely correspond to a signal $x(t)$ if the ROC is not specified. That is, two different signals may have identical Laplace transforms, but different ROCs. We demonstrate this property in the next example.

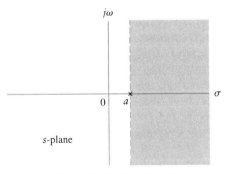

FIGURE 6.4   The ROC for $x(t) = e^{at}u(t)$ is depicted by the shaded region. A pole is located at $s = a$.

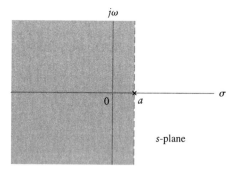

FIGURE 6.5   The ROC for $y(t) = -e^{at}u(-t)$ is depicted by the shaded region. A pole is located at $s = a$.

EXAMPLE 6.2 LAPLACE TRANSFORM OF AN ANTICAUSAL EXPONENTIAL SIGNAL   An anticausal signal is zero for $t > 0$. Determine the Laplace transform and ROC for the anticausal signal

$$y(t) = -e^{at}u(-t).$$

Solution: Using $y(t) = -e^{at}u(-t)$ in place of $x(t)$ in Eq. (6.5), we obtain

$$
\begin{aligned}
Y(s) &= \int_{-\infty}^{\infty} -e^{at}u(-t)e^{-st}\, dt \\
&= -\int_{-\infty}^{0} e^{-(s-a)t}\, dt \\
&= \frac{1}{s-a}e^{-(s-a)t}\Big|_{-\infty}^{0} \\
&= \frac{1}{s-a}, \quad \mathrm{Re}(s) < a.
\end{aligned}
\tag{6.9}
$$

The ROC and the location of the pole at $s = a$ are depicted in Fig. 6.5.   ■

     Examples 6.1 and 6.2 reveal that the Laplace transforms $X(s)$ and $Y(s)$ are equal, even though the signals $x(t)$ and $y(t)$ are clearly different. However, the ROCs of the two signals are different. This ambiguity occurs in general with signals that are one sided. To see why, let $x(t) = g(t)u(t)$ and $y(t) = -g(t)u(-t)$. We may thus write

$$
\begin{aligned}
X(s) &= \int_{0}^{\infty} g(t)e^{-st}\, dt \\
&= G(s, \infty) - G(s, 0),
\end{aligned}
$$

where

$$G(s, t) = \int g(t)e^{-st}\, dt.$$

Next, we write

$$Y(s) = -\int_{-\infty}^{0} g(t)e^{-st}\, dt$$

$$= \int_{0}^{-\infty} g(t)e^{-st}\, dt$$

$$= G(s, -\infty) - G(s, 0).$$

We see that $X(s) = Y(s)$ whenever $G(s, \infty) = G(s, -\infty)$. In Examples 6.1 and 6.2, we have $G(s, -\infty) = G(s, \infty) = 0$. The values of s for which the integral represented by $G(s, \infty)$ converges differ from those for which the integral represented by $G(s, -\infty)$ converges, and thus the ROCs are different. The ROC must be specified for the Laplace transform to be unique.

▶ **Problem 6.1**   Determine the Laplace transform and ROC of the following signals:

(a) $x(t) = u(t - 5)$
(b) $x(t) = e^{5t}u(-t + 3)$

Answers:

(a)

$$X(s) = \frac{e^{-5s}}{s}, \quad \text{Re}(s) > 0$$

(b)

$$X(s) = -\frac{e^{-3(s-5)}}{s - 5}, \quad \text{Re}(s) < 5 \qquad\blacktriangleleft$$

▶ **Problem 6.2**   Determine the Laplace transform, ROC, and locations of poles and zeros of $X(s)$ for the following signals:

(a) $x(t) = e^{j\omega_o t}u(t)$
(b) $x(t) = \sin(3t)u(t)$
(c) $x(t) = e^{-2t}u(t) + e^{-3t}u(t)$

Answers:

(a)

$$X(s) = \frac{1}{s - j\omega_o}, \quad \text{Re}(s) > 0$$

There is a pole at $s = j\omega_o$

(b)

$$X(s) = \frac{3}{s^2 + 9}, \quad \text{Re}(s) > 0$$

There are poles at $s = \pm j3$

(c)

$$X(s) = \frac{2s + 5}{s^2 + 5s + 6}, \quad \text{Re}(s) > -3$$

There is a zero at $s = -5/2$ and poles at $s = -2$ and $s = -3$ ◀

6.3 *The Unilateral Laplace Transform*

There are many applications of Laplace transforms in which it is reasonable to assume that the signals involved are causal—that is, zero for times $t < 0$. For example, if we apply an input that is zero for time $t < 0$ to a causal system, the output will also be zero for $t < 0$. Also, the choice of time origin is arbitrary in many problems. Thus, time $t = 0$ is often chosen as the time at which an input is presented to the system, and the behavior of the system for time $t \geq 0$ is of interest. In such problems, it is advantageous to define the unilateral or one-sided Laplace transform, which is based only on the nonnegative-time $(t \geq 0)$ portions of a signal. By working with causal signals, we remove the ambiguity inherent in the bilateral transform and thus do not need to consider the ROC. Also, the differentiation property for the unilateral Laplace transform may be used to analyze the behavior of a causal system described by a differential equation with initial conditions. Indeed, this is the most common use for the unilateral transform in engineering applications.

The *unilateral Laplace transform* of a signal $x(t)$ is defined by

$$X(s) = \int_{0^-}^{\infty} x(t)e^{-st}\,dt. \tag{6.10}$$

The lower limit of 0^- implies that we do include discontinuities and impulses that occur at $t = 0$ in the integral. Hence, $X(s)$ depends on $x(t)$ for $t \geq 0$. Since the inverse Laplace transform given by Eq. (6.6) depends only on $X(s)$, the inverse unilateral transform is still given by that equation. We shall denote the relationship between $X(s)$ and $x(t)$ as

$$x(t) \xleftrightarrow{\;\mathcal{L}_u\;} X(s),$$

where the subscript u in $\mathcal{L}_u$ denotes the unilateral transform. Naturally, the unilateral and bilateral Laplace transforms are equivalent for signals that are zero for times $t < 0$. For example, repeating Example 6.1, but this time using the definition of the unilateral Laplace transform given in Eq. (6.10), we find that

$$e^{at}u(t) \xleftrightarrow{\;\mathcal{L}_u\;} \frac{1}{s - a} \tag{6.11}$$

is equivalent to

$$e^{at}u(t) \xleftrightarrow{\;\mathcal{L}\;} \frac{1}{s - a} \quad \text{with ROC } \mathrm{Re}\{s\} > a.$$

▶ **Problem 6.3**   Determine the unilateral Laplace transforms of the following signals:
(a) $x(t) = u(t)$
(b) $x(t) = u(t + 3)$
(c) $x(t) = u(t - 3)$

Answers:
(a) $X(s) = 1/s$
(b) $X(s) = 1/s$
(c) $X(s) = e^{-3s}/s$ ◀

6.4 Properties of the Unilateral Laplace Transform

The properties of the Laplace transform are similar to those of the Fourier transform; hence, we simply state many of them. (Proofs of some are given in the problems at the end of the chapter.) The properties described in this section specifically apply to the unilateral Laplace transform. The unilateral and bilateral transforms have many properties in common, although there are important differences, discussed in Section 6.8. In the properties discussed next, we assume that

$$x(t) \xleftarrow{\mathcal{L}_u} X(s)$$

and

$$y(t) \xleftarrow{\mathcal{L}_u} Y(s).$$

Linearity

$$ax(t) + by(t) \xleftarrow{\mathcal{L}_u} aX(s) + bY(s). \qquad (6.12)$$

The linearity of the Laplace transform follows from its definition as an integral and the fact that integration is a linear operation.

Scaling

$$x(at) \xleftarrow{\mathcal{L}_u} \frac{1}{a} X\left(\frac{s}{a}\right) \quad \text{for } a > 0. \qquad (6.13)$$

Scaling in time introduces the inverse scaling in s.

Time Shift

$$x(t - \tau) \xleftarrow{\mathcal{L}_u} e^{-s\tau} X(s) \quad \text{for all } \tau \text{ such that } x(t - \tau)u(t) = x(t - \tau)u(t - \tau). \qquad (6.14)$$

A shift of τ in time corresponds to multiplication of the Laplace transform by the complex exponential $e^{-s\tau}$. The restriction on the shift arises because the unilateral transform is defined solely in terms of the nonnegative-time portions of the signal. Hence, this property applies only if the shift does not move a nonzero $t \geq 0$ component of the signal to $t < 0$, as depicted in Fig. 6.6(a), or does not move a nonzero $t < 0$ portion of the signal to $t \geq 0$, as depicted in Fig. 6.6(b). The time-shift property is most commonly applied to causal signals $x(t)$ with shifts $\tau > 0$, in which case the shift restriction is always satisfied.

s-Domain Shift

$$e^{s_o t} x(t) \xleftarrow{\mathcal{L}_u} X(s - s_o). \qquad (6.15)$$

Multiplication by a complex exponential in time introduces a shift in complex frequency s into the Laplace transform.

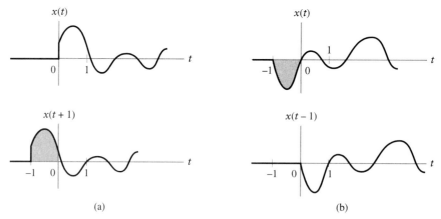

(a)   (b)

FIGURE 6.6   Time shifts for which the unilateral Laplace transform time-shift property does not apply. (a) A nonzero portion of $x(t)$ that occurs at times $t \geq 0$ is shifted to times $t < 0$. (b) A nonzero portion of $x(t)$ that occurs at times $t < 0$ is shifted to times $t \geq 0$.

Convolution

$$\boxed{x(t) * y(t) \xleftrightarrow{\;\mathcal{L}_u\;} X(s)Y(s).}$$ (6.16)

Convolution in time corresponds to multiplication of Laplace transforms. This property only applies when $x(t) = 0$ and $y(t) = 0$ for $t < 0$.

Differentiation in the s-Domain

$$\boxed{-tx(t) \xleftrightarrow{\;\mathcal{L}_u\;} \frac{d}{ds}X(s).}$$ (6.17)

Differentiation in the s-domain corresponds to multiplication by $-t$ in the time domain.

EXAMPLE 6.3 APPLYING PROPERTIES   Find the unilateral Laplace transform of

$$x(t) = (-e^{3t}u(t)) * (tu(t)).$$

Solution: Using Eq. (6.11), we have

$$-e^{3t}u(t) \xleftrightarrow{\;\mathcal{L}_u\;} \frac{-1}{s-3}$$

and

$$u(t) \xleftrightarrow{\;\mathcal{L}_u\;} \frac{1}{s}.$$

Applying the s-domain differentiation property given by Eq. (6.17), we obtain

$$tu(t) \xleftrightarrow{\;\mathcal{L}_u\;} \frac{1}{s^2}.$$

Now we use the convolution property given by Eq. (6.16) to get

$$x(t) = (e^{3t}u(t)) * (tu(t)) \xleftrightarrow{\;\mathcal{L}_u\;} X(s) = \frac{-1}{s^2(s-3)}.$$

■

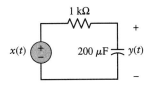

FIGURE 6.7 *RC* circuit for Examples 6.4 and 6.10. Note that $RC = 0.2$ s.

EXAMPLE 6.4 RC FILTER OUTPUT   Find the Laplace transform of the output of the *RC* circuit depicted in Fig. 6.7 for the input $x(t) = te^{2t}u(t)$.

Solution: The impulse response of the *RC* circuit was obtained in Example 1.21 as

$$h(t) = \frac{1}{RC}e^{-t/(RC)}u(t).$$

We apply the convolution property (Eq. (6.16)) to obtain the Laplace transform of the output $y(t)$ as the product of the Laplace transforms of the input $x(t)$ and impulse response $h(t)$: $Y(s) = H(s)X(s)$. Using $RC = 0.2$ s and Eq. (6.11), we obtain

$$h(t) \xleftrightarrow{\;\mathcal{L}_u\;} \frac{5}{s+5}.$$

Next, we use the *s*-domain differentiation property given by Eq. (6.17) to write

$$X(s) = \frac{1}{(s-2)^2},$$

and we conclude that

$$Y(s) = \frac{5}{(s-2)^2(s+5)}.$$

Note that Fourier methods are not applicable to this particular problem because the FT of the input signal $x(t)$ does not exist, since $x(t)$ is not absolutely integrable. ∎

▶ **Problem 6.4**   Find the unilateral Laplace transform of the following signals:

(a) $x(t) = e^{-t}(t-2)u(t-2)$
(b) $x(t) = t^2 e^{-2t}u(t)$
(c) $x(t) = tu(t) - (t-1)u(t-1) - (t-2)u(t-2) + (t-3)u(t-3)$
(d) $x(t) = e^{-t}u(t) * \cos(t-2)u(t-2)$

Answers:

(a)
$$X(s) = \frac{e^{-2(s+1)}}{(s+1)^2}$$

(b)
$$X(s) = \frac{2}{(s+2)^3}$$

(c)
$$X(s) = (1 - e^{-s} - e^{-2s} + e^{-3s})/s^2$$

(d)
$$X(s) = \frac{e^{-2s}s}{(s+1)(s^2+1)}$$

◀

Differentiation in the Time Domain

Suppose that the Laplace transform of $x(t)$ exists, and consider the unilateral Laplace transform of $dx(t)/dt$. By definition,

$$\frac{d}{dt}x(t) \xleftrightarrow{\mathcal{L}_u} \int_{0^-}^{\infty} \left(\frac{d}{dt}x(t)\right)e^{-st}dt.$$

Integrating by parts, we obtain

$$\frac{d}{dt}x(t) \xleftrightarrow{\mathcal{L}_u} x(t)e^{-st}\Big|_{0^-}^{\infty} + s\int_{0^-}^{\infty} x(t)e^{-st}\,dt.$$

Since $X(s)$ exists, it follows that $x(t)e^{-st}$ approaches zero as t approaches infinity; thus, $x(t)e^{-st}\big|_{t=\infty} = 0$. Furthermore, the integral corresponds to the definition of the unilateral Laplace transform in Eq. (6.10), so we have

$$\boxed{\frac{d}{dt}x(t) \xleftrightarrow{\mathcal{L}_u} sX(s) - x(0^-).} \tag{6.18}$$

EXAMPLE 6.5 VERIFYING THE DIFFERENTIATION PROPERTY   Let $x(t) = e^{at}u(t)$. Find the Laplace transform of $dx(t)/dt$ by direct calculation and by using Eq. (6.18).

Solution: Apply the product rule for differentiation to obtain the derivative of $x(t), t > 0^-$:

$$\frac{d}{dt}e^{at}u(t) = ae^{at}u(t) + \delta(t).$$

The unilateral Laplace transform of $ae^{at}u(t)$ is a times the unilateral Laplace transform of $e^{at}u(t)$; so, using Eq. (6.11) and $\delta(t) \xleftrightarrow{\mathcal{L}_u} 1$, we have

$$\frac{d}{dt}x(t) = ae^{at}u(t) + \delta(t) \xleftrightarrow{\mathcal{L}_u} \frac{a}{s-a} + 1 = \frac{s}{s-a}.$$

Next, let us rederive this result, using the differentiation property given in Eq. (6.18). From that equation,

$$\frac{d}{dt}x(t) \xleftrightarrow{\mathcal{L}_u} sX(s) - x(0^-) = \frac{s}{s-a}.$$ ■

The general form for the differentiation property is

$$\boxed{\frac{d^n}{dt^n}x(t) \xleftrightarrow{\mathcal{L}_u} s^nX(s) - \frac{d^{n-1}}{dt^{n-1}}x(t)\Big|_{t=0^-} - s\frac{d^{n-2}}{dt^{n-2}}x(t)\Big|_{t=0^-} \\ - \cdots - s^{n-2}\frac{d}{dt}x(t)\Big|_{t=0^-} - s^{n-1}x(0^-).} \tag{6.19}$$

Integration Property

$$\boxed{\int_{-\infty}^{t} x(\tau)d\tau \xleftrightarrow{\mathcal{L}_u} \frac{x^{(-1)}(0^-)}{s} + \frac{X(s)}{s},} \tag{6.20}$$

where

$$x^{(-1)}(0^-) = \int_{-\infty}^{0^-} x(\tau)d\tau$$

is the area under $x(t)$ from $t = -\infty$ to $t = 0^-$.

▶ **Problem 6.5**   Use the integration property to show that the unilateral Laplace transform of

$$tu(t) = \int_{-\infty}^{t} u(\tau)d\tau$$

is given by $1/s^2$.   ◀

Initial- and Final-Value Theorems

The initial- and final-value theorems allow us to determine the initial value, $x(0^+)$, and the final value, $x(\infty)$, of $x(t)$ directly from $X(s)$. These theorems are most often used to evaluate either the initial or final values of a system output without explicitly determining the entire time response of the system. The initial-value theorem states that

$$\lim_{s\to\infty} sX(s) = x(0^+). \tag{6.21}$$

The initial-value theorem does not apply to rational functions $X(s)$ in which the order of the numerator polynomial is greater than or equal to that of the denominator polynomial order. The final-value theorem states that

$$\lim_{s\to 0} sX(s) = x(\infty). \tag{6.22}$$

The final-value theorem applies only if all the poles of $X(s)$ are in the left half of the s-plane, with at most a single pole at $s = 0$.

EXAMPLE 6.6 APPLYING THE INITIAL- AND FINAL-VALUE THEOREMS   Determine the initial and final values of a signal $x(t)$ whose unilateral Laplace transform is

$$X(s) = \frac{7s + 10}{s(s + 2)}.$$

Solution: We may apply the initial-value theorem, Eq. (6.21), to obtain

$$x(0^+) = \lim_{s\to\infty} s\frac{7s + 10}{s(s + 2)}$$
$$= \lim_{s\to\infty} \frac{7s + 10}{s + 2}$$
$$= 7.$$

The final value theorem, Eq. (6.22), is applicable also, since $X(s)$ has only a single pole at $s = 0$ and the remaining poles are in the left half of the s-plane. We have

$$x(\infty) = \lim_{s\to 0} s\frac{7s + 10}{s(s + 2)}$$
$$= \lim_{s\to 0} \frac{7s + 10}{s + 2}$$
$$= 5.$$

The reader may verify these results by showing that $X(s)$ is the Laplace transform of $x(t) = 5u(t) + 2e^{-2t}u(t)$.   ■

▶ **Problem 6.6**   Find the initial and final values of the time-domain signal $x(t)$ corresponding to the following Laplace transforms:

(a)
$$X(s) = e^{-5s}\left(\frac{-2}{s(s+2)}\right)$$

(b)
$$X(s) = \frac{2s+3}{s^2 + 5s + 6}$$

Answers:

(a) $x(0^+) = 0$ and $x(\infty) = -1$
(b) $x(0^+) = 2$ and $x(\infty) = 0$ ◀

6.5 Inversion of the Unilateral Laplace Transform

Direct inversion of the Laplace transform using Eq. (6.6) requires an understanding of contour integration, which is beyond the scope of this book. Instead, we shall determine inverse Laplace transforms using the one-to-one relationship between a signal and its unilateral Laplace transform. Given knowledge of several basic transform pairs and the Laplace transform properties, we are able to invert a very large class of Laplace transforms in this manner. A table of basic Laplace transform pairs is given in Appendix D.1.

In the study of LTI systems described by integro-differential equations, we frequently encounter Laplace transforms that are a ratio of polynomials in s. In this case, the inverse transform is obtained by expressing $X(s)$ as a sum of terms for which we already know the time function, using a partial-fraction expansion. Suppose

$$X(s) = \frac{B(s)}{A(s)}$$
$$= \frac{b_M s^M + b_{M-1}s^{M-1} + \cdots + b_1 s + b_0}{s^N + a_{N-1}s^{N-1} + \cdots + a_1 s + a_0}.$$

If $X(s)$ is an improper rational function, (i.e, $M \geq N$), then we may use long division to express $X(s)$ in the form

$$X(s) = \sum_{k=0}^{M-N} c_k s^k + \widetilde{X}(s),$$

where

$$\widetilde{X}(s) = \frac{\widetilde{B}(s)}{A(s)}.$$

The numerator polynomial $\widetilde{B}(s)$ now has order one less than that of the denominator polynomial, and the partial-fraction expansion method is used to determine the inverse transform of $\widetilde{X}(s)$. Given that the impulse and its derivatives are zero at $t = 0^-$, we find the inverse transform of the terms in the sum component of $X(s)$, $\sum_{k=0}^{M-N} c_k s^k$, using the pair $\delta(t) \xleftrightarrow{\mathcal{L}_u} 1$ and the differentiation property given by Eq. (6.18). We obtain

$$\sum_{k=0}^{M-N} c_k \delta^{(k)}(t) \xleftrightarrow{\mathcal{L}_u} \sum_{k=0}^{M-N} c_k s^k,$$

where $\delta^{(k)}(t)$ denotes the kth derivative of the impulse $\delta(t)$.

Now we factor the denominator polynomial as a product of pole factors to obtain

$$\widetilde{X}(s) = \frac{b_P s^P + b_{P-1} s^{P-1} + \cdots + b_1 s + b_0}{\prod_{k=1}^{N}(s - d_k)},$$

where $P < N$. If all the poles d_k are distinct, then, using a partial-fraction expansion, we may rewrite $\widetilde{X}(s)$ as a sum of simple terms:

$$\widetilde{X}(s) = \sum_{k=1}^{N} \frac{A_k}{s - d_k}.$$

Here, the A_k are determined by using the method of residues or by solving a system of linear equations as described in Appendix B. The inverse Laplace transform of each term in the sum may now be found from Eq. (6.11), resulting in the pair

$$\boxed{A_k e^{d_k t} u(t) \xleftrightarrow{\ \mathcal{L}_u\ } \frac{A_k}{s - d_k}.} \qquad (6.23)$$

If a pole d_i is repeated r times, then there are r terms in the partial-fraction expansion associated with that pole, namely,

$$\frac{A_{i_1}}{s - d_i}, \frac{A_{i_2}}{(s - d_i)^2}, \ldots, \frac{A_{i_r}}{(s - d_i)^r}.$$

The inverse Laplace transform of each term is found using Eqs. (6.23) and (6.17) to obtain

$$\boxed{\frac{A t^{n-1}}{(n - 1)!} e^{d_k t} u(t) \xleftrightarrow{\ \mathcal{L}_u\ } \frac{A}{(s - d_k)^n}.} \qquad (6.24)$$

EXAMPLE 6.7 INVERSION BY PARTIAL-FRACTION EXPANSION Find the inverse Laplace transform of

$$X(s) = \frac{3s + 4}{(s + 1)(s + 2)^2}.$$

Solution: We use a partial-fraction expansion of $X(s)$ to write

$$X(s) = \frac{A_1}{s + 1} + \frac{A_2}{s + 2} + \frac{A_3}{(s + 2)^2}.$$

Solving for A_1, A_2, and A_3 by the method of residues, we obtain

$$X(s) = \frac{1}{s + 1} - \frac{1}{s + 2} + \frac{2}{(s + 2)^2}.$$

Using Eqs. (6.23) and (6.24), we may construct $x(t)$ from the inverse Laplace transform of each term in the partial-fraction expansion as follows:

▶ The pole of the first term is at $s = -1$, so

$$e^{-t} u(t) \xleftrightarrow{\ \mathcal{L}_u\ } \frac{1}{s + 1}.$$

▶ The second term has a pole at $s = -2$; thus,

$$-e^{2t}u(t) \xleftrightarrow{\mathcal{L}_u} -\frac{1}{s+2}.$$

▶ The double pole in the last term is also at $s = -2$; hence,

$$2te^{-2t}u(t) \xleftrightarrow{\mathcal{L}_u} \frac{2}{(s+2)^2}.$$

Combining these three terms, we obtain

$$x(t) = e^{-t}u(t) - e^{-2t}u(t) + 2te^{-2t}u(t).$$ ■

EXAMPLE 6.8 INVERTING AN IMPROPER RATIONAL LAPLACE TRANSFORM   Find the inverse unilateral Laplace transform of

$$X(s) = \frac{2s^3 - 9s^2 + 4s + 10}{s^2 - 3s - 4}.$$

Solution: We use long division to express $X(s)$ as the sum of a proper rational function and a polynomial in s:

$$
\begin{array}{r}
2s - 3 \\
s^2 - 3s - 4 \overline{)2s^3 - 9s^2 + 4s + 10} \\
\underline{2s^3 - 6s^2 - 8s } \\
-3s^2 + 12s + 10 \\
\underline{-3s^2 + 9s + 12} \\
3s - 2
\end{array}
$$

Thus, we may write

$$X(s) = 2s - 3 + \frac{3s - 2}{s^2 - 3s - 4}.$$

Using a partial-fraction expansion to expand the rational function, we obtain

$$X(s) = 2s - 3 + \frac{1}{s+1} + \frac{2}{s-4}.$$

Term-by-term inversion of $X(s)$ yields

$$x(t) = 2\delta^{(1)}(t) - 3\delta(t) + e^{-t}u(t) + 2e^{4t}u(t).$$ ■

**S & S
Solutions**

▶ **Problem 6.7**   Find the inverse Laplace transforms of the following functions:

(a)
$$X(s) = \frac{-5s - 7}{(s+1)(s-1)(s+2)}$$

(b)
$$X(s) = \frac{s}{s^2 + 5s + 6}$$

(c)
$$X(s) = \frac{s^2 + s - 3}{s^2 + 3s + 2}$$

Answers:

(a)
$$x(t) = e^{-t}u(t) - 2e^{t}u(t) + e^{-2t}u(t)$$

(b)
$$x(t) = -2e^{-2t}u(t) + 3e^{-3t}u(t)$$

(c)
$$x(t) = \delta(t) + e^{-2t}u(t) - 3e^{-t}u(t) \qquad \blacktriangleleft$$

The partial-fraction expansion procedure is applicable to either real or complex poles. A complex pole usually results in complex-valued expansion coefficients and a complex exponential function of time. If the coefficients in the denominator polynomial are real, then all the complex poles occur in complex-conjugate pairs. In cases where $X(s)$ has real-valued coefficients and thus corresponds to a real-valued time signal, we may simplify the algebra by combining complex-conjugate poles in the partial-fraction expansion in such a way as to ensure real-valued expansion coefficients and a real-valued inverse transform. This is accomplished by combining all pairs of complex-conjugate poles into quadratic terms with real coefficients. The inverse Laplace transforms of these quadratic terms are exponentially damped sinusoids.

Suppose, then, that $\alpha + j\omega_0$ and $\alpha - j\omega_0$ make up a pair of complex-conjugate poles. The first-order terms associated with these two poles in the partial-fraction expansion are written as

$$\frac{A_1}{s - \alpha - j\omega_o} + \frac{A_2}{s - \alpha + j\omega_o}.$$

In order for this sum to represent a real-valued signal, A_1 and A_2 must be complex conjugates of each other. Hence, we may replace these two terms with the single quadratic term

$$\frac{B_1 s + B_2}{(s - \alpha - j\omega_o)(s - \alpha + j\omega_o)} = \frac{B_1 s + B_2}{(s - \alpha)^2 + \omega_o^2},$$

where both B_1 and B_2 are real valued. We then solve for B_1 and B_2 and factor the result into the sum of two quadratic terms for which the inverse Laplace transforms are known. That is, we write

$$\frac{B_1 s + B_2}{(s - \alpha)^2 + \omega_o^2} = \frac{C_1(s - \alpha)}{(s - \alpha)^2 + \omega_o^2} + \frac{C_2 \omega_o}{(s - \alpha)^2 + \omega_o^2},$$

where $C_1 = B_1$ and $C_2 = (B_1\alpha + B_2)/\omega_o$. The inverse Laplace transform of the first term is given by

$$C_1 e^{\alpha t} \cos(\omega_o t)u(t) \xleftarrow{\mathcal{L}_u} \frac{C_1(s - \alpha)}{(s - \alpha)^2 + \omega_o^2}. \qquad (6.25)$$

Likewise, the inverse Laplace transform of the second term is obtained from the pair

$$C_2 e^{\alpha t} \sin(\omega_o t)u(t) \xleftarrow{\mathcal{L}_u} \frac{C_2 \omega_o}{(s - \alpha)^2 + \omega_o^2}. \qquad (6.26)$$

The next example illustrates this approach.

EXAMPLE 6.9 INVERSE LAPLACE TRANSFORM FOR COMPLEX-CONJUGATE POLES   Find the inverse Laplace transform of

$$X(s) = \frac{4s^2 + 6}{s^3 + s^2 - 2}.$$

Solution: There are three poles in $X(s)$. By trial and error, we find that $s = 1$ is a pole. We factor $s - 1$ out of $s^3 + s^2 - 2$ to obtain $s^2 + 2s + 2 = 0$ as the equation defining the remaining two poles. Finding the roots of this quadratic equation gives the complex-conjugate poles $s = -1 \pm j$.

We may write the quadratic equation $s^2 + 2s + 2$ in terms of the perfect square $(s^2 + 2s + 1) + 1 = (s + 1)^2 + 1$, so the partial-fraction expansion for $X(s)$ takes the form

$$X(s) = \frac{A}{s - 1} + \frac{B_1 s + B_2}{(s + 1)^2 + 1} \tag{6.27}$$

The expansion coefficient A is easily obtained by the method of residues. That is, we multiply both sides of Eq. (6.27) by $(s - 1)$ and evaluate at $s = 1$ to obtain

$$A = X(s)(s - 1)|_{s=1}$$
$$= \left.\frac{4s^2 + 6}{(s + 1)^2 + 1}\right|_{s=1}$$
$$= 2.$$

The remaining expansion coefficients B_1 and B_2 are obtained by placing both terms on the right-hand side of Eq. (6.27) over a common denominator and equating the numerator of the result to the numerator of $X(s)$. We thus write

$$4s^2 + 6 = 2((s + 1)^2 + 1) + (B_1 s + B_2)(s - 1)$$
$$= (2 + B_1)s^2 + (4 - B_1 + B_2)s + (4 - B_2).$$

Equating coefficients of s^2 gives $B_1 = 2$, and equating coefficients of s^0 gives $B_2 = -2$. Hence,

$$X(s) = \frac{2}{s - 1} + \frac{2s - 2}{(s + 1)^2 + 1}$$
$$= \frac{2}{s - 1} + 2\frac{s + 1}{(s + 1)^2 + 1} - 4\frac{1}{(s + 1)^2 + 1}.$$

In arriving at the second equation, we have factored $2s - 2$ into $2(s + 1) - 4$. Now we take the inverse Laplace transform of each term, using Eqs. (6.23), (6.25), and (6.26). Putting these results together, we obtain

$$x(t) = 2e^t u(t) + 2e^{-t}\cos(t)u(t) - 4e^{-t}\sin(t)u(t). \qquad ■$$

▶ **Problem 6.8**   Find the inverse Laplace transform of

(a)

$$X(s) = \frac{3s + 2}{s^2 + 4s + 5}$$

(b)

$$X(s) = \frac{s^2 + s - 2}{s^3 + 3s^2 + 5s + 3}$$

Answers:

(a)
$$x(t) = 3e^{-2t}\cos(t)u(t) - 4e^{-2t}\sin(t)u(t)$$

(b)
$$x(t) = -e^{-t}u(t) + 2e^{-t}\cos(\sqrt{2}t)u(t) - \frac{1}{\sqrt{2}}e^{-t}\sin(\sqrt{2}t)u(t) \qquad \blacktriangleleft$$

The poles of $X(s)$ determine the inherent characteristics of the signal $x(t)$. A complex pole at $s = d_k$ results in a complex exponential term of the form $e^{d_k t}u(t)$. Letting $d_k = \sigma_k + j\omega_k$, we may write this term as $e^{\sigma_k t}e^{j\omega_k t}u(t)$. Hence, the real part of the pole determines the exponential damping factor σ_k, and the imaginary part determines the sinusoidal frequency ω_k. The rate of decay of this term increases as $\mathrm{Re}\{d_k\}$ becomes more negative. The rate of oscillation is proportional to $|\mathrm{Im}\{d_k\}|$. With these properties in mind, we can infer much about the characteristics of a signal from the locations of its poles in the s-plane.

6.6 Solving Differential Equations with Initial Conditions

The primary application of the unilateral Laplace transform in systems analysis is solving differential equations with nonzero initial conditions. The initial conditions are incorporated into the solution as the values of the signal and its derivatives that occur at time zero in the differentiation property given by Eq. (6.19). This is illustrated by way of an example.

S & S
Solutions

EXAMPLE 6.10 RC CIRCUIT ANALYSIS   Use the Laplace transform to find the voltage across the capacitor, $y(t)$, for the RC circuit depicted in Fig. 6.7 in response to the applied voltage $x(t) = (3/5)e^{-2t}u(t)$ and initial condition $y(0^-) = -2$.

Solution: Using Kirchhoff's voltage law, we may describe the behavior of the circuit in Fig. 6.7 by the differential equation

$$\frac{d}{dt}y(t) + \frac{1}{RC}y(t) = \frac{1}{RC}x(t).$$

Letting $RC = 0.2$ s, we obtain

$$\frac{d}{dt}y(t) + 5y(t) = 5x(t).$$

Now we take the unilateral Laplace transform of each side of the differential equation and apply the differentiation property of Eq. (6.18), yielding

$$sY(s) - y(0^-) + 5Y(s) = 5X(s).$$

Solving for $Y(s)$, we get

$$Y(s) = \frac{1}{s+5}[5X(s) + y(0^-)].$$

Next, we use $x(t) \xrightarrow{\mathcal{L}_u} X(s) = \dfrac{3/5}{s+2}$ and the initial condition $y(0^-) = -2$, obtaining

$$Y(s) = \frac{3}{(s+2)(s+5)} + \frac{-2}{s+5}.$$

Expanding $Y(s)$ in partial fractions results in

$$Y(s) = \frac{1}{s+2} + \frac{-1}{s+5} + \frac{-2}{s+5}$$

$$= \frac{1}{s+2} - \frac{3}{s+5},$$

and taking the inverse unilateral Laplace transform yields the voltage across the capacitor:

$$y(t) = e^{-2t}u(t) - 3e^{-5t}u(t).$$

Care must be exercised in evaluating $y(t)$ at $t = 0$, as explained under Note 5 in Further Reading. ∎

The Laplace transform method for solving differential equations offers a clear separation between the natural response of the system to initial conditions and the forced response of the system associated with the input. Taking the unilateral Laplace transform of both sides of the general differential equation

$$\frac{d^N}{dt^N}y(t) + a_{N-1}\frac{d^{N-1}}{dt^{N-1}}y(t) + \cdots + a_1\frac{d}{dt}y(t) + a_0 y(t) =$$

$$b_M\frac{d^M}{dt^M}x(t) + b_{M-1}\frac{d^{M-1}}{dt^{M-1}}x(t) + \cdots + b_1\frac{d}{dt}x(t) + b_0 x(t),$$

we obtain

$$A(s)Y(s) - C(s) = B(s)X(s),$$

where

$$A(s) = s^N + a_{N-1}s^{N-1} + \cdots + a_1 s + a_0,$$

$$B(s) = b_M s^M + b_{M-1}s^{M-1} + \cdots + b_1 s + b_0,$$

$$C(s) = \sum_{k=1}^{N}\sum_{l=0}^{k-1} a_k s^{k-1-l}\frac{d^l}{dt^l}y(t)\bigg|_{t=0^-},$$

and we have assumed that the input is zero for $t < 0$.

We note that $C(s) = 0$ if all the initial conditions on $y(t)$ are zero and $B(s)X(s) = 0$ if the input $x(t)$ is zero. Now, we separate the effects of the initial conditions on $y(t)$ and the input to write

$$Y(s) = \frac{B(s)X(s)}{A(s)} + \frac{C(s)}{A(s)}$$

$$= Y^{(f)}(s) + Y^{(n)}(s),$$

where

$$Y^{(f)}(s) = \frac{B(s)X(s)}{A(s)} \quad \text{and} \quad Y^{(n)}(s) = \frac{C(s)}{A(s)}.$$

The term $Y^{(f)}(s)$ represents the component of the response associated entirely with the input, or the *forced response*, of the system. The same term represents the output when the initial conditions are zero. The term $Y^{(n)}(s)$ represents the component of the output due entirely to the initial conditions, or the *natural response*, of the system. This term represents the system output when the input is zero.

EXAMPLE 6.11 FINDING THE FORCED AND NATURAL RESPONSES   Use the unilateral Laplace transform to determine the output of a system represented by the differential equation

$$\frac{d^2}{dt^2}y(t) + 5\frac{d}{dt}y(t) + 6y(t) = \frac{d}{dt}x(t) + 6x(t)$$

in response to the input $x(t) = u(t)$. Assume that the initial conditions on the system are

$$y(0^-) = 1 \text{ and } \frac{d}{dt}y(t)\big|_{t=0^-} = 2. \tag{6.28}$$

Identify the forced response of the system, $y^{(f)}(t)$, and the natural response $y^{(n)}(t)$.

Solution: Using the differentiation property in Eq. (6.19) and taking the unilateral Laplace transform of both sides of the differential equation, we obtain

$$(s^2 + 5s + 6)Y(s) - \frac{d}{dt}y(t)\big|_{t=0^-} - sy(0^-) - 5y(0^-) = (s + 6)X(s).$$

Solving for $Y(s)$, we get

$$Y(s) = \frac{(s + 6)X(s)}{s^2 + 5s + 6} + \frac{sy(0^-) + \frac{d}{dt}y(t)\big|_{t=0^-} + 5y(0^-)}{s^2 + 5s + 6}.$$

The first term is associated with the forced response of the system, $Y^{(f)}(s)$. The second term corresponds to the natural response, $Y^{(n)}(s)$. Using $X(s) = 1/s$ and the initial conditions Eq. (6.28), we obtain

$$Y^{(f)}(s) = \frac{s + 6}{s(s + 2)(s + 3)}$$

and

$$Y^{(n)}(s) = \frac{s + 7}{(s + 2)(s + 3)}.$$

Partial fraction expansions of both terms yields

$$Y^{(f)}(s) = \frac{1}{s} + \frac{-2}{s + 2} + \frac{1}{s + 3}$$

and

$$Y^{(n)}(s) = \frac{5}{s + 2} + \frac{-4}{s + 3}.$$

Next, taking the inverse unilateral Laplace transforms of $Y^{(f)}(s)$ and $Y^{(n)}(s)$, we obtain

$$y^{(f)}(t) = u(t) - 2e^{-2t}u(t) + e^{-3t}u(t)$$

and

$$y^{(n)}(t) = 5e^{-2t}u(t) - 4e^{-3t}u(t).$$

The output of the system is $y(t) = y^{(f)}(t) + y^{(n)}(t)$. Figures 6.8(a), (b), and (c) depict the forced response, natural response, and system output, respectively.   ■

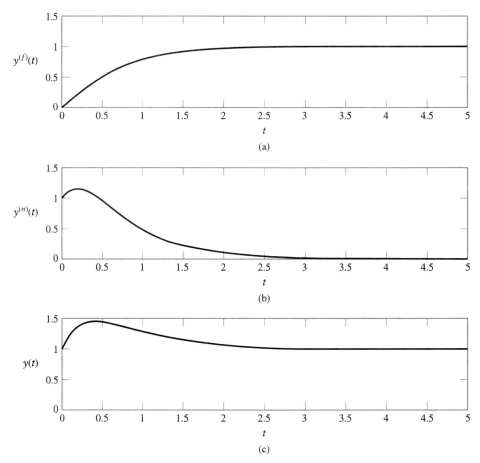

FIGURE 6.8   The solution to Example 6.11. (a) Forced response of the system, $y^{(f)}(t)$. (b) Natural response of the system, $y^{(n)}(t)$. (c) Overall system output.

S & S
Solutions

EXAMPLE 6.12 MEMS ACCELEROMETER: FORCED AND NATURAL RESPONSE   The MEMS accelerometer introduced in Section 1.10 is governed by the differential equation

$$\frac{d^2}{dt^2}y(t) + \frac{\omega_n}{Q}\frac{d}{dt}y(t) + \omega_n^2 y(t) = x(t),$$

where $x(t)$ is the external acceleration and $y(t)$ is the position of the proof mass. Find the forced and natural responses if $\omega_n = 10{,}000$ rad/s and $Q = 1/2$, assuming that the initial position of the mass is $y(0^-) = -2 \times 10^{-7}$ m, the initial velocity is $\frac{d}{dt}y(t)|_{t=0^-} = 0$, and the input is $x(t) = 20[u(t) - u(t - 3 \times 10^{-4})]$ m/s^2.

Solution: We take the unilateral Laplace transform of both sides of the differential equation and rearrange terms to identify

$$Y^{(f)}(s) = \frac{X(s)}{s^2 + 20{,}000s + (10{,}000)^2}$$

and

$$Y^{(n)}(s) = \frac{(s + 20{,}000)y(0^-) + \frac{d}{dt}y(t)\Big|_{t=0^-}}{s^2 + 20{,}000s + (10{,}000)^2}.$$

Using the prescribed initial conditions and a partial-fraction expansion of the resulting $y^{(n)}(s)$, we obtain

$$Y^{(n)}(s) = \frac{-2 \times 10^{-7}(s + 20{,}000)}{(s + 10{,}000)^2}$$

$$= \frac{-2 \times 10^{-7}}{s + 10{,}000} + \frac{-2 \times 10^{-3}}{(s + 10{,}000)^2}.$$

Now we take the inverse unilateral transform of $Y^{(n)}(s)$ to obtain the natural response:

$$y^{(n)}(t) = -2 \times 10^{-7}e^{-10{,}000t}u(t) - 2 \times 10^{-3}te^{-10{,}000t}u(t).$$

Next, we use $X(s) = 20(1 - e^{-3\times 10^{-4}s})/s$ to obtain the Laplace transform of the forced response:

$$Y^{(f)}(s) = (1 - e^{-3\times 10^{-4}s})\frac{20}{s(s^2 + 20{,}000s + 10{,}000^2)}.$$

Performing a partial-fraction expansion of $Y^{(f)}(s)$ yields

$$Y^{(f)}(s) = (1 - e^{-3\times 10^{-4}s})\frac{20}{10^8}\left[\frac{1}{s} - \frac{1}{s + 10{,}000} - \frac{10{,}000}{(s + 10{,}000)^2}\right].$$

The term $e^{-3\times 10^{-4}s}$ introduces a time delay of 3×10^{-4} s to each term in the partial fraction expansion. Taking the inverse unilateral Laplace transform, we obtain

$$y^{(f)}(t) = \frac{20}{10^8}[u(t) - u(t - 3 \times 10^{-4}) - e^{-10{,}000t}u(t) + e^{-(10{,}000t-3)}u(t - 3 \times 10^{-4})$$

$$- 10{,}000te^{-10{,}000t}u(t) + (10{,}000t - 3)e^{-(10{,}000t-3)}u(t - 3 \times 10^{-4})].$$

The natural and forced responses are depicted in Fig. 6.9. This MEMS has a low Q-factor. Thus, we expect the response of the system to be heavily damped, an expectation that is borne out by the results obtained for $y^{(n)}(t)$ and $y^{(f)}(t)$. ∎

S & S
Solutions

▶ **Problem 6.9** Determine the forced and natural responses of the systems described by the following differential equations with the specified input and initial conditions:

(a)
$$\frac{d}{dt}y(t) + 3y(t) = 4x(t), \quad x(t) = \cos(2t)u(t), \quad y(0^-) = -2$$

(b)
$$\frac{d^2}{dt^2}y(t) + 4y(t) = 8x(t), \quad x(t) = u(t), \quad y(0^-) = 1, \quad \frac{d}{dt}y(t)\Big|_{t=0^-} = 2$$

Answers:

(a)
$$y^{(f)}(t) = -\frac{12}{13}e^{-3t}u(t) + \frac{12}{13}\cos(2t)u(t) + \frac{8}{13}\sin(2t)u(t)$$

$$y^{(n)}(t) = -2e^{-3t}u(t)$$

(b)
$$y^{(f)}(t) = 2u(t) - 2\cos(2t)u(t)$$
$$y^{(n)}(t) = \cos(2t)u(t) + \sin(2t)u(t)$$

◀

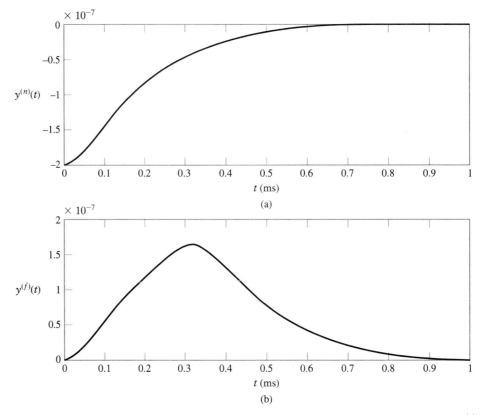

FIGURE 6.9   MEMS accelerometer responses for Example 6.12. (a) Natural response $y^{(n)}(t)$. (b) Forced response $y^{(f)}(t)$.

The natural response of the system is obtained from a partial-fraction expansion using the poles of $\frac{C(s)}{A(s)}$, which are the roots of $A(s)$. Each pole factor contributes a term of the form e^{pt} where p is the corresponding root of $A(s)$. For this reason, these roots are sometimes termed the *natural frequencies* of the system. The natural frequencies provide valuable information about the system characteristics. If the system is stable, then the natural frequencies must have negative real parts; that is, they must lie in the left half of the s-plane. The distance of the real part of the natural frequencies to the left of the $j\omega$ axis determines how fast the system responds, since it determines how fast the corresponding term in the natural response decays to zero. The imaginary part of the natural frequency determines the frequency of oscillation for the corresponding term in the natural response. As the magnitude of the imaginary part increases, the oscillation frequency increases.

6.7 *Laplace Transform Methods in Circuit Analysis*

The differentiation and integration properties may also be used to transform circuits involving capacitive and inductive elements so that the circuits may be solved directly in terms of Laplace transforms, rather than by first writing the differential equation in the time domain. This is accomplished by replacing resistive, capacitive, and inductive elements by their Laplace transform equivalents.

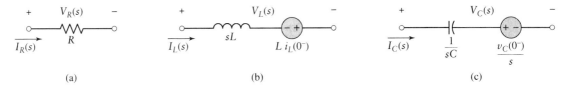

(a) (b) (c)

FIGURE 6.10 Laplace transform circuit models for use with Kirchhoff's voltage law. (a) Resistor. (b) Inductor with initial current $i_L(0^-)$. (c) Capacitor with initial voltage $v_C(0^-)$.

A resistance R with corresponding voltage $v_R(t)$ and current $i_R(t)$ satisfies the relation

$$v_R(t) = Ri_R(t).$$

Transforming this equation, we write

$$V_R(s) = RI_R(s), \tag{6.29}$$

which is represented by the transformed resistor element of Fig. 6.10(a). Next, we consider an inductor, for which

$$v_L(t) = L\frac{d}{dt}i_L(t).$$

Transforming this relationship and using the differentiation property given in Eq. (6.19) yields

$$V_L(s) = sLI_L(s) - Li_L(0^-). \tag{6.30}$$

This relationship is represented by the transformed inductor element of Fig. 6.10(b). Lastly, consider a capacitor, which may be described by

$$v_C(t) = \frac{1}{C}\int_{0^-}^{t} i_C(\tau)d\tau + v_C(0^-).$$

Transforming this relationship and using the integration property given in Eq. (6.20), we obtain

$$V_C(s) = \frac{1}{sC}I_C(s) + \frac{v_C(0^-)}{s}. \tag{6.31}$$

Figure 6.10 (c) depicts the transformed capacitor element described by Eq. (6.31).

The circuit models corresponding to Eqs. (6.29), (6.30), and (6.31) are most useful when one is applying Kirchhoff's voltage law to solve a circuit. If Kirchhoff's current law is to be used, then it is more convenient to rewrite Eqs. (6.29), (6.30), and (6.31) to express current as a function of voltage. This results in the transformed circuit elements depicted in Fig. 6.11. The next example illustrates the Laplace transform method for solving an electrical circuit.

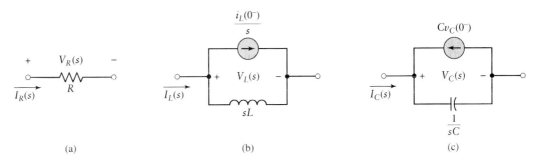

(a) (b) (c)

FIGURE 6.11 Laplace transform circuit models for use with Kirchhoff's current law. (a) Resistor. (b) Inductor with initial current $i_L(0^-)$. (c) Capacitor with initial voltage $v_C(0^-)$.

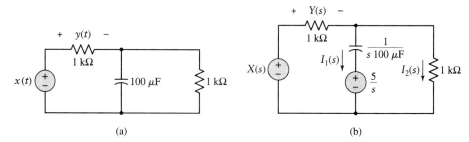

FIGURE 6.12   Electrical circuit for Example 6.13. (a) Original circuit. (b) Transformed circuit.

EXAMPLE 6.13 SOLVING A SECOND-ORDER CIRCUIT   Use Laplace transform circuit models to determine the voltage $y(t)$ in the circuit of Fig. 6.12(a) for an applied voltage $x(t) = 3e^{-10t}u(t)$ V. The voltage across the capacitor at time $t = 0^-$ is 5 V.

Solution: The transformed circuit is drawn in Fig. 6.12(b), with symbols $I_1(s)$ and $I_2(s)$ representing the current through each branch. Using Kirchhoff's laws, we write the following equations to describe the circuit:

$$Y(s) = 1000(I_1(s) + I_2(s));$$

$$X(s) = Y(s) + \frac{1}{s(10^{-4})}I_1(s) + \frac{5}{s};$$

$$X(s) = Y(s) + 1000I_2(s).$$

Combining these three equations to eliminate $I_1(s)$ and $I_2(s)$ yields

$$Y(s) = X(s)\frac{s + 10}{s + 20} - \frac{5}{s + 20}.$$

Using $X(s) = 3/(s + 10)$, we obtain

$$Y(s) = \frac{-2}{s + 20}$$

and thus conclude that

$$y(t) = -2e^{-20t}u(t) \text{ V.}   \blacksquare$$

The natural and forced responses of a circuit are easily determined by using the transformed circuit representation. The natural response is obtained by setting the voltage or current source associated with the input equal to zero. In this case, the only voltage or current sources in the transformed circuit are those associated with the initial conditions in the transformed capacitor and inductor circuit models. The forced response due to the input is obtained by setting the initial conditions equal to zero, which eliminates the voltage or current sources present in the transformed capacitor and inductor circuit models.

▶ **Problem 6.10**   Use the Laplace transform circuit representation to obtain the natural and forced responses for the *RC* circuit depicted in Fig. 6.7, assuming that $x(t) = (3/5)e^{-2t}u(t)$ and $y(0^-) = -2$.

Answer:

$$y^{(n)}(t) = -2e^{-5t}u(t)$$
$$y^{(f)}(t) = e^{-2t}u(t) - e^{-5t}u(t)   ◀$$

6.8 *Properties of the Bilateral Laplace Transform*

The bilateral Laplace transform involves the values of the signal $x(t)$ for both $t \geq 0$ and $t < 0$ and is given by

$$x(t) \xleftrightarrow{\mathcal{L}} X(s) = \int_{-\infty}^{\infty} x(t)e^{-st}\, dt.$$

Hence, the bilateral Laplace transform is well suited to problems involving noncausal signals and systems, applications studied in subsequent sections. In this section, we note important differences between unilateral and bilateral Laplace transform properties.

The properties of linearity, scaling, s-domain shift, convolution, and differentiation in the s-domain are identical for the bilateral and unilateral Laplace transforms, although the operations associated by these properties may change the region of convergence (ROC). The effect of each of these operations on the ROC is given in the table of Laplace transform properties in Appendix D.2.

To illustrate the change in ROC that may occur, consider the linearity property. If $x(t) \xleftrightarrow{\mathcal{L}} X(s)$ with ROC R_x and $y(t) \xleftrightarrow{\mathcal{L}} Y(s)$ with ROC R_y, then $ax(t) + by(t) \xleftrightarrow{\mathcal{L}} aX(s) + bY(s)$ with ROC at least $R_x \cap R_y$, where the symbol $\cap$ indicates intersection. Usually, the ROC for a sum of signals is just the intersection of the individual ROCs. The ROC may be larger than the intersection of the individual ROCs if a pole and a zero cancel in the sum $aX(s) + bY(s)$. The effect of pole–zero cancellation on the ROC is illustrated in the next example.

EXAMPLE 6.14 EFFECT OF POLE–ZERO CANCELLATION ON THE ROC   Suppose

$$x(t) = e^{-2t}u(t) \xleftrightarrow{\mathcal{L}} X(s) = \frac{1}{s + 2}, \quad \text{with ROC } \mathrm{Re}(s) > -2$$

and

$$y(t) = e^{-2t}u(t) - e^{-3t}u(t) \xleftrightarrow{\mathcal{L}} Y(s) = \frac{1}{(s + 2)(s + 3)} \text{ with ROC } \mathrm{Re}(s) > -2.$$

The s-plane representations of the ROCs are shown in Fig. 6.13. The intersection of the ROCs is $\mathrm{Re}(s) > -2$. However, if we choose $a = 1$ and $b = -1$, then the difference $x(t) - y(t) = e^{-3t}u(t)$ has ROC $\mathrm{Re}(s) > -3$, which is larger than the intersection of the ROCs. Here, the subtraction eliminates the signal $e^{-2t}u(t)$ in the time domain; consequently, the ROC is enlarged. This corresponds to a pole–zero cancellation in the s-domain, since

$$
\begin{aligned}
X(s) - Y(s) &= \frac{1}{s + 2} - \frac{1}{(s + 2)(s + 3)} \\
&= \frac{(s + 3) - 1}{(s + 2)(s + 3)} \\
&= \frac{(s + 2)}{(s + 2)(s + 3)}.
\end{aligned}
$$

The zero of $(X(s) - Y(s))$ located at $s = -2$ cancels the pole at $s = -2$, so we have

$$X(s) - Y(s) = \frac{1}{s + 3}.$$ ∎

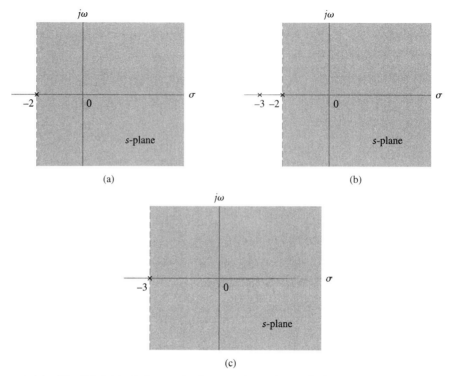

FIGURE 6.13   The ROC (shaded region) of a sum of signals may be larger than the intersection of individual ROCs when pole–zero cancellation occurs. (a) ROC for $x(t) = e^{-2t}u(t)$; (b) ROC for $y(t) = e^{-2t}u(t) - e^{-3t}u(t)$; (c) ROC for $x(t) - y(t)$.

If the intersection of the ROCs is the empty set and pole–zero cancellation does not occur, then the Laplace transform of $ax(t) + by(t)$ does not exist. Note that the ROC for the convolution of two signals may also be larger than the intersection of the individual ROCs if pole–zero cancellation occurs.

The bilateral Laplace transform properties involving time shifts, differentiation in the time domain, and integration with respect to time differ slightly from their unilateral counterparts. We state them without proof as follows:

Time Shift

$$\boxed{x(t - \tau) \xleftrightarrow{\ \mathcal{L}\ } e^{-s\tau}X(s).} \tag{6.32}$$

The restriction on the shift that is present in the unilateral property given by Eq. (6.14) is removed because the bilateral Laplace transform is evaluated over both positive and negative values of time. Note that the ROC is unchanged by a time shift.

Differentiation in the Time Domain

$$\boxed{\frac{d}{dt}x(t) \xleftrightarrow{\ \mathcal{L}\ } sX(s), \quad \text{with ROC at least } R_x,} \tag{6.33}$$

where R_x is the ROC associated with $X(s)$. Differentiation in time corresponds to multiplication by s. The ROC associated with $sX(s)$ may be larger than R_x if $X(s)$ has a single pole at $s = 0$ on the ROC boundary. Multiplication by s, corresponding to differentiation, cancels this pole and therefore eliminates the dc component in $x(t)$.

EXAMPLE 6.15   USING THE BILATERAL TIME-SHIFT AND DIFFERENTIATION PROPERTIES
Find the Laplace transform of

$$x(t) = \frac{d^2}{dt^2}(e^{-3(t-2)}u(t-2)).$$

Solution: We know from Example 6.1 that

$$e^{-3t}u(t) \overset{\mathcal{L}}{\longleftrightarrow} \frac{1}{s+3}, \quad \text{with ROC } \text{Re}(s) > -3.$$

The time-shift property given by Eq. (6.32) implies that

$$e^{-3(t-2)}u(t-2) \overset{\mathcal{L}}{\longleftrightarrow} \frac{1}{s+3}e^{-2s}, \quad \text{with ROC } \text{Re}(s) > -3.$$

Now we apply the time-differentiation property given by Eq. (6.33) twice, as shown by

$$x(t) = \frac{d^2}{dt^2}(e^{-3(t-2)}u(t-2)) \overset{\mathcal{L}}{\longleftrightarrow} X(s) = \frac{s^2}{s+3}e^{-2s}, \quad \text{with ROC } \text{Re}(s) > -3. \quad \blacksquare$$

Integration with Respect to Time

$$\boxed{\int_{-\infty}^{t} x(\tau)d\tau \overset{\mathcal{L}}{\longleftrightarrow} \frac{X(s)}{s}, \quad \text{with ROC } R_x \cap \text{Re}(s) > 0.} \qquad (6.34)$$

Integration corresponds to division by s. Since this introduces a pole at $s = 0$ and we are integrating to the right, the ROC must lie to the right of $s = 0$.

The initial- and final-value theorems apply to the bilateral Laplace transform, with the additional restriction that $x(t) = 0$ for $t < 0$.

**S & S
Solutions**

▶ **Problem 6.11**   Determine the bilateral Laplace transform and the corresponding ROC for each of the following signals:

(a)
$$x(t) = e^{-t}\frac{d}{dt}(e^{-(t+1)}u(t+1))$$

(b)
$$x(t) = \int_{-\infty}^{t} e^{2\tau}\sin(\tau)u(-\tau)d\tau$$

Answers:

(a)
$$X(s) = \frac{(s+1)e^{s+1}}{s+2}, \quad \text{Re}(s) > -2$$

(b)
$$X(s) = \frac{-1}{s((s-2)^2+1)}, \quad 0 < \text{Re}(s) < 2 \qquad \blacktriangleleft$$

6.9 Properties of the Region of Convergence

In Section 6.2, we discovered that the bilateral Laplace transform is not unique, unless the ROC is specified. In this section, we show how the ROC is related to the characteristics of a signal $x(t)$. We develop these properties using intuitive arguments rather than rigorous proofs. Once we know the ROC properties, we can often identify the ROC from knowledge of the Laplace transform $X(s)$ and limited knowledge of the characteristics of $x(t)$.

First, we note that the ROC cannot contain any poles. If the Laplace transform converges, then $X(s)$ is finite over the entire ROC. Suppose d is a pole of $X(s)$. This implies that $X(d) = \pm\infty$, so the Laplace transform does not converge at d. Thus, $s = d$ cannot lie in the ROC.

Next, convergence of the bilateral Laplace transform for a signal $x(t)$ implies that

$$I(\sigma) = \int_{-\infty}^{\infty} |x(t)| e^{-\sigma t} dt < \infty$$

for some values of σ. The set of σ for which this integral is finite determines the ROC of the bilateral Laplace transform of $x(t)$. The quantity σ is the real part of s, so the ROC depends only on that part; the imaginary component of s does not affect convergence. This implies that the ROC consists of strips parallel to the $j\omega$-axis in the s-plane.

Suppose $x(t)$ is a finite-duration signal; that is, $x(t) = 0$ for $t < a$ and $t > b$. If we can find a finite bounding constant A such that $|x(t)| \leq A$, then

$$I(\sigma) \leq \int_a^b A e^{-\sigma t} dt$$

$$= \begin{cases} \dfrac{-A}{\sigma} [e^{-\sigma t}|_a^b, & \sigma \neq 0 \\ A(b - a), & \sigma = 0 \end{cases}.$$

In this case, we see that $I(\sigma)$ is finite for all finite values of σ, and we conclude that the ROC for a finite-duration signal includes the entire s-plane.

Now we separate $I(\sigma)$ into positive- and negative-time sections; that is,

$$I(\sigma) = I_-(\sigma) + I_+(\sigma),$$

where

$$I_-(\sigma) = \int_{-\infty}^{0} |x(t)| e^{-\sigma t} dt$$

and

$$I_+(\sigma) = \int_0^{\infty} |x(t)| e^{-\sigma t} dt.$$

In order for $I(\sigma)$ to be finite, both of these integrals must be finite. This implies that $|x(t)|$ must be bounded in some sense.

Suppose we can bound $|x(t)|$ for both positive and negative t by finding the smallest constants $A > 0$ and σ_p such that

$$|x(t)| \leq A e^{\sigma_p t}, \quad t > 0,$$

and the largest constant σ_n such that

$$|x(t)| \le Ae^{\sigma_n t}, \quad t < 0.$$

A signal $x(t)$ that satisfies these bounds is said to be of *exponential order*. The bounds imply that $|x(t)|$ grows no faster than $e^{\sigma_p t}$ for positive t and $e^{\sigma_n t}$ for negative t. There are signals that are not of exponential order, such as e^{t^2} or t^{3t}, but such signals generally do not arise in the study of physical systems.

Using the exponential order bounds on $|x(t)|$, we may write

$$I_-(\sigma) \le A \int_{-\infty}^{0} e^{(\sigma_n - \sigma)t}\, dt$$

$$= \frac{A}{\sigma_n - \sigma}[e^{(\sigma_n-\sigma)t}|_{-\infty}^{0}]$$

and

$$I_+(\sigma) \le A \int_{0}^{\infty} e^{(\sigma_p - \sigma)t}\, dt$$

$$= \frac{A}{\sigma_p - \sigma}[e^{(\sigma_p-\sigma)t}|_{0}^{\infty}].$$

We note that $I_-(\sigma)$ is finite whenever $\sigma < \sigma_n$ and $I_+(\sigma)$ is finite whenever $\sigma > \sigma_p$. The quantity $I(\sigma)$ is finite at values σ for which both $I_-(\sigma)$ and $I_+(\sigma)$ are finite. Hence, the Laplace transform converges for $\sigma_p < \sigma < \sigma_n$. Note that if $\sigma_p > \sigma_n$, then there are no values of σ for which the bilateral Laplace transform converges.

We may draw the following conclusions from the analysis just presented: Define a *left-sided signal* as a signal for which $x(t) = 0$ for $t > b$, a *right-sided signal* as a signal for which $x(t) = 0$ for $t < a$, and a *two-sided signal* as a signal that is infinite in extent in both directions. Note that a and b are arbitrary constants. If $x(t)$ is of exponential order, then

▸ The ROC of a left-sided signal is of the form $\sigma < \sigma_n$.
▸ The ROC of a right-sided signal is of the form $\sigma > \sigma_p$.
▸ The ROC of a two-sided signal is of the form $\sigma_p < \sigma < \sigma_n$.

Each of these cases is illustrated in Fig. 6.14.

Exponential signals of the form Ae^{at} are frequently encountered in physical problems. In this case, there is a clear relationship between the ROC and the signal. Specifically, the real part of one or more poles determines the ROC boundaries σ_n and σ_p. Suppose we have the right-sided signal, $x(t) = e^{at}u(t)$, where, in general, a is complex. This signal is of exponential order, with the smallest exponential bounding signal $e^{\mathrm{Re}(a)t}$. Hence, $\sigma_p = \mathrm{Re}(a)$, and the ROC is $\sigma > \mathrm{Re}(a)$. The bilateral Laplace transform of $x(t)$ has a pole at $s = a$, so the ROC is the region of the s-plane that lies to the right of the pole. Likewise, if $x(t) = e^{at}u(-t)$, then the ROC is $\sigma < \mathrm{Re}(a)$; that is, the ROC is the region of the s-plane to the left of the pole. If a signal $x(t)$ consists of a sum of exponentials, then the ROC is the intersection of the ROCs associated with each term in the sum. This property is demonstrated in the next example.

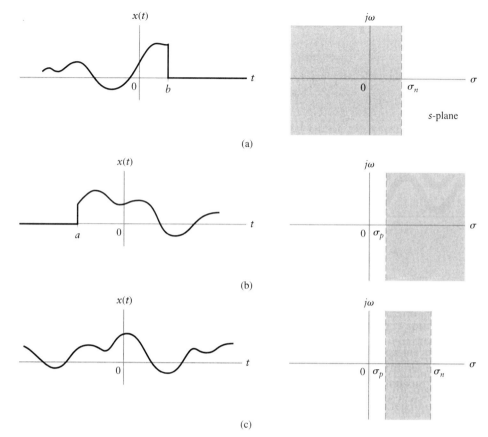

FIGURE 6.14   Relationship between the time extent of a signal and the ROC, shown as the shaded region. (a) A left-sided signal has ROC to the left of a vertical line in the *s*-plane. (b) A right-sided signal has ROC to the right of a vertical line in the *s*-plane. (c) A two-sided signal has ROC given by a vertical strip of finite width in the *s*-plane.

EXAMPLE 6.16   ROC OF A SUM OF EXPONENTIALS   Consider the two signals

$$x_1(t) = e^{-2t}u(t) + e^{-t}u(-t)$$

and

$$x_2(t) = e^{-t}u(t) + e^{-2t}u(-t).$$

Identify the ROC associated with the bilateral Laplace transform of each signal.

Solution: We check the absolute integrability of $|x_1(t)|e^{-\sigma t}$ by writing

$$
\begin{aligned}
I_1(\sigma) &= \int_{-\infty}^{\infty} |x_1(t)|e^{-\sigma t}\, dt \\
&= \int_{-\infty}^{0} e^{-(1+\sigma)t}\, dt + \int_{0}^{\infty} e^{-(2+\sigma)t}\, dt \\
&= \frac{-1}{1+\sigma}\big[e^{-(1+\sigma)t}\big|_{-\infty}^{0} + \frac{-1}{2+\sigma}\big[e^{-(2+\sigma)t}\big|_{0}^{\infty}.
\end{aligned}
$$

The first term converges for $\sigma < -1$, while the second term converges for $\sigma > -2$. Hence, both terms converge for $-2 < \sigma < -1$. This is the intersection of the ROC for each term.

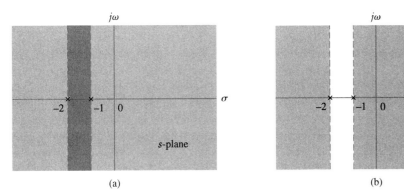

FIGURE 6.15   ROCs for signals in Example 6.16. (a) The shaded regions denote the ROCs of each individual term, $e^{-2t}u(t)$ and $e^{-t}u(-t)$. The doubly shaded region is the intersection of the individual ROCs and represents the ROC of the sum. (b) The shaded regions represent the individual ROCs of $e^{-2t}u(-t)$ and $e^{-t}u(t)$. In this case there is no intersection, and the Laplace transform of the sum does not converge for any value of s.

The ROC for each term and the intersection of the ROCs, which is shown as the doubly shaded region, are depicted in Fig. 6.15(a). The reader may verify that the Laplace transform of $x_1(t)$ is

$$X_1(s) = \frac{1}{s + 2} + \frac{-1}{s + 1}$$

$$= \frac{-1}{(s + 1)(s + 2)},$$

which has poles at $s = -1$ and $s = -2$. We see that the ROC associated with $X_1(s)$ is the strip of the s-plane located between the poles.

For the second signal, $x_2(t)$, we have

$$I_1(\sigma) = \int_{-\infty}^{\infty} |x_2(t)| e^{-\sigma t}\, dt$$

$$= \int_{-\infty}^{0} e^{-(2+\sigma)t}\, dt + \int_{0}^{\infty} e^{-(1+\sigma)t}\, dt$$

$$= \frac{-1}{2 + \sigma} \left[e^{-(2+\sigma)t}\big|_{-\infty}^{0} + \frac{-1}{1 + \sigma}\left[e^{-(1+\sigma)t}\big|_{0}^{\infty}\right.\right.$$

The first term converges for $\sigma < -2$ and the second term for $\sigma > -1$. Here, there is no value of σ for which both terms converge, so the intersection is empty. Hence, there are no values of s for which $X_2(s)$ converges, as illustrated in Fig. 6.15(b). Thus, the bilateral Laplace transform of $x_2(t)$ does not exist.   ■

▶ **Problem 6.12**   Describe the ROC of the signal

$$x(t) = e^{-b|t|}$$

for the two cases $b > 0$ and $b < 0$.

Answer: For $b > 0$, the ROC is the region $-b < \sigma < b$. For $b < 0$, the ROC is the empty set   ◀

6.10 *Inversion of the Bilateral Laplace Transform*

As in the unilateral case discussed in Section 6.5, we consider the inversion of bilateral Laplace transforms that are expressed as ratios of polynomials in s. The primary difference between the inversions of bilateral and unilateral Laplace transforms is that we must use the ROC to determine a unique inverse transform in the bilateral case.

Suppose we wish to invert the ratio of polynomials in s given by

$$X(s) = \frac{B(s)}{A(s)}$$
$$= \frac{b_M s^M + b_{M-1} s^{M-1} + \cdots + b_1 s + b_0}{s^N + a_{N-1} s^{N-1} + \cdots + a_1 s + a_0}.$$

As in the unilateral case, if $M \geq N$, then we use long division to express

$$X(s) = \sum_{k=0}^{M-N} c_k s^k + \widetilde{X}(s),$$

where

$$\widetilde{X}(s) = \frac{\widetilde{B}(s)}{A(s)}$$

is expressed as a partial-fraction expansion in terms of nonrepeated poles; that is,

$$\widetilde{X}(s) = \sum_{k=1}^{N} \frac{A_k}{s - d_k}.$$

We have

$$\sum_{k=0}^{M-N} c_k \delta^{(k)}(t) \xleftrightarrow{\quad\mathcal{L}\quad} \sum_{k=0}^{M-N} c_k s^k,$$

where $\delta^{(k)}(t)$ denotes the kth derivative of the impulse $\delta(t)$. Note that the ROC of $\widetilde{X}(s)$ is the same as the ROC of $X(s)$ because the Laplace transform of the impulse and its derivatives converge everywhere in the s-plane.

In the bilateral case, there are two possibilities for the inverse Laplace transform of each term in the partial-fraction expansion of $\widetilde{X}(s)$: We may use either the right-sided transform pair

$$A_k e^{d_k t} u(t) \xleftrightarrow{\quad\mathcal{L}\quad} \frac{A_k}{s - d_k}, \quad \text{with ROC } \text{Re}(s) > d_k, \qquad (6.35)$$

or the left-sided transform pair

$$-A_k e^{d_k t} u(-t) \xleftrightarrow{\quad\mathcal{L}\quad} \frac{A_k}{s - d_k}, \quad \text{with ROC } \text{Re}(s) < d_k. \qquad (6.36)$$

The ROC associated with $X(s)$ determines whether the left-sided or right-sided inverse transform is chosen. Recall that the ROC of a right-sided exponential signal lies to the right of the pole, while the ROC of a left-sided exponential signal lies to the left of the pole.

The linearity property states that the ROC of $X(s)$ is the intersection of the ROCs of the individual terms in the partial-fraction expansion. In order to find the inverse transform of each term, we must infer the ROC of each term from the given ROC of $X(s)$. This is easily accomplished by comparing the location of each pole with the ROC of $X(s)$. If the ROC of $X(s)$ lies to the left of a particular pole, we choose the left-sided inverse Laplace transform for that pole. If the ROC of $X(s)$ lies to the right of a particular pole, we choose the right-sided inverse Laplace transform for that pole. This procedure is illustrated in the next example.

EXAMPLE 6.17 INVERTING A PROPER RATIONAL LAPLACE TRANSFORM   Find the inverse bilateral Laplace transform of

$$X(s) = \frac{-5s - 7}{(s + 1)(s - 1)(s + 2)}, \quad \text{with ROC } -1 < \text{Re}(s) < 1.$$

Solution: Use the partial-fraction expansion

$$X(s) = \frac{1}{s + 1} - \frac{2}{s - 1} + \frac{1}{s + 2}.$$

The ROC and the locations of the poles are depicted in Fig. 6.16. We find the inverse Laplace transform of each term, using the relationship between the locations of poles and the ROC:

▶ The pole of the first term is at $s = -1$. The ROC lies to the right of this pole, so Eq. (6.35) is applicable, and we choose the right-sided inverse Laplace transform,

$$e^{-t}u(t) \xleftrightarrow{\mathcal{L}} \frac{1}{s + 1}.$$

▶ The second term has a pole at $s = 1$. Here, the ROC is to the left of the pole, so Eq. (6.36) is applicable, and we choose the left-sided inverse Laplace transform,

$$2e^{t}u(-t) \xleftrightarrow{\mathcal{L}} -\frac{2}{s - 1}.$$

▶ The pole in the last term is at $s = -2$. The ROC is to the right of this pole, so from Eq. (6.35) we choose the right-sided inverse Laplace transform,

$$e^{-2t}u(t) \xleftrightarrow{\mathcal{L}} \frac{1}{s + 2}.$$

Combining these three terms, we obtain

$$x(t) = e^{-t}u(t) + 2e^{t}u(-t) + e^{-2t}u(t). \qquad \blacksquare$$

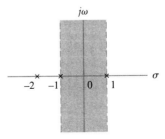

FIGURE 6.16   Poles and ROC for Example 6.17.

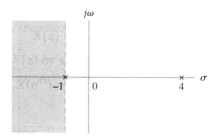

FIGURE 6.17   Poles and ROC for Example 6.18.

▶ **Problem 6.13**   Repeat the previous example if the ROC is $-2 < \mathrm{Re}(s) < -1$.

Answer:

$$x(t) = -e^{-t}u(-t) + 2e^{t}u(-t) + e^{-2t}u(t)$$ ◀

EXAMPLE 6.18 INVERTING AN IMPROPER RATIONAL LAPLACE TRANSFORM   Find the inverse bilateral Laplace transform of

$$X(s) = \frac{2s^3 - 9s^2 + 4s + 10}{s^2 - 3s - 4}, \quad \text{with ROC } \mathrm{Re}(s) < -1.$$

Solution: Use the result of Example 6.8 to expand $X(s)$:

$$X(s) = 2s - 3 + \frac{1}{s+1} + \frac{2}{s-4}, \quad \mathrm{Re}(s) < -1.$$

The locations of the poles and the ROC are shown in Fig. 6.17. The ROC is to the left of both poles, so applying Eq. (6.36), we choose left-sided inverse Laplace transforms and obtain

$$x(t) = 2\delta^{(1)}(t) - 3\delta(t) - e^{-t}u(-t) - 2e^{4t}u(-t).$$ ■

▶ **Problem 6.14**   Find the inverse Laplace transform of

$$X(s) = \frac{s^4 + 3s^3 - 4s^2 + 5s + 5}{s^2 + 3s - 4}, \quad \text{with ROC } -4 < \mathrm{Re}(s) < 1.$$

Answer:

$$x(t) = \delta^{(2)}(t) - 2e^{t}u(-t) + 3e^{-4t}u(t)$$ ◀

The relationship between the locations of poles and the ROC in the s-plane also determines the inverse transform for the other terms that can occur in a partial-fraction expansion. For example, using Eq. (6.24), the inverse bilateral Laplace transform of the term

$$\frac{A}{(s - d_k)^n}$$

is given by the right-sided signal

$$\frac{At^{n-1}}{(n-1)!}e^{d_k t}u(t)$$

if the ROC lies to the right of the poles. If the ROC lies to the left of the poles, then the inverse Laplace transform is

$$\frac{-At^{n-1}}{(n-1)!}e^{d_k t}u(-t).$$

Similarly, the inverse bilateral Laplace transform of the term

$$\frac{C_1(s-\alpha)}{(s-\alpha)^2 + \omega_o^2}$$

is the right-sided signal

$$C_1 e^{\alpha t}\cos(\omega_o t)u(t)$$

if the ROC lies to the right of the poles at $s = \alpha \pm j\omega_o$. If the ROC lies to the left of the poles at $s = \alpha \pm j\omega_o$, then the inverse Laplace transform is

$$-C_1 e^{\alpha t}\cos(\omega_o t)u(-t).$$

● **S & S**
Solutions

▶ **Problem 6.15**   Find the inverse bilateral Laplace transform of

$$X(s) = \frac{4s^2 + 6}{s^3 + s^2 - 2}, \quad \text{with ROC } -1 < \text{Re}(s) < 1.$$

Answer:

$$x(t) = -2e^t u(-t) + 2e^{-t}\cos(t)u(t) - 4e^{-t}\sin(t)u(t) \qquad ◀$$

Note that we may determine a unique inverse bilateral Laplace transform by using knowledge other than the ROC. The most common form of other knowledge is that of causality, stability, or the existence of the Fourier transform.

▹ If the signal is known to be causal, then we choose the right-sided inverse transform for each term. This is the approach followed with the unilateral Laplace transform.

▹ A stable signal is absolutely integrable and thus has a Fourier transform. Hence, stability and the existence of the Fourier transform are equivalent conditions. In both of these cases, the ROC includes the $j\omega$-axis in the s-plane, or $\text{Re}(s) = 0$. The inverse Laplace transform is obtained by comparing the locations of poles with the $j\omega$-axis. If a pole lies to the left of the $j\omega$-axis, then the right-sided inverse transform is chosen. If the pole lies to the right of the $j\omega$-axis, then the left-sided inverse transform is chosen.

▶ **Problem 6.16**   Find the inverse Laplace transform of

$$X(s) = \frac{4s^2 + 15s + 8}{(s+2)^2(s-1)},$$

assuming that (a) $x(t)$ is causal and (b) the Fourier transform of $x(t)$ exists.

Answers:

(a) $x(t) = e^{-2t}u(t) + 2te^{-2t}u(t) + 3e^t u(t)$

(b) $x(t) = e^{-2t}u(t) + 2te^{-2t}u(t) - 3e^t u(-t)$ ◀

6.11 *The Transfer Function*

The transfer function of an LTI system was defined in Eq. (6.2) as the Laplace transform of the impulse response. Recall that the output of an LTI system is related to the input in terms of the impulse response via the convolution

$$y(t) = h(t) * x(t).$$

In general, this equation applies to $h(t)$ and $x(t)$, regardless of whether they are causal or noncausal. Hence, if we take the bilateral Laplace transform of both sides of this equation and use the convolution property, then we have

$$Y(s) = H(s)X(s). \tag{6.37}$$

The Laplace transform of the system output is equal to the product of the transfer function and the Laplace transform of the input. Hence, the transfer function of an LTI system provides yet another description of the input–output behavior of the system.

Note that Eq. (6.37) implies that

$$\boxed{H(s) = \frac{Y(s)}{X(s)}.} \tag{6.38}$$

That is, the transfer function is the ratio of the Laplace transform of the output signal to the Laplace transform of the input signal. This definition applies at values of s for which $X(s)$ is nonzero.

■ 6.11.1 THE TRANSFER FUNCTION AND DIFFERENTIAL-EQUATION SYSTEM DESCRIPTION

The transfer function may be related directly to the differential-equation description of an LTI system by using the bilateral Laplace transform. Recall that the relationship between the input and output of an Nth-order LTI system is described by the differential equation

$$\sum_{k=0}^{N} a_k \frac{d^k}{dt^k} y(t) = \sum_{k=0}^{M} b_k \frac{d^k}{dt^k} x(t).$$

In Section 6.2, we showed that the input e^{st} is an eigenfunction of the LTI system, with the corresponding eigenvalue equal to the transfer function $H(s)$. That is, if $x(t) = e^{st}$, then $y(t) = e^{st}H(s)$. Substitution of e^{st} for $x(t)$ and $e^{st}H(s)$ for $y(t)$ into the differential equation gives

$$\left(\sum_{k=0}^{N} a_k \frac{d^k}{dt^k}\{e^{st}\}\right)H(s) = \sum_{k=0}^{M} b_k \frac{d^k}{dt^k}\{e^{st}\}.$$

We now use the relationship

$$\frac{d^k}{dt^k}\{e^{st}\} = s^k e^{st}$$

and solve for $H(s)$ to obtain

$$\boxed{H(s) = \frac{\sum_{k=0}^{M} b_k s^k}{\sum_{k=0}^{N} a_k s^k}.} \tag{6.39}$$

$H(s)$ is a ratio of polynomials in s and is thus termed a *rational transfer function*. The coefficient of s^k in the numerator polynomial corresponds to the coefficient b_k of the kth derivative of $x(t)$. The coefficient of s^k in the denominator polynomial corresponds to the coefficient a_k of the kth derivative of $y(t)$. Hence, we may obtain the transfer function of an LTI system from the differential-equation description of the system. Conversely, we may determine the differential-equation description of a system from its transfer function.

EXAMPLE 6.19 TRANSFER FUNCTION OF A SECOND-ORDER SYSTEM   Find the transfer function of the LTI system described by the differential equation

$$\frac{d^2}{dt^2}y(t) + 3\frac{d}{dt}y(t) + 2y(t) = 2\frac{d}{dt}x(t) - 3x(t).$$

Solution: Apply Eq. (6.39) to obtain

$$H(s) = \frac{2s - 3}{s^2 + 3s + 2}.$$
■

The poles and zeros of a rational transfer function offer much insight into LTI system characteristics, as we shall see in the sections that follow. Recall from Section 6.2.5 that the transfer function is expressed in pole–zero form by factoring the numerator and denominator polynomials in Eq. (6.39) as follows:

$$\boxed{H(s) = \frac{\widetilde{b}\prod_{k=1}^{M}(s - c_k)}{\prod_{k=1}^{N}(s - d_k)}.} \tag{6.40}$$

where c_k and d_k are the zeros and poles of the system, respectively. Knowledge of the poles, zeros, and gain factor $\widetilde{b} = b_M/a_N$ completely determines the transfer function $H(s)$ and thus offers yet another description of an LTI system. Note that the poles of the system are the roots of the characteristic equation, as defined in Section 2.10.

▶ **Problem 6.17**   Find the transfer functions of the systems described by the following differential equations:

(a)

$$\frac{d^2}{dt^2}y(t) + 2\frac{d}{dt}y(t) + y(t) = \frac{d}{dt}x(t) - 2x(t)$$

(b)

$$\frac{d^3}{dt^3}y(t) - \frac{d^2}{dt^2}y(t) + 3y(t) = 4\frac{d}{dt}x(t)$$

Answers:

(a)

$$H(s) = \frac{s - 2}{s^2 + 2s + 1}$$

(b)

$$H(s) = \frac{4s}{s^3 - s^2 + 3}$$
◀

▶ **Problem 6.18**   Find a differential-equation description of the systems described by the following transfer functions:

(a)

$$H(s) = \frac{s^2 - 2}{s^3 - 3s + 1}$$

(b)

$$H(s) = \frac{2(s + 1)(s - 1)}{s(s + 2)(s + 1)}$$

Answers:

(a)

$$\frac{d^3}{dt^3}y(t) - 3\frac{d}{dt}y(t) + y(t) = \frac{d^2}{dt^2}x(t) - 2x(t)$$

(b)

$$\frac{d^3}{dt^3}y(t) + 3\frac{d^2}{dt^2}y(t) + 2\frac{d}{dt}y(t) = 2\frac{d^2}{dt^2}x(t) - 2x(t) \qquad ◀$$

EXAMPLE 6.20 TRANSFER FUNCTION OF AN ELECTROMECHANICAL SYSTEM   An electromechanical system consisting of a dc motor and a load is depicted in Fig. 6.18(a). The input is the applied voltage $x(t)$ and output is the angular position of the load, $y(t)$. The rotational inertia of the load is given by J. Under ideal circumstances, the torque produced by the motor is directly proportional to the input current; that is,

$$\tau(t) = K_1 i(t),$$

where K_1 is a constant. Rotation of the motor results in a back electromotive force $v(t)$ that is proportional to the angular velocity, or

$$v(t) = K_2 \frac{d}{dt} y(t), \qquad (6.41)$$

where K_2 is another constant. The circuit diagram in Fig. 6.18(b) depicts the relationship between the input current $i(t)$, applied voltage $x(t)$, back electromotive force $v(t)$, and armature resistance R. Express the transfer function of this system in pole–zero form.

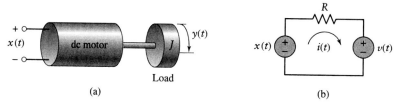

(a)   (b)

FIGURE 6.18   (a) Electromechanical system in which a motor is used to position a load. (b) Circuit diagram relating applied voltage to back electromotive force, armature resistance, and input current. Note that $v(t) = K_2 \, dy(t)/dt$.

Solution: By definition, the torque experienced by the load is given by the product of the rotational inertia and the angular acceleration. Equating the torque produced by the motor and that experienced by the load results in the relationship

$$J\frac{d^2}{dt^2}y(t) = K_1 i(t). \tag{6.42}$$

Application of Ohm's law to the circuit in Fig. 6.18(b) indicates that the current is expressed in terms of the input and back electromotive force by the relationship

$$i(t) = \frac{1}{R}[x(t) - v(t)].$$

Hence, using this expression for $i(t)$ in Eq. (6.42), we have

$$J\frac{d^2}{dt^2}y(t) = \frac{K_1}{R}[x(t) - v(t)].$$

Next, we express $v(t)$ in terms of the angular velocity, using Eq. (6.41) to obtain the differential equation relating the applied voltage to position. The result is

$$J\frac{d^2}{dt^2}y(t) + \frac{K_1 K_2}{R}\frac{d}{dt}y(t) = \frac{K_1}{R}x(t).$$

Application of Eq. (6.39) implies that the transfer function is given by

$$H(s) = \frac{\dfrac{K_1}{R}}{Js^2 + \dfrac{K_1 K_2}{R}s}.$$

We express $H(s)$ in pole–zero form as

$$H(s) = \frac{\dfrac{K_1}{RJ}}{s\left(s + \dfrac{K_1 K_2}{RJ}\right)}.$$

Hence, this system has a pole at $s = 0$ and another one at $s = -K_1 K_2/(RJ)$. ∎

6.12 *Causality and Stability*

The impulse response is the inverse Laplace transform of the transfer function. In order to obtain a unique inverse transform, we must know the ROC or have other knowledge of the impulse response. The differential-equation description of a system does not contain this information. Hence, to obtain the impulse response, we must have additional knowledge of the system characteristics. The relationships between the poles, zeros, and system characteristics can provide this additional knowledge.

The impulse response of a causal system is zero for $t < 0$. Therefore, if we know that a system is causal, the impulse response is determined from the transfer function by using the right-sided inverse Laplace transforms. A system pole at $s = d_k$ in the left half of the s-plane $[\text{Re}(d_k) < 0]$ contributes an exponentially decaying term to the impulse response, while a pole in the right half of the s-plane $[\text{Re}(d_k) > 0]$ contributes an increasing exponential term to the impulse response. These relationships are illustrated in Fig. 6.19.

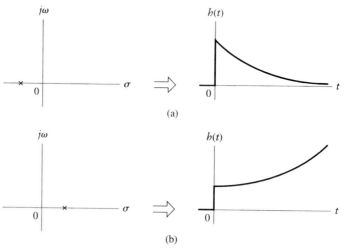

FIGURE 6.19 The relationship between the locations of poles and the impulse response in a causal system. (a) A pole in the left half of the s-plane corresponds to an exponentially decaying impulse response. (b) A pole in the right half of the s-plane corresponds to an exponentially increasing impulse response. The system is unstable in case (b).

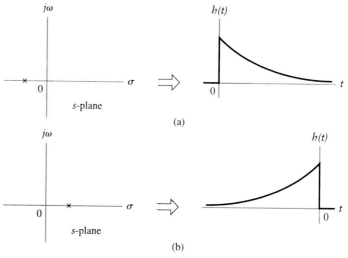

FIGURE 6.20 The relationship between the locations of poles and the impulse response in a stable system. (a) A pole in the left half of the s-plane corresponds to a right-sided impulse response. (b) A pole in the right-half of the s-plane corresponds to a left-sided impulse response. In case (b), the system is noncausal.

Alternatively, if we know that a system is stable, then the impulse response is absolutely integrable. This implies that the Fourier transform exists, and thus the ROC includes the $j\omega$-axis in the s-plane. Such knowledge is sufficient to uniquely determine the inverse Laplace transform of the transfer function. A pole of the system transfer function that is in the right half of the s-plane contributes a left-sided decaying exponential term to the impulse response, while a pole in the left half of the s-plane contributes a right-sided decaying exponential term to the impulse response, as illustrated in Fig. 6.20. Note that a sta-

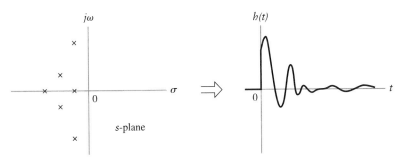

FIGURE 6.21   A system that is both stable and causal must have a transfer function with all of its poles in the left half of the *s*-plane, as shown here.

ble impulse response cannot contain any increasing exponential terms, since an increasing exponential is not absolutely integrable.

Now suppose a system is known to be both causal and stable. Then a pole that is in the left half of the *s*-plane contributes a right-sided decaying exponential term to the impulse response. We cannot have a pole in the right half of the *s*-plane, however, because a pole in the right half will contribute either a left-sided decaying exponential that is not causal or a right-sided increasing exponential that results in an unstable impulse response. That is, the inverse Laplace transform of a pole in the right half of the *s*-plane is either stable or causal, but cannot be both stable and causal. Systems that are stable and causal must have all their poles in the left half of the *s*-plane. Such a system is illustrated in Fig. 6.21.

EXAMPLE 6.21 INVERSE LAPLACE TRANSFORM WITH STABILITY AND CAUSALITY CONSTRAINTS   A system has the transfer function

$$H(s) = \frac{2}{s+3} + \frac{1}{s-2}.$$

Find the impulse response, (a) assuming that the system is stable and (b) assuming that the system is causal. Can this system be both stable and causal?

Solution: This system has poles at $s = -3$ and $s = 2$. (a) If the system is stable, then the pole at $s = -3$ contributes a right-sided term to the impulse response, while the pole at $s = 2$ contributes a left-sided term. We thus have

$$h(t) = 2e^{-3t}u(t) - e^{2t}u(-t).$$

(b) If the system is causal, then both poles must contribute right-sided terms to the impulse response, and we have

$$h(t) = 2e^{-3t}u(t) + e^{2t}u(t).$$

Note that this system is not stable, since the term $e^{2t}u(t)$ is not absolutely integrable. In fact, the system cannot be both stable and causal because the pole at $s = 2$ is in the right half of the *s*-plane.   ∎

526 CHAPTER 6 ▪ THE LAPLACE TRANSFORM

▶ **Problem 6.19** For the following systems described by differential equations, find the impulse response, assuming that the system is (i) stable and (ii) causal:

(a)

$$\frac{d^2}{dt^2}y(t) + 5\frac{d}{dt}y(t) + 6y(t) = \frac{d^2}{dt^2}x(t) + 8\frac{d}{dt}x(t) + 13x(t)$$

(b)

$$\frac{d^2}{dt^2}y(t) - 2\frac{d}{dt}y(t) + 10y(t) = x(t) + 2\frac{d}{dt}x(t)$$

Answers:

(a) (i) and (ii): $h(t) = 2e^{-3t}u(t) + e^{-2t}u(t) + \delta(t)$

(b) (i) $h(t) = -2e^t\cos(3t)u(-t) - e^t\sin(3t)u(-t)$

    (ii) $h(t) = 2e^t\cos(3t)u(t) + e^t\sin(3t)u(t)$ ◀

▪ 6.12.1 INVERSE SYSTEMS

Given an LTI system with impulse response $h(t)$, the impulse response of the inverse system, $h^{\text{inv}}(t)$, satisfies the condition (see Section 2.7.4)

$$h^{\text{inv}}(t) * h(t) = \delta(t).$$

If we take the Laplace transform of both sides of this equation, we find that the inverse system transfer function $H^{\text{inv}}(s)$ satisfies

$$H^{\text{inv}}(s)H(s) = 1,$$

or

$$H^{\text{inv}}(s) = \frac{1}{H(s)}.$$

The inverse system transfer function is therefore the inverse of the transfer function of the original system. If $H(s)$ is written in pole–zero form, as in Eq. (6.40), then we have

$$H^{\text{inv}}(s) = \frac{\prod_{k=1}^{N}(s - d_k)}{\widetilde{b}\prod_{k=1}^{M}(s - c_k)}. \tag{6.43}$$

The zeros of the inverse system are the poles of $H(s)$, and the poles of the inverse system are the zeros of $H(s)$. We conclude that any system with a rational transfer function has an inverse system.

Often, we are interested in inverse systems that are both stable and causal. In this section, we previously concluded that a stable, causal system must have all of its poles in the left half of the s-plane. Since the poles of the inverse system $H^{\text{inv}}(s)$ are the zeros of $H(s)$, a stable and causal inverse system exists only if all of the zeros of $H(s)$ are in the left half of the s-plane. A system whose transfer function $H(s)$ has all of its poles and zeros in the left half of the s-plane is said to be *minimum phase*. A nonminimum-phase system cannot have a stable and causal inverse system, as it has zeros in the right half of the s-plane.

One important property of a minimum-phase system is the unique relationship between the magnitude and phase response. That is, the phase response of a minimum phase system can be uniquely determined from the magnitude response and vice versa.

EXAMPLE 6.22 FINDING AN INVERSE SYSTEM   Consider an LTI system described by the differential equation

$$\frac{d}{dt}y(t) + 3y(t) = \frac{d^2}{dt^2}x(t) + \frac{d}{dt}x(t) - 2x(t).$$

Find the transfer function of the inverse system. Does a stable and causal inverse system exist?

Solution: First, we find the system transfer function $H(s)$ by taking the Laplace transform of both sides of the given differential equation, obtaining

$$Y(s)(s + 3) = X(s)(s^2 + s - 2).$$

Hence, the transfer function of the system is

$$H(s) = \frac{Y(s)}{X(s)}$$
$$= \frac{s^2 + s - 2}{s + 3},$$

and the inverse system has the transfer function

$$H^{inv}(s) = \frac{1}{H(s)}$$
$$= \frac{s + 3}{s^2 + s - 2}$$
$$= \frac{s + 3}{(s - 1)(s + 2)}.$$

The inverse system has poles at $s = 1$ and $s = -2$. The pole at $s = 1$ is in the right half of the s-plane. Therefore, the inverse system represented by $H^{inv}(s)$ cannot be both stable and causal. ∎

▶ **Problem 6.20**   Consider a system with impulse response

$$h(t) = \delta(t) + e^{-3t}u(t) + 2e^{-t}u(t).$$

Find the transfer function of the inverse system. Does a stable and causal inverse system exist?

Answer:

$$H^{inv}(s) = \frac{s^2 + 4s + 3}{(s + 2)(s + 5)}$$

A stable and causal system does exist ◀

▶ **Problem 6.21**   Consider the following transfer functions:

(a)

$$H(s) = \frac{s^2 - 2s - 3}{(s + 2)(s^2 + 4s + 5)}$$

(b)

$$H(s) = \frac{s^2 + 2s + 1}{(s^2 + 3s + 2)(s^2 + s - 2)}$$

(i) Determine whether the systems described by these transfer functions can be both stable and causal.

(ii) Determine whether a stable and causal inverse system exists.

Answers:

(a) (i) stable and causal; (ii) inverse system cannot be both stable and causal

(b) (i) cannot be both stable and causal; (ii) inverse system is stable and causal ◀

6.13 *Determining the Frequency Response from Poles and Zeros*

The locations of the poles and zeros in the s-plane provide insight into the frequency response of a system. Recall that the frequency response is obtained from the transfer function by substituting $j\omega$ for s—that is, by evaluating the transfer function along the $j\omega$-axis in the s-plane. This operation assumes that the $j\omega$ axis lies in the ROC. Substituting $s = j\omega$ into Eq. (6.40) yields

$$H(j\omega) = \frac{\tilde{b}\prod_{k=1}^{M}(j\omega - c_k)}{\prod_{k=1}^{N}(j\omega - d_k)}. \tag{6.44}$$

We will examine both the magnitude and phase of $H(j\omega)$, using a graphical technique for determining the frequency response and also employing the *Bode diagram* or *Bode plot* approach. The Bode diagram displays the system magnitude response in dB and the phase response in degrees as a function of the logarithm of frequency. The concepts learned from constructing Bode diagrams are helpful in developing engineering intuition regarding the effect of pole–zero placement on the system frequency response, and Bode diagrams find extensive use in the design of control systems, as discussed in Chapter 9. Both methods construct the overall frequency response by appropriately combining the frequency response of each pole and zero.

■ 6.13.1 GRAPHICAL EVALUATION OF THE FREQUENCY RESPONSE

We begin with the magnitude response at some fixed value of ω, say, ω_o, and write

$$|H(j\omega_o)| = \frac{|\tilde{b}|\prod_{k=1}^{M}|j\omega_o - c_k|}{\prod_{k=1}^{N}|j\omega_o - d_k|}.$$

This expression involves a ratio of products of terms having the form $|j\omega_o - g|$, where g is either a pole or a zero. The zero contributions are in the numerator, while the pole contributions are in the denominator. The factor $(j\omega_o - g)$ is a complex number that may be represented in the s-plane as a vector from the point g to the point $j\omega_o$, as illustrated in Fig. 6.22. The length of this vector is $|j\omega_o - g|$. By examining the length of the vector as ω_o changes, we may assess the contribution of each pole or zero to the overall magnitude response.

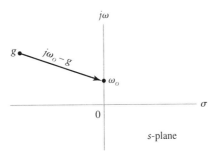

FIGURE 6.22   The quantity $j\omega_o - g$ shown as a vector from g to $j\omega_o$ in the s-plane.

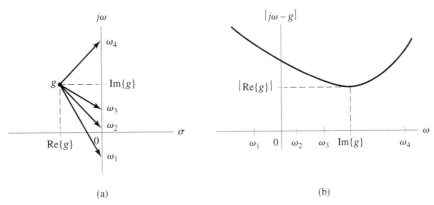

(a)

(b)

FIGURE 6.23   The function $|j\omega - g|$ corresponds to the lengths of vectors from g to the $j\omega$-axis in the s-plane. (a) Vectors from g to $j\omega$ for several frequencies. (b) $|j\omega - g|$ as a function of $j\omega$.

Figure 6.23(a) depicts the vector $j\omega - g$ for several different values of ω, and Fig. 6.23(b) depicts $|j\omega - g|$ as a continuous function of frequency. Note that when $\omega = \text{Im}\{g\}$, $|j\omega - g| = |\text{Re}\{g\}|$. Hence, if g is close to the $j\omega$-axis ($\text{Re}\{g\} \approx 0$), then $|j\omega - g|$ will become very small for $\omega = \text{Im}\{g\}$. Also, if g is close to the $j\omega$-axis, then the most rapid change in $|j\omega - g|$ occurs at frequencies closest to g.

If g represents a zero, then $|j\omega - g|$ contributes to the numerator of $|H(j\omega)|$. Accordingly, at frequencies close to a zero, $|H(j\omega)|$ tends to decrease. How far $|H(j\omega)|$ decreases depends on how close the zero is to the $j\omega$-axis. If the zero is on the $j\omega$-axis, then $|H(j\omega)|$ goes to zero at the frequency corresponding to the zero location. At frequencies far from a zero (i.e., when $|\omega| \gg \text{Re}\{g\}$), $|j\omega - g|$ is approximately equal to $|\omega|$. The component of the magnitude response due to a zero is illustrated in Fig. 6.24(a). In contrast, if g corresponds to a pole, then $|j\omega - g|$ contributes to the denominator of $|H(j\omega)|$; thus, when $|j\omega - g|$ decreases, $|H(j\omega)|$ increases. How far $|H(j\omega)|$ increases depends on how close the pole is to the $j\omega$-axis. A pole that is close to the $j\omega$-axis will result in a large peak in $|H(j\omega)|$. The component of the magnitude response associated with a pole is illustrated in Fig. 6.24(b). Zeros near the $j\omega$-axis tend to pull the response magnitude down, while poles near the $j\omega$-axis tend to push the response magnitude up. Note that a pole cannot lie on the $j\omega$-axis, since we have assumed that the ROC includes the $j\omega$-axis.

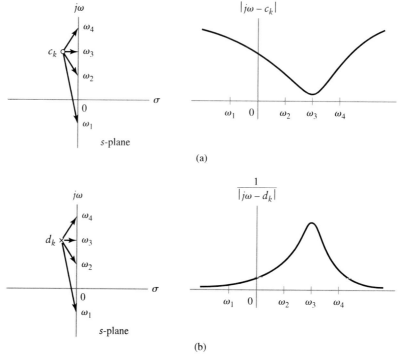

FIGURE 6.24   Components of the magnitude response. (a) Magnitude response associated with a zero. (b) Magnitude response associated with a pole.

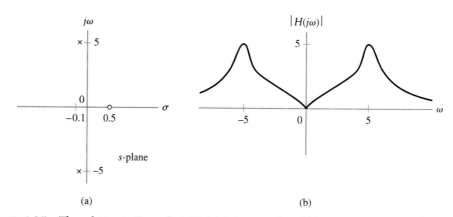

FIGURE 6.25   The solution to Example 6.23. (a) Pole–zero plot. (b) Approximate magnitude response.

EXAMPLE 6.23 GRAPHICAL DERIVATION OF MAGNITUDE RESPONSE   Sketch the magnitude response of the LTI system having the transfer function

$$H(s) = \frac{(s - 0.5)}{(s + 0.1 - 5j)(s + 0.1 + 5j)}.$$

Solution: The system has a zero at $s = 0.5$ and poles at $s = -0.1 \pm 5j$, as depicted in Fig. 6.25(a). Hence, the zero causes the response to decrease near $\omega = 0$ while the pole causes it to increase near $\omega = \pm 5$. At $\omega = 0$, we have

$$|H(j0)| = \frac{0.5}{|0.1 - 5j||0.1 + 5j|}$$

$$\approx \frac{0.5}{5^2}$$

$$= 0.02.$$

At $\omega = 5$, we have

$$|H(j5)| = \frac{|j5 - 0.5|}{|0.1||j10 + 0.1|}$$

$$\approx \frac{5}{0.1(10)}$$

$$= 5.$$

For $\omega \gg 5$, the length of the vector from $j\omega$ to one of the poles is approximately equal to the length of the vector from $j\omega$ to the zero, so the zero is canceled by one of the poles. The distance from $j\omega$ to the remaining pole increases as the frequency increases; thus, the magnitude response goes to zero. The magnitude response is sketched in Fig. 6.25(b). ∎

The phase of $H(j\omega)$ may also be evaluated in terms of the phase associated with each pole and zero. Using Eq. (6.44), we may evaluate

$$\arg\{H(j\omega)\} = \arg\{\widetilde{b}\} + \sum_{k=1}^{M} \arg\{j\omega - c_k\} - \sum_{k=1}^{N} \arg\{j\omega - d_k\}. \qquad (6.45)$$

In this case, the phase of $H(j\omega)$ is the sum of the phase angles due to all the zeros, minus the sum of the phase angles due to all the poles. The first term, $\arg\{\widetilde{b}\}$, is independent of frequency. The phase associated with each zero and pole is evaluated when $\omega = \omega_o$ by considering a term of the form $\arg\{j\omega_o - g\}$. This is the angle of a vector pointing from g to $j\omega_o$ in the s-plane. The angle of the vector is measured relative to the horizontal line through g, as illustrated in Fig. 6.26. By examining the phase of this vector as ω changes, we may assess the contribution of each pole or zero to the overall phase response.

Figure 6.27(a) depicts the phase of $j\omega - g$ for several different frequencies, and Fig. 6.27(b) illustrates the phase as a continuous function of frequency. We assume g represents a zero. Note that since g is in the left half of the s-plane, the phase is $-\pi/2$ for ω large and negative, increasing to zero when $\omega = \text{Im}\{g\}$, and increasing further to $\pi/2$ for ω large and positive. If g is in the right half of the s-plane, then the phase begins at $-\pi/2$ for ω large and negative, decreases to $-\pi$ when $\omega = \text{Im}\{g\}$, and then decreases to $-3\pi/2$ for ω large and positive. If g is close to the $j\omega$-axis, then the change from $-\pi/2$ to

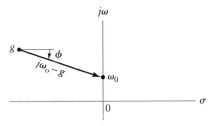

FIGURE 6.26 The quantity $j\omega_o - g$ shown as a vector from g to $j\omega_o$ in the s-plane. The phase angle of the vector is ϕ, defined with respect to a horizontal line through g.

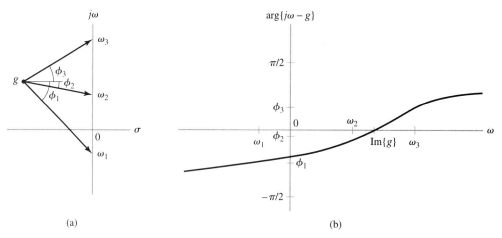

FIGURE 6.27   The phase angle of $j\omega - g$. (a) Vectors from g to $j\omega$ for several different values of ω. (b) Plot of $\arg\{j\omega - g\}$ as a continuous function of ω.

$\pi/2$ (or $-3\pi/2$) occurs rapidly in the vicinity of $\omega = \text{Im}\{g\}$. If g corresponds to a pole, then the contribution of g to the phase of $H(j\omega)$ is the negative of that described.

EXAMPLE 6.24 GRAPHICAL DERIVATION OF PHASE RESPONSE   Sketch the phase response of an LTI system described by the transfer function

$$H(s) = \frac{(s - 0.5)}{(s + 0.1 - 5j)(s + 0.1 + 5j)}.$$

Solution: The locations of the poles and zeros of this system in the s-plane are depicted in Fig. 6.25(a). The phase response associated with the zero at $s = 0.5$ is illustrated in Fig. 6.28(a), the phase response associated with the pole at $s = -0.1 + j5$ is shown in Fig. 6.28(b), and that associated with the pole at $s = -0.1 - j5$ is presented in Fig. 6.28(c). The phase response of the system is obtained by subtracting the phase contributions of the poles from that of the zero. The result is shown in Fig. 6.28(d).   ■

▶ **Problem 6.22**   Sketch the magnitude response and phase response of an LTI system with the transfer function

$$H(s) = \frac{-2}{(s + 0.2)(s^2 + 2s + 5)}.$$

Answer: Poles are at $s = -0.2$ and $s = -1 \pm j2$. [See Fig. 6.29 (a) and (b).]   ◀

■ **6.13.2 BODE DIAGRAMS**

Assume for the moment that, for an LTI system, all poles and zeros are real. The Bode diagram of the system is obtained by expressing the magnitude response of Eq. (6.44) in dB as

$$|H(j\omega)|_{\text{dB}} = 20 \log_{10}|K| + \sum_{k=1}^{M} 20 \log_{10}\left|1 - \frac{j\omega}{c_k}\right| - \sum_{k=1}^{N} 20 \log_{10}\left|1 - \frac{j\omega}{d_k}\right| \quad (6.46)$$

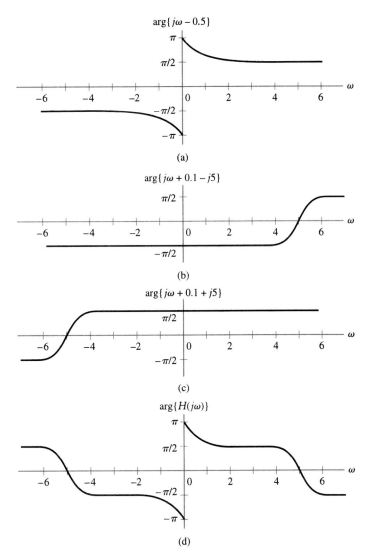

FIGURE 6.28 The phase response of the system in Example 6.24. (a) Phase of the zero at $s = 0.5$. (b) Phase of the pole at $s = -0.1 + j5$. (c) Phase of the pole at $s = -0.1 - j5$. (d) Phase response of the system.

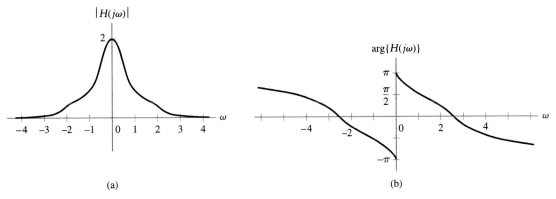

FIGURE 6.29 Solution to Problem 6.22.

and the phase response as

$$\arg\{H(j\omega)\} = \arg K + \sum_{k=1}^{M} \arg\left(1 - \frac{j\omega}{c_k}\right) - \sum_{k=1}^{N} \arg\left(1 - \frac{j\omega}{d_k}\right). \qquad (6.47)$$

In Eqs. (6.46) and (6.47), the gain factor is given by

$$K = \frac{\widetilde{b}\prod_{k=1}^{M}(-c_k)}{\prod_{k=1}^{N}(-d_k)}.$$

Hence, in computing the magnitude response $|H(j\omega)|_{\text{dB}}$, the product and division factors in Eq. (6.44) are associated with additions and subtractions, respectively. Moreover, the individual contributions of the zero and pole factors to the phase response, $\arg\{H(j\omega)\}$, also involve additions and subtractions. The computation of $H(j\omega)$ for varying ω is thereby made relatively easy. The intuitive appeal of the Bode diagram comes from the fact that the computation of $|H(j\omega)|_{\text{dB}}$ and $\arg\{H(j\omega)\}$ may be readily approximated by straight-line segments obtained by summing the straight-line-segment approximations associated with each pole or zero factor.

Consider the case of a pole factor $(1 - j\omega/d_0)$ for which $d_0 = -\omega_b$, some real number. The contribution of this pole factor to the gain component $|H(j\omega)|_{\text{dB}}$ is written as

$$-20\log_{10}\left|1 + \frac{j\omega}{\omega_b}\right| = -10\log_{10}\left(1 + \frac{\omega^2}{\omega_b^2}\right). \qquad (6.48)$$

We may obtain asymptotic approximations of this contribution by considering both very small and very large values of ω, compared with ω_b, as follows:

▶ *Low-frequency asymptote.* For $\omega \ll \omega_b$, Eq. (6.48) approximates to

$$-20\log_{10}\left|1 + \frac{j\omega}{\omega_b}\right| \approx -20\log_{10}(1) = 0\text{ dB},$$

which consists of the 0-dB line.

▶ *High-frequency asymptote.* For $\omega \gg \omega_b$, Eq. (6.48) approximates to

$$-20\log_{10}\left|1 + \frac{j\omega}{\omega_b}\right| \approx -10\log_{10}\left|\frac{\omega}{\omega_b}\right|^2 = -20\log_{10}\left|\frac{\omega}{\omega_b}\right|,$$

which represents a straight line with a slope of -20 dB/decade.

These two asymptotes intersect at $\omega = \omega_b$. Accordingly, the contribution of the pole factor $(1 + j\omega/\omega_b)$ to $|H(j\omega)|_{\text{dB}}$ may be approximated by a pair of straight-line segments, as illustrated in Fig. 6.30(a). The intersection frequency ω_b is called a *corner* or *break frequency* of the Bode diagram. The figure also includes the actual magnitude characteristics of a simple pole factor. The approximation error (i.e., the difference between the actual magnitude characteristic and the approximate form) attains its maximum value of 3 dB at the corner frequency ω_b. The table shown in the figure presents a listing of the approximation errors for a logarithmically spaced set of frequencies normalized with respect to ω_b. Note that the magnitude characteristic of a zero is the negative of that of a pole. Thus, the high-frequency asymptote for a zero has a slope of 20 dB/decade.

The phase response of the simple pole factor is defined by

$$-\arg\{1 + j\omega/\omega_b\} = -\arctan\left(\frac{\omega}{\omega_b}\right),$$

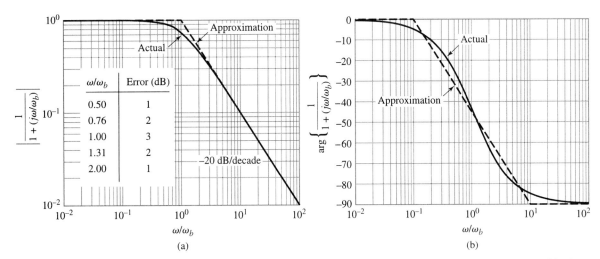

ω/ω_b	Error (dB)
0.50	1
0.76	2
1.00	3
1.31	2
2.00	1

FIGURE 6.30   Bode diagram for first-order pole factor: $1/(1 + s/\omega_b)$. (a) Gain response. (b) Phase response.

which is plotted exactly in Fig. 6.30(b), as the solid line. The dashed line depicts a piecewise linear approximation to the phase response. The phase response of a zero is the negative of the phase response of a pole. Recall from Section 6.12.1 that a minimum-phase system has all its poles and zeros in the left-half of the s-plane. Note that if a pole or zero is in the left-half of the s-plane, then $\omega_b > 0$ and the magnitude and phase response are uniquely related through ω_b. This is why there is a unique relationship between the magnitude and phase response of a minimum-phase system.

We can now see the practical merit of the Bode diagram: By using the approximations described for simple pole or zero factors of the transfer function, we can quickly sketch $|H(j\omega)|_{dB}$. The next example illustrates how to combine the individual factors to obtain $|H(j\omega)|_{dB}$.

EXAMPLE 6.25 BODE DIAGRAM CONSTRUCTION   Sketch the magnitude and phase response as a Bode diagram for the LTI system described by the transfer function

$$H(s) = \frac{5(s + 10)}{(s + 1)(s + 50)}.$$

Solution: First, we express

$$H(j\omega) = \frac{\left(1 + \dfrac{j\omega}{10}\right)}{(1 + j\omega)\left(1 + \dfrac{j\omega}{50}\right)},$$

from which we identify two pole corner frequencies of $\omega = 1$ and $\omega = 50$ and a single zero corner frequency of $\omega = 10$. The asymptotic approximation of each pole and zero are depicted in Fig. 6.31(a). The sum of the asymptotes approximates $|H(j\omega)|_{dB}$, as shown in Fig. 6.31(b). Note that for $\omega > 10$, the high-frequency asymptotes of the zero and the pole with corner frequency $\omega = 1$ cancel. Similarly, Fig. 6.31(c) depicts the asymptotic approximation of the phase of each pole and the zero. The sum of the asymptotes approximates $\arg\{H(j\omega)\}$, as shown in Fig. 6.31(d). ∎

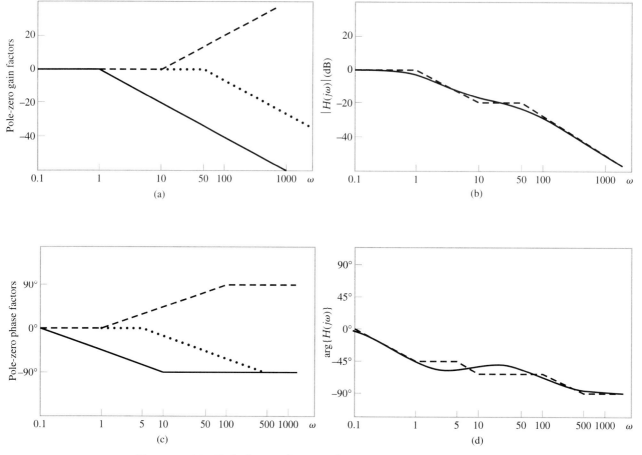

FIGURE 6.31   Bode diagram for Example 6.25. (a) Gain response of pole at $s = -1$ (solid line), zero at $s = -10$ (dashed line), and pole at $s = -50$ (dotted line). (b) Actual gain response (solid line) and asymptotic approximation (dashed line). (c) Phase response of pole at $s = -1$ (solid line), zero at $s = -10$ (dashed line), and pole at $s = -50$ (dotted line). (d) Actual phase response (solid line) and asymptotic approximation (dashed line).

Thus far, we have assumed that poles and zeros are real valued. Pairs of complex-conjugate poles or zeros are grouped into real-valued quadratic factors. For example, a quadratic pole factor is expressed as

$$Q(s) = \frac{1}{1 + 2(\zeta/\omega_n)s + \left(\dfrac{s}{\omega_n}\right)^2}.$$

The poles of $Q(s)$ are located at

$$s = -\zeta\omega_n \pm j\omega_n\sqrt{1 - \zeta^2},$$

where we have assumed that $\zeta \leq 1$. Expressing $Q(s)$ in this form simplifies the determination of the Bode diagram. (A physical interpretation of ζ and ω_n is given in Section 9.10

in the context of the characteristics of second-order all-pole systems.) Substituting $s = j\omega$ into the expression for $Q(s)$ yields

$$Q(j\omega) = \frac{1}{1 - (\omega/\omega_n)^2 + j2\zeta\omega/\omega_n},$$

and thus the magnitude of $Q(j\omega)$ in decibels is given by

$$|Q(j\omega)|_{\text{dB}} = -20\log_{10}[(1 - (\omega/\omega_n)^2)^2 + 4\zeta^2(\omega/\omega_n)^2]^{1/2}, \qquad (6.49)$$

while the phase of $Q(j\omega)$ is given by

$$\arg\{Q(j\omega)\} = -\arctan\left(\frac{2\zeta(\omega/\omega_n)}{1 - (\omega/\omega_n)^2}\right). \qquad (6.50)$$

For $\omega \ll \omega_n$, Eq. (6.49) is approximated as

$$|Q(j\omega)|_{\text{dB}} \approx -20\log_{10}(1) = 0 \text{ dB};$$

for $\omega \gg \omega_n$, it is approximated as

$$|Q(j\omega)|_{\text{dB}} \approx -20\log_{10}\left(\frac{\omega}{\omega_n}\right)^2 = -40\log_{10}\left(\frac{\omega}{\omega_n}\right).$$

Hence, the gain component $|Q(j\omega)|_{\text{dB}}$ may be approximated by a pair of straight-line segments, one represented by the 0-dB line and the other having a slope of -40 dB/decade, as shown in Fig. 6.32. The two asymptotes intersect at $\omega = \omega_n$, which is referred to as the *corner frequency* of the quadratic factor. However, unlike the case of a simple pole factor, the actual magnitude of the quadratic pole factor may differ markedly from its asymptotic approximation, depending on how small the factor ζ is, compared with unity. Figure 6.33(a) shows the exact plot of $|Q(j\omega)|_{\text{dB}}$ for three different values of ζ in the range $0 < \zeta < 1$. The difference between the exact curve and the asymptotic approximation of $|Q(j\omega)|_{\text{dB}}$ defines the approximation error. Evaluating Eq. (6.49) at $\omega = \omega_n$ and noting that the corresponding value of the asymptotic approximation is 0 dB, we find that the value of the error at $\omega = \omega_n$ is given by

$$(\text{Error})_{\omega=\omega_n} = -20\log_{10}(2\zeta) \text{ dB}.$$

This error is zero for $\zeta = 0.5$, positive for $\zeta < 0.5$, and negative for $\zeta > 0.5$.

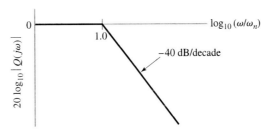

FIGURE 6.32   Asymptotic approximation to $20\log_{10}|Q(j\omega)|$, where

$$Q(s) = \frac{1}{1 + (2\zeta/\omega_n)s + s^2/\omega_n^2}.$$

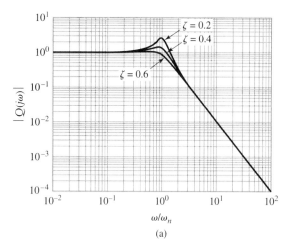
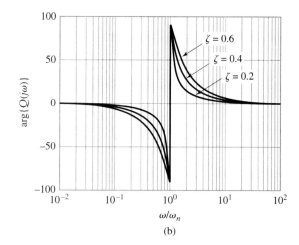

FIGURE 6.33   Bode diagram of second-order pole factor

$$Q(s) = \frac{1}{1 + (2\zeta/\omega_n)s + s^2/\omega_n^2}$$

for varying ζ: (a) Gain response. (b) Phase response.

The phase component of $Q(j\omega)$ is given by Eq. (6.50). Figure 6.33(b) shows exact plots of arg $\{Q(j\omega)\}$ for the same values of ζ used in Fig. 6.33(a). At $\omega = \omega_n$, we have

$$\arg\{Q(j\omega_n)\} = -90 \text{ degrees.}$$

Note the change in the algebraic sign of $Q(j\omega)$ at $\omega = \omega_n$, which introduces a 180-degree change in the phase.

EXAMPLE 6.26 BODE DIAGRAM FOR AN ELECTROMECHANICAL SYSTEM   The combination of dc motor and load depicted in Fig. 6.18 has transfer function

$$H(s) = \frac{\dfrac{K_1}{RJ}}{s\left(s + \dfrac{K_1 K_2}{RJ}\right)},$$

as derived in Example 6.20. Sketch the magnitude and phase responses as a Bode diagram, assuming that $\frac{K_1}{RJ} = 100$ and $\frac{K_1 K_2}{RJ} = 50$.

Solution: First we use $s = j\omega$ to write

$$H(j\omega) = \frac{2}{j\omega\left(1 + \dfrac{j\omega}{50}\right)},$$

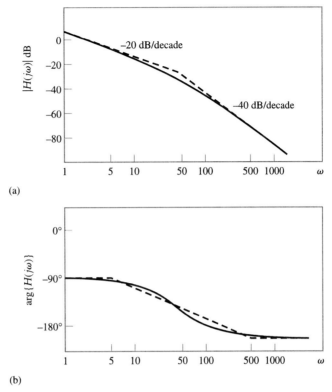

(a)

(b)

FIGURE 6.34 Bode diagram for electromechanical system in Example 6.20. (a) Actual magnitude response (solid line) and asymptotic approximation (dashed line). (b) Actual phase response (solid line) and asymptotic approximation (dashed line).

from which we identify a pole corner frequency at $\omega = 50$. The $j\omega$ term in the denominator contributes a line of slope -20 dB per decade to the magnitude response and a -90 degree factor to the phase. Figure 6.34 depicts the magnitude and phase responses as a Bode diagram. ∎

▶ **Problem 6.23** Sketch the asymptotic approximation to the gain and phase components of the Bode diagram for the following system transfer functions:

(a) $H(s) = \dfrac{8s + 40}{s(s + 20)}$

(b) $H(s) = \dfrac{10}{(s + 1)(s + 2)(s + 10)}$

Answer: See Fig. 6.35 ◀

▶ **Problem 6.24** Sketch the asymptotic approximation to the gain component of the Bode diagram for a system with transfer function $H(s) = \dfrac{100(s + 1)}{s(s^2 + 20s + 100)}$.

Answer: See Fig. 6.36 ◀

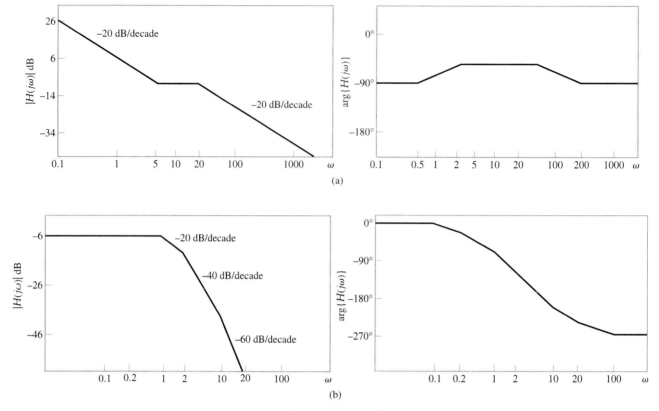

FIGURE 6.35   Solution to Problem 6.23.

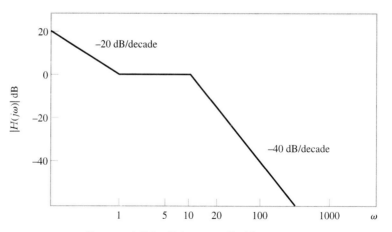

FIGURE 6.36   Solution to Problem 6.24.

6.14 *Exploring Concepts with MATLAB*

The MATLAB Control System Toolbox contains numerous commands that are useful for working with Laplace transforms and continuous-time LTI systems described in terms of transfer functions, poles and zeros, or state variables.

■ 6.14.1 POLES AND ZEROS

The command `r = roots(a)` finds the roots of a polynomial described by the vector **a** and thus may be used to determine the zeros and poles of a Laplace transform expressed as a ratio of polynomials in *s*. The elements of the vector **a** correspond to descending powers of *s*.

For example, we may find the poles and zeros of the Laplace transform in Example 6.9, namely,

$$X(s) = \frac{4s^2 + 6}{s^3 + s^2 - 2},$$

using the following commands:

```
>> z = roots([4, 0, 6])
z =

      0 + 1.2247i
      0 - 1.2247i
>> p = roots([1, 1, 0, -2])
p =

    -1.0000 + 1.0000i
    -1.0000 - 1.0000i
     1.0000
```

Hence, we identify zeros at $s = \pm j1.2247$, a pole at $s = 1$, and a pair of complex-conjugate poles at $s = -1 \pm j$. (Recall that MATLAB has `i = ` $\sqrt{-1}$.)

The command `poly(r)` uses the poles or zeros specified in the vector r to determine the coefficients of the corresponding polynomial.

■ 6.14.2 PARTIAL-FRACTION EXPANSIONS

The `residue` command finds the partial-fraction expansion of a ratio of two polynomials. The syntax is `[r,p,k] = residue(b,a),` where b represents the numerator polynomial, a represents the denominator polynomial, r represents the coefficients or residues of the partial-fraction expansion, p represents the poles, and k is a vector describing any terms in powers of *s*. If the order of the numerator is less than that of the denominator, then k is an empty matrix.

To illustrate the use of the residue command, we find the partial-fraction expansion of the Laplace transform considered in Example 6.7, viz.,

$$X(s) = \frac{3s + 4}{(s + 1)(s + 2)^2} = \frac{3s + 4}{s^3 + 5s^2 + 8s + 4},$$

using the following commands:

```
>> [r,p,k] = residue([3, 4], [1, 5, 8, 4])

r =

      -1.0000
       2.0000
       1.0000

p =

      -2.0000
      -2.0000
      -1.0000

k =
      []
```

Hence, the residue `r(1)` = −1 corresponds to the pole at $s = -2$ given by `p(1)`, the residue `r(2)` = 2 corresponds to the double pole at $s = -2$ given by `p(2)`, and the residue `r(3)` = 1 corresponds to the pole at $s = -1$ given by `p(3)`. The partial-fraction expansion is therefore

$$X(s) = \frac{-1}{s+2} + \frac{2}{(s+2)^2} + \frac{1}{s+1}.$$

This result agrees with that of Example 6.7.

▶ **Problem 6.25**   Use `residue` to solve Problem 6.7.   ◀

■ 6.14.3 RELATING SYSTEM DESCRIPTIONS

Recall that a system may be described in terms of a differential equation, a transfer function, poles and zeros, a frequency response, or state variables. The Control System Toolbox contains routines for relating the transfer function, pole–zero, and state-variable representations of LTI systems. All of these routines are based on LTI objects that represent the different forms of the system description. State-space objects are defined with the MATLAB command `ss`, as discussed in Section 2.14. The command `H = tf(b, a)` creates an LTI object `H` representing a transfer function with numerator and denominator polynomials defined by the coefficients in `b` and `a`, ordered in descending powers of s. The command `H = zpk(z, p, k)` creates an LTI object representing the pole–zero-gain form of system description. The zeros and poles are described by the vectors `z` and `p`, respectively, and the gain is represented by the scalar `k`.

The commands `ss`, `tf`, and `zpk` also convert among models when applied to an LTI object of a different form. For example, if `syszpk` is an LTI object representing a system in zero–pole-gain form, then the command `sysss = ss(syszpk)` generates a state-space object `sysss` representing the same system.

The commands `tzero(sys)` and `pole(sys)` find the zeros and poles of the LTI object `sys`, while `pzmap(sys)` produces a pole–zero plot. Additional commands that apply directly to LTI objects include `freqresp` for determining the frequency response, `bode` for determining the Bode plot, `step` for determining the step response, and `lsim` for simulating the system output in response to a specified input.

Consider a system containing zeros at $s = 0$ and $s = \pm j10$ and poles at $s = -0.5 \pm j5$, $s = -3$, and $s = -4$ with gain 2. We may determine the transfer function representation of this system, plot the locations of the poles and zeros in the s-plane, and plot the system's magnitude response using the following MATLAB commands:

```
>> z = [0, j*10, -j*10];   p = [-0.5+j*5, -0.5-j*5, -3,
    -4];    k = 2;
>> syszpk = zpk(z, p, k)
Zero/pole/gain
      2 s  (s ^ 2 + 100)
- - - - - - - - - - - - - - - - - - - - - - -
(s+4)  (s+3)  (s ^ 2 + s + 25.25)

>> systf = tf(syszpk)       % convert to transfer
                               function form
Transfer function:
      2 s ^ 3 + 200s
- - - - - - - - - - - - - - - - - - - - - - -
s ^ 4 + 7  s ^ 3 + 44.25  s ^ 2 + 188.8  s + 303

>> pzmap(systf)             % generate pole-zero plot
>> w = [0:499]*20/500;       % Frequencies from 0 to
                               20 rad/sec
>> H = freqresp(systf,w);
>> Hmag = abs(squeeze(H));      plot(w,Hmag)
```

Figure 6.37 depicts the pole–zero plot resulting from these commands, while Fig. 6.38 illustrates the magnitude response of the system for $0 \leq \omega < 20$. Note that the magnitude

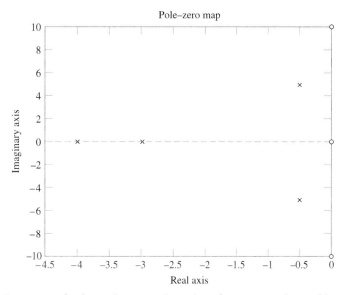

FIGURE 6.37 Locations of poles and zeros in the s-plane for a system obtained by using MATLAB.

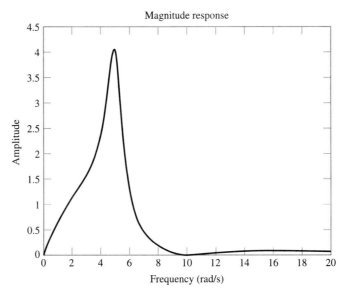

FIGURE 6.38   Magnitude response for a system obtained by using MATLAB.

response is zero at frequencies corresponding to the locations of zeros on the $j\omega$-axis, at $\omega = 0$ and $\omega = 10$. Similarly, the magnitude response is large at the frequency corresponding to the location of the pole near the $j\omega$-axis, at $\omega = 5$.

The Bode diagram for the system in Example 6.25 is obtained with the use of MATLAB's `bode` command as follows:

```
>> z = [-10];     p = [-1, -50];     k = 5;
>> sys = zpk(z, p, k);
>> bode(sys)
```

The result is shown in Fig. 6.39.

6.15 *Summary*

The Laplace transform represents continuous-time signals as weighted superpositions of complex exponentials, which are more general signals than complex sinusoids (which they include as a special case). Correspondingly, the Laplace transform represents a more general class of signals than does the Fourier transform, including signals that are not absolutely integrable. Hence, we may use the Laplace transform to analyze signals and LTI systems that are not stable. The transfer function is the Laplace transform of the impulse response and offers another description for the input–output characteristics of an LTI system. The Laplace transform converts the convolution of time signals to multiplication of Laplace transforms, so the Laplace transform of an LTI system output is the product of the Laplace transform of the input and the transfer function. The locations of the poles and zeros of a transfer function in the s-plane offer yet another characterization of an LTI system, providing information regarding the system's stability, causality, invertibility, and frequency response.

Complex exponentials are parameterized by a complex variable s. The Laplace transform is a function of s and is represented in a complex plane termed the s-plane. The Fouri-

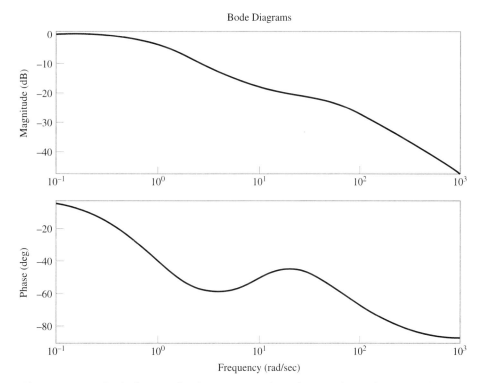

Bode Diagrams

FIGURE 6.39   Bode diagram for the system in Example 6.25 obtained using MATLAB.

er transform is obtained by evaluating the Laplace transform on the $j\omega$-axis—that is, by setting $s = j\omega$. The properties of the Laplace transform are analogous to those of the Fourier transform. The frequency response of an LTI system is obtained from the transfer function by setting $s = j\omega$. The Bode diagram uses the poles and zeros of an LTI system to depict the magnitude response in dB and the phase response as logarithmic functions of frequency.

The unilateral, or one-sided, Laplace transform applies to causal signals and provides a convenient tool for solving system problems involving differential equations with initial conditions. The bilateral Laplace transform applies to two-sided signals; it is not unique unless the ROC is specified. The relative positions of the ROC and the Laplace transform poles in the s-plane determine whether a signal is left-sided, right-sided, or two-sided.

The Fourier and Laplace transforms have many similarities and can often be used interchangeably, yet they have decidedly different roles in signal and system analysis. The Laplace transform is most often used in the transient and stability analysis of systems. Problems involving this type of analysis occur frequently in control system applications, where we are interested in the manner in which the system output tracks a desired system output. The system poles provide essential information about an LTI system's stability and transient response characteristics, while the unilateral transform can be used to obtain an LTI system's response, given the input and initial conditions. There is no Fourier transform counterpart with these capabilities. The role of the Laplace transform in the transient and stability analysis of control systems is further explored in Chapter 9.

In contrast, the Fourier transform is usually employed as a signal representation tool and in solving system problems in which steady-state characteristics are of interest. The Fourier transform is easier to visualize and use than the Laplace transform in such problems because it is a function of the real-valued frequency ω, whereas the Laplace transform is a function of the complex frequency $s = \sigma + j\omega$. Examples of the signal representation and steady-state system analysis problems to which the Fourier transform is applied were given in Chapters 4 and 5 and are further developed in the context of filtering in Chapter 8.

FURTHER READING

1. The Laplace transform is named after Pierre-Simon de Laplace (1749–1827), who studied a wide variety of natural phenomena, including hydrodynamics, the propagation of sound, heat, the tides, and the liquid state of matter, although the majority of his life was devoted to celestial mechanics. Laplace presented complete analytic solutions to the mechanical problems associated with the solar system in a series of five volumes titled *Mécanique céleste*. A famous passage in another of his books asserted that the future of the world is determined entirely by the past—that if one possessed knowledge of the "state" of the world at any instant, then one could predict the future. Although Laplace made many important discoveries in mathematics, his primary interest was the study of nature. Mathematics was simply a means to that end.

2. The following texts are devoted specifically to the Laplace transform and applications thereof:

    ▶ Holbrook, J. G., *Laplace Transforms for Electronic Engineers*, 2nd ed. (Pergamon Press, 1966)
    ▶ Kuhfittig, P. K. F., *Introduction to the Laplace Transform* (Plenum Press, 1978)
    ▶ Thomson, W. T., *Laplace Transformation*, 2nd ed. (Prentice-Hall, 1960)

3. The text

    ▶ Bode, H. W., *Network Analysis and Feedback Amplifier Design* (Van Nostrand, 1947)
    is a classic, presenting a detailed analysis of the unique relationships that exist between the magnitude and phase responses of minimum-phase systems. The Bode diagram is named in honor of H. W. Bode.

4. The technique of contour integration, used to evaluate the inverse Laplace transform in Eq. (6.4), is usually studied in mathematics under the topic of complex variables. An introductory treatment of contour integration is given in:

    ▶ Brown, J. and R. Churchill, *Complex Variables and Applications,* (McGraw-Hill, 1996)

5. Solutions to differential equations obtained using the unilateral Laplace transform do not apply for $t < 0$. Care must be exercised in evaluating the solution at the point $t = 0$, since at this instant the step function is discontinuous and the amplitude of any impulses in the solution is undefined. We have taken the conventional approach to deal with this difficult issue and are not aware of any way to address it at an introductory level that is technically precise and correct. To resolve the problem in a rigorous manner requires the use of generalized functional analysis, which is beyond the scope of this book.

ADDITIONAL PROBLEMS

6.26 A signal $x(t)$ has the indicated Laplace transform $X(s)$. Plot the poles and zeros in the s-plane and determine the Fourier transform of $x(t)$ without inverting $X(s)$.

(a) $X(s) = \dfrac{s^2 + 1}{s^2 + 5s + 6}$

(b) $X(s) = \dfrac{s^2 - 1}{s^2 + s + 1}$

(c) $X(s) = \dfrac{1}{s - 4} + \dfrac{2}{s - 2}$

6.27 Determine the bilateral Laplace transform and ROC for the following signals:

(a) $x(t) = e^{-t}u(t + 2)$

(b) $x(t) = u(-t + 3)$

(c) $x(t) = \delta(t + 1)$

(d) $x(t) = \sin(t)u(t)$

6.28 Determine the unilateral Laplace transform of the following signals, using the defining equation:

(a) $x(t) = u(t - 2)$

(b) $x(t) = u(t + 2)$

(c) $x(t) = e^{-2t}u(t + 1)$

(d) $x(t) = e^{2t}u(-t + 2)$

(e) $x(t) = \sin(\omega_o t)$

(f) $x(t) = u(t) - u(t - 2)$

(g) $x(t) = \begin{cases} \sin(\pi t), & 0 < t < 1 \\ 0, & \text{otherwise} \end{cases}$

6.29 Use the basic Laplace transforms and the Laplace transform properties given in Tables D.1 and D.2 to determine the unilateral Laplace transform of the following signals:

(a) $x(t) = \dfrac{d}{dt}\{te^{-t}u(t)\}$

(b) $x(t) = tu(t) * \cos(2\pi t)u(t)$

(c) $x(t) = t^3 u(t)$

(d) $x(t) = u(t - 1) * e^{-2t}u(t - 1)$

(e) $x(t) = \int_0^t e^{-3\tau}\cos(2\tau)d\tau$

(f) $x(t) = t\dfrac{d}{dt}(e^{-t}\cos(t)u(t))$

6.30 Use the basic Laplace transforms and the Laplace transform properties given in Tables D.1 and D.2 to determine the time signals corresponding to the following unilateral Laplace transforms:

(a) $X(s) = \left(\dfrac{1}{s + 2}\right)\left(\dfrac{1}{s + 3}\right)$

(b) $X(s) = e^{-2s}\dfrac{d}{ds}\left(\dfrac{1}{(s + 1)^2}\right)$

(c) $X(s) = \dfrac{1}{(2s + 1)^2 + 4}$

(d) $X(s) = s\dfrac{d^2}{ds^2}\left(\dfrac{1}{s^2 + 9}\right) + \dfrac{1}{s + 3}$

6.31 Given the transform pair $\cos(2t)u(t) \xleftrightarrow{\mathcal{L}_u} X(s)$, determine the time signals corresponding to the following Laplace transforms:

(a) $(s + 1)X(s)$

(b) $X(3s)$

(c) $X(s + 2)$

(d) $s^{-2}X(s)$

(e) $\dfrac{d}{ds}(e^{-3s}X(s))$

6.32 Given the transform pair $x(t) \xleftrightarrow{\mathcal{L}_u} \dfrac{2s}{s^2 + 2}$, where $x(t) = 0$ for $t < 0$, determine the Laplace transform of the following time signals:

(a) $x(3t)$

(b) $x(t - 2)$

(c) $x(t) * \dfrac{d}{dt}x(t)$

(d) $e^{-t}x(t)$

(e) $2tx(t)$

(f) $\int_0^t x(3\tau)d\tau$

6.33 Use the s-domain shift property and the transform pair $e^{-at}u(t) \xleftrightarrow{\mathcal{L}_u} \dfrac{1}{s + a}$ to derive the unilateral Laplace transform of $x(t) = e^{-at}\cos(\omega_1 t)u(t)$.

6.34 Prove the following properties of the unilateral Laplace transform:

(a) Linearity

(b) Scaling

(c) Time shift

(d) s-domain shift

(e) Convolution

(f) Differentiation in the s-domain

6.35 Determine the initial value $x(0^+)$, given the following Laplace transforms $X(s)$:

(a) $X(s) = \dfrac{1}{s^2 + 5s - 2}$

(b) $X(s) = \dfrac{s + 2}{s^2 + 2s - 3}$

(c) $X(s) = e^{-2s}\dfrac{6s^2 + s}{s^2 + 2s - 2}$

6.36 Determine the final value $x(\infty)$, given the following Laplace transforms $X(s)$:

(a) $X(s) = \dfrac{2s^2 + 3}{s^2 + 5s + 1}$

(b) $X(s) = \dfrac{s + 2}{s^3 + 2s^2 + s}$

(c) $X(s) = e^{-3s}\dfrac{2s^2 + 1}{s(s + 2)^2}$

6.37 Use the method of partial fractions to find the time signals corresponding to the following unilateral Laplace transforms:

(a) $X(s) = \dfrac{s + 3}{s^2 + 3s + 2}$

(b) $X(s) = \dfrac{2s^2 + 10s + 11}{s^2 + 5s + 6}$

(c) $X(s) = \dfrac{2s - 1}{s^2 + 2s + 1}$

(d) $X(s) = \dfrac{5s + 4}{s^3 + 3s^2 + 2s}$

(e) $X(s) = \dfrac{s^2 - 3}{(s + 2)(s^2 + 2s + 1)}$

(f) $X(s) = \dfrac{3s + 2}{s^2 + 2s + 10}$

(g) $X(s) = \dfrac{4s^2 + 8s + 10}{(s + 2)(s^2 + 2s + 5)}$

(h) $X(s) = \dfrac{3s^2 + 10s + 10}{(s + 2)(s^2 + 6s + 10)}$

(i) $X(s) = \dfrac{2s^2 + 11s + 16 + e^{-2s}}{s^2 + 5s + 6}$

6.38 Determine the forced and natural responses for the LTI systems described by the following differential equations with the specified input and initial conditions:

(a) $\dfrac{d}{dt}y(t) + 10y(t) = 10x(t), \quad y(0^-) = 1,$

$x(t) = u(t)$

(b) $\dfrac{d^2}{dt^2}y(t) + 5\dfrac{d}{dt}y(t) + 6y(t)$

$= -4x(t) - 3\dfrac{d}{dt}x(t),$

$y(0^-) = -1, \dfrac{d}{dt}y(t)\bigg|_{t=0^-} = 5, x(t) = e^{-t}u(t)$

(c) $\dfrac{d^2}{dt^2}y(t) + y(t) = 8x(t), y(0^-) = 0,$

$\dfrac{d}{dt}y(t)\bigg|_{t=0^-} = 2, x(t) = e^{-t}u(t)$

(d) $\dfrac{d^2}{dt^2}y(t) + 2\dfrac{d}{dt}y(t) + 5y(t) = \dfrac{d}{dt}x(t),$

$y(0^-) = 2, \dfrac{d}{dt}y(t)\bigg|_{t=0^-} = 0, x(t) = u(t)$

6.39 Use Laplace transform circuit models to determine the current $y(t)$ in the circuit of Fig. P6.39, assuming normalized values $R = 1\Omega$ and $L = \frac{1}{2}H$ for the specified inputs. The current through the inductor at time $t = 0^-$ is 2 A.

(a) $x(t) = e^{-t}u(t)$

(b) $x(t) = \cos(t)u(t)$

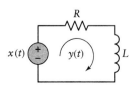

FIGURE P6.39

6.40 The circuit in Fig. P6.40 represents a system with input $x(t)$ and output $y(t)$. Determine the forced response and the natural response of this system under the specified conditions.

(a) $R = 3\Omega, L = 1\,H, C = \frac{1}{2}F, x(t) = u(t),$ the current through the inductor at $t = 0^-$ is 2 A, and the voltage across the capacitor at $t = 0^-$ is 1 V.

(b) $R = 2\Omega, L = 1\,H, C = \frac{1}{5}F, x(t) = u(t),$ the current through the inductor at $t = 0^-$ is 2 A, and the voltage across the capacitor at $t = 0^-$ is 1 V.

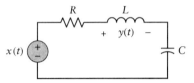

FIGURE P6.40

6.41 Determine the bilateral Laplace transform and the corresponding ROC for the following signals:

(a) $x(t) = e^{-t/2}u(t) + e^{-t}u(t) + e^t u(-t)$

(b) $x(t) = e^t \cos(2t)u(-t) + e^{-t}u(t) + e^{t/2}u(t)$

(c) $x(t) = e^{3t+6}u(t + 3)$

(d) $x(t) = \cos(3t)u(-t) * e^{-t}u(t)$

(e) $x(t) = e^t \sin(2t + 4)u(t + 2)$

(f) $x(t) = e^t \dfrac{d}{dt}(e^{-2t}u(-t))$

6.42 Use the tables of transforms and properties to determine the time signals that correspond to the following bilateral Laplace transforms:

(a) $X(s) = e^{5s}\dfrac{1}{s + 2}$ with ROC $Re(s) < -2$

(b) $X(s) = \dfrac{d^2}{ds^2}\left(\dfrac{1}{s - 3}\right)$ with ROC $Re(s) > 3$

(c) $X(s) = s\left(\dfrac{1}{s^2} - \dfrac{e^{-s}}{s^2} - \dfrac{e^{-2s}}{s}\right)$ with ROC $Re(s) < 0$

(d) $X(s) = s^{-2}\dfrac{d}{ds}\left(\dfrac{e^{-3s}}{s}\right)$ with ROC $Re(s) > 0$

6.43 Use the method of partial fractions to determine the time signals corresponding to the following bilateral Laplace transforms:

(a) $X(s) = \dfrac{-s - 4}{s^2 + 3s + 2}$

   (i)   with ROC $Re(s) < -2$
   (ii) with ROC $Re(s) > -1$
   (iii) with ROC $-2 < Re(s) < -1$

(b) $X(s) = \dfrac{4s^2 + 8s + 10}{(s + 2)(s^2 + 2s + 5)}$

(i) with ROC $\text{Re}(s) < -2$

(ii) with ROC $\text{Re}(s) > -1$

(iii) with ROC $-2 < \text{Re}(s) < -1$

(c) $X(s) = \dfrac{5s + 4}{s^2 + 2s + 1}$

(i) with ROC $\text{Re}(s) < -1$

(ii) with ROC $\text{Re}(s) > -1$

(d) $X(s) = \dfrac{2s^2 + 2s - 2}{s^2 - 1}$

(i) with ROC $\text{Re}(s) < -1$

(ii) with ROC $\text{Re}(s) > 1$

(iii) with ROC $-1 < \text{Re}(s) < 1$

6.44 Consider the RC circuit depicted in Fig. P6.44.

(a) Find the transfer function, assuming that $y_1(t)$ is the output. Plot the poles and zeros, and characterize the system as low pass, high pass, or band pass.

(b) Repeat Part (a), assuming that $y_2(t)$ is the system output.

(c) Find the impulse responses for the systems in Parts (a) and (b).

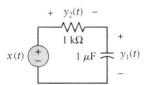

FIGURE P6.44

6.45 A system has the indicated transfer function $H(s)$. Determine the impulse response, assuming (i) that the system is causal and (ii) that the system is stable.

(a) $H(s) = \dfrac{2s^2 + 2s - 2}{s^2 - 1}$

(b) $H(s) = \dfrac{2s - 1}{s^2 + 2s + 1}$

(c) $H(s) = \dfrac{s^2 + 5s - 9}{(s + 1)(s^2 - 2s + 10)}$

(d) $H(s) = e^{-5s} + \dfrac{2}{s - 2}$

6.46 A stable system has the indicated input $x(t)$ and output $y(t)$. Use Laplace transforms to determine the transfer function and impulse response of the system.

(a) $x(t) = e^{-t}u(t), y(t) = e^{-2t}\cos(t)u(t)$

(b) $x(t) = e^{-2t}u(t), y(t) = -2e^{-t}u(t) + 2e^{-3t}u(t)$

6.47 The relationship between the input $x(t)$ and output $y(t)$ of a causal system is described by the indicated differential equation. Use Laplace transforms to determine the transfer function and impulse response of the system.

(a) $\dfrac{d}{dt}y(t) + 10y(t) = 10x(t)$

(b) $\dfrac{d^2}{dt^2}y(t) + 5\dfrac{d}{dt}y(t) + 6y(t) = x(t) + \dfrac{d}{dt}x(t)$

(c) $\dfrac{d^2}{dt^2}y(t) - \dfrac{d}{dt}y(t) - 2y(t)$
$$= -4x(t) + 5\dfrac{d}{dt}x(t)$$

6.48 Determine a differential equation description for a system with the following transfer function:

(a) $H(s) = \dfrac{1}{s(s + 3)}$

(b) $H(s) = \dfrac{6s}{s^2 - 2s + 8}$

(c) $H(s) = \dfrac{2(s - 2)}{(s + 1)^2(s + 3)}$

6.49 (a) Use the time-differentiation property to show that the transfer function of an LTI system is expressed in terms of the state-variable description as
$$H(s) = \mathbf{c}(s\mathbf{I} - \mathbf{A})^{-1}\mathbf{b} + D.$$

(b) Determine the transfer function, impulse response, and differential-equation descriptions of a stable LTI system represented by the following state-variable descriptions:

(i) $\mathbf{A} = \begin{bmatrix} -1 & 1 \\ 0 & -2 \end{bmatrix}$, $\mathbf{b} = \begin{bmatrix} 3 \\ -1 \end{bmatrix}$,
$\mathbf{c} = \begin{bmatrix} 1 & 2 \end{bmatrix}$, $D = \begin{bmatrix} 0 \end{bmatrix}$

(ii) $\mathbf{A} = \begin{bmatrix} 1 & 2 \\ 1 & -6 \end{bmatrix}$, $\mathbf{b} = \begin{bmatrix} 1 \\ 2 \end{bmatrix}$,
$\mathbf{c} = \begin{bmatrix} 0 & 1 \end{bmatrix}$, $D = \begin{bmatrix} 0 \end{bmatrix}$

6.50 Determine (i) whether the systems described by the following transfer functions are both stable and causal and (ii) whether a stable and causal inverse system exists:

(a) $H(s) = \dfrac{(s + 1)(s + 2)}{(s + 1)(s^2 + 2s + 10)}$

(b) $H(s) = \dfrac{s^2 + 2s - 3}{(s + 3)(s^2 + 2s + 5)}$

(c) $H(s) = \dfrac{s^2 - 3s + 2}{(s + 2)(s^2 - 2s + 8)}$

(d) $H(s) = \dfrac{s^2 + 2s}{(s^2 + 3s - 2)(s^2 + s + 2)}$

6.51 The relationship between the input $x(t)$ and output $y(t)$ of a system is described by the differential equation

$$\frac{d^2}{dt^2}y(t) + \frac{d}{dt}y(t) + 5y(t)$$
$$= \frac{d^2}{dt^2}x(t) - 2\frac{d}{dt}x(t) + x(t).$$

(a) Does this system have a stable and causal inverse? Why or why not?

(b) Find a differential-equation description of the inverse system.

6.52 A stable, causal system has a rational transfer function $H(s)$. The system satisfies the following conditions:

(i) The impulse response $h(t)$ is real valued;

(ii) $H(s)$ has exactly two zeros, one of which is at $s = 1 + j$;

(iii) The signal $\frac{d^2}{dt^2}h(t) + 3\frac{d}{dt}h(t) + 2h(t)$ contains an impulse and doublet of unknown strengths and a unit amplitude step. Find $H(s)$.

6.53 Sketch the magnitude response of the systems described by the following transfer functions, using the relationship between the pole and zero locations and the $j\omega$-axis in the s-plane:

(a) $H(s) = \dfrac{s}{s^2 + 2s + 101}$

(b) $H(s) = \dfrac{s^2 + 16}{s + 1}$

(c) $H(s) = \dfrac{s - 1}{s + 1}$

6.54 Sketch the phase response of the systems described by the following transfer functions, using the relationship between the locations of the poles and zeros and the $j\omega$-axis in the s-plane:

(a) $H(s) = \dfrac{s - 1}{s + 2}$

(b) $H(s) = \dfrac{s + 1}{s + 2}$

(c) $H(s) = \dfrac{1}{s^2 + 2s + 17}$

(d) $H(s) = s^2$

6.55 Sketch the Bode diagrams for the systems described by the following transfer functions:

(a) $H(s) = \dfrac{50}{(s + 1)(s + 10)}$

(b) $H(s) = \dfrac{20(s + 1)}{s^2(s + 10)}$

(c) $H(s) = \dfrac{5}{(s + 1)^3}$

(d) $H(s) = \dfrac{s + 2}{s^2 + s + 100}$

(e) $H(s) = \dfrac{s + 2}{s^2 + 10s + 100}$

6.56 The output of a multipath system $y(t)$ may be expressed in terms of the input $x(t)$ as

$$y(t) = x(t) + ax(t - T_{\text{diff}}),$$

where a and T_{diff} respectively represent the relative strength and time delay of the second path.

(a) Find the transfer function of the multipath system.

(b) Express the transfer function of the inverse system as an infinite sum, using the formula for summing a geometric series.

(c) Determine the impulse response of the inverse system. What condition must be satisfied for the inverse system to be both stable and causal?

(d) Find a stable inverse system, assuming that the condition determined in Part (c) is violated.

6.57 In Section 2.12, we derived block diagram descriptions of systems represented by linear constant-coefficient differential equations by rewriting the differential equation as an integral equation. Consider the second-order system described by the integral equation

$$y(t) = -a_1 y^{(1)}(t) - a_0 y^{(2)}(t) + b_2 x(t)$$
$$+ b_1 x^{(1)}(t) + b_0 x^{(2)}(t).$$

Recall that $v^{(n)}(t)$ is the n-fold integral of $v(t)$ with respect to time. Use the integration property to take the Laplace transform of the integral equation and derive the direct form I and II block diagrams for the transfer function of this system.

ADVANCED PROBLEMS

6.58 Prove the initial-value theorem by assuming that $x(t) = 0$ for $t < 0$ and taking the Laplace transform of the Taylor series expansion of $x(t)$ about $t = 0^+$.

6.59 The system with impulse response $h(t)$ is causal and stable and has a rational transfer function. Identify the conditions on the transfer function so that each of the following systems with impulse response $g(t)$ is stable and causal:

(a) $g(t) = \frac{d}{dt}h(t)$

(b) $g(t) = \int_{-\infty}^{t} h(\tau)d\tau$

6.60 Use the continuous-time representation $x_\delta(t)$ for the discrete-time signal $x[n]$ introduced in Section 4.4 to determine the Laplace transforms of the following discrete-time signals:

(a) $x[n] = \begin{cases} 1, & -2 \le n \le 2 \\ 0, & \text{otherwise} \end{cases}$

(b) $x[n] = (1/2)^n u[n]$

(c) $x[n] = e^{-2t}u(t)|_{t=nT}$

6.61 The autocorrelation function for a signal $x(t)$ is defined as

$$r(t) = \int_{-\infty}^{\infty} x(\tau)x(t + \tau)d\tau.$$

(a) Write $r(t) = x(t) * h(t)$. Express $h(t)$ in terms of $x(t)$. The system with impulse response $h(t)$ is called a *matched filter* for $x(t)$.

(b) Use the result from Part (a) to find the Laplace transform of $r(t)$.

(c) If $x(t)$ is real and $X(s)$ has two poles, one of which is located at $s = \sigma_p + j\omega_p$, determine the locations of all the poles of $R(s)$.

6.62 Suppose a system has M poles at $d_k = \alpha_k + j\beta_k$ and M zeros at $c_k = -\alpha_k + j\beta_k$. That is, the poles and zeros are symmetric about the $j\omega$-axis.

(a) Show that the magnitude response of any system that satisfies this symmetry condition is unity. Such a system is termed an *all-pass* system, since it passes all frequencies with unit gain.

(b) Evaluate the phase response of a single real pole–zero pair; that is, sketch the phase response of $\dfrac{s - \alpha}{s + \alpha}$, where $\alpha > 0$.

6.63 Consider the nonminimum-phase system described by the transfer function

$$H(s) = \frac{(s + 2)(s - 1)}{(s + 4)(s + 3)(s + 5)}.$$

(a) Does this system have a stable and causal inverse system?

(b) Express $H(s)$ as the product of a minimum-phase system $H_{\min}(s)$ and an all-pass system $H_{\text{ap}}(s)$ containing a single pole and zero. (See Problem 6.62 for the definition of an all-pass system.)

(c) Let $H_{\min}^{\text{inv}}(s)$ be the inverse system of $H_{\min}(s)$. Find $H_{\min}^{\text{inv}}(s)$. Can it be both stable and causal?

(d) Sketch the magnitude response and phase response of the system $H(s)H_{\min}^{\text{inv}}(s)$.

(e) Generalize your results from Parts (b) and (c) to an arbitrary nonminimum-phase system $H(s)$, and determine the magnitude response of the system $H(s)H_{\min}^{\text{inv}}(s)$.

6.64 An Nth-order low-pass *Butterworth* filter has squared magnitude response

$$|H(j\omega)|^2 = \frac{1}{1 + (j\omega/j\omega_c)^{2N}}.$$

The Butterworth filter is said to be maximally flat, because the first $2N$ derivatives of $|H(j\omega)|^2$ are zero at $\omega = 0$. The cutoff frequency, defined as the value for which $|H(j\omega)|^2 = 1/2$, is $\omega = \omega_c$. Assuming that the impulse response is real, the conjugate symmetry property of the Fourier transform may be used to write $|H(j\omega)|^2 = H(j\omega)H^*(j\omega) = H(j\omega)H(-j\omega)$. Noting that $H(s)|_{s=j\omega} = H(j\omega)$, we conclude that the Laplace transform of the Butterworth filter is characterized by the equation

$$H(s)H(-s) = \frac{1}{1 + (s/j\omega_c)^{2N}}.$$

(a) Find the poles and zeros of $H(s)H(-s)$, and sketch them in the s-plane.

(b) Choose the poles and zeros of $H(s)$ so that the impulse response is both stable and causal. Note that if s_p is a pole or zero of $H(s)$, then $-s_p$ is a pole or zero of $H(-s)$.

(c) Note that $H(s)H(-s)|_{s=0} = 1$. Find $H(s)$ for $N = 1$ and $N = 2$.

(d) Find the third-order differential equation that describes a Butterworth filter with cutoff frequency $\omega_c = 1$.

6.65 It is often convenient to change the cutoff frequency of a filter or to change a low-pass filter to a high-pass filter. Consider a system described by the transfer function

$$H(s) = \frac{1}{(s + 1)(s^2 + s + 1)}.$$

(a) Find the poles and zeros and sketch the magnitude response of this system. Determine whether the system is low pass or high pass, and find the cutoff frequency (the value of ω for which $|H(j\omega)| = 1/\sqrt{2}$).

(b) Perform the transformation of variables in which s is replaced by $s/10$ in $H(s)$. Repeat Part (a) for the transformed system.

(c) Perform the transformation of variables in which s is replaced by $1/s$ in $H(s)$. Repeat Part (a) for the transformed system.

(d) Find the transformation that converts $H(s)$ to a high-pass system with cutoff frequency $\omega = 100$.

COMPUTER EXPERIMENTS

6.66 Use the MATLAB command `roots` to determine the poles and zeros of the following systems:

(a) $H(s) = \dfrac{s^2 + 2}{s^3 + 2s^2 - s + 1}$

(b) $H(s) = \dfrac{s^3 + 1}{s^4 + 2s^2 + 1}$

(c) $H(s) = \dfrac{4s^2 + 8s + 10}{2s^3 + 8s^2 + 18s + 20}$

6.67 Use the MATLAB command `pzmap` to plot the poles and zeros of the following systems:

(a) $H(s) = \dfrac{s^3 + 1}{s^4 + 2s^2 + 1}$

(b) $\mathbf{A} = \begin{bmatrix} 1 & 2 \\ 1 & -6 \end{bmatrix}, \quad \mathbf{b} = \begin{bmatrix} 1 \\ 2 \end{bmatrix},$

$\mathbf{c} = \begin{bmatrix} 0 & 1 \end{bmatrix}, \quad D = \lfloor 0 \rfloor$

6.68 Use the MATLAB command `freqresp` to evaluate and plot the magnitude and phase responses for Examples 6.23 and 6.24, respectively.

6.69 Use the MATLAB command `freqresp` to evaluate and plot the magnitude and phase responses for each part of Problem 6.53.

6.70 Use your knowledge of the effect of poles and zeros on the magnitude response, to design systems having the specified magnitude response. Place poles and zeros in the s-plane, and evaluate the corresponding magnitude response using the MATLAB command `freqresp`. Repeat this process until you find poles and zeros that satisfy the specifications.

(a) Design a high-pass filter with two poles and two zeros that satisfies $|H(j0)| = 0$, $0.8 \le |H(j\omega)| \le 1.2$, for $|\omega| > 100\pi$ and that has real-valued coefficients.

(b) Design a low-pass filter with real-valued coefficients that satisfies $0.8 \le |H(j\omega)| \le 1.2$ for $|\omega| < \pi$ and $|H(j\omega)| < 0.1$ for $|\omega| > 10\pi$.

6.71 Use the MATLAB command `bode` to find the Bode diagrams for the systems in Problem 6.55.

6.72 Use the MATLAB command `ss` to find state-variable descriptions of the systems in Problem 6.48.

6.73 Use the MATLAB command `tf` to find transfer function descriptions of the systems in Problem 6.49.

7 Representing Signals by Using Discrete-Time Complex Exponentials: The z-Transform

7.1 Introduction

In this chapter, we generalize the complex sinusoidal representation of a discrete-time signal offered by the DTFT to a representation in terms of complex exponential signals that is termed the *z-transform*, the discrete-time counterpart to the Laplace transform. By using this more general representation of discrete-time signals, we are able to obtain a broader characterization of discrete-time LTI systems and their interaction with signals than is possible with the DTFT. For example, the DTFT can be applied solely to stable LTI systems, since the DTFT exists only if the impulse response is absolutely summable. In contrast, the z-transform of the impulse response exists for unstable LTI systems, and thus the z-transform can be used to study a much broader class of discrete-time LTI systems and signals.

As in continuous time, we shall see that discrete-time complex exponentials are eigenfunctions of LTI systems. This characteristic endues the z-transform with a powerful set of properties for use in analyzing signals and systems. Many of these properties parallel those of the DTFT; for example, convolution of time signals corresponds to multiplication of z-transforms. Hence, the output of an LTI system is obtained by multiplying the z-transform of the input by the z-transform of the impulse response. We define the z-transform of the impulse response as the transfer function of a system. The transfer function generalizes the frequency response characterization of a system's input–output behavior and offers new insights into system characteristics.

The primary roles of the z-transform in engineering practice are the study of system characteristics and the derivation of computational structures for implementing discrete-time systems on computers. The unilateral z-transform is also used to solve difference equations subject to initial conditions. We shall explore such problems in this chapter as we study the z-transform.

7.2 The z-Transform

We will derive the z-transform by examining the effect of applying a complex exponential input to an LTI system. Let $z = re^{j\Omega}$ be a complex number with magnitude r and angle Ω. The signal $x[n] = z^n$ is a complex exponential signal. We use $z = re^{j\Omega}$ to write

$$x[n] = r^n \cos(\Omega n) + jr^n \sin(\Omega n). \qquad (7.1)$$

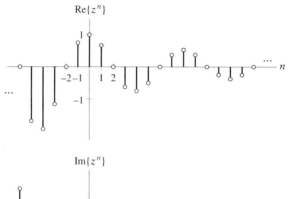

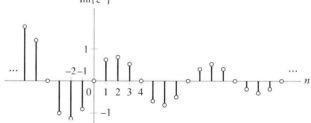

FIGURE 7.1   Real and imaginary parts of the signal z^n.

As illustrated in Fig. 7.1, the real part of $x[n]$ is an exponentially damped cosine and the imaginary part is an exponentially damped sine. The positive number r determines the damping factor and Ω is the sinusoidal frequency. Note that $x[n]$ is a complex sinusoid if $r = 1$.

Consider applying $x[n]$ to an LTI system with impulse response $h[n]$. The system output is given by

$$
\begin{aligned}
y[n] &= H\{x[n]\} \\
&= h[n] * x[n] \\
&= \sum_{k=-\infty}^{\infty} h[k]x[n-k].
\end{aligned}
$$

We use $x[n] = z^n$ to obtain

$$
\begin{aligned}
y[n] &= \sum_{k=-\infty}^{\infty} h[k]z^{n-k} \\
&= z^n \left(\sum_{k=-\infty}^{\infty} h[k]z^{-k} \right).
\end{aligned}
$$

We define the *transfer function*

$$
\boxed{H(z) = \sum_{k=-\infty}^{\infty} h[k]z^{-k}} \tag{7.2}
$$

so that we may write

$$
H\{z^n\} = H(z)z^n.
$$

This equation has the form of an eigenrelation, where z^n is the eigenfunction and $H(z)$ is the eigenvalue. The action of an LTI system on an input z^n is equivalent to multiplication of the input by the complex number $H(z)$. If we express $H(z)$ in polar form as $H(z) = |H(z)|e^{j\phi(z)}$, then the system output is written as

$$
y[n] = |H(z)|e^{j\phi(z)}z^n.
$$

Using $z = re^{j\Omega}$ and applying Euler's formula, we obtain

$$y[n] = |H(re^{j\Omega})|r^n \cos(\Omega n + \phi(re^{j\Omega})) + j|H(re^{j\Omega})|r^n \sin(\Omega n + \phi(re^{j\Omega})).$$

Comparing $y[n]$ with $x[n]$ in Eq. (7.1), we see that the system modifies the amplitude of the input by $|H(re^{j\Omega})|$ and shifts the phase of the sinusoidal components by $\phi(re^{j\Omega})$.

We now seek to represent arbitrary signals as a weighted superposition of the eigenfunctions z^n. Substituting $z = re^{j\Omega}$ into Eq. (7.2) yields

$$H(re^{j\Omega}) = \sum_{n=-\infty}^{\infty} h[n](re^{j\Omega})^{-n}$$

$$= \sum_{n=-\infty}^{\infty} (h[n]r^{-n})e^{-j\Omega n}.$$

We see that $H(re^{j\Omega})$ corresponds to the DTFT of a signal $h[n]r^{-n}$. Hence, the inverse DTFT of $H(re^{j\Omega})$ must be $h[n]r^{-n}$, so we may write

$$h[n]r^{-n} = \frac{1}{2\pi} \int_{-\pi}^{\pi} H(re^{j\Omega})e^{j\Omega n} d\Omega.$$

Multiplying this result by r^n gives

$$h[n] = \frac{r^n}{2\pi} \int_{-\pi}^{\pi} H(re^{j\Omega})e^{j\Omega n} d\Omega$$

$$= \frac{1}{2\pi} \int_{-\pi}^{\pi} H(re^{j\Omega})(re^{j\Omega})^n d\Omega.$$

We may convert the preceding equation to an integral over z by substituting $re^{j\Omega} = z$. Since the integration is performed only over Ω, r may be considered a constant, and we have $dz = jre^{j\Omega}d\Omega$. Accordingly, we identify $d\Omega = \frac{1}{j}z^{-1} dz$. Lastly, consider the limits on the integral. As Ω goes from $-\pi$ to π, z traverses a circle of radius r in a counterclockwise direction. Thus, we write

$$h[n] = \frac{1}{2\pi j} \oint H(z)z^{n-1}dz, \tag{7.3}$$

where the symbol $\oint$ denotes integration around a circle of radius $|z| = r$ in a counterclockwise direction. Equation (7.2) indicates how to determine $H(z)$ from $h[n]$, while Eq. (7.3) expresses $h[n]$ as a function of $H(z)$. We say that the transfer function $H(z)$ is the z-transform of the impulse response $h[n]$.

More generally, the *z-transform* of an arbitrary signal $x[n]$ is

$$X(z) = \sum_{n=-\infty}^{\infty} x[n]z^{-n}, \tag{7.4}$$

and the *inverse z-transform* is

$$x[n] = \frac{1}{2\pi j} \oint X(z)z^{n-1} dz. \tag{7.5}$$

We express the relationship between $x[n]$ and $X(z)$ with the notation

$$x[n] \overset{z}{\longleftrightarrow} X(z).$$

Note that Eq. (7.5) expresses the signal $x[n]$ as a weighted superposition of complex exponentials z^n. The weights are $(1/2\pi j)X(z)z^{-1}\,dz$. In practice, we will not evaluate this integral directly, since that would require knowledge of complex-variable theory. Instead, we will evaluate inverse z-transforms by inspection, using the one-to-one relationship between $x[n]$ and $X(z)$.

■ 7.2.1 CONVERGENCE

The z-transform exists when the infinite sum in Eq. (7.4) converges. A necessary condition for convergence is absolute summability of $x[n]z^{-n}$. Since $|x[n]z^{-n}| = |x[n]r^{-n}|$, we must have

$$\sum_{n=-\infty}^{\infty} |x[n]r^{-n}| < \infty.$$

The range of r for which this condition is satisfied is termed the *region of convergence* (ROC) of the z-transform.

The z-transform exists for signals that do not have a DTFT. Recall that existence of the DTFT requires absolute summability of $x[n]$. By limiting ourselves to restricted values of r, we ensure that $x[n]r^{-n}$ is absolutely summable, even though $x[n]$ is not. For example, the DTFT of $x[n] = \alpha^n u[n]$ does not exist for $|\alpha| > 1$, since $x[n]$ is then an increasing exponential signal, as illustrated in Fig. 7.2(a). However, if $r > \alpha$, then r^{-n}, depicted in Fig. 7.2(b), decays faster than $x[n]$ grows. Hence, the signal $x[n]r^{-n}$, depicted in Fig. 7.2(c),

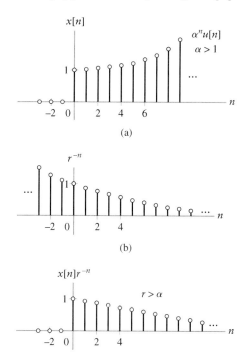

(a)

(b)

(c)

FIGURE 7.2   Illustration of a signal that has a z-transform, but does not have a DTFT. (a) An increasing exponential signal for which the DTFT does not exist. (b) The attenuating factor r^{-n} associated with the z-transform. (c) The modified signal $x[n]r^{-n}$ is absolutely summable, provided that $r > \alpha$, and thus the z-transform of $x[n]$ exists.

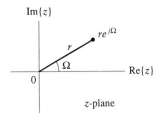

FIGURE 7.3   The z-plane. A point $z = re^{j\Omega}$ is located at a distance r from the origin and an angle Ω relative to the real axis.

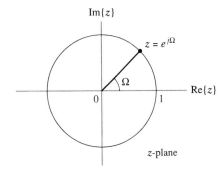

FIGURE 7.4   The unit circle, $z = e^{j\Omega}$, in the z-plane.

is absolutely summable, and the z-transform exists. The ability to work with signals that do not have a DTFT is a significant advantage offered by the z-transform.

■ 7.2.2 THE z-PLANE

It is convenient to represent the complex number z as a location in a complex plane termed the *z-plane* and depicted graphically in Fig. 7.3. The point $z = re^{j\Omega}$ is located at a distance r from the origin and an angle Ω from the positive real axis. Note that if $x[n]$ is absolutely summable, then the DTFT is obtained from the z-transform by setting $r = 1$, or substituting $z = e^{j\Omega}$ into Eq. (7.4). That is,

$$X(e^{j\Omega}) = X(z)|_{z=e^{j\Omega}}. \tag{7.6}$$

The equation $z = e^{j\Omega}$ describes a circle of unit radius centered on the origin in the z-plane, as illustrated in Fig. 7.4. This contour is termed the *unit circle* in the z-plane. The frequency Ω in the DTFT corresponds to the point on the unit circle at an angle of Ω with respect to the positive real axis. As the discrete-time frequency Ω goes from $-\pi$ to π, we take one trip around the unit circle. We say that the DTFT corresponds to the z-transform evaluated on the unit circle.

EXAMPLE 7.1 THE z-TRANSFORM AND THE DTFT   Determine the z-transform of the signal

$$x[n] = \begin{cases} 1, & n = -1 \\ 2, & n = 0 \\ -1, & n = 1 \\ 1, & n = 2 \\ 0, & \text{otherwise} \end{cases} .$$

Use the z-transform to determine the DTFT of $x[n]$.

Solution: We substitute the prescribed $x[n]$ into Eq. (7.4) to obtain

$$X(z) = z + 2 - z^{-1} + z^{-2}.$$

We obtain the DTFT from $X(z)$ by substituting $z = e^{j\Omega}$:

$$X(e^{j\Omega}) = e^{j\Omega} + 2 - e^{-j\Omega} + e^{-j2\Omega}.$$

■

■ 7.2.3 POLES AND ZEROS

The most commonly encountered form of the z-transform in engineering applications is a ratio of two polynomials in z^{-1}, as shown by the rational function

$$X(z) = \frac{b_0 + b_1 z^{-1} + \cdots + b_M z^{-M}}{a_0 + a_1 z^{-1} + \cdots + a_N z^{-N}}.$$

It is useful to rewrite $X(z)$ as a product of terms involving the roots of the numerator and denominator polynomials; that is,

$$X(z) = \frac{\widetilde{b} \prod_{k=1}^{M}(1 - c_k z^{-1})}{\prod_{k=1}^{N}(1 - d_k z^{-1})},$$

where $\widetilde{b} = b_0/a_0$. The c_k are the roots of the numerator polynomial and are termed the *zeros* of $X(z)$. The d_k are the roots of the denominator polynomial and are termed the *poles* of $X(z)$. Locations of zeros are denoted with the " $\bigcirc$ " symbol and locations of poles with the "$\times$" symbol in the z-plane. The locations of poles and zeros completely specify $X(z)$, except for the gain factor $\widetilde{b}$.

EXAMPLE 7.2 z-TRANSFORM OF A CAUSAL EXPONENTIAL SIGNAL   Determine the z-transform of the signal

$$x[n] = \alpha^n u[n].$$

Depict the ROC and the locations of poles and zeros of $X(z)$ in the z-plane.

Solution: Substituting $x[n] = \alpha^n u[n]$ into Eq. (7.4) yields

$$X(z) = \sum_{n=-\infty}^{\infty} \alpha^n u[n] z^{-n}$$

$$= \sum_{n=0}^{\infty} \left(\frac{\alpha}{z}\right)^n.$$

This is a geometric series of infinite length in the ratio α/z; the sum converges, provided that $|\alpha/z| < 1$, or $|z| > |\alpha|$. Hence,

$$X(z) = \frac{1}{1 - \alpha z^{-1}}, \qquad |z| > |\alpha|$$

$$= \frac{z}{z - \alpha}, \qquad |z| > |\alpha|.$$

(7.7)

There is thus a pole at $z = \alpha$ and a zero at $z = 0$, as illustrated in Fig. 7.5. The ROC is depicted as the shaded region of the z-plane.

■

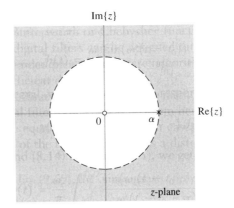

FIGURE 7.5 Locations of poles and zeros of $x[n] = \alpha^n u[n]$ in the z-plane. The ROC is the shaded area.

As with the Laplace transform, the expression for $X(z)$ does not correspond to a unique time signal, unless the ROC is specified. This means that two different time signals may have identical z-transforms, but different ROC's, as demonstrated in the next example.

EXAMPLE 7.3 Z-TRANSFORM OF AN ANTICAUSAL EXPONENTIAL SIGNAL Determine the z-transform of the signal

$$y[n] = -\alpha^n u[-n - 1].$$

Depict the ROC and the locations of poles and zeros of $X(z)$ in the z-plane.

Solution: We substitute $y[n] = -\alpha^n u[-n - 1]$ into Eq. (7.4) and write

$$Y(z) = \sum_{n=-\infty}^{\infty} -\alpha^n u[-n - 1]z^{-n}$$

$$= -\sum_{n=-\infty}^{-1} \left(\frac{\alpha}{z}\right)^n$$

$$= -\sum_{k=1}^{\infty} \left(\frac{z}{\alpha}\right)^k$$

$$= 1 - \sum_{k=0}^{\infty} \left(\frac{z}{\alpha}\right)^k.$$

The sum converges, provided that $|z/\alpha| < 1$, or $|z| < |\alpha|$. Hence,

$$Y(z) = 1 - \frac{1}{1 - z\alpha^{-1}}, \qquad |z| < |\alpha|,$$

$$= \frac{z}{z - \alpha}, \qquad |z| < |\alpha|.$$

(7.8)

The ROC and the locations of poles and zeros are depicted in Fig. 7.6. ∎

Note that $Y(z)$ in Eq. (7.8) is identical to $X(z)$ in Eq. (7.7), even though the time signals are quite different. Only the ROC differentiates the two transforms. We must know the ROC to determine the correct inverse z-transform. This ambiguity is a general feature of signals that are one sided.

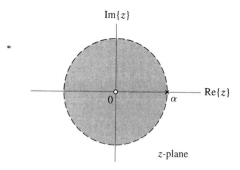

FIGURE 7.6 ROC and locations of poles and zeros of $x[n] = -\alpha^n u[-n-1]$ in the z-plane.

EXAMPLE 7.4 z-TRANSFORM OF A TWO-SIDED SIGNAL   Determine the z-transform of

$$x[n] = -u[-n-1] + \left(\frac{1}{2}\right)^n u[n].$$

Depict the ROC and the locations of poles and zeros of $X(z)$ in the z-plane.

Solution: Substituting for $x[n]$ in Eq. (7.4), we obtain

$$X(z) = \sum_{n=-\infty}^{\infty} \left(\frac{1}{2}\right)^n u[n]z^{-n} - u[-n-1]z^{-n}$$

$$= \sum_{n=0}^{\infty} \left(\frac{1}{2z}\right)^n - \sum_{n=-\infty}^{-1} \left(\frac{1}{z}\right)^n$$

$$= \sum_{n=0}^{\infty} \left(\frac{1}{2z}\right)^n + 1 - \sum_{k=0}^{\infty} z^k.$$

Both sums must converge in order for $X(z)$ to converge. This implies that we must have $|z| > 1/2$ and $|z| < 1$. Hence,

$$X(z) = \frac{1}{1 - \frac{1}{2}z^{-1}} + 1 - \frac{1}{1-z}, \qquad 1/2 < |z| < 1$$

$$= \frac{z\left(2z - \frac{3}{2}\right)}{\left(z - \frac{1}{2}\right)(z-1)}, \qquad 1/2 < |z| < 1.$$

The ROC and the locations of the poles and zeros are depicted in Fig. 7.7. In this case, the ROC is a ring in the z-plane.   ■

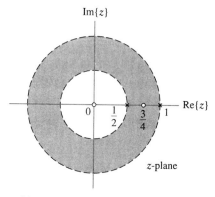

FIGURE 7.7 ROC and locations of poles and zeros in the z-plane for Example 7.4.

▶ **Problem 7.1**   Determine the z-transform, the ROC, and the locations of poles and zeros of $X(z)$ for the following signals:

(a)
$$x[n] = \left(\frac{1}{2}\right)^n u[n] + \left(\frac{-1}{3}\right)^n u[n]$$

(b)
$$x[n] = -\left(\frac{1}{2}\right)^n u[-n-1] - \left(\frac{-1}{3}\right)^n u[-n-1]$$

(c)
$$x[n] = -\left(\frac{3}{4}\right)^n u[-n-1] + \left(\frac{-1}{3}\right)^n u[n]$$

(d)
$$x[n] = e^{j\Omega_o n}u[n]$$

Answers:

(a)
$$X(z) = \frac{z\left(2z - \frac{1}{6}\right)}{\left(z - \frac{1}{2}\right)\left(z + \frac{1}{3}\right)}, \qquad |z| > 1/2$$

Poles are at $z = 1/2$ and $z = -1/3$, and zeros are at $z = 0$ and $z = 1/12$

(b)
$$X(z) = \frac{z\left(2z - \frac{1}{6}\right)}{\left(z - \frac{1}{2}\right)\left(z + \frac{1}{3}\right)}, \qquad |z| < 1/3$$

Poles are at $z = 1/2$ and $z = -1/3$, and zeros are at $z = 0$ and $z = 1/12$

(c)
$$X(z) = \frac{z(2z - 5/12)}{(z - 3/4)(z + 1/3)}, \qquad 1/3 < |z| < 3/4$$

Poles are at $z = 3/4$ and $z = -1/3$, and zeros are at $z = 0$ and $z = 5/24$

(d)
$$X(z) = \frac{z}{z - e^{j\Omega_o}}, \qquad |z| > 1$$

A pole is at $z = e^{j\Omega_o}$, and a zero is at $z = 0$ ◀

7.3 *Properties of the Region of Convergence*

The basic properties of the ROC are examined in this section. In particular, we show how the ROC is related to the characteristics of a signal $x[n]$. Given properties of the ROC, we can often identify the ROC from $X(z)$ and limited knowledge of the characteristics of $x[n]$. The relationship between the ROC and the characteristics of the time-domain signal is used to find inverse z-transforms in Section 7.5. The results presented here are derived with the use of intuitive arguments rather than rigorous proofs.

First, we note that the ROC cannot contain any poles. This is because the ROC is defined as the set of all z for which the z-transform converges. Hence, $X(z)$ must be finite for all z in the ROC. If d is a pole, then $|X(d)| = \infty$, and the z-transform does not converge at the pole. Thus, the pole cannot lie in the ROC.

Second, the ROC for a finite-duration signal includes the entire z-plane, except possibly $z = 0$ or $|z| = \infty$ (or both). To see this, suppose $x[n]$ is nonzero only on the interval $n_1 \le n \le n_2$. We have

$$X(z) = \sum_{n=n_1}^{n_2} x[n]z^{-n}.$$

This sum will converge, provided that each of its terms is finite. If a signal has any nonzero causal components ($n_2 > 0$), then the expression for $X(z)$ will have a term involving z^{-1}, and thus the ROC cannot include $z = 0$. If a signal is noncausal, a power of ($n_1 < 0$), then the expression for $X(z)$ will have a term involving a power of z, and thus the ROC cannot include $|z| = \infty$. Conversely, if $n_2 \le 0$, then the ROC will include $z = 0$. If a signal has no nonzero noncausal components ($n_1 \ge 0$), then the ROC will include $|z| = \infty$. This line of reasoning also indicates that $x[n] = c\delta[n]$ is the only signal whose ROC is the entire z-plane.

Now consider infinite-duration signals. The condition for convergence is $|X(z)| < \infty$. We may thus write

$$|X(z)| = \left| \sum_{n=-\infty}^{\infty} x[n]z^{-n} \right|$$
$$\le \sum_{n=-\infty}^{\infty} |x[n]z^{-n}|$$
$$= \sum_{n=-\infty}^{\infty} |x[n]||z|^{-n}.$$

The second line follows from the fact that the magnitude of a sum of complex numbers is less than or equal to the sum of the individual magnitudes. We obtain the third line by noting that the magnitude of a product equals the product of the magnitudes. Splitting the infinite sum into negative- and positive-time portions, we define

$$I_-(z) = \sum_{n=-\infty}^{-1} |x[n]||z|^{-n}$$

and

$$I_+(z) = \sum_{n=0}^{\infty} |x[n]||z|^{-n}.$$

We note that $|X(z)| \le I_-(z) + I_+(z)$. If both $I_-(z)$ and $I_+(z)$ are finite, then $|X(z)|$ is guaranteed to be finite, too. This clearly requires that $|x[n]|$ be bounded in some way.

Suppose we can bound $|x[n]|$ by finding the smallest positive constants A_-, A_+, r_-, and r_+ such that

$$|x[n]| \le A_-(r_-)^n, \qquad n < 0 \tag{7.9}$$

and

$$|x[n]| \le A_+(r_+)^n, \qquad n \ge 0. \tag{7.10}$$

A signal that satisfies these two bounds grows no faster than $(r_+)^n$ for positive n and $(r_-)^n$ for negative n. While we can construct signals that do not satisfy these bounds, such as a^{n^2}, these signals do not generally occur in engineering problems.

If the bound given in Eq. (7.9) is satisfied, then

$$I_-(z) \leq A_- \sum_{n=-\infty}^{-1} (r_-)^n |z|^{-n}$$

$$= A_- \sum_{n=-\infty}^{-1} \left(\frac{r_-}{|z|}\right)^n$$

$$= A_- \sum_{k=1}^{\infty} \left(\frac{|z|}{r_-}\right)^k$$

where we have substituted $k = -n$ in the third line. The last sum converges if and only if $|z| \leq r_-$. Now consider the positive-time portion. If the bound given in Eq. (7.10) is satisfied, then

$$I_+(z) \leq A_+ \sum_{n=0}^{\infty} (r_+)^n |z|^{-n}$$

$$= A_+ \sum_{n=0}^{\infty} \left(\frac{r_+}{|z|}\right)^n.$$

This sum converges if and only if $|z| > r_+$. Hence, if $r_+ < |z| < r_-$, then both $I_+(z)$ and $I_-(z)$ converge and $|X(z)|$ also converges. Note that if $r_+ > r_-$, then there are no values of z for which convergence is guaranteed.

Now define a *left-sided signal* as a signal for which $x[n] = 0$ for $n \geq 0$, a *right-sided signal* as a signal for which $x[n] = 0$ for $n < 0$, and a *two-sided signal* as a signal that has infinite duration in both the positive and negative directions. Then, for signals $x[n]$ that satisfy the exponential bounds of Eqs. (7.9) and (7.10), we have the following conclusions:

▸ The ROC of a right-sided signal is of the form $|z| > r_+$.
▸ The ROC of a left-sided signal is of the form $|z| < r_-$.
▸ The ROC of a two-sided signal is of the form $r_+ < |z| < r_-$.

Each of these cases is illustrated in Fig. 7.8.

One-sided exponential signals are encountered frequently in engineering problems because we are often interested in the behavior of a signal either before or after a given time instant. With such signals, the magnitude of one or more poles determines the ROC boundaries r_- and r_+. Suppose we have a right-sided signal $x[n] = \alpha^n u[n]$, where α is, in general, complex. The z-transform of $x[n]$ has a pole at $z = \alpha$ and the ROC is $|z| > |\alpha|$. Hence, the ROC is the region of the z-plane with a radius greater than the radius of the pole. Likewise, if $x[n]$ is the left-sided signal $x[n] = \alpha^n u[-n - 1]$, then the ROC is $|z| < |\alpha|$, the region of the z-plane with radius less than the radius of the pole. If a signal consists of a sum of exponentials, then the ROC is the intersection of the ROCs associated with each term; it will have a radius greater than that of the pole of largest radius associated with right-sided terms and a radius less than that of the pole of smallest radius associated with left-sided terms. Some of these properties of the ROC are illustrated in the next example.

EXAMPLE 7.5 ROCs OF TWO-SIDED SIGNALS   Identify the ROC associated with the z-transform for each of the following signals:

$$x[n] = (-1/2)^n u[-n] + 2(1/4)^n u[n];$$

$$y[n] = (-1/2)^n u[n] + 2(1/4)^n u[n];$$

$$w[n] = (-1/2)^n u[-n] + 2(1/4)^n u[-n].$$

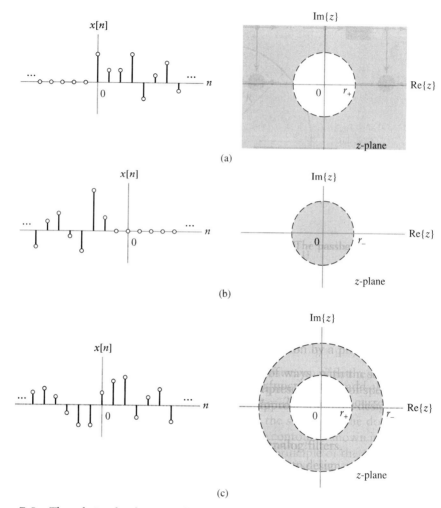

FIGURE 7.8 The relationship between the ROC and the time extent of a signal. (a) A right-sided signal has an ROC of the form $|z| > r_+$. (b) A left-sided signal has an ROC of the form $|z| < r_-$. (c) A two-sided signal has an ROC of the form $r_+ < |z| < r_-$.

Solution: Beginning with $x[n]$, we use Eq. (7.4) to write

$$X(z) = \sum_{n=-\infty}^{0} \left(\frac{-1}{2z}\right)^n + 2\sum_{n=0}^{\infty} \left(\frac{1}{4z}\right)^n$$

$$= \sum_{k=0}^{\infty} (-2z)^k + 2\sum_{n=0}^{\infty} \left(\frac{1}{4z}\right)^n.$$

The first series converges for $|z| < \frac{1}{2}$, while the second converges for $|z| > \frac{1}{4}$. Both series must converge for $X(z)$ to converge, so the ROC is $\frac{1}{4} < |z| < \frac{1}{2}$. The ROC of this two-sided signal is depicted in Fig. 7.9(a). Summing the two geometric series, we obtain

$$X(z) = \frac{1}{1 + 2z} + \frac{2z}{z - \frac{1}{4}},$$

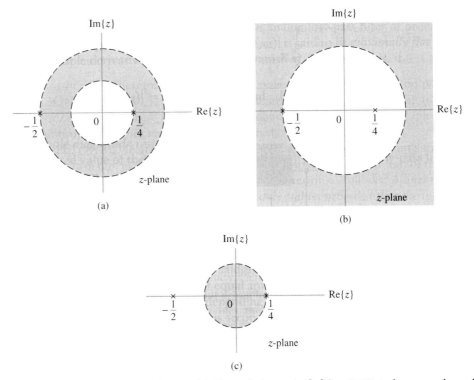

FIGURE 7.9 ROCs for Example 7.5. (a) Two-sided signal $x[n]$ has ROC in between the poles. (b) Right-sided signal $y[n]$ has ROC outside of the circle containing the pole of largest magnitude. (c) Left-sided signal $w[n]$ has ROC inside the circle containing the pole of smallest magnitude.

which has poles at $z = -1/2$ and $z = 1/4$. Note that the ROC is the ring-shaped region located between the poles.

Next, $y[n]$ is a right-sided signal, and again using the definition of the z-transform given by Eq. (7.4), we write

$$Y(z) = \sum_{n=0}^{\infty} \left(\frac{-1}{2z} \right)^n + 2 \sum_{n=0}^{\infty} \left(\frac{1}{4z} \right)^n.$$

The first series converges for $|z| > 1/2$, while the second converges for $|z| > 1/4$. Hence, the combined ROC is $|z| > 1/2$, as depicted in Fig. 7.9(b). In this case, we write

$$Y(z) = \frac{z}{z + \frac{1}{2}} + \frac{2z}{z - \frac{1}{4}},$$

for which the poles are again at $z = -1/2$ and $z = 1/4$. The ROC is outside a circle containing the pole of largest radius, $z = -1/2$.

The last signal, $w[n]$, is left sided and has z-transform given by

$$W(z) = \sum_{n=-\infty}^{0} \left(\frac{-1}{2z} \right)^n + 2 \sum_{n=-\infty}^{0} \left(\frac{1}{4z} \right)^n$$

$$= \sum_{k=0}^{\infty} (-2z)^k + 2 \sum_{k=0}^{\infty} (4z)^k.$$

Here, the first series converges for $|z| < 1/2$, while the second series converges for $|z| < 1/4$, giving a combined ROC of $|z| < 1/4$, as depicted in Fig. 7.9(c). In this case, we have

$$W(z) = \frac{1}{1 + 2z} + \frac{2}{1 - 4z},$$

where the poles are at $z = -1/2$ and $z = 1/4$. The ROC is inside a circle containing the pole of smallest radius, $z = 1/4$.

This example illustrates that the ROC of a two-sided signal is a ring, the ROC of a right-sided signal is the exterior of a circle, and the ROC of a left-sided signal is the interior of a circle. In each case, the poles define the boundaries of the ROC.   ■

▶ **Problem 7.2**   Determine the z-transform and ROC for the two-sided signal

$$x[n] = \alpha^{|n|},$$

assuming that $|\alpha| < 1$. Repeat for $|\alpha| > 1$.

Answer: For $|\alpha| < 1$,

$$X(z) = \frac{z}{z - \alpha} - \frac{z}{z - 1/\alpha}, \qquad |\alpha| < |z| < 1/|\alpha|$$

For $|\alpha| > 1$, the ROC is the empty set   ◀

7.4 *Properties of the z-Transform*

Most properties of the z-transform are analogous to those of the DTFT. Hence, in this section we state the properties and defer the proofs to problems. We assume that

$$x[n] \xleftrightarrow{\ z\ } X(z), \quad \text{with ROC } R_x$$

and

$$y[n] \xleftrightarrow{\ z\ } Y(z), \quad \text{with ROC } R_y.$$

The ROC is changed by certain operations. In the previous section, we established that the general form of the ROC is a ring in the z-plane. Thus, the effect of an operation on the ROC is described by a change in the radii of the ROC boundaries.

Linearity

The linearity property states that the z-transform of a sum of signals is just the sum of the individual z-transforms. That is,

$$\boxed{ax[n] + by[n] \xleftrightarrow{\ z\ } aX(z) + bY(z), \quad \text{with ROC at least } R_x \cap R_y.} \qquad (7.11)$$

The ROC is the intersection of the individual ROCs because the z-transform of the sum is valid only wherever both $X(z)$ and $Y(z)$ converge. The ROC can be larger than the intersection if one or more terms in $x[n]$ or $y[n]$ cancel each other in the sum. In the z-plane, this corresponds to a zero canceling a pole that defines one of the ROC boundaries. This phenomenon is illustrated in the next example.

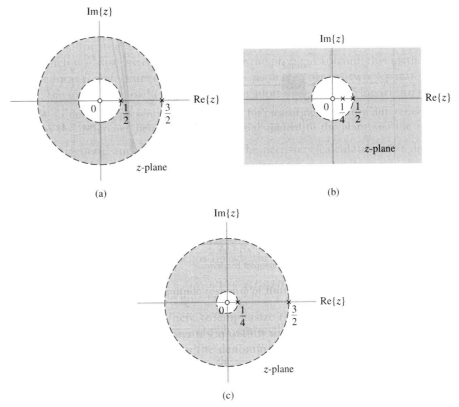

FIGURE 7.10   ROCs for Example 7.6. (a) ROC and pole–zero plot for $X(z)$. (b) ROC and pole–zero plot for $Y(z)$. (c) ROC and pole–zero plot for $a(X(z) + Y(z))$.

EXAMPLE 7.6   POLE–ZERO CANCELLATION   Suppose

$$x[n] = \left(\frac{1}{2}\right)^n u[n] - \left(\frac{3}{2}\right)^n u[-n-1] \xleftrightarrow{\ z\ } X(z) = \frac{-z}{\left(z - \frac{1}{2}\right)\left(z - \frac{3}{2}\right)},$$

$$\text{with ROC } \frac{1}{2} < |z| < \frac{3}{2}$$

and

$$y[n] = \left(\frac{1}{4}\right)^n u[n] - \left(\frac{1}{2}\right)^n u[n] \xleftrightarrow{\ z\ } Y(z) = \frac{-\frac{1}{4}z}{\left(z - \frac{1}{4}\right)\left(z - \frac{1}{2}\right)}, \quad \text{with ROC } |z| > \frac{1}{2}.$$

Evaluate the z-transform of $ax[n] + by[n]$.

Solution: The pole–zero plots and ROCs for $x[n]$ and $y[n]$ are depicted in Figs. 7.10(a) and (b), respectively. The linearity property given by Eq. (7.11) indicates that

$$ax[n] + by[n] \xleftrightarrow{\ z\ } a\frac{-z}{\left(z - \frac{1}{2}\right)\left(z - \frac{3}{2}\right)} + b\frac{-\frac{1}{4}z}{\left(z - \frac{1}{4}\right)\left(z - \frac{1}{2}\right)}.$$

In general, the ROC is the intersection of individual ROCs, or $\frac{1}{2} < |z| < \frac{3}{2}$ in this example, which corresponds to the ROC depicted in Fig. 7.10(a). Note, however, what happens when $a = b$: We have

$$ax[n] + ay[n] = a\left(-\left(\frac{3}{2}\right)^n u[-n-1] + \left(\frac{1}{4}\right)^n u[n]\right),$$

and we see that the term $\left(\frac{1}{2}\right)^n u[n]$ has been canceled in the time-domain signal. The ROC is now easily verified to be $\frac{1}{4} < |z| < \frac{3}{2}$, as shown in Fig. 7.10(c). This ROC is larger than the intersection of the individual ROCs, because the term $\left(\frac{1}{2}\right)^n u[n]$ is no longer present. Combining z-transforms and using the linearity property gives

$$aX(z) + aY(z) = a\left(\frac{-z}{\left(z-\frac{1}{2}\right)\left(z-\frac{3}{2}\right)} + \frac{-\frac{1}{4}z}{\left(z-\frac{1}{4}\right)\left(z-\frac{1}{2}\right)}\right)$$

$$= a\frac{-\frac{1}{4}z\left(z-\frac{3}{2}\right) - z\left(z-\frac{1}{4}\right)}{\left(z-\frac{1}{4}\right)\left(z-\frac{1}{2}\right)\left(z-\frac{3}{2}\right)}$$

$$= a\frac{-\frac{5}{4}z\left(z-\frac{1}{2}\right)}{\left(z-\frac{1}{4}\right)\left(z-\frac{1}{2}\right)\left(z-\frac{3}{2}\right)}.$$

The zero at $z = \frac{1}{2}$ cancels the pole at $z = \frac{1}{2}$, so we have

$$aX(z) + aY(z) = a\frac{-\frac{5}{4}z}{\left(z-\frac{1}{4}\right)\left(z-\frac{3}{2}\right)}.$$

Hence, cancellation of the $\left(\frac{1}{2}\right)^n u[n]$ term in the time domain corresponds to cancellation of the pole at $z = \frac{1}{2}$ by a zero in the z-domain. This pole defined the ROC boundary, so the ROC enlarges when the pole is removed. ■

Time Reversal

$$x[-n] \xleftrightarrow{z} X\left(\frac{1}{z}\right), \quad \text{with ROC } \frac{1}{R_x}. \tag{7.12}$$

Time reversal, or reflection, corresponds to replacing z by z^{-1}. Hence, if R_x is of the form $a < |z| < b$, the ROC of the reflected signal is $a < 1/|z| < b$, or $1/b < |z| < 1/a$.

Time Shift

$$x[n - n_o] \xleftrightarrow{z} z^{-n_o}X(z), \quad \text{with ROC } R_x, \text{except possibly } z = 0 \text{ or } |z| = \infty. \tag{7.13}$$

Multiplication by z^{-n_o} introduces a pole of order n_o at $z = 0$ if $n_o > 0$. In this case, the ROC cannot include $z = 0$, even if R_x does include $z = 0$, unless $X(z)$ has a zero of at least order n_o at $z = 0$ that cancels all of the new poles. If $n_o < 0$, then multiplication by z^{-n_o} introduces n_o poles at infinity. If these poles are not canceled by zeros at infinity in $X(z)$, then the ROC of $z^{-n_o}X(z)$ cannot include $|z| = \infty$.

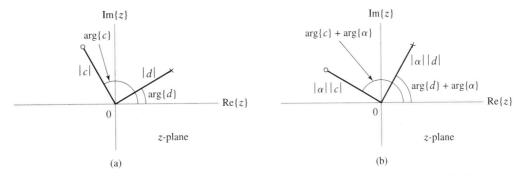

FIGURE 7.11   The effect of multiplication by α^n on the poles and zeros of a transfer function. (a) Locations of pole d and zero c of $X(z)$. (b) Locations of pole and zero of $X(z/\alpha)$.

Multiplication by an Exponential Sequence

Let α be a complex number. Then

$$\alpha^n x[n] \xleftrightarrow{\ z\ } X\!\left(\frac{z}{\alpha}\right), \quad \text{with ROC } |\alpha| R_x. \tag{7.14}$$

The notation $|\alpha| R_x$ implies that the ROC boundaries are multiplied by $|\alpha|$. If R_x is $a < |z| < b$, then the new ROC is $|\alpha| a < |z| < |\alpha| b$. If $X(z)$ contains a factor $1 - dz^{-1}$ in the denominator, so that d is a pole, then $X\!\left(\frac{z}{\alpha}\right)$ has a factor $1 - \alpha d z^{-1}$ in the denominator and thus has a pole at αd. Similarly, if c is a zero of $X(z)$, then $X\!\left(\frac{z}{\alpha}\right)$ has a zero at αc. This indicates that the poles and zeros of $X(z)$ have their radii changed by $|\alpha|$, and their angles are changed by $\arg\{\alpha\}$. (See Fig. 7.11.) If α has unit magnitude, then the radius is unchanged; if α is a positive real number, then the angle is unchanged.

Convolution

$$x[n] * y[n] \xleftrightarrow{\ z\ } X(z)Y(z), \quad \text{with ROC at least } R_x \cap R_y. \tag{7.15}$$

Convolution of time-domain signals corresponds to multiplication of z-transforms. As with the linearity property, the ROC may be larger than the intersection of R_x and R_y if a pole–zero cancellation occurs in the product $X(z)Y(z)$.

Differentiation in the z-Domain

$$nx[n] \xleftrightarrow{\ z\ } -z\frac{d}{dz}X(z), \quad \text{with ROC } R_x. \tag{7.16}$$

Multiplication by n in the time domain corresponds to differentiation with respect to z and multiplication of the result by $-z$ in the z-domain. This operation does not change the ROC.

EXAMPLE 7.7 APPLYING MULTIPLE PROPERTIES   Find the z-transform of the signal

$$x[n] = \left(n\left(\frac{-1}{2}\right)^n u[n] \right) * \left(\frac{1}{4}\right)^{-n} u[-n].$$

Solution: First we find the z-transform of $w[n] = n\left(\frac{-1}{2}\right)^n u[n]$. We know from Example 7.2 that

$$\left(\frac{-1}{2}\right)^n u[n] \xleftarrow{\quad z \quad} \frac{z}{z + \frac{1}{2}}, \quad \text{with ROC } |z| > \frac{1}{2}.$$

Thus, the z-domain differentiation property of Eq. (7.16) implies that

$$w[n] = n\left(\frac{-1}{2}\right)^n u[n] \xleftarrow{\quad z \quad} W(z) = -z\frac{d}{dz}\left(\frac{z}{z + \frac{1}{2}}\right), \quad \text{with ROC } |z| > \frac{1}{2}$$

$$= -z\left(\frac{z + \frac{1}{2} - z}{\left(z + \frac{1}{2}\right)^2}\right)$$

$$= \frac{-\frac{1}{2}z}{\left(z + \frac{1}{2}\right)^2}, \quad \text{with ROC } |z| > \frac{1}{2}.$$

Next, we find the z-transform of $y[n] = \left(\frac{1}{4}\right)^{-n} u[-n]$. We do this by applying the time-reversal property given in Eq. (7.12) to the result of Example 7.2. Noting that

$$\left(\frac{1}{4}\right)^n u[n] \xleftarrow{\quad z \quad} \frac{z}{z - \frac{1}{4}}, \quad \text{with ROC } |z| > \frac{1}{4},$$

we see that Eq. (7.12) implies that

$$y[n] \xleftarrow{\quad z \quad} Y(z) = \frac{\frac{1}{z}}{\frac{1}{z} - \frac{1}{4}}, \quad \text{with ROC } \frac{1}{|z|} > \frac{1}{4}$$

$$= \frac{-4}{z - 4}, \quad \text{with ROC } |z| < 4.$$

Last, we apply the convolution property given in Eq. (7.15) to obtain $X(z)$ and thus write

$$x[n] = w[n] * y[n] \xleftarrow{\quad z \quad} X(z) = W(z)Y(z), \quad \text{with ROC } R_w \cap R_y$$

$$= \frac{2z}{(z - 4)\left(z + \frac{1}{2}\right)^2}, \quad \text{with ROC } \frac{1}{2} < |z| < 4. \quad ■$$

EXAMPLE 7.8 Z-TRANSFORM OF AN EXPONENTIALLY DAMPED COSINE   Use the properties of linearity and multiplication by a complex exponential to find the z-transform of

$$x[n] = a^n \cos(\Omega_o n) u[n],$$

where a is real and positive.

Solution: First we note from Example 7.2 that $y[n] = a^n u[n]$ has the z-transform

$$Y(z) = \frac{1}{1 - az^{-1}}, \quad \text{with ROC } |z| > a.$$

Now we rewrite $x[n]$ as the sum

$$x[n] = \frac{1}{2}e^{j\Omega_o n}y[n] + \frac{1}{2}e^{-j\Omega_o n}y[n]$$

and apply the property of multiplication by a complex exponential given in Eq. (7.14) to each term, obtaining

$$X(z) = \frac{1}{2}Y(e^{-j\Omega_o}z) + \frac{1}{2}Y(e^{j\Omega_o}z), \quad \text{with ROC } |z| > a$$

$$= \frac{1}{2}\frac{1}{1 - ae^{j\Omega_o}z^{-1}} + \frac{1}{2}\frac{1}{1 - ae^{-j\Omega_o}z^{-1}}$$

$$= \frac{1}{2}\left(\frac{1 - ae^{-j\Omega_o}z^{-1} + 1 - ae^{j\Omega_o}z^{-1}}{(1 - ae^{j\Omega_o}z^{-1})(1 - ae^{-j\Omega_o}z^{-1})}\right)$$

$$= \frac{1 - a\cos(\Omega_o)z^{-1}}{1 - 2a\cos(\Omega_o)z^{-1} + a^2z^{-2}}, \quad \text{with ROC } |z| > a. \qquad \blacksquare$$

▶ **Problem 7.3**   Find the z-transform of the following signals:

(a)
$$x[n] = u[n-2] * (2/3)^n u[n]$$

(b)
$$x[n] = \sin(\pi n/8 - \pi/4)u[n-2]$$

(c)
$$x[n] = (n-1)(1/2)^n u[n-1] * (1/3)^n u[n+1]$$

(d)
$$x[n] = (2)^n u[-n-3]$$

Answers:

(a)
$$X(z) = \frac{1}{(z-1)(z-2/3)}, \quad \text{with ROC } |z| > 1$$

(b)
$$X(z) = \frac{z^{-1}\sin(\pi/8)}{z^2 - 2z\cos(\pi/8) + 1}, \quad \text{with ROC } |z| > 1$$

(c)
$$X(z) = \frac{3/4\, z^2}{(z - 1/3)(z - 1/2)^2}, \quad \text{with ROC } |z| > 1/2$$

(d)
$$X(z) = \frac{-z^3}{4(z-2)}, \quad \text{with ROC } |z| < 2 \qquad \blacktriangleleft$$

7.5 *Inversion of the z-Transform*

We now turn our attention to the problem of recovering a time-domain signal from its z-transform. Direct evaluation of the inversion integral defined in Eq. (7.5) requires an understanding of complex variable theory, which is beyond the scope of this book. Hence, two alternative methods for determining inverse z-transforms are presented. The method of partial fractions uses knowledge of several basic z-transform pairs and the z-transform properties to invert a large class of z-transforms. The approach also relies on an important property of the ROC: A right-sided time signal has an ROC that lies outside the pole radius, while a left-sided time signal has an ROC that lies inside the pole radius. The second inversion method expresses $X(z)$ as a power series in z^{-1} of the form Eq. (7.4), so that the values of the signal may be determined by inspection.

■ 7.5.1 PARTIAL-FRACTION EXPANSIONS

In the study of LTI systems, we often encounter z-transforms that are a rational function of z^{-1}. Let

$$
\begin{aligned}
X(z) &= \frac{B(z)}{A(z)} \\
&= \frac{b_0 + b_1 z^{-1} + \cdots + b_M z^{-M}}{a_0 + a_1 z^{-1} + \cdots + a_N z^{-N}},
\end{aligned} \tag{7.17}
$$

and assume that $M < N$. If $M \geq N$, then we may use long division to express $X(z)$ in the form

$$
X(z) = \sum_{k=0}^{M-N} f_k z^{-k} + \frac{\widetilde{B}(z)}{A(z)}.
$$

The numerator polynomial $\widetilde{B}(z)$ now has order one less than that of the denominator polynomial, and the partial-fraction expansion is applied to determine the inverse transform of $\widetilde{B}(z)/A(z)$. The inverse z-transform of the terms in the sum are obtained from the pair $1 \overset{z}{\longleftrightarrow} \delta[n]$ and the time-shift property.

In some problems, $X(z)$ may be expressed as a ratio of polynomials in z rather than z^{-1}. In this case, we may use the partial-fraction expansion method described here if we first convert $X(z)$ to a ratio of polynomials in z^{-1} as described by Eq. (7.17). This conversion is accomplished by factoring out the highest power of z present in the numerator from the numerator polynomial and the term with the highest power of z present in the denominator, an operation which ensures that the remainder has the form described by Eq. (7.17). For example, if

$$
X(z) = \frac{2z^2 - 2z + 10}{3z^3 - 6z + 9},
$$

then we factor z^2 from the numerator and $3z^3$ from the denominator and thus write

$$
\begin{aligned}
X(z) &= \frac{z^2}{3z^3}\left(\frac{2 - 2z^{-1} + 10z^{-2}}{1 - 2z^{-2} + 3z^{-3}} \right) \\
&= \frac{1}{3}z^{-1}\left(\frac{2 - 2z^{-1} + 10z^{-2}}{1 - 2z^{-2} + 3z^{-3}} \right).
\end{aligned}
$$

The partial-fraction expansion is applied to the term in parentheses, and the factor $(1/3)z^{-1}$ is incorporated later, using the time-shift property given in Eq. (7.13).

The partial-fraction expansion of Eq. (7.17) is obtained by factoring the denominator polynomial into a product of first-order terms. The result is

$$X(z) = \frac{b_0 + b_1 z^{-1} + \cdots + b_M z^{-M}}{a_0 \prod_{k=1}^{N}(1 - d_k z^{-1})},$$

where the d_k are the poles of $X(z)$. If none of the poles are repeated, then, using the partial-fraction expansion, we may rewrite $X(z)$ as a sum of first-order terms:

$$\boxed{X(z) = \sum_{k=1}^{N} \frac{A_k}{1 - d_k z^{-1}}.}$$

Depending on the ROC, the inverse z-transform associated with each term is then determined by using the appropriate transform pair. We obtain

$$\boxed{A_k(d_k)^n u[n] \xleftrightarrow{\ z\ } \frac{A_k}{1 - d_k z^{-1}}, \quad \text{with ROC } |z| > d_k}$$

or

$$\boxed{-A_k(d_k)^n u[-n-1] \xleftrightarrow{\ z\ } \frac{A_k}{1 - d_k z^{-1}}, \quad \text{with ROC } |z| < d_k.}$$

For each term, the relationship between the ROC associated with $X(z)$ and each pole determines whether the right-sided or left-sided inverse transform is chosen.

If a pole d_i is repeated r times, then there are r terms in the partial-fraction expansion associated with that pole:

$$\frac{A_{i_1}}{1 - d_i z^{-1}}, \frac{A_{i_2}}{(1 - d_i z^{-1})^2}, \dots, \frac{A_{i_r}}{(1 - d_i z^{-1})^r}.$$

Again, the ROC of $X(z)$ determines whether the right- or left-sided inverse transform is chosen. If the ROC is of the form $|z| > d_i$, the right-sided inverse z-transform is chosen:

$$A\frac{(n+1)\cdots(n+m-1)}{(m-1)!}(d_i)^n u[n] \xleftrightarrow{\ z\ } \frac{A}{(1 - d_i z^{-1})^m}, \quad \text{with ROC } |z| > d_i.$$

If, instead, the ROC is of the form $|z| < d_i$, then the left-sided inverse z-transform is chosen:

$$-A\frac{(n+1)\cdots(n+m-1)}{(m-1)!}(d_i)^n u[-n-1] \xleftrightarrow{\ z\ } \frac{A}{(1 - d_i z^{-1})^m}, \quad \text{with ROC } |z| < d_i.$$

The linearity property of Eq. (7.11) indicates that the ROC of $X(z)$ is the intersection of the ROCs associated with the individual terms in the partial-fraction expansion. In order to choose the correct inverse transform, we must infer the ROC of each term from the ROC of $X(z)$. This is accomplished by comparing the location of each pole with the ROC of $X(z)$. If the ROC of $X(z)$ has a radius greater than that of the pole associated with a given term, we choose the right-sided inverse transform. If the ROC of $X(z)$ has a radius less than that of the pole, we choose the left-sided inverse transform for that term. The next example illustrates this procedure.

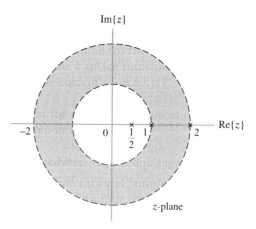

FIGURE 7.12   Locations of poles and ROC for Example 7.9.

EXAMPLE 7.9 INVERSION BY PARTIAL-FRACTION EXPANSION   Find the inverse z-transform of

$$X(z) = \frac{1 - z^{-1} + z^{-2}}{\left(1 - \frac{1}{2}z^{-1}\right)(1 - 2z^{-1})(1 - z^{-1})}, \quad \text{with ROC } 1 < |z| < 2.$$

Solution: We use a partial-fraction expansion to write

$$X(z) = \frac{A_1}{1 - \frac{1}{2}z^{-1}} + \frac{A_2}{1 - 2z^{-1}} + \frac{A_3}{1 - z^{-1}}.$$

Solving for A_1, A_2, and A_3 gives

$$X(z) = \frac{1}{1 - \frac{1}{2}z^{-1}} + \frac{2}{1 - 2z^{-1}} - \frac{2}{1 - z^{-1}}.$$

Now we find the inverse z-transform of each term, using the relationship between the locations of the poles and the ROC of $X(z)$, each of which is depicted in Fig. 7.12. The figure shows that the ROC has a radius greater than the pole at $z = \frac{1}{2}$, so this term has the right-sided inverse transform

$$\left(\frac{1}{2}\right)^n u[n] \overset{z}{\longleftrightarrow} \frac{1}{1 - \frac{1}{2}z^{-1}}.$$

The ROC also has a radius less than the pole at $z = 2$, so this term has the left-sided inverse transform

$$-2(2)^n u[-n - 1] \overset{z}{\longleftrightarrow} \frac{2}{1 - 2z^{-1}}.$$

Finally, the ROC has a radius greater than the pole at $z = 1$, so this term has the right-sided inverse z-transform

$$-2u[n] \overset{z}{\longleftrightarrow} -\frac{2}{1 - z^{-1}}.$$

Combining the individual terms gives

$$x[n] = \left(\frac{1}{2}\right)^n u[n] - 2(2)^n u[-n - 1] - 2u[n]. \qquad ■$$

▶ **Problem 7.4**   Repeat Example (7.9) for the following ROCs:

(a) $\frac{1}{2} < |z| < 1$

(b) $|z| < 1/2$

Answers:

(a)

$$x[n] = \left(\frac{1}{2}\right)^n u[n] - 2(2)^n u[-n-1] + 2u[-n-1]$$

(b)

$$x[n] = -\left(\frac{1}{2}\right)^n u[-n-1] - 2(2)^n u[-n-1] + 2u[-n-1]$$ ◀

EXAMPLE 7.10   INVERSION OF AN IMPROPER RATIONAL FUNCTION   Find the inverse z-transform of

$$X(z) = \frac{z^3 - 10z^2 - 4z + 4}{2z^2 - 2z - 4}, \quad \text{with ROC } |z| < 1.$$

Solution: The poles at $z = -1$ and $z = 2$ are found by determining the roots of the denominator polynomial. The ROC and pole locations in the z-plane are depicted in Fig. 7.13. We convert $X(z)$ into a ratio of polynomials in z^{-1} in accordance with Eq. (7.17). We do this by factoring z^3 from the numerator and $2z^2$ from the denominator, yielding

$$X(z) = \frac{1}{2}z\left(\frac{1 - 10z^{-1} - 4z^{-2} + 4z^{-3}}{1 - z^{-1} - 2z^{-2}}\right).$$

The factor $\frac{1}{2}z$ is easily incorporated later by using the time-shift property, so we focus on the ratio of polynomials in parentheses. Using long division to reduce the order of the numerator polynomial, we have

$$
\begin{array}{r}
-2z^{-1} + 3 \\
-2z^{-2} - z^{-1} + 1 \overline{)\ 4z^{-3} - 4z^{-2} - 10z^{-1} + 1} \\
4z^{-3} + 2z^{-2} -\ 2z^{-1} \\
\hline
-6z^{-2} -\ 8z^{-1} + 1 \\
-6z^{-2} -\ 3z^{-1} + 3 \\
\hline
-\ 5z^{-1} - 2
\end{array}
$$

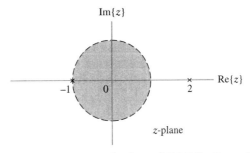

FIGURE 7.13   Locations of poles and ROC for Example 7.10.

Thus, we may write

$$\frac{1 - 10z^{-1} - 4z^{-2} + 4z^{-3}}{1 - z^{-1} - 2z^{-2}} = -2z^{-1} + 3 + \frac{-5z^{-1} - 2}{1 - z^{-1} - 2z^{-2}}$$

$$= -2z^{-1} + 3 + \frac{-5z^{-1} - 2}{(1 + z^{-1})(1 - 2z^{-1})}.$$

Next, using a partial-fraction expansion, we have

$$\frac{-5z^{-1} - 2}{(1 + z^{-1})(1 - 2z^{-1})} = \frac{1}{1 + z^{-1}} - \frac{3}{1 - 2z^{-1}}$$

and thus define

$$X(z) = \frac{1}{2}zW(z), \tag{7.18}$$

where

$$W(z) = -2z^{-1} + 3 + \frac{1}{1 + z^{-1}} - \frac{3}{1 - 2z^{-1}}, \quad \text{with ROC } |z| < 1.$$

The ROC has a smaller radius than either pole, as shown in Fig. 7.13, so the inverse z-transform of $W(z)$ is

$$w[n] = -2\delta[n - 1] + 3\delta[n] - (-1)^n u[-n - 1] + 3(2)^n u[-n - 1].$$

Finally, we apply the time-shift property (Eq. (7.13)) to Eq. (7.18) to obtain

$$x[n] = \frac{1}{2}w[n + 1]$$

and thus

$$x[n] = -\delta[n] + \frac{3}{2}\delta[n + 1] - \frac{1}{2}(-1)^{n+1}u[-n - 2] + 3(2)^n u[-n - 2]. \quad ■$$

▶ **Problem 7.5**   Find the time-domain signals corresponding to the following z-transforms:

(a)

$$X(z) = \frac{(1/4)z^{-1}}{(1 - (1/2)z^{-1})(1 - (1/4)z^{-1})}, \quad \text{with ROC } 1/4 < |z| < 1/2$$

(b)

$$X(z) = \frac{16z^2 - 2z + 1}{8z^2 + 2z - 1}, \quad \text{with ROC } |z| > \frac{1}{2}$$

(c)

$$X(z) = \frac{2z^3 + 2z^2 + 3z + 1}{2z^4 + 3z^3 + z^2}, \quad \text{with ROC } |z| > 1$$

Answers:

(a)

$$x[n] = -(1/4)^n u[n] - (1/2)^n u[-n - 1]$$

(b)

$$x[n] = -\delta[n] + \left(\frac{1}{4}\right)^n u[n] + 2\left(\frac{-1}{2}\right)^n u[n]$$

(c)

$$x[n] = \delta[n - 2] + 2(-1)^{n-1}u[n - 1] - (-1/2)^{n-1}u[n - 1] \qquad ◀$$

The method of partial fractions also applies when the poles are complex valued. In this case, the expansion coefficients will generally be complex valued. However, if the coefficients in $X(z)$ are real valued, then the expansion coefficients corresponding to complex-conjugate poles will be complex conjugates of each other.

Note that information other than the ROC can be used to establish a unique inverse transform. For example, causality, stability, or the existence of the DTFT is sufficient to determine the inverse transform.

▶ If a signal is known to be causal, then right-sided inverse transforms are chosen.

▶ If a signal is stable, then it is absolutely summable and has a DTFT. Hence, stability and the existence of the DTFT are equivalent conditions. In both cases, the ROC includes the unit circle in the z-plane, $|z| = 1$. The inverse z-transform is determined by comparing the locations of the poles with the unit circle. If a pole is inside the unit circle, then the right-sided inverse z-transform is chosen; if a pole is outside the unit circle, then the left-sided inverse z-transform is chosen.

▶ **Problem 7.6**   Find the inverse z-transform of

$$X(z) = \frac{1}{1 - \frac{1}{2}z^{-1}} + \frac{2}{1 - 2z^{-1}},$$

assuming that (a) the signal is causal and (b) the signal has a DTFT.

Answers:

(a)

$$x[n] = \left(\frac{1}{2}\right)^n u[n] + 2(2)^n u[n]$$

(b)

$$x[n] = \left(\frac{1}{2}\right)^n u[n] - 2(2)^n u[-n-1] \qquad ◀$$

▥ 7.5.2 Power Series Expansion

We now seek to express $X(z)$ as a power series in z^{-1} or z of the form defined in Eq. (7.4). The values of the signal $x[n]$ are then given by the coefficient associated with z^{-n}. This inversion method is limited to signals that are one sided—that is, discrete-time signals with ROCs of the form $|z| < a$ or $|z| > a$. If the ROC is $|z| > a$, then we express $X(z)$ as a power series in z^{-1}, so that we obtain a right-sided signal. If the ROC is $|z| < a$, then we express $X(z)$ as a power series in z and obtain a left-sided inverse transform.

EXAMPLE 7.11   Inversion by Means of Long Division   Find the inverse z-transform of

$$X(z) = \frac{2 + z^{-1}}{1 - \frac{1}{2}z^{-1}}, \quad \text{with ROC } |z| > \frac{1}{2},$$

using a power series expansion.

Solution: We use long division to write $X(z)$ as a power series in z^{-1}, since the ROC indicates that $x[n]$ is right sided. We have

$$
\begin{array}{r}
2 + 2z^{-1} + z^{-2} + \tfrac{1}{2}z^{-3} + \cdots \\[2pt]
1 - \tfrac{1}{2}z^{-1} \,)\overline{2 + z^{-1}} \\[2pt]
\underline{2 - z^{-1}} \\[2pt]
2z^{-1} \\[2pt]
\underline{2z^{-1} - z^{-2}} \\[2pt]
z^{-2} \\[2pt]
\underline{z^{-2} - \tfrac{1}{2}z^{-3}} \\[2pt]
\tfrac{1}{2}z^{-3}
\end{array}
$$

That is,

$$
X(z) = 2 + 2z^{-1} + z^{-2} + \frac{1}{2}z^{-3} + \cdots.
$$

Thus, comparing $X(z)$ with Eq. (7.4), we obtain

$$
x[n] = 2\delta[n] + 2\delta[n-1] + \delta[n-2] + \tfrac{1}{2}\delta[n-3] + \cdots.
$$

If the ROC is changed to $|z| < \tfrac{1}{2}$, then we expand $X(z)$ as a power series in z:

$$
\begin{array}{r}
-2 - 8z - 16z^2 - 32z^3 + \cdots \\[2pt]
-\tfrac{1}{2}z^{-1} + 1 \,)\overline{z^{-1} + 2 } \\[2pt]
\underline{z^{-1} - 2} \\[2pt]
4 \\[2pt]
\underline{4 - 8z} \\[2pt]
8z \\[2pt]
\underline{8z - 16z^2} \\[2pt]
16z^2
\end{array}
$$

That is,

$$
X(z) = -2 - 8z - 16z^2 - 32z^3 - \cdots.
$$

In this case, we therefore have

$$
x[n] = -2\delta[n] - 8\delta[n+1] - 16\delta[n+2] - 32\delta[n+3] - \cdots. \qquad ■
$$

Long division may be used to obtain the power series whenever $X(z)$ is a ratio of polynomials and long division is simple to perform. However, long division may not lead to a closed-form expression for $x[n]$.

An advantage of the power series approach is the ability to find inverse z-transforms for signals that are not a ratio of polynomials in z. This is illustrated in the next example.

EXAMPLE 7.12 INVERSION VIA A POWER SERIES EXPANSION   Find the inverse z-transform of

$$
X(z) = e^{z^2}, \quad \text{with ROC all } z \text{ except } |z| = \infty.
$$

Solution: Using the power series representation for e^a, viz.,

$$
e^a = \sum_{k=0}^{\infty} \frac{a^k}{k!},
$$

we write

$$X(z) = \sum_{k=0}^{\infty} \frac{(z^2)^k}{k!}$$

$$= \sum_{k=0}^{\infty} \frac{z^{2k}}{k!}.$$

Thus,

$$x[n] = \begin{cases} 0, & n > 0 \text{ or } n \text{ odd} \\ \dfrac{1}{\left(\frac{-n}{2}\right)!}, & \text{otherwise} \end{cases} .$$

$\blacksquare$

▶ **Problem 7.7**   Use the power series approach to find the inverse z-transform of $X(z) = \cos(2z^{-1})$, with ROC all z except $z = 0$.

Answer:

$$x[n] = \begin{cases} 0, & n < 0 \text{ or } n \text{ odd} \\ (-1)^{n/2}2^n/(n!), & \text{otherwise} \end{cases}$$

◀

7.6 *The Transfer Function*

In this section, we examine the relationship between the transfer function and input—output descriptions of LTI discrete-time systems. In Section 7.2, we defined the transfer function as the z-transform of the impulse response. The output $y[n]$ of an LTI system may be expressed as the convolution of the impulse response $h[n]$ and the input $x[n]$:

$$y[n] = h[n] * x[n].$$

If we take the z-transform of both sides of this equation, then we may express the output $Y(z)$ as the product of the transfer function $H(z)$ and the transformed input $X(z)$:

$$Y(z) = H(z)X(z). \tag{7.19}$$

Thus, the z-transform has converted convolution of time sequences into multiplication of z-transforms, and we see that the transfer function offers yet another description of the input–output characteristics of a system.

Note that Eq. (7.19) implies that the transfer function may also be viewed as the ratio of the z-transform of the ouput to that of the input; that is,

$$\boxed{H(z) = \frac{Y(z)}{X(z)}.} \tag{7.20}$$

This definition applies at all z in the ROCs of $X(z)$ and $Y(z)$ for which $X(z)$ is nonzero.

The impulse response is the inverse z-transform of the transfer function. In order to uniquely determine the impulse response from the transfer function, we must know the ROC. If the ROC is not known, then other system characteristics, such as stability or causality, must be known in order to uniquely determine the impulse response.

EXAMPLE 7.13 SYSTEM IDENTIFICATION   The problem of finding the system description from knowledge of the input and output is known as system identification. Find the transfer function and impulse response of a causal LTI system if the input to the system is

$$x[n] = (-1/3)^n u[n]$$

and the output is

$$y[n] = 3(-1)^n u[n] + (1/3)^n u[n].$$

Solution: The z-transforms of the input and output are respectively given by

$$X(z) = \frac{1}{(1 + (1/3)z^{-1})}, \quad \text{with ROC } |z| > 1/3,$$

and

$$Y(z) = \frac{3}{1 + z^{-1}} + \frac{1}{1 - (1/3)z^{-1}}$$

$$= \frac{4}{(1 + z^{-1})(1 - (1/3)z^{-1})}, \quad \text{with ROC } |z| > 1.$$

We apply Eq. (7.20) to obtain the transfer function:

$$H(z) = \frac{4(1 + (1/3)z^{-1})}{(1 + z^{-1})(1 - (1/3)z^{-1})}, \quad \text{with ROC } |z| > 1.$$

The impulse response of the system is obtained by finding the inverse z-transform of $H(z)$. Applying a partial fraction expansion to $H(z)$ yields

$$H(z) = \frac{2}{1 + z^{-1}} + \frac{2}{1 - (1/3)z^{-1}}, \quad \text{with ROC } |z| > 1.$$

The impulse response is thus given by

$$h[n] = 2(-1)^n u[n] + 2(1/3)^n u[n]. \qquad ■$$

▶ **Problem 7.8**   An LTI system has impulse response $h[n] = (1/2)^n u[n]$. Determine the input to the system if the output is given by $y[n] = (1/2)^n u[n] + (-1/2)^n u[n]$.

Answer:

$$x[n] = 2(-1/2)^n u[n] \qquad ◀$$

■ 7.6.1 RELATING THE TRANSFER FUNCTION AND THE DIFFERENCE EQUATION

The transfer function may be obtained directly from the difference-equation description of an LTI system. Recall that an Nth-order difference equation relates the input $x[n]$ to the output $y[n]$:

$$\sum_{k=0}^{N} a_k y[n - k] = \sum_{k=0}^{M} b_k x[n - k].$$

In Section 7.2, we showed that the transfer function $H(z)$ is an eigenvalue of the system associated with the eigenfunction z^n. That is, if $x[n] = z^n$, then the output of an LTI system is $y[n] = z^n H(z)$. Substituting $x[n-k] = z^{n-k}$ and $y[n-k] = z^{n-k}H(z)$ into the difference equation gives the relationship

$$z^n \sum_{k=0}^{N} a_k z^{-k} H(z) = z^n \sum_{k=0}^{M} b_k z^{-k}.$$

We may now solve for $H(z)$:

$$H(z) = \frac{\sum_{k=0}^{M} b_k z^{-k}}{\sum_{k=0}^{N} a_k z^{-k}}. \tag{7.21}$$

The transfer function of an LTI system described by a difference equation is a ratio of polynomials in z^{-1} and is thus termed a *rational transfer function*. The coefficient of z^{-k} in the numerator polynomial is the coefficient associated with $x[n-k]$ in the difference equation. The coefficient of z^{-k} in the denominator polynomial is the coefficient associated with $y[n-k]$ in the difference equation. This correspondence allows us not only to find the transfer function, given the difference equation, but also to find a difference-equation description for a system, given a rational transfer function.

EXAMPLE 7.14 FINDING THE TRANSFER FUNCTION AND IMPULSE RESPONSE   Determine the transfer function and the impulse response for the causal LTI system described by the difference equation

$$y[n] - (1/4)y[n-1] - (3/8)y[n-2] = -x[n] + 2x[n-1].$$

Solution: We obtain the transfer function by applying Eq. (7.21):

$$H(z) = \frac{-1 + 2z^{-1}}{1 - (1/4)z^{-1} - (3/8)z^{-2}}.$$

The impulse response is found by identifying the inverse z-transform of $H(z)$. Applying a partial-fraction expansion to $H(z)$ gives

$$H(z) = \frac{-2}{1 + (1/2)z^{-1}} + \frac{1}{1 - (3/4)z^{-1}}.$$

The system is causal, so we choose the right-sided inverse z-transform for each term to obtain the following impulse response:

$$h[n] = -2(-1/2)^n u[n] + (3/4)^n u[n].$$ ∎

EXAMPLE 7.15 FINDING A DIFFERENCE-EQUATION DESCRIPTION   Find the difference-equation description of an LTI system with transfer function

$$H(z) = \frac{5z + 2}{z^2 + 3z + 2}.$$

Solution: We rewrite $H(z)$ as a ratio of polynomials in z^{-1}. Dividing both the numerator and denominator by z^2, we obtain

$$H(z) = \frac{5z^{-1} + 2z^{-2}}{1 + 3z^{-1} + 2z^{-2}}.$$

Comparing this transfer function with Eq. (7.21), we conclude that $M = 2$, $N = 2$, $b_0 = 0$, $b_1 = 5$, $b_2 = 2$, $a_0 = 1$, $a_1 = 3$, and $a_2 = 2$. Hence, this system is described by the difference equation

$$y[n] + 3y[n-1] + 2y[n-2] = 5x[n-1] + 2x[n-2].$$ ■

▶ **Problem 7.9**   Determine the transfer function and a difference-equation representation of an LTI system described by the impulse response

$$h[n] = (1/3)^n u[n] + (1/2)^{n-2} u[n-1].$$

Answer:

$$H(z) = \frac{1 + (3/2)z^{-1} - (2/3)z^{-2}}{1 - (5/6)z^{-1} + (1/6)z^{-2}};$$

$$y[n] - (5/6)y[n-1] + (1/6)y[n-2] = x[n] + (3/2)x[n-1] - (2/3)x[n-2]$$ ◀

The poles and zeros of a rational transfer function offer much insight into LTI system characteristics, as we shall see in the sections that follow. The transfer function is expressed in pole–zero form by factoring the numerator and denominator polynomials in Eq. (7.21). To proceed we write

$$H(z) = \frac{\tilde{b}\prod_{k=1}^{M}(1 - c_k z^{-1})}{\prod_{k=1}^{N}(1 - d_k z^{-1})}, \tag{7.22}$$

where the c_k and the d_k are, respectively, the zeros and poles of the system and $\tilde{b} = b_0/a_0$ is the gain factor. This form assumes that there are no poles or zeros at $z = 0$. A pth-order pole at $z = 0$ occurs when $b_0 = b_1 = \cdots = b_{p-1} = 0$, while an lth order zero at $z = 0$ occurs when $a_0 = a_1 = \cdots = a_{l-1} = 0$. In this case, we write

$$H(z) = \frac{\tilde{b}z^{-p}\prod_{k=1}^{M-p}(1 - c_k z^{-1})}{z^{-l}\prod_{k=1}^{N-l}(1 - d_k z^{-1})}, \tag{7.23}$$

where $\tilde{b} = b_p/a_l$. The system in Example 7.15 had a first-order pole at $z = 0$. The system's poles, zeros, and gain factor $\tilde{b}$ uniquely determine the transfer function and thus provide another description for the input–output behavior of the system. Note that the poles of the system are the roots of the characteristic equation defined in Section 2.10.

7.7 Causality and Stability

The impulse response of a causal LTI system is zero for $n < 0$. Therefore, the impulse response of a causal LTI system is determined from the transfer function by using right-sided inverse transforms. A pole that is inside the unit circle in the z-plane (i.e., $|d_k| < 1$) contributes an exponentially decaying term to the impulse response, while a pole that is outside the unit circle (i.e., $|d_k| > 1$) contributes an exponentially increasing term. These relationships are illustrated in Fig. 7.14. A pole on the unit circle contributes a complex sinusoid.

Alternatively, if a system is stable, then the impulse response is absolutely summable and the DTFT of the impulse response exists. It follows that the ROC must include the unit circle in the z-plane. Hence, the relationship between the location of a pole and the unit circle determines the component of the impulse response associated with that pole. A pole inside the unit circle contributes a right-sided decaying exponential term to the impulse response, while a pole outside the unit circle contributes a left-sided decaying exponential term to the impulse response, as depicted in Fig. 7.15. Note that a stable impulse response cannot contain any increasing exponential or sinusoidal terms, since then the impulse response is not absolutely summable.

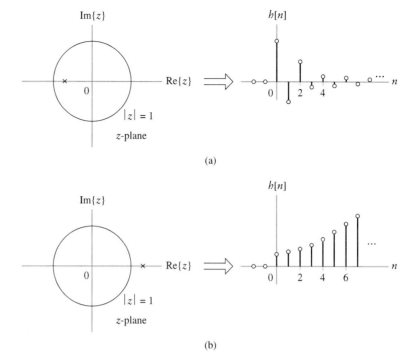

(a)

(b)

FIGURE 7.14   The relationship between the location of a pole and the impulse response characteristics for a causal system. (a) A pole inside the unit circle contributes an exponentially decaying
term to the impulse response. (b) A pole outside the unit circle contributes an exponentially increasing term to the impulse response.

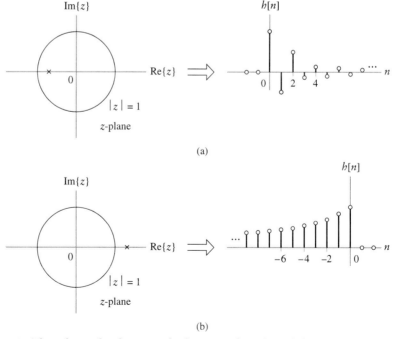

(a)

(b)

FIGURE 7.15   The relationship between the location of a pole and the impulse response characteristics for a stable system. (a) A pole inside the unit circle contributes a right-sided term to the impulse response. (b) A pole outside the unit circle contributes a left-sided term to the impulse response.

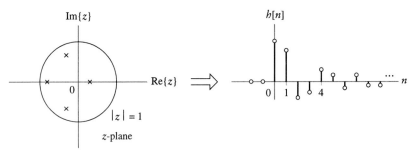

FIGURE 7.16   A system that is both stable and causal must have all its poles inside the unit circle in the z-plane, as illustrated here.

LTI systems that are both stable and causal must have all their poles inside the unit circle. A pole that is inside the unit circle in the z-plane contributes a right-sided, or causal, decaying exponential term to the impulse response. We cannot have a pole outside the unit circle, since the inverse transform of a pole located outside the circle will contribute either a right-sided increasing exponential term, which is not stable, or a left-sided decaying exponential term, which is not causal. Also, a pole on the unit circle contributes a complex sinusoidal term, which is not stable. An example of a stable and causal LTI system is depicted in Fig. 7.16.

EXAMPLE 7.16 CAUSALITY AND STABILITY   An LTI system has the transfer function

$$H(z) = \frac{2}{1 - 0.9e^{j\frac{\pi}{4}}z^{-1}} + \frac{2}{1 - 0.9e^{-j\frac{\pi}{4}}z^{-1}} + \frac{3}{1 + 2z^{-1}}.$$

Find the impulse response, assuming that the system is (a) stable or (b) causal. Can this system be both stable and causal?

Solution: The given system has poles at $z = 0.9e^{j\frac{\pi}{4}}$, $z = 0.9e^{-j\frac{\pi}{4}}$, and $z = -2$, depicted in Fig. 7.17. If the system is stable, then the ROC includes the unit circle. The two poles inside the unit circle contribute right-sided terms to the impulse response, while the pole outside the unit circle contributes a left-sided term. Hence, for case (a),

$$h[n] = 2\left(0.9e^{j\frac{\pi}{4}}\right)^n u[n] + 2\left(0.9e^{-j\frac{\pi}{4}}\right)^n u[n] - 3(-2)^n u[-n - 1]$$

$$= 4(0.9)^n \cos\left(\frac{\pi}{4}n\right)u[n] - 3(-2)^n u[-n - 1].$$

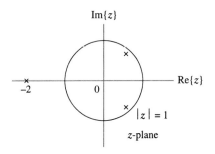

FIGURE 7.17   Locations of poles in the z-plane for the system in Example 7.16.

If the system is assumed causal, then all poles contribute right-sided terms to the impulse response, so for case (b), we have

$$h[n] = 2\left(0.9e^{j\frac{\pi}{4}}\right)^n u[n] + 2\left(0.9e^{-j\frac{\pi}{4}}\right)^n u[n] + 3(-2)^n u[n]$$

$$= 4(0.9)^n \cos\left(\frac{\pi}{4}n\right)u[n] + 3(-2)^n u[n].$$

Note that this LTI system cannot be both stable and causal, since there is a pole outside the unit circle.   ∎

EXAMPLE 7.17 FIRST-ORDER RECURSIVE SYSTEM: INVESTMENT COMPUTATION   In Example 2.5, we showed that the first-order recursive equation

$$y[n] - \rho y[n - 1] = x[n]$$

may be used to describe the value $y[n]$ of an investment by setting $\rho = 1 + r/100$, where r is the interest rate per period, expressed in percent. Find the transfer function of this system and determine whether it can be both stable and causal.

Solution: The transfer function is determined by substituting into Eq. (7.21) to obtain

$$H(z) = \frac{1}{1 - \rho z^{-1}}.$$

This LTI system cannot be both stable and causal, because the pole at $z = \rho$ is outside the unit circle.   ∎

▶ **Problem 7.10**   A stable and causal LTI system is described by the difference equation

$$y[n] + \frac{1}{4}y[n - 1] - \frac{1}{8}y[n - 2] = -2x[n] + \frac{5}{4}x[n - 1].$$

Find the system impulse response.

Answer: The system impulse response is

$$h[n] = \left(\frac{1}{4}\right)^n u[n] - 3\left(-\frac{1}{2}\right)^n u[n]$$

◀

▪ 7.7.1 INVERSE SYSTEMS

Recall from Section 2.7.4 that the impulse response of an inverse system, $h^{inv}[n]$, satisfies

$$h^{inv}[n] * h[n] = \delta[n],$$

where $h[n]$ is the impulse response of the system to be inverted. Taking the z-transform of both sides of this equation, we find that the transfer function of the inverse system must satisfy

$$H^{inv}(z)H(z) = 1.$$

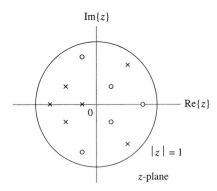

FIGURE 7.18 A system that has a causal and stable inverse must have all its poles and zeros inside the unit circle, as illustrated here.

That is,

$$H^{\text{inv}}(z) = \frac{1}{H(z)}.$$

Thus, the transfer function of an LTI inverse system is the inverse of the transfer function of the system that we desire to invert. If $H(z)$ is written in the pole–zero form shown in Eq. (7.23), then

$$H^{\text{inv}}(z) = \frac{z^{-l}\prod_{k=1}^{N-l}(1 - d_k z^{-1})}{\widetilde{b}z^{-p}\prod_{k=1}^{M-p}(1 - c_k z^{-1})}. \tag{7.24}$$

The zeros of $H(z)$ are the poles of $H^{\text{inv}}(z)$, and the poles of $H(z)$ are the zeros of $H^{\text{inv}}(z)$. Any system described by a rational transfer function has an inverse system of this form.

We are often interested in inverse systems that are both stable and causal, so we can implement a system $H^{\text{inv}}(z)$ that reverses the distortion introduced by $H(z)$ to a signal of interest. $H^{\text{inv}}(z)$ is both stable and causal if all of its poles are inside the unit circle. Since the poles of $H^{\text{inv}}(z)$ are the zeros of $H(z)$, we conclude that a stable and causal inverse of an LTI system $H(z)$ exists if and only if all the zeros of $H(z)$ are inside the unit circle. If $H(z)$ has any zeros outside the unit circle, then a stable and causal inverse system does not exist. A system with all its poles and zeros inside the unit circle, as illustrated in Fig. 7.18, is termed a *minimum-phase* system. As with continuous-time minimum-phase systems, there is a unique relationship between the magnitude and phase responses of a discrete-time minimum-phase system. That is, the phase response of a minimum-phase system is uniquely determined by the magnitude response. Alternatively, the magnitude response of a minimum-phase system is uniquely determined by the phase response.

EXAMPLE 7.18 A STABLE AND CAUSAL INVERSE SYSTEM   An LTI system is described by the difference equation

$$y[n] - y[n-1] + \frac{1}{4}y[n-2] = x[n] + \frac{1}{4}x[n-1] - \frac{1}{8}x[n-2].$$

Find the transfer function of the inverse system. Does a stable and causal LTI inverse system exist?

Solution: We find the transfer function of the given system by applying Eq. (7.21) to obtain

$$H(z) = \frac{1 + \frac{1}{4}z^{-1} - \frac{1}{8}z^{-2}}{1 - z^{-1} + \frac{1}{4}z^{-2}}$$

$$= \frac{\left(1 - \frac{1}{4}z^{-1}\right)\left(1 + \frac{1}{2}z^{-1}\right)}{\left(1 - \frac{1}{2}z^{-1}\right)^2}.$$

The inverse system then has the transfer function

$$H^{\text{inv}}(z) = \frac{\left(1 - \frac{1}{2}z^{-1}\right)^2}{\left(1 - \frac{1}{4}z^{-1}\right)\left(1 + \frac{1}{2}z^{-1}\right)}.$$

The poles of the inverse system are at $z = \frac{1}{4}$ and $z = -\frac{1}{2}$. Both of these poles are inside the unit circle, and therefore the inverse system can be both stable and causal. Note that this system is also minimum phase, since the double zero at $z = 1/2$ is located inside the unit circle. ∎

EXAMPLE 7.19 MULTIPATH COMMUNICATION CHANNEL: INVERSE SYSTEM Recall from Section 1.10 that a discrete-time LTI model for a two-path communication channel is

$$y[n] = x[n] + ax[n - 1].$$

Find the transfer function and difference-equation description of the inverse system. What must the parameter a satisfy for the inverse system to be stable and causal?

Solution: We use Eq. (7.21) to obtain the transfer function of the multipath system:

$$H(z) = 1 + az^{-1}.$$

Hence, the inverse system has transfer function

$$H^{\text{inv}}(z) = \frac{1}{H(z)} = \frac{1}{1 + az^{-1}},$$

which satisfies the difference-equation description

$$y[n] + ay[n - 1] = x[n].$$

The inverse system is both stable and causal when $|a| < 1$. ∎

▶ **Problem 7.11** An LTI system has the impulse response

$$h[n] = 2\delta[n] + \frac{5}{2}\left(\frac{1}{2}\right)^n u[n] - \frac{7}{2}\left(\frac{-1}{4}\right)^n u[n].$$

Find the transfer function of the inverse system. Does a stable and causal inverse system exist?

Answer:

$$H^{\text{inv}}(z) = \frac{\left(1 - \frac{1}{2}z^{-1}\right)\left(1 + \frac{1}{4}z^{-1}\right)}{\left(1 - \frac{1}{8}z^{-1}\right)(1 + 2z^{-1})}$$

The inverse system cannot be both stable and causal ◀

▶ **Problem 7.12**   Determine whether each of the following LTI systems is (i) causal and stable and (ii) minimum phase.

(a)

$$H(z) = \frac{1 + 2z^{-1}}{1 + (14/8)z^{-1} + (49/64)z^{-2}}.$$

(b)

$$y[n] - (6/5)y[n - 1] - (16/25)y[n - 2] = 2x[n] + x[n - 1].$$

Answers:

(a) (i) stable and causal; (ii) non minimum phase

(b) (i) not stable and causal; (ii) non minimum phase   ◀

7.8 Determining the Frequency Response from Poles and Zeros

We now explore the relationship between the locations of poles and zeros in the z-plane and the frequency response of the system. Recall that the frequency response is obtained from the transfer function by substituting $e^{j\Omega}$ for z in $H(z)$. That is, the frequency response corresponds to the transfer function evaluated on the unit circle in the z-plane. This assumes that the ROC includes the unit circle. Substituting $z = e^{j\Omega}$ into Eq. (7.23) gives

$$H(e^{j\Omega}) = \frac{\widetilde{b}e^{-jp\Omega} \prod_{k=1}^{M-p}(1 - c_k e^{-j\Omega})}{e^{-jl\Omega} \prod_{k=1}^{N-l}(1 - d_k e^{-j\Omega})}.$$

We rewrite $H(e^{j\Omega})$ in terms of positive powers of $e^{j\Omega}$ by multiplying both the numerator and denominator by $e^{jN\Omega}$ to obtain

$$H(e^{j\Omega}) = \frac{\widetilde{b}e^{j(N-M)\Omega} \prod_{k=1}^{M-p}(e^{j\Omega} - c_k)}{\prod_{k=1}^{N-l}(e^{j\Omega} - d_k)}. \tag{7.25}$$

We shall examine both the magnitude and phase of $H(e^{j\Omega})$, using Eq. (7.25).

The magnitude of $H(e^{j\Omega})$ at some fixed value of Ω, say, Ω_o, is defined by

$$|H(e^{j\Omega_o})| = \frac{|\widetilde{b}| \prod_{k=1}^{M-p}|e^{j\Omega_o} - c_k|}{\prod_{k=1}^{N-l}|e^{j\Omega_o} - d_k|}.$$

This expression involves a ratio of products of terms of the form $|e^{j\Omega_o} - g|$, where g represents either a pole or a zero. The terms involving zeros are in the numerator, while those involving poles are in the denominator. If we use vectors to represent complex numbers in the z-plane, then $e^{j\Omega_o}$ is a vector from the origin to the point $e^{j\Omega_o}$ and g is a vector from the origin to g. Hence, $e^{j\Omega_o} - g$ is represented as a vector from the point g to the point $e^{j\Omega_o}$, as illustrated in Fig. 7.19. The length of this vector is $|e^{j\Omega_o} - g|$. We assess the contribution of each pole and zero to the overall frequency response by examining $|e^{j\Omega_o} - g|$ as Ω_o changes.

Figure 7.20(a) depicts the vector $e^{j\Omega} - g$ for several different values of Ω, while Fig. 7.20(b) depicts $|e^{j\Omega} - g|$ as a continuous function of frequency. Note that if $\Omega = \arg\{g\}$, then $|e^{j\Omega} - g|$ attains its minimum value of $1 - |g|$ when g is inside the unit circle and takes on the value $|g| - 1$ when g is outside the unit circle. Hence, if g is close to the unit circle ($|g| \approx 1$), then $|e^{j\Omega} - g|$ becomes very small when $\Omega = \arg\{g\}$.

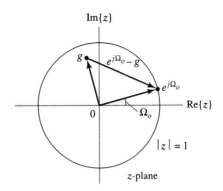

FIGURE 7.19 Vector interpretation of $e^{j\Omega_o} - g$ in the z-plane.

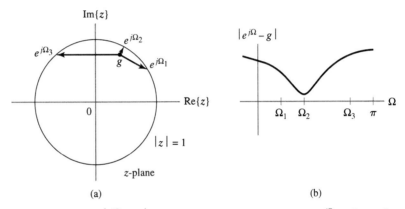

(a)   (b)

FIGURE 7.20 The quantity $|e^{j\Omega} - g|$ is the length of a vector from g to $e^{j\Omega}$ in the z-plane. (a) Vectors from g to $e^{j\Omega}$ at several frequencies. (b) The function $|e^{j\Omega} - g|$.

If g represents a zero, then $|e^{j\Omega} - g|$ contributes to the numerator of $|H(e^{j\Omega})|$. Thus, at frequencies near $\arg\{g\}$, $|H(e^{j\Omega})|$ tends to have a minimum. How far $|H(e^{j\Omega})|$ decreases depends on how close the zero is to the unit circle; if the zero is on the unit circle, then $|H(e^{j\Omega})|$ goes to zero at the frequency corresponding to the zero. On the other hand, if g represents a pole, then $|e^{j\Omega} - g|$ contributes to the denominator of $|H(e^{j\Omega})|$. When $|e^{j\Omega} - g|$ decreases, $|H(e^{j\Omega})|$ will increase, with the size of the increase dependent on how far the pole is from the unit circle. A pole that is very close to the unit circle will cause a large peak in $|H(e^{j\Omega})|$ at the frequency corresponding to the phase angle of the pole. Hence, zeros tend to pull the frequency response magnitude down, while poles tend to push it up.

S & S
Solutions

EXAMPLE 7.20 MULTIPATH COMMUNICATION CHANNEL: MAGNITUDE RESPONSE In Example 7.19, the transfer function of the discrete-time model for a two-path communication system is found to be

$$H(z) = 1 + az^{-1}.$$

Sketch the magnitude response of this system and the corresponding inverse system for $a = 0.5e^{j\pi/4}$, $a = 0.8e^{j\pi/4}$, and $a = 0.95e^{j\pi/4}$.

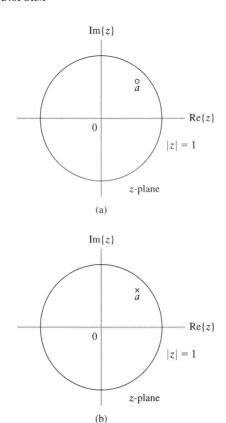

FIGURE 7.21   (a) Location of zero for multipath channel. (b) Location of pole for inverse of the multipath channel.

Solution: The multipath channel has a single zero at $z = a$, while the inverse system has a single pole at $z = a$, as shown in Figs. 7.21(a) and (b). The magnitude responses are sketched in Figs. 7.22(a)–(c). Both the minimum of $|H(e^{j\Omega})|$ and maximum of $|H^{\text{inv}}(e^{j\Omega})|$ occur at the frequency corresponding to the angle of the zero of $H(z)$, namely, $\Omega = \pi/4$. The minimum of $|H(e^{j\Omega})|$ is $1 - |a|$. Hence, as $|a|$ approaches unity, the channel magnitude response at $\Omega = \pi/4$ approaches zero and the two-path channel suppresses any components of the input having frequency $\Omega = \pi/4$. The inverse system maximum occurs at $\Omega = \pi/4$ and is given by $1/(1 - |a|)$. Thus, as $|a|$ approaches unity, the magnitude response of the inverse system approaches infinity. If the multipath channel eliminates the component of the input at frequency $\Omega = \pi/4$, the inverse system cannot restore this component to its original value. Large values of gain in the inverse system are generally undesirable, since noise in the received signal would then be amplified. Furthermore, the inverse system is highly sensitive to small changes in a as $|a|$ approaches unity.   ■

EXAMPLE 7.21 MAGNITUDE RESPONSE FROM POLES AND ZEROS   Sketch the magnitude response for an LTI system having the transfer function

$$H(z) = \frac{1 + z^{-1}}{\left(1 - 0.9e^{j\frac{\pi}{4}}z^{-1}\right)\left(1 - 0.9e^{-j\frac{\pi}{4}}z^{-1}\right)}.$$

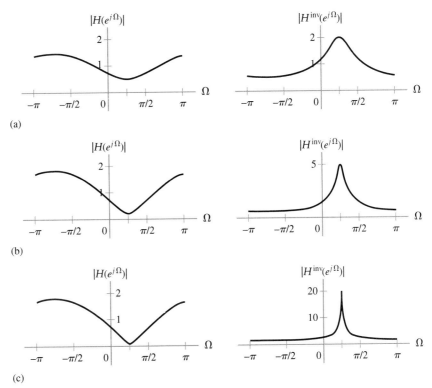

FIGURE 7.22   Magnitude response of multipath channel (left panel) and inverse system (right panel). (a) $a = 0.5e^{j\pi/4}$. (b) $a = 0.8e^{j\pi/4}$. (c) $a = 0.95e^{j\pi/4}$.

Solution: The system has a zero at $z = -1$ and poles at $z = 0.9e^{j\frac{\pi}{4}}$ and $z = 0.9e^{-j\frac{\pi}{4}}$, as depicted in Fig. 7.23(a). Hence, the magnitude response will be zero at $\Omega = \pi$ and will be large at $\Omega = \pm\frac{\pi}{4}$, because the poles are close to the unit circle. Figures 7.23 (b)–(d) depict the component of the magnitude response associated with the zero and each pole. Multiplication of these contributions gives the overall magnitude response sketched in Fig. 7.23(e). ∎

▶ **Problem 7.13**   Sketch the magnitude response of the LTI system with transfer function

$$H(z) = \frac{z - 1}{z + 0.9}.$$

Answer: See Fig. 7.24   ◀

The phase of $H(e^{j\Omega})$ may also be evaluated in terms of the phase associated with each pole and zero. Using Eq. (7.25), we obtain

$$\arg\{H(e^{j\Omega})\} = \arg\{\tilde{b}\} + (N - M)\Omega + \sum_{k=1}^{M-p} \arg\{e^{j\Omega} - c_k\} - \sum_{k=1}^{N-l} \arg\{e^{j\Omega} - d_k\}.$$

The phase of $H(e^{j\Omega})$ involves the sum of the phase angles due to each zero minus the phase angle due to each pole. The first term, $\arg\{\tilde{b}\}$, is independent of frequency. The phase

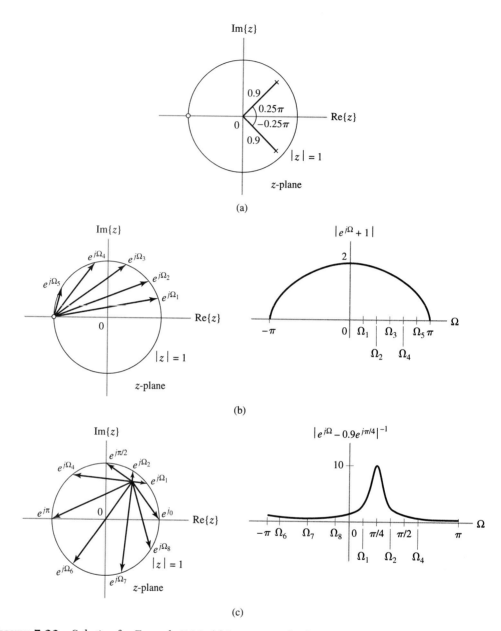

FIGURE 7.23 Solution for Example 7.21. (a) Locations of poles and zeros in the z-plane. (b) The component of the magnitude response associated with a zero is given by the length of a vector from the zero to $e^{j\Omega}$. (c) The component of the magnitude response associated with the pole at $z = 0.9\, e^{j\frac{\pi}{4}}$ is the inverse of the length of a vector from the pole to $e^{j\Omega}$. (Parts (d) and (e) on following page).

associated with each zero and pole is evaluated by considering a term of the form $\arg\{e^{j\Omega} - g\}$. This is the angle associated with a vector pointing from g to $e^{j\Omega}$. The angle of this vector is measured with respect to a horizontal line passing through g, as illustrated in Fig. 7.25. The contribution of any pole or zero to the overall phase response is determined by the angle of the $e^{j\Omega} - g$ vector as the frequency changes.

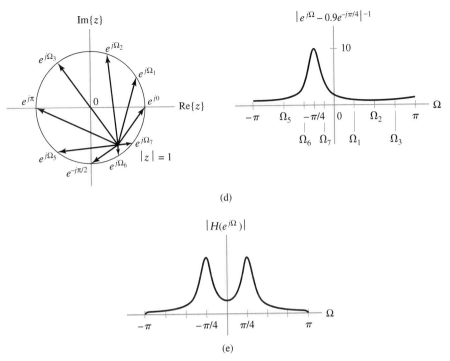

(d)

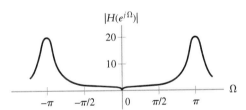

(e)

FIGURE 7.23 Continued (d) The component of the magnitude response associated with the pole at $z = 0.9\,e^{-j\frac{\pi}{4}}$ is the inverse of the length of a vector from the pole to $e^{j\Omega}$. (e) The system magnitude response is the product of the response in parts (b)–(d).

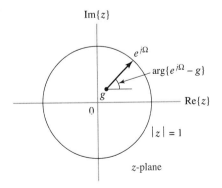

FIGURE 7.24 Solution to Problem 7.13.

FIGURE 7.25 The quantity $\arg\{e^{j\Omega} - g\}$ is the angle of the vector from g to $e^{j\Omega}$ with respect to a horizontal line through g, as shown here.

Exact evaluation of the frequency response is best performed numerically. However, we can often obtain a rough approximation from the locations of the poles and zeros as discussed here, and thus develop an insight into the nature of the frequency response. Asymptotic approximations analogous to those of the Bode diagram for continuous-time systems introduced in Chapter 6 are not used with discrete-time systems, because the frequency range is limited to $-\pi < \Omega \leq \pi$.

7.9 Computational Structures for Implementing Discrete-Time LTI Systems

Discrete-time LTI systems lend themselves to implementation on a computer. In order to write the computer program that determines the system output from the input, we must first specify the order in which each computation is to be performed. The z-transform is often used to develop such computational structures for implementing discrete-time systems that have a given transfer function. Recall from Chapter 2 that there are many different block diagram implementations corresponding to a system with a given input–output characteristic. The freedom to choose between alternative implementations can be used to optimize some criteria associated with the computation, such as the number of numerical operations or the sensitivity of the system to numerical rounding of computations. A detailed study of such issues is beyond the scope of this book; here we illustrate the role of the z-transform in obtaining alternative computational structures.

Several block diagrams for implementing systems described by difference equations were derived in Section 2.12. These block diagrams consist of time-shift operations, denoted by the operator S, multiplication by constants, and summing junctions. We may represent rational transfer function descriptions of systems with analogous block diagrams by taking the z-transform of the block diagram representing the difference equation. The time-shift operator corresponds to multiplication by z^{-1} in the z-domain. Scalar multiplication and addition are linear operations and are thus not modified by taking the z-transform. Hence, the block diagrams representing rational transfer functions use z^{-1} in place of the time-shift operators. For example, the block diagram depicted in Fig. 2.33 represents a system described by the difference equation

$$y[n] + a_1 y[n-1] + a_2 y[n-2] = b_0 x[n] + b_1 x[n-1] + b_2 x[n-2]. \quad (7.26)$$

Taking the z-transform of this difference equation gives

$$(1 + a_1 z^{-1} + a_2 z^{-2})Y(z) = (b_0 + b_1 z^{-1} + b_2 z^{-2})X(z).$$

The block diagram depicted in Fig. 7.26 implements the foregoing relationship, and it is obtained by replacing the shift operators in Fig. 2.33 with z^{-1}. The transfer function of the system in Fig. 7.26 is given by

$$\begin{aligned} H(z) &= \frac{Y(z)}{X(z)} \\ &= \frac{b_0 + b_1 z^{-1} + b_2 z^{-2}}{1 + a_1 z^{-1} + a_2 z^{-2}}. \end{aligned} \quad (7.27)$$

The direct form II representation of an LTI system was derived in Section 2.12 by writing the difference equation described by Eq. (7.26) as two coupled difference equations involving an intermediate signal $f[n]$. We may also derive the direct form II representation

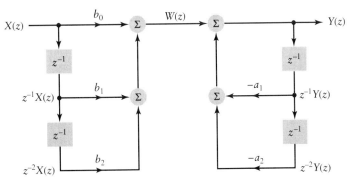

FIGURE 7.26 Block diagram of the transfer function corresponding to Fig. 2.33.

directly from the system transfer function. The transfer function of the system described by Eq. (7.26) is given by $H(z)$ in Eq. (7.27). Now suppose we write $H(z) = H_1(z)H_2(z)$, where

$$H_1(z) = b_0 + b_1 z^{-1} + b_2 z^{-2}$$

and

$$H_2(z) = \frac{1}{1 + a_1 z^{-1} + a_2 z^{-2}}.$$

The direct form II implementation for $H(z)$ is obtained by writing

$$Y(z) = H_1(z)F(z), \tag{7.28}$$

where

$$F(z) = H_2(z)X(z). \tag{7.29}$$

The block diagram depicted in Fig. 7.27(a) implements Eqs. (7.28) and (7.29). The z^{-1} blocks in $H_1(z)$ and $H_2(z)$ generate identical quantities and thus may be combined to obtain the direct form II block diagram depicted in Fig. 7.27(b).

The pole–zero form of the transfer function leads to two alternative system implementations: the cascade and parallel forms. In these forms, the transfer function is represented as an interconnection of lower order transfer functions, or sections. In the cascade form, we write

$$H(z) = \prod_{i=1}^{P} H_i(z),$$

where the $H_i(z)$ contain distinct subsets of the poles and zeros of $H(z)$. Usually, one or two of the poles and zeros of $H(z)$ are assigned to each $H_i(z)$. We say that the system is represented as a cascade of first- or second-order sections in this case. Poles and zeros that occur in complex-conjugate pairs are usually placed into the same section so that the coefficients of the section are real valued. In the parallel form, we use a partial-fraction expansion to write

$$H(z) = \sum_{i=1}^{P} H_i(z),$$

where each $H_i(z)$ contains a distinct set of the poles of $H(z)$. Here again, one or two poles usually are assigned to each section, and we say that the system is represented by a parallel connection of first- or second-order sections. The next example and problem illustrate both the parallel and cascade forms.

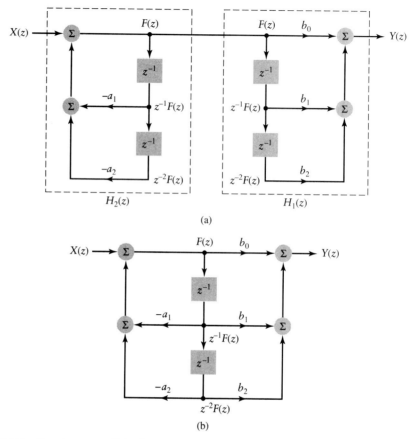

FIGURE 7.27 Development of the direct form II representation of an LTI system. (a) Representation of the transfer function $H(z)$ as $H_2(z)H_1(z)$. (b) Direct form II implementation of the transfer function $H(z)$ obtained from (a) by collapsing the two sets of z^{-1} blocks.

EXAMPLE 7.22 CASCADE IMPLEMENTATION   Consider the system represented by the transfer function

$$H(z) = \frac{(1 + jz^{-1})(1 - jz^{-1})(1 + z^{-1})}{\left(1 - \frac{1}{2}e^{j\frac{\pi}{4}}z^{-1}\right)\left(1 - \frac{1}{2}e^{-j\frac{\pi}{4}}z^{-1}\right)\left(1 - \frac{3}{4}e^{j\frac{\pi}{8}}z^{-1}\right)\left(1 - \frac{3}{4}e^{-j\frac{\pi}{8}}z^{-1}\right)}.$$

Depict the cascade form for this system, using real-valued second-order sections. Assume that each second-order section is implemented as a direct form II representation.

Solution: We combine complex-conjugate poles and zeros into the sections, obtaining

$$H_1(z) = \frac{1 + z^{-2}}{1 - \cos\left(\frac{\pi}{4}\right)z^{-1} + \frac{1}{4}z^{-2}}$$

and

$$H_2(z) = \frac{1 + z^{-1}}{1 - \frac{3}{2}\cos\left(\frac{\pi}{8}\right)z^{-1} + \frac{9}{16}z^{-2}}.$$

The block diagram corresponding to $H_1(z)H_2(z)$ is depicted in Fig. 7.28. Note that this solution is not unique, since we could have interchanged the order of $H_1(z)$ and $H_2(z)$ or interchanged the pairing of poles and zeros.   ■

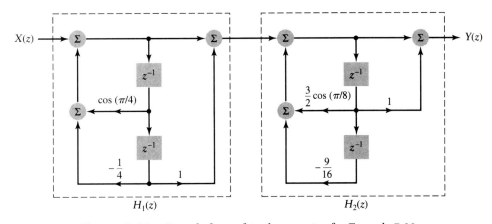

FIGURE 7.28 Cascade form of implementation for Example 7.22.

▶ **Problem 7.14** Depict the parallel-form representation of the transfer function

$$H(z) = \frac{4 - \frac{1}{2}z^{-1} - \frac{1}{2}z^{-2}}{\left(1 - \frac{1}{2}z^{-1}\right)\left(1 + \frac{1}{2}z^{-1}\right)\left(1 - \frac{1}{4}z^{-1}\right)},$$

using first-order sections implemented as a direct form II representation.

Answer:

$$H(z) = \frac{1}{1 - \frac{1}{2}z^{-1}} + \frac{1}{1 + \frac{1}{2}z^{-1}} + \frac{2}{1 - \frac{1}{4}z^{-1}}$$

(See Fig. 7.29)   ◀

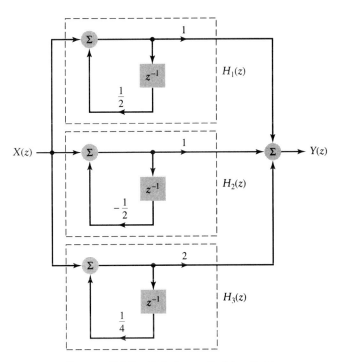

FIGURE 7.29 Solution to Problem 7.14.

7.10 *The Unilateral z-Transform*

The *unilateral*, or *one-sided*, *z-transform* is evaluated by using the portion of a signal associated with nonnegative values of the time index ($n \geq 0$). This form of the z-transform is appropriate for problems involving causal signals and LTI systems. It is reasonable to assume causality in many applications of z-transforms. For example, we are often interested in the response of a causal LTI system to an input signal. The choice of time origin is usually arbitrary, so we may choose $n = 0$ as the time at which the input is applied and then study the response for times $n \geq 0$. There are several advantages to using the unilateral transform in such problems, the chief two of which are that we do not need to use ROCs and, perhaps most important, the unilateral transform allows us to study LTI systems described by difference equations with initial conditions.

▪ 7.10.1 DEFINITION AND PROPERTIES

The unilateral z-transform of a signal $x[n]$ is defined as

$$X(z) = \sum_{n=0}^{\infty} x[n]z^{-n}, \tag{7.30}$$

which depends only on $x[n]$ for $n \geq 0$. The inverse z-transform may be obtained by evaluating Eq. (7.5) for $n \geq 0$. We denote the relationship between $x[n]$ and $X(z)$ as

$$x[n] \xleftrightarrow{\;z_u\;} X(z).$$

The unilateral and bilateral z-transforms are equivalent for causal signals. For example,

$$\alpha^n u[n] \xleftrightarrow{\;z_u\;} \frac{1}{1 - \alpha z^{-1}}$$

and

$$a^n \cos(\Omega_o n)u[n] \xleftrightarrow{\;z_u\;} \frac{1 - a\cos(\Omega_o)z^{-1}}{1 - 2a\cos(\Omega_o)z^{-1} + a^2 z^{-2}}.$$

It is straightforward to show that the unilateral z-transform satisfies the same properties as the bilateral z-transform, with one important exception: the time-shift property. In order to develop this property as it applies to the unilateral z-transform, let $w[n] = x[n-1]$. Now, from Eq. (7.30), we have

$$X(z) = \sum_{n=0}^{\infty} x[n]z^{-n}.$$

The unilateral z-transform of $w[n]$ is defined similarly as

$$W(z) = \sum_{n=0}^{\infty} w[n]z^{-n}.$$

We express $W(z)$ as a function of $X(z)$. Substituting $w[n] = x[n-1]$, we obtain

$$\begin{aligned}
W(z) &= \sum_{n=0}^{\infty} x[n-1]z^{-n} \\
&= x[-1] + \sum_{n=1}^{\infty} x[n-1]z^{-n} \\
&= x[-1] + \sum_{m=0}^{\infty} x[m]z^{-(m+1)} \\
&= x[-1] + z^{-1}\sum_{m=0}^{\infty} x[m]z^{-m} \\
&= x[-1] + z^{-1}X(z).
\end{aligned}$$

Hence, a one-unit time-shift results in multiplication by z^{-1} and addition of the constant $x[-1]$. We obtain the time-shift property for delays greater than unity in an identical manner. If

$$x[n] \xleftrightarrow{\;z_u\;} X(z),$$

then

$$\boxed{\begin{aligned} x[n-k] \xleftrightarrow{\;z_u\;} x[-k] + \\ x[-k+1]z^{-1} + \cdots + x[-1]z^{-k+1} + z^{-k}X(z) \quad \text{for} \quad k>0. \end{aligned}} \tag{7.31}$$

In the case of a time advance, the time-shift property changes somewhat. Here, we obtain

$$x[n+k] \xleftrightarrow{\;z_u\;} -x[0]z^k - x[1]z^{k-1} - \cdots - x[k-1]z + z^k X(z) \quad \text{for} \quad k>0. \tag{7.32}$$

Both time-shift properties correspond to the bilateral time-shift property, with additional terms that account for values of the sequence that are shifted into or out of the nonnegative time portion of the signal.

■ 7.10.2 SOLVING DIFFERENCE EQUATIONS WITH INITIAL CONDITIONS

The primary application of the unilateral z-transform is in solving difference equations subject to nonzero initial conditions. The difference equation is solved by taking the unilateral z-transform of both sides, using algebra to obtain the z-transform of the solution, and then inverse z-transforming. The initial conditions are incorporated naturally into the problem as a consequence of the time-shift property given in Eq. (7.31).

Consider taking the unilateral z-transform of both sides of the difference equation

$$\sum_{k=0}^{N} a_k y[n-k] = \sum_{k=0}^{M} b_k x[n-k].$$

We may write the z-transform as

$$A(z)Y(z) + C(z) = B(z)X(z),$$

where

$$A(z) = \sum_{k=0}^{N} a_k z^{-k},$$

$$B(z) = \sum_{k=0}^{M} b_k z^{-k},$$

and

$$C(z) = \sum_{m=0}^{N-1} \sum_{k=m+1}^{N} a_k y[-k+m] z^{-m}.$$

Here, we have assumed that $x[n]$ is causal, so that $x[n-k] \xleftrightarrow{\;z_u\;} z^{-k}X(z)$. The term $C(z)$ depends on the N initial conditions $y[-1], y[-2], \ldots, y[-N]$ and the a_k. $C(z)$ is zero if all the initial conditions are zero. Solving for $Y(z)$ yields

$$Y(z) = \frac{B(z)}{A(z)}X(z) - \frac{C(z)}{A(z)}.$$

The output is the sum of the forced response due to the input, represented by $\frac{B(z)}{A(z)}X(z)$, and the natural response induced by the initial conditions, represented by $\frac{C(z)}{A(z)}$. Since $C(z)$ is a polynomial, the poles of the natural response are the roots of $A(z)$, which are also the poles of the transfer function. Hence, the form of the natural response depends only on the poles of the system, which are the roots of the characteristic equation defined in Section 2.10. Note that if the system is stable, then the poles must lie inside the unit circle.

EXAMPLE 7.23 FIRST-ORDER RECURSIVE SYSTEM: INVESTMENT COMPUTATION Recall from Example 2.5 that the growth in an asset due to compound interest is described by the first-order difference equation

$$y[n] - \rho y[n-1] = x[n],$$

where $\rho = 1 + r/100$, r is the interest rate per period in percent, and $y[n]$ represents the balance after the deposit or withdrawal represented by $x[n]$. Assume that a bank account has an initial balance of \$10,000 and earns 6% annual interest compounded monthly. Starting in the first month of the second year, the owner withdraws \$100 per month from the account at the beginning of each month. Determine the balance at the start of each month (following any withdrawals) and how many months it will take for the account balance to reach zero.

Solution: We take the unilateral z-transform of both sides of the difference equation and use the time-shift property of Eq. (7.31) to obtain

$$Y(z) - \rho(y[-1] + z^{-1}Y(z)) = X(z).$$

Now we rearrange this equation to determine $Y(z)$. We have

$$(1 - \rho z^{-1})Y(z) = X(z) + \rho y[-1],$$

or

$$Y(z) = \frac{X(z)}{1 - \rho z^{-1}} + \frac{\rho y[-1]}{1 - \rho z^{-1}}.$$

Note that $Y(z)$ is given as the sum of two terms: one that depends on the input and another that depends on the initial condition. The input-dependent term represents the forced response of the system; the initial-condition term represents the natural response of the system.

The initial balance of \$10,000 at the start of the first month is the initial condition $y[-1]$, and there is an offset of two between the time index n and the month index. That is, $y[n]$ represents the balance in the account at the start of the $n + 2$nd month. We have $\rho = 1 + \frac{6/12}{100} = 1.005$. Since the owner withdraws \$100 per month at the start of month 13 ($n = 11$), we may express the input to the system as $x[n] = -100u[n - 11]$. Thus,

$$X(z) = \frac{-100z^{-11}}{1 - z^{-1}},$$

and we have

$$Y(z) = \frac{-100z^{-11}}{(1 - z^{-1})(1 - 1.005z^{-1})} + \frac{1.005(10,000)}{1 - 1.005z^{-1}}.$$

Now we perform a partial-fraction expansion on the first term of $Y(z)$, obtaining

$$Y(z) = \frac{20,000z^{-11}}{1 - z^{-1}} - \frac{20,000z^{-11}}{1 - 1.005z^{-1}} + \frac{10,050}{1 - 1.005z^{-1}}.$$

The monthly account balance is obtained by inverse z-transforming $Y(z)$, resulting in

$$y[n] = 20,000u[n - 11] - 20,000(1.005)^{n-11}u[n - 11] + 10,050(1.005)^n u[n].$$

The last term, $10,050(1.005)^n u[n]$ is the natural response associated with the initial balance, while the first two terms represent the forced response associated with the withdrawals. The account balance, natural response, and forced response for the first 60 months are illustrated in Fig. 7.30 as a function of the month, not n. The account balance reaches zero during the withdrawal at the start of month 163. ∎

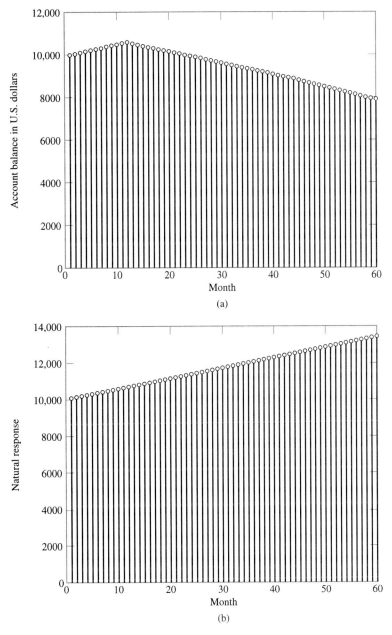

(a)

(b)

FIGURE 7.30 Solution to Example 7.23, depicted as a function of the month. (a) Account balance at the start of each month following possible withdrawal. (b) Natural response.

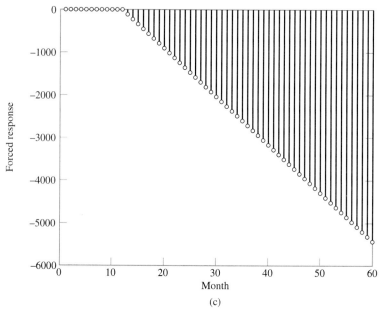

FIGURE 7.30   Continued (c) Forced response.

▶ **Problem 7.15**   Determine the forced response $y^{(f)}[n]$, natural response $y^{(n)}[n]$, and output $y[n]$ of the system described by the difference equation

$$y[n] + 3y[n-1] = x[n] + x[n-1]$$

if the input is $x[n] = \left(\frac{1}{2}\right)^n u[n]$ and $y[-1] = 2$ is the initial condition.

Answer:

$$y^{(f)}[n] = \frac{4}{7}(-3)^n u[n] + \frac{3}{7}\left(\frac{1}{2}\right)^n u[n]$$

$$y^{(n)}[n] = -6(-3)^n u[n]$$

$$y[n] = y^{(f)}[n] + y^{(n)}[n]$$ ◀

7.11 *Exploring Concepts with MATLAB*

The MATLAB Signal Processing Toolbox contains routines for working with *z*-transforms.

■ 7.11.1 POLES AND ZEROS

The poles and zeros of an LTI system may be determined by applying `roots` to the respective polynomial. For example, to find the roots of $1 + 4z^{-1} + 3z^{-2}$, we give the command `roots([1, 4, 3])`. The poles and zeros may be displayed in the *z*-plane by using `zplane(b, a)`. If b and a are row vectors, then `zplane` finds the roots of the numerator and denominator polynomials represented by b and a, respectively, before finding the poles and zeros and displaying them. If b and a are column vectors, then `zplane` assumes that b and a contain the locations of the zeros and poles, respectively, and displays them directly.

■ 7.11.2 INVERSION OF THE z-TRANSFORM

The `residuez` command computes partial-fraction expansions for z-transforms expressed as a ratio of two polynomials in z^{-1}. The syntax is `[r, p, k] = residuez(b, a)`, where `b` and `a` are vectors representing the numerator and denominator polynomial coefficients, ordered in descending powers of z. The vector `r` represents the partial-fraction expansion coefficents corresponding to the poles given in `p`. The vector `k` contains the coefficients associated with powers of z^{-1} that result from long division when the order of the numerator equals or exceeds that of the denominator.

For example, we may use MATLAB to find the partial-fraction expansion for the z-transform given in Example 7.10:

$$X(z) = \frac{z^3 - 10z^2 - 4z + 4}{2z^2 - 2z - 4}.$$

Since `residuez` assumes that the numerator and denominator polynomials are expressed in powers of z^{-1}, we first write $X(z) = zY(z)$, where

$$Y(z) = \frac{1 - 10z^{-1} - 4z^{-2} + 4z^{-3}}{2 - 2z^{-1} - 4z^{-2}}.$$

Now we use `residuez` to find the partial-fraction expansion for $Y(z)$ as follows:

```
>> [r, p, k] = residuez([1, -10, -4, 4], [2, -2, -4])

r =
    -1.5000
     0.5000
p =
     2
    -1

k =
    1.5000    -1.0000
```

This implies a partial-fraction expansion of the form

$$Y(z) = \frac{-1.5}{1 - 2z^{-1}} + \frac{0.5}{1 + z^{-1}} + 1.5 - z^{-1},$$

which, as expected, corresponds to $\frac{1}{2}W(z)$ in Example 7.10.

▶ **Problem 7.16** Solve Problem 7.5, using MATLAB and the `residuez` command. ◀

■ 7.11.3 TRANSFORM ANALYSIS OF LTI SYSTEMS

Recall that the difference equation, transfer function, poles and zeros, frequency response, and state-variable description offer different, yet equivalent, representations of the input–output characteristics of an LTI system. The MATLAB Signal Processing Toolbox contains several routines for converting between different LTI system descriptions. If `b` and `a` contain the coefficients of the transfer function numerator and denominator polynomials, respectively, ordered in descending powers of z, then `tf2ss(b, a)` determines a state-variable description of the system and `tf2zp(b, a)` determines the pole–zero-gain description of the system.

Similarly, `zp2ss` and `zp2tf` convert from pole–zero-gain descriptions to state-variable and transfer-function descriptions, respectively, while `ss2tf` and `ss2zp`, respectively, convert from state-variable description to transfer function and pole–zero-gain forms. As noted in Section 3.19, the frequency response of a system described by a difference equation is evaluated from the transfer function with the use of `freqz`.

Consider an LTI system with transfer function

$$H(z) = \frac{0.094(1 + 4z^{-1} + 6z^{-2} + 4z^{-3} + z^{-4})}{1 + 0.4860z^{-2} + 0.0177z^{-4}}. \tag{7.33}$$

We may depict the poles and zeros of $H(z)$ in the z-plane and plot the system's magnitude response with the following commands:

```
>> b = .094*[1, 4, 6, 4, 1];
>> a = [1, 0, 0.486, 0, 0.0177];
>> zplane(b, a)
>> [H,w] = freqz(b, a, 250);
>> plot(w,abs(H))
```

Figure 7.31 indicates that this system has a zero of multiplicity four at $z = -1$ and four poles on the imaginary axis. The magnitude response is depicted in Fig. 7.32. Note that the zeros at $z = -1$ force the magnitude response to be small at high frequencies.

■ 7.11.4 COMPUTATIONAL STRUCTURES FOR IMPLEMENTING DISCRETE-TIME LTI SYSTEMS

One useful means for implementing a discrete-time LTI system is as a cascade of second-order sections. The MATLAB Signal Processing Toolbox contains routines for converting a state-variable or pole–zero-gain description of a system to a cascade connection of second-order sections. This is accomplished by using `ss2sos` and `zp2sos`. The syntax for `zp2sos` is `sos = zp2sos(z, p, k)`, where z and p are vectors containing zeros and poles, respectively, and k is the gain. The matrix `sos` is L by 6, where each row contains the coefficients of the transfer function for that section. The first three elements of the row contain the numerator coefficients, while the last three elements contain the denomi-

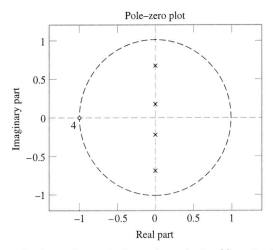

FIGURE 7.31   Location of poles and zeros in the z-plane obtained by using MATLAB. The number "4" near the zero at $z = -1$ indicates that there are 4 zeros at this location.

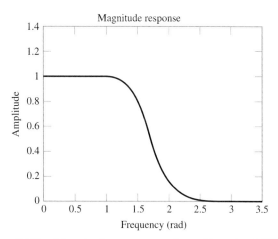

FIGURE 7.32   Magnitude response evaluated by using MATLAB.

nator coefficients. The commands `sos2zp`, `sos2ss`, and `sos2tf` convert from a cascade of second-order sections to pole–zero-gain, state-variable, and transfer-function descriptions.

Suppose we repeat Example 7.22, using MATLAB to obtain a representation of the system as a cascade of second-order sections. The transfer function is given in pole–zero-gain form:

$$H(z) = \frac{(1 + jz^{-1})(1 - jz^{-1})(1 + z^{-1})}{\left(1 - \frac{1}{2}e^{j\frac{\pi}{4}}z^{-1}\right)\left(1 - \frac{1}{2}e^{-j\frac{\pi}{4}}z^{-1}\right)\left(1 - \frac{3}{4}e^{j\frac{\pi}{8}}z^{-1}\right)\left(1 - \frac{3}{4}e^{-j\frac{\pi}{8}}z^{-1}\right)}.$$

The system has zeros at $z = \pm j$ and $z = -1$, while the poles are at $z = \frac{1}{2}e^{\pm j\frac{\pi}{4}}$ and $z = \frac{3}{4}e^{\pm j\frac{\pi}{8}}$. We employ `zp2sos` to convert from pole–zero-gain form to second-order sections as follows:

```
>> z = [ -1, -j, j];
>> p = [ 0.5*exp(j*pi/4), 0.5*exp(-j*pi/4),
         0.75*exp(j*pi/8), 0.75exp(-j*pi/8) ];
>>k = 1;
>> sos = zp2sos(z, p, k)

sos =

      0.2706    0.2706         0    1.0000   -0.7071    0.2500
      3.6955         0    3.6955    1.0000   -1.3858    0.5625
```

Hence, the system is described as a cascade of second-order sections given by

$$F_1(z) = \frac{0.2706 + 0.2706z^{-1}}{1 - 0.7071z^{-1} + 0.25z^{-2}} \quad \text{and} \quad F_2(z) = \frac{3.6955 + 3.6955z^{-2}}{1 - 1.3858z^{-1} + 0.5625z^{-2}}.$$

Note that this solution differs from that of Example 7.22 in that the pairing of zeros and poles are interchanged. A scaling factor is also introduced into each section by `zp2sos`. The overall gain is unchanged, however, since the product of the scaling factors is unity. The procedures employed by `zp2sos` for scaling and pairing poles with zeros are chosen to minimize the effect of numerical errors when such systems are implemented with fixed-point arithmetic.

7.12 Summary

The z-transform represents discrete-time signals as a weighted superposition of complex exponentials, a more general signal class than complex sinusoids, so the z-transform can represent a broader class of discrete-time signals than the DTFT, including signals that are not absolutely summable. Thus, we may use the z-transform to analyze discrete-time signals and LTI systems that are not stable. The transfer function of a discrete-time LTI system is the z-transform of its impulse response. The transfer function offers another description of the input–output characteristics of an LTI system. The z-transform converts convolution of time signals into multiplication of z-transforms, so the z-transform of a system's output is the product of the z-transform of the input and the system's transfer function.

A complex exponential is described by a complex number. Hence, the z-transform is a function of a complex variable z represented in the complex plane. The DTFT is obtained by evaluating the z-transform on the unit circle, $|z| = 1$, by setting $z = e^{j\Omega}$. The properties of the z-transform are analogous to those of the DTFT. The ROC defines the values of z for which the z-transform converges. The ROC must be specified in order to have a unique relationship between the time signal and its z-transform. The relative locations of the ROC and z-transform poles determine whether the corresponding time signal is right sided, left sided, or both. The locations of z-transform's poles and zeros offer another representation of the input–output characteristics of an LTI system, providing information regarding the system's stability, causality, invertibility, and frequency response.

The z-transform and DTFT have many common features. However, they have distinct roles in signal and system analysis. The z-transform is generally used to study LTI system characteristics such as stability and causality, to develop computational structures for implementing discrete-time systems, and in the design of digital filters, the subject of Chapter 8. The z-transform is also used for transient and stability analysis of sampled-data control systems, a topic we visit in Chapter 9. The unilateral z-transform applies to causal signals and offers a convenient tool for solving problems associated with LTI systems defined by difference equations with nonzero initial conditions. None of these problems are addressable with the DTFT. Instead, the DTFT is usually used as a tool for representing signals and to study the steady-state characteristics of LTI systems, as we illustrated in Chapters 3 and 4. In these problems, the DTFT is easier to visualize than the z-transform, since it is a function of the real-valued frequency Ω, while the z-transform is a function of a complex number $z = re^{j\Omega}$.

FURTHER READING

1. The following text is devoted entirely to z-transforms:
   ▶ Vich, R., *Z Transform Theory and Applications* (D. Reidel Publishing, 1987)

2. The z-transform is also discussed in most texts on signal processing, including the following:
   ▶ Proakis, J. G., and D. G. Manolakis, *Digital Signal Processing: Principles, Algorithms and Applications*, 3rd ed. (Prentice Hall, 1995)
   ▶ Oppenheim, A. V., R. W. Schafer, and J. R. Buck, *Discrete Time Signal Processing*, 2nd ed. (Prentice Hall, 1999)

   The book by Oppenheim et al. discusses the relationship between the magnitude and phase responses of minimum-phase discrete-time systems.

3. Evaluation of the inverse z-transform using Eq. (7.5) is discussed in:

▸ Oppenheim, A. V., R. W. Schafer, and J. R. Buck, *op. cit.*

and an introductory treatment of the techniques involved in contour integration is given in:

▸ Brown, J., and R. Churchill, *Complex Variables and Applications,* (McGraw-Hill, 1996)

4. A thorough, yet advanced treatment of computational structures for implementing discrete-time LTI systems is contained in

▸ Roberts, R. A., and C. T. Mullis, *Digital Signal Processing,* (Addison-Wesley, 1987)

ADDITIONAL PROBLEMS

7.17 Determine the z-transform and ROC for the following time signals:

(a) $x[n] = \delta[n - k], \quad k > 0$

(b) $x[n] = \delta[n + k], \quad k > 0$

(c) $x[n] = u[n]$

(d) $x[n] = \left(\frac{1}{4}\right)^n (u[n] - u[n - 5])$

(e) $x[n] = \left(\frac{1}{4}\right)^n u[-n]$

(f) $x[n] = 3^n u[-n - 1]$

(g) $x[n] = \left(\frac{2}{3}\right)^{|n|}$

(h) $x[n] = \left(\frac{1}{2}\right)^n u[n] + \left(\frac{1}{4}\right)^n u[-n - 1]$

Sketch the ROC, poles, and zeros in the z-plane.

7.18 Given each of the following z-transforms, determine whether the DTFT of the corresponding time signal exists without determining the signal, and identify the DTFT in those cases where it does exist:

(a) $X(z) = \dfrac{5}{1 + \frac{1}{3}z^{-1}}, \quad |z| > \frac{1}{3}$

(b) $X(z) = \dfrac{5}{1 + \frac{1}{3}z^{-1}}, \quad |z| < \frac{1}{3}$

(c) $X(z) = \dfrac{z^{-1}}{\left(1 - \frac{1}{2}z^{-1}\right)(1 + 3z^{-1})}, \quad |z| < \frac{1}{2}$

(d) $X(z) = \dfrac{z^{-1}}{\left(1 - \frac{1}{2}z^{-1}\right)(1 + 3z^{-1})}, \quad \frac{1}{2} < |z| < 3$

7.19 The locations of the poles and zeros of $X(z)$ are depicted in the z-plane in the following figures:

(a) Fig. P7.19(a)

(b) Fig. P7.19(b)

(c) Fig. P7.19(c)

In each case, identify all valid ROCs for $X(z)$, and specify whether the time signal corresponding to each ROC, is right sided, left sided or two-sided.

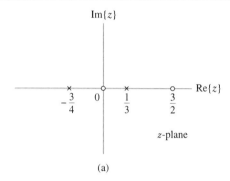

(a)

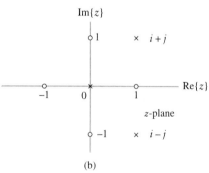

(b)

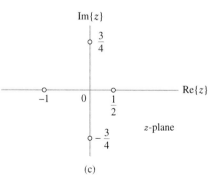

(c)

FIGURE P7.19

7.20 Use the tables of z-transforms and the z-transform properties given in Appendix E to determine the z-transforms of the following signals:

(a) $x[n] = \left(\frac{1}{2}\right)^n u[n] * 2^n u[-n - 1]$

(b) $x[n] = n\left(\left(\frac{1}{2}\right)^n u[n] * \left(\frac{1}{4}\right)^n u[n-2]\right)$

(c) $x[n] = u[-n]$

(d) $x[n] = n\sin(\frac{\pi}{2}n)u[-n]$

(e) $x[n] = 3^{n-2}u[n] * \cos(\frac{\pi}{6}n + \pi/3)u[n]$

7.21 Given the z-transform pair $x[n] \xleftrightarrow{\;z\;} \frac{z^2}{z^2-16}$, with ROC $|z| < 4$, use the z-transform properties to determine the z-transform of the following signals:

(a) $y[n] = x[n-2]$

(b) $y[n] = (1/2)^n x[n]$

(c) $y[n] = x[-n] * x[n]$

(d) $y[n] = nx[n]$

(e) $y[n] = x[n+1] + x[n-1]$

(f) $y[n] = x[n] * x[n-3]$

7.22 Given the z-transform pair $n^2 3^n u[n] \xleftrightarrow{\;z\;} X(z)$,

S&S
Solutions

use the z-transform properties to determine the time-domain signals corresponding to the following z transforms:

(a) $Y(z) = X(2z)$

(b) $Y(z) = X(z^{-1})$

(c) $Y(z) = \frac{d}{dz}X(z)$

(d) $Y(z) = \frac{z^2 - z^{-2}}{2}X(z)$

(e) $Y(z) = [X(z)]^2$

7.23 Prove the following z-transform properties:

(a) time reversal

(b) time shift

(c) multiplication by exponential sequence

(d) convolution

(e) differentiation in the z-domain

7.24 Use the method of partial fractions to obtain the time-domain signals corresponding to the following z-transforms:

(a) $X(z) = \frac{1 + \frac{7}{6}z^{-1}}{\left(1 - \frac{1}{2}z^{-1}\right)\left(1 + \frac{1}{3}z^{-1}\right)}$, $|z| > \frac{1}{2}$

(b) $X(z) = \frac{1 + \frac{7}{6}z^{-1}}{\left(1 - \frac{1}{2}z^{-1}\right)\left(1 + \frac{1}{3}z^{-1}\right)}$, $|z| < \frac{1}{3}$

(c) $X(z) = \frac{1 + \frac{7}{6}z^{-1}}{\left(1 - \frac{1}{2}z^{-1}\right)\left(1 + \frac{1}{3}z^{-1}\right)}$, $\frac{1}{3} < |z| < \frac{1}{2}$

(d) $X(z) = \frac{z^2 - 3z}{z^2 + \frac{3}{2}z - 1}$, $\frac{1}{2} < |z| < 2$

(e) $X(z) = \frac{3z^2 - \frac{1}{4}z}{z^2 - 16}$, $|z| > 4$

(f) $X(z) = \frac{z^3 + z^2 + \frac{3}{2}z + \frac{1}{2}}{z^3 + \frac{3}{2}z^2 + \frac{1}{2}z}$, $|z| < \frac{1}{2}$

(g) $X(z) = \frac{2z^4 - 2z^3 - 2z^2}{z^2 - 1}$, $|z| > 1$

7.25 Determine the time-domain signals corresponding to the following z-transforms:

(a) $X(z) = 1 + 2z^{-6} + 4z^{-8}$, $|z| > 0$

(b) $X(z) = \sum_{k=5}^{10} \frac{1}{k}z^{-k}$, $|z| > 0$

(c) $X(z) = (1 + z^{-1})^3$, $|z| > 0$

(d) $X(z) = z^6 + z^2 + 3 + 2z^{-3} + z^{-4}$, $|z| > 0$

7.26 Use the following clues to determine the signals $x[n]$ and rational z-transforms $X(z)$:

(a) $X(z)$ has poles at $z = 1/2$ and $z = -1$, $x[1] = 1$, $x[-1] = 1$, and the ROC includes the point $z = 3/4$.

(b) $x[n]$ is right sided, $X(z)$ has a single pole, $x[0] - 2$, and $x[2] = 1/2$.

(c) $x[n]$ is two sided, $X(z)$ has one pole at $z = 1/4$, $x[-1] = 1$, $x[-3] = 1/4$, and $X(1) = 11/3$.

7.27 Determine the impulse response corresponding to

S&S
Solutions

the following transfer functions if (i) the system is stable or (ii) the system is causal:

(a) $H(z) = \frac{2 - \frac{3}{2}z^{-1}}{\left(1 - 2z^{-1}\right)\left(1 + \frac{1}{2}z^{-1}\right)}$

(b) $H(z) = \frac{5z^2}{z^2 - z - 6}$

(c) $H(z) = \frac{4z}{z^2 - \frac{1}{4}z + \frac{1}{16}}$

7.28 Use a power series expansion to determine the time-domain signal corresponding to the following z-transforms:

(a) $X(z) = \frac{1}{1 - \frac{1}{4}z^{-2}}$, $|z| > \frac{1}{2}$

(b) $X(z) = \frac{1}{1 - \frac{1}{4}z^{-2}}$, $|z| < \frac{1}{2}$

(c) $X(z) = \cos(z^{-3})$, $|z| > 0$

(d) $X(z) = \ln(1 + z^{-1})$, $|z| > 0$

7.29 A causal system has input $x[n]$ and output $y[n]$. Use the transfer function to determine the impulse response of this system.

(a) $x[n] = \delta[n] + \frac{1}{4}\delta[n-1] - \frac{1}{8}\delta[n-2]$,
    $y[n] = \delta[n] - \frac{3}{4}\delta[n-1]$

(b) $x[n] = (-3)^n u[n]$,
    $y[n] = 4(2)^n u[n] - \left(\frac{1}{2}\right)^n u[n]$

7.30 A system has impulse response $h[n] = \left(\frac{1}{2}\right)^n u[n]$. Determine the input to the system if the output is given by

(a) $y[n] = 2\delta[n-4]$

(b) $y[n] = \frac{1}{3}u[n] + \frac{2}{3}\left(\frac{-1}{2}\right)^n u[n]$

7.31 Determine (i) transfer function and (ii) impulse response representations of the causal systems described by the following difference equations:

(a) $y[n] - \frac{1}{2}y[n-1] = 2x[n-1]$

(b) $y[n] = x[n] - x[n-2]$
$+ x[n-4] - x[n-6]$

(c) $y[n] - \frac{4}{5}y[n-1] - \frac{16}{25}y[n-2]$
$= 2x[n] + x[n-1]$

7.32 Determine (i) transfer function and (ii) difference-equation representations of the systems with the following impulse responses:

(a) $h[n] = 3\left(\frac{1}{4}\right)^n u[n-1]$

(b) $h[n] = \left(\frac{1}{3}\right)^n u[n] + \left(\frac{1}{2}\right)^{n-2} u[n-1]$

(c) $h[n] = 2\left(\frac{2}{3}\right)^n u[n-1]$
$+ \left(\frac{1}{4}\right)^n \left[\cos\left(\frac{\pi}{6}n\right) - 2\sin\left(\frac{\pi}{6}n\right)\right]u[n]$

(d) $h[n] = \delta[n] - \delta[n-5]$

7.33 (a) Take the z-transform of the state-update equation (2.62), utilizing the time-shift property Eq. (7.13) to obtain

$$\widetilde{\mathbf{q}}(z) = (z\mathbf{I} - \mathbf{A})^{-1}\mathbf{b}X(z),$$

where

$$\widetilde{\mathbf{q}}(z) = \begin{bmatrix} Q_1(z) \\ Q_2(z) \\ \vdots \\ Q_N(z) \end{bmatrix}$$

is the z-transform of $\mathbf{q}[n]$. Use this result to show that the transfer function of an LTI system is expressed in terms of the state-variable description as

$$H(z) = \mathbf{c}(z\mathbf{I} - \mathbf{A})^{-1}\mathbf{b} + D.$$

(b) Determine transfer function and difference-equation representations for the systems described by the following state-variable descriptions, and then plot the locations of the poles and zeros in the z-plane:

(i) $\mathbf{A} = \begin{bmatrix} -\frac{1}{2} & 0 \\ 0 & \frac{1}{2} \end{bmatrix}$, $\mathbf{b} = \begin{bmatrix} 0 \\ 2 \end{bmatrix}$,
$\mathbf{c} = \begin{bmatrix} 1 & -1 \end{bmatrix}$, $D = [1]$

(ii) $\mathbf{A} = \begin{bmatrix} \frac{1}{2} & -\frac{1}{2} \\ -\frac{1}{2} & -\frac{1}{4} \end{bmatrix}$, $\mathbf{b} = \begin{bmatrix} 1 \\ 0 \end{bmatrix}$,
$\mathbf{c} = \begin{bmatrix} 2 & 1 \end{bmatrix}$, $D = [0]$

(iii) $\mathbf{A} = \begin{bmatrix} -\frac{1}{4} & \frac{1}{8} \\ -\frac{7}{2} & \frac{3}{4} \end{bmatrix}$, $\mathbf{b} = \begin{bmatrix} 2 \\ 2 \end{bmatrix}$,
$\mathbf{c} = \begin{bmatrix} 0 & 1 \end{bmatrix}$, $D = [0]$

7.34 Determine whether each of the following systems is (i) causal and stable and (ii) minimum phase.

(a) $H(z) = \dfrac{2z + 3}{z^2 + z - \frac{5}{16}}$

(b) $y[n] - y[n-1] - \frac{1}{4}y[n-2]$
$= 3x[n] - 2x[n-1]$

(c) $y[n] - 2y[n-2] = x[n] - \frac{1}{2}x[n-1]$

7.35 For each of the following systems, identify the transfer function of the inverse system, and determine whether the inverse system can be both causal and stable:

(a) $H(z) = \dfrac{1 - 8z^{-1} + 16z^{-2}}{1 - \frac{1}{2}z^{-1} + \frac{1}{4}z^{-2}}$

(b) $H(z) = \dfrac{z^2 - \frac{81}{100}}{z^2 - 1}$

(c) $h[n] = 10\left(\frac{-1}{2}\right)^n u[n] - 9\left(\frac{-1}{4}\right)^n u[n]$

(d) $h[n] = 24\left(\frac{1}{2}\right)^n u[n-1] - 30\left(\frac{1}{3}\right)^n u[n-1]$

(e) $y[n] - \frac{1}{4}y[n-2] = 6x[n]$
$- 7x[n-1] + 3x[n-2]$

(f) $y[n] - \frac{1}{2}y[n-1] = x[n]$

7.36 A system described by a rational transfer function $H(z)$ has the following properties: (1) the system is causal; (2) $h[n]$ is real; (3) $H(z)$ has a pole at $z = j/2$ and exactly one zero; (4) the inverse system has two zeros; (5) $\sum_{n=0}^{\infty} h[n]2^{-n} = 0$; and (6) $h[0] = 1$.

(a) Is this system stable?

(b) Is the inverse system both stable and causal?

(c) Find $h[n]$.

(d) Find the transfer function of the inverse system.

7.37 Use the graphical method to sketch the magnitude response of the systems having the following transfer functions:

(a) $H(z) = \dfrac{z^{-2}}{1 + \frac{49}{64}z^{-2}}$

(b) $H(z) = \dfrac{1 + z^{-1} + z^{-2}}{3}$

(c) $H(z) = \dfrac{1 + z^{-1}}{1 + (18/10)\cos\left(\frac{\pi}{4}\right)z^{-1} + (81/100)z^{-2}}$

7.38 Draw block diagram implementations of the following systems as a cascade of second-order sections with real-valued coefficients:

(a) $H(z) = \dfrac{\left(1 - \frac{1}{4}e^{j\frac{\pi}{4}}z^{-1}\right)\left(1 - \frac{1}{4}e^{-j\frac{\pi}{4}}z^{-1}\right)}{\left(1 - \frac{1}{2}e^{j\frac{\pi}{3}}z^{-1}\right)\left(1 - \frac{1}{2}e^{-j\frac{\pi}{3}}z^{-1}\right)}$.

$\dfrac{\left(1 + \frac{1}{4}e^{j\frac{\pi}{8}}z^{-1}\right)\left(1 + \frac{1}{4}e^{-j\frac{\pi}{8}}z^{-1}\right)}{\left(1 - \frac{3}{4}e^{j\frac{7\pi}{8}}z^{-1}\right)\left(1 - \frac{3}{4}e^{-j\frac{7\pi}{8}}z^{-1}\right)}$

(b) $H(z) = \dfrac{(1 + 2z^{-1})^2\left(1 - \frac{1}{2}e^{j\frac{\pi}{2}}z^{-1}\right)}{\left(1 - \frac{3}{8}z^{-1}\right)\left(1 - \frac{3}{8}e^{j\frac{\pi}{3}}z^{-1}\right)}$.

$\dfrac{\left(1 - \frac{1}{2}e^{-j\frac{\pi}{2}}z^{-1}\right)}{\left(1 - \frac{3}{8}e^{-j\frac{\pi}{3}}z^{-1}\right)\left(1 + \frac{3}{4}z^{-1}\right)}$

7.39 Draw block diagram implementations of the following systems as a parallel combination of second-order sections with real-valued coefficients:

(a) $h[n] = 2\left(\frac{1}{2}\right)^n u[n] + \left(\frac{j}{2}\right)^n u[n] +$

$\left(\frac{-j}{2}\right)^n u[n] + \left(\frac{-1}{2}\right)^n u[n]$

(b) $h[n] = 2\left(\frac{1}{2}e^{j\frac{\pi}{4}}\right)^n u[n] + \left(\frac{1}{4}e^{j\frac{\pi}{3}}\right)^n u[n] +$

$\left(\frac{1}{4}e^{-j\frac{\pi}{3}}\right)^n u[n] + 2\left(\frac{1}{2}e^{-j\frac{\pi}{4}}\right)^n u[n]$

7.40 Determine the transfer function of the system depicted in Fig. P7.40.

7.41 Let $x[n] = u[n + 4]$.

(a) Determine the unilateral z-transform of $x[n]$.

(b) Use the unilateral z-transform time-shift property and the result of (a) to determine the unilateral z-transform of $w[n] = x[n - 2]$.

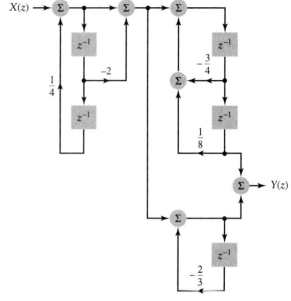

FIGURE P7.40

7.42 Use the unilateral z-transform to determine the forced response, the natural response, and the complete response of the systems described by the following difference equations with the given inputs and initial conditions:

(a) $y[n] - \frac{1}{3}y[n - 1] = 2x[n]$, $y[-1] = 1$,
$x[n] = \left(\frac{-1}{2}\right)^n u[n]$

(b) $y[n] - \frac{1}{9}y[n - 2] = x[n - 1]$, $y[-1] = 1$,
$y[-2] = 0$, $x[n] = 2u[n]$

(c) $y[n] - \frac{1}{4}y[n - 1] - \frac{1}{8}y[n - 2]$
$\qquad\qquad = x[n] + x[n - 1]$,
$y[-1] = 1$, $y[-2] = -1$, $x[n] = 3^n u[n]$

ADVANCED PROBLEMS

7.43 Use the z-transform of $u[n]$ and the property of differentiation in the z-domain to derive the formula for evaluating the sum

$$\sum_{n=0}^{\infty} n^2 a^n,$$

assuming that $|a| < 1$.

7.44 A continuous-time signal $y(t)$ satisfies the first-order differential equation

$$\frac{d}{dt}y(t) + 2y(t) = x(t).$$

Approximate the derivative as $(y(nT_s) - y((n-1)T_s))/T_s$ and show that the sampled signal $y[n] = y(nT_s)$

satisfies the first-order difference equation

$$y[n] + \alpha y[n - 1] = v[n].$$

Express α and $v[n]$ in terms of T_s and $x[n] = x(nT_s)$.

7.45 The autocorrelation signal for a real-valued causal signal $x[n]$ is defined as

$$r_x[n] = \sum_{l=0}^{\infty} x[l]x[n + l].$$

Assume that the z-transform of $r_x[n]$ converges for some values of z.

Find $x[n]$ if

$$R_x(z) = \frac{1}{(1 - \alpha z^{-1})(1 - \alpha z)},$$

where $|\alpha| < 1$.

7.46 The cross-correlation of two real-valued signals $x[n]$ and $y[n]$ is expressed as

$$r_{xy}[n] = \sum_{l=-\infty}^{\infty} x[l]y[n + l].$$

(a) Express $r_{xy}[n]$ as a convolution of two sequences.
(b) Find the z-transform of $r_{xy}[n]$ as a function of the z-transforms of $x[n]$ and $y[n]$.

7.47 A signal with rational z-transform has even symmetry; that is, $x[n] = x[-n]$.
(a) What constraints must the poles of such a signal satisfy?
(b) Show that the z-transform corresponds to the impulse response of a stable system if and only if $\sum_{n=-\infty}^{\infty} |x[n]| < \infty$.
(c) Suppose

$$X(z) = \frac{2 - (17/4)z^{-1}}{(1 - (1/4)z^{-1})(1 - 4z^{-1})}.$$

Determine the ROC and find $x[n]$.

7.48 Consider an LTI system with transfer function

$$H(z) = \frac{1 - a^*z}{z - a}, \quad |a| < 1.$$

Here, the pole and zero are a conjugate reciprocal pair.
(a) Sketch a pole–zero plot for this system in the z-plane.
(b) Use the graphical method to show that the magnitude response of the system is unity for all frequencies. A system with this characteristic is termed an *all-pass* system.
(c) Use the graphical method to sketch the phase response of the system for $a = \frac{1}{2}$.
(d) Use the result from (b) to prove that any system with a transfer function of the form

$$H(z) = \prod_{k=1}^{P} \frac{1 - a_k^*z}{z - a_k}, \quad |a_k| < 1,$$

corresponds to a stable and causal all-pass system.
(e) Can a stable and causal all-pass system also be minimum phase? Explain.

7.49 Let:

$$H(z) = F(z)(z - a) \quad \text{and} \quad G(z) = F(z)(1 - az),$$

where $0 < a < 1$ is real.
(a) Show that $|G(e^{j\Omega})| = |H(e^{j\Omega})|$.
(b) Show that $g[n] = h[n] * v[n]$, where

$$V(z) = \frac{z^{-1} - a}{1 - az^{-1}}.$$

$V(z)$ is thus the transfer function of an all-pass system. (See Problem 7.48.)
(c) One definition of the average delay introduced by a causal system is the normalized first moment

$$d = \frac{\sum_{k=0}^{\infty} k v^2[k]}{\sum_{k=0}^{\infty} v^2[k]}.$$

Calculate the average delay introduced by the all-pass system $V(z)$.

7.50 The transfer function of an LTI system is expressed as

$$H(z) = \frac{b_0 \prod_{k=1}^{M}(1 - c_k z^{-1})}{\prod_{k=1}^{N}(1 - d_k z^{-1})},$$

where $|d_k| < 1$, $k = 1, 2, \ldots, N$; $|c_k| < 1$, $k = 1, 2, \ldots, M - 1$; and $|c_M| > 1$.

(a) Show that $H(z)$ can be factored into the form $H(z) = H_{min}(z)H_{ap}(z)$, where $H_{min}(z)$ is minimum phase and $H_{ap}(z)$ is all pass. (See Problem 7.48.)
(b) Find a minimum-phase equalizer with transfer function $H_{eq}(z)$ chosen so that $|H(e^{j\Omega})H_{eq}(e^{j\Omega})| = 1$, and determine the transfer function of the cascade $H(z)H_{eq}(z)$.

7.51 A very useful structure for implementing nonrecursive systems is the so-called lattice stucture. The lattice is constructed as a cascade of two-input, two-output sections of the form depicted in Fig. P7.51(a). An Mth-order lattice structure is depicted in Fig. P7.51(b).

(a) Find the transfer function of a second-order ($M = 2$) lattice having $c_1 = \frac{1}{2}$ and $c_2 = -\frac{1}{4}$.
(b) Determine the relationship between the transfer function and the lattice structure by examining the effect of adding a section on the transfer function, as depicted in Fig. P7.51(c).

Here, we have defined $H_i(z)$ as the transfer function between the input and the output of the lower branch in the ith section and $\tilde{H}_i(z)$ as the transfer function between the input and the output of the upper branch in the ith section. Write the relationship

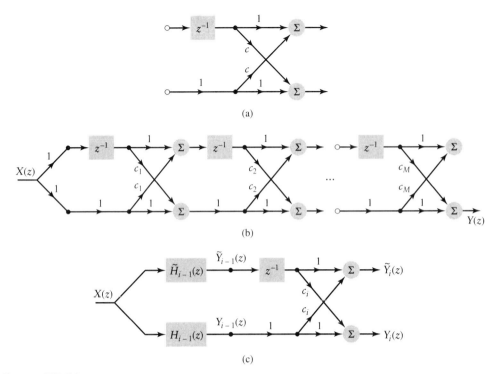

FIGURE P7.51

between the transfer functions to the $(i-1)$th and ith stages as

$$\begin{bmatrix} \widetilde{H}_i(z) \\ H_i(z) \end{bmatrix} = \mathbf{T}(z) \begin{bmatrix} \widetilde{H}_{i-1}(z) \\ H_{i-1}(z) \end{bmatrix},$$

where $\mathbf{T}(z)$ is a two-by-two matrix. Express $\mathbf{T}(z)$ in terms of c_i and z^{-1}.

(c) Use induction to prove that $\widetilde{H}_i(z) = z^{-i}H_i(z^{-1})$.

(d) Show that the coefficient of z^{-i} in $H_i(z)$ is given by c_i.

(e) By combining the results of (b)–(d), we may derive an algorithm for finding the c_i required by the lattice structure to implement an arbitrary order-M nonrecursive transfer function $H(z)$. Start with $i = M$, so that $H_M(z) = H(z)$. The result of (d) implies that c_M is the coefficient of z^{-M} in $H(z)$. By decreasing i, continue this algorithm to find the remaining c_i. *Hint:* Use the result of (b) to find a two-by-two matrix $\mathbf{A}(z)$ such that

$$\begin{bmatrix} \widetilde{H}_{i-1}(z) \\ H_{i-1}(z) \end{bmatrix} = \mathbf{A}(z) \begin{bmatrix} \widetilde{H}_i(z) \\ H_i(z) \end{bmatrix}.$$

7.52 Causal filters always have a nonzero phase response. One technique for attaining a zero phase response

from a causal filter involves filtering the signal twice, once in the forward direction and the second time in the reverse direction. We may describe this operation in terms of the input $x[n]$ and filter impulse response $h[n]$ as follows: Let $y_1[n] = x[n] * h[n]$ represent filtering the signal in the forward direction. Now filter $y_1[n]$ backwards to obtain $y_2[n] = y_1[-n] * h[n]$. The output is then given by reversing $y_2[n]$ to obtain $y[n] = y_2[-n]$.

(a) Show that this set of operations is represented equivalently by a filter with impulse response $h_o[n]$ as $y[n] = x[n] * h_o[n]$, and express $h_o[n]$ in terms of $h[n]$.

(b) Show that $h_o[n]$ is an even signal and that the phase response of any system with an even impulse response is zero.

(c) For every pole or zero at $z = \beta$ in $h[n]$, show that $h_o[n]$ has a pair of poles or zeros at $z = \beta$ and $z = \frac{1}{\beta}$.

7.53 The present value of a loan with interest compounded monthly may be described in terms of the first-order difference equation

$$y[n] = \rho y[n-1] - x[n],$$

where $\rho = \left(1 + \frac{r/12}{100}\right)$, r is the annual interest rate expressed as a percent, $x[n]$ is the payment credited at the end of the nth month, and $y[n]$ is the loan balance at the beginning of the $(n + 1)$th month. The beginning loan balance is the initial condition $y[-1]$. If uniform payments of c are made for L consecutive months, then $x[n] = c\{u[n] - u[n - L]\}$.

(a) Use the unilateral z-transform to show that

$$Y(z) = \frac{y[-1]\rho - c\sum_{n=0}^{L-1} z^{-n}}{1 - \rho z^{-1}}$$

Hint: Use long division to demonstrate that

$$\frac{1 - z^{-L}}{1 - z^{-1}} = \sum_{n=0}^{L-1} z^{-n}.$$

(b) Show that $z = \rho$ must be a zero of $Y(z)$ if the loan is to have zero balance after L payments.

(c) Find the monthly payment c as a function of the initial loan value $y[-1]$ and the interest rate r, assuming that the loan has zero balance after L payments.

COMPUTER EXPERIMENTS

7.54 Use the MATLAB command `zplane` to obtain a pole–zero plot for the following systems:

(a) $H(z) = \dfrac{1 + z^{-2}}{2 + z^{-1} - \frac{1}{2}z^{-2} + \frac{1}{4}z^{-3}}$

(b) $H(z) = \dfrac{1 + z^{-1} + \frac{3}{2}z^{-2} + \frac{1}{2}z^{-3}}{1 + \frac{3}{2}z^{-1} + \frac{1}{2}z^{-2}}$

7.55 Use the MATLAB command `residuez` to obtain the partial-fraction expansions required to solve Problem 7.24(d)–(g).

7.56 Use the MATLAB command `tf2ss` to find state-variable descriptions of the systems in Problem 7.27.

7.57 Use the MATLAB command `ss2tf` to find the transfer functions in Problem 7.33.

7.58 Use the MATLAB command `zplane` to solve Problems 7.35(a) and (b).

7.59 Use the MATLAB command `freqz` to evaluate and plot the magnitude and phase response of the system given in Example 7.21.

7.60 Use the MATLAB command `freqz` to evaluate and plot the magnitude and phase response of the systems given in Problem 7.37.

7.61 Use the MATLAB commands `filter` and `filtic` to plot the loan balance at the start of each month $n = 0, 1, \ldots L + 1$ in Problem 7.53. Assume that $y[-1] = \$10,000$, $L = 60$, $r = 10\%$ and the monthly payment is chosen to bring the loan balance to zero after 60 payments.

7.62 Use the MATLAB command `zp2sos` to determine a cascade connection of second-order sections for implementing the systems in Problem 7.38.

7.63 A causal discrete-time LTI system has the transfer function

(a) Use the locations of poles and zeros to sketch the magnitude response.

(b) Use the MATLAB commands `zp2tf` and `freqz` to evaluate and plot the magnitude and phase response.

(c) Use the MATLAB command `zp2sos` to obtain a representation of this filter as a cascade of two second-order sections with real-valued coefficients.

(d) Use the MATLAB command `freqz` to evaluate and plot the magnitude response of each section in part (c).

(e) Use the MATLAB command `filter` to determine the impulse response of the system by obtaining the output for an input $x[n] = \delta[n]$.

(f) Use the MATLAB command `filter` to determine the system output for the input

$$x[n] = \left(1 + \cos\left(\frac{\pi}{4}n\right) + \cos\left(\frac{\pi}{2}n\right)\right.$$
$$\left. + \cos\left(\frac{3\pi}{4}n\right) + \cos(\pi n)\right)u[n].$$

Plot the first 250 points of the input and output.

$$H(z) = \frac{0.0976(z - 1)^2(z + 1)^2}{(z - 0.3575 - j0.5889)(z - 0.3575 + j0.5889)(z - 0.7686 - j0.3338)(z - 0.7686 + j0.3338)}.$$

8 | Application to Filters and Equalizers

8.1 *Introduction*

In Chapters 3–5, we made use of *filters* as functional blocks to suppress spurious signals by exploiting the fact that the frequency content of these signals is separated from the frequency content of wanted signals. Equalization was discussed in Chapters 2–4 in the context of inverse systems, and particularly the theme example on multipath channels; in Chapter 5, we made use of *equalizers* as functional blocks to compensate for distortion that arises when a signal is transmitted through a physical system such as a telephone channel. The treatments of both filters and equalizers presented in those chapters were from a system-theoretic viewpoint. Now that we have the Laplace and *z*-transforms at our disposal, we are ready to describe procedures for the *design* of these two important functional blocks.

We begin the discussion by considering the issue of *distortionless transmission*, which is basic to the study of linear filters and equalizers. This leads naturally to a discussion of an idealized framework for filtering, which, in turn, provides the basis for the design of practical filters. The design of a filter can be accomplished by using continuous-time concepts, in which case we speak of *analog filters*. Alternatively, the design can be accomplished by using discrete-time concepts, in which case we speak of *digital filters*. Analog and digital filters have their own advantages and disadvantages. Both of these types of filter are discussed in this chapter. The topic of equalization is covered toward the end of the chapter.

8.2 *Conditions for Distortionless Transmission*

Consider a continuous-time LTI system with impulse response $h(t)$. Equivalently, the system may be described in terms of its frequency response $H(j\omega)$, defined as the Fourier transform of $h(t)$. Let a signal $x(t)$ with Fourier transform $X(j\omega)$ be applied to the input of the system. Let the signal $y(t)$ with Fourier transform $Y(j\omega)$ denote the output of the system. We wish to know conditions for *distortionless transmission* through the system. By "distortionless transmission" we mean that the output signal of the system is an exact replica of the input signal, except, possibly, for two minor modifications:

- ▶ A scaling of amplitude
- ▶ A constant time delay

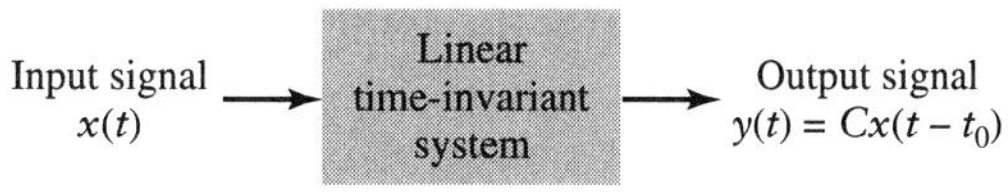

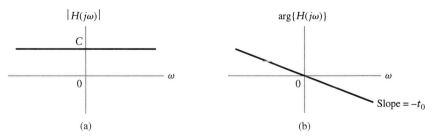

FIGURE 8.1   Time-domain condition for distortionless transmission of a signal through a linear time-invariant system.

On this basis, we say that a signal $x(t)$ is transmitted through the system without distortion if the output signal $y(t)$ is defined by (see Fig. 8.1)

$$y(t) = Cx(t - t_0), \tag{8.1}$$

where the constant C accounts for a change in amplitude and the constant t_0 accounts for a delay in transmission.

Applying the Fourier transform to Eq. (8.1) and using the time-shifting property of the Fourier transform described in Table 3.7, we get

$$Y(j\omega) = CX(j\omega)e^{-j\omega t_0}. \tag{8.2}$$

The frequency response of a distortionless LTI system is therefore

$$\begin{aligned} H(j\omega) &= \frac{Y(j\omega)}{X(j\omega)} \\ &= Ce^{-j\omega t_0}. \end{aligned} \tag{8.3}$$

Correspondingly, the impulse response of the system is given by

$$h(t) = C\delta(t - t_0). \tag{8.4}$$

Equations (8.3) and (8.4) describe the frequency-domain and time-domain conditions, respectively, that an LTI system has to satisfy for distortionless transmission. From a practical viewpoint, Eq. (8.3) is the more revealing of the two, indicating that, in order to achieve the distortionless transmission of a signal with some finite frequency content through a continuous-time LTI system, the frequency response of the system must satisfy two conditions:

1. The magnitude response $|H(j\omega)|$ must be constant for all frequencies of interest; that is, we must have

$$\boxed{|H(j\omega)| = C} \tag{8.5}$$

   for some constant C.

2. For the same frequencies of interest, the phase response $\arg\{H(\omega)\}$ must be linear in frequency, with slope $-t_0$ and intercept zero; that is, we must have

$$\boxed{\arg\{H(j\omega)\} = -\omega t_0.} \tag{8.6}$$

These two conditions are illustrated in Figs. 8.2(a) and (b), respectively.

FIGURE 8.2   Frequency response for distortionless transmission through a linear time-invariant system. (a) Magnitude response. (b) Phase response.

Consider next the case of a discrete-time LTI system with transfer function $H(e^{j\Omega})$. Following a procedure similar to that just described, we may show that the conditions for distortionless transmission through such a system are as follows:

1. The magnitude response $|H(e^{j\Omega})|$ is constant for all frequencies of interest; that is,

$$\boxed{|H(e^{j\Omega})| = C,}$$
(8.7)

   where C is a constant.

2. For the same frequencies of interest, the phase response $\arg\{H(e^{j\Omega})\}$ is linear in frequency; that is,

$$\boxed{\arg\{H(e^{j\Omega})\} = -\Omega n_0,}$$
(8.8)

   where n_0 accounts for delay in transmission through the discrete-time LTI system.

▶ **Problem 8.1**   Using the impulse response of Eq. (8.4) in the convolution integral, show that the input–output relation of a distortionless system is as given in Eq. (8.1). ◀

EXAMPLE 8.1 PHASE RESPONSE FOR DISTORTIONLESS TRANSMISSION   Suppose that the condition of Eq. (8.6) on the phase response $\arg\{H(j\omega)\}$ for distortionless transmission is modified by adding a constant phase angle equal to a positive or negative integer multiple of π radians (i.e., 180°). What is the effect of this modification?

Solution: We begin by rewriting Eq. (8.6) as

$$\arg\{H(j\omega)\} = \omega t_0 + k\pi,$$

where k is an integer. Correspondingly, the frequency response of the system given in Eq. (8.3) takes the new form

$$H(j\omega) = Ce^{-j(\omega t_0 + k\pi)}.$$

But

$$e^{+jk\pi} = \begin{cases} -1, & k = \pm1, \pm3, \ldots \\ +1, & k = 0, \pm2, \pm4, \ldots \end{cases}.$$

Therefore,

$$H(j\omega) = \pm Ce^{-j\omega t_0},$$

which is of exactly the same form as Eq. (8.3), except for a possible change in the algebraic sign of the scaling factor C. We conclude that the conditions for distortionless transmission through a linear time-invariant system remain unchanged when the phase response of the system is changed by a constant amount equal to a positive or negative integer multiple of 180°. ■

8.3 *Ideal Low-Pass Filters*

Typically, the spectral content of an information-bearing signal occupies a frequency band of some finite extent. For example, the spectral content of a speech signal essential for telephonic communication lies in the frequency band from 300 to 3100 Hz. To extract the essential information content of a speech signal for such an application, we need a frequency-selective system—that is, a filter which limits the spectrum of the signal to the desired band of frequencies. Indeed, filters are basic to the study of signals and systems, in the sense that every system used to process signals contains a filter of some kind in its composition.

As noted in Chapter 3, the frequency response of a filter is characterized by a *passband* and a *stopband*, which are separated by a *transition band*, also known as a *guard band*. Signals with frequencies inside the passband are transmitted with little or no distortion, whereas those with frequencies inside the stopband are effectively rejected. The filter may thus be of the low-pass, high-pass, band-pass, or band-stop type, depending on whether it transmits low, high, intermediate, or all but intermediate frequencies, respectively.

Consider, then, an *ideal* low-pass filter, which transmits all the low frequencies inside the passband without any distortion and rejects all the high frequencies inside the stopband. The transition from the passband to the stopband is assumed to occupy zero width. Insofar as low-pass filtering is concerned, the primary interest is in the faithful transmission of an information-bearing signal whose spectral content is confined to some frequency band defined by $0 \leq \omega \leq \omega_c$. Accordingly, in such an application, the conditions for distortionless transmission need be satisfied only inside the passband of the filter, as illustrated in Fig. 8.3. Specifically, the frequency response of an ideal low-pass filter with cutoff frequency ω_c is defined by

$$H(j\omega) = \begin{cases} e^{-j\omega t_0}, & |\omega| \leq \omega_c \\ 0, & |\omega| > \omega_c \end{cases} \tag{8.9}$$

where, for convenience of presentation, we have set the constant $C = 1$. For a finite delay t_0, the ideal low-pass filter is noncausal, which is confirmed next by examining the impulse response $h(t)$ of the filter.

To evaluate $h(t)$, we take the inverse Fourier transform of Eq. (8.9), obtaining

$$\begin{aligned} h(t) &= \frac{1}{2\pi} \int_{-\omega_c}^{\omega_c} e^{j\omega(t-t_0)} \, d\omega \\ &= \frac{1}{2\pi} \frac{e^{j\omega(t-t_0)}}{j(t-t_0)} \bigg|_{\omega=-\omega_c}^{\omega_c} \\ &= \frac{\sin(\omega_c(t-t_0))}{\pi(t-t_0)}. \end{aligned} \tag{8.10}$$

Recall the definition of the sinc function given by Eq. (3.24):

$$\text{sinc}(\omega t) = \frac{\sin(\pi \omega t)}{\pi \omega t}. \tag{8.11}$$

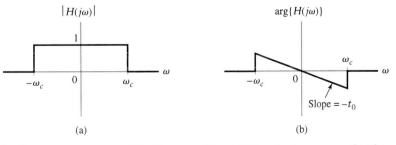

(a)   (b)

FIGURE 8.3   Frequency response of ideal low-pass filter. (a) Magnitude response. (b) Phase response.

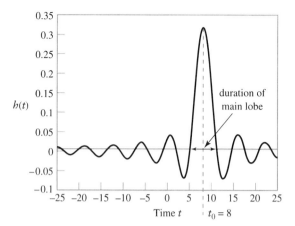

FIGURE 8.4 Time-shifted form of the sinc function, representing the impulse response of an ideal (but noncausal) low-pass filter for $\omega_c = 1$ and $t_0 = 8$.

Accordingly, we may rewrite Eq. (8.10) in the compact form

$$h(t) = \frac{\omega_c}{\pi} \operatorname{sinc}\left(\frac{\omega_c}{\pi}(t - t_0) \right). \tag{8.12}$$

This impulse response has a peak amplitude of ω_c/π, centered at time t_0, as shown in Fig. 8.4 for $\omega_c = 1$ and $t_0 = 8$. The duration of the mainlobe of the impulse response is $2\pi/\omega_c$, and the rise time from the zero at the beginning of the mainlobe to the peak is π/ω_c. We see from the figure that, for any finite value of t_0, including $t_0 = 0$, there is some response from the filter before the time $t = 0$ at which the unit impulse is applied to the input of the filter. This response confirms that the ideal low-pass filter is noncausal.

Despite its noncausal nature, the ideal low-pass filter is a useful theoretical concept. In particular, it provides the framework for the design of practical (i.e., causal) filters.

■ 8.3.1 TRANSMISSION OF A RECTANGULAR PULSE THROUGH AN IDEAL LOW-PASS FILTER

A rectangular pulse plays a key role in digital communications. For example, for the electrical representation of a binary sequence transmitted through a channel, we may use the following protocols:

▶ Transmit a rectangular pulse for symbol 1.

▶ Switch off the pulse for symbol 0.

Consider, then, a rectangular pulse

$$x(t) = \begin{cases} 1, & |t| \le \dfrac{T_0}{2} \\ 0, & |t| > \dfrac{T_0}{2} \end{cases} \tag{8.13}$$

of unit amplitude and duration T_0. This pulse is applied to a communication channel modeled as an ideal low-pass filter whose frequency response is defined by Eq. (8.9). The issue of interest is that of determining the response of the channel to the pulse input.

The impulse response of the filter representing the channel is given by Eq. (8.12), which we may rewrite as

$$h(t) = \frac{\omega_c}{\pi} \frac{\sin(\omega_c(t - t_0))}{\omega_c(t - t_0)}. \tag{8.14}$$

Using the convolution integral, we may express the response of the filter as

$$y(t) = \int_{-\infty}^{\infty} x(\tau)h(t - \tau)\, d\tau. \tag{8.15}$$

Substituting Eqs. (8.13) and (8.14) into Eq. (8.15), we get

$$y(t) = \frac{\omega_c}{\pi} \int_{-T_0/2}^{T_0/2} \frac{\sin(\omega_c(t - t_0 - \tau))}{\omega_c(t - t_0 - \tau)}\, d\tau.$$

Let

$$\lambda = \omega_c(t - t_0 - \tau).$$

Then, changing the variable of integration from τ to λ, we may rewrite $y(t)$ as

$$
\begin{aligned}
y(t) &= \frac{1}{\pi} \int_b^a \frac{\sin \lambda}{\lambda}\, d\lambda \\
&= \frac{1}{\pi} \left[\int_0^a \frac{\sin \lambda}{\lambda}\, d\lambda - \int_0^b \frac{\sin \lambda}{\lambda}\, d\lambda \right],
\end{aligned}
\tag{8.16}
$$

where the limits of integration, a and b, are defined by

$$a = \omega_c\left(t - t_0 + \frac{T_0}{2} \right) \tag{8.17}$$

and

$$b = \omega_c\left(t - t_0 - \frac{T_0}{2} \right). \tag{8.18}$$

To rewrite Eq. (8.16) in a compact form, we introduce the *sine integral*, defined by

$$\boxed{\; \mathrm{Si}(u) = \int_0^u \frac{\sin \lambda}{\lambda}\, d\lambda. \;} \tag{8.19}$$

The sine integral cannot be evaluated in closed form in terms of elementary functions, but it can be integrated by using a power series. Its plot is shown in Fig. 8.5. From this figure, we see that

▸ The sine integral $\mathrm{Si}(u)$ has odd symmetry about the origin $u = 0$;
▸ it has maxima and minima at multiples of π; and
▸ it approaches the limiting value of $\pm\pi/2$ for large values of $|u|$.

Using the definition of the sine integral in Eq. (8.19), we may rewrite the response $y(t)$ defined in Eq. (8.16) in the compact form

$$\boxed{\; y(t) = \frac{1}{\pi}[\mathrm{Si}(a) - \mathrm{Si}(b)], \;} \tag{8.20}$$

where a and b are themselves defined in Eqs. (8.17) and (8.18), respectively.

Figure 8.6 depicts the response $y(t)$ for three different values of the cutoff frequency ω_c, assuming that the pulse duration $T_0 = 1$ s and the transmission delay t_0 is zero. In each case,

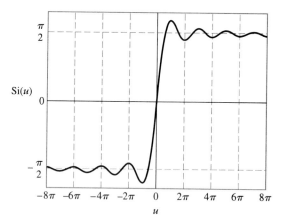

FIGURE 8.5   Sine integral.

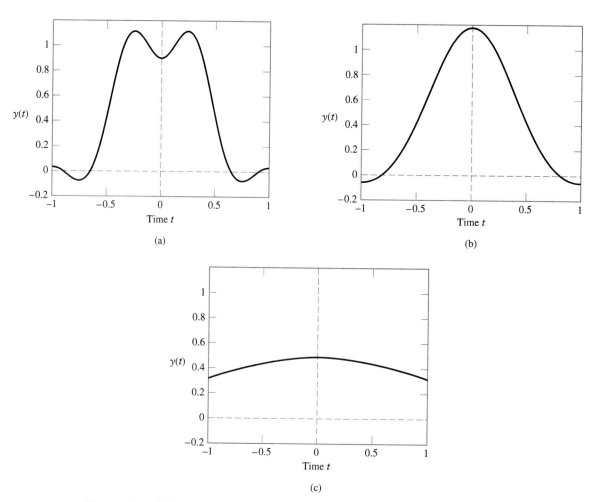

FIGURE 8.6   Pulse response of an ideal low-pass filter for input pulse of duration $T_0 = 1$ s and varying filter cutoff frequency ω_c: (a) $\omega_c = 4\pi$ rad/s; (b) $\omega_c = 2\pi$ rad/s; and (c) $\omega_c = 0.4\pi$ rad/s.

we see that the response $y(t)$ is symmetric about $t = 0$. We further observe that the shape of the response $y(t)$ is markedly dependent on the cutoff frequency. In particular, we note the following points:

1. When ω_c is larger than $2\pi/T_0$, as in Fig. 8.6(a), the response $y(t)$ has approximately the same duration as the rectangular pulse $x(t)$ applied to the filter input. However, it differs from $x(t)$ in two major respects:

   ▸ Unlike the input $x(t)$, the response $y(t)$ has nonzero rise and fall times that are inversely proportional to the cutoff frequency ω_c.

   ▸ The response $y(t)$ exhibits *ringing* at both the leading and trailing edges.

2. When $\omega_c = 2\pi/T_0$, as in Fig. 8.6(b), the response $y(t)$ is recognizable as a pulse. However, the rise and fall times of $y(t)$ are now significant compared with the duration of the input rectangular pulse $x(t)$.

3. When the cutoff frequency ω_c is smaller than $2\pi/T_0$, as in Fig. 8.6(c), the response $y(t)$ is a grossly distorted version of the input $x(t)$.

These observations point to the inverse relationship that exists between the two parameters: (1) the duration of the rectangular input pulse applied to an ideal low-pass filter and (2) the cutoff frequency of the filter. This inverse relationship is a manifestation of the constancy of the time–bandwidth product discussed in Chapter 3. From a practical perspective, the inverse relationship between pulse duration and filter cutoff frequency has a simple interpretation, as illustrated here, in the context of digital communications: If the requirement is merely that of recognizing that the response of a low-pass channel is due to the transmission of the symbol 1, represented by a rectangular pulse of duration T_0, it is adequate to set the cutoff frequency of the channel at $\omega_c = 2\pi/T_0$.

EXAMPLE 8.2 OVERSHOOT FOR INCREASING CUTOFF FREQUENCY The response $y(t)$ shown in Fig. 8.6(a), corresponding to a cutoff frequency $\omega_c = 4\pi/T_0$ for $T_0 = 1$ s, exhibits an overshoot of approximately 9%. Investigate what happens to this overshoot when the cutoff frequency ω_c is allowed to approach infinity.

Solution: In Figs. 8.7(a) and (b), we show the pulse response of the ideal low-pass filter for cutoff frequency $\omega_c = 10\pi/T_0$ and $\omega_c = 40\pi/T_0$. The two graphs illustrate that the overshoot remains approximately equal to 9% in a manner that is practically independent of how large the cutoff frequency ω_c is. This result is, in fact, another manifestation of the *Gibbs phenomenon* discussed in Chapter 3.

To provide an analytic proof of what the graphs illustrate, we observe from Fig. 8.5 that the sine integral $\text{Si}(u)$ defined in Eq. (8.19) oscillates at a frequency of $1/(2\pi)$. The implication of this observation is that the filter response $y(t)$ will oscillate at a frequency equal to $\omega_c/(2\pi)$, where ω_c is the cutoff frequency of the filter. The filter response $y(t)$ has its first maximum at

$$t_{\max} = \frac{T_0}{2} - \frac{\pi}{\omega_c}. \tag{8.21}$$

Correspondingly, the integration limits a and b defined in Eqs. (8.17) and (8.18) take on the following values (assuming that $t_0 = 0$):

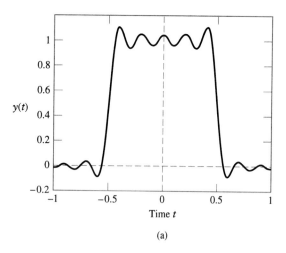

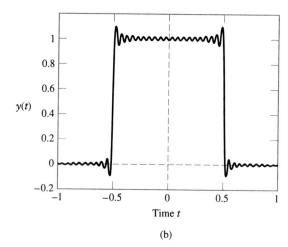

FIGURE 8.7   The Gibbs phenomenon, exemplified by the pulse response of an ideal low-pass filter. The overshoot remains essentially the same despite a significant increase in the cutoff frequency ω_c: (a) $\omega_c T_0 = 10\pi$ rad and (b) $\omega_c T_0 = 40\pi$ rad. The pulse duration T_0 is maintained constant at 1 s.

$$
\begin{aligned}
a_{\max} &= \omega_c \left(t_{\max} + \frac{T_0}{2} \right) \\
&= \omega_c \left(\frac{T_0}{2} - \frac{\pi}{\omega_c} + \frac{T_0}{2} \right) \\
&= \omega_c T_0 - \pi;
\end{aligned}
\tag{8.22}
$$

$$
\begin{aligned}
b_{\max} &= \omega_c \left(t_{\max} - \frac{T_0}{2} \right) \\
&= \omega_c \left(\frac{T_0}{2} - \frac{\pi}{\omega_c} - \frac{T_0}{2} \right) \\
&= -\pi.
\end{aligned}
\tag{8.23}
$$

Substituting Eqs. (8.22) and (8.23) into (8.20) yields

$$
\begin{aligned}
y(t_{\max}) &= \frac{1}{\pi} [\mathrm{Si}(a_{\max}) - \mathrm{Si}(b_{\max})] \\
&= \frac{1}{\pi} [\mathrm{Si}(\omega_c T_0 - \pi) - \mathrm{Si}(-\pi)] \\
&= \frac{1}{\pi} [\mathrm{Si}(\omega_c T_0 - \pi) + \mathrm{Si}(\pi)].
\end{aligned}
\tag{8.24}
$$

Let

$$
\mathrm{Si}(\omega_c T_0 - \pi) = \frac{\pi}{2} (1 + \Delta),
\tag{8.25}
$$

where Δ is the absolute value of the deviation in the value of $\mathrm{Si}(\omega_c T_0 - \pi)$, expressed as a fraction of the final value $+\pi/2$. The maximum value of $\mathrm{Si}(u)$ occurs at $u_{\max} = \pi$ and is equal to 1.852, which we may write as $(1.179)(\pi/2)$; that is,

$$
\mathrm{Si}(\pi) = (1.179) \left(\frac{\pi}{2} \right).
$$

Hence, we may rewrite Eq. (8.24) as

$$y(t_{\max}) = \tfrac{1}{2}(1.179 + 1 + \Delta)$$
$$= 1.09 + \tfrac{1}{2}\Delta. \tag{8.26}$$

Viewing ω_c as a measure of the filter's bandwidth, we note from Fig. 8.5 that for a time–bandwidth product $\omega_c T_0$ large compared with unity, the fractional deviation Δ has a very small value. We may thus write the approximation

$$y(t_{\max}) \simeq 1.09 \quad \text{for } \omega_c \gg 2\pi/T_0, \tag{8.27}$$

which shows that the overshoot in the filter response is approximately 9%, a result that is practically independent of the cutoff frequency ω_c. ∎

8.4 *Design of Filters*

The low-pass filter with frequency response shown in Fig. 8.3 is "ideal" in that it passes all frequency components lying inside the passband with no distortion and rejects all frequency components lying inside the stopband, and the transition from the passband to the stopband is abrupt. Recall that these characteristics result in a nonimplementable filter. Therefore, from a practical perspective, the prudent approach is to tolerate an acceptable level of distortion by permitting prescribed "deviations" from these ideal conditions, as described here for the case of continuous-time or analog filters:

▶ Inside the passband, the magnitude response of the filter should lie between 1 and $1 - \epsilon$; that is,

$$1 - \epsilon \leq |H(j\omega)| \leq 1 \quad \text{for } 0 \leq |\omega| \leq \omega_p, \tag{8.28}$$

where ω_p is the *passband cutoff frequency* and ϵ is a *tolerance parameter*.

▶ Inside the stopband, the magnitude response of the filter should not exceed δ; that is,

$$|H(j\omega)| \leq \delta \quad \text{for } |\omega| \geq \omega_s, \tag{8.29}$$

where ω_s is the *stopband cutoff frequency* and δ is another tolerance parameter. (The parameter δ used here should not be confused with the symbol for the unit impulse.)

▶ The transition bandwidth has a finite width equal to $\omega_s - \omega_p$.

The tolerance diagram of Fig. 8.8 presents a portrayal of these filter specifications. Analogous specifications are used for discrete-time filters, with the added provision that the response is always 2π periodic in Ω. So long as these specifications meet the goal for the filtering problem at hand and the filter design is accomplished at a reasonable cost, the job is satisfactorily done. Indeed, this is the very nature of engineering design.

The specifications just described favor an approach that focuses on the design of the filter based on its frequency response rather than its impulse response. This is in recognition of the fact that the application of a filter usually involves the separation of signals on the basis of their frequency content.

Having formulated a set of specifications describing the desired properties of the frequency-selective filter, we set forth two distinct steps involved in the design of the filter, pursued in the following order:

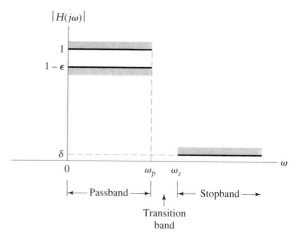

FIGURE 8.8   Tolerance diagram of a practical low-pass filter: The passband, transition band, and stopband are shown for positive frequencies.

1. The *approximation* of a prescribed frequency response (i.e., magnitude response, phase response, or both) by a rational transfer function which represents a system that is both causal and stable.

2. The *realization* of the approximating transfer function by a physical system.

Both of these steps can be implemented in a variety of ways, with the result that there is no unique solution to the filter design problem for a prescribed set of specifications.

Nevertheless, we may mention three different approaches to the design of analog and digital filters, as summarized here:

1. *Analog approach*, which applies to the class of analog filters.

2. *Analog-to-digital approach*, where the motivation is to design a digital filter by building on what we know about an analog filter design.

3. *Direct digital approach*, which applies to the class of digital filters.

In what follows, the basic ideas of these approaches, are presented and illustrated with different design examples.

8.5 *Approximating Functions*

The choice of a transfer function for solving the approximation problem is the transition step from a set of design specifications to the realization of the transfer function by means of a specific filter structure. Accordingly, this is the most fundamental step in filter design, because the choice of the transfer function determines the performance of the filter. At the outset, however, it must be emphasized that there is no unique solution to the approximation problem. Rather, we have a set of possible solutions, each with its own distinctive properties.

Basically, the approximation problem is an *optimization problem* that can be solved only in the context of a specific *criterion of optimality*. In other words, before we proceed to solve the approximation problem, we have to specify a criterion of optimality in an implicit or explicit sense. Moreover, the choice of that criterion uniquely determines the solution. Two optimality criteria commonly used in filter design are as follows:

1. *Maximally flat magnitude response.*
Let $|H(j\omega)|$ denote the magnitude response of an analog low-pass filter of order k, where K is an integer. Then the magnitude response $|H(j\omega)|$ is said to be *maximally flat* at the origin if its multiple derivatives with respect to ω vanish at $\omega = 0$—that is, if

$$\frac{\partial^K}{\partial\omega^K}|H(j\omega)| = 0 \quad \text{at } \omega = 0 \quad \text{and} \quad k = 1, 2, \ldots, 2K - 1.$$

2. *Equiripple magnitude response.*
Let the squared value of the magnitude response $|H(j\omega)|$ of an analog low-pass filter be expressed in the form

$$|H(j\omega)|^2 = \frac{1}{1 + \gamma^2 F^2(\omega)},$$

where γ is related to the passband tolerance parameter ϵ and $F(\omega)$ is some function of ω. Then the magnitude response $|H(j\omega)|$ is said to be *equiripple in the passband* if $F^2(\omega)$ oscillates between maxima and minima of equal amplitude over the entire passband. Here we must distinguish between two cases, depending on whether the filter order K is odd or even. We illustrate the formulation of this second optimality criterion for two cases, $K = 3$ and $K = 4$, as follows (see Fig. 8.9):

   Case (a): $K = 3$ and $\omega_c = 1$

   (i) $F^2(\omega) = 0$ if $\omega = 0, \pm\omega_a$.
   (ii) $F^2(\omega) = 1$ if $\omega = \pm\omega_b, \pm1$.

   (iii) $\dfrac{\partial}{\partial\omega}F^2(\omega) = 0$ if $\omega = 0, \pm\omega_b, \pm\omega_a$,

      where $0 < \omega_b < \omega_a < 1$.

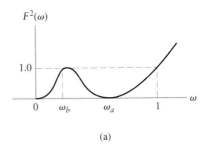

(a)

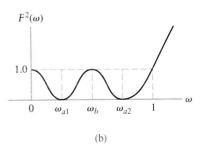

(b)

FIGURE 8.9   Two different forms of function $F^2(\omega)$: (a) $K = 3$. (b) $K = 4$.

Case (b): $K = 4$, and $\omega_c = 1$

(i) $F(\omega) = 0$ if $\omega = \pm\omega_{a1}, \pm\omega_{a2}$.

(ii) $F^2(\omega) = 1$ if $\omega = 0, \pm\omega_b, \pm1$.

(iii) $\dfrac{\partial}{\partial\omega}F^2(\omega) = 0$ if $\omega = 0, \pm\omega_{a1}, \pm\omega_b, \pm\omega_{a2}$,

where $0 < \omega_{a1} < \omega_b < \omega_{a2} < 1$.

The two optimality criteria described under points 1 and 2 are satisfied by two classes of filters known as Butterworth filters and Chebyshev filters, respectively. They are both described in what follows.

■ 8.5.1 BUTTERWORTH FILTERS

A Butterworth function of order K is defined by

$$|H(j\omega)|^2 = \frac{1}{1 + \left(\dfrac{\omega}{\omega_c}\right)^{2K}}, \qquad K = 1, 2, 3, \ldots, \tag{8.30}$$

and a filter so designed is referred to as a *Butterworth filter of order K*.

The approximating function of Eq. (8.30) satisfies the requirement that $|H(j\omega)|$ is an even function of ω. The parameter ω_c is the *cutoff frequency* of the filter. For prescribed values of tolerance parameters ϵ and δ defined in Fig. 8.8, we readily find from Eq. (8.30) that the passband and stopband cutoff frequencies are, respectively,

$$\omega_p = \omega_c\left(\frac{\epsilon}{1-\epsilon}\right)^{1/(2K)} \tag{8.31}$$

and

$$\omega_s = \omega_c\left(\frac{1-\delta}{\delta}\right)^{1/(2K)}. \tag{8.32}$$

The squared magnitude response $|H(j\omega)|^2$ obtained by using the approximating function of Eq. (8.30) is plotted in Fig. 8.10 for four different values of filter order K as a function of the normalized frequency ω/ω_c. All these curves pass through the half-power point at $\omega = \omega_c$.

A Butterworth function is monotonic throughout the passband and stopband. In particular, in the vicinity of $\omega = 0$, we may expand the magnitude of $H(j\omega)$ as a power series:

$$|H(j\omega)| = 1 - \frac{1}{2}\left(\frac{\omega}{\omega_c}\right)^{2K} + \frac{3}{8}\left(\frac{\omega}{\omega_c}\right)^{4K} - \frac{5}{16}\left(\frac{\omega}{\omega_c}\right)^{6K} + \cdots. \tag{8.33}$$

This equation implies that the first $2K - 1$ derivatives of $|H(j\omega)|$ with respect to ω are zero at the origin. It follows that the Butterworth function is indeed *maximally flat* at $\omega = 0$.

To design an analog filter, we need to know the transfer function $H(s)$, expressed as a function of the complex variable s. Given the Butterworth function $|H(j\omega)|^2$, how do we find the corresponding transfer function $H(s)$? To address this issue, we put $j\omega = s$ and recognize that

$$H(s)H(-s)|_{s=j\omega} = |H(j\omega)|^2. \tag{8.34}$$

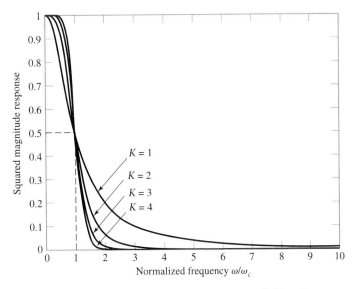

FIGURE 8.10 Squared magnitude response of Butterworth filter for varying orders.

S & S

Solutions

Hence, setting $\omega = s/j$, we may rewrite Eq. (8.30) in the equivalent form

$$H(s)H(-s) = \frac{1}{1 + \left(\dfrac{s}{j\omega_c}\right)^{2K}}. \tag{8.35}$$

The roots of the denominator polynomial are located at the following points in the s plane:

$$\begin{aligned} s &= j\omega_c(-1)^{1/(2K)} \\ &= \omega_c e^{j\pi(2k+1)/(2K)} \quad \text{for } k = 0, 1, \ldots, 2K - 1. \end{aligned} \tag{8.36}$$

That is, the poles of $H(s)H(-s)$ form symmetrical patterns on a circle of radius ω_c, as illustrated in Fig. 8.11 for $K = 3$ and $K = 4$. Note that, for any K, none of the poles fall on the imaginary axis of the s-plane.

Which of these $2K$ poles belong to $H(s)$? To answer this fundamental question, we recall from Chapter 6 that, for the transfer function $H(s)$ to represent a stable and causal filter, all of its poles must lie in the left half of the s-plane. Accordingly, those K poles of $H(s)H(-s)$ which lie in the left half of the s-plane are allocated to $H(s)$, and the remaining right-half poles are allocated to $H(-s)$. So, when $H(s)$ is stable, $H(-s)$ is unstable.

EXAMPLE 8.3 BUTTERWORTH LOW-PASS FILTER OF ORDER 3 Determine the transfer function of a Butterworth filter of the low-pass type of order $K = 3$. Assume that the 3-dB cutoff frequency $\omega_c = 1$.

Solution: For filter order $K = 3$, the $2K = 6$ poles of $H(s)H(-s)$ are located on a circle of unit radius with angular spacing $60°$, as shown in Fig. 8.11(a). Hence, allocating the left-half plane poles to $H(s)$, we may define them as

$$s = -\frac{1}{2} + j\frac{\sqrt{3}}{2},$$

$$s = -1,$$

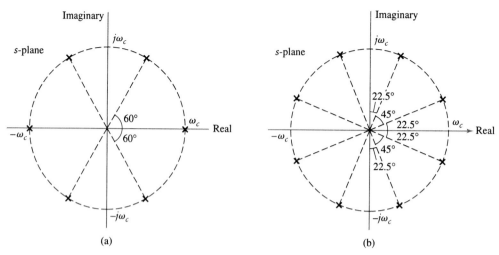

(a)   (b)

FIGURE 8.11   Distribution of poles of $H(s)H(-s)$ in the s-plane for two different filter orders: (a) $K = 3$ and (b) $K = 4$, for which the total number of poles is 6 and 8, respectively.

S & S
Solutions

and

$$s = -\frac{1}{2} - j\frac{\sqrt{3}}{2}.$$

The transfer function of a Butterworth filter of order 3 is therefore

$$H(s) = \frac{1}{(s + 1)\left(s + \frac{1}{2} - j\frac{\sqrt{3}}{2}\right)\left(s + \frac{1}{2} + j\frac{\sqrt{3}}{2}\right)}$$

$$= \frac{1}{(s + 1)(s^2 + s + 1)} \tag{8.37}$$

$$= \frac{1}{s^3 + 2s^2 + 2s + 1}.$$ ■

▶ **Problem 8.2**   How is the transfer function of Eq. (8.37) modified for a Butterworth filter of order 3 and cutoff frequency ω_c?

Answer:

$$H(s) = \frac{1}{\left(\frac{s}{\omega_c}\right)^3 + 2\left(\frac{s}{\omega_c}\right)^2 + 2\left(\frac{s}{\omega_c}\right) + 1}$$ ◀

S & S
Solutions

▶ **Problem 8.3**   Find the transfer function of a Butterworth filter with cutoff frequency $\omega_c = 1$ and filter order (a) $K = 1$ and (b) $K = 2$.

Answers:

(a) $H(s) = \dfrac{1}{s + 1}$

(b) $H(s) = \dfrac{1}{s^2 + \sqrt{2}\,s + 1}$ ◀

TABLE 8.1   Summary of Butterworth Filter Transfer Functions.

$$H(s) = \frac{1}{Q(s)}$$

Filter Order K	Polynomial $Q(s)$
1	$s + 1$
2	$s^2 + \sqrt{2}\,s + 1$
3	$s^3 + 2s^2 + 2s + 1$
4	$s^4 + 2.6131s^3 + 3.4142s^2 + 2.6131s + 1$
5	$s^5 + 3.2361s^4 + 5.2361s^3 + 5.2361s^2 + 3.2361s + 1$
6	$s^6 + 3.8637s^5 + 7.4641s^4 + 9.1416s^3 + 7.4641s^2 + 3.8637s + 1$

Table 8.1 presents a summary of the transfer functions of Butterworth filters of cutoff frequency $\omega_c = 1$ for up to and including filter order $K = 6$.

■ 8.5.2 CHEBYSHEV FILTERS

The tolerance diagram of Fig. 8.8 calls for an approximating function that lies between 1 and $1 - \epsilon$ inside the passband range $0 \le \omega \le \omega_p$. The Butterworth function meets this requirement, but concentrates its approximating ability near $\omega = 0$. For a given filter order, we can obtain a filter with a reduced transition bandwidth by using an approximating function that exhibits an *equiripple* characteristic in the passband (i.e., it oscillates uniformly between 1 and $1 - \epsilon$ for $0 \le \omega \le \omega_p$), as illustrated in Figs. 8.12(a) and (b) for $K = 3, 4$, respectively, and 0.5-dB ripple in the passband. The magnitude responses plotted here satisfy the equiripple criteria described earlier for K odd and K even, respectively. Approximating functions with an equiripple magnitude response are known collectively

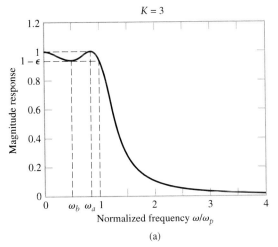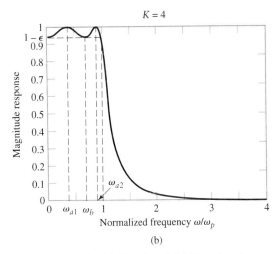

FIGURE 8.12   Magnitude response of Chebyshev filter for order (a) $K = 3$ and (b) $K = 4$ and passband ripple = 0.5 dB. The frequencies ω_b and ω_a in case (a) and the frequencies ω_{a1}, and ω_b, and ω_{a2} in case (b) are defined in accordance with the optimality criteria for equiripple amplitude response.

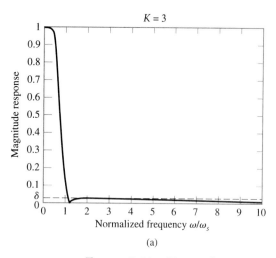

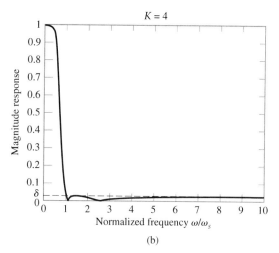

FIGURE 8.13   Magnitude response of inverse Chebyshev filter for order (a) $K = 3$ and (b) $K = 4$ and stopband ripple = 30 dB.

as *Chebyshev functions*. A filter designed on this basis is called a *Chebyshev filter*. The poles of a transfer function $H(s)$ pertaining to a Chebyshev filter lie on an ellipse in the s-plane in a manner closely related to those of the corresponding Butterworth filter.

The Chebyshev functions shown in Fig. 8.12 exhibit a monotonic behavior in the stopband. Alternatively, we may use another class of Chebyshev functions that exhibit a monotonic response in the passband, but an equiripple response in the stopband, as illustrated in Figs. 8.13(a) and (b) for $K = 3$, 4, respectively, and 30-dB stopband ripple. A filter designed on this basis is called an *inverse Chebyshev filter*. Unlike a Chebyshev filter, the transfer function of an inverse Chebyshev filter has zeros on the $j\omega$-axis of the s-plane.

The ideas embodied in Chebyshev and inverse Chebyshev filters can be combined to further reduce the transition bandwidth by making the approximating function equiripple in both the passband and the stopband. Such an approximating function is called an *elliptic function*, and a filter resulting from its use is called an *elliptic filter*. An elliptic filter is optimum in the sense that, for a prescribed set of design specifications, the width of the transition band is the smallest that we can achieve. This permits the smallest possible separation between the passband and stopband of the filter. From the standpoint of analysis, however, determining the transfer function $H(s)$ is simplest for a Butterworth filter and most challenging for an elliptic filter. The elliptic filter is able to achieve its optimum behavior by virtue of the fact that its transfer function $H(s)$ has finite zeros in the s-plane, the number of which is uniquely determined by filter order K. In contrast, the transfer function $H(s)$ of a Butterworth filter or that of a Chebyshev filter has all of its zeros located at $s = \infty$.

8.6 *Frequency Transformations*

Up to this point, we have considered the issue of solving the approximation problem for low-pass filters. In that context, it is common practice to speak of a low-pass "prototype" filter, by which we mean a low-pass filter whose cutoff frequency ω_c is normalized to unity. Given that we have found the transfer function of a low-pass prototype filter, we may use it to derive the transfer function of a low-pass filter with an arbitrary cutoff frequency, a high-pass, a band-pass, or a band-stop filter by means of an appropriate transformation of the independent variable. Such a transformation has no effect on the tolerances within which the ideal characteristic of interest is approximated. In Problem 8.2, we considered low-pass to low-pass transformation. In what follows, we consider two other *frequency transformations*: low pass to high pass and low pass to band pass. Other frequency transformations follow the principles described herein.

■ 8.6.1 Low-Pass to High-Pass Transformation

The points $s = 0$ and $s = \infty$ in the s-plane are of particular interest here. In the case of a low-pass filter, $s = 0$ defines the midpoint of the passband (defined for both positive and negative frequencies), and $s \to \infty$ defines the vicinity where the transfer function of the filter behaves asymptotically. The roles of these two points are interchanged in a high-pass filter. Accordingly, the low-pass to high-pass transformation is described by

$$s \to \frac{\omega_c}{s}, \tag{8.38}$$

where ω_c is the desired cutoff frequency of the high-pass filter. This notation implies that we replace s in the transfer function of the low-pass prototype with ω_c/s to obtain the transfer function of the corresponding high-pass filter with cutoff frequency ω_c.

To be more precise, let $(s - d_j)$ denote a pole factor of the transfer function $H(s)$ of a low-pass prototype. Using the formula (8.38), we may thus write

$$\boxed{\frac{1}{s - d_j} \to \frac{-s/d_j}{s - D_j},} \tag{8.39}$$

where $D_j = \omega_c/d_j$. The transformation equation (8.39) results in a zero at $s = 0$ and a pole at $s = D_j$ for a pole at $s = d_j$ in the original transfer function $H(s)$.

EXAMPLE 8.4 THIRD-ORDER BUTTERWORTH HIGH-PASS FILTER   Equation (8.37) defines the transfer function of a Butterworth low-pass filter of order 3 and unity cutoff frequency. Determine the transfer function of the corresponding high-pass filter with cutoff frequency $\omega_c = 1$.

Solution: Applying the frequency transformation equation (8.38) to the low-pass transfer function of Eq. (8.37) yields the transfer function of the corresponding high-pass filter with $\omega_c = 1$:

$$H(s) = \frac{1}{\left(\dfrac{1}{s} + 1\right)\left(\dfrac{1}{s^2} + \dfrac{1}{s} + 1\right)}$$

$$= \frac{s^3}{(s + 1)(s^2 + s + 1)}.$$

■

▶ **Problem 8.4**   Given the transfer function

$$H(s) = \frac{1}{s^2 + \sqrt{2}\,s + 1}$$

pertaining to a second-order Butterworth low-pass filter with unity cutoff frequency, find the transfer function of the corresponding high-pass filter with cutoff frequency ω_c.

Answer:

$$\frac{s^2}{s^2 + \sqrt{2}\,\omega_c s + \omega_c^2}$$ ◄

▶ **Problem 8.5** Let $(s - c_j)$ denote a zero factor in the transfer function of a low-pass prototype. How is this factor transformed by the use of Eq. (8.38)?

Answer:

$$\frac{s - C_j}{-s/c_j}, \qquad \text{where } C_j = \omega_c/c_j$$ ◄

■ 8.6.2 LOW-PASS TO BAND-PASS TRANSFORMATION

Consider next the transformation of a low-pass prototype filter into a band-pass filter. By definition, a band-pass filter rejects both low- and high-frequency components and passes a certain band of frequencies somewhere between them. Thus, the frequency response $H(j\omega)$ of a band-pass filter has the following properties:

1. $H(j\omega) = 0$ at both $\omega = 0$ and $\omega = \infty$.
2. $|H(j\omega)| \simeq 1$ for a frequency band centered on ω_0, the midband frequency of the filter.

Accordingly, we want to create a transfer function with zeros at $s = 0$ and $s = \infty$ and poles near $s = \pm j\omega_0$ on the $j\omega$-axis in the s-plane. A low-pass to band-pass transformation that meets these requirements is described by

$$s \to \frac{s^2 + \omega_0^2}{Bs}, \tag{8.40}$$

where ω_0 is the midband frequency and B is the bandwidth of the band-pass filter. Both ω_0 and B are measured in radians per second. According to Eq. (8.40), the point $s = 0$ for the low-pass prototype filter is transformed into $s = \pm j\omega_0$ for the band-pass filter, and the point $s = \infty$ for the low-pass prototype filter is transformed into $s = 0$ and $s = \infty$ for the band-pass filter.

Thus, a pole factor $(s - d_j)$ in the transfer function of a low-pass prototype filter is transformed as follows:

$$\frac{1}{s - d_j} \to \frac{Bs}{(s - p_1)(s - p_2)}. \tag{8.41}$$

Note that the poles p_1 and p_2 are defined by

$$p_1, p_2 = \tfrac{1}{2}\left(Bd_j \pm \sqrt{B^2 d_j^2 - 4\omega_0^2}\right). \tag{8.42}$$

An important point to observe is that the frequency transformations described in Eqs. (8.38) and (8.40) are *reactance functions*. By a "reactance function," we mean the driving-point impedance of a network composed entirely of inductors and capacitors. Indeed, we may generalize this result by saying that all frequency transformations, regardless of complications in the passband specifications of interest, are in the form of reactance functions.

▶ **Problem 8.6** Consider a low-pass filter whose transfer function is

$$H(s) = \frac{1}{s + 1}.$$

Find the transfer function of the corresponding band-pass filter with midband frequency $\omega_0 = 1$ and bandwidth $B = 0.1$.

Answer:

$$\frac{0.1s}{s^2 + 0.1s + 1} \qquad \blacktriangleleft$$

▶ **Problem 8.7** Here, we revisit the phase response of a band-pass channel considered in Example 5.8, namely,

$$\phi(\omega) = -\tan^{-1}\left(\frac{\omega^2 - \omega_c^2}{\omega\omega_c}\right),$$

where ω_c is the carrier frequency of the modulated signal applied to the filter. Show that $\phi(\omega)$ is the phase response of the filter obtained by applying the low-pass to band-pass transformation to the Butterworth low-pass filter of order 1:

$$H(s) = \frac{1}{s + 1}. \qquad \blacktriangleleft$$

Having familiarized ourselves with the notion of approximating functions and the fundamental role of low-pass prototype filters, we consider the implementation of passive analog filters in the next section, followed by the design of digital filters in Section 8.8.

8.7 Passive Filters

A filter is said to be *passive* when its composition is made up entirely of passive circuit elements (i.e., inductors, capacitors, and resistors). However, the design of highly frequency-selective passive filters is based exclusively on reactive elements (i.e., inductors and capacitors). Resistive elements enter the design only as source resistance or load resistance. The order K of the filter is usually determined by the number of reactive elements the filter contains.

Figure 8.14(a) shows a low-pass Butterworth filter of order $K = 1$ and 3-dB cutoff frequency $\omega_c = 1$. The filter is driven from an ideal current source. The resistance $R_l = 1\Omega$ represents the load resistance. The capacitor $C = 1F$ represents the only reactive element of the filter.

Figure 8.14(b) shows a lowpass Butterworth filter of order $K = 3$ and 3-dB cutoff frequency $\omega_c = 1$. As with the previous configuration, the filter is driven from a current source

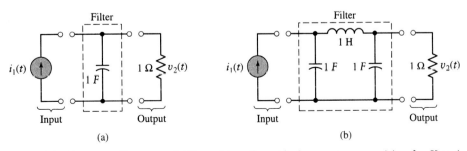

(a)   (b)

FIGURE 8.14 Low-pass Butterworth filters driven from ideal current source: (a) order $K = 1$ and (b) order $K = 3$.

and $R_l = 1\Omega$ represents the load resistance. In this case, the filter is made up of three reactive elements: two equal shunt capacitors and a single series inductor.

Note that in both Figs. 8.14(a) and (b) the transfer function $H(s)$ is in the form of a transfer impedance, defined by the Laplace transform of the output voltage $v_2(t)$, divided by the Laplace transform of the current source $i_1(t)$.

▶ **Problem 8.8**   Show that the transfer function of the filter in Fig. 8.14(b) is equal to the Butterworth function given in Eq. (8.37).   ◀

▶ **Problem 8.9**   The passive filters depicted in Fig. 8.14 have impulse response of infinite duration. Justify this statement.   ◀

The determination of the elements of a filter, starting from a particular transfer function $H(s)$, is referred to as *network synthesis*. It encompasses a number of highly advanced procedures that are beyond the scope of this text. Indeed, passive filters occupied a dominant role in the design of communication and other systems for several decades, until the advent of active filters and digital filters in the 1960s. Active filters (using operational amplifiers) are discussed in Chapter 9; digital filters are discussed next.

8.8 *Digital Filters*

A digital filter uses *computation* to implement the filtering action that is to be performed on a continuous-time signal. Figure 8.15 shows a block diagram of the operations involved in such an approach to design a frequency-selective filter; the ideas behind these operations were discussed in Section 4.7. The block labeled "analog-to-digital (A/D) converter" is used to convert the continuous-time signal $x(t)$ into a corresponding sequence $x[n]$ of numbers. The digital filter processes the sequence of numbers $x[n]$ on a sample-by-sample basis to produce a new sequence of numbers, $y[n]$, which is then converted into the corresponding continuous-time signal by the digital-to-analog (D/A) converter. Finally, the reconstruction (low-pass) filter at the output of the system produces a continuous-time signal $y(t)$, representing the filtered version of the original input signal $x(t)$.

Two important points should be carefully noted in the study of digital filters:

1. The underlying design procedures are usually based on the use of an analog or infinite-precision model for the samples of input data and all internal calculations; this is done in order to take advantage of well-understood discrete-time, but continuous-amplitude, mathematics. The resulting *discrete-time filter* provides the designer with a theoretical framework for the task at hand.

2. When the discrete-time filter is implemented in digital form for practical use, as depicted in Fig. 8.15, the input data and internal calculations are all quantized to a finite precision. In so doing, *round-off errors* are introduced into the operation of the

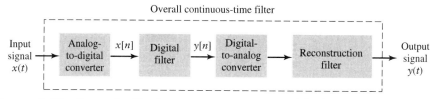

FIGURE 8.15   System for filtering a continuous-time signal, built around a digital filter.

digital filter, causing its performance to deviate from that of the theoretical discrete-time filter from which it is derived.

In this section, we confine ourselves to matters relating to point 1. Although, in light of this point, the filters considered herein should in reality be referred to as discrete-time filters, we will refer to them as digital filters to conform to commonly used terminology.

Analog filters, exemplified by the passive filters discussed in Section 8.7, are characterized by an impulse response of infinite duration. (See Problem 8.9.) In contrast, there are two classes of digital filters, depending on the duration of the impulse response:

1. *Finite-duration impulse response (FIR) digital filters*, the operation of which is governed by linear constant-coefficient difference equations of a nonrecursive nature. The transfer function of an FIR digital filter is a polynomial in z^{-1}. Consequently, FIR digital filters exhibit three important properties:
   ▸ They have finite memory, and therefore, any transient start-up is of limited duration.
   ▸ They are always BIBO stable.
   ▸ They can realize a desired magnitude response with an exactly linear phase response (i.e., with no phase distortion), as explained subsequently.

2. *Infinite-duration impulse response (IIR) digital filters*, whose input–output characteristics are governed by linear constant-coefficient difference equations of a recursive nature. The transfer function of an IIR digital filter is a rational function in z^{-1}. Consequently, for a prescribed frequency response, the use of an IIR digital filter generally results in a shorter filter length than does the use of the corresponding FIR digital filter. However, this improvement is achieved at the expense of phase distortion and a transient start-up that is not limited to a finite time interval.

In what follows, examples of both FIR and IIR digital filters are discussed.

8.9 *FIR Digital Filters*

An inherent property of FIR digital filters is that they can realize a frequency response with *linear phase*. Recognizing that a linear phase response corresponds to a constant delay, we can greatly simplify the approximation problem in the design of FIR digital filters. Specifically, the design simplifies to that of approximating a desired magnitude response.

Let $h[n]$ denote the impulse response of an FIR digital filter, defined as the inverse discrete-time Fourier transform of the frequency response $H(e^{j\Omega})$. Let M denote the filter order, corresponding to a filter length of $M + 1$. To design the filter, we are required to determine the filter coefficients $h[n]$, $n = 0, 1, \ldots, M$, so that the actual frequency response of the filter, namely, $H(e^{j\Omega})$, provides a good approximation to a desired frequency response $H_d(e^{j\Omega})$ over the frequency interval $-\pi < \Omega \leq \pi$. As a measure of the goodness of this approximation, we define the *mean-square error*

$$E = \frac{1}{2\pi} \int_{-\pi}^{\pi} |H_d(e^{j\Omega}) - H(e^{j\Omega})|^2 \, d\Omega. \tag{8.43}$$

Let $h_d[n]$ denote the inverse discrete-time Fourier transform of $H_d(e^{j\Omega})$. Then, invoking Parseval's theorem from Section 3.16, we may redefine the error measure in the equivalent form

$$E = \sum_{n=-\infty}^{\infty} |h_d[n] - h[n]|^2. \tag{8.44}$$

The only adjustable parameters in this equation are the filter coefficients $h[n]$. Accordingly, the error measure is minimized by setting

$$h[n] = \begin{cases} h_d[n], & 0 \leq n \leq M \\ 0, & \text{otherwise} \end{cases}. \tag{8.45}$$

Equation (8.45) is equivalent to the use of a *rectangular window* defined by

$$w[n] = \begin{cases} 1, & 0 \leq n \leq M \\ 0, & \text{otherwise} \end{cases}. \tag{8.46}$$

We may therefore rewrite Eq. (8.45) in the equivalent form

$$h[n] = w[n]h_d[n]. \tag{8.47}$$

It is for this reason that the design of an FIR filter based on Eq. (8.45) is called the *window method*. The mean-square error resulting from the use of the window method is

$$E = \sum_{n=-\infty}^{-1} h_d^2[n] + \sum_{n=M+1}^{\infty} h_d^2[n].$$

Since the multiplication of two discrete-time sequences is equivalent to the convolution of their DTFTs, we may express the frequency response of the FIR filter with an impulse response $h[n]$, as given by:

$$\begin{aligned} H(e^{j\Omega}) &= \sum_{n=0}^{M} h[n]e^{-jn\Omega} \\ &= \frac{1}{2\pi} \int_{-\pi}^{\pi} W(e^{j\Lambda})H_d(e^{j(\Omega-\Lambda)}) \, d\Lambda. \end{aligned} \tag{8.48}$$

The function

$$W(e^{j\Omega}) = \frac{\sin[\Omega(M+1)/2]}{\sin(\Omega/2)} e^{-jM\Omega/2}, \qquad -\pi < \Omega \leq \pi, \tag{8.49}$$

is the frequency response of the rectangular window $w[n]$. In Fig. 8.16, we have plotted the magnitude response $|W(e^{j\Omega})|$ of the rectangular window for filter order $M = 12$. For

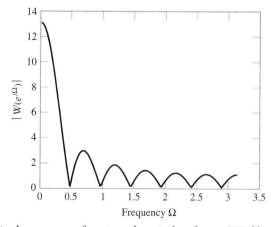

FIGURE 8.16 Magnitude response of rectangular window for an FIR filter of order $M = 12$, depicted on $0 \leq \Omega \leq \pi$.

the actual frequency response $H(e^{j\Omega})$ of the FIR digital filter to equal the ideal frequency response $H_d(e^{j\Omega})$, one period of the function $W(e^{j\Omega})$ must consist of a single unit impulse located at $\Omega = 0$. The frequency response $W(e^{j\Omega})$ of the rectangular window $w[n]$ can only approximate this ideal condition in an oscillatory manner.

The *mainlobe* of a window $w[n]$ is defined as the frequency band between the first zero crossings of its magnitude response $|W(e^{j\Omega})|$ on either side of the origin. The parts of the magnitude response that lie on either side of the mainlobe are referred to as *sidelobes*. The width of the mainlobe and amplitudes of the sidelobes provide measures of the extent to which the frequency response $W(e^{j\Omega})$ deviates from an impulse function located at $\Omega = 0$.

▶ **Problem 8.10**   Referring to Fig. 8.16 describing the frequency response of a rectangular window, verify that the width of the mainlobe is

$$\Delta\Omega_{\text{mainlobe}} = \frac{4\pi}{M + 1},$$

where M is the order of the filter.   ◀

▶ **Problem 8.11**   Referring again to the frequency response of Fig. 8.16, verify that, for a rectangular window, (a) all the sidelobes have a common width equal to $2\pi/(M + 1)$ and (b) the first sidelobes have a peak amplitude that is 13 dB below that of the mainlobe.   ◀

Recall from the discussion presented in Chapter 3 that the convolution of $H_d(e^{j\Omega})$ with $W(e^{j\Omega})$ described in Eq. (8.48) results in an oscillatory approximation of the desired frequency response $H_d(e^{j\Omega})$ by the frequency response $H(e^{j\Omega})$ of the FIR filter. The oscillations, a consequence of the sidelobes in $|W(e^{j\Omega})|$, may be reduced by using a different window with smaller sidelobes. A practical window commonly used for this purpose is the *Hamming window*, defined by

$$w[n] = \begin{cases} 0.54 - 0.46 \cos\left(\dfrac{2\pi n}{M}\right), & 0 \le n \le M \\[2mm] 0, & \text{otherwise} \end{cases}. \tag{8.50}$$

In the case when M is an even integer, $w[n]$ becomes *symmetric* about the point $n = M/2$. In Fig. 8.17, we have plotted $w[n]$ for the Hamming window with $M = 12$.

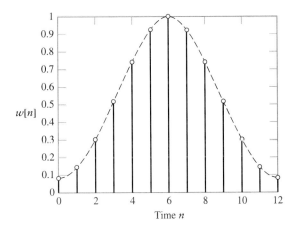

FIGURE 8.17   Impulse response of Hamming window of order $M = 12$.

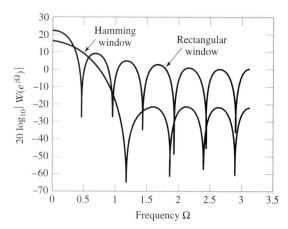

FIGURE 8.18   Comparison of magnitude responses of rectangular and Hamming windows of order $M = 12$, plotted in decibels.

To further compare the frequency response of the Hamming window with that of the rectangular window, we have chosen to plot $20 \log_{10}|W(e^{j\Omega})|$ for these two windows in Fig. 8.18 for $M = 12$. From this figure, we may make two important observations:

▶ The mainlobe of the rectangular window is less than half the width of the mainlobe of the Hamming window.

▶ The sidelobes of the Hamming window, relative to the mainlobe, are greatly reduced compared with those of the rectangular window. Specifically, the peak amplitude of the first sidelobe of the rectangular window is only about 13 dB below that of the main-lobe, whereas the corresponding value for the Hamming window is about 40 dB below.

It is because of the latter property that the Hamming window reduces oscillations in the frequency response of an FIR digital filter, as illustrated in the next two examples. However, there is a price to be paid for this improvement, namely, a wider transition band.

In order to obtain the best possible approximation of the desired response, the window must preserve as much of the energy in $h_d[n]$ as possible. Since the windows are symmetric about $n = M/2$, for M even, we desire to concentrate the maximum values of $h_d[n]$ about $n = M/2$. This is accomplished by choosing the phase response $\arg\{H_d(e^{j\Omega})\}$ to be linear, with zero intercept and a slope equal to $-M/2$. This point is illustrated in the next example.

EXAMPLE 8.5   COMPARISON OF RECTANGULAR AND HAMMING WINDOWS   Consider the desired frequency response

$$H_d(e^{j\Omega}) = \begin{cases} e^{-jM\Omega/2}, & |\Omega| \le \Omega_c \\ 0, & \Omega_c < |\Omega| \le \pi \end{cases} \tag{8.51}$$

which represents the frequency response of an ideal low-pass filter with a linear phase. Investigate the frequency response of an FIR digital filter of length $M = 12$, using (a) a rectangular window and (b) a Hamming window. Assume that $\Omega_c = 0.2\pi$ radians.

Solution: The desired response is

$$h_d[n] = \frac{1}{2\pi} \int_{-\pi}^{\pi} H_d(e^{j\Omega})e^{jn\Omega}\, d\Omega$$

$$= \frac{1}{2\pi} \int_{-\Omega_c}^{\Omega_c} e^{j\Omega(n-M/2)}\, d\Omega. \tag{8.52}$$

Invoking the definition of the sinc function, we may express $h_d[n]$ in the compact form

$$h_d[n] = \frac{\Omega_c}{\pi} \operatorname{sinc}\left[\frac{\Omega_c}{\pi}\left(n - \frac{M}{2}\right)\right], \qquad -\infty < n < \infty. \tag{8.53}$$

This impulse response is symmetric about $n = M/2$, for M even, at which point we have

$$h_d\left[\frac{M}{2}\right] = \frac{\Omega_c}{\pi}. \tag{8.54}$$

(a) *Rectangular window.* For the case of a rectangular window, the use of Eq. (8.47) yields

$$h[n] = \begin{cases} \dfrac{\Omega_c}{\pi} \operatorname{sinc}\left[\dfrac{\Omega_c}{\pi}\left(n - \dfrac{M}{2}\right)\right], & 0 \le n \le M \\[2mm] 0, & \text{otherwise} \end{cases}, \tag{8.55}$$

the value of which is given in the second column of Table 8.2 for $\Omega_c = 0.2\pi$ and $M = 12$. The corresponding magnitude response $|H(e^{j\Omega})|$ is plotted in Fig. 8.19. The oscillations in $|H(e^{j\Omega})|$ due to windowing the ideal impulse response are evident at frequencies greater than $\Omega_c = 0.2\pi$.

(b) *Hamming window.* For the case of a Hamming window, the use of Eqs. (8.50) and (8.53) yields

$$h[n] = \begin{cases} \dfrac{\Omega_c}{\pi} \operatorname{sinc}\left[\dfrac{\Omega_c}{\pi}\left(n - \dfrac{M}{2}\right)\right]\left(0.54 - 0.46\cos\left(2\pi\dfrac{n}{M}\right)\right), & 0 \le n \le M \\[2mm] 0, & \text{otherwise} \end{cases}, \tag{8.56}$$

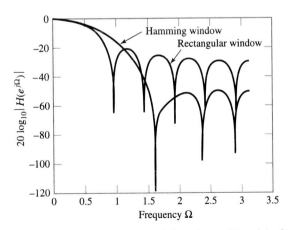

FIGURE 8.19   Comparison of magnitude responses (plotted on a dB scale) of two low-pass FIR digital filters of order $M = 12$ each, one filter using the rectangular window and the other using the Hamming window.

the value of which is given in the last column of Table 8.2 for $\Omega_c = 0.2\pi$ and $M = 12$. The corresponding magnitude response, $|H(e^{j\Omega})|$, is plotted in Fig. 8.19. We see that the oscillations due to windowing have been greatly reduced in amplitude. However, this improvement has been achieved at the expense of a wider transition band compared with that attained by using a rectangular window.

Note that the filter coefficients in the table have been *scaled*, so that the magnitude response of the filter at $\Omega = 0$ is exactly unity after windowing. This explains the deviation of the coefficient $h[M/2]$ from the theoretical value $\Omega_c/\pi = 0.2$. ■

The structure of an FIR digital filter for implementing either window is shown in Fig. 8.20. The filter coefficients for the two windows are, of course, different, taking the respective values given in Table 8.2.

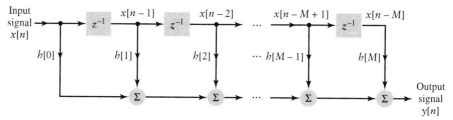

FIGURE 8.20   Structure for implementing an FIR digital filter.

TABLE 8.2   *Filter Coefficients of Rectangular and Hamming Windows for Low-pass Filter ($\Omega_c = 0.2\pi$ and $M = 12$).*

	$h[n]$	
n	Rectangular window	Hamming window
0	−0.0281	−0.0027
1	0.0000	0.0000
2	0.0421	0.0158
3	0.0909	0.0594
4	0.1364	0.1271
5	0.1686	0.1914
6	0.1802	0.2180
7	0.1686	0.1914
8	0.1364	0.1271
9	0.0909	0.0594
10	0.0421	0.0158
11	0.0000	0.0000
12	−0.0281	−0.0027

EXAMPLE 8.6 DISCRETE-TIME DIFFERENTIATOR   In Section 1.10, we discussed ⎨
of a simple *RC* circuit of the high-pass type as an approximate differentiator. In the
rent example, we address the use of an FIR digital filter as the basis for designing a mo
accurate differentiator. Specifically, consider a discrete-time differentiator, the frequency
response of which is defined by

$$H_d(e^{j\Omega}) = j\Omega e^{-jM\Omega/2}, \qquad -\pi < \Omega \le \pi. \tag{8.57}$$

Design an FIR digital filter that approximates this desired frequency response for $M = 12$,
using (a) a rectangular window and (b) a Hamming window.

Solution: The desired impulse response is

$$\begin{aligned} h_d[n] &= \frac{1}{2\pi} \int_{-\pi}^{\pi} H_d(e^{j\Omega}) e^{jn\Omega} \, d\Omega \\ &= \frac{1}{2\pi} \int_{-\pi}^{\pi} j\Omega e^{j\Omega(n-M/2)} \, d\Omega. \end{aligned} \tag{8.58}$$

Integrating by parts, we get

$$h_d[n] = \frac{\cos[\pi(n - M/2)]}{(n - M/2)} - \frac{\sin[\pi(n - M/2)]}{\pi(n - M/2)^2}, \qquad -\infty < n < \infty. \tag{8.59}$$

(a) *Rectangular window.* Multiplying the impulse response of Eq. (8.59) by the rectangular window of Eq. (8.46), we get

$$h_d[n] = \begin{cases} \dfrac{\cos[\pi(n - M/2)]}{(n - M/2)} - \dfrac{\sin[\pi(n - M/2)]}{\pi(n - M/2)^2}, & 0 \le n \le M \\ 0, & \text{otherwise} \end{cases}. \tag{8.60}$$

This impulse response is antisymmetric in that $h[M - n] = -h[n]$. Also, for M even,
$h[n]$ is zero at $n = M/2$; see Problem 8.12. The value of $h[n]$ is given in the second
column of Table 8.3 for $M = 12$. The table clearly demonstrates the antisymmetric
property of $h[n]$. The corresponding magnitude response $|H(e^{j\Omega})|$ is plotted in
Fig. 8.21(a). The oscillatory deviations from the ideal frequency response are mani-
festations of windowing the ideal impulse response in Eq. (8.59).

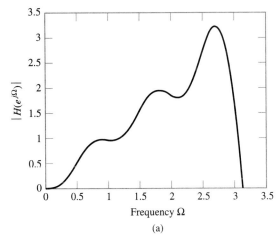

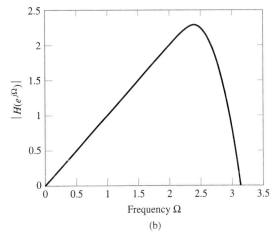

(a)   (b)

FIGURE 8.21   Magnitude response of FIR digital filter as differentiator, designed using (a) a rectangular window and (b) a Hamming window. In both cases, the filter order M is 12.

TABLE 8.3 *Filter Coefficients of Rectangular and Hamming Windows for a Differentiator.*

	$h[n]$	
n	Rectangular window	Hamming window
0	−0.1667	−0.0133
1	0.2000	0.0283
2	−0.2500	−0.0775
3	0.3333	0.1800
4	−0.5000	−0.3850
5	1.0000	0.9384
6	0	0
7	−1.0000	−0.9384
8	0.5000	0.3850
9	−0.3333	−0.1800
10	0.2500	0.0775
11	−0.2000	−0.0283
12	0.1667	0.0133

(b) *Hamming window.* Multiplying the impulse response $h_d[n]$ of Eq. (8.59) by the Hamming window of Eq. (8.50), we get the impulse response $h[n]$ given in the last column of Table 8.3. The corresponding magnitude response $|H(e^{j\Omega})|$ is plotted in Fig. 8.21(b). Comparing this response with that of Fig. 8.21(a), we see that the oscillations have been greatly reduced in amplitude, but the bandwidth over which $|H(e^{j\Omega})|$ is linear with Ω also has been reduced, yielding less usable bandwidth for the operation of differentiation. ∎

Note that many other windows besides the Hamming window allow different trade-offs between mainlobe width and sidelobe height.

S & S
Solutions

▶ **Problem 8.12** Starting with Eq. (8.58), derive the formula for the impulse response $h_d[n]$ given in Eq. (8.59), and show that $h_d[M/2] = 0$. ◀

■ **8.9.1 FILTERING OF SPEECH SIGNALS**

The preprocessing of speech signals is fundamental to many applications, such as the digital transmission and storage of speech, automatic speech recognition, and automatic speaker recognition systems. FIR digital filters are well suited for the preprocessing of speech signals, for two important reasons:

1. In speech-processing applications, it is essential to maintain precise time alignment. The *exact* linear phase property inherent in an FIR digital filter caters to this requirement in a natural way.

2. The approximation problem in filter design is greatly simplified by the exact linear phase property of an FIR digital filter. In particular, in not having to deal with delay (phase) distortion, our only concern is that of approximating a desired magnitude response.

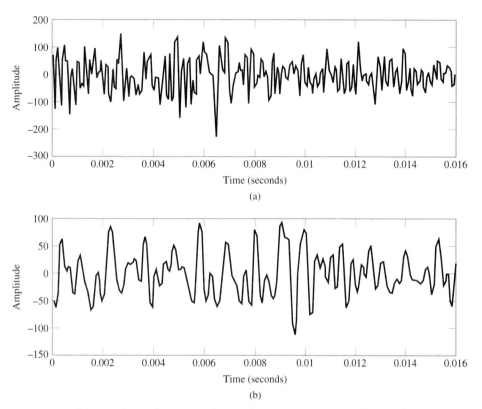

FIGURE 8.22 (a) Waveform of raw speech signal, containing an abundance of high-frequency noise. (b) Waveform of speech signal after passing it through a low-pass FIR digital filter of order $M = 98$ and cutoff frequency $f_c = 3.1 \times 10^3$ Hz.

However, there is a price to be paid for achieving these two desirable features: To design an FIR digital filter with a sharp cutoff characteristic, the length of the filter has to be large, producing an impulse response with a long duration.

In this subsection, we will illustrate the use of an FIR digital filter for the preprocessing of a real-life speech signal, so that it would be suitable for transmission over a telephone channel. Figure 8.22(a) shows a short portion of the waveform of a speech signal produced by a female speaker saying the phrase, "This was easy for us." The original sampling rate of this speech signal was 16 kHz, and the total number of samples contained in the whole sequence was 27,751.

Before transmission, the speech signal is applied to an FIR digital low-pass filter with the following specifications:

length of filter, $M + 1 = 99$;
symmetric about midpoint to obtain a linear phase response;
cutoff frequency $f_c = \omega_c/2\pi = 3.1 \times 10^3$ rad/s.

The design of the filter was based on the window method, using the *Hanning* or *raised-cosine window*, which is not to be confused with the Hamming window. This new window is defined by

$$w[n] = \begin{cases} \dfrac{1}{2}\left[1 - \cos\left(\dfrac{2\pi n}{M}\right)\right], & 0 \le n \le M \\ 0, & \text{otherwise} \end{cases}. \qquad (8.61)$$

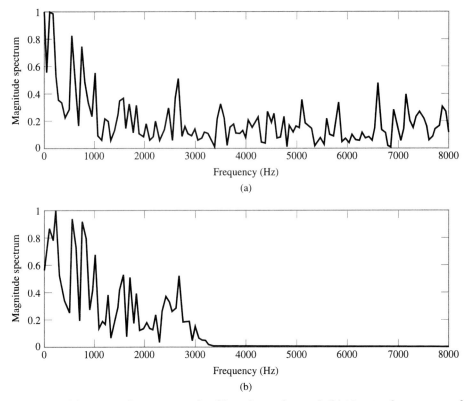

FIGURE 8.23   (a) Magnitude spectrum of unfiltered speech signal. (b) Magnitude spectrum of unfiltered speech signal. Note the sharp cutoff of the spectrum around 3100 Hz.

The Hanning window goes to zero, with zero slope at the edges of the window (i.e., $n = 0$ and $n = M$).

   Figure 8.23 shows the magnitude spectra of the speech signal before and after filtering. In the both cases, the FFT algorithm was used to perform the computation. In comparing the magnitude spectrum of the filtered signal shown in Fig. 8.23(b) with that of the unfiltered signal shown in Fig. 8.23(a), we clearly see the effect of the relatively sharp cutoff produced by the FIR low-pass filter around 3.1 kHz.

   In listening to the unfiltered and filtered versions of the speech signal, the following observations were made:

   1. The unfiltered speech signal was harsh, with an abundance of high-frequency noise such as clicks, pops, and hissing sounds.

   2. The filtered signal, in contrast, was found to be much softer, smoother, and natural sounding.

The essence of these observations can be confirmed by examining 16 milliseconds of the speech waveforms and their spectra, shown in Figs. 8.22 and 8.23, respectively.

   As mentioned previously, the original speech signal was sampled at the rate of 16 kHz, which corresponds to a sampling interval $T_s = 62.5\ \mu$s. The structure used to implement the FIR filter was similar to that described in Fig. 8.20. The filter order was chosen to be $M = 98$, so as to provide a frequency response with a fairly steep transition from the passband into the stopband. Hence, in passing the speech signal through this filter with $M + 1 = 99$ coefficients, a delay of

$$T_s\left(\frac{M}{2}\right) = 62.5 \times 49 = 3.0625 \text{ ms}$$

is introduced into the filtered speech signal. This time delay is clearly discernible in comparing the waveform of that signal in Fig. 8.22(b) with that of the raw speech signal in Fig. 8.22(a).

8.10 *IIR Digital Filters*

Various techniques have been developed for the design of IIR digital filters. In this section, we describe a popular method for converting analog transfer functions to digital transfer functions. The method is based on the *bilinear transform*, which provides a unique mapping between points in the s-plane and those in the z-plane.

The bilinear transform is defined by

$$s = \left(\frac{2}{T_s}\right)\left(\frac{z-1}{z+1}\right), \tag{8.62}$$

where T_s is the implied sampling interval associated with conversion from the s-domain to the z-domain. To simplify matters, we shall set $T_s = 2$ henceforth. The resulting filter design is independent of the actual choice of T_s. Let $H_a(s)$ denote the transfer function of an analog (continuous-time) filter. Then the transfer function of the corresponding digital filter is obtained by substituting the bilinear transformation of Eq. (8.62) into $H_a(s)$, yielding

$$H(z) = H_a(s)\big|_{s=((z-1)/(z+1))}. \tag{8.63}$$

What can we say about the properties of the transfer function $H(z)$ derived from Eq. (8.63)? To answer this question, we rewrite Eq. (8.62) in the form

$$z = \frac{1+s}{1-s}$$

with $T_s = 2$. Putting $s = \sigma + j\omega$ in this equation, we may express the complex variable z in the polar form

$$z = re^{j\theta},$$

where the radius and angle are defined, respectively, by

$$\begin{aligned} r &= |z| \\ &= \left[\frac{(1+\sigma)^2 + \omega^2}{(1-\sigma)^2 + \omega^2}\right]^{1/2} \end{aligned} \tag{8.64}$$

and

$$\begin{aligned} \theta &= \arg\{z\} \\ &= \tan^{-1}\left(\frac{\omega}{1+\sigma}\right) + \tan^{-1}\left(\frac{\omega}{1-\sigma}\right). \end{aligned} \tag{8.65}$$

From Eqs. (8.64) and (8.65), we readily see that

- ▶ $r < 1$ for $\sigma < 0$.
- ▶ $r = 1$ for $\sigma = 0$.
- ▶ $r > 1$ for $\sigma > 0$.
- ▶ $\theta = 2\tan^{-1}(\omega)$ for $\sigma = 0$.

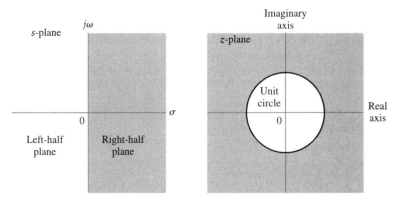

FIGURE 8.24   Illustration of the properties of the bilinear transform. The left half of the s-plane (shown on the left) is mapped onto the interior of the unit circle in the z-plane (shown on the right). Likewise, the right half of the s-plane is mapped onto the exterior of the unit circle in the z-plane; the two corresponding regions are shown shaded.

Accordingly, we may state the properties of the bilinear transform as follows:

1. The left-half of the s-plane is mapped onto the interior of the unit circle in the z-plane.
2. The entire $j\omega$-axis of the s-plane is mapped onto one complete revolution of the unit circle in the z-plane.
3. The right-half of the s-plane is mapped onto the exterior of the unit circle in the z-plane.

These properties are illustrated in Fig. 8.24.

An immediate implication of Property 1 is that, if the analog filter represented by the transfer function $H_a(s)$ is stable and causal, then the digital filter derived from it by using the bilinear transform given by Eq. (8.62) is guaranteed to be stable and causal also. Since the bilinear transform has real coefficients, it follows that $H(z)$ will have real coefficients if $H_a(s)$ has real coefficients. Hence, the transfer function $H(z)$ resulting from the use of Eq. (8.63) is indeed physically realizable.

▶ **Problem 8.13**   What do the points $s = 0$ and $s = \pm j\infty$ in the s-plane map onto in the z-plane, using the bilinear transform?

Answer: $s = 0$ is mapped onto $z = +1$. The points $s = j\infty$ and $s = -j\infty$ are mapped onto just above and just below $z = -1$, respectively.   ◀

For $\sigma = 0$ and $\theta = \Omega$, Eq. (8.65) reduces to

$$\Omega = 2 \tan^{-1}(\omega), \tag{8.66}$$

which is plotted in Fig. 8.25 for $\omega > 0$. Note that Eq. (8.66) has odd symmetry. The infinitely long range of frequency variations $-\infty < \omega < \infty$ for an analog (continuous-time) filter is nonlinearly compressed into the finite frequency range $-\pi < \Omega < \pi$ for a digital (discrete-time) filter. This form of nonlinear distortion is known as *warping*. In the design of frequency-selective filters in which the emphasis is on the approximation of a piecewise magnitude response, we must compensate for this nonlinear distortion by *prewarping* the design specifications of the analog filter. Specifically, the critical frequencies (i.e., the prescribed passband cutoff and stopband cutoff frequencies) are prewarped in accordance with the formula

$$\omega = \tan\left(\frac{\Omega}{2}\right), \tag{8.67}$$

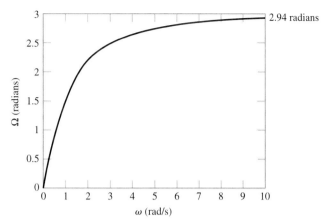

FIGURE 8.25   Graphical plot of the relation between the frequency Ω pertaining to the discrete-time domain and the frequency ω pertaining to the continuous-time domain: $\Omega = 2\tan^{-1}(\omega)$.

which is the inverse of Eq. (8.66). To illustrate the prewarping procedure, let Ω_k', $k = 1, 2, \ldots$, denote the critical frequencies that a digital filter is required to realize. Before applying the bilinear transform, the corresponding critical frequencies of the continuous-time filter are prewarped by using Eq. (8.67) to obtain

$$\omega_k = \tan\left(\frac{\Omega_k'}{2}\right), \qquad k = 1, 2, \ldots. \tag{8.68}$$

Then, when the bilinear transform is applied to the transfer function of the analog filter designed using the prewarped frequencies in Eq. (8.68), we find from Eq. (8.66) that

$$\Omega_k = \Omega_k', \qquad k = 1, 2, \ldots. \tag{8.69}$$

That is, the prewarping procedure ensures that the digital filter will meet the prescribed design specifications exactly.

EXAMPLE 8.7   DESIGN OF DIGITAL IIR LOW-PASS FILTER BASED ON A BUTTERWORTH RESPONSE   Using an analog filter with a Butterworth response of order 3, design a digital IIR low-pass filter with a 3-dB cutoff frequency $\Omega_c = 0.2\pi$.

Solution: The prewarping formula of Eq. (8.68) indicates that the cutoff frequency of the analog filter should be

$$\omega_c = \tan(0.1\pi) = 0.3249.$$

Adapting Eq. (8.37) to the problem at hand in light of Problem 8.2, we find that the transfer function of the analog filter is

$$H_a(s) = \frac{1}{\left(\dfrac{s}{\omega_c} + 1\right)\left(\dfrac{s^2}{\omega_c^2} + \dfrac{s}{\omega_c} + 1\right)}$$

$$= \frac{0.0343}{(s + 0.3249)(s^2 + 0.3249s + 0.1056)}. \tag{8.70}$$

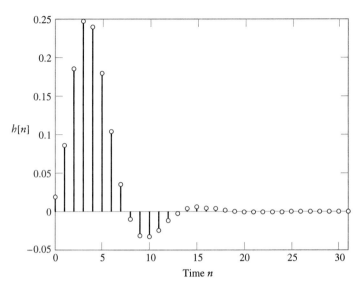

FIGURE 8.26   Impulse response of digital IIR low-pass filter with Butterworth response of order 3 and 3-dB cutoff frequency $\Omega_c = 0.2\pi$.

Hence, using Eq. (8.63), we get

$$H(z) = \frac{0.0181(z + 1)^3}{(z - 0.50953)(z^2 - 1.2505z + 0.39812)}. \tag{8.71}$$

Figure 8.26 shows the impulse response $h[n]$ of the filter [i.e., the inverse z-transform of the $H(z)$ given in Eq. (8.71)].

In Section 7.9, we discussed different computational structures (i.e., cascade and parallel forms) for implementing discrete-time systems. In light of the material covered therein, we readily see that the transfer function of Eq. (8.71) can be realized by using a cascade of two sections, as shown in Fig. 8.27. The section resulting from the bilinear transformation of the simple pole factor $((s/\omega_c) + 1)$ in $H_a(s)$ is referred to as a *first-order section*. Similarly, the section resulting from the bilinear transformation of the quadratic pole fac-

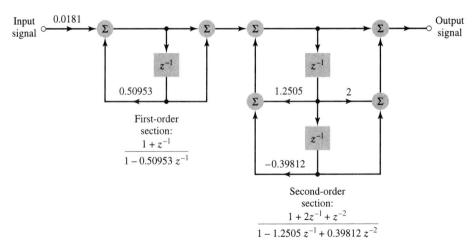

FIGURE 8.27   Cascade implementation of IIR low-pass digital filter, made up of a first-order section followed by a second-order section.

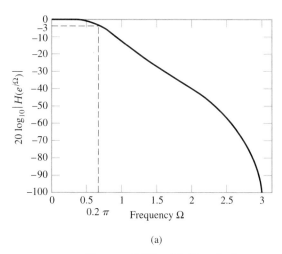 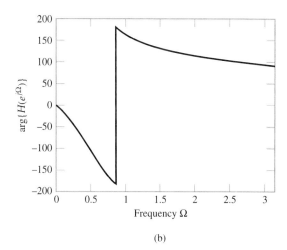

(a)   (b)

FIGURE 8.28 (a) Magnitude response of the IIR low-pass digital filter characterized by the impulse response shown in Fig. 8.26, plotted in decibels. (b) Phase response of the filter.

tor $((s/\omega_c)^2 + (s/\omega_c) + 1)$ in $H_a(s)$ is referred to as a *second-order section*. Indeed, this result may be generalized to say that the application of the bilinear transform to $H_a(s)$ in factored form results in a realization of $H(z)$ that consists of a cascade of first-order and second-order sections. From a practical point of view, this kind of structure for implementing a digital filter has intuitive appeal.

Putting $z = e^{j\Omega}$ in Eq. (8.71) and plotting $H(e^{j\Omega})$ versus Ω, we get the magnitude and phase responses shown in Fig. 8.28. We see that the passband of the filter extends up to 0.2π, as prescribed.   ■

8.11 *Linear Distortion*

In practice, the conditions for distortionless transmission described in Section 8.2 can be satisfied only approximately; the material presented in the previous sections testifies to this statement. That is to say, there is always a certain amount of distortion present in the output signal of a physical LTI system, be it of a continuous-time or discrete-time type, due to deviation in the frequency response of the system from the ideal conditions described in Eqs. (8.5) and (8.6). In particular, we may distinguish two components of linear distortion produced by transmitting a signal through an LTI system:

1. *Amplitude distortion.* When the magnitude response of the system is not constant inside the frequency band of interest, the frequency components of the input signal are transmitted through the system with different amounts of gain or attenuation. This effect is called *amplitude distortion.* The most common form of amplitude distortion is excess gain or attenuation at one or both ends of the frequency band of interest.

2. *Phase distortion.* The second form of linear distortion arises when the phase response of the system is not linear with frequency inside the frequency band of interest. If the input signal is divided into a set of components, each one of which occupies a narrow band of frequencies, we find that each such component is subject to a different delay in passing through the system, with the result that the output signal emerges with a waveform different from that of the input signal. This form of linear distortion is called *phase* or *delay distortion.*

We emphasize the distinction between a constant delay and a constant phase shift. In the case of a continuous-time LTI system, a constant delay means a linear phase response (i.e., $\arg\{H(j\omega)\} = -t_0\omega$, where t_0 is the constant delay). In contrast, a constant phase shift means that $\arg\{H(j\omega)\}$ equals some constant for all ω. These two conditions have different implications. Constant delay is a requirement for distortionless transmission; constant phase shift causes the signal to be distorted.

An LTI system that suffers from linear distortion is said to be *dispersive*, in that the frequency components of the input signal emerge with amplitude or phase characteristics that are different from those of the original input signal after it is transmitted through the system. The telephone channel is an example of a dispersive system.

▶ **Problem 8.14 Multipath Propagation Channel** In Section 1.10, we introduced the discrete-time model

$$y[n] = x[n] + ax[n-1]$$

as the descriptor of a multipath propagation channel. In general, the model parameter a can be real or complex, so long as $|a| < 1$.

(a) What form of distortion is introduced by this channel? Justify your answer.

(b) Determine the transfer function of an FIR filter of order 4 designed to equalize the channel; here, it is assumed that a is small enough to ignore higher order terms.

Answers:

(a) Both amplitude distortion and phase distortion are introduced by the channel

(b) The transfer function of the FIR equalizer is

$$H(z) = 1 - az^{-1} + a^2z^{-2} - a^3z^{-3} + a^4z^{-4}$$ ◀

8.12 *Equalization*

To compensate for linear distortion, we may use a network known as an *equalizer* connected in cascade with the system in question, as illustrated in Fig. 8.29. The equalizer is designed in such a way that, inside the frequency band of interest, the overall magnitude and phase responses of this cascade connection approximate the conditions for distortionless transmission to within prescribed limits.

Consider, for example, a communication channel with frequency response $H_c(j\omega)$. Let an equalizer of frequency response $H_{eq}(j\omega)$ be connected in cascade with the channel, as in Fig. 8.29. Then the overall frequency response of this combination is equal to $H_c(j\omega)H_{eq}(j\omega)$. For overall transmission through the cascade connection to be distortionless, we require that

$$H_c(j\omega)H_{eq}(j\omega) = e^{-j\omega t_0}, \tag{8.72}$$

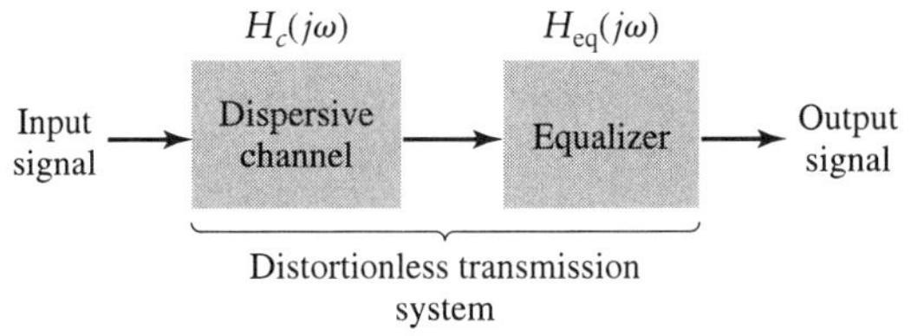

FIGURE 8.29 Cascade connection of a dispersive (LTI) channel and an equalizer for distortionless transmission.

where t_0 is a constant time delay. [See Eq. (8.3); for convenience of presentation, we have set the scaling factor C equal to unity.] Ideally, therefore, the frequency response of the equalizer is *inversely related* to that of the channel, according to the formula

$$H_{eq}(j\omega) = \frac{e^{-j\omega t_0}}{H_c(j\omega)}. \tag{8.73}$$

In practice, the equalizer is designed such that its frequency response approximates the ideal value of Eq. (8.73) closely enough for the linear distortion to be reduced to a satisfactory level.

The frequency response $H_{eq}(j\omega)$ of the equalizer in Eq. (8.73) is formulated in continuous time. Although it is indeed possible to design the equalizer with an analog filter, the preferred method is to do the design in discrete time, using a digital filter. With a discrete-time approach, the channel output is sampled prior to equalization. Depending on the application, the equalizer output may be converted back to a continuous-time signal or left in discrete-time form.

A system that is well suited for equalization is the FIR digital filter, also referred to as a *tapped-delay-line equalizer*. The structure of such a filter is depicted in Fig. 8.20. Since the channel frequency response is represented in terms of the Fourier transform, we shall employ the Fourier transform representation for the FIR filter frequency response. If the sampling interval equals T_s seconds, then we see from Eq. (4.18) that the equalizer frequency response is

$$H_{\delta,eq}(j\omega) = \sum_{n=0}^{M} h[n] \exp(-jn\omega T_s). \tag{8.74}$$

The subscript δ in $H_{\delta,eq}(j\omega)$ is intended to distinguish H from its continuous-time counterpart $H_{eq}(j\omega)$. For convenience of analysis, it is assumed that the number of filter coefficients $M + 1$ in the equalizer is odd (i.e., M is even).

The goal of equalizer design is to determine the filter coefficients $h[0], h[1], \ldots, h[M]$, so that $H_{\delta,eq}(j\omega)$ approximates $H_{eq}(j\omega)$ in Eq. (8.73) over a frequency band of interest, say, $-\omega_c \leq \omega \leq \omega_c$. Note that $H_{\delta,eq}(j\omega)$ is periodic, with one period occupying the frequency range $-\pi/T_s \leq \omega \leq \pi/T_s$. Hence, we choose $T_s = \pi/\omega_c$, so that one period of $H_{\delta,eq}(j\omega)$ corresponds to the frequency band of interest. Let

$$H_d(j\omega) = \begin{cases} e^{-j\omega t_0}/H_c(j\omega), & -\omega_c \leq \omega \leq \omega_c \\ 0, & \text{otherwise} \end{cases} \tag{8.75}$$

be the frequency response we seek to approximate with $H_{\delta,eq}(j\omega)$. We accomplish this task by using a variation of the window method of FIR filter design, as summarized in Procedure 8.1.

PROCEDURE 8.1 *Summary of Window Method for the Design of an Equalizer.*

Start with a specified order M, assumed to be an even integer. Then, for a given sampling interval T_s, proceed as follows:

1. Set the constant time delay $t_0 = (M/2)/T_s$.

2. Take the inverse Fourier transform of $H_d(j\omega)$ to obtain a desired impulse response $h_d(t)$.

3. Set $h[n] = w[n]h_d(nT_s)$, where $w[n]$ is a window of length $(M + 1)$. Note that the sampling operation does not cause aliasing of the desired response in the band $-\omega_c \leq \omega \leq \omega_c$, since we chose $T_s = \pi/\omega_c$.

Typically, $H_c(j\omega)$ is given numerically in terms of its magnitude and phase, in which case numerical integration is used to evaluate $h_d(nT_s)$. The number of terms, $M + 1$, is chosen just big enough to produce a satisfactory approximation to $H_d(j\omega)$.

EXAMPLE 8.8 DESIGN OF AN EQUALIZER FOR A FIRST-ORDER BUTTERWORTH CHANNEL
Consider a simple channel whose frequency response is described by the first-order Butterworth response

$$H_c(j\omega) = \frac{1}{1 + j\omega/\pi}.$$

Design an FIR filter with 13 coefficients (i.e., $M = 12$) for equalizing this channel over the frequency band $-\pi \le \omega \le \pi$. Ignore the effect of channel noise.

Solution: In this example, the channel equalization problem is simple enough for us to solve without having to resort to the use of numerical integration. With $\omega_c = \pi$, the sampling interval is $T_s = 1$ s. Now, from Eq. (8.75), we have

$$H_d(j\omega) - \begin{cases} \left(1 + \dfrac{j\omega}{\pi}\right)e^{-j6\omega}, & -\pi \le \omega \le \pi \\ 0, & \text{otherwise} \end{cases}.$$

The nonzero part of the frequency response $H_d(j\omega)$ consists of the sum of two terms: unity and $j\omega/\pi$, except for a linear phase term. These two terms are approximated as follows:

▶ The term $j\omega/\pi$ represents a scaled form of differentiation. The design of a differentiator using an FIR filter was discussed in Example 8.6. Indeed, evaluating the inverse Fourier transform of $j\omega/\pi$ and then setting $t = n$ for a sampling interval $T_s = 1$ s, we get Eq. (8.59), scaled by $1/\pi$. Thus, using the result obtained in that example, which incorporated the Hamming window of length 13, and scaling it by $1/\pi$, we get the values listed in the second column of Table 8.4.

TABLE 8.4   *Filter Coefficients for Example 8.8 on Equalization.*

n	Hamming-Windowed Inverse Fourier Transform of $j\omega/\pi$	Hamming-Windowed Inverse Fourier Transform of 1	$h_d[n]$
0	−0.0042	0	−0.0042
1	0.0090	0	0.0090
2	−0.0247	0	−0.0247
3	0.0573	0	0.0573
4	−0.1225	0	−0.1225
5	0.2987	0	0.2987
6	0	1	1.0000
7	−0.2987	0	−0.2987
8	0.1225	0	0.1225
9	−0.0573	0	−0.0573
10	0.0247	0	0.0247
11	−0.0090	0	−0.0090
12	0.0042	0	0.0042

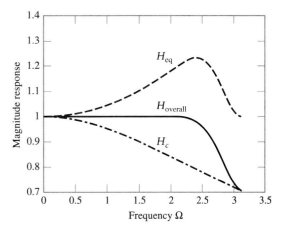

FIGURE 8.30   Magnitude response of Butterworth channel of order 1: dashed and dotted ($-\cdot-\cdot$) curve. Magnitude response of FIR equalizer of order $M = 12$: dashed ($--$) curve. Magnitude response of equalized channel: solid curve. The flat region of the overall (equalized) magnitude response is extended up to about $\Omega = 2.5$.

> ► The inverse Fourier transform of the unity term is sinc(t). Setting $t = nT_s = n$ and weighting it with the Hamming window of length 13, we get the set of values listed in column 3 of the table.

Adding these two sets of values, we get the Hamming-windowed FIR filter coefficients for the equalizer listed in the last column of the table. Note that this filter is antisymmetric about the midpoint $n = 6$.

Figure 8.30 superposes of the magnitude responses of the channel, the FIR equalizer, and the equalized channel. The responses are plotted for the band $0 \leq \Omega \leq \pi$. From the figure, we see that the magnitude response of the equalized channel is essentially flat over the band $0 \leq \Omega \leq 2.5$. In other words, an FIR filter with $M = 12$ equalizes a channel with a finite-order Butterworth response of cutoff frequency π for a large portion of its passband, since

$$\frac{2.5}{\pi} \approx 0.8 \text{ (i.e., 80 percent).} \qquad \blacksquare$$

8.13 *Exploring Concepts with MATLAB*

In this chapter, we studied the design of linear filters and equalizers. While these two systems act on input signals of their own, their purposes are entirely different. The goal of filtering is to produce an output signal with a specified frequency content. An equalizer, by contrast, is used to compensate for some form of linear distortion contained in the input signal.

The MATLAB Signal Processing Toolbox possesses a rich collection of functions that are tailor made for the analysis and design of linear filters and equalizers. In this section, we explore the use of some of those functions as tools for consolidating ideas and design procedures described in previous sections.

■ 8.13.1 TRANSMISSION OF A RECTANGULAR PULSE THROUGH AN IDEAL LOW-PASS FILTER

In Section 8.3.1, we studied the response of an ideal low-pass filter to an input rectangular pulse. This response, denoted by $y(t)$, is given in terms of the sine integral by Eq. (8.20); that is,

$$y(t) = \frac{1}{\pi}[\text{Si}(a) - \text{Si}(b)],$$

where

$$a = \omega_c\left(t - t_0 + \frac{T_0}{2}\right)$$

and

$$b = \omega_c\left(t - t_0 - \frac{T_0}{2}\right),$$

in which T_0 is the pulse duration, ω_c is the cutoff frequency of the filter, and t_0 is the transmission delay through the filter. For convenience of presentation, we set $t_0 = 0$.

The expression for $y(t)$ shows that the sine integral $\text{Si}(u)$, defined in Eq. (8.19), plays a dominant role in determining the response of an ideal low-pass filter to a rectangular pulse input. Unfortunately, there is no analytic solution of this integral. We therefore have to resort to the use of numerical integration for its evaluation. A common procedure for numerical integration is to compute an estimate of the area under the integrand between the limits of integration. Such a procedure is referred to as a *quadrature technique*. The MATLAB function

```
quad('function_name', a,b)
```

returns the area under the integrand between the limits of integration, `a` and `b`. The function `quad` uses a form of Simpson's rule in which the integrand is uniformly sampled across `[a,b]`. For the sine integral plotted in Fig. 8.5, we used the commands

```
>> x = -20:.1:20;
>> For u = 1:length(x),
       z(u) = quad('sincnopi', 0,x(u));
   end
```

which incorporate the M-file called `'sincnopi.m'` described as follows:

```
function y = sincnopi(w)
y = ones(size(w));
i = find(w);
y(i) = sin(w(i))./w(i);
```

Returning to the issue at hand, we produce the MATLAB code for computing the pulse response $y(t)$ as follows:

```
function [y]=sin_pr(wc, r)
%r is a user-specified resolution parameter
T=1;
to=0;            % transmission delay = 0
t=-T*1.01:r:T*1.01;
```

```
ta=wc*(t-to+T/2);
tb=wc*(t-to-T/2);
for q=1:length(ta),
z1(q)=quad('sincnopi',0,ta(q));
end
for q=1:length(tb),
z2(q)=quad('sincnopi',0,tb(q));
end
plot(t,(z1-z2)/pi)
axis(gca,'YLim',[-.2 1.2])
axis(gca,'XLim',[-1 1])
```

■ 8.13.2 FIR DIGITAL FILTERS

The MATLAB Signal Processing Toolbox has two types of routines, namely, `fir1` and `fir2`, for designing FIR filters based on the window methods. The functions of these routines are summarized here:

1. The command

    ```
    b=fir1(M,wc)
    ```

    designs an Mth-order low-pass digital filter and returns the filter coefficients in vector `b` of length `M+1`. The cutoff frequency `wc` is normalized so that it lies in the interval $[0, 1]$, with 1 corresponding to one-half the sampling rate, or $\Omega = \pi$ in discrete-time frequency. By default, `fir1` uses a Hamming window; it also allows the use of several other windows, including the rectangular and Hanning windows. (In MATLAB, the rectangular window is referred to as the `boxcar`.) The use of a desired window can be specified with an optional trailing argument. For example, `fir1(M,wc,boxcar(M+1))` uses a rectangular window. Note that, by default, the filter is *scaled* so that the center of the first passband has a magnitude of exactly unity after windowing.

2. `fir2` designs an FIR filter with arbitrary frequency response. The command

    ```
    b=fir2(M,F,K)
    ```

    designs a filter of order `M` with frequency response specified by vectors `F` and `K`. The vector `F` specifies frequency points in the range $[0, 1]$, where 1 corresponds to one-half the sampling rate, or $\Omega = \pi$. The vector `K` is a vector containing the desired magnitude response at the points specified in `F`. The vectors `F` and `K` must have the same length. As with `fir1`, by default `fir2` uses a Hamming window; other windows can be specified with an optional trailing argument.

`fir1` was used to design the FIR digital filters considered in Examples 8.5 and 8.6. In particular, in Example 8.5 we studied the window method for the design of a low-pass filter of order `M = 12`, using (a) a rectangular (boxcar) window and (b) a Hamming window. We employed the following MATLAB commands for designing these filters and evaluating their frequency responses:

(a) *Rectangular window*

```
>> b=fir1(12,0.2,boxcar(13));
>> [H,w]=freqz(b,1,512);
>> db=20*log10(abs(H));
>> plot(w,db);
```

(b) *Hamming window*

```
>> b=fir1(12,0.2,hamming(13));
>> [H,w]=freqz(b,1,512);
>> db=20*log10(abs(H));
>> plot(w,db)
```

In Example 8.6, we studied the design of a discrete-time differentiator whose frequency response is defined by

$$H_d(e^{j\Omega}) = j\Omega e^{-j M\Omega/2}.$$

Here again, we examined the use of a rectangular window and a Hamming window as the basis for filter design. The respective MATLAB commands for designing these filters are as follows:

```
>> taps=13;  M=taps-1;  %M - filter order
>> n=0:M;  f=n-M/2;
>> a = cos(pi*f) ./ f;  % integration by parts eq.8.59
>> b = sin(pi*f) ./ (pi*f.^2);
>> h=a-b;  % impulse response for rectangular windowing
>> k=isnan(h);  h(k)=0;  % get rid of not a number
>> [H,w]=freqz(h,1,512,2*pi);
>> hh=hamming(taps)'.*h;            % apply Hamming window
>> [HH,w]=freqz(hh,1,512,2*pi);
>> figure (i); clf;
>> plot (w,abs(H)); hold on;
>> plot (w,abs(HH),'f'); hold off;
```

■ 8.13.3 PROCESSING OF SPEECH SIGNALS

The filtering of speech signals was used as an illustration in Section 8.9. The filter considered therein was an FIR low-pass filter designed by using the Hanning window. Insight into the effect of filtering was achieved by comparing the spectra of the raw and filtered speech signals. Since the speech data represent a continuous-time signal, the Fourier transform is the appropriate Fourier representation. We will approximate the Fourier transform by evaluating the discrete-time Fourier series of a finite-duration sampled section of speech, using the `fft` command as discussed in Chapter 4. Thus, the MATLAB commands for studying the effect of filtering on speech signals are as follows:

```
clear
load spk_sam
%Note there are two speech vectors loaded here: tst
    and tst1.
>> speech=tst1;
>> b=fir1(98,3000/8000,hanning(99));
>> filt_sp=filter(b,1,speech);
>> f=0:8000/127:8000;
>> subplot(2,1,1);
>> spect=fft(speech,256);
>> plot(f,abs(spect(1:128))/max(abs(spect(1:128))));
    subplot(2,1,2)
>> filt_spect=fft(filt_sp,256);
>> plot(f,abs(filt_spect(1:128))/
    max(abs(filt_spect(1:128))));
```

■ 8.13.4 IIR Digital Filters

In Example 8.7, we used an analog filter as the basis for the design of an IIR low-pass filter with cutoff frequency Ω_c. It is a simple matter to design such a digital filter with the use of the Signal Processing Toolbox. For the problem posed in Example 8.8, the requirement is to design an IIR digital low-pass filter with a Butterworth response of order 3.

The MATLAB command

```
[b,a]=butter(K,w)
```

designs a low-pass digital IIR filter with a Butterworth response of order `K` and returns the coefficients of the transfer function's numerator and denominator polynomials in vectors `b` and `a`, respectively, of length $K + 1$. The cutoff frequency `w` of the filter must be normalized so that it lies in the interval $[0, 1]$, with 1 corresponding to $\Omega = \pi$.

Thus, the commands for designing the IIR digital filter in Example 8.7 and evaluating its frequency response are as follows:

```
>> [b,a]=butter(3,0.2);
>> [H,w]=freqz(b,a,512);
>> mag=20*log10(abs(H));
>> plot(w,mag)
>> phi=angle(H);
>> phi=(180/pi)*phi;  % convert from radians to degrees
>> plot(w,phi)
```

▶ **Problem 8.15**   In the experiment on double sideband-suppressed carrier modulation described in Section 5.11, we used a Butterworth low-pass digital filter with the following specifications:

Filter order 3

Cutoff frequency 0.125 Hz

Sampling rate 10 Hz

Use the MATLAB command `butter` to design this filter. Plot the frequency response of the filter, and show that it satisfies the specifications of the afore-mentioned experiment. ◀

■ 8.13.5 Equalization

In Example 8.8, we considered the design of an FIR digital filter to equalize a channel with frequency response

$$H_c(j\omega) = \frac{1}{1 + (j\omega/\pi)}.$$

The desired frequency response of the equalizer is

$$H_d(j\omega) = \begin{cases} \left(1 + \dfrac{j\omega}{\pi}\right)e^{-jM\omega/2}, & -\pi < \omega \leq \pi, \\ 0, & \text{otherwise} \end{cases},$$

where $M + 1$ is the length of the equalizer. Following the procedure described in Example 8.8, we note that the equalizer consists of two components connected in parallel: an ideal low-pass filter and a differentiator. Assuming a Hamming window of length $M + 1$, we may build on the MATLAB commands used for Examples 8.1 and 8.6. The corresponding set of commands for designing the equalizer and evaluating its frequency response may thus be formulated as follows:

```
>> clear;clc;
>> taps=13;
>> M=taps-1;
>> n=0:M;
>> f=n-M/2;
>> a=cos(pi*f)./f; %Integration by parts eq.8.59
>> b=sin(pi*f)./(pi*f.^2);
>> h=a-b; %Impulse resp. of window
>> k=isnan(h); h(k)=0; %Get rid of not a number

   %Response of Equalizer
>> hh=(hamming(taps)'.*h)/pi;
>> k=fftshift(ifft(ones(taps,1))).*hamming(taps);
>> [Heq,w]=freqz(hh+k',1,512,2*pi);

   %Response of Channel
>> den=sqrt(1+(w/pi).^2);
>> Hchan=1./den;

   %Response of Equalized Channel
>> Hcheq=Heq.*Hchan;

   %Plot
>> figure(1);clf
   hold on
   plot(w,abs(Heq),'b--')
   plot(w,abs(Hchan),'g-.')
   plot(w,abs(Hcheq),'r')
   hold off
   axis([0 3.5 0.7 1:4])
   legend('Equalizer','Channel','Equalized Channel')
```

8.14 *Summary*

In this chapter, we discussed procedures for the design of two important building blocks of systems and of processing signals: linear filters and equalizers. Later, using MATLAB we explored these procedures. The purpose of a filter is to separate signals on the basis of their frequency content. The purpose of an equalizer is to compensate for linear distortion produced when signals are transmitted through a dispersive channel.

Frequency-selective analog filters may be realized by using inductors and capacitors. The resulting networks are referred to as passive filters; their design is based on continuous-time ideas. Alternatively, we may use digital filters whose design is based on discrete-time concepts. Digital filters are of two kinds: finite-duration impulse response (FIR) and infinite-duration impulse response (IIR).

FIR digital filters are characterized by finite memory and BIBO stability; they can be designed to have a linear phase response. IIR digital filters have infinite memory; they are therefore able to realize a prescribed magnitude response with a shorter filter length than is possible with FIR filters.

For the design of FIR digital filters, we may use the window method, wherein a window is used to provide a trade-off between transition bandwidth and passband/stopband ripple. For the design of IIR digital filters, we may start with a suitable continuous-time transfer function (e.g., a Butterworth or Chebyshev function) and then apply the bilinear transform. Both of these digital filters can be designed directly from the prescribed specifications, using computer-aided procedures. Here, algorithmic computational complexity is traded off for a more efficient design.

Turning finally to the issue of equalization, the method most commonly used in practice involves an FIR digital filter. The central problem here is to evaluate the filter coefficients such that when the equalizer is connected in cascade with, say, a communication channel, the combination of the two approximates a distortionless filter.

FURTHER READING

1. The classic texts for the synthesis of passive filters include the following works:
   ▶ Guillemin, E. A., *Synthesis of Passive Networks* (Wiley, 1957)
   ▶ Tuttle, D. F. Jr., *Network Synthesis* (Wiley, 1958)
   ▶ Weinberg, L., *Network Analysis and Synthesis* (McGraw-Hill, 1962)

2. The Hamming window and the Hanning window (also referred to as the Hann window) are named after their respective originators: Richard W. Hamming and Julius von Hann. The term "Hanning" window was introduced in
   ▶ Blackman, R. B., and J. W. Tukey, *The Measurement of Power Spectra* (Dover Publications, 1958)

   A discussion of the window method for the design of FIR digital filters would be incomplete without mentioning the *Kaiser window*, named after James F. Kaiser. This window is defined in terms of an adjustable parameter, denoted by a, that controls the trade-off between mainlobe width and sidelobe level. When a goes to zero, the Kaiser window becomes simply the rectangular window. For a succinct description of the Kaiser window, see
   ▶ Kaiser, J. F., "Nonrecursive digital filter design using the I_0–sinh window function," *Selected Papers in Digital Signal Processing, II*, edited by the Digital Signal Processing Committee, IEEE Acoustics, Speech, and Signal Processing Society, pp. 123–126 (IEEE Press, 1975)

3. Digital filters were first described in the following books:
   ▶ Gold, B., and C. M. Rader, *Digital Processing of Signals* (McGraw-Hill, 1969)
   ▶ Kuo, F., and J. F. Kaiser, eds., *System Analysis by Digital Computer* (Wiley, 1966)

4. For an advanced treatment of digital filters, see the following books:
   ▶ Antoniou, A., *Digital Filters: Analysis, Design, and Applications*, 2nd ed. (McGraw-Hill, 1993)
   ▶ Mitra, S. K., *Digital Signal Processing: A Computer-Based Approach* (McGraw-Hill, 1998)
   ▶ Oppenheim, A. V., R. W. Schafer, and J. R. Buck, *Discrete-Time Signal Processing*, 2nd ed. (Prentice-Hall, 1999)
   ▶ Parks, T. W., and C. S. Burrus, *Digital Filter Design* (Wiley, 1987)
   ▶ Rabiner, L. R., and B. Gold, *Theory and Application of Digital Signal Processing* (Prentice-Hall, 1975)

5. For books on speech processing using digital filter techniques, see
   ▶ Rabiner, L. R., and R. W. Schafer, *Digital Processing of Speech Signals* (Prentice-Hall, 1978)
   ▶ Deller, J., J. G. Proakis, and J. H. L. Hanson, *Discrete-Time Processing of Speech Signals* (Prentice-Hall, 1993)

6. For a discussion of equalization, see the classic book

▶ Lucky, R. W., J. Salz, and E. J. Weldon, Jr., *Principles of Data Communication* (McGraw-Hill, 1968)

Equalization is also discussed in the following books:

▶ Haykin, S., *Communications Systems*, 4th ed. (Wiley, 2001)
▶ Proakis, J. G., *Digital Communications*, 3rd ed. (McGraw-Hill, 1995)

ADDITIONAL PROBLEMS

8.16 A rectangular pulse of 1-μs duration is transmitted through a low-pass channel. Suggest a small enough value for the cutoff frequency of the channel, such that the pulse is recognizable at the output of the filter.

8.17 Derive Eqs. (8.31) and (8.32), defining the passband and stopband cutoff frequencies of a Butterworth filter of order K.

8.18 Consider a Butterworth low-pass filter of order $N = 5$ and cutoff frequency $\omega_c = 1$.

(a) Find the $2K$ poles of $H(s)H(-s)$.
(b) Determine $H(s)$.

8.19 Show that, for a Butterworth low-pass filter, the following properties are satisfied:

(a) The transfer function $H(s)$ has a pole at $s = -\omega_c$ for K odd.
(b) All the poles of the transfer function $H(s)$ appear in complex-conjugate form for K even.

8.20 The denominator polynomial of the transfer function of a Butterworth low-pass prototype filter of order $K = 5$ is defined by

$$(s + 1)(s^2 + 0.618s + 1)(s^2 + 1.618s + 1).$$

Find the transfer function of the corresponding high-pass filter with cutoff frequency $\omega_c = 1$. Plot the magnitude response of the filter.

8.21 Consider again the low-pass prototype filter of order $K = 5$ described in Problem 8.20. The requirement is to modify the cutoff frequency of the filter to some arbitrary value ω_c. Find the transfer function of this filter.

8.22 For the low-pass transfer function $H(s)$ specified in Example 8.3, find the transfer function of the corresponding bandpass filter with midband frequency $\omega_0 = 1$ and bandwidth $B = 0.1$. Plot the magnitude response of the filter.

8.23 The low-pass Butterworth filters shown in Fig. 8.14 are driven by current sources. Construct the low-pass structures that are equivalent to those of Fig. 8.14, but that are driven by voltage sources.

8.24 Filter specifications call for the design of low-pass Butterworth filters based on the prototype structures shown in Figs. 8.14(a) and 8.14(b). The specifications are as follows:

Cutoff frequency, $f_c = 100\,\text{kHz}$;

Load resistance $R_l = 10\,\text{k}\Omega$.

Determine the required values of the reactive elements of the filter.

8.25 An FIR digital filter is required to have a zero at $z = 1$ for its transfer function $H(z)$. Find the condition that the impulse response $h[n]$ of the filter must satisfy for this requirement to be satisfied.

8.26 In Section 8.9, we presented one procedure for deriving an FIR digital filter, using the window method. In this problem, we want to proceed in two steps, as follows:

(a) Define

$$h[n] = \begin{cases} h_d[n], & -M/2 \le n \le M/2 \\ 0, & \text{otherwise} \end{cases},$$

where $h_d[n]$ is the desired impulse response corresponding to a frequency response with zero phase. The phase response of $h[n]$ is also zero.

(b) Having determined $h[n]$, shift it to the right by $M/2$ samples. This second step makes the filter causal.

Show that this procedure is equivalent to that described in Section 8.9.

8.27 Equation (8.64) and (8.65) pertain to the bilinear transform

$$s = \frac{z - 1}{z + 1}.$$

How are these equations modified for

$$s = \frac{2}{T_s}\left(\frac{z - 1}{z + 1}\right),$$

where T_s is the sampling interval?

8.28 In Section 1.10, we discussed a first-order recursive discrete-time filter defined by the transfer function

$$H(z) = \frac{1}{1 - \rho z^{-1}},$$

where the coefficient ρ is positive and limited to $0 < \rho < 1$. In the current problem, we consider the use of this filter for designing the discrete-time equivalent of an ideal integrator and its approximation.

(a) For the limiting case of $\rho = 1$, plot the frequency response of the filter for $-\pi < \omega \leq \pi$, and compare it with the frequency response of an ideal integrator. In particular, determine the frequency range over which the recursive discrete-time filter does not deviate from the ideal integrator by more than 1%.

(b) To ensure stability of the recursive discrete-time filter, the coefficient ρ must be less than unity. Repeat the calculations of Part (a) for $\rho = 0.99$.

8.29 Figure 8.27 shows a cascade realization of the digital IIR filter specified in Eq. (8.71). Formulate a direct form II for realizing this transfer function.

8.30 The multipath propagation channel of a wireless communications environment, discussed in Section 1.10, involves three paths from the transmitter to the receiver:

▶ a direct path.
▶ an indirect path via a small reflector introducing a differential delay of 10 μs and signal gain of 0.1.
▶ a second indirect path via a large reflector introducing a differential delay of 15 μs and signal gain of 0.2.

The differential delays and the attenuations are measured with respect to the direct path.

(a) Formulate a discrete-time equation that relates the received signal to the transmitted signal, ignoring the presence of noise in the received signal.

(b) Find the structure of an IIR equalizer and identify its coefficients. Is this equalizer stable?

(c) Suppose the equalizer is to be implemented in the form of an FIR filter. Determine the coefficients of this second equalizer, ignoring all coefficients smaller than 1%.

ADVANCED PROBLEMS

8.31 Suppose that, for a given signal $x(t)$, the integrated value of the signal over an interval T_0 is required. The relevant integral is

$$y(t) = \int_{t-T_0}^{t} x(\tau)\, d\tau.$$

(a) Show that $y(t)$ can be obtained by transmitting $x(t)$ through a filter with transfer function given by

$$H(j\omega) = T_0 \operatorname{sinc}(\omega T_0/2\pi) \exp(-j\omega T_0/2).$$

(b) Assuming the use of an ideal low-pass filter, determine the filter output at time $t = T_0$ due to a step function applied to the filter at $t = 0$. Compare the result with the corresponding output of the ideal integrator.
[*Note:* Si$(\pi) = 1.85$ and Si$(\infty) = \pi/2$.]

8.32 A low-pass prototype filter is to be transformed into a bandstop filter with midband rejection frequency ω_0. Suggest a suitable frequency transformation.

8.33 An FIR digital filter has a total of $M + 1$ coefficients, where M is an even integer. The impulse response of the filter is symmetric with respect to the $(M/2)$th point; that is,

$$h[n] = h[M - n], \quad 0 \leq n \leq M.$$

(a) Find the magnitude response of the filter.

(b) Show that this filter has a linear phase response.

Show that there are no restrictions on the frequency response $H(e^{j\Omega})$ at both $\Omega = 0$ and $\Omega = \pi$. This filter is labeled type I.

8.34 Suppose that in Problem 8.33 the $M + 1$ coefficients of the FIR digital filter satisfy the antisymmetry condition with respect to the $(M/2)$th point; that is,

$$h[n] = -h[M - n], \quad 0 \leq n \leq \frac{M}{2} - 1.$$

In this case, show that the frequency response $H(e^{j\Omega})$ of the filter must satisfy the conditions $H(e^{j0}) = 0$ and $H(e^{j\pi}) = 0$. Also, show that the filter has a linear phase response. This filter is labeled type III.

8.35 In Problems 8.33 and 8.34, the filter order M is an even integer. In the current problem and the next one, the filter order M is an odd integer. Suppose the impulse response $h[n]$ of the filter is symmetric about the noninteger point $n = M/2$. Let

$$b[k] = 2h[(M + 1)/2 - k],$$
$$k = 1, 2, \ldots, (M + 1)/2.$$

Find the frequency response $H(e^{j\Omega})$ of the filter in terms of $b[k]$. That is, show that

(a) The phase response of the filter is linear.

(b) There is no restriction on $H(e^{j0})$, but $H(e^{j\pi}) = 0$.

The filter considered in this problem is labeled type II.

8.36 Continuing with Problem 8.35 involving an FIR digital filter of order M that is an odd integer, suppose that the impulse response $h[n]$ of the filter is antisymmetric about the noninteger point $n = M/2$. Let

$$c[k] = 2h[(M + 1)/2 - k],$$
$$k = 1, 2, \ldots, (M + 1)/2.$$

Find the frequency response $H(e^{j\Omega})$ of the filter in terms of $c[k]$. That is, show that:

(a) The phase response of the filter is linear.

(b) $H(e^{j0}) = 0$, but there is no restriction on $H(e^{j\pi})$.

The filter considered in this problem is labeled type IV.

8.37 Equation (8.59) defines the impulse response $h_d[n]$ of an FIR digital filter used as a differentiator with a rectangular window. Show that $h_d[n]$ is antisymmetric; that is, show that

$$h[n - M] = -h[n], \quad 0 \le n \le M/2 - 1,$$

where M is the order of the filter, assumed to be even. In light of Problem 8.34, what can you say about the frequency response of this particular differentiator? Check your answer against the magnitude responses shown in Fig. 8.21.

8.38 It is possible for a digital IIR filter to be unstable. How can such a condition arise? Assuming that a bilinear transform is used, where would some of the poles of the corresponding analog transfer function have to lie for instability to occur?

8.39 In Section 8.10, we described a bilinear transform method for designing IIR digital filters. Here, we

consider another method, called the *method of impulse invariance*, for digital filter design. In this procedure for transforming a continuous-time (analog) filter into a discrete-time (digital) filter, the impulse response $h[n]$ of the discrete-time filter is chosen as equally spaced samples of the continuous-time filter's impulse response $h_a(t)$; that is,

$$h[n] = T_s h_a(nT_s)$$

where T_s is the sampling interval. Let

$$H_a(s) = \sum_{k=1}^{N} \frac{A_k}{s - d_k}$$

denote the transfer function of the continuous-time filter. Show that the transfer function of the corresponding discrete-time filter obtained by using the method of impulse invariance is given by

$$H(z) = \sum_{k=1}^{N} \frac{T_s A_k}{1 - e^{d_k T_s} z^{-1}}.$$

8.40 Equation (8.73) defines the frequency response of an equalizer for dealing with linear distortion produced by a continuous-time LTI system. Formulate the corresponding relation for an equalizer used to deal with linear distortion produced by a discrete-time LTI system.

8.41 Consider a tapped-delay-line equalizer whose frequency response is specified in Eq. (8.74). In theory, this equalizer can compensate for any linear distortion simply by making the number of coefficients, $M + 1$, large enough. What is the penalty for making M large? Justify your answer.

COMPUTER EXPERIMENTS

8.42 Design an FIR digital low-pass filter with a total of 23 coefficients. Use a Hamming window for the design. The cutoff frequency of the filter is $\omega_c = \pi/3$ for sampling interval $T_s = 15$.

(a) Plot the impulse response of the filter.

(b) Plot the magnitude response of the filter.

8.43 Design a differentiator using an FIR digital filter of order $M = 100$. For this design, use (a) a rectangular window and (b) a Hamming window. In each case, plot the impulse response and magnitude response of the filter.

8.44 You are given a data sequence with sampling rate of $2\pi \times 8000$ rad/s. A low-pass digital IIR filter is required for processing this data sequence to meet the following specifications:

cutoff frequency $\omega_c = 2\pi \times 800$ rad/s;

attenuation at $2\pi \times 1200$ rad/s = 15 dB

(a) Assuming a Butterworth response, determine a suitable value for the filter order K.

(b) Using the bilinear transform, design the filter.

(c) Plot the magnitude response and phase response of the filter.

8.45 Design a high-pass digital IIR filter with Butterworth response. The filter specifications are as follows: filter order $K = 5$, cutoff frequency $\omega_c = 0.6$, and sampling interval $T_s = 15$.

8.46 Consider a channel whose frequency response is described by the second-order Butterworth response

$$H_c(j\omega) = \frac{1}{(1 + j\omega/\pi)^2}.$$

Design an FIR filter with 95 coefficients for equalizing this channel over the frequency band $-\pi < \omega \le \pi$.

9 Application to Linear Feedback Systems

9.1 Introduction

Feedback is a concept of profound engineering importance. The need for feedback arises in the design of power amplifiers, operational amplifiers, digital filters, and control systems, just to mention a few applications. In all these applications, feedback is introduced into the design of the system with a specific purpose in mind: Improving the linear behavior of the system, reducing the sensitivity of the gain of the system to variations in the values of certain parameters, and reducing the effect of external disturbances on the operation of the system. However, these practical benefits are achieved at the cost of a more complicated system behavior. Also, the feedback system may become unstable, unless special precautions are taken in its design.

We begin the study of linear feedback systems in this chapter by describing some basic feedback concepts that provide the motivation for two important applications: operational amplifiers and feedback control systems, discussed in that order. The next topic is the stability problem, which features prominently in the study of feedback systems. Two approaches are taken here, one based on pole locations of a feedback system and the other based on the frequency response of the system. One other important topic, covered toward the end of the chapter, is that of a sampled-data system, which is a feedback control system that uses a computer for control. The study of sampled-data systems is important not only in an engineering context, but also in a theoretical one: It combines the use of the z-transform and the Laplace transform under one umbrella.

9.2 What Is Feedback?

Let us define *feedback* as *the return of a fraction of the output signal of a system to its input, thereby forming a loop of dependencies among signals around the system*. However, it can be argued that the presence or absence of feedback in a system is more a matter of viewpoint than that of physical reality. This simple, yet profound, statement is illustrated by way of the theme example on recursive discrete-time filters, presented earlier in Section 1.10.

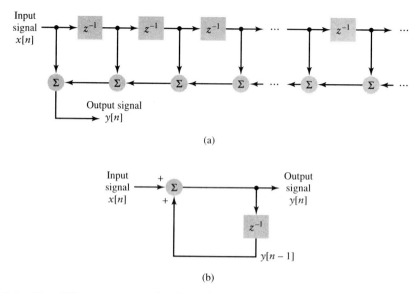

FIGURE 9.1   Two different structures for the realization of an accumulator. (a) Feed forward struc ture of infinite order. (b) First-order recusive structure.

To be specific, consider an *accumulator*, whose function is to add all previous values of a discrete-time input signal, namely, $x[n - 1], x[n - 2], \ldots$, to its current value $x[n]$, to produce the output signal

$$y[n] = \sum_{k=0}^{\infty} x[n - k].$$

According to this input–output description, the accumulator may be realized by a feed-forward system of infinite order, as depicted in Fig. 9.1(a). Clearly, there is *no* feedback in such a realization.

Recall from Chapter 1 that the accumulator may also be implemented as the first-order recursive discrete-time filter shown in Fig. 9.1(b). The presence of feedback is clearly visible in this second realization of the accumulator, in that we have a feedback loop consisting of two components:

▶ A *memory unit*, represented by z^{-1}, which acts on the current output $y[n]$ to supply the past output $y[n - 1]$.

▶ An *addet*, which adds the past (delayed) output $y[n - 1]$ to the current input $x[n]$ to produce $y[n]$.

The two structures of Fig. 9.1 provide two entirely different methods of realizing an accumulator. They are, however, equivalent in that they are indistinguishable from each other in terms of input–output behavior. In particular, they both have exactly the same impulse response, of infinite duration. Yet, one structure has no feedback, while the other is a simple example of a feedback system.

To further illustrate the fact that feedback is a matter of viewpoint, consider the simple parallel *RC* circuit shown in Fig. 9.2(a). In physical terms, we do not normally

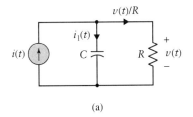

(a)

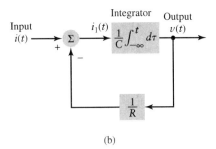

(b)

FIGURE 9.2   (a) Simple parallel RC circuit driven by current source $i(t)$. (b) Block diagram of the RC circuit, formulated in terms of two variables: the current $i_1(t)$ through the capacitor C and the voltage $v(t)$ across C. This figure clearly displays the presence of a feedback loop, even though there is no physical evidence of feedback in the RC circuit itself.

think of this circuit as an example of a feedback system. Yet its mathematical formulation in terms of the current $i_1(t)$ through the capacitor C, given by the equation (see Problem 1.92)

$$i_1(t) = i(t) - \frac{1}{RC} \int_{-\infty}^{t} i_1(\tau) \, d\tau,$$

clearly reveals the presence of a feedback loop, as depicted in Fig. 9.2(b).

The accumulator of Fig. 9.1 and the parallel RC circuit of Fig. 9.2 are presented here merely to illustrate that feedback is indeed a matter of viewpoint, depending on how the input–output behavior of the system is formulated.

Our primary interest in this chapter is the study of LTI systems whose block diagrams, by virtue of the underlying philosophy of their design, exhibit feedback loops—hence the reference to these systems as *linear feedback systems*. The accumulator, implemented in the recursive form shown in Fig. 9.1(b), is one example of such a system.

The motivation for the study of linear feedback systems is twofold:

1. Practical benefits of engineering importance, resulting directly from the application of feedback, are achieved.

2. An understanding of the stability problem ensures that the feedback system is stable under all operating conditions.

The rest of the chapter focuses on these two important issues. The discussion of basic feedback concepts begins with a continuous-time perspective, and lays down the framework for the study of linear feedback systems in general.

9.3 Basic Feedback Concepts

Figure 9.3(a) shows the block diagram of a feedback system in its most basic form. The system consists of three components connected together to form a single feedback loop:

▶ a *plant*, which acts on an *error signal* $e(t)$ to produce the output signal $y(t)$;

▶ a *sensor*, which measures the output signal $y(t)$ to produce a *feedback signal* $r(t)$;

▶ a *comparator*, which calculates the difference between the externally applied input (reference) signal $x(t)$ and the feedback signal $r(t)$ to produce the error signal

$$e(t) = x(t) - r(t). \tag{9.1}$$

The terminology used here pertains more closely to a control system, but can readily be adapted to deal with a feedback amplifier.

In what follows, we assume that the plant dynamics and sensor dynamics in Fig. 9.3(a) are each modeled as LTI systems. Given the time-domain descriptions of both systems, we may proceed to relate the output signal $y(t)$ to the input signal $x(t)$. However, we find it more convenient to work with Laplace transforms and do the formulation in the s-domain, as described in Fig. 9.3(b). Let $X(s)$, $Y(s)$, $R(s)$, and $E(s)$ denote the Laplace transforms of $x(t)$, $y(t)$, $r(t)$, and $e(t)$, respectively. We may then transform Eq. (9.1) into the equivalent form

$$E(s) = X(s) - R(s). \tag{9.2}$$

Let $G(s)$ denote the transfer function of the plant and $H(s)$ denote the transfer function of the sensor. Then by definition, we may write

$$G(s) = \frac{Y(s)}{E(s)} \tag{9.3}$$

and

$$H(s) = \frac{R(s)}{Y(s)}. \tag{9.4}$$

Using Eq. (9.3) to eliminate $E(s)$ and Eq. (9.4) to eliminate $R(s)$ from Eq. (9.2), we get

$$\frac{Y(s)}{G(s)} = X(s) - H(s)Y(s).$$

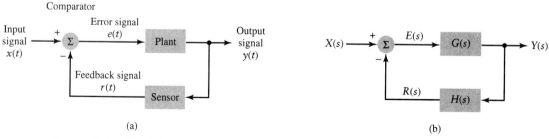

(a)   (b)

FIGURE 9.3 Block diagram representations of a single-loop feedback system: (a) time-domain representation and (b) s-domain representation.

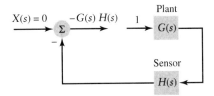

FIGURE 9.4   *s*-domain representation of a scheme for measuring the return difference $F(s)$, which is defined as the difference between the unit signal applied to the plant and the return signal $G(s)H(s)$.

Collecting terms and solving for the ratio $Y(s)/X(s)$, we find that the *closed-loop transfer function* of the feedback system in Fig. 9.4 is

$$
\begin{aligned}
T(s) &= \frac{Y(s)}{X(s)} \\
&= \frac{G(s)}{1 + G(s)H(s)}.
\end{aligned}
\tag{9.5}
$$

The term "closed-loop" is used here to emphasize the fact that there is a closed signal-transmission loop around which signals may flow in the system.

   The quantity $1 + G(s)H(s)$ in the denominator of Eq. (9.5) provides a measure of the feedback acting around $G(s)$. For a physical interpretation of this quantity, examine the configuration of Fig. 9.4, in which we have made two changes to the feedback system of Fig. 9.3(b):

▶ The input signal $x(t)$, and therefore $X(s)$, is reduced to zero.
▶ The feedback loop around $G(s)$ is opened.

Suppose that a test signal with unit Laplace transform is applied to $G(s)$ (i.e., the plant) shown in the figure. Then the signal returned to the other open end of the loop is $-G(s)H(s)$. The difference between the unit test signal and the returned signal is equal to $1 + G(s)H(s)$, a quantity called the *return difference*. Denoting this quantity by $F(s)$, we may thus write

$$
F(s) = 1 + G(s)H(s).
\tag{9.6}
$$

   The product term $G(s)H(s)$ is called the *loop transfer function* of the system. It is simply the transfer function of the plant and the sensor connected in cascade, as shown in Fig. 9.5. This configuration is the same as that of Fig. 9.4 with the comparator removed. Denoting the loop transfer function by $L(s)$, we may thus write

$$
L(s) = G(s)H(s)
\tag{9.7}
$$

FIGURE 9.5   *s*-domain representation of a scheme for measuring the loop transfer function $L(s)$.

and so relate the return difference $F(s)$ to $L(s)$ by the formula

$$\boxed{F(s) = 1 + L(s).}$$

(9.8)

In what follows, we use $G(s)H(s)$ and $L(s)$ interchangeably when referring to the loop transfer function.

■ 9.3.1 NEGATIVE AND POSITIVE FEEDBACK

Consider an operating range of frequencies for which G and H, pertaining, respectively, to the plant and sensor, may be treated as essentially independent of the complex frequency s. In such a situation, the feedback in Fig. 9.3(a) is said to be *negative*. When the comparator is replaced with an adder, the feedback is said to be *positive*.

The terms, *negative* and *positive*, however, are of limited value. We say this because, in the general setting depicted in Fig. 9.3(b), the loop transfer function $G(s)H(s)$ is dependent on the complex frequency s. For $s = j\omega$, we find that $G(j\omega)H(j\omega)$ has a phase that varies with the frequency ω. When the phase of $G(j\omega)H(j\omega)$ is zero, the situation in Fig. 9.3(b) corresponds to negative feedback. When the phase of $G(j\omega)H(j\omega)$ is 180°, the same configuration behaves like a positive feedback system. Thus, for the single-loop feedback system of Fig. 9.3(b), there will be different frequency bands for which the feedback is alternately negative and positive. Care must therefore be exercised in the use of the terms *negative feedback* and *positive feedback*.

9.4 *Sensitivity Analysis*

A primary motivation for the use of feedback is to reduce the sensitivity of the closed-loop transfer function of the system in Fig. 9.3 to changes in the transfer function of the plant. For the purpose of this discussion, we ignore the dependencies on the complex frequency s in Eq. (9.5) and treat G and H as "constant" parameters. We may thus write

$$T = \frac{G}{1 + GH}.$$

(9.9)

In Eq. (9.9), we refer to G as the *gain* of the plant and to T as the *closed-loop gain* of the feedback system.

Suppose now that the gain G is changed by a small amount ΔG. Then, differentiating Eq. (9.9) with respect to G, we find that the corresponding change in T is

$$\Delta T = \frac{\partial T}{\partial G}\Delta G$$
$$= \frac{1}{(1 + GH)^2}\Delta G.$$

(9.10)

The *sensitivity* of T with respect to changes in G is formally defined by

$$\boxed{S_G^T = \frac{\Delta T/T}{\Delta G/G}.}$$

(9.11)

In words, the sensitivity of T with respect to G is the percentage change in T divided by the percentage change in G. Using Eqs. (9.5) and (9.10) in Eq. (9.11) yields

$$
\boxed{
\begin{aligned}
S_G^T &= \frac{1}{1 + GH} \\
&= \frac{1}{F},
\end{aligned}
}
\tag{9.12}
$$

which shows that the sensitivity of T with respect to G is equal to the reciprocal of the return difference F.

With the availability of two degrees of freedom, represented by the parameters G and H pertaining to the plant and sensor, respectively, the use of feedback permits a system designer to simultaneously realize prescribed values for the closed-loop gain T and sensitivity S_G^T. This is achieved through the use of Eqs. (9.9) and (9.12), respectively.

▶ **Problem 9.1**   To make the sensitivity S_G^T small compared with unity, the loop gain GH must be large compared with unity. What are the approximate values of the closed-loop gain T and sensitivity S_G^T under this condition?

Answers:

$$
T \simeq \frac{1}{H} \quad \text{and} \quad S_G^T \simeq \frac{1}{GH} \qquad ◀
$$

EXAMPLE 9.1 FEEDBACK AMPLIFIER   Consider a single-loop feedback amplifier, the block diagram of which is shown in Fig. 9.6. The system consists of a linear amplifier and a feedback network made up of positive resistors only. The amplifier has gain A, and the feedback network feeds a controllable fraction β of the output signal back to the input. Suppose that the gain $A = 1000$.

(a) Determine the value of β that will result in a closed-loop gain $T = 10$.

(b) Suppose that the gain A changes by 10%. What is the corresponding percentage change in the closed-loop gain T?

Solution:

(a) For the problem at hand, the plant and sensor are represented by the amplifier and feedback network, respectively. We may thus put $G = A$ and $H = \beta$ and rewrite Eq. (9.9) in the form

$$
T = \frac{A}{1 + \beta A}.
$$

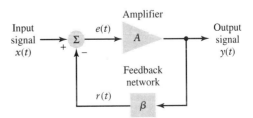

FIGURE 9.6   Block diagram of single-loop feedback amplifier.

Solving for β, we get

$$\beta = \frac{1}{A}\left(\frac{A}{T} - 1\right).$$

With $A = 1000$ and $T = 10$, we obtain

$$\beta = \frac{1}{1000}\left(\frac{1000}{10} - 1\right) = 0.099.$$

(b) From Eq. (9.12), the sensitivity of the closed-loop gain T with respect to A is

$$S_A^T = \frac{1}{1 + \beta A}$$

$$= \frac{1}{1 + 0.099 \times 1000} = \frac{1}{100}.$$

Hence, with a 10% change in A, the corresponding percentage change in T is

$$\Delta T = S_A^T \frac{\Delta A}{A}$$

$$= \frac{1}{100} \times 10\% = 0.1\%,$$

which indicates that, for this example, the feedback amplifier of Fig. 9.6 is relatively insensitive to variations in the gain A of the internal amplifier. ■

9.5 *Effect of Feedback on Disturbance or Noise*

The use of feedback has another beneficial effect on a system's performance: It reduces the effect of a disturbance or noise generated inside the feedback loop. To see how this works, consider the single-loop feedback system depicted in Fig. 9.7. This system differs from the basic configuration of Fig. 9.3 in two respects: G and H are both treated as constant parameters, and the system includes a disturbance signal denoted by ν inside the loop. Since the system is linear, we may use the principle of superposition to calculate the effects of the externally applied input signal x and the disturbance signal ν separately and then add the results:

1. We set the disturbance signal ν equal to zero. Then the closed-loop gain of the system with the signal x as the input is equal to $G/(1 + GH)$. Hence, the output signal resulting from x acting alone is

$$y|_{\nu=0} = \frac{G}{1 + GH}x.$$

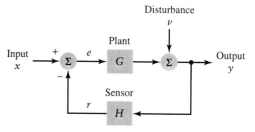

FIGURE 9.7   Block diagram of a single-loop feedback system that includes a disturbance inside the loop.

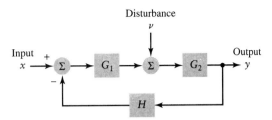

FIGURE 9.8   Feedback system for Problem 9.2.

2. We set the input signal x equal to zero. Then, with the disturbance v acting as the only external signal acting on the system, the pertinent closed-loop gain of the system is $1/(1 + GH)$. Correspondingly, the output of the system so produced is given by

$$y|_{x=0} = \frac{1}{1 + GH}v.$$

Adding these two contributions, we obtain the output due to the combined action of x and v:

$$y = \frac{G}{1 + GH}x + \frac{1}{1 + GH}v. \tag{9.13}$$

Here, the first term represents the desired output and the second term represents the unwanted output. Equation (9.13) clearly shows that the use of feedback in Fig. 9.7 has the effect of reducing the disturbance v by the factor $1 + GH$ (i.e., the return difference F).

▶ **Problem 9.2**   Consider the system configuration of Fig. 9.8. Determine the effect produced by the disturbance signal v acting alone.

Answer:

$$y|_{x=0} = \frac{G_2}{1 + G_1 G_2 H}v \qquad\qquad ◀$$

9.6 *Distortion Analysis*

Nonlinearity arises in a physical system whenever it is driven outside its linear range of operation. We may improve the linearity of such a system by applying feedback around it. To investigate this important effect, we may proceed in one of two ways:

▶ The output of the system is expressed as a nonlinear function of the input, and a pure sine wave is used as the input signal.

▶ The input to the system is expressed as a nonlinear function of the output.

The latter approach may seem strange at first sight; however, it is more general in formulation and provides a more intuitively satisfying description of how feedback affects the nonlinear behavior of a system. It is therefore the approach that is pursued in what follows.

Consider then a feedback system in which the dependence of the error e on the system output y is represented by

$$e = a_1 y + a_2 y^2, \tag{9.14}$$

where a_1 and a_2 are constants. The linear term $a_1 y$ represents the desired behavior of the plant, and the parabolic term $a_2 y^2$ accounts for its deviation from linearity. Let the parameter H determine the fraction of the plant output y that is fed back to the input. With x denoting the input applied to the feedback system, we may thus write

$$e = x - Hy. \tag{9.15}$$

Eliminating e between Eqs. (9.14) and (9.15) and rearranging terms, we get

$$x = (a_1 + H)y + a_2 y^2.$$

Differentiating x with respect to y yields

$$\begin{aligned}\frac{dx}{dy} &= a_1 + H + 2a_2 y \\ &= (a_1 + H)\left(1 + \frac{2a_2}{a_1 + H}y\right),\end{aligned} \tag{9.16}$$

which holds in the presence of feedback.

In the absence of feedback, the plant operates by itself, as shown by

$$x = a_1 y + a_2 y^2, \tag{9.17}$$

which is a rewrite of Eq. (9.14) with the input x used in place of the error e. Differentiating Eq. (9.17) with respect to y yields

$$\begin{aligned}\frac{dx}{dy} &= a_1 + 2a_2 y \\ &= a_1\left(1 + \frac{2a_2}{a_1}y\right).\end{aligned} \tag{9.18}$$

The derivatives in Eqs. (9.16) and (9.18) have both been *normalized* to their respective linear terms to make for a fair comparison between them. In the presence of feedback, the term $2a_2 y/(a_1 + H)$ in Eq. (9.16) provides a measure of *distortion* due to the parabolic term $a_2 y^2$ in the input–output relationship of the plant. The corresponding measure of distortion in the absence of feedback is represented by the term $2a_2 y/a_1$ in Eq. (9.18). Accordingly, the application of feedback has reduced the distortion due to deviation of the plant from linearity by the factor

$$D = \frac{2a_2 y/(a_1 + H)}{2a_2 y/a_1} = \frac{a_1}{a_1 + H}.$$

From Eq. (9.17), we readily see that the coefficient a_1 is the reciprocal of the gain G of the plant. Hence, we may rewrite the foregoing result as

$$\boxed{D = \frac{1/G}{(1/G) + H} = \frac{1}{1 + GH} = \frac{1}{F}}, \tag{9.19}$$

which shows that the distortion is reduced by a factor equal to the return difference F.

▶ **Problem 9.3**   Suppose the nonlinear relation of Eq. (9.14) defining the error e in terms of the output y, is expanded to include a cubic term; that is,

$$e = a_1 y + a_2 y^2 + a_3 y^3.$$

Show that the application of feedback also reduces the effect of distortion due to this cubic term by a factor equal to the return difference F.   ◀

9.7 *Summarizing Remarks on Feedback*

■ 9.7.1 BENEFITS OF FEEDBACK

From the analysis presented in Sections 9.4 through 9.6, we now see t⎺⎽ ⎽
ference *F* plays a central role in the study of feedback systems, in three important respects:

1. Control of sensitivity
2. Control of the effect of an internal disturbance
3. Control of distortion in a nonlinear system

With regard to point 1, the application of feedback to a plant reduces the sensitivity of the closed-loop gain of the feedback system to parameter variations in the plant by a factor equal to *F*. In respect of point 2, the transmission of a disturbance from some point inside the loop of the feedback system to the closed-loop output of the system is also reduced by a factor equal to *F*. Finally, as regards point 3, distortion due to nonlinear effects in the plant is again reduced by a factor equal to *F*. These improvements in overall system performance resulting from the application of feedback are of immense engineering importance.

■ 9.7.2 COST OF FEEDBACK

Naturally, there are attendant costs to the benefits gained from the application of feedback to a control system:

- ► *Increased complexity.* The application of feedback to a control system requires the addition of new components. Thus, there is the cost of increased system complexity.

- ► *Reduced gain.* In the absence of feedback, the transfer function of a plant is $G(s)$. When feedback is applied to the plant, the transfer function of the system is modified to $G(s)/F(s)$, where $F(s)$ is the return difference. Since, the benefits of feedback are now realized only when $F(s)$ is greater than unity, it follows that the application of feedback results in reduced gain.

- ► *Possible instability.* Often, an open-loop system (i.e., the plant operating on its own) is stable. However, when feedback is applied to the system, there is a real possibility that the closed-loop system may become unstable. To guard against such a possibility, we have to take precautionary measures in the design of the feedback control system.

In general, the advantages of feedback outweigh the disadvantages. It is therefore necessary that we account for the increased complexity in designing a control system and pay particular attention to the stability problem. The stability problem will occupy our attention from Sections 9.11 through 9.16.

9.8 *Operational Amplifiers*

An important application of feedback is in operational amplifiers. An *operational amplifier*, or an *op amp*, as it is often referred to, provides the basis for realizing a transfer function with prescribed poles and zeros in a relatively straightforward manner. Ordinarily, an op amp has two input terminals, one inverting and the other noninverting, and an output terminal. Figure 9.9(a) shows the conventional symbol used to represent an operational amplifier; only the principal signal terminals are included in this symbol.

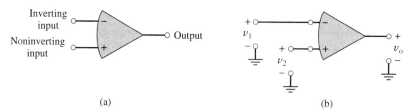

(a)   (b)

FIGURE 9.9   (a) Conventional symbol for operational amplifier. (b) Operational amplifier with input and output voltages.

The *ideal model* for an operational amplifier encompasses four assumptions (refer to Fig. 9.9(b) for the input and output signals):

1. The op amp acts as a voltage-controlled voltage source described by the input–output relation

$$v_o = A(v_2 - v_1), \qquad (9.20)$$

where v_1 and v_2 are the signals applied to the inverting and noninverting input terminals, respectively, and v_o is the output signal. All these signals are measured in volts.

2. The open-loop voltage gain A has a constant value that is very large compared with unity, which means that, for a finite output signal v_o, we must have $v_1 \simeq v_2$. This property is referred to as *virtual ground*.

3. The impedance between the two input terminals is infinitely large, and so is the impedance between each one of them and the ground, which means that the input terminal currents are zero.

4. The output impedance is zero.

Typically, the operational amplifier is not used in an open-loop fashion. Rather, it is normally used as the amplifier component of a feedback circuit in which the feedback controls the closed-loop transfer function of the circuit. Figure 9.10 shows one such circuit, where the noninverting input terminal of the operational amplifier is grounded and the impedances $Z_1(s)$ and $Z_2(s)$ represent the input element and feedback element of the circuit, respectively. Let $V_{in}(s)$ and $V_{out}(s)$ denote the Laplace transforms of the input and output voltage signals, respectively. Then, using the ideal model to describe the operational amplifier, we may in a corresponding way construct the model shown in Fig. 9.11 for the feedback circuit of Fig. 9.10. The following condition may be derived from properties 2 and 3 of the ideal operational amplifier:

$$\frac{V_{in}(s)}{Z_1(s)} \simeq -\frac{V_{out}(s)}{Z_2(s)}.$$

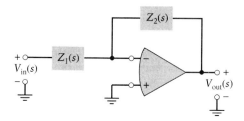

FIGURE 9.10   Operational amplifier embedded in a single-loop feedback circuit.

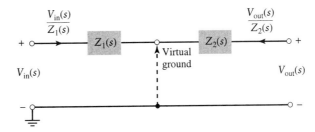

FIGURE 9.11 Ideal model for the feedback circuit of Fig. 9.10.

The closed-loop transfer function of the feedback circuit in Fig. 9.10 is therefore

$$
\begin{aligned}
T(s) &= \frac{V_{\text{out}}(s)}{V_{\text{in}}(s)} \\
&\simeq -\frac{Z_2(s)}{Z_1(s)}.
\end{aligned}
\tag{9.21}
$$

We derived this result without recourse to the feedback theory developed in Section 9.3. How, then, do we interpret the result in light of the general feedback formula of Eq. (9.5)? To answer this question, we have to understand the way in which feedback manifests itself in the operational amplifier circuit of Fig. 9.10. The feedback element $Z_2(s)$ is connected in parallel to the amplifier at both its input and output ports. This would therefore suggest the use of currents as the basis for representing the input signal $x(t)$ and feedback signal $r(t)$. The application of feedback in the system of Fig. 9.10 has the effect of making the input impedance measured looking into the operational amplifier small compared with both $Z_1(s)$ and $Z_2(s)$, but nevertheless of some finite value. Let $Z_{\text{in}}(s)$ denote this input impedance. We may then use current signals to represent the Laplace transforms of the current signals $x(t)$ and $r(t)$ in terms of the Laplace transforms of the voltage signals $v_{\text{in}}(t)$ and $v_{\text{out}}(t)$, respectively:

$$
X(s) = \frac{V_{\text{in}}(s)}{Z_1(s)};
\tag{9.22}
$$

$$
R(s) = -\frac{V_{\text{out}}(s)}{Z_2(s)}.
\tag{9.23}
$$

The error signal $e(t)$, defined as the difference between $x(t)$ and $r(t)$, is applied across the input terminals of the operational amplifier to produce an output voltage equal to $v_{\text{out}}(t)$. With $e(t)$, represented by the Laplace transform $E(s)$, viewed as a current signal, we may invoke the following considerations:

▶ A generalization of Ohm's law, according to which the voltage produced across the input terminals of the operational amplifier is $Z_{\text{in}}(s)E(s)$, where $Z_{\text{in}}(s)$ is the input impedance.

▶ A voltage gain equal to $-A$.

We may thus express the Laplace transform of the voltage $y(t)$ produced across the output terminals of the operational amplifier as

$$
\begin{aligned}
Y(s) &= V_{\text{out}}(s) \\
&= -AZ_{\text{in}}(s)E(s).
\end{aligned}
\tag{9.24}
$$

By definition (see Eq. (9.3)), the transfer function of the operational amplifier (viewed as the plant) is

$$G(s) = \frac{Y(s)}{E(s)}.$$

For the problem at hand, from Eq. (9.24), it follows that

$$G(s) = -AZ_{in}(s). \tag{9.25}$$

From the definition of Eq. (9.4) we recall that the transfer function of the feedback path is

$$H(s) = \frac{R(s)}{Y(s)}.$$

Hence, with $V_{out}(s) = Y(s)$, it follows from Eq. (9.23) that

$$H(s) = -\frac{1}{Z_2(s)}. \tag{9.26}$$

Using Eqs. (9.22) and (9.24), we may now reformulate the feedback circuit of Fig. 9.10 as depicted in Fig. 9.12, where $G(s)$ and $H(s)$ are defined by Eqs. (9.25) and (9.26), respectively. Figure 9.12 is configured in the same way as the basic feedback system shown in Fig. 9.3(b).

From Fig. 9.12, we readily find that

$$\frac{Y(s)}{X(s)} = \frac{G(s)}{1 + G(s)H(s)}$$

$$= \frac{-AZ_{in}(s)}{1 + \frac{AZ_{in}(s)}{Z_2(s)}}.$$

In light of Eq. (9.22) and the first line of Eq. (9.24), we may rewrite this result in the equivalent form

$$\frac{V_{out}(s)}{V_{in}(s)} = \frac{-AZ_{in}(s)}{Z_1(s)\left[1 + \frac{AZ_{in}(s)}{Z_2(s)}\right]}. \tag{9.27}$$

Since, for an operational amplifier, the gain A is very large compared with unity, we may approximate Eq. (9.27) as

$$\frac{V_{out}(s)}{V_{in}(s)} \simeq -\frac{Z_2(s)}{Z_1(s)},$$

which is the result that we derived earlier.

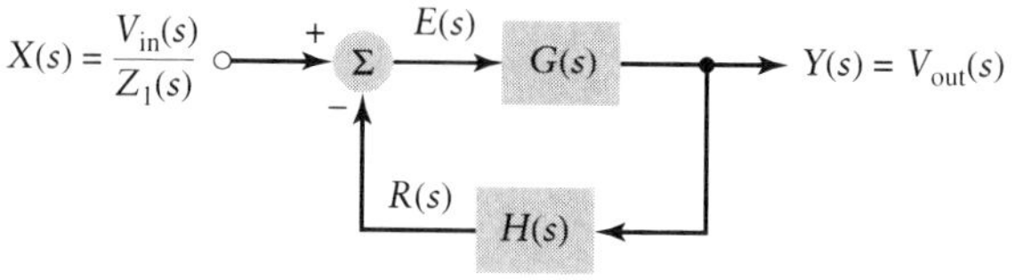

FIGURE 9.12   Reformulation of the feedback circuit of Fig. 9.10 so that it corresponds to the basic feedback system of Fig. 9.3(b).

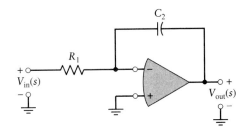

FIGURE 9.13   Operational amplifier circuit used as an integrator in Example 9.2.

EXAMPLE 9.2   INTEGRATOR   In Section 1.10, we discussed the use of a simple RC circuit as an approximate realization of an ideal integrator. In this example, we improve on the realization of an integrator in a significant way through the use of an operational amplifier. To be specific, consider the operational amplifier circuit of Fig. 9.13, in which the input element is a resistor R_1 and the feedback element is a capacitor C_2. Show that this circuit operates as an integrator.

Solution: The impedances are

$$Z_1(s) = R_1$$

and

$$Z_2(s) = \frac{1}{sC_2}.$$

Thus, substituting these values into Eq. (9.21), we get

$$T(s) \simeq -\frac{1}{sC_2R_1},$$

which shows that the closed-loop transfer function of Fig. 9.13 has a pole at the origin. Since division by the complex variable s corresponds to integration in time, we conclude that this circuit performs integration on the input signal.   ■

S & S
Solutions

▶ **Problem 9.4**   The circuit elements of the integrator in Fig. 9.13 have the values $R_1 = 100\,\text{k}\Omega$ and $C_2 = 1.0\,\mu\text{F}$. The initial value of the output voltage is $v_{\text{out}}(0)$. Determine the output voltage $v_{\text{out}}(t)$ for varying time t.

Answer: $v_{\text{out}}(t) \simeq -\displaystyle\int_0^t 10v_{\text{in}}(\tau)\,d\tau + v_{\text{out}}(0)$, where time $t > 0$ and is measured in seconds   ◀

EXAMPLE 9.3   ANOTHER OPERATIONAL AMPLIFIER CIRCUIT WITH RC ELEMENTS   Consider the operational amplifier circuit of Fig. 9.14. Determine the closed-loop transfer function of this circuit.

Solution: The input element is the parallel combination of resistor R_1 and capacitor C_1; hence,

$$Z_1(s) = \frac{R_1}{1 + sC_1R_1}.$$

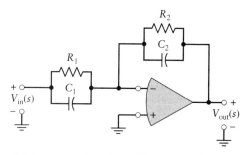

FIGURE 9.14   Operational amplifier circuit for Example 9.3.

The feedback element is the parallel combination of resistor R_2 and capacitor C_2; hence,

$$Z_2(s) = \frac{R_2}{1 + sC_2R_2}.$$

Substituting these expressions into Eq. (9.21) yields the closed-loop transfer function

$$T(s) \simeq -\frac{R_2}{R_1}\frac{1 + sC_1R_1}{1 + sC_2R_2},$$

which has a zero at $s = -1/C_1R_1$ and a pole at $s = -1/C_2R_2$.   ▪

▶ **Problem 9.5 Differentiator**   The operational amplifier circuit of Fig. 9.14 includes a differentiator as a special case.

(a) How is the differentiator realized?

(b) How does the differentiator realized in this way differ from the approximate RC differentiator discussed in Section 1.10?

Answers:

(a) $R_1 = \infty$ and $C_2 = 0$

(b) $T(s) \simeq -sC_1R_2$, which, except for a minus sign, represents a differentiator far more accurate than the passive high-pass RC circuit   ◀

▪ 9.8.1 ACTIVE FILTERS

In Chapter 8, we discussed procedures for the design of passive filters and digital filters. We may also design filters by using operational amplifiers; filters synthesized in this way are referred to as *active filters*.

   In particular, by cascading different versions of the basic circuit of Fig. 9.14, it is possible to synthesize an overall transfer function with arbitrary real poles and arbitrary real zeros. Indeed, with more elaborate forms of the impedances $Z_1(s)$ and $Z_2(s)$ in Fig. 9.10, we can realize a transfer function with arbitrary complex poles and zeros.

   Compared to passive LC filters, active filters offer an advantage by eliminating the need for using inductors. Compared to digital filters, active filters offer the advantages of continuous-time operation and reduced complexity. However, active filters lack the computing power and flexibility offered by digital filters.

9.9 *Control Systems*

Consider a *plant* that is controllable. The function of a control system in such a facility is to *obtain accurate control over the plant, so that the output of the plant remains close to a target (desired) response.* This is accomplished through proper modification of the plant input. We may identify two basic types of control system:

- ▶ *open-loop control*, in which the modification to the plant input is derived directly from the target response;
- ▶ *closed-loop control*, in which feedback is used around the plant.

In both cases, the target response acts as the input to the control system. Let us examine these two types of control system in the order given.

■ 9.9.1 OPEN-LOOP CONTROL

Figure 9.15(a) shows the block diagram of an open-loop control system. The plant dynamics are represented by the transfer function $G(s)$. The controller, represented by the transfer function $H(s)$, acts on the target response $y_d(t)$ to produce the desired control signal $c(t)$. The disturbance $\nu(t)$ is included to account for noise and distortion produced at the output of the plant. The configuration shown in Fig. 9.15(b) depicts the error $e(t)$ as the difference between the target response $y_d(t)$ and the actual output $y(t)$ of the system; that is,

$$e(t) = y_d(t) - y(t). \tag{9.28}$$

Let $Y_d(s)$, $Y(s)$, and $E(s)$ denote the Laplace transforms of $y_d(t)$, $y(t)$, and $e(t)$, respectively. Then we may rewrite Eq. (9.28) in the s-domain as

$$E(s) = Y_d(s) - Y(s). \tag{9.29}$$

From Fig. 9.15(a), we also readily find that

$$Y(s) = G(s)H(s)Y_d(s) + N(s), \tag{9.30}$$

where $N(s)$ is the Laplace transform of the disturbance signal $\nu(t)$. Eliminating $Y(s)$ between Eqs. (9.29) and (9.30) yields

$$E(s) = [1 - G(s)H(s)]Y_d(s) - N(s). \tag{9.31}$$

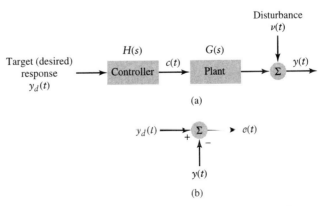

(a)

(b)

FIGURE 9.15   (a) Block diagram of open-loop control system and (b) configuration for calculation of the error signal $e(t)$.

The error $e(t)$ is minimized by setting

$$1 - G(s)H(s) = 0.$$

For this condition to be satisfied, the controller must act as the inverse of the plant; that is,

$$H(s) = \frac{1}{G(s)}. \tag{9.32}$$

From Fig. 9.15(a), we see that with $y_d(t) = 0$, the plant output $y(t)$ is equal to $v(t)$. Therefore, the best that an open-loop control system can do is to leave the disturbance $v(t)$ unchanged.

The overall transfer function of the system (in the absence of the disturbance $v(t)$) is simply

$$T(s) = \frac{Y(s)}{Y_d(s)} \tag{9.33}$$
$$= G(s)H(s).$$

Ignoring the dependence on s, and assuming that H does not change, the sensitivity of T with respect to changes in G is therefore

$$S_G^T = \frac{\Delta T/T}{\Delta G/G}$$
$$= \frac{H\,\Delta G/(GH)}{\Delta G/G} \tag{9.34}$$
$$= 1.$$

The implication of $S_G^T = 1$ is that a percentage change in G is translated into an equal percentage change in T.

The conclusion to be drawn from this analysis is that an open-loop control system leaves both the sensitivity and the effect of a disturbance unchanged.

■ 9.9.2 CLOSED-LOOP CONTROL

Consider next the closed-loop control system shown in Fig. 9.16. As before, the plant and controller are represented by the transfer functions $G(s)$ and $H(s)$, respectively. The controller or compensator in the forward path preceding the plant is the only "free" part of the system that is available for adjustment by the system designer. Accordingly, this closed-loop control system is referred to as a *single-degree-of-freedom (1-DOF) structure*.

To simplify matters, Fig. 9.16 assumes that the sensor (measuring the output signal to produce a feedback signal) is perfect. That is, the transfer function of the sensor is unity, and noise produced by the sensor is zero. Under this assumption, the actual output $y(t)$ of the plant is fed back directly to the input of the system. This system is therefore said to be

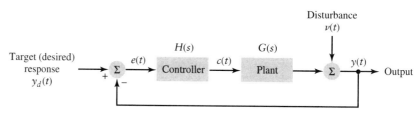

FIGURE 9.16   Control system with unity feedback.

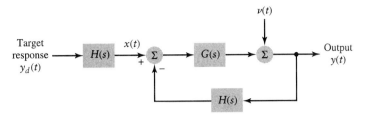

FIGURE 9.17   Reformulation of the feedback control system of Fig. 9.16.

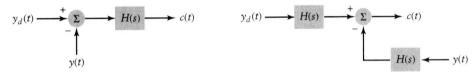

FIGURE 9.18   A pair of equivalent block diagrams used to change Fig. 9.16 into the equivalent form shown in Fig. 9.17.

a *unity-feedback system.* The controller is actuated by the "measured" error $e(t)$, defined as the difference between the target (desired) response $y_d(t)$ (acting as the input) and the feedback (output) signal $y(t)$.

For the purpose of analysis, we may recast the closed-loop control system of Fig. 9.16 into the equivalent form shown in Fig. 9.17. Here we have made use of the equivalence between the two block diagrams shown in Fig. 9.18. Except for the block labeled $H(s)$ at the input end, the single-loop feedback system shown in Fig. 9.17 is of exactly the same form as that of Fig. 9.3(b). By transforming the original closed-loop control system of Fig. 9.16 into the equivalent form shown in Fig. 9.17, we may make full use of the results developed in Section 9.4. Specifically, we note from Fig. 9.17 that

$$X(s) = H(s)Y_d(s). \qquad (9.35)$$

Hence, using Eq. (9.35) in Eq. (9.5), we readily find that the closed-loop transfer function of the 1-DOF system of Fig. 9.16 is given by

$$
\begin{aligned}
T(s) &= \frac{Y(s)}{Y_d(s)} \\
&= \frac{Y(s)}{X(s)} \cdot \frac{X(s)}{Y_d(s)} \\
&= \frac{G(s)H(s)}{1 + G(s)H(s)}.
\end{aligned}
\qquad (9.36)
$$

Assuming that $G(s)H(s)$ is large compared with unity for all values of s that are of interest, we see that Eq. (9.36) reduces to

$$T(s) \simeq 1.$$

That is, with the disturbance $\nu(t) = 0$, we have

$$y(t) \simeq y_d(t). \qquad (9.37)$$

It is therefore desirable to have a large loop gain $G(s)H(s)$. Under this condition, the system of Fig. 9.16 has the potential to achieve the desired goal of accurate control, exemplified by the actual output $y(t)$ of the system closely approximating the target response $y_d(t)$.

There are other good reasons for using a large loop gain. Specifically, in light of the results summarized in Section 9.7, we may state that

▶ The sensitivity of the closed-loop control system $T(s)$ is reduced by a factor equal to the return difference

$$F(s) = 1 + G(s)H(s).$$

▶ The disturbance $v(t)$ inside the feedback loop is reduced by the same factor $F(s)$.
▶ The effect of distortion due to nonlinear behavior of the plant is also reduced by $F(s)$.

9.10 *Transient Response of Low-Order Systems*

To set the stage for the presentation of material on the stability analysis of feedback control systems, we find it informative to examine the transient response of first-order and second-order systems. Although feedback control systems of such low order are indeed rare in practice, their transient analysis forms the basis for a better understanding of higher order systems.

■ 9.10.1 FIRST-ORDER SYSTEM

Using the notation of Chapter 6, we define the transfer function of a first-order system by

$$T(s) = \frac{b_0}{s + a_0}.$$

In order to give physical meaning to the coefficients of the transfer function $T(s)$, we find it more convenient to rewrite it in the *standard form*

$$T(s) = \frac{T(0)}{\tau s + 1}, \tag{9.38}$$

where $T(0) = b_0/a_0$ and $\tau = 1/a_0$. The parameter $T(0)$ is the gain of the system at $s = 0$. The parameter τ is measured in units of time and is therefore referred to as the *time constant* of the system. According to Eq. (9.38), the single pole of $T(s)$ is located at $s = -1/\tau$.

For a step input (i.e., $Y_d(s) = 1/s$), the response of the system has the Laplace transform

$$Y(s) = \frac{T(0)}{s(\tau s + 1)}. \tag{9.39}$$

Expanding $Y(s)$ in partial fractions, using the table of Laplace transform pairs in Appendix D, and assuming that $T(0) = 1$, we find that the step response of the system is

$$y(t) = (1 - e^{-t/\tau})u(t), \tag{9.40}$$

which is plotted in Fig. 9.19. At $t = \tau$, the response $y(t)$ reaches 63.21% of its final value—hence the "time-constant" terminology.

■ 9.10.2 SECOND-ORDER SYSTEM

Again using the notation of Chapter 6, we define the transfer function of a second-order system by

$$T(s) = \frac{b_0}{s^2 + a_1 s + a_0}.$$

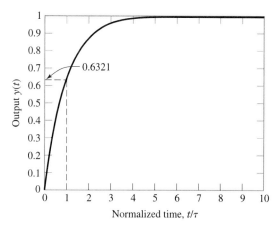

FIGURE 9.19 Transient response of first-order system, plotted against the normalized time t/τ, where τ is the time constant of the system. It is assumed that $T(0) = 1$.

However, as in the first-order case, we find it more convenient to reformulate $T(s)$ so that its coefficients have physical meaning. Specifically, we redefine $T(s)$ in *standard form* as

$$T(s) = \frac{T(0)\omega_n^2}{s^2 + 2\zeta\omega_n s + \omega_n^2}, \tag{9.41}$$

where $T(0) = b_0/a_0$, $\omega_n^2 = a_0$, and $2\zeta\omega_n = a_1$. The parameter $T(0)$ is the gain of the system at $s = 0$. The dimensionless parameter ζ is called the *damping ratio*, and ω_n is called the *undamped frequency* of the system. These three parameters characterize the system in their own individual ways. The poles of the system are located at

$$s = -\zeta\omega_n \pm j\omega_n\sqrt{1 - \zeta^2}. \tag{9.42}$$

For a step input, the response of the system has the Laplace transform

$$Y(s) = \frac{T(0)\omega_n^2}{s(s^2 + 2\zeta\omega_n s + \omega_n^2)}.$$

For the moment, we assume that the poles of $T(s)$ are complex with negative real parts, which implies that $0 < \zeta < 1$. Then, assuming that $T(0) = 1$, expanding $Y(s)$ in partial fractions, and using the table of Laplace transform pairs in Appendix D, we may express the step response of the system as the exponentially damped sinusoidal signal

$$y(t) = \left[1 - \frac{1}{\sqrt{1 - \zeta^2}}e^{-\zeta\omega_n t}\sin\left(\omega_n\sqrt{1 - \zeta^2}\,t + \tan^{-1}\left(\frac{\sqrt{1 - \zeta^2}}{\zeta}\right)\right)\right]u(t). \tag{9.43}$$

The *time constant* of the exponentially damped sinusoid is defined by

$$\tau = \frac{1}{\zeta\omega_n}, \tag{9.44}$$

which is measured in seconds. The frequency of the exponentially damped sinusoid is $\omega_n\sqrt{1 - \zeta^2}$.

Depending on the value of ζ, we may now formally identify three regimes of operation, keeping in mind that the undamped frequency ω_n is always positive:

1. $0 < \zeta < 1$. In this case, the two poles of $T(s)$ constitute a complex-conjugate pair, and the step response of the system is defined by Eq. (9.43). The system is said to be *underdamped*.

2. $\zeta > 1$. In this second case, the two poles of $T(s)$ are real. The step response now involves two exponential functions and is given by

$$y(t) = (1 + k_1 e^{-t/\tau_1} + k_2 e^{-t/\tau_2})u(t), \qquad (9.45)$$

where the time constants are

$$\tau_1 = \frac{1}{\zeta\omega_n - \omega_n\sqrt{\zeta^2 - 1}}$$

and

$$\tau_2 = \frac{1}{\zeta\omega_n + \omega_n\sqrt{\zeta^2 - 1}}$$

and the scaling factors are

$$k_1 = \frac{1}{2}\left(1 + \frac{\zeta}{\sqrt{\zeta^2 - 1}}\right)$$

and

$$k_2 = \frac{1}{2}\left(1 - \frac{\zeta}{\sqrt{\zeta^2 - 1}}\right).$$

Thus, for $\zeta > 1$, the system is said to be *overdamped*.

3. $\zeta = 1$. In this final case, the two poles are coincident at $s = -\omega_n$, and the step response of the system is defined by

$$y(t) = (1 - e^{-t/\tau} - te^{-t/\tau})u(t), \qquad (9.46)$$

where $\tau = 1/\omega_n$ is the only time constant of the system. The system is said to be *critically damped*.

Figure 9.20 shows the step response $y(t)$ plotted against time t for $\omega_n = 1$ and three different damping ratios: $\zeta = 2$, $\zeta = 1$, and $\zeta = 0.1$. These three values of ζ correspond to regimes 2, 3, and 1, respectively.

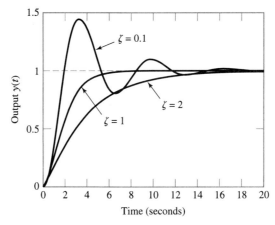

FIGURE 9.20   Transient response of the second-order system with $T(0) = 1$ and $\omega_n = 1$ for three different values of the damping ratio ζ: overdamped ($\zeta = 2$), critically damped ($\zeta = 1$), and underdamped ($\zeta = 0.1$).

With the foregoing material on first-order and second-order systems at our disposal, we are ready to resume our study of feedback control systems.

▶ **Problem 9.6**   Using the table of Laplace transform pairs in Appendix D, derive the following:

  (a) The step response of an underdamped system, as defined in Eq. (9.43)
  (b) The step response of an overdamped system, as defined in Eq. (9.45)
  (c) The step response of a critically damped system, as defined in Eq. (9.46)   ◀

9.11 *The Stability Problem*

In Sections 9.4 to 9.6, we showed that a large loop gain $G(s)H(s)$ is required to make the closed-loop transfer function $T(s)$ of a feedback system less sensitive to variations in the values of parameters, mitigate the effects of disturbance or noise, and reduce nonlinear distortion. Indeed, based on the findings presented there, it would be tempting to propose the following recipe for improving the performance of a feedback system: Make the loop gain $G(s)H(s)$ of the system as large as possible in the passband of the system. Unfortunately, the utility of this simple recipe is limited by a stability problem that is known to arise in feedback systems under certain conditions: If the number of poles contained in $G(s)H(s)$ is three or higher, then the system becomes more prone to instability and therefore more difficult to control as the loop gain is increased. In the design of a feedback system, the task is therefore not only to meet the various performance requirements imposed on the system for satisfactory operation inside a prescribed passband, but also to ensure that the system is stable and remains stable under all possible operating conditions.

The stability of a feedback system, like that of any other LTI system, is completely determined by the location of the system's poles or natural frequencies in the *s*-plane. The *natural frequencies* of a linear feedback system with closed-loop transfer function $T(s)$ are defined as the roots of the *characteristic equation*

$$A(s) = 0, \qquad (9.47)$$

where $A(s)$ is the denominator polynomial of $T(s)$. *The feedback system is stable if the roots of this characteristic equation are all confined to the left half of the s-plane.*

It would therefore seem appropriate for us to begin a detailed study of the stability problem by discussing how the natural frequencies of a feedback system are modified by the application of feedback. We now examine this issue by using three simple feedback systems.

■ 9.11.1 FIRST-ORDER FEEDBACK SYSTEM

Consider a first-order feedback system with unity feedback. The loop transfer function of the system is defined by

$$G(s)H(s) = \frac{K}{\tau_0 s + 1}, \qquad (9.48)$$

where τ_0 is the *open-loop time constant* of the system and K is an adjustable loop gain. The loop transfer function $G(s)H(s)$ has a single pole at $s = -1/\tau_0$. Using Eq. (9.48) in Eq. (9.36), we find that the closed-loop transfer function of the system is

$$T(s) = \frac{G(s)H(s)}{1 + G(s)H(s)}$$

$$= \frac{K}{\tau_0 s + K + 1}.$$

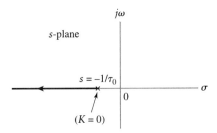

Figure 9.21   Effect of feedback, with increasing K, on the locations of the single pole of a first-order system.

The characteristic equation of the system is therefore

$$\tau_0 s + K + 1 = 0, \tag{9.49}$$

which has a single root at $s = -(K + 1)/\tau_0$. As K is increased, this root moves along the real axis of the s-plane, tracing the locus shown in Fig. 9.21. Indeed, it remains confined to the left half of the s-plane for $K > -1$. We may therefore state that the first-order feedback system with a loop transfer function described by Eq. (9.48) is stable for all $K > 1$.

■ 9.11.2 Second-Order Feedback System

Consider next a specific second-order feedback system with unity feedback. The loop transfer function of the system is defined by

$$G(s)H(s) = \frac{K}{s(\tau s + 1)}, \tag{9.50}$$

where K is an adjustable loop gain measured in rad/s and $G(s)H(s)$ has simple poles at $s = 0$ and $s = -1/\tau$. Using Eq. (9.50) in Eq. (9.36), we find that the closed-loop transfer function of the system is

$$\begin{aligned} T(s) &= \frac{G(s)H(s)}{1 + G(s)H(s)} \\ &= \frac{K}{\tau s^2 + s + K}. \end{aligned}$$

The characteristic equation of the system is therefore

$$\tau s^2 + s + K = 0. \tag{9.51}$$

This is a quadratic equation in s with a pair of roots defined by

$$s = -\frac{1}{2\tau} \pm \sqrt{\frac{1}{4\tau^2} - \frac{K}{\tau}}. \tag{9.52}$$

Figure 9.22 shows the locus traced by the two roots of Eq. (9.52) as the loop gain K is varied, starting from zero. We see that for $K = 0$, the characteristic equation has a root at $s = 0$ and another at $s = -1/\tau$. As K is increased, the two roots move toward each other along the real axis, until they meet at $s = -1/2\tau$ for $K = 1/4\tau$. When K is increased further, the two roots separate from each other along a line parallel to the $j\omega$-axis and that passes through the point $s = -1/2\tau$. This point, called the *breakaway point*, is where the root loci break away from the real axis of the s-plane.

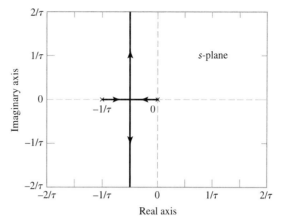

FIGURE 9.22   Effect of feedback, with increasing K, on the locations of the two poles of a second-order system. The loop transfer function has poles at $s = 0$ and $s = -1/\tau$.

When there is no feedback applied to the system (i.e., when $K = 0$), the characteristic equation of the system has a root at $s = 0$, and the system is therefore on the verge of instability. When K is assigned a value greater than zero, the two roots of the characteristic equation are both confined to the left half of the s-plane. It follows that the second-order feedback system with a loop transfer function described by Eq. (9.50) is stable for all positive values of K.

S & S
Solutions

▶ **Problem 9.7**   Refer to the second-order feedback system of Eq. (9.50), and identify the values of K that result in the following forms of step response for the system: (a) underdamped, (b) overdamped, and (c) critically damped.

Answers: (a) $K > 0.25/\tau$ (b) $K < 0.25/\tau$ (c) $K = 0.25/\tau$ ◀

▶ **Problem 9.8**   For the case when the loop gain K is large enough to produce an underdamped step response, show that the damping ratio and natural frequency of the second-order feedback system are respectively defined in terms of the loop gain K and time constant τ as

$$\zeta = \frac{1}{2\sqrt{\tau K}} \quad \text{and} \quad \omega_n = \sqrt{\frac{K}{\tau}} \qquad ◀$$

▶ **Problem 9.9**   The characteristic equation of a second-order feedback system may, in general, be written in the form

$$s^2 + as + b = 0$$

Show that such a system is stable, provided that the coefficients a and b are both positive.   ◀

■ 9.11.3 THIRD-ORDER FEEDBACK SYSTEM

From the analysis just presented, we see that first-order and second-order feedback systems do not pose a stability problem. In both cases, the feedback system is stable for all positive values of the loop gain K. To probe further into the stability problem, we now consider a third-order feedback system whose loop transfer function is described by

$$G(s)H(s) = \frac{K}{(s + 1)^3}. \tag{9.53}$$

TABLE 9.1   *Roots of the Characteristic Equation $s^3 + 3s^2 + 3s + K + 1 = 0$.*

K	Roots
0	Third-order root at $s = -1$
5	$s = -2.71$ $s = -0.1450 \pm j1.4809$
10	$s = -3.1544$ $s = 0.0772 \pm j1.8658$

Correspondingly, the closed-loop transfer function of the system is

$$T(s) = \frac{G(s)H(s)}{1 + G(s)H(s)}$$

$$= \frac{K}{s^3 + 3s^2 + 3s + K + 1}.$$

The characteristic equation of the system is therefore

$$s^3 + 3s^2 + 3s + K + 1 = 0. \tag{9.54}$$

This cubic characteristic equation is more difficult to handle than the lower order characteristic equations (9.49) and (9.51). So we resort to the use of a computer in order to gain some insight into how variations in the loop gain K affect the stability of the system.

Table 9.1 presents the roots of the characteristic equation (9.54) for three different values of K. For $K = 0$, we have a third-order root at $s = -1$. For $K = 5$, the characteristic equation has a simple root and a pair of complex-conjugate roots, all of which have negative real parts (i.e., they are located in the left half of the s-plane). Hence, for $K = 5$, the system is stable. For $K = 10$, the pair of complex-conjugate roots moves into the right half of the s-plane, and the system is therefore unstable. Thus, in the case of a third-order feedback system with a loop transfer function described by Eq. (9.53), the loop gain K has a profound influence on the stability of the system.

The majority of feedback systems used in practice are of order 3 or higher. The stability of such systems is therefore a problem of paramount importance. Much of the material presented in the rest of this chapter is devoted to a study of this problem.

9.12 *Routh–Hurwitz Criterion*

The *Routh–Hurwitz criterion* provides a simple procedure for ascertaining whether all the roots of a polynomial $A(s)$ have negative real parts (i.e., lie in the left-half of the s-plane), *without* having to compute the roots of $A(s)$. Let the polynomial $A(s)$ be expressed in the expanded form

$$A(s) = a_n s^n + a_{n-1}s^{n-1} + \cdots + a_1 s + a_0, \tag{9.55}$$

where $a_n \neq 0$. The procedure begins by arranging all the coefficients of $A(s)$ in the form of two rows as follows:

Row n: a_n a_{n-2} a_{n-4} $\cdots$
Row $n-1$: a_{n-1} a_{n-3} a_{n-5} $\cdots$

If the order n of polynomial $A(s)$ is even, and therefore coefficient a_0 belongs to row n, then a zero is placed under a_0 in row $n - 1$. The next step is to construct row $n - 2$ by using the entries of rows n and $n - 1$ in accordance with the following formula:

$$\text{Row } n - 2: \quad \frac{a_{n-1}a_{n-2} - a_n a_{n-3}}{a_{n-1}} \quad \frac{a_{n-1}a_{n-4} - a_n a_{n-5}}{a_{n-1}} \quad \cdots .$$

Note that the entries in this row have determinantlike quantities for their numerators. That is, $a_{n-1}a_{n-2} - a_n a_{n-3}$ corresponds to the negative of the determinant of the two-by-two matrix

$$\begin{bmatrix} a_n & a_{n-2} \\ a_{n-1} & a_{n-3} \end{bmatrix}.$$

A similar formulation applies to the numerators of the other entries in row $n - 2$. Next, the entries of rows $n - 1$ and $n - 2$ are used to construct row $n - 3$, following a procedure similar to that just described, and the process is continued until we reach row 0. The resulting array of $(n + 1)$ rows is called the *Routh array*.

We may now state the Routh–Hurwitz criterion: *All the roots of the polynomial $A(s)$ lie in the left half of the s-plane if all the entries in the leftmost column of the Routh array are nonzero and have the same sign. If sign changes are encountered in scanning the leftmost column, the number of such changes is the number of roots of $A(s)$ in the right half of the s-plane.*

EXAMPLE 9.4 FOURTH-ORDER FEEDBACK SYSTEM   The characteristic polynomial of a fourth-order feedback system is given by

$$A(s) = s^4 + 3s^3 + 7s^2 + 3s + 10.$$

Construct the Routh array of the system, and determine whether the system is stable.

Solution: Constructing the Routh array for $n = 4$, we obtain the following:

Row 4:	1	7	10
Row 3:	3	3	0
Row 2:	$\frac{3 \times 7 - 3 \times 1}{3} = 6$	$\frac{3 \times 10 - 0 \times 1}{3} = 10$	0
Row 1:	$\frac{6 \times 3 - 10 \times 3}{6} = -2$	0	0
Row 0:	$\frac{-2 \times 10 - 0 \times 6}{-2} = 10$	0	0

There are two sign changes in the entries in the leftmost column of the Routh array. We therefore conclude that (1) the system is unstable and (2) the characteristic equation of the system has two roots in the right half of the s-plane. ■

The Routh–Hurwitz criterion may be used to determine the critical value of the loop gain K for which the polynomial $A(s)$ has a pair of roots on the $j\omega$-axis of the s-plane by exploiting a special case of the criterion. If $A(s)$ has a pair of roots on the $j\omega$-axis, the Routh–Hurwitz test terminates prematurely in that an entire (always odd numbered) row of zeros is encountered in constructing the Routh array. When this happens, the feedback system is said to be *on the verge of instability*. The critical value of K is deduced from the entries of the particular row in question. The corresponding pair of roots on the $j\omega$-axis is found in the auxiliary polynomial formed from the entries of the preceding row, as illustrated in the next example.

EXAMPLE 9.5 THIRD-ORDER FEEDBACK SYSTEM   Consider again a third-order feedback system whose loop transfer function $L(s) = G(s)H(s)$ is defined by Eq. (9.53); that is,

$$L(s) = \frac{K}{(s + 1)^3}.$$

Find (a) the value of K for which the system is on the verge of instability and (b) the corresponding pair of roots on the $j\omega$-axis of the s-plane.

Solution: The characteristic polynomial of the system is defined by

$$A(s) = (s + 1)^3 + K$$
$$= s^3 + 3s^2 + 3s + 1 + K$$

Constructing the Routh array, we obtain the following:

Row 3:	1	3
Row 2:	3	$1 + K$
Row 1:	$\dfrac{9 - (1 + K)}{3}$	0
Row 0:	$1 + K$	0

(a) For the only nonzero entry of row 1 to become zero, we require that

$$9 - (1 + K) = 0,$$

which yields $K = 8$.

(b) For this value of K, the auxiliary polynomial is obtained from row 2. We have

$$3s^2 + 9 = 0,$$

which has a pair of roots at $s = \pm j\sqrt{3}$. This result is readily checked by putting $K = 8$ into the expression for $A(s)$, in which case we may express $A(s)$ in the factored form

$$A(s) = (s^2 + 3)(s + 3).$$ ■

▶ **Problem 9.10**   Consider a linear feedback system with loop transfer function

$$L(s) = \frac{0.2K(s + 5)}{(s + 1)^3}.$$

Find (a) the critical value of the loop gain K for which the system is on the verge of instability and (b) the corresponding pair of roots on the $j\omega$-axis of the s-plane.

Answers: (a) $K = 20$ (b) $s = \pm j\sqrt{7}$ ◀

■ 9.12.1 SINUSOIDAL OSCILLATORS

In the design of *sinusoidal oscillators*, feedback is applied to an amplifier with the specific objective of making the system *unstable*. In such an application, the oscillator consists of an amplifier and a frequency-determining network, forming a closed-loop feedback system. The amplifier sets the necessary condition for oscillation. To avoid distorting the output signal, the degree of nonlinearity in the amplifier is maintained at a very low level. In the next example, we show how the Routh–Hurwitz criterion may be used for such an application.

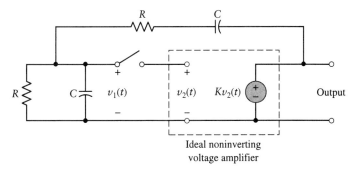

FIGURE 9.23   *RC* audio oscillator.

EXAMPLE 9.6 *RC* OSCILLATOR   Figure 9.23 shows the simplified circuit diagram of an *RC audio* oscillator. Determine the frequency of oscillation and the condition for oscillation.

Solution: With the switch open, and in accordance with the terminology of Fig. 9.4, we find that the loop transfer function is

$$L(s) = -\frac{V_1(s)}{V_2(s)}$$

$$= -\frac{K\left(\dfrac{R}{sC}\right)\Big/\left(R + \dfrac{1}{sC}\right)}{\left(\dfrac{R}{sC}\Big/\left(R + \dfrac{1}{sC}\right)\right) + \left(R + \dfrac{1}{sC}\right)}$$

$$= -\frac{KRCs}{(RCs)^2 + 3(RCs) + 1}.$$

The characteristic equation of the feedback circuit is therefore

$$(RCs)^2 + (3 - K)RCs + 1 = 0.$$

A quadratic characteristic equation is simple enough for us to determine the condition for instability without having to set up the Routh array. For the problem at hand, we see that when the switch is closed, the circuit will be on the verge of instability, provided that the voltage gain K of the amplifier is 3. The natural frequencies of the circuit will then lie on the $j\omega$-axis at $s = \pm j/(RC)$. In practice, the gain K is chosen to be slightly larger than 3 so that the two roots of the characteristic equation lie just to the right of the $j\omega$-axis. This is done in order to make sure that the oscillator is self-starting. As the oscillations build up in amplitude, a resistive component of the amplifier (not shown in the figure) is modified slightly, helping to stabilize the gain K at the desired value of 3.   ∎

▶ **Problem 9.11**   The element values in the oscillator circuit of Fig. 9.23 are $R = 100\,\text{k}\Omega$ and $C = 0.01\,\mu\text{F}$. Find the oscillation frequency.

Answer: 159.15 Hz   ◀

9.13 *Root Locus Method*

The *root locus method* is an analytical tool for the design of a linear feedback system, with emphasis on the locations of the poles of the system's closed-loop transfer function. Recall that the poles of a system's transfer function determine its transient response. Hence, by knowing the locations of the closed-loop poles, we can deduce considerable information about the transient response of the feedback system. The method derives its name from the fact that a "root locus" is the geometric path or locus traced out by the roots of the system's characteristic equation in the s-plane as some parameter (usually, but not necessarily, the loop gain) is varied from zero to infinity. Such a root locus is exemplified by the plots shown in Fig. 9.21 for a first-order feedback system and Fig. 9.22 for a second-order feedback system.

In a general setting, construction of the root locus begins with the loop transfer function of the system, expressed in factored form as

$$\boxed{\begin{aligned} L(s) &= G(s)H(s) \\ &= K\frac{\prod_{i=1}^{M}(1 - s/c_i)}{\prod_{j=1}^{N}(1 - s/d_j)}, \end{aligned}}\tag{9.56}$$

where K is the loop gain and d_j and c_i are the poles and zeros of $L(s)$, respectively. These poles and zeros are fixed numbers, independent of K. In a linear feedback system, they may be determined directly from the block diagram of the system, since the system is usually made up of a cascade connection of first- and second-order components.

Traditionally, the term "root locus" refers to a situation in which the loop gain is nonnegative—that is, $0 \leq K \leq \infty$. This is the case treated in what follows.

■ 9.13.1 ROOT LOCUS CRITERIA

Let the numerator and denominator polynomials of the loop transfer function $L(s)$ be defined by

$$P(s) = \prod_{i=1}^{M}\left(1 - \frac{s}{c_i}\right)\tag{9.57}$$

and

$$Q(s) = \prod_{j=1}^{N}\left(1 - \frac{s}{d_j}\right).\tag{9.58}$$

The characteristic equation of the system is defined by

$$A(s) = Q(s) + KP(s) = 0.\tag{9.59}$$

Equivalently, we may write the characteristic equation as

$$\boxed{L(s) = K\frac{P(s)}{Q(s)} = -1.}\tag{9.60}$$

Since the variable $s = \sigma + j\omega$ is complex valued, we may express the polynomial $P(s)$ in terms of its magnitude and phase components as

$$P(s) = |P(s)|e^{j\arg\{P(s)\}},\tag{9.61}$$

where

$$|P(s)| = \prod_{i=1}^{M} \left| 1 - \frac{s}{c_i} \right| \tag{9.62}$$

and

$$\arg\{P(s)\} = \sum_{i=1}^{M} \arg\left\{ 1 - \frac{s}{c_i} \right\}. \tag{9.63}$$

Similarly, the polynomial $Q(s)$ may be expressed in terms of its magnitude and phase components as

$$Q(s) = |Q(s)| e^{j \arg\{Q(s)\}}, \tag{9.64}$$

where

$$|Q(s)| = \prod_{j=1}^{N} \left| 1 - \frac{s}{d_j} \right| \tag{9.65}$$

and

$$\arg\{Q(s)\} = \sum_{j=1}^{N} \arg\left\{ 1 - \frac{s}{d_j} \right\}. \tag{9.66}$$

Substituting Eqs. (9.62), (9.63), (9.65), and (9.66) into Eq. (9.60), we may readily establish two basic criteria for a root locus (assuming that K is nonnegative):

1. *Angle criterion.* For a point s_l to lie on a root locus, the angle criterion

$$\arg\{P(s)\} - \arg\{Q(s)\} = (2k + 1)\pi, \qquad k = 0, \pm1, \pm2, \ldots, \tag{9.67}$$

must be satisfied for $s = s_l$. The angles $\arg\{Q(s)\}$ and $\arg\{P(s)\}$ are themselves determined by the angles of the pole and zero factors of $L(s)$, as in Eqs. (9.66) and (9.63).

2. *Magnitude criterion.* Once a root locus is constructed, the value of the loop gain K corresponding to the point s_l is determined from the magnitude criterion

$$K = \frac{|Q(s)|}{|P(s)|}, \tag{9.68}$$

evaluated at $s = s_l$. The magnitudes $|Q(s)|$ and $|P(s)|$ are themselves determined by the magnitudes of the pole and zero factors of $L(s)$, as in Eqs. (9.65) and (9.62).

To illustrate the use of the angle and magnitude criteria for the construction of root loci, consider the loop transfer function

$$L(s) = \frac{K(1 - s/c)}{s(1 - s/d)(1 - s/d^*)},$$

which has a zero at $s = c$, a simple pole at $s = 0$, and a pair of complex-conjugate poles at $s = d, d^*$. Select an arbitrary trial point g in the s-plane, and construct vectors from the poles and zeros of $L(s)$ to that point, as depicted in Fig. 9.24. For the angle criterion of Eq. (9.67) and the magnitude criterion of Eq. (9.68) to be both satisfied by the choice of point g, we should find that

$$\theta_{z_1} - \theta_{p_1} - \theta_{p_2} - \theta_{p_3} = (2k + 1)\pi, \qquad k = 0, \pm1, \ldots,$$

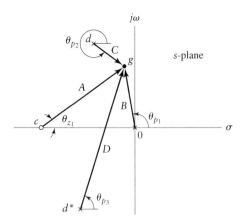

FIGURE 9.24   Illustrating the angle criterion of Eq. (9.67) and the magnitude criterion of Eq. (9.68) for the loop transfer function

$$L(s) = \frac{K(1 - s/c)}{s(1 - s/d)(1 - s/d^*)}.$$

The various angles and magnitudes of vectors drawn from the poles and zeros of $L(s)$ to point g in the complex s-plane are defined as follows:

$$\theta_{z_1} = \arg\left\{1 - \frac{g}{c}\right\}, \qquad A = \left|1 - \frac{g}{c}\right|$$

$$\theta_{p_1} = \arg\{g\}, \qquad B = |g|$$

$$\theta_{p_2} = \arg\left\{1 - \frac{g}{d}\right\}, \qquad C = \left|1 - \frac{g}{d}\right|$$

$$\theta_{p_3} = \arg\left\{1 - \frac{g}{d^*}\right\}, \qquad D = \left|1 - \frac{g}{d^*}\right|$$

and

$$K = \frac{BCD}{A},$$

where the angles and lengths of the vectors are as defined in the figure.

▪ 9.13.2 PROPERTIES OF THE ROOT LOCUS

Given the poles and zeros of the loop transfer function as described in Eq. (9.56), we may construct an approximate form of the root locus of a linear feedback system by exploiting some basic properties of the root locus:

Property 1. *The root locus has a number of branches equal to N or M, whichever is greater.* A *branch* of the root locus refers to the locus of one of the roots of the characteristic equation $A(s) = 0$ as K varies from zero to infinity. Property 1 follows from Eq. (9.59), bearing in mind that the polynomials $P(s)$ and $Q(s)$ are themselves defined by Eqs. (9.57) and (9.58).

Property 2. *The root locus starts at the poles of the loop transfer function.* For $K = 0$, the characteristic equation, given by Eq. (9.59), reduces to

$$Q(s) = 0.$$

The roots of this equation are the same as the poles of the loop transfer function $L(s)$, given by Eq. (9.56), which proves that Property 2 holds.

Property 3. *The root locus terminates on the zeros of the loop transfer function, including those zeros which lie at infinity.*
As K approaches infinity, the characteristic equation, given by Eq. (9.59), reduces to

$$P(s) = 0.$$

The roots of this equation are the same as the zeros of the loop transfer function $L(s)$, which proves that Property 3 holds.

Property 4. *The root locus is symmetrical about the real axis of the s-plane.*
Either the poles and zeros of the loop transfer function $L(s)$ are real, or else they occur in complex-conjugate pairs. The roots of the characteristic equation given by Eq. (9.59), must therefore be real or complex-conjugate pairs, from which Property 4 follows immediately.

Property 5. *As the loop gain K approaches infinity, the branches of the root locus tend to straight-line asymptotes with angles given by*

$$\theta_k = \frac{(2k + 1)\pi}{N - M}, \qquad k = 0, 1, 2, \ldots, |N - M| - 1. \tag{9.69}$$

The asymptotes intersect at a common point on the real axis of the s-plane, the location of which is defined by

$$\sigma_0 = \frac{\sum_{j=1}^{N} d_j - \sum_{i=1}^{M} c_i}{N - M}. \tag{9.70}$$

That is,

$$\sigma_0 = \frac{(\text{sum of finite poles}) - (\text{sum of finite zeros})}{(\text{number of finite poles}) - (\text{number of finite zeros})}.$$

The intersection point $s = \sigma_0$ is called the *centroid* of the root locus.

▶ **Problem 9.12**   The loop transfer function of a linear feedback system is defined by

$$L(s) = \frac{0.2K(s + 5)}{(s + 1)^3}.$$

Find (a) the asymptotes of the root locus of the system and (b) the centroid of the root locus.
Answers: (a) $\theta = 90°, 270°$ (b) $\sigma_0 = 1$ ◀

Property 6. *The intersection points of the root locus with the imaginary axis of the s-plane, and the corresponding values of loop gain K, may be determined from the Routh–Hurwitz criterion.*
This property was discussed in Section 9.12.

Property 7. *The breakaway points, where the branches of the root locus intersect, must satisfy the condition*

$$\frac{d}{ds}\left(\frac{1}{L(s)}\right) = 0, \tag{9.71}$$

where $L(s)$ is the loop transfer function.

Equation (9.71) is a necessary, but not sufficient, condition for a breakaway point. In other words, all breakaway points satisfy Eq. (9.71), but not all solutions of this equation are breakaway points.

EXAMPLE 9.7 SECOND-ORDER FEEDBACK SYSTEM   Consider again the second-order feedback system of Eq. (9.50), assuming that $\tau = 1$. The loop transfer function of the system is

$$L(s) = \frac{K}{s(1 + s)}.$$

Find the breakaway point of the root locus of this system.

Solution: The use of Eq. (9.71) yields

$$\frac{d}{ds}[s(1 + s)] = 0.$$

That is,

$$1 + 2s = 0,$$

from which we readily see that the breakaway point is at $s = -\frac{1}{2}$. This agrees with the result displayed in Fig. 9.22 for $\tau = 1$.   ∎

The seven properties just described are usually adequate to construct a reasonably accurate root locus, starting from the factored form of the loop transfer function of a linear feedback system. The next two examples illustrate how this is done.

EXAMPLE 9.8 LINEAR FEEDBACK AMPLIFIER   Consider a linear feedback amplifier involving three transistor stages. The loop transfer function of the amplifier is defined by

$$L(s) = \frac{6K}{(s + 1)(s + 2)(s + 3)}.$$

Sketch the root locus of this feedback amplifier.

Solution: The loop transfer function $L(s)$ has poles at $s = -1$, $s = -2$, and $s = -3$. All three zeros of $L(s)$ occur at infinity. Thus, the root locus has three branches that start at the aforementioned poles and terminate at infinity.

From Eq. (9.69), we find that the angles made by the three asymptotes are 60°, 180°, and 300°. Moreover, the intersection point of these asymptotes (i.e., the centroid of the root locus) is obtained from Eq. (9.70) as

$$\sigma_0 = \frac{-1 - 2 - 3}{3} = -2.$$

The asymptotes are depicted in Fig. 9.25.

To find the intersection points of the root locus with the imaginary axis of the s-plane, we first form the characteristic polynomial, using Eq. (9.59):

$$A(s) = (s + 1)(s + 2)(s + 3) + 6K$$
$$= s^3 + 6s^2 + 11s + 6(K + 1).$$

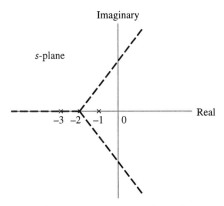

FIGURE 9.25   Diagram showing the intersection point (i.e., the centroid of the root locus) of the three asymptotes for the feedback system of Example 9.8.

Next, we construct the Routh array:

Row 3:	1	11
Row 2:	6	$6(K + 1)$
Row 1:	$\dfrac{66 - 6(K + 1)}{6}$	0
Row 0:	$6(K + 1)$	0

Setting the only nonzero entry of row 1 equal to zero, in accordance with Property 6, we find that the critical value of K for which the system is on the verge of instability is

$$K = 10.$$

Using row 2 to construct the auxiliary polynomial with $K = 10$, we write

$$6s^2 + 66 = 0.$$

Hence, the intersection points of the root locus with the imaginary axis are at $s = \pm j\sqrt{11}$. Finally, using Eq. (9.71), we find that the breakaway point must satisfy the condition

$$\frac{d}{ds}[(s + 1)(s + 2)(s + 3)] = 0.$$

That is,

$$3s^2 + 12s + 11 = 0.$$

The roots of this quadratic equation are

$$s = -1.423 \quad \text{and} \quad s = -2.577.$$

Examining the real-axis segments of the root locus, we infer from Fig. 9.25 that the first point ($s = -1.423$) is on the root locus and is therefore a breakaway point, but the second point ($s = -2.577$) is not on the root locus. Moreover, for $s = -1.423$, the use of Eq. (9.60) yields

$$K = (|1 - 1.423| \times |2 - 1.423| \times |3 - 1.423|)/6$$
$$= 0.0641.$$

Finally, putting all of these results together, we may sketch the root locus of the feedback amplifier as shown in Fig. 9.26. ∎

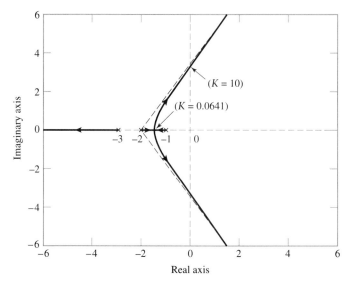

FIGURE 9.26   Root locus of third-order feedback system with loop transfer function

$$L(s) = \frac{6K}{(s + 1)(s + 2)(s + 3)}.$$

EXAMPLE 9.9 UNITY FEEDBACK SYSTEM   Consider the unity-feedback control system of Fig. 9.27. The plant is unstable, with a transfer function defined by

$$G(s) = \frac{0.5K}{(s + 5)(s - 4)}.$$

The controller has a transfer function defined by

$$H(s) = \frac{(s + 2)(s + 5)}{s(s + 12)}.$$

Sketch the root locus of the system, and determine the values of K for which the system is stable.

Solution: The plant has two poles, one at $s = -5$ and the other at $s = 4$. The latter pole, inside the right-half of the s-plane, is responsible for instability of the plant. The controller has a pair of zeros at $s = -2$ and $s = -5$ and a pair of poles at $s = 0$ and $s = -12$. When the controller is connected in cascade with the plant, a *pole–zero cancellation* takes place, yielding the loop transfer function

$$L(s) = G(s)H(s)$$
$$= \frac{0.5K(s + 2)}{s(s + 12)(s - 4)}.$$

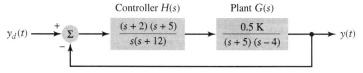

FIGURE 9.27   Unity-feedback system for Example 9.9.

The root locus has three branches. One branch starts at the pole $s = -12$ and terminates at the zero $s = -2$. The other two branches start at the poles $s = 0$ and $s = 4$ and terminate at infinity.

With $L(s)$ having three poles and one finite zero, we find from Eq. (9.69) that the root locus has two asymptotes, defined by $\theta = 90°$ and $270°$. The centroid of the root locus is obtained from Eq. (9.70) and is

$$\sigma_0 = \frac{(-12 + 0 + 4) - (-2)}{3 - 1}$$

$$= -3.$$

Next, on the basis of Eq. (9.59), the characteristic polynomial of the feedback system is

$$A(s) = s^3 + 8s^2 + (0.5K - 48)s + K.$$

Constructing the Routh array, we obtain the following:

Row 3:	1	$0.5K - 48$
Row 2:	8	K
Row 1:	$\dfrac{8(0.5K - 48) - K}{8}$	0
Row 0:	K	0

Setting the only nonzero entry of row 1 to zero, we get

$$8(0.5K - 48) - K = 0,$$

which yields the critical value of the loop gain K, namely,

$$K = 128.$$

Next, using the entries of row 2 with $K = 128$, we get the auxiliary polynomial,

$$8s^2 + 128 = 0,$$

which has roots at $s = \pm j4$. Thus, the root locus intersects the imaginary axis of the s-plane at $s = \pm j4$, and the corresponding value of K is 128.

Finally, applying Eq. (9.71), we find that the breakaway point of the root locus must satisfy the condition

$$\frac{d}{ds}\left(\frac{s(s + 12)(s - 4)}{0.5K(s + 2)}\right) = 0;$$

that is,

$$s^3 + 7s^2 + 16s - 48 = 0.$$

Using the computer, we find that this cubic equation has a single real root at $s = 1.6083$. The corresponding value of K is 29.01.

Putting these results together, we may construct the root locus shown in Fig. 9.28. Here, we see that the feedback system is unstable for $0 \le K \le 128$. When $K > 128$, all three roots of the characteristic equation become confined to the left half of the s-plane. Thus, the application of feedback has the beneficial effect of *stabilizing* an unstable plant, provided that the loop gain is large enough.   ■

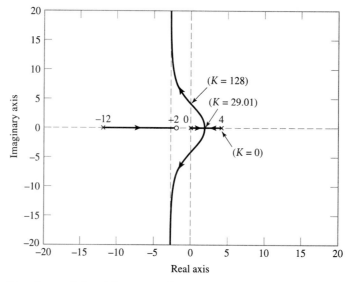

S & S

Solutions

FIGURE 9.28   Root locus of closed-loop control system with loop transfer function

$$L(s) = \frac{0.5K(s + 2)}{s(s - 4)(s + 12)}.$$

▶ **Problem 9.13**   How is the root locus modified if the pole at $s = 4$ of the loop transfer function $L(s)$ in Example 9.9 is replaced with the pole $s = -4$ in the left half plane?

Answer: The new root locus has three branches. One branch starts at the pole $s = 0$ and terminates at the zero $s = -2$. The other two branches start at the poles $s = -4$ and $s = -12$, move towards each other, intersecting at $s = -7.6308$, and then separate from each other; their asymptotes intersect at $s = -7$. The feedback system is stable for all $K > 0$.   ◀

9.14 *Nyquist Stability Criterion*

The root locus method provides information on the roots of the characteristic equation of a linear feedback system (i.e., the poles of the system's closed-loop transfer function) as the loop gain is varied. This information may, in turn, be used to assess not only the stability of the system, but also matters relating to its transient response, as discussed in Section 9.10. For the method to work, we require knowledge of the poles and zeros of the system's loop transfer function. However, in certain situations, this requirement may be difficult to meet. For example, it could be that the only way of assessing the stability of a feedback system is by experimental means, or the feedback loop may include a time delay, in which case the loop transfer function is not a rational function. In such situations, we may look to the Nyquist criterion as an alternative method for evaluating the stability of the system. In any event, the Nyquist criterion is important enough to be considered in its own right.

The *Nyquist stability criterion* is a frequency-domain method that is based on a plot (in polar coordinates) of the loop transfer function $L(s)$ for $s = j\omega$. The criterion has three desirable features that make it a useful tool for the analysis and design of a linear feedback system:

1. It provides information on the absolute stability of the system, the degree of stability, and how to stabilize the system if it is unstable.

2. It provides information on the frequency-domain response of the system.

3. It can be used to study the stability of a linear feedback system with a time delay, which may arise due to the presence of distributed components.

A limitation of the Nyquist criterion, however, is that, unlike the root locus technique, it does not give the exact location of the roots of the system's characteristic equation. Also, a word of caution is in order: Derivation of the Nyquist stability criterion is intellectually more demanding than the material presented hitherto on the stability problem.

■ 9.14.1 ENCLOSURES AND ENCIRCLEMENTS

To prepare the way for a statement of the Nyquist stability criterion, we need to understand what is meant by the terms "enclosure" and "encirclement," which arise in the context of contour mapping. Toward that end, consider some function $F(s)$ of the complex variable s. We are accustomed to representing matters relating to s in a complex plane referred to as the s-plane. Since the function $F(s)$ is complex valued, it is represented in a complex plane of its own, hereafter referred to as the F-plane. Let C denote a *closed contour* traversed by the complex variable s in the s-plane. A contour is said to be *closed* if it terminates onto itself and does not intersect itself as it is transversed by the complex variable s. Let Γ denote the corresponding contour traversed by the function $F(s)$ in the F-plane. If $F(s)$ is a *single-valued* function of s, then Γ is also a closed contour. The customary practice is to traverse the contour C in a counterclockwise direction, as indicated in Fig. 9.29(a). Two different situations may arise in the F-plane:

▶ The interior of contour C in the s-plane is mapped onto the interior of contour Γ in the F-plane, as illustrated in Fig. 9.29(b). In this case, contour Γ is traversed in the counterclockwise direction (i.e., in the same direction as contour C is traversed).

▶ The interior of contour C in the s-plane is mapped onto the exterior of contour Γ in the F-plane, as illustrated in Fig. 9.29(c). In this second case, contour Γ is traversed in the clockwise direction (i.e., in the opposite direction to that in which contour C is traversed).

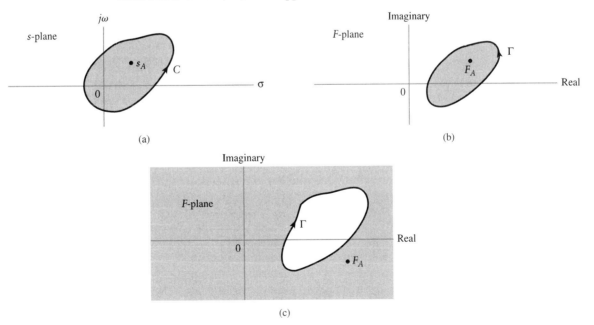

FIGURE 9.29 (a) Contour C traversed in counterclockwise direction in s-plane. (b) and (c) Two possible ways in which contour C is mapped onto the F-plane, with point $F_A = F(s_A)$.

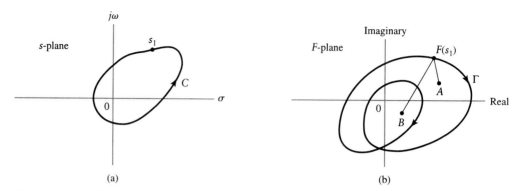

FIGURE 9.30   Illustration of the definition of encirclement. As point s_1 traverses contour C in the s-plane in the counterclockwise direction, as shown in (a), point A is encircled by contour Γ only once and point B is encircled twice, both in the clockwise direction in the F-plane, as shown in (b).

On the basis of this figure, we may offer the following definition: *A region or point in a plane is said to be "enclosed" by a closed contour if the region or point is mapped inside that contour traversed in the counterclockwise direction.* For example, point s_A inside contour C in Fig. 9.29(a) is mapped onto point $F_A = F(s_A)$ inside contour Γ in Fig. 9.29(b), but outside contour Γ in Fig. 9.29(c). Thus, point F_A is enclosed by Γ in Fig. 9.29(b), but not in Fig. 9.29(c).

The notion of enclosure as defined herein should be carefully distinguished from that of encirclement. For the latter, we may offer the following definition: *A point is said to be encircled by a closed contour if it lies inside the contour.* It is possible for a point of interest in the F-plane to be encircled more than once in a positive or negative direction. In particular, the contour Γ in the F-plane makes a total of *m positive encirclements* of a point A if the phasor (i.e., the line drawn from point A to a moving point $F(s_1)$ on the contour Γ) rotates through $2\pi m$ in a counterclockwise direction as the point s_1 traverses contour C in the s-plane once in the same counterclockwise direction. Thus, in the situation described in Fig. 9.30, we find that as the point s_1 traverses the contour C in the s-plane once in the counterclockwise direction, point A is encircled by contour Γ in the F-plane only once, whereas point B is encircled by contour Γ twice, both in the clockwise direction. Thus, in the case of point A we have $m = -1$, and in the case of point B we have $m = -2$.

▶ **Problem 9.14**   Consider the situations described in Fig. 9.31. How many times are points A and B encircled by the locus Γ in this figure?

Answer: For point A the number of encirclements is 2, and for point B it is 1   ◀

■ 9.14.2 PRINCIPLE OF THE ARGUMENT

Assume that a function $F(s)$ is a single-valued rational function of the complex variable s that satisfies the following two requirements:

1. $F(s)$ is *analytic* in the interior of a closed contour C in the s-plane, except at a finite number of poles. The requirement of analyticity means that at every point $s = s_0$ inside the contour C, excluding the points at which the poles are located, $F(s)$ has a derivative at $s = s_0$ and at every point in the neighborhood of s_0.
2. $F(s)$ has neither poles nor zeros on the contour C.

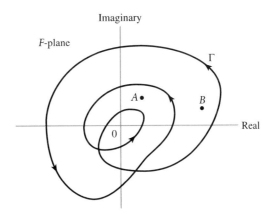

FIGURE 9.31　　Diagram for Problem 9.14.

We may then state the *principle of the argument* in complex-variable theory as

$$\frac{1}{2\pi}\arg\{F(s)\}_C = Z - P, \tag{9.72}$$

where $\arg\{F(s)\}_C$ is the change in the argument (angle) of the function $F(s)$ as the contour C is traversed once in the counterclockwise direction and Z and P are the number of zeros and poles, respectively, of the function $F(s)$ inside the contour C. Note that the change in the magnitude of $F(s)$ as s moves on the contour C once is zero, because $F(s)$ is single valued and the contour C is closed; hence, $\arg\{F(s)\}_C$ is the only term representing the change in $F(s)$ on the left-hand side of Eq. (9.72) as s traverses the contour C once. Suppose now that the origin in the F-plane is encircled a total of m times as the contour C is traversed once in the counterclockwise direction. We may then write

$$\arg\{F(s)\}_C = 2\pi m, \tag{9.73}$$

in light of which Eq. (9.72) reduces to

$$m = Z - P. \tag{9.74}$$

As mentioned previously, m may be positive or negative. Accordingly, we may identify three distinct cases, given that the contour C is traversed in the s-plane once in the counterclockwise direction:

1. $Z > P$, in which case the contour Γ encircles the origin of the F-plane m times in the counterclockwise direction.
2. $Z = P$, in which case the origin of the F-plane is not encircled by the contour Γ.
3. $Z < P$, in which case the contour Γ encircles the origin of the F-plane m times in the clockwise direction.

9.14.3　NYQUIST CONTOUR

We are now equipped with the tools we need to return to the issue at hand: the evaluation of the stability of a linear feedback system. From Eq. (9.8), we know that the characteristic equation of such a system is defined in terms of its loop transfer function $L(s) = G(s)H(s)$ as

$$1 + L(s) = 0,$$

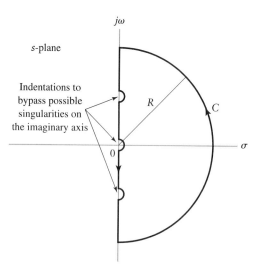

FIGURE 9.32   Nyquist contour.

or, equivalently,

$$F(s) = 0, \tag{9.75}$$

where $F(s)$ is the return difference. With $F(s)$ as the function of interest, the Nyquist stability criterion is basically an application of the principle of the argument, described as follows: Determine the number of roots of the characteristic equation, given by Eq. (9.75), that lie in the right half of the s-plane. With this part of the s-plane as the domain of interest, we may solve the stability problem by considering the contour C shown in Fig. 9.32, which is constructed so as to satisfy the requirements of the principle of the argument:

▸ The semicircle has a radius R that tends to infinity; hence, the contour C encompasses the entire right half of the s-plane as $R \to \infty$.

▸ The small semicircles shown along the imaginary axis are included to bypass the singularities (i.e., poles and zeros) of $F(s)$ that are located at the centers of the semicircles. This ensures that the return difference $F(s)$ has no poles or zeros on the contour C.

The contour C shown in the figure is referred to as the *Nyquist contour.*

Let Γ be the closed contour traced by the return difference $F(s)$ in the F-plane as the Nyquist contour C of Fig. 9.32 is traversed once in the s-plane in the counterclockwise direction. If Z is the (unknown) number of the zeros of $F(s)$ in the right half of the s-plane, then, from Eq. (9.74), we readily see that

$$Z = m + P, \tag{9.76}$$

where P is the number of poles of $F(s)$ in the right half of the s-plane and m is the net number of counterclockwise encirclements of the origin in the F-plane by the contour Γ. Recognizing that the zeros of $F(s)$ are the same as the roots of the system's characteristic equation, we may now formally state the Nyquist stability criterion as follows: *A linear feedback system is absolutely stable, provided that its characteristic equation has no roots in the right half of the s-plane or on the $j\omega$-axis—that is, provided that*

$$m + P = 0. \tag{9.77}$$

The Nyquist stability criterion may be simplified for a large class of linear feedback systems. By definition, the return difference $F(s)$ is related to the loop transfer function $L(s)$ by Eq. (9.8), reproduced here for convenience of presentation,

$$F(s) = 1 + L(s). \tag{9.78}$$

The poles of $F(s)$ are therefore the same as the poles of $L(s)$. If $L(s)$ has no poles in the right half of the s-plane (i.e., if the system is stable in the absence of feedback), then $P = 0$, and Eq. (9.77) reduces to $m = 0$. That is, the feedback system is absolutely stable, provided that the contour Γ does not encircle the origin in the F-plane.

From Eq. (9.78), we also note that the origin in the F-plane corresponds to the point $(-1, 0)$ in the L-plane. For the case when $L(s)$ has no poles in the right half of the s-plane, we may therefore reformulate the Nyquist stability criterion as follows: *A linear feedback system with loop transfer function $L(s)$ is absolutely stable, provided that the locus traced by $L(s)$ in the L-plane does not encircle the point $(-1, 0)$ as s traverses the Nyquist contour once in the s-plane.* The point $(-1, 0)$ in the L-plane is called the *critical point* of the feedback system.

Typically, the loop transfer function $L(s)$ has more poles than zeros, which means that $L(s)$ approaches zero as s approaches infinity. Hence, the contribution of the semicircular part of the Nyquist contour C to the $L(s)$ locus approaches zero as the radius R approaches infinity. In other words, the $L(s)$ locus reduces simply to a plot of $L(j\omega)$ for $-\infty < \omega < \infty$ (i.e., the values of s on the imaginary axis of the s-plane). It is also helpful to view the locus as a *polar* plot of $L(j\omega)$ for varying ω, with $|L(j\omega)|$ denoting the magnitude and $\arg\{L(j\omega)\}$ denoting the phase angle. The resulting plot is called the *Nyquist locus* or *Nyquist diagram*.

Construction of the Nyquist locus is simplified by recognizing that

$$\left|L(-j\omega)\right| = \left|L(j\omega)\right|$$

and

$$\arg\{L(-j\omega)\} = -\arg\{L(j\omega)\}.$$

Accordingly, it is necessary to plot the Nyquist locus only for positive frequencies $0 \le \omega < \infty$. The locus for negative frequencies is inserted simply by reflecting the locus for positive frequencies about the real axis of the L-plane, as illustrated in Fig. 9.33 for a system whose loop transfer function has a pole at $s = 0$. Figure 9.33(a) represents a stable

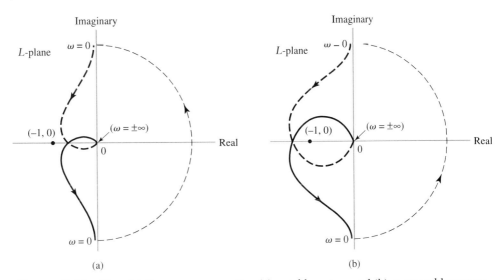

FIGURE 9.33   Nyquist diagrams representing (a) a stable system and (b) an unstable system.

system, whereas Fig. 9.33(b) represents an unstable system whose characteristic equation has two roots in the right-half plane and for which the Nyquist locus encircles the critical point $(-1, 0)$ twice in the counterclockwise direction. Note in Fig. 9.33 that both Nyquist loci start at $\omega = \infty$ and terminate at $\omega = 0$, so as to be consistent with the fact that the Nyquist contour in Fig. 9.32 is traversed in the counterclockwise direction.

EXAMPLE 9.10 LINEAR FEEDBACK AMPLIFIER   Using the Nyquist stability criterion, investigate the stability of the three-stage transistor feedback amplifier examined in Example 9.8. Putting $s = j\omega$ in $L(s)$, we get the loop frequency response

$$L(j\omega) = \frac{6K}{(j\omega + 1)(j\omega + 2)(j\omega + 3)}.$$

Show that the amplifier is stable with $K = 6$.

Solution: With $K = 6$, the magnitude and phase of $L(j\omega)$ are given by

$$|L(j\omega)| = \frac{36}{(\omega^2 + 1)^{1/2}(\omega^2 + 4)^{1/2}(\omega^2 + 9)^{1/2}}$$

and

$$\arg\{L(j\omega)\} = -\tan^{-1}(\omega) - \tan^{-1}\left(\frac{\omega}{2}\right) - \tan^{-1}\left(\frac{\omega}{3}\right).$$

Figure 9.34 shows a plot of the Nyquist contour, which is seen not to encircle the critical point $(-1, 0)$. The amplifier is therefore stable.   ■

▶ **Problem 9.15**   Consider a feedback amplifier described by the loop frequency response

$$L(j\omega) = \frac{K}{(1 + j\omega)^3}.$$

Using the Nyquist stability criterion, show that the amplifier is on the verge of instability for $K = 8$.   ◀

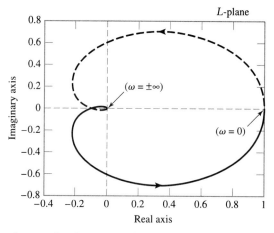

FIGURE 9.34   Nyquist diagram for three-stage feedback amplifier with loop frequency response

$$L(j\omega) = \frac{6K}{(j\omega + 1)(j\omega + 2)(j\omega + 3)} \text{ with } K = 6.$$

9.15 *Bode Diagram*

Another method for studying the stability of a linear feedback system is based on the *Bode diagram*, which was discussed in Chapter 6. For the problem at hand, this method involves plotting the loop transfer function $L(s)$ for $s = j\omega$ in the form of two separate graphs. In one graph, the magnitude of $L(j\omega)$ in decibels is plotted against the logarithm of ω. In the other graph, the phase of $L(j\omega)$ in degrees is plotted against the logarithm of ω.

The attractive feature of the Bode diagram is twofold:

1. The relative ease and speed with which the necessary calculations for different frequencies can be performed make the Bode diagram a useful design tool.

2. The concepts learned from the Bode diagram are very helpful in developing engineering intuition regarding the effect of pole–zero placement on the frequency response $L(j\omega)$.

The intuitive appeal of the Bode diagram comes from the fact that the computation of $|L(j\omega)|_{dB}$ may readily be approximated by straight-line segments. As shown in Section 6.13, the form of the approximation depends on whether the pole or zero factor in question is a simple or quadratic factor:

▶ The contribution of a simple pole factor $(1 + s/\sigma_0)$ to the gain response $|L(j\omega)|_{dB}$ is approximated by a *low-frequency asymptote* consisting simply of the 0-dB line and a *high-frequency asymptote* represented by a straight line with a slope of -20 dB/decade. The two asymptotes intersect at $\omega = \sigma_0$, which is called the *corner* or *break frequency*. The approximation error—that is, the difference between the actual gain response and its approximate form—attains its maximum value of 3 dB at the corner frequency σ_0.

▶ The contribution of a quadratic pole factor $1 + 2\zeta(s/\omega_n) + (s/\omega_n)^2$, consisting of a pair of complex-conjugate poles with the damping factor $\zeta < 1$, to the gain response $|L(j\omega)|_{dB}$ is a pair of asymptotes. One asymptote is represented by the 0-dB line, and the other has a slope of -40 dB/decade. The two asymptotes intersect at the natural frequency $\omega = \omega_n$. However, unlike the case of a simple pole factor, the actual contribution of a quadratic pole factor may differ markedly from its asymptotic approximation, depending on how close the damping factor ζ is to unity. The error is zero for $\zeta = 0.5$, positive for $\zeta < 0.5$, and negative for $\zeta > 0.5$.

The next example illustrates the computation of the Bode diagram for a third-order loop transfer function.

EXAMPLE 9.11   LINEAR FEEDBACK AMPLIFIER (CONTINUED)   Consider the three-stage feedback amplifier with loop frequency response

$$L(j\omega) = \frac{6K}{(j\omega + 1)(j\omega + 2)(j\omega + 3)}$$

$$= \frac{K}{(1 + j\omega)(1 + j\omega/2)(1 + j\omega/3)}.$$

Construct the Bode diagram for $K = 6$.

Solution: The numerator in the second line of $L(j\omega)$ is a constant equal to 6 for $K = 6$. Expressed in decibels, this numerator contributes a constant gain equal to

$$20 \log_{10} 6 = 15.56 \text{ dB}.$$

The denominator is made up of three simple pole factors with corner frequencies equal to 1, 2, and 3 rad/s. Putting the contributions of the numerator and denominator terms together, we get the straight-line approximation to the gain component of $L(j\omega)$ shown in Fig. 9.35.

Figures 9.36(a) and (b) show the exact gain and phase components of $L(j\omega)$, respectively. (The new terms included in the figure are explained in the next subsection.) ■

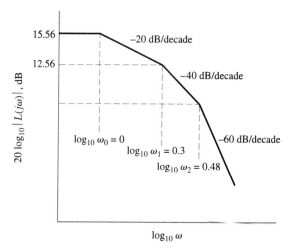

FIGURE 9.35   Straight-line approximation to the gain component of Bode diagram for open-loop response

$$L(j\omega) = \frac{6K}{(j\omega + 1)(j\omega + 2)(j\omega + 3)} \text{ for } K = 6.$$

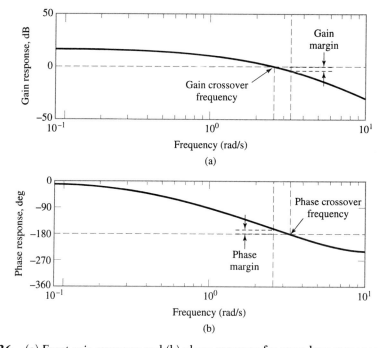

FIGURE 9.36   (a) Exact gain response and (b) phase response for open-loop response

$$L(j\omega) = \frac{6K}{(j\omega + 1)(j\omega + 2)(j\omega + 3)} \text{ for } K = 6.$$

■ 9.15.1 RELATIVE STABILITY OF A FEEDBACK SYSTEM

Now that we have familiarized ourselves with the construction of the Bode diagram, we are ready to consider its use in studying the stability problem. The *relative stability* of a feedback system is determined by how close a plot of the loop transfer function $L(s)$ of the system is to the critical point $L(s) = -1$ for $s = j\omega$. With the Bode diagram consisting of two graphs, one pertaining to $20 \log_{10}|L(j\omega)|$ and the other to $\arg\{L(j\omega)\}$, there are two commonly used measures of relative stability, as illustrated in Fig. 9.37.

The first of these two measures is the *gain margin*, expressed in decibels. For a stable feedback system, the gain margin is defined as the number of decibels by which $20 \log_{10}|L(j\omega)|$ must be changed to bring the system to the verge of instability. Assume that when the phase angle of the loop frequency response $L(j\omega)$ equals $-180°$, its magnitude $|L(j\omega)|$ equals $1/K_m$, where $K_m > 1$. Then the quantity $20 \log_{10} K_m$ is equal to the gain margin of the system, as indicated in Fig. 9.37(a). The frequency ω_p at which $\arg\{L(j\omega_p)\} = -180°$ is called the *phase crossover frequency*.

The second measure of relative stability is the *phase margin*, expressed in degrees. Again for a stable feedback system, the phase margin is defined as the magnitude of the minimum angle by which $\arg\{L(j\omega)\}$ must be changed in order to intersect the critical point $L(j\omega) = -1$. Assume that when the magnitude $|L(j\omega)|$ equals unity, the phase angle $\arg\{L(j\omega)\}$ equals $-180° + \phi_m$. The angle ϕ_m is called the phase margin of the system, as indicated in Fig. 9.37(b). The frequency ω_g at which $|L(j\omega_g)| = 1$ is called the *gain crossover frequency*.

On the basis of these definitions, we can make two observations regarding the stability of a feedback system:

1. For a stable feedback system, both the gain margin and phase margin must be positive. By implication, the phase crossover frequency must be larger than the gain crossover frequency.

2. The system is unstable if the gain margin is negative or the phase margin is negative.

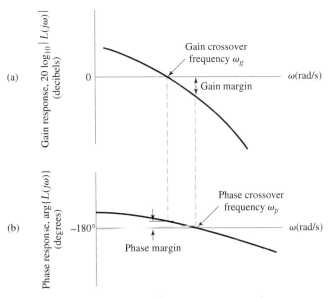

FIGURE 9.37 Illustration of the definitions of (a) gain margin and gain crossover frequency and (b) phase margin and phase crossover frequency.

EXAMPLE 9.12 LINEAR FEEDBACK AMPLIFIER (CONTINUED)   Calculate the gain and phase margins for the loop frequency response of Example 9.11 for $K = 6$.

Solution: Figure 9.36 includes the locations of the gain and phase crossover frequencies:

$$\omega_p = \text{phase crossover frequency} = 3.317 \, \text{rad/s};$$
$$\omega_g = \text{gain crossover frequency} = 2.59 \, \text{rad/s}.$$

With $\omega_p > \omega_g$, we have further confirmation that the three-stage feedback amplifier described by the loop frequency response $L(j\omega)$ of Examples 9.10 and 9.11 is stable for $K = 6$.

At $\omega = \omega_p$, we have, by definition, $\arg\{L(j\omega_p)\} = -180°$. At this frequency, we find from Fig. 9.36(a) that

$$20 \log_{10}|L(j\omega_p)| = -4.437 \, \text{dB}.$$

The gain margin is therefore equal to 4.437 dB.

At $\omega = \omega_g$, we have, by definition, $|L(j\omega_g)| = 1$. At this frequency, we find from Fig. 9.36(b) that

$$\arg\{L(j\omega_p)\} = -162.01°.$$

The phase margin is therefore equal to

$$180 - 162.01 = 17.99°.$$

These stability margins are included in the Bode diagram of Fig. 9.36.   ■

■ 9.15.2 RELATION BETWEEN THE BODE DIAGRAM AND NYQUIST CRITERION

The Bode diagram discussed in this section and the Nyquist diagram discussed in the previous section are frequency-domain techniques that offer different perspectives on the stability of a linear feedback system. The Bode diagram consists of two separate graphs, one for displaying the gain response and the other for displaying the phase response. By contrast, the Nyquist diagram combines the magnitude and phase responses in a single polar plot.

The Bode diagram illustrates the frequency response of the system. It uses straight-line approximations that can be sketched with little effort, thereby providing an easy-to-use method for assessing the absolute stability and relative stability of the system. Accordingly, a great deal of insight can be derived from using the Bode diagram to design a feedback system by frequency-domain techniques.

The Nyquist criterion is important for two reasons:

1. It provides the theoretical basis for using the loop frequency response to determine the stability of a closed-loop system.

2. It may be used to assess stability from experimental data describing the system.

The Nyquist criterion is the ultimate test for stability, in the sense that any determination of stability may be misleading unless it is used in conjunction with the Nyquist criterion. This is particularly so when the system is *conditionally stable*, which means that the system goes through stable and unstable conditions as the loop gain is varied. Such a phenomenon is illustrated in Fig. 9.38, where we see that there are two phase crossover frequencies, namely, ω_{p1} and ω_{p2}. For $\omega_{p1} \leq \omega \leq \omega_{p2}$, the magnitude response $|L(j\omega)|$ is greater than unity. Moreover, the gain crossover frequency ω_g is greater than both ω_{p1} and ω_{p2}. Based on these superficial observations, it would be tempting to conclude that a closed-loop feedback system represented by Fig. 9.38 is unstable. In reality, however, the

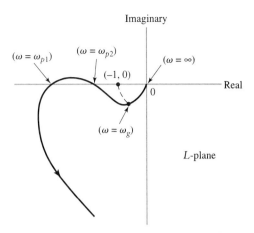

FIGURE 9.38   Nyquist diagram illustrating the notion of conditional stability.

system is stable, since the Nyquist locus shown therein does *not* encircle the critical point $(-1, 0)$.

A closed-loop system characterized by a Nyquist locus such as that shown in Fig. 9.38 is said to be *conditionally stable* because a reduced loop gain or an increased loop gain will make the system unstable.

▶ **Problem 9.16**   Verify that the Nyquist locus shown in Fig. 9.38 does not encircle the critical point $(-1, 0)$.   ◀

9.16 *Sampled-Data Systems*

In the treatment of feedback control systems discussed thus far, we have assumed that the whole system behaves in a continuous-time fashion. However, in many applications of control theory, a digital computer is included as an integral part of the control system. Examples of digital control of dynamic systems include such important applications as aircraft autopilots, mass-transit vehicles, oil refineries, and papermaking machines. A distinct advantage of using a digital computer for control is increased *flexibility* of the control program and better decision making.

The use of a digital computer to calculate the control action for a continuous-time system introduces two effects: sampling and quantization. Sampling is made necessary by virtue of the fact that a digital computer can manipulate only discrete-time signals. Thus, samples are taken from physical signals such as position or velocity and are then used in the computer to calculate the appropriate control. As for quantization, it arises because the digital computer operates with finite arithmetic. The computer takes in numbers, stores them, performs calculations on them, and then returns them with some finite accuracy. In other words, quantization introduces round-off errors into the calculations performed by the computer. In this section, we confine our attention to the effects of sampling in feedback control systems.

Feedback control systems using digital computers are "hybrid" systems, in the sense that continuous-time signals appear in some places and discrete-time signals appear in other places. Such systems are commonly referred to as *sampled-data systems*. Their hybrid nature makes the analysis of sampled-data systems somewhat less straightforward than that of a purely continuous-time system or a purely discrete-time system, since it requires the combined use of both continuous-time and discrete-time analysis methods.

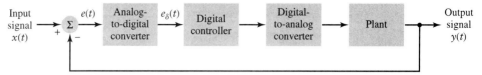

FIGURE 9.39   Block diagram of sampled-data feedback control system, which includes both discrete-time and continuous-time components.

■ 9.16.1 SYSTEM DESCRIPTION

Consider, for example, the feedback control system of Fig. 9.39, in which the digital computer (controller) performs the controlling action. The analog-to-digital (A/D) converter, at the front end of the system, acts on the continuous-time error signal and converts it into a stream of numbers for processing in the computer. The control calculated by the computer is a second stream of numbers, which is converted by the digital-to-analog (D/A) converter back into a continuous-time signal applied to the plant.

For the purpose of analysis, the various components of the sampled-data system of Fig. 9.39 are modeled as follows:

1. *A/D converter.* This component is represented simply by an impulse sampler. Let $e(t)$ denote the error signal, defined as the difference between the system input $x(t)$ and system output $y(t)$. Let $e[n] = e(nT_s)$ be the samples of $e(t)$, where T_s is the sampling period. Recall from Chapter 4 that the discrete-time signal $e[n]$ can be represented by the continuous-time signal

$$e_\delta(t) = \sum_{n=-\infty}^{\infty} e[n]\delta(t - nT_s). \tag{9.79}$$

2. *Digital controller.* The computer program responsible for the control is viewed as a difference equation, whose input–output effect is represented by the z-transform $D(z)$ or, equivalently, the impulse response $d[n]$:

$$D(z) = \sum_{n=-\infty}^{\infty} d[n]z^{-n}. \tag{9.80}$$

Alternatively, we may represent the computer program by the continuous-time transfer function $D_\delta(s)$, where s is the complex frequency in the Laplace transform. This representation follows from the continuous-time representation of the signal $d[n]$ given by

$$d_\delta(t) = \sum_{n=-\infty}^{\infty} d[n]\delta(t - nT_s).$$

Taking the Laplace transform of $d_\delta(t)$ gives

$$D_\delta(s) = \sum_{n=-\infty}^{\infty} d[n]e^{-nsT_s}. \tag{9.81}$$

From Eqs. (9.80) and (9.81), we see that, given the transfer function $D_\delta(s)$, we may determine the corresponding z-transform $D(z)$ by letting $z = e^{sT_s}$:

$$\boxed{D(z) = D_\delta(s)|_{e^{sT_s}=z}.} \tag{9.82}$$

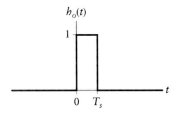

Figure 9.40   Impulse response of zero-order hold.

Conversely, given $D(z)$, we may determine $D_\delta(s)$ by writing

$$D_\delta(s) = D(z)|_{z=e^{sT_s}}. \tag{9.83}$$

The inverse z-transform of $D(z)$ is a sequence of numbers whose individual values are *equal to* the impulse response $d[n]$. In contrast, the inverse Laplace transform of $D_\delta(s)$ is a sequence of impulses whose individual strengths are *weighted by* the impulse response $d[n]$. Note also that $D_\delta(s)$ is periodic in s, with a period equal to $2\pi/T_s$.

3. *D/A converter.* A commonly used type of D/A converter is the *zero-order hold*, which simply holds the amplitude of an incoming sample constant for the entire sampling period, until the next sample arrives. The impulse response of the zero-order hold, denoted by $h_o(t)$, may thus be described as shown in Fig. 9.40 (see Section 4.6); that is,

$$h_o(t) = \begin{cases} 1, & 0 < t < T_s \\ 0, & \text{otherwise} \end{cases}.$$

The transfer function of the zero-order hold is therefore

$$\begin{aligned} H_o(s) &= \int_0^{T_s} e^{-st}\, dt \\ &= \frac{1 - e^{-sT_s}}{s}. \end{aligned} \tag{9.84}$$

4. *Plant.* The plant operates on the continuous-time control delivered by the zero-order hold to produce the overall system output. The plant, as usual, is represented by the transfer function $G(s)$.

On the basis of these representations, we may model the digital control system of Fig. 9.39 as depicted in Fig. 9.41.

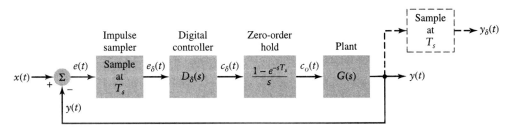

Figure 9.41   Model of sampled-data feedback control system shown in Fig. 9.39.

■ **9.16.2 PROPERTIES OF LAPLACE TRANSFORMS OF SAMPLED SIGNALS**

To prepare the way for determining the closed-loop transfer function of the sampled-data system modeled in Fig. 9.41, we need to introduce some properties of the Laplace transforms of sampled signals. Let $a_\delta(t)$ denote the impulse-sampled version of a continuous-time signal $a(t)$; that is,

$$a_\delta(t) = \sum_{n=-\infty}^{\infty} a(nT_s)\delta(t - nT_s).$$

Let $A_\delta(s)$ denote the Laplace transform of $a_\delta(t)$. (In the control literature, $a^*(t)$ and $A^*(s)$ are commonly used to denote the impulse-sampled version of $a(t)$ and its Laplace transform, respectively, so $A^*(s)$ is referred to as a *starred transform*. We have not used this terminology, largely because the asterisk is used to denote complex conjugation in this book.) The Laplace transform $A_\delta(s)$ has two important properties that follow from the material on impulse sampling presented in Chapter 4:

1. *The Laplace transform $A_\delta(s)$ of a sampled signal $a_\delta(t)$ is periodic in the complex variable s with period $j\omega_s$, where $\omega_s = 2\pi/T_s$, and T_s is the sampling period.* This property follows directly from Eq. (4.23). Specifically, using s in place of $j\omega$ in that equation, we may write

$$\boxed{A_\delta(s) = \frac{1}{T_s} \sum_{k=-\infty}^{\infty} A(s - jk\omega_s),}$$ (9.85)

from which we readily find that

$$A_\delta(s) = A_\delta(s + j\omega_s).$$ (9.86)

2. *If the Laplace transform $A(s)$ of the original continuous-time signal $a(t)$ has a pole at $s = s_1$, then the Laplace transform $A_\delta(s)$ of the sampled signal $a_\delta(t)$ has poles at $s = s_1 + jm\omega_s$, where $m = 0, \pm1, \pm2, \dots$.* This property follows directly from Eq. (9.85) by rewriting it in the expanded form

$$A_\delta(s) = \frac{1}{T_s}[A(s) + A(s + j\omega_s) + A(s - j\omega_s) + A(s + j2\omega_s) + A(s - j2\omega_s) + \cdots].$$

Here, we clearly see that if $A(s)$ has a pole at $s = s_1$, then each term of the form $A(s - jm\omega_s)$ contributes a pole at $s = s_1 + jm\omega_s$, because

$$A(s - jm\omega_s)|_{s=s_1+jm\omega_s} = A(s_1), \qquad m = 0, \pm1, \pm2, \dots.$$

Property 2 of $A_\delta(s)$ is illustrated in Fig. 9.42.

Examining Eq. (9.85), we see that, because of the summation involving terms of the form $A(s - jk\omega_s)$, both the poles and zeros of $A(s)$ contribute to the zeros of $A_\delta(s)$. Accordingly, no statement equivalent to Property 2 can be made regarding the zeros of $A_\delta(s)$. Nevertheless, we can say that the zeros of $A_\delta(s)$ exhibit periodicity with period $j\omega_s$, as illustrated in Fig. 9.42.

Thus far in this subsection, we have discussed only discrete-time signals. However, in a sampled-data system we have a mixture of continuous-time and discrete-time signals. The issue we discuss next concerns such a situation. Suppose we have a signal $l(t)$ that is the result of convolving a discrete-time signal $a_\delta(t)$ with a continuous-time signal $b(t)$; that is,

$$l(t) = a_\delta(t) * b(t).$$

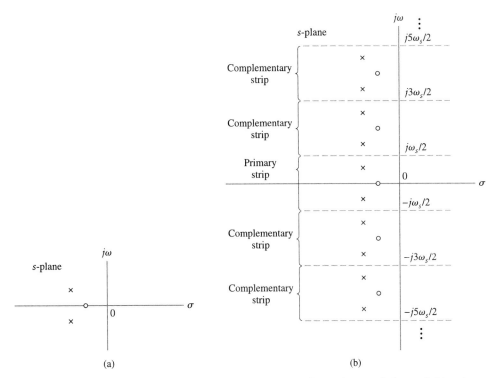

FIGURE 9.42 Illustration of Property 2 of the Laplace transform of a sampled signal. (a) Pole–zero map of $A(s)$. (b) Pole–zero map of

$$A_\delta(s) = \frac{1}{T_s} \sum_{k=-\infty}^{\infty} A(s - jk\omega_s),$$

where ω_s is the sampling frequency.

We now sample $l(t)$ at the same rate as $a_\delta(t)$ and so write

$$l_\delta(t) = [a_\delta(t) * b(t)]_\delta.$$

Transforming this relation into the complex s-domain, we may write, equivalently,

$$L_\delta(s) = [A_\delta(s)B(s)]_\delta,$$

where $a_\delta(t) \xleftrightarrow{\mathscr{L}} A_\delta(s)$, $b(t) \xleftrightarrow{\mathscr{L}} B(s)$, and $l_\delta(t) \xleftrightarrow{\mathscr{L}} L_\delta(s)$. Adapting Eq. (9.85) to this new situation, we have

$$L_\delta(s) = \frac{1}{T_s} \sum_{k=-\infty}^{\infty} A_\delta(s - jk\omega_s)B(s - jk\omega_s), \qquad (9.87)$$

where, as before, $\omega_s = 2\pi/T_s$. However, by definition, the Laplace transform $A_\delta(s)$ is periodic in s with period $j\omega_s$. It follows that

$$A_\delta(s - jk\omega_s) = A_\delta(s) \quad \text{for } k = 0, \pm 1, \pm 2, \ldots.$$

Hence we may simplify Eq. (9.87) to

$$\boxed{\begin{aligned} L_\delta(s) &= A_\delta(s) \cdot \frac{1}{T_s} \sum_{k=-\infty}^{\infty} B(s - jk\omega_s) \\ &= A_\delta(s)B_\delta(s), \end{aligned}} \qquad (9.88)$$

where $b_\delta(t) \overset{\mathscr{L}}{\longleftrightarrow} B_\delta(s)$ and $b_\delta(t)$ is the impulse-sampled version of $b(t)$; that is,

$$B_\delta(s) = \frac{1}{T_s} \sum_{k=-\infty}^{\infty} B(s - jk\omega_s).$$

In light of Eq. (9.88), we may now state another property of impulse sampling: *If the Laplace transform of a signal to be sampled at the rate $1/T_s$ is the product of a Laplace transform that is already periodic in s with period $j\omega_s = j2\pi/T_s$ and another Laplace transform that is not, then the periodic Laplace transform comes out as a factor of the result.*

■ 9.16.3 CLOSED-LOOP TRANSFER FUNCTION

Returning to the issue at hand, namely, that of determining the closed-loop transfer function of the sampled-data system in Fig. 9.39, we note that each one of the functional blocks in the model of Fig. 9.41, except for the sampler, is characterized by a transfer function of its own. Unfortunately, a sampler does not have a transfer function, which complicates the determination of the closed-loop transfer function of a sampled-data system. To get around this problem, we commute the sampling operator with the summer and so reformulate the model of Fig. 9.41 in the equivalent form shown in Fig. 9.43, where the signals entering into the analysis are now all represented by their respective Laplace transforms. The usual approach in sampled-data systems analysis is to relate the sampled version of the input, $X_\delta(s)$, to a sampled version of the output, $Y_\delta(s)$. That is, we analyze the closed-loop transfer function $T_\delta(s)$ contained in the dashed box in Fig. 9.43. This approach describes the behavior of the plant output $y(t)$ *at the instants of sampling*, but provides no information on how the output varies between those instants.

In Fig. 9.43, the transfer function of the zero-order hold has been split into two parts. One part, represented by $(1 - e^{-sT_s})$, has been integrated with the transfer function $D_\delta(s)$ of the digital controller. The other part, represented by $1/s$, has been integrated with the transfer function $G(s)$ of the plant. In so doing, we now have only two kinds of transforms to think about in the model of Fig. 9.43:

▶ transforms of continuous-time quantities, represented by the Laplace transform $y(t) \overset{\mathscr{L}}{\longleftrightarrow} Y(s)$ and the transfer function

$$B(s) = \frac{G(s)}{s}; \qquad (9.89)$$

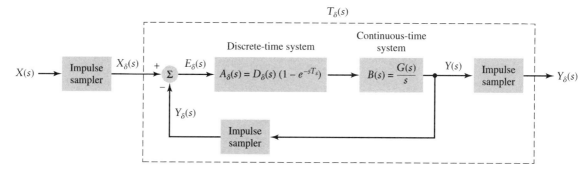

FIGURE 9.43   Block diagram of sampled-data system obtained by reformulating the model of Fig. 9.41. $X(s)$ is the Laplace transform of the input $x(t)$, and $Y_\delta(s)$ is the Laplace transform of the sampled signal $y_\delta(t)$ appearing at the output of the dashed sampler in Fig. 9.41.

▶ transforms of discrete-time quantities, represented by the Laplace transforms $x_\delta(t) \xleftrightarrow{\mathscr{L}} X_\delta(s)$, $e_\delta(t) \xleftrightarrow{\mathscr{L}} E_\delta(s)$, and $y_\delta(t) \xleftrightarrow{\mathscr{L}} Y_\delta(s)$ and the transfer function

$$A_\delta(s) = D_\delta(s)(1 - e^{-sT_s}). \tag{9.90}$$

We are now ready to describe a straightforward procedure for the analysis of sampled-data systems:

1. Write cause-and-effect equations, using Laplace transforms to obtain the closed-loop transfer function $T_\delta(s)$.

2. Convert $T_\delta(s)$ to a discrete-time transfer function $T(z)$.

3. Use z-plane analysis tools, such as the root locus method, to assess the system's stability and performance.

Although we have described this procedure in the context of the sampled-data system shown in Fig. 9.39 containing a single sampler, the procedure generalizes to a sampled-data system containing any number of samplers.

Looking at Fig. 9.43, we may readily set up the cause-and-effect equations

$$E_\delta(s) = X_\delta(s) - Y_\delta(s) \tag{9.91}$$

and

$$Y(s) = A_\delta(s)B(s)E_\delta(s), \tag{9.92}$$

where $B(s)$ and $A_\delta(s)$ are defined by Eqs. (9.89) and (9.90), respectively. The impulse sampler applied to $y(t)$, depicted as the dashed output unit in Fig. 9.41, has the same sampling period T_s and is synchronous with the impulse sampler at the front end of the system. Thus, sampling $y(t)$ in this manner, we may rewrite Eq. (9.92) in the sampled form

$$\begin{aligned} Y_\delta(s) &= A_\delta(s)B_\delta(s)E_\delta(s) \\ &= L_\delta(s)E_\delta(s), \end{aligned} \tag{9.93}$$

where $B_\delta(s)$ is the sampled form of $B(s)$ and $L_\delta(s)$ is defined by Eq. (9.88). Solving Eqs. (9.91) and (9.93) for the ratio $Y_\delta(s)/X_\delta(s)$, we may express the closed-loop transfer function of the sampled-data system of Fig. 9.41 as

$$\begin{aligned} T_\delta(s) &= \frac{Y_\delta(s)}{X_\delta(s)} \\ &= \frac{L_\delta(s)}{1 + L_\delta(s)}. \end{aligned} \tag{9.94}$$

Finally, adapting Eq. (9.82) to our present situation, we may rewrite Eq. (9.94) in terms of the z-transform as

$$\boxed{T(z) = \frac{L(z)}{1 + L(z)}} \tag{9.95}$$

where

$$L(z) = L_\delta(s)\big|_{e^{sT_s} = z}$$

and

$$T(z) = T_\delta(s)\big|_{e^{sT_s}=z}.$$

As stated previously, Eq. (9.95) defines the transfer function $T(z)$ between the sampled input of the original sampled-data system in Fig. 9.39 and the plant output $y(t)$, measured only at the sampling instants.

EXAMPLE 9.13 CALCULATION OF CLOSED-LOOP TRANSFER FUNCTION   In the sampled-data system of Fig. 9.39, the transfer function of the plant is

$$G(s) = \frac{a_0}{s + a_0},$$

and the z-transform of the digital controller (computer program) is

$$D(z) = \frac{K}{1 - z^{-1}}.$$

Determine the closed-loop transfer function $T(z)$ of the system.

Solution: Consider first $B(s) = G(s)/s$, expressed in partial fractions as

$$B(s) = \frac{a_0}{s(s + a_0)}$$
$$= \frac{1}{s} - \frac{1}{s + a_0}.$$

The inverse Laplace transform of $B(s)$ is

$$b(t) = \mathcal{L}^{-1}[B(s)] = (1 - e^{-a_0 t})u(t).$$

Hence, adapting the definition of Eq. (9.81) for the problem at hand, we have (see Note 8 under Further Reading)

$$B_\delta(s) = \sum_{n=-\infty}^{\infty} b[n]e^{-snT_s}$$
$$= \sum_{n=0}^{\infty} (1 - e^{-a_0 nT_s})e^{-snT_s}$$
$$= \sum_{n=0}^{\infty} e^{-snT_s} - \sum_{n=0}^{\infty} e^{-(s+a_0)nT_s}$$
$$= \frac{1}{1 - e^{-sT_s}} - \frac{1}{1 - e^{-a_0 T_s}e^{-sT_s}}$$
$$= \frac{(1 - e^{-a_0 T_s})e^{-sT_s}}{(1 - e^{-sT_s})(1 - e^{-a_0 T_s}e^{-sT_s})}.$$

For convergence, we have to restrict our analysis to values of s for which both $\left|e^{-sT_s}\right|$ and $\left|e^{-T_s(s+a_0)}\right|$ are less than unity. Next, applying Eq. (9.83) to the given z-transform $D(z)$, we get

$$D_\delta(s) = \frac{K}{1 - e^{-sT_s}},$$

the use of which in Eq. (9.90) yields

$$A_\delta(s) = K.$$

Hence, using the results obtained for $A_\delta(s)$ and $B_\delta(s)$ in Eq. (9.88), we obtain

$$L_\delta(s) = \frac{K(1 - e^{-a_0 T_s})e^{-sT_s}}{(1 - e^{-sT_s})(1 - e^{-a_0 T_s}e^{-sT_s})}.$$

Finally, setting $e^{-sT_s} = z^{-1}$, we get the z-transform

$$L(z) = \frac{K(1 - e^{-a_0 T_s})z^{-1}}{(1 - z^{-1})(1 - e^{-a_0 T_s}z^{-1})}$$

$$= \frac{K(1 - e^{-a_0 T_s})z}{(z - 1)(z - e^{-a_0 T_s})},$$

which has a zero at the origin, a pole at $z = e^{-a_0 T_s}$ inside the unit circle, and a pole at $z = 1$ on the unit circle in the z-plane.   ∎

▶ **Problem 9.17** The transfer function of a plant is

$$G(s) = \frac{1}{(s + 1)(s + 2)}.$$

Determine $(G(s)/s)_\delta$.

Answer: $\left(\dfrac{G(s)}{s}\right)_\delta = \dfrac{\frac{1}{2}}{1 - e^{-sT_s}} - \dfrac{1}{1 - e^{-T_s}e^{-sT_s}} + \dfrac{\frac{1}{2}}{1 - e^{-2T_s}e^{-sT_s}}$ ◀

■ 9.16.4 STABILITY

The stability problem in a sampled-data system is different from its continuous-time counterpart, because we are performing our analysis in the z-plane instead of the s-plane. The stability domain for a continuous-time system is represented by the left half of the s-plane; the stability domain for a sampled-data system is represented by the interior of the unit circle in the z-plane.

Referring to Eq. (9.95), we see that the stability of the sampled-data system of Fig. 9.39 is determined by the poles of the closed-loop transfer function $T(z)$ or, equivalently, the roots of the characteristic equation:

$$1 + L(z) = 0.$$

Subtracting 1 from both sides yields

$$\boxed{L(z) = -1.}$$ (9.96)

The significant point to note about this equation is that it has the same mathematical form as the corresponding equation for the continuous-time feedback system described in Eq. (9.60). Accordingly, the mechanics of constructing the root locus in the z-plane are exactly the same as the mechanics of constructing the root locus in the s-plane. In other words, all the properties of the s-plane root locus described in Section 9.13 carry over to the z-plane root locus. The only point of difference is that in order for the sampled-data feedback system to be stable, all the roots of the characteristic equation (9.96) must be confined to the interior of the unit circle in the z-plane.

In a similar way, the principle of the argument used to derive the Nyquist criterion in Section 9.14 applies to the z-plane as well as the s-plane. This time, however, the imaginary axis of the s-plane is replaced by the unit circle in the z-plane, and all the poles of the closed-loop transfer function $T(z)$ are required to be inside the unit circle.

EXAMPLE 9.14   ROOT LOCUS OF SECOND-ORDER SAMPLED-DATA SYSTEM
Continuing with Example 9.13, assume that $e^{-a_0 T_s} = \frac{1}{2}$. Then

$$L(z) = \frac{\frac{1}{2}Kz}{(z - 1)\left(z - \frac{1}{2}\right)}.$$

Construct the z-plane root locus of the system.

Solution: The characteristic equation of the system is

$$(z - 1)\left(z - \frac{1}{2}\right) + \frac{1}{2}Kz = 0;$$

that is,

$$z^2 + \frac{1}{2}(K - 3)z + \frac{1}{2} = 0.$$

This is a quadratic equation in z; its two roots are given by

$$z = -\frac{1}{4}(K - 3) \pm \frac{1}{4}\sqrt{K^2 - 6K + 1}.$$

The root locus of the system is shown in Fig. 9.44, where we note the following:

▶ Starting with $K = 0$, the breakaway point of the root locus occurs at $z = 1/\sqrt{2} \simeq 0.707$ for $K = 3 - 2\sqrt{2} \simeq 0.172$.

▶ For $K = 3 + 2\sqrt{2} \simeq 5.828$, the root locus again intersects the real axis of the z-plane, but this time at $z = -1/\sqrt{2} \simeq -0.707$.

▶ For $0.172 \le K \le 5.828$, the roots of the characteristic equation trace a circle centered on the origin of the z-plane and with radius equal to $1/\sqrt{2}$.

▶ For $K > 5.828$, the two roots start separating from each other, with one root moving toward the zero at the origin and the other root moving toward infinity.

▶ For $K = 6$, the two roots of the characteristic equation move to $z = -\frac{1}{2}$ and $z = -1$. Thus, for this value of K, the system is on the verge of instability, and for $K > 6$, the system becomes unstable.   ■

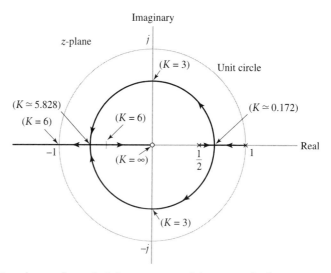

FIGURE 9.44   Root locus of sampled-data system with loop transfer function

$$L(z) = \frac{\frac{1}{2}Kz}{(z - 1)\left(z - \frac{1}{2}\right)}.$$

9.17 *Exploring Concepts with MATLAB*

The issue of stability is of paramount importance in the study of feedback systems. In dealing with these systems, we are given the (open) loop transfer function of the system, denoted by $L(s)$, and we are required to determine the closed-loop stability of the system. In the material presented in this chapter, two basic methods for the study of this problem have been presented:

1. Root locus method.
2. Nyquist stability criterion.

The MATLAB Control System Toolbox is designed to explore these two methods in a computationally efficient manner.

■ 9.17.1 CLOSED-LOOP POLES OF FEEDBACK SYSTEM

Let the loop transfer function $L(s)$ be expressed as the ratio of two polynomials in s; that is,

$$L(s) = K\frac{P(s)}{Q(s)},$$

where K is a scaling factor. The characteristic equation of the feedback system is defined by

$$1 + L(s) = 0,$$

or, equivalently,

$$Q(s) + KP(s) = 0.$$

The roots of this equation define the poles of the closed-loop transfer function of the feedback system. To extract these roots, we use the command `roots` introduced in Section 6.14.

This command was used to compute the results presented in Table 9.1, detailing the roots of the characteristic equation of a third-order feedback system, namely,

$$s^3 + 3s^2 + 3s + K + 1 = 0,$$

for $K = 0, 5$, and 10. For example, for $K = 10$, we have

```
>> sys = [1, 3, 3, 11];
>> roots(sys)
ans =
    -3.1544
     0.0772 + 1.8658i
     0.0772 - 1.8658i
```

Now suppose we want to calculate the natural frequencies and damping factors pertaining to the closed-loop poles of the third-order feedback system for $K = 10$. For this system, we write and get

```
>> sys = [1, 3, 3, 11];
>> damp(sys)
   Eigenvalue            Damping        Freq. (rad/s)
   0.0772 + 1.8658i     -0.0414         1.8674
   0.0772 - 1.8658i     -0.0414         1.8674
   -3.1544               1.000          3.1544
```

The values returned in the first column are the roots of the characteristic equation. The column `Eigenvalue` is merely a reflection of the way in which this part of the calculation is performed.

A related issue of interest is that of calculating the damping factors corresponding to the poles of the closed-loop transfer function of the system or the roots of the characteristic equation. This calculation is easily accomplished on MATLAB by using the command

```
[Wn, z] = damp(sys)
```

which returns vectors `Wn` and `z` containing the natural frequencies and damping factors of the feedback system, respectively.

■ 9.17.2 ROOT LOCUS DIAGRAM

Constructing the root locus of a feedback system requires that we calculate and plot the locus of the roots of the characteristic equation

$$Q(s) + KP(s) = 0$$

for varying K. This task is easily accomplished by using the MATLAB command

```
rlocus(tf(num, den))
```

where `num` and `den` denote the coefficients of the numerator polynomial $P(s)$ and denominator polynomial $Q(s)$, respectively, in descending powers of s. Indeed, this command was used to generate the results plotted in Figs. 9.22, 9.26, and 9.28. For example, the root locus of Fig. 9.28 pertains to the loop transfer function

$$L(s) = \frac{0.5K(s + 2)}{s(s - 4)(s + 12)}$$
$$= \frac{K(0.5s + 1.0)}{s^3 + 8s^2 - 48s}.$$

The root locus is computed and plotted by using the following commands:

```
>> num = [.5, 1];
>> den = [1, 8, -48, 0];
>> rlocus(tf(num, den))
```

▶ **Problem 9.18**   Use the command `rlocus` to plot the root locus of a feedback system having the loop transfer function

$$L(s) = \frac{K}{(s + 1)^3}.$$

Answer: The breakaway point is -1. The system is on the verge of instability for $K = 8$, for which the closed loop poles of the system are at $s = -3$ and $s = \pm 1.7321j$ ◀

Another useful command is `rlocfind`, which finds the value of the scaling factor K required to realize a specified set of roots on the root locus. To illustrate the use of this command, consider again the root locus of Fig. 9.28 and issue the following commands:

```
>> num = [.5, 1];
>> den = [1, 8, -48, 0];
>> rlocus(tf(num, den));
>> K = rlocfind(num, den)
Select a point in the graphics window
```

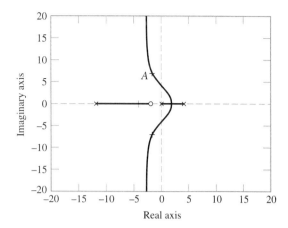

FIGURE 9.45 Root locus diagram illustrating the application of MATLAB command `rlocfind`.

We then respond by placing the cursor at point A, representing the location of the root in the top left-hand quadrant, say, a symbol "+" as indicated in Fig. 9.45. Upon clicking on this point, MATLAB responds as follows:

```
selected point =
    -1.6166 + 6.393i
K =
    213.68
```

■ 9.17.3 NYQUIST STABILITY CRITERION

Constructing the Nyquist diagram involves making a polar plot of the loop frequency response $L(j\omega)$, which is obtained from the loop transfer function $L(s)$ by putting $s = j\omega$. The frequency ω is varied over the range $-\infty < \omega < \infty$. To proceed with the construction, we first express $L(j\omega)$ as the ratio of two polynomials in descending powers of $j\omega$:

$$L(j\omega) = \frac{p'_M(j\omega)^M + p'_{M-1}(j\omega)^{M-1} + \cdots + p'_1(j\omega) + p'_0}{q_N(j\omega)^N + q_{N-1}(j\omega)^{N-1} + \cdots + q_1(j\omega) + q_0}.$$

Here, $p'_i = Kp_i$ for $i = M, M-1, \ldots, 1, 0$. Let `num` and `den` denote the numerator and denominator coefficients of $L(j\omega)$, respectively. We may then construct the Nyquist diagram by using the MATLAB command

```
nyquist(tf(num, den))
```

The results displayed in Fig. 9.34 for Example 9.10 were obtained with this MATLAB command. For that example, we have

$$L(j\omega) = \frac{36}{(j\omega)^3 + 6(j\omega)^2 + 11(j\omega) + 6}.$$

To compute the Nyquist diagram, we therefore write

```
>> num = [36];
>> den = [1, 6, 11, 6];
>> nyquist(tf(num, den))
```

▶ **Problem 9.19**   Using the command `nyquist`, plot the Nyquist diagram for the feedback system defined by

$$L(j\omega) = \frac{6}{(1 + j\omega)^3}.$$

Determine whether the system is stable

Answer: The system is stable   ◀

■ 9.17.4 BODE DIAGRAM

The Bode diagram for a linear feedback system consists of two graphs. In one graph, the loop gain response $20 \log_{10}|L(j\omega)|$ is plotted against the logarithm of ω. In the other graph, the loop phase response $\arg\{L(j\omega)\}$ is plotted against the logarithm of ω. With the given loop frequency response expressed in the form

$$L(j\omega) = \frac{p'_M(j\omega)^M + p'_{M-1}(j\omega)^{M-1} + \cdots + p'_1(j\omega) + p'_0}{q_N(j\omega)^N + q_{N-1}(j\omega)^{N-1} + \cdots + q_1(j\omega) + q_0},$$

we first set up the vectors `num` and `den` to represent the coefficients of the numerator and denominator polynomials of $L(j\omega)$, respectively. The Bode diagram for $L(j\omega)$ may then be easily constructed by using the MATLAB command

```
margin(tf(num, den))
```

This command calculates the gain margin, phase margin, and associated crossover frequencies from frequency response data. The result also includes plots of both the loop gain and phase responses.

The preceding command was used to compute the results presented in Fig. 9.36 for Example 9.11. For that example, we have

$$L(j\omega) = \frac{36}{(j\omega)^3 + 6(j\omega)^2 + 11(j\omega) + 6}.$$

The commands for computing the Bode diagram, including the stability margins, are as follows:

```
>> num = [36];
>> den = [1, 6, 11, 6];
>> margin(tf(num, den))
```

▶ **Problem 9.20**   Compute the Bode diagram, the stability margins, and the associated crossover frequencies for the loop frequency response

$$L(j\omega) = \frac{6}{(1 + j\omega)^3}.$$

Answers:

> Gain margin = 2.499 dB
>
> Phase margin = 10.17°
>
> Phase crossover frequency = 1.7321
>
> Gain crossover frequency = 1.5172   ◀

9.18 *Summary*

In this chapter, we discussed the concept of feedback, which is of fundamental importance to the study of feedback amplifiers and control systems. The application of feedback has beneficial effects of engineering importance:

▶ It reduces the sensitivity of the closed-loop gain of a system with respect to changes in the gain of a plant inside the loop.

▶ It reduces the effect of a disturbance generated inside the loop.

▶ It reduces nonlinear distortion due to deviation of the plant from a linear behavior.

Indeed, these improvements get better as the amount of feedback, measured by the return difference, is increased.

However, feedback is like a double-edged sword that can become harmful if it is used improperly. In particular, it is possible for a feedback system to become unstable, unless special precautions are taken. Stability features prominently in the study of feedback systems. There are two fundamentally different methods for assessing the stability of linear feedback systems:

1. The root locus method, a transform-domain method, which is related to the transient response of the closed-loop system.

2. The Nyquist stability criterion, a frequency-domain method, which is related to the open-loop frequency response of the system.

From an engineering perspective, it is not enough to ensure that a feedback system is stable. Rather, the design of the system must include an adequate *margin* of stability to guard against variations in parameters due to external factors. The root locus technique and the Nyquist stability criterion cater to this requirement in their own particular ways.

The Nyquist stability criterion may itself be pursued by using one of two presentations:

1. *Nyquist diagram (locus).* In this method of presentation, the open-loop frequency response of the system is plotted in polar form, with attention focusing on whether the critical point $(-1, 0)$ is encircled or not.

2. *Bode diagram.* In this method, the open-loop frequency response of the system is presented as a combination of two graphs. One graph plots the loop gain response, and the other graph plots the loop phase response. In an unconditionally stable system, we should find that the gain crossover frequency is smaller than the phase crossover frequency.

In a sense, the root locus method and the Nyquist stability criterion (represented by the Nyquist diagram or the Bode diagram) complement each other: the root locus method highlights the stability problem in the time domain, whereas the Nyquist criterion highlights the stability problem in the frequency domain.

FURTHER READING

1. According to the book

   ▶ Waldhauer, F. D., *Feedback* (Wiley, 1982), p. 3,

   the development of feedback theory may be traced to a journey on the Lackawanna Ferry between Hoboken, New Jersey, and Manhattan, New York, on the morning of August 1927. On that day, Harold S. Black, a member of the technical staff at Bell Telephone Laboratories, in Murray Hill, New Jersey, was a passenger on the ferry on his way to

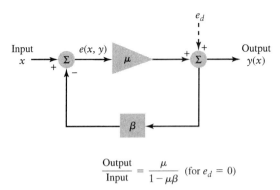

$$\frac{\text{Output}}{\text{Input}} = \frac{\mu}{1 - \mu\beta} \quad \text{(for } e_d = 0\text{)}$$

FIGURE 9.46 Depictions of Harold Black's original feedback diagram and equation.

work. By that time, he had been working for some six years on the problem of reducing nonlinear distortion in amplifiers for use in telephone transmission repeaters. On a blank space in his copy of *The New York Times*, he drew the diagram and the equation shown in Fig. 9.46, and with that figure, the language with which to talk about feedback systems was established. In a related context, see the classic paper

▸ Black, H. S., "Stabilized feedback amplifiers," *Bell System Technical Journal*, vol. 13, 1934, pp. 1–18

Two other members of Bell Telephone Laboratories, Harry Nyquist and Hendrick W. Bode, made significant contributions to the development of modern feedback theory; for their classic works, see

▸ Nyquist, H., "Regeneration theory," *Bell System Technical Journal*, vol. 11, 1932, pp. 126–147

▸ Bode, H. W., *Network Analysis and Feedback Amplifier Design* (Van Nostrand, 1945)

2. For a short history of control systems, see

▸ Dorf, R. C., and R. H. Bishop, *Modern Control Systems*, 9th ed. (Prentice-Hall, 2001)

▸ Phillips, C. L., and R. D. Harbor, *Feedback Control Systems*, 4th ed. (Prentice Hall, 1996)

3. For a complete treatment of automatic control systems, see the following books:

▸ Belanger, P. R., *Control Engineering: A Modern Approach* (Saunders, 1995)

▸ Dorf, R. C., and R. H. Bishop, *op. cit.*

▸ Kuo, B. C., *Automatic Control Systems*, 7th ed. (Prentice Hall, 1995)

▸ Palm, W. J. III, *Control Systems Engineering* (Wiley, 1986)

▸ Phillips, C. L., and R. D. Harbor, *op. cit.*

These books cover both continuous-time and discrete-time aspects of control systems. They also present detailed system design procedures.

4. Feedback amplifiers are discussed in the following books:

▸ Siebert, W. McC., *Circuits, Signals, and Systems* (MIT Press, 1986)

▸ Waldhauer, F. D., *op. cit.*

5. For a discussion of operational amplifiers and their applications, see

▸ Kennedy, E. J., *Operational Amplifier Circuits* (Holt, Rinehart, and Winston, 1988)

▸ Wait, J. V., L. P. Huelsman, and G. A. Korn, *Introduction to Operational Amplifier Theory and Applications*, 2d ed. (McGraw-Hill, 1992)

6. For a proof of Property 5, embodying Eqs. (9.69) and (9.71), see

▸ Truxal, J. G., *Control System Synthesis* (McGraw-Hill, 1955), pp. 227–228

7. For a discussion of the practical issues involved in the operation of D/A converters, basic to the construction of sample-data systems, see the following article:

▶ Hendriks, P. "Specifying communication DACs," *IEEE Spectrum*, vol. 34, pp. 58–69, July 1997

8. Evaluation of $B_\delta(s)$ from $b(t)$ in Example 9.13 requires that we uniformly sample $b(t)$ at the rate $1/T_s$, which forces us to assume a value for the unit-step function $u(t)$ at time $t = 0$. For convenience of presentation, we made the choice $u(0) = 1$ in the second line of the equation defining $B_\delta(s)$. In a sense, it may be argued that this choice continues on the presentation made in the Further Reading section of Chapter 6 under Note 5.

ADDITIONAL PROBLEMS

9.21 A transistor amplifier has a gain of 2500. Feedback is applied around the amplifier by using a network that returns a fraction $\beta = 0.01$ of the amplifier output to the input.

(a) Calculate the closed-loop gain of the feedback amplifier.

(b) The gain of the transistor amplifier changes by 10% due to external factors. Calculate the corresponding change in the closed-loop gain of the feedback amplifier.

9.22 Figure P9.22 shows the block diagram of a position-control system. The preamplifier has a gain G_a. The gain of the motor and load combination (i.e., the plant) is G_p. The sensor in the feedback path returns a fraction H of the motor output to the input of the system.

(a) Determine the closed-loop gain T of the feedback system.

(b) Determine the sensitivity of T with respect to changes in G_p.

(c) Assuming that $H = 1$ and, nominally, $G_p = 1.5$, what is the value of G_a that would make the sensitivity of T with respect to changes in G_p equal to 1%?

radar antenna, respectively. The controller has gain G_c, and the plant (made up of motor, gears, and an antenna pedestal) is represented by G_p. To improve the system's performance, "local" feedback via the sensor H is applied around the plant. In addition, the system uses unity feedback, as indicated in the figure. The purpose of the system is to drive the antenna so as to track the target with sufficient accuracy. Determine the closed-loop gain of the system.

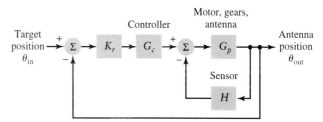

FIGURE P9.23

9.24 Figure P9.24 shows the circuit diagram of an inverting op amp circuit. The op amp is modeled as having infinite input impedance, zero output impedance, and infinite voltage gain. Evaluate the transfer function $V_2(s)/V_1(s)$ of this circuit.

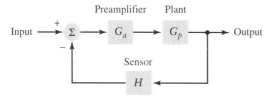

FIGURE P9.22

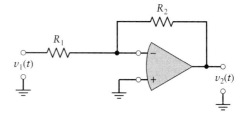

FIGURE P9.24

9.23 Figure P9.23 shows the simplified block diagram of a radar tracking system. The radar is represented by K_r, denoting some gain; θ_{in} and θ_{out} denote the angular positions of the target being tracked and the

9.25 Figure P9.25 shows a practical differentiator that uses an op amp. Assume that the op amp is ideal, having infinite input impedance, zero output impedance, and infinite gain.

(a) Determine the transfer function of this circuit.

(b) What is the range of frequencies for which the circuit acts like an ideal differentiator?

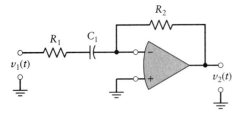

FIGURE P9.25

9.26 Figure P9.26 shows a control system with unity feedback. Determine the Laplace transform of the error signal $e(t)$ for (i) a unit-step input, (ii) a unit-ramp input, and (iii) a unit-parabolic input. Do these calculations for each of the following cases:

(a) $G(s) = \dfrac{15}{(s + 1)(s + 3)}$

(b) $G(s) = \dfrac{5}{s(s + 1)(s + 4)}$

(c) $G(s) = \dfrac{5(s + 1)}{s^2(s + 3)}$

(d) $G(s) = \dfrac{5(s + 1)(s + 2)}{s^2(s + 3)}$

You may assume that, in each case, the closed-loop transfer function of the system is stable.

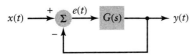

FIGURE P9.26

9.27 Using the Routh–Hurwitz criterion, demonstrate the stability of the closed-loop transfer function of the control system of Fig. P9.26 for all four cases specified in Problem 9.26.

9.28 Use the Routh–Hurwitz criterion to determine the number of roots in the left half, on the imaginary axis, and in the right half of the s-plane for each of the following characteristic equations:

(a) $s^4 + 2s^2 + 1 = 0$

(b) $s^4 + s^3 + s + 0.5 = 0$

(c) $s^4 + 2s^3 + 3s^2 + 2s + 4 = 0$

9.29 Using the Routh–Hurwitz criterion, find the range of the parameter K for which the characteristic equation

$$s^3 + s^2 + s + K = 0$$

represents a stable system.

9.30 (a) The characteristic equation of a third-order feedback system is defined, in general, by

$$A(s) = a_3 s^3 + a_2 s^2 + a_1 s + a_0 = 0.$$

Using the Routh–Hurwitz criterion, determine the conditions that the coefficients a_0, a_1, a_2, and a_3 must satisfy for the system to be stable.

(b) Revisit Problem 9.29 in light of the result obtained in part (a).

9.31 The loop transfer function of a feedback control system is defined by

$$L(s) = \frac{K}{s(s^2 + s + 2)}, \qquad K > 0.$$

The system uses unity feedback.

(a) Sketch the root locus of the system for varying K.

(b) What is the value of K for which the system is on the verge of instability?

9.32 Consider a control system with unity feedback whose loop transfer function is given by

$$L(s) = \frac{K}{s(s + 1)}.$$

Plot the root locus of the system for the following values of gain factor:

(a) $K = 0.1$

(b) $K = 0.25$

(c) $K = 2.5$

9.33 Consider a feedback system for which the loop transfer function is

$$L(s) = \frac{K(s + 0.5)}{(s + 1)^4}.$$

Plot the root locus of the system and determine the values of gain K for which the feedback system is stable.

9.34 Consider a three-stage feedback amplifier for which the loop transfer function is

$$L(s) = \frac{K}{(s + 1)^2(s + 5)}.$$

(a) Using the root locus, investigate the stability of this system for varying K. Repeat the investigation using the Nyquist criterion.

(b) Determine the values of gain K for which the feedback amplifier is stable.

9.35 The loop transfer function of a unity-feedback control system is defined by

$$L(s) = \frac{K}{s(s + 1)}.$$

Using the Nyquist criterion, investigate the stability of this system for varying K. Show that the system is stable for all $K > 0$.

9.36 A unity-feedback control system has the loop transfer function

$$L(s) = \frac{K}{s^2(s + 1)}.$$

Using the Nyquist criterion, show that the system is unstable for all gains $K > 0$. Also, verify your answer by using the Routh–Hurwitz criterion.

9.37 The loop transfer function of a unity-feedback system is defined by

$$L(s) = \frac{K}{s(s + 1)(s + 2)}.$$

(a) Using the Nyquist stability criterion, show that the system is stable for $0 < K < 6$. Also, verify your answer by using the Routh–Hurwitz criterion.

(b) For $K = 2$, determine the gain margin in decibels and the phase margin in degrees.

(c) A phase margin of 20° is required. What value of K is needed to attain this requirement? What is the corresponding value of the gain margin?

9.38 Figure 9.37 illustrates the definitions of gain margin and phase margin, using the Bode diagram. Illustrate the definitions of these two measures of relative stability, using the Nyquist diagram.

9.39 (a) Construct the Bode diagram for the loop frequency response

$$L(j\omega) = \frac{K}{(j\omega + 1)(j\omega + 2)(j\omega + 3)}$$

for $K = 7$, 8, 9, 10, and 11. Show that the three-stage feedback amplifier characterized by this loop frequency response is stable for $K = 7$, 8, and 9; is on the verge of instability for $K = 10$; and is unstable for $K = 11$.

(b) Calculate the gain and phase margins of the feedback amplifier for $K = 7$, 8, and 9.

9.40 Sketch the Bode diagram for each of the following loop transfer functions:

(a) $L(s) = \dfrac{50}{(s + 1)(s + 2)}$

(b) $L(s) = \dfrac{10}{(s + 1)(s + 2)(s + 5)}$

(c) $L(s) = \dfrac{5}{(s + 1)^3}$

(d) $L(s) = \dfrac{10(s + 0.5)}{(s + 1)(s + 2)(s + 5)}$

9.41 Investigate the stability performance of the system described in Example 9.9 for varying gain K. This time, however, use the Bode diagram to do the investigation.

9.42 Consider the sampled-data system shown in Fig. P9.42. Express the z-transform of the sampled output $y(t)$ as a function of the z-transform of the sampled input $x(t)$.

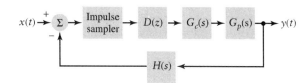

FIGURE P9.42

9.43 Figure P9.43 shows the block diagram of a satellite control system that uses digital control. The transfer function of the digital controller is defined by

$$D(z) = K\left(1.5 + \frac{z - 1}{z}\right).$$

Find the closed-loop transfer function of the system, assuming that the sampling period $T_s = 0.1$ s.

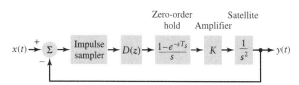

FIGURE P9.43

9.44 Figure P9.44 shows the block diagram of a sampled-data system.

(a) Determine the closed-loop transfer function of the system for a sampling period $T_s = 0.1$ s.

(b) Repeat the problem for $T_s = 0.05$ s.

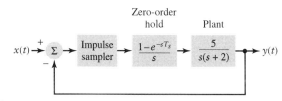

FIGURE P9.44

ADVANCED PROBLEMS

9.45 Consider the linear feedback system of Fig. 9.3(b), and let the forward component be a tuned amplifier (i.e., plant) defined by the transfer function

$$G(s) = \frac{A}{1 + Q\left(\dfrac{s}{\omega_0} + \dfrac{\omega_0}{s}\right)}$$

and the feedback component (i.e., sensor) be

$$H(s) = \beta,$$

where the forward gain A and feedback factor β are both positive and the quality factor Q and resonant frequency ω_0 are both fixed. Assume that Q is very large compared with unity, in which case the open-loop bandwidth equals ω_0/Q.

(a) Determine the closed-loop gain of the system

$$T(s) = \frac{G(s)}{1 + G(s)H(s)},$$

and plot the root locus of the system for increasing loop gain $L(0) = \beta A$.

(b) Show that the closed-loop Q-factor is reduced by the factor $1 + \beta A$, or equivalently, the closed-loop bandwidth is increased by the same factor. (*Note:* The bandwidth is defined as the difference between the two frequencies at which the magnitude response is reduced to $1/\sqrt{2}$ of its value at $\omega = \omega_0$.)

9.46 *Phase-locked loop.* Figure P9.46 shows the linearized block diagram of a phase-locked loop.

(a) Show that the closed-loop transfer of the system is

$$\frac{V(s)}{\Phi_1(s)} = \frac{(s/K_v)L(s)}{1 + L(s)},$$

where K_v is a constant and $\Phi_1(s)$ and $V(s)$ are the Laplace transforms of $\phi_1(t)$ and $v(t)$. The loop transfer function is itself defined by

$$L(s) = K_0 \frac{H(s)}{s},$$

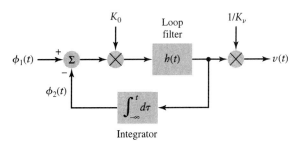

FIGURE P9.46

where K_0 is the gain factor and $H(s)$ is the transfer function of the loop filter.

(b) Specify the condition under which the phase-locked loop acts as an ideal differentiator. Under this condition, define the output voltage $v(t)$ in terms of the phase angle $\phi_1(t)$ acting as input.

9.47 *Steady-state error specifications.* The steady-state error of a feedback control system is defined as the value of the error signal $e(t)$ as time t approaches infinity. Denoting this error by ε_{ss}, we may write

$$\varepsilon_{ss} = \lim_{t \to \infty} e(t).$$

(a) Using the final-value theorem of Laplace transform theory described in Eq. (6.22), and referring to the feedback control system of Fig. 9.16, show that

$$\varepsilon_{ss} = \lim_{s \to 0} \frac{s Y_d(s)}{1 + G(s)H(s)},$$

where $Y_d(s)$ is the Laplace transform of the target response $y_d(t)$.

(b) In general, we may write

$$G(s)H(s) = \frac{P(s)}{s^p Q_1(s)},$$

where neither the polynomial $P(s)$ nor $Q_1(s)$ has a zero at $s = 0$. Since $1/s$ is the transfer function of an integrator, the order p is the number of integrators in the feedback loop, so p is referred to as the *type* of the feedback control system. Derive the following formulas for varying p:

(i) For a step input $y_d(t) = u(t)$,

$$\varepsilon_{ss} = \frac{1}{1 + K_p},$$

where

$$K_p = \lim_{s \to 0} \frac{P(s)}{s^p Q_1(s)}$$

is the *position error constant*. What is the value of ε_{ss} for $p = 0, 1, 2, \ldots$?

(ii) For a ramp input $y_d(t) = t\,u(t)$,

$$\varepsilon_{ss} = \frac{1}{K_v},$$

where

$$K_v = \lim_{s \to 0} \frac{P(s)}{s^{p-1} Q_1(s)}$$

is the *velocity error constant*. What is the value of ε_{ss} for $p = 0, 1, 2, \ldots$?

(iii) For a parabolic input $y_d(t) = (t^2/2)u(t)$,

$$\varepsilon_{ss} = \frac{1}{K_a},$$

where

$$K_a = \lim_{s \to 0} \frac{P(s)}{s^{p-2}Q_1(s)}$$

is the *acceleration error constant*. What is the value of ε_{ss} for $p = 0, 1, 2, \dots$?

(c) List your results as a table, summarizing the steady-state errors according to the type of system.

(d) Determine the steady-state errors for the feedback control system of Fig. P9.47.

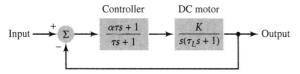

FIGURE P9.47

9.48 Figure P9.48 shows the block diagram of a feedback control system of type 1. (The system type is defined in Problem 9.47.) Determine the damping ratio, natural frequency, and time constant of this system for $K = 20$.

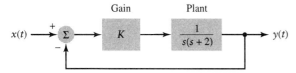

FIGURE P9.48

9.49 The block diagram of a feedback control system that uses a *proportional (P) controller* is shown in Fig. P9.49. This form of compensation is employed when satisfactory performance is attainable merely by setting the constant K_P. For the plant specified in the figure, determine the value of K_P needed to realize a natural (undamped) frequency $\omega_n = 2$. What are the corresponding values of (a) the damping factor and (b) the time constant of the system?

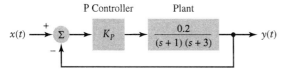

FIGURE P9.49

9.50 Figure P9.50 shows the block diagram of a feedback control system that uses a *proportional-plus-integral (PI) controller*. This form of controller, characterized by the parameters K_P and K_I, is employed to improve the steady-state error of the system by increasing the system type by 1. (The system type is defined in Problem 9.47.) Let $K_I/K_P = 0.1$. Plot the root locus of the system for varying K_P. Find the value of K_P needed to place a pole of the closed-loop transfer function for the system at $s = -5$. What is the steady-state error of the system for a unit-ramp input?

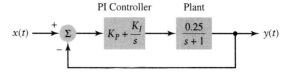

FIGURE P9.50

9.51 Figure P9.51 shows the block diagram of a feedback control system that uses a *proportional-plus-derivative (PD) controller*. This form of compensation, characterized by the parameters K_P and K_D, is employed to improve the transient response of the system. Let $K_P/K_D = 4$. Plot the root locus of the system for varying K_D. Determine the value of K_D needed to place a pair of poles of the closed-loop transfer function of the system at $s = -2 \pm j2$.

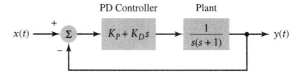

FIGURE P9.51

9.52 Consider again the PI and PD controllers of Problems 9.50 and 9.51, respectively. We may make the following statements in the context of their frequency responses:

(a) The PI controller is a *phase-lag* element, in that it adds a negative contribution to the angle criterion of the root locus.

(b) The PD controller is a *phase-lead* element, in that it adds a positive contribution to the angle criterion of the root locus.

Justify the validity of these two statements.

9.53 Figure P9.53 shows the block diagram of a control system that uses a popular compensator known as the *proportional-plus-integral-plus-derivative (PID) controller*. The parameters K_P, K_I, and K_D of the controller are chosen to introduce a pair of complex-conjugate zeros at $s = -1 \pm j2$ into the loop transfer function of the system. Plot the root locus of the system for increasing K_D. Determine the range of values of K_D for the system to remain stable.

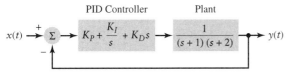

FIGURE P9.53

9.54 Figure P9.54 shows an *inverted pendulum* that moves in a vertical plane on a cart. The cart itself moves along a horizontal axis under the influence of a force, applied to keep the pendulum vertical. The transfer function of the inverted pendulum on a cart, viewed as the plant, is given by

$$G(s) = \frac{(s + 3.1)(s - 3.1)}{s^2(s + 4.4)(s - 4.4)}.$$

Assuming the use of a proportional controller in a manner similar to that described in Problem 9.49, is the use of such a controller sufficient to stabilize the system? Justify your answer. How would you stabilize the system?

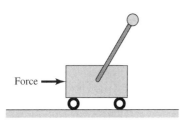

FIGURE P9.54

9.55 *Time-domain specifications.* In describing the step response of a feedback control system, we have two conflicting criteria: the swiftness of the response and the closeness of the response to some desired response. Swiftness is measured in terms of the rise time and the peak time. Closeness to the desired response is measured in terms of the percentage overshoot and the settling time. These four quantities are defined as follows:

▸ The *rise time* is the time taken by the step response $y(t)$ to rise from 10% to 90% of its final value $y(\infty)$.

▸ The *peak time* is the time taken by the step response to reach the overshoot maximum value y_{max}.

▸ The *percentage overshoot* is $(y_{max} - y(\infty))/y(\infty)$, expressed as a percentage.

▸ The *selling time* is the time required for the step response to settle within $\pm\delta\%$ of the final value $y(\infty)$, where δ is a user-specified parameter.

Consider an undamped second-order system with damping factor ζ and undamped frequency ω_n, as defined in Eq. (9.41). Using the specifications given in Fig. P9.55, we may postulate the following formulas:

1. Rise time $T_r \approx \dfrac{1}{\omega_n}(0.60 + 2.16\zeta)$.

2. Peak time $T_p = \dfrac{\pi}{\omega_n\sqrt{1 - \zeta^2}}$.

3. Percentage overshoot P.O. $= 100e^{-\pi\zeta/\sqrt{1-\zeta^2}}$.

4. Settling time $T_{settling} \approx \dfrac{4.6}{\zeta\omega_n}$.

Formulas 1 and 4 are approximations, as explicit formulas are difficult to obtain; formulas 2 and 3 are exact.

(a) Using Eq. (9.43), derive the exact formulas for the peak time T_p and the percentage overshoot P.O.

(b) Use computer simulations to justify the approximate formula for rise time T_r and settling time $T_{settling}$. Do this for the damping factor $\zeta = 0.1, 0.2, \dots, 0.9$ (in increments of 0.1) and $\delta = 1$.

9.56 *Reduced-order models.* In practice, we often find that the poles and zeros of the closed-loop transfer function $T(s)$ of a feedback system are grouped in the complex s-plane roughly in the manner illustrated in Fig. P9.56. Depending on how close the poles and zeros are to the $j\omega$-axis, we may identify two groupings:

▸ *Dominant poles and zeros*—namely, those poles at zeros of $T(s)$ which lie close to the $j\omega$-axis. They are said to be dominant because they exert a profound influence on the frequency response of the system.

▸ *Insignificant poles and zeros*—that is, those poles and zeros of $T(s)$ which are far removed from the $j\omega$-axis. They are said to be insignificant because they have relatively little influence on the frequency response of the system.

Given that we have a high-order feedback system whose closed-loop response transfer function $T(s)$ fits the picture portrayed in Fig. P9.56, we may approximate the system by a reduced-order model simply by retaining the dominant poles and zeros of $T(s)$. Such an endeavor is motivated by two considerations: Low-order models are simple and therefore appealing in system analysis and design, and they are less demanding in computational terms. Consider again the linear feedback amplifier of Example 9.8 with constant $K = 8$, and do the following:

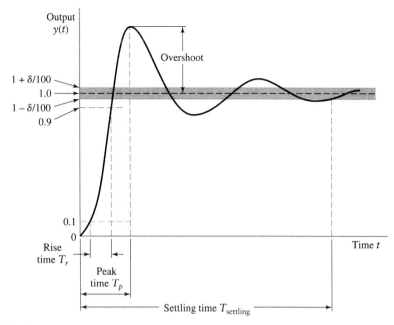

FIGURE P9.55

(a) Using the computer, find the roots of the characteristic equation of the system.

(b) The complex-conjugate roots found in Part (a) constitute the dominant poles of the system. Using these poles, approximate the system with a second-order model. Here you must make sure that the transfer function $T'(s)$ of the second-order model is properly scaled relative to the original $T(s)$, such that $T'(0) = T(0)$.

(c) Compute the step response of the second-order model, and show that it is fairly close to the step response of the original third-order feedback amplifier.

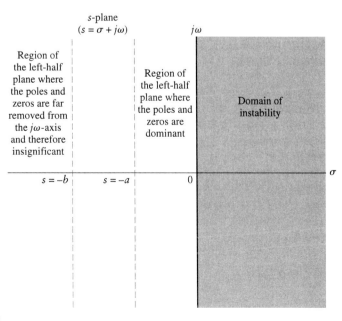

FIGURE P9.56

9.57 In this problem, we revisit Problem 9.56, involving the approximation of a third-order feedback amplifier with a second-order model. In that problem, we used the step response as the basis for assessing the quality of the approximation. In the current problem, we use the Bode diagram as the basis for assessing the quality of the approximation. Specifically, plot the Bode diagram for the reduced-order model

$$T'(s) = \frac{8.3832}{s^2 + 0.2740s + 9.4308},$$

and compare the diagram with the Bode diagram for the original system,

$$T(s) = \frac{48}{s^3 + 6s^2 + 11s + 54}.$$

Comment on your results.

9.58 *Relation between phase margin and damping factor.* The guidelines used in the classical approach to the design of a linear feedback control system are usually derived from the analysis of second-order system dynamics, which is justified on the following grounds: First, when the loop gain is large, the closed-loop transfer function of the system develops a pair of dominant complex-conjugate poles. Second, a second-order model provides an adequate approximation to the system. (See the discussion of reduced-order models in Problem 9.56.) Consider, then, a second-order system whose loop transfer function is given by

$$L(s) = \frac{K}{s(\tau s + 1)},$$

which was studied in Section 9.11.2. Hence, do the following for the case when K is large enough to produce an underdamped step response:

(a) Using the results of Problem 9.8, express the loop transfer function in terms of the damping factor ζ and natural frequency ω_n as

$$L(s) = \frac{\omega_n^2}{s(s + 2\zeta\omega_n)}.$$

(b) By definition, the gain crossover frequency ω_g is determined from the relation

$$|L(j\omega_g)| = 1.$$

Applying this definition to the loop transfer function $L(s)$ of Part (a), show that

$$\omega_g = \tan^{-1}\left(\sqrt{\sqrt{4\zeta^4 + 1} - 2\zeta^2}\right).$$

Next, show that the phase margin, measured in degrees, is given by

$$\phi_m = \tan^{-1}\left(\frac{2\zeta}{\sqrt{\sqrt{4\zeta^4 + 1} - 2\zeta^2}}\right).$$

(c) Using the exact formula found in Part (b), plot ϕ_m versus ζ for values of the damping factor in the range $0 \le \zeta \le 0.6$. For this range, show that ζ can be linearly related to ϕ_m by the approximate formula

$$\zeta \approx \frac{\phi_m}{100}, \qquad 0 \le \zeta \le 0.6.$$

(d) Given ω_g and ϕ_m, discuss how the results of Parts (b) and (c) could be used to determine the rise time, peak time, percentage overshoot, and settling time as descriptors of the step response of the system. For this discussion, you may refer to the results of Problem 9.55.

COMPUTER EXPERIMENTS

9.59 Consider again the third-order feedback system studied in Problem 9.18. Use the MATLAB command **rlocfind** to determine the value of the scaling factor K in the loop transfer function

$$L(s) = \frac{K}{(s + 1)^3}$$

that satisfies the following requirement: The complex-conjugate closed-loop poles of the feedback system have a damping factor equal to 0.5. What is the corresponding value of the undamped frequency?

9.60 In Problem 9.31, we considered a unity-feedback system with loop transfer function

$$L(s) = \frac{K}{s(s^2 + s + 2)}, \quad K > 0.$$

Use the following MATLAB commands to evaluate the stability of the system:

(a) **rlocus**, for constructing the root locus diagram of the system.

(b) `rlocfind`, for determining the value of K for which the complex-conjugate closed-loop poles of the system have a damping factor of about 0.707.

(c) `margin`, for evaluating the gain and phase margins of the system for $K = 1.5$.

9.61 The loop transfer function of a feedback system is defined by

$$L(s) = \frac{K(s - 1)}{(s + 1)(s^2 + s + 1)}.$$

This transfer function includes an all-pass component represented by $(s - 1)/(s + 1)$, which is so called because it passes all frequencies with no amplitude distortion.

(a) Use the MATLAB command `rlocus` to construct the root locus of the system. Next, use the command `rlocfind` to determine the value of K for which the system is on the verge of instability. Check the value so obtained by using the Routh–Hurwitz criterion.

(b) Use the command `nyquist` to plot the Nyquist diagram of the system for $K = 0.8$; confirm that the system is stable.

(c) For $K = 0.8$, use the command `margin` to assess the stability margins of the system.

9.62 The loop transfer function of a feedback system is defined by

$$L(s) = \frac{K(s + 1)}{s^4 + 5s^3 + 6s^2 + 2s - 8}.$$

This system is stable only when K lies inside a certain range $K_{\min} < K < K_{\max}$.

(a) Use the MATLAB command `rlocus` to plot the root locus of the system.

(b) Find the critical limits of stability, $K_{\min}$ and $K_{\max}$, by using the command `rlocfind`.

(c) For K lying midway between $K_{\min}$ and $K_{\max}$, determine the stability margins of the system, using the command `margin`.

(d) For the value of K employed in Part (c), confirm the stability of the system by using the command `nyquist` to plot the Nyquist diagram.

9.63 (a) Construct the root locus of the feedback system described in Problem 9.13, and compare the result with those of Figs. 9.22 and 9.28.

(b) Construct the Nyquist diagram of the system. Hence demostate its stability for all $K > 0$.

9.64 In this problem, we study the design of a unity-feedback control system that uses a *phase-lead compensator* to improve the transient response of the system. The transfer function of this compensator is defined by

$$G_c(s) = \frac{\alpha \tau s + 1}{\tau s + 1}, \qquad \alpha > 1.$$

The phase-lead compensator is called that because it introduces a phase advance into the loop frequency response of the system. The loop transfer function of the uncompensated system is defined by

$$L(s) = \frac{K}{s(s + 1)}.$$

The lead compensator is connected in cascade with the open-loop system, resulting in the modified loop transfer function

$$L_c(s) = G_c(s)L(s).$$

(a) For $K = 1$, determine the damping factor and the undamped frequency of the closed-loop poles of the uncompensated system.

(b) Consider next the compensated system. Suppose that the transient specifications require that we have closed-loop poles with a damping factor $\zeta = 0.5$ and an undamped frequency $\omega_n = 2$. Assuming that $\alpha = 10$, show that the angle criterion (pertaining to the construction of the root locus) is satisfied by choosing the time constant $\tau = 0.027$ for the lead compensator.

(c) Use the MATLAB command `rlocfind` to confirm that a phase-lead compensator with $\alpha = 10$ and $\tau = 0.027$ does indeed satisfy the desired transient response specifications.

9.65 Consider a unity-feedback control system that uses the cascade connection of a plant and a controller, as shown in Fig. 9.16. The transfer function of the plant is

$$G(s) = \frac{10}{s(0.2s + 1)}.$$

The transfer function of the controller is defined by

$$H(s) = K\left(\frac{\alpha \tau s + 1}{\tau s + 1}\right), \qquad \alpha < 1.$$

The controller must be designed to meet the following requirements:

(a) The steady-state error to a ramp input of unit slope should be 0.1.

(b) The overshoot of the step response should not exceed 10%.

(c) The 5% settling time of the step response should be less than 2 s.

Refer to Problems 9.47 and 9.55 for definitions of the pertinent terms.

  Carry out the design, using (a) the frequency-domain approach based on the Bode diagram and (b) the time-domain approach, employing the root locus method. Having designed the controller by either approach, construct an operational amplifier circuit for its implementation.

9.66 Throughout this chapter, we have treated the adjustable scale factor K in the loop transfer function

$$L(s) = K \frac{P(s)}{Q(s)}$$

as a positive number. In this last problem, we use MATLAB to explore feedback systems for which K is negative.

(a) Use the `rlocus` command to plot the root locus of the loop transfer function

$$L(s) = \frac{K(s - 1)}{(s + 1)(s^2 + s + 1)},$$

where K is negative. Now, using the `rlocfind` command, show that this feedback system is on the verge of instability for $K = -1.0$. Verify the result with the Routh–Hurwitz criterion.

(b) Use the `rlocus` command to show that a feedback system with loop transfer function

$$L(s) = \frac{K(s + 1)}{s^4 + 5s^3 + 6s^2 + 2s - 8}$$

is unstable for all $K < 0$.

10 | Epilogue

10.1 *Introduction*

In the material covered in Chapters 1 through 9, we have presented an introductory treatment of signals and systems, with an emphasis on fundamental issues and their applications to three areas: digital filters, communications systems, and feedback systems. Insofar as signals are concerned, we placed particular emphasis on *Fourier analysis* as a method for their representation. Fourier theory is an essential part of the signal-processing practitioner's kit of tools. Basically, it enables us to transform a signal described in the time domain into an equivalent representation in the frequency domain, subject to certain conditions imposed on the signal. Most importantly, the transformation is one to one, in that there is no loss of information as we go back and forth from one domain to the other. As for the analysis of systems, we restricted our attention primarily to a special class known as *LTI systems*, whose characterization satisfies two distinct properties: linearity and time invariance. The motivation for invoking these properties is to make system analysis mathematically tractable.

Fourier theory presupposes that the signal under study is stationary. However, many of the real-life signals encountered in practice are nonstationary. A signal is said to be *nonstationary* if its intrinsic properties vary with time. For example, a speech signal is nonstationary. Other examples of nonstationary signals are the time series representing the fluctuations in stock prices observed at the various capital markets around the world, the received signal of a radar system monitoring variations in prevalent weather conditions, and the received signal of a radio telescope listening to radio emissions from the galaxies around us.

Turning next to the LTI model, we find that many physical systems do indeed permit the use of such a model. Nevertheless, strictly speaking, a physical system may depart from the idealized LTI model due to the presence of nonlinear components or time-varying parameters, depending on the conditions under which the system is operated.

To deal with the practical realities of nonstationary signals and nonlinear and time-varying systems, we need new tools. With that in mind, the purpose of this concluding chapter is to provide brief expositions of the following topics, in the order shown:

▶ Speech signals: An example of nonstationarity
▶ Time-frequency analysis of nonstationary signals
▶ Nonlinear systems
▶ Adaptive filters

In these examinations, the reader is presented with a more realistic assessment of the world of signals and systems than is portrayed in Chapters 1 through 9.

10.2 *Speech Signals: An Example of Nonstationarity*

As already mentioned, a speech signal is a nonstationary signal, in that its intrinsic characteristics vary with time. In this section, we endeavor to explain why indeed that is so. We have picked speech signals for a discussion of nonstationarity because of their ubiquity in our daily lives.

A simple model of the speech production process is given by a form of filtering in which a sound source excites a vocal tract filter. The vocal tract is then modeled as a tube of nonuniform cross-sectional area, beginning at the glottis (i.e., the opening between the vocal cords) and ending at the lips, as outlined by the dashed lines in Fig. 10.1. The figure shows a photograph of a sagittal-plane X-ray of a person's vocal system. Depending on the mode of excitation provided by the source, the sounds constituting a speech signal may be classified into two distinct types:

▶ *Voiced sounds*, for which the source of excitation is pulselike and periodic. In this case, the speech signal is produced by forcing air (from the lungs) through the glottis with the vocal cords vibrating in a relaxed manner. An example of a voiced sound is /e/ in *eve*. (The symbol / / is commonly used to denote a phoneme, a basic linguistic unit.)

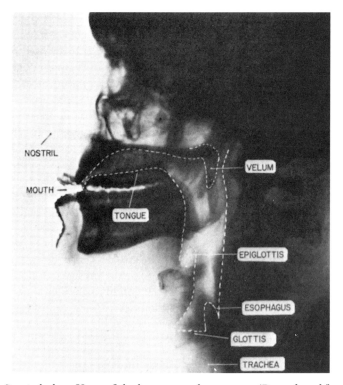

FIGURE 10.1   Saggital-plane X-ray of the human vocal apparatus. (Reproduced from J. L. Flanagan et al., "Speech coding," *IEEE Transactions in Communications*, vol. COM-27, pp. 710–737, 1979; courtesy of the IEEE.)

▶ *Unvoiced sounds*, for which the source of excitation is noiselike (i.e., random). In this second case, the speech signal is produced by forming a constriction in the vocal tract toward its mouth end and forcing a continuous stream of air through the constriction at a high velocity. An example of an unvoiced sound is /f/ in *f*ish.

Voiced sounds produced during the utterance of vowels are characterized by quasi periodicity, low-frequency content, and a large amplitude. In contrast, unvoiced sounds, or fricatives, are characterized by randomness, high-frequency content, and a relatively low amplitude. The transition in time between voiced and unvoiced sounds is gradual (on the order of tens of milliseconds). Thus, recognizing that a typical speech signal contains many voiced and unvoiced sounds strung together in a manner that depends on what is being spoken, we can readily appreciate that a speech signal is indeed a nonstationary signal.

10.3   *Time–Frequency Analysis*

Fourier theory is valid only for stationary signals. For the analysis of a nonstationary signal, the preferred method is to use a description of the signal that involves both time and frequency. As the name implies, *time–frequency analysis* maps a signal (i.e., a one-dimensional function of time) onto an image (i.e., a two-dimensional function of time and frequency) that displays the signal's spectral components as a function of time. In conceptual terms, we may think of this mapping as a *time-varying spectral representation* of the signal. This representation is analogous to a musical score, with time and frequency representing the two principal axes. The values of the time–frequency representation of the signal provide an indication of the specific times at which certain spectral components of the signal are observed.

Basically, there are two classes of time–frequency representations of signals: *linear* and *quadratic*. In this section, we concern ourselves with linear representations only; specifically, we present brief expositions of the short-time Fourier transform and the wavelet transform, in that order.

■ 10.3.1 Orthonormal Bases of Functions

In one way or another, the formulation of time–frequency analysis builds on a two-parameter family of basis functions, denoted by $\psi_{\tau,\alpha}(t)$. The subscript parameter τ denotes delay time. The subscript parameter α depends on the particular type of time-frequency analysis being considered. In the short-time Fourier transform, α equals the frequency ω; in the continuous wavelet transform, α equals the scale parameter a that governs the frequency content. For practical reasons, we usually choose the basis function $\psi_{\tau,\alpha}(t)$ so that it is well concentrated in both the time and frequency domains.

Furthermore, expanding on the exposition of Fourier theory presented in Chapter 3, we note that it is desirable that the basis functions $\psi_{\tau,\alpha}(t)$ form an *orthonormal set*. There are two requirements for orthonormality:

1. *Normalization*, which means that the energy of the basis function $\psi_{\tau,\alpha}(t)$ is unity.

2. *Orthogonality*, which means that the inner product

$$\int_{-\infty}^{\infty} \psi_{\tau,\alpha}(t)\psi_{\tau',\alpha'}^{*}(t)\, dt = 0. \tag{10.1}$$

The asterisk is included in the integrand to account for the possibility that the basis function is complex valued. The parameters τ, α, and τ', α' are chosen from a restricted set of possible values in order to satisfy the orthogonality condition of Eq. (10.1).

(The issue of orthogonality is explored in Problems 3.85 and 3.100, with Problem 3.100 focusing on the basis-function expansion.)

In the formulation of the wavelet transform, it turns out that there is enough freedom to use orthonormal basis functions; unfortunately, this is not the case for the short-time Fourier transform.

■ 10.3.2 SHORT-TIME FOURIER TRANSFORM

Let $x(t)$ denote a signal of interest, and let $w(t)$ denote a *window function* of limited temporal extent; $w(t)$ may be complex valued, hence the use of an asterisk for complex conjugation. We thus define a *modified signal* given by

$$\boxed{x_\tau(t) = x(t)w^*(t - \tau),}$$ (10.2)

where τ is a delay parameter. The modified signal $x_\tau(t)$ is a function of two time variables:

- ▶ The running time t
- ▶ The fixed time delay τ, in which we are interested

As illustrated in Fig. 10.2, the window function is chosen in such a way that we may write

$$x_\tau(t) \approx \begin{cases} x(t) & \text{for } t \text{ close to } \tau \\ 0 & \text{for } t \text{ far away from } \tau \end{cases}.$$ (10.3)

In words, the original signal $x(t)$ is essentially unchanged as a result of windowing for values of the running time t around the delay time τ; for values of t far away from τ, the signal is practically suppressed by the window function. Thus, for all practical purposes, the modified signal $x_\tau(t)$ may be treated as a stationary signal, thereby permitting the application of the standard Fourier theory.

With this background, we may now apply the Fourier transform given in Eq. (3.31) to the modified signal $x_\tau(t)$. Accordingly, the short-time Fourier transform (STFT) of a nonstationary signal $x(t)$ is formally defined as

$$\boxed{\begin{aligned} X_\tau(j\omega) &= \int_{-\infty}^{\infty} x_\tau(t)e^{-j\omega t}\, dt \\ &= \int_{-\infty}^{\infty} x(t)w^*(t - \tau)e^{-j\omega t}\, dt. \end{aligned}}$$ (10.4)

The subscript τ in $X_\tau(j\omega)$ is included to remind us that the STFT naturally depends on the value assigned to the delay parameter τ, thereby distinguishing it from the standard Fourier transform $X(j\omega)$.

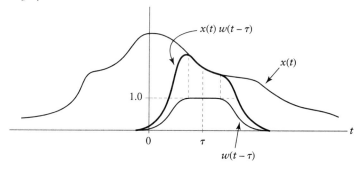

FIGURE 10.2   Result of multiplying a signal $x(t)$ by a window function $w(t)$ delayed in time by τ.

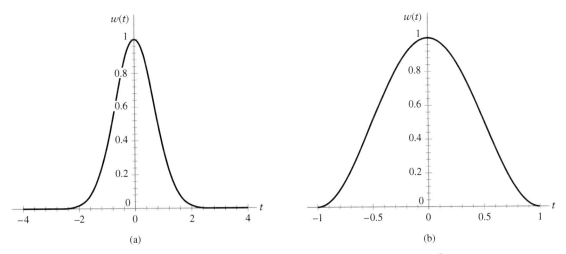

FIGURE 10.3   (a) Gaussian window. (b) Hanning window.

Clearly, $X_\tau(j\omega)$ is linear in the signal $x(t)$. The parameter ω plays a role similar to that of angular frequency in the ordinary Fourier transform. For a given $x(t)$, the result obtained by computing $X_\tau(j\omega)$ is dependent on the choice of the window $w(t)$. In the literature on time-frequency analysis, the short-time Fourier transform is usually denoted by $X(\tau, \omega)$; we have used $X_\tau(j\omega)$ here to be consistent with the terminology used in this book.

Many different shapes of window are used in practice. Typically, they are symmetric, unimodal, and smooth; two examples are a *Gaussian window*, as illustrated in Fig. 10.3(a), and a single period of a *Hanning window* (i.e., a raised-cosine window), as illustrated in Fig. 10.3(b). [See Eq. (8.61) for a definition of the Hanning window in discrete time.] The STFT using the Gaussian window is often called the *Gabor transform*.

In mathematical terms, the integral of Eq. (10.4) represents the inner (scalar) product of the signal $x(t)$ with a two-parameter family of basis functions, which is denoted by

$$\psi_{\tau,\omega}(t) = w(t - \tau)e^{j\omega t}. \tag{10.5}$$

The complex-valued basis function $\psi_{\tau,\omega}(t)$ varies with τ and ω, the time localization and the frequency of $\psi_{\tau,\omega}(t)$, respectively. Assuming that the window function $w(t)$ is real valued, the real and imaginary parts of $\psi_{\tau,\omega}(t)$ consist of an envelope function shifted in time by τ and filled in with a quadrature pair of sinusoidal waves, as illustrated in Figs. 10.4(a) and 10.4(b). It is important to note that, in general, it is difficult to find orthonormal basis functions based on the method of construction defined in Eq. (10.5).

Many of the properties of the Fourier transform are carried over to the STFT. In particular, the following two signal-preserving properties are noteworthy:

1. *The STFT preserves time shifts, except for a linear modulation;* that is, if $X_\tau(j\omega)$ is the STFT of a signal $x(t)$, then the STFT of the time-shifted signal $x(t - t_0)$ is equal to $X_{\tau-t_0}(j\omega)e^{-j\omega t_0}$.

2. *The STFT preserves frequency shifts;* that is, if $X_\tau(j\omega)$ is the STFT of a signal $x(t)$, then the STFT of the modulated signal $x(t)e^{j\omega_0 t}$ is equal to $X_\tau(j\omega - j\omega_0)$.

An issue of major concern in using the STFT is that of *time-frequency resolution.* To be specific, consider a pair of purely sinusoidal signals whose angular frequencies are spaced $\Delta\omega$ rad/s apart. The smallest value of $\Delta\omega$ for which the two signals are resolvable is called the *frequency resolution.* The duration of the window $w(t)$ is called the *time resolution,*

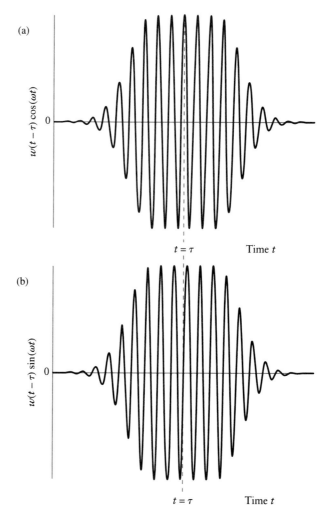

FIGURE 10.4 Real and imaginary parts of the complex-valued basis function $\psi_{\tau,\omega}(t)$, assuming that the window $w(t)$ is real-valued.

denoted by $\Delta\tau$. The frequency resolution $\Delta\omega$ and time resolution $\Delta\tau$ are inversely related according to the inequality

$$\Delta\tau\Delta\omega \geq \tfrac{1}{2}, \tag{10.6}$$

which is a manifestation of the duality property of the STFT, inherited from the Fourier transform. This relationship is referred to as the *uncertainty principle*, a term borrowed by analogy with statistical quantum mechanics; it was discussed in Section 3.17 in the context of the time–bandwidth product. The best that we can do is to satisfy Eq. (10.6) with the equality sign, which we can do with the use of a Gaussian window. Consequently, the time–frequency resolution capability of the STFT is *fixed* over the entire time–frequency plane. This point is illustrated in Fig. 10.5(a), in which the time–frequency plane is partitioned into *tiles* of the same shape and size. Figure 10.5(b) displays the real parts of the associated basis functions of the STFT; all have exactly the same duration, but different frequencies.

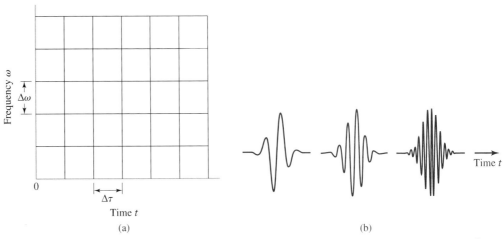

FIGURE 10.5   (a) Uniform tiling of the time–frequency plane by the short-time Fourier transform. (b) Real parts of associated basis functions for different frequency slots.

The squared magnitude of the STFT of a signal $x(t)$ is referred to as the *spectrogram* of the signal and is defined by

$$|X_\tau(j\omega)|^2 = \left| \int_{-\infty}^{\infty} x(t)w^*(t - \tau)e^{-j\omega t}\, dt \right|^2. \tag{10.7}$$

The spectrogram represents a simple, yet powerful extension of classical Fourier theory. In physical terms, it provides a measure of the energy of the signal in the time–frequency plane.

■ 10.3.3 SPECTROGRAMS OF SPEECH SIGNALS

Figure 10.6 shows the spectrograms of prefiltered and postfiltered versions of a speech signal that were displayed in Figs. 8.21(a) and (b). The spectrograms were computed using a raised-cosine window, 256 samples long. The gray scale of a particular pattern in the two spectrograms is indicative of the energy of the signal in that pattern. The gray scale code (in order of decreasing energy) is that black is the highest, followed by gray, and then white.

The following observations pertaining to the characteristics of speech signals can be made from the spectrograms of Fig. 10.6:

▶ The resonant frequencies of the vocal tract are represented by the dark areas of the spectrograms; these resonant frequencies are called *formants*.

▶ In the voiced regions, the striations appear darker and horizontal, as the energy is concentrated in narrow frequency bands, representing the harmonics of the glottal (excitation) pulse train.

▶ The unvoiced sounds have lower amplitudes because their energy is less than for voiced sounds and is distributed over a broader band of frequencies.

The sharp horizontal boundary at about 3.1 kHz between regions of significant energy and low (almost zero) energy seen in the spectrogram of Fig. 10.6(b), representing the filtered speech signal, is due to the action of the FIR digital low-pass filter.

(a)

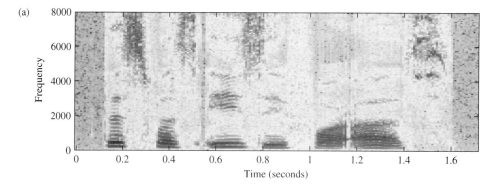

(b)

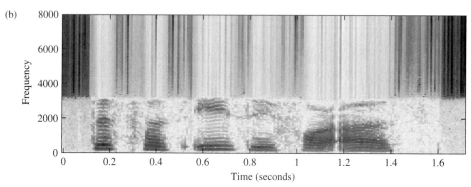

FIGURE 10.6   Spectrograms of speech signals. (a) Noisy version of the speech signal produced by a female speaker saying the phrase "This was easy for us." (b) Filtered version of the speech signal. (See Note 6 under Further Reading.)

■ 10.3.4 WAVELET TRANSFORM

To overcome the time–frequency resolution limitation of the STFT, we need a form of mapping that has the ability to trade off time resolution for frequency resolution and vice versa. One such method, known as the *wavelet transform* (WT), acts like a "mathematical microscope," in the sense that different parts of the signal may be examined by adjusting the "focus."

Wavelet analysis is based on a two-parameter family of basis functions denoted by

$$\psi_{\tau,a}(t) = |a|^{-1/2}\psi\left(\frac{t-\tau}{a}\right), \tag{10.8}$$

where a is a *nonzero scale factor* (also referred to as a *dilation parameter*) and τ is a time delay. The basis functions $\psi_{\tau,a}(t)$ for varying τ and a are called *wavelets* and are obtained by shifting and scaling a *mother wavelet* denoted by $\psi(t)$. By definition, the Fourier transform of the mother wavelet is

$$\Psi(j\omega) = \int_{-\infty}^{\infty} \psi(t)e^{-j\omega t}\, dt. \tag{10.9}$$

We assume that $\psi(t)$ satisfies the *admissability condition*, written in terms of the magnitude spectrum $|\Psi(j\omega)|$ as

$$C_\psi = \int_{-\infty}^{\infty} \frac{|\Psi(j\omega)|^2}{|\omega|}\, d\omega < \infty. \tag{10.10}$$

Typically, the magnitude spectrum $|\Psi(j\omega)|$ decreases sufficiently with increasing ω, so that the admissability condition reduces to a simpler requirement on the mother wavelet itself, namely,

$$\int_{-\infty}^{\infty} \psi(t)\,dt = \Psi(j0) = 0, \tag{10.11}$$

which means that $\psi(t)$ has at least some oscillations. Equation (10.11) indicates that the Fourier transform $\Psi(j\omega)$ is zero at the origin. With the magnitude spectrum $|\Psi(j\omega)|$ decreasing at high frequencies, it follows that the mother wavelet $\psi(t)$ has a band-pass characteristic. Furthermore, normalizing the mother wavelet so that it has unit energy, we may use Parseval's theorem, Eq. (3.62), to write

$$\int_{-\infty}^{\infty} |\psi(t)|^2\,dt = \frac{1}{2\pi}\int_{-\infty}^{\infty} |\Psi(j\omega)|^2\,d\omega = 1. \tag{10.12}$$

Consequently, using the time-shifting and time-scaling properties of the Fourier transform given in Table 3.7 and Eq. (3.60), respectively, we find that the wavelet $\psi_{\tau,a}(t)$ also has unity energy; that is,

$$\int_{-\infty}^{\infty} |\psi_{\tau,a}(t)|^2\,dt = 1 \qquad \text{for all } a \neq 0 \text{ and all } \tau. \tag{10.13}$$

Given a nonstationary signal $x(t)$, the WT is now formally defined as the inner product of the wavelet $\psi_{\tau,a}(t)$ and the signal $x(t)$:

$$\boxed{\begin{aligned} W_x(\tau, a) &= \int_{-\infty}^{\infty} x(t)\psi_{\tau,a}^*(t)\,dt \\ &= |a|^{-1/2}\int_{-\infty}^{\infty} x(t)\psi^*\!\left(\frac{t-\tau}{a}\right)dt. \end{aligned}} \tag{10.14}$$

Like the Fourier transform, the WT is invertible; that is, the original signal $x(t)$ can be recovered from $W_x(\tau, a)$ without loss of information by using the synthesis formula

$$x(t) = \frac{1}{a^2 C_\psi}\int_{-\infty}^{\infty}\int_{-\infty}^{\infty} W_x(\tau, a)\psi_{\tau,a}(t)\,da\,d\tau, \tag{10.15}$$

where C_ψ is defined by Eq. (10.10). Equation (10.15) is called the *resolution of the identity*; it states that the signal $x(t)$ can be expressed as a superposition of shifted and dilated wavelets.

In wavelet analysis, the basis function $\psi_{\tau,a}(t)$ is an oscillating function; there is therefore no need to use sines and cosines (waves) as in Fourier analysis. More specifically, the basis function $e^{j\omega t}$ in Fourier analysis oscillates forever; in contrast, the basis function $\psi_{\tau,a}(t)$ in wavelet analysis is localized in time, lasting for only a few cycles. The delay parameter τ gives the position of the wavelet $\psi_{\tau,a}(t)$, while the scale factor a governs its frequency content. For $|a| \ll 1$, the wavelet $\psi_{\tau,a}(t)$ is a highly concentrated and shrunken version of the mother wavelet $\psi(t)$, with frequency content concentrated mostly in the high-frequency range. By contrast, for $|a| \gg 1$, the wavelet $\psi_{\tau,a}(t)$ is very much spread out and has mostly low frequency content. Figure 10.7(a) shows the *Haar wavelet*, the simplest example of a wavelet. The *Daubechies wavelet*, shown in Fig. 10.7(b), is a more sophisticated example. Both of these wavelets have *compact support* in time, meaning that the wavelets have finite duration. The Daubechies wavelet has length $N = 12$ and is therefore less local than the Haar wavelet, which has length $N = 1$. However, the Daubechies wavelet is continuous and has better frequency resolution than the Haar wavelet.

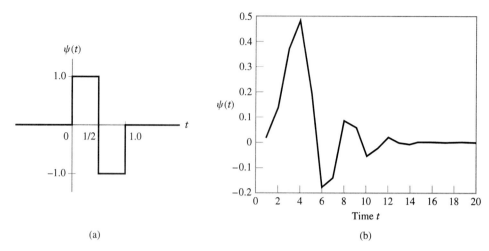

FIGURE 10.7   (a) Haar wavelet. (b) Daubechies wavelet.

The Haar and Daubechies wavelets are each orthonormal; that is, in both cases, the mother wavelet $\psi(t)$ satisfies the two requirements of orthonormality: unit energy and orthogonality of $\psi(t)$ with respect to all its dilations and translations in accordance with Eq. (10.1). In particular, the parameters τ, α and τ', α' are chosen from discrete sets. The most common example has τ (and τ') selected from the set $\{k2^{-j}; k \text{ and } j \text{ integer}\}$, while α (and α') is selected from the set $\{2^{-j}; j \text{ integer}\}$. (The integer j used here is not to be confused with $j = \sqrt{-1}$.) On this basis, it is relatively straightforward to verify the orthonormality of the Haar wavelets:

1. From Fig. 10.7(a), we readily see that

$$\int_{-\infty}^{\infty} |\psi(t)|^2 \, dt = \int_{0}^{1/2} (1)^2 \, dt + \int_{1/2}^{1} (-1)^2 \, dt \qquad (10.16)$$
$$= 1.$$

2. The Haar wavelet basis, containing all its dilations and translations, may be expressed by $2^{j/2}\psi(2^j t - k)$, where j and k are integers (positive, negative, or zero); this expression is obtained by setting $a = 2^{-j}$ and $\tau = k2^{-j}$ in Eq. (10.8). Orthogonality is assured by virtue of the fact that

$$\text{inner product} = \int_{-\infty}^{\infty} \psi(2^j t - k)\psi(2^l t - m) \, dt = 0, \quad \text{for} \quad l \neq j \quad \text{or} \quad m \neq k.$$
$$(10.17)$$

Note that since the Haar wavelet is real valued, there is no need for complex conjugation. Equation (10.17) is easily verified for the first few Haar wavelets, with reference to Fig. 10.8:

$$\begin{pmatrix} j = 0, & k = 0 \\ l = 1, & m = 0 \end{pmatrix}, \quad \text{inner product} = \int_{-\infty}^{\infty} \psi(t)\psi(2t)dt = 0;$$

$$\begin{pmatrix} j = 0, & k = 0 \\ l = 1, & m = 1 \end{pmatrix}, \quad \text{inner product} = \int_{-\infty}^{\infty} \psi(t)\psi(2t - 1)dt = 0;$$

$$\begin{pmatrix} j = 1, & k = 0 \\ l = 1, & m = 1 \end{pmatrix}, \quad \text{inner product} = \int_{-\infty}^{\infty} \psi(2t)\psi(2t - 1)dt = 0.$$

From Figs. 10.8(a) and 10.8(b), we see that when $\psi(t) = 1$, the dilated $\psi(2t)$ assumes the values $+1$ and -1, making the first integral zero. Likewise, from Figs. 10.8(a) and 10.8(c), we see that when $\psi(t) = -1$, the dilated and translated $\psi(2t - 1)$ assumes the values $+1$ and -1, making the second integral zero. The third integral is zero for a different reason:

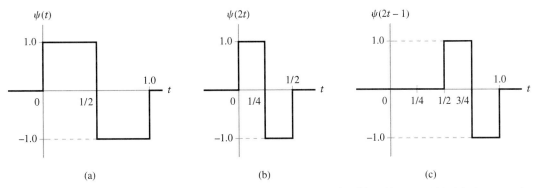

FIGURE 10.8   (a) Haar mother wavelet $\psi(t)$. (b) Haar wavelet dilated by $a = 1/2$. (c) Haar wavelet dilated by $1/2$ and translated by $\tau = 1/2$.

The functions $\psi(2t)$ and $\psi(2t - 1)$ do not overlap, as is observed from Figs. 10.8(b) and 10.8(c), where we see that when one of these two functions is zero, the other one is nonzero. This pattern of behavior holds for all other dilations by 2^{-j} and translations by k, thereby justifying the orthogonality condition of Eq. (10.17).

In windowed Fourier analysis, the goal is to measure the local frequency content of the signal. In wavelet analysis, by contrast, we measure the similarity between the signal $x(t)$ and the wavelet $\psi_{\tau, a}(t)$ for varying τ and a. The dilations by a result in several magnifications of the signal, with distinct resolutions.

Just as the STFT has signal-preserving properties of its own, so does the WT:

1. *The WT preserves time shifts;* that is, if $W_x(\tau, a)$ is the WT of a signal $x(t)$, then $W_x(\tau - t_0, a)$ is the WT of the time-shifted signal $x(t - t_0)$.

2. *The WT preserves time scaling;* that is, if $W_x(\tau, a)$ is the WT of a signal $x(t)$, then the WT of the time-scaled signal $|a_0|^{1/2} x(a_0 t)$ is equal to $W_x(a_0 \tau, a a_0)$.

However, unlike the STFT, the WT does not preserve frequency shifts.

As mentioned previously, the mother wavelet $\psi(t)$ can be any band-pass function. To establish a connection with the modulated window in the STFT, we choose

$$\psi(t) = w(t)e^{j\omega_0 t}. \tag{10.18}$$

The window $w(t)$ is typically a low-pass function. Thus, Eq. (10.18) describes the mother wavelet $\psi(t)$ as a complex, linearly modulated signal whose frequency content is concentrated essentially around its own carrier frequency ω_0; note, however, this particular mother wavelet does not lead to an orthonormal set. Let ω denote the carrier frequency of an analyzing wavelet $\psi_{\tau, a}(t)$. Then the scale factor a of $\psi_{\tau, a}(t)$ is inversely related to the carrier frequency ω; that is,

$$a = \frac{\omega_0}{\omega}. \tag{10.19}$$

Since, by definition, a wavelet is a scaled version of the same prototype, it follows that

$$\frac{\Delta \omega}{\omega} = Q, \tag{10.20}$$

where $\Delta \omega$ is the frequency resolution of the analyzing wavelet $\psi_{\tau, a}(t)$ and Q is a constant. Choosing the window $w(t)$ to be a Gaussian function, and therefore using Eq. (10.6) with the equality sign, we may express the time resolution of the wavelet $\psi_{\tau, a}(t)$ as

$$\Delta \tau = \frac{1}{2 \Delta \omega} \tag{10.21}$$

$$= \frac{1}{2Q\omega}.$$

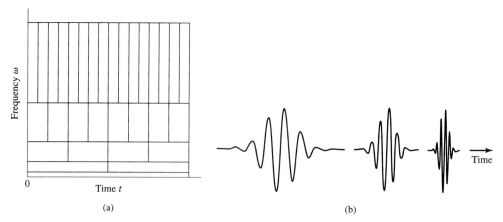

FIGURE 10.9 (a) Partitioning of the time–frequency plane by the wavelet transform. (b) Real parts of associated basis functions.

In light of Eqs. (10.19) and (10.20), we may now formally state the time–frequency resolution properties of the WT:

1. The time resolution $\Delta\tau$ varies inversely with the carrier frequency ω of the analyzing wavelet $\psi_{\tau,a}(t)$; hence, $\Delta\tau$ can be made arbitrarily small at high frequencies.
2. The frequency resolution $\Delta\omega$ varies linearly with the carrier frequency ω of the analyzing wavelet $\psi_{\tau,a}(t)$; hence, $\Delta\omega$ can be made arbitrarily small at low frequencies.

Thus, the WT is well suited for the analysis of nonstationary signals containing high-frequency transients superimposed on longer-lived low-frequency components.

In Section 10.3.2, we stated that the STFT has a fixed resolution, as illustrated in Fig. 10.5(a). In contrast, the WT has a multiresolution capability, as illustrated in Fig. 10.9(a). In the latter figure, we see that the WT partitions the time–frequency plane into tiles of the same area, but with varying widths and heights that depend on the carrier frequency ω of the analyzing wavelet $\psi_{\tau,a}(t)$. Thus, unlike the STFT, the WT provides a trade-off between time and frequency resolutions, which are represented by the widths and heights of the tiles, respectively (i.e., narrower widths and heights correspond to better resolution). Figure 10.9(b) displays the real parts of the basis functions of the WT. Here, we see that every time the basis function is compressed by a factor of, say, 2, its carrier frequency is increased by the same factor.

The WT performs a time-scale analysis of its own. Thus, its squared magnitude is called a *scalogram*, defined by

$$\left| W_x(\tau, a) \right|^2 = \frac{1}{a} \left| \int_{-\infty}^{\infty} x(t)\psi^* \left(\frac{t - \tau}{a} \right) dt \right|^2. \tag{10.22}$$

The scalogram represents a distribution of the energy of the signal in the time-scale plane.

■ 10.3.5 IMAGE COMPRESSION USING THE WAVELET TRANSFORM

The transmission of an image over a communication channel can be made more efficient by compressing the image at the transmitting end of the system and reconstructing the original image at the receiving end. This combination of signal-processing operations is called *image compression*. Basically, there are two types of image compression:

1. *Lossless compression*, which operates by removing the *redundant* information contained in the image. Lossless compression is completely reversible, in that the original image can be reconstructed exactly.

2. *Lossy compression*, which involves the loss of some information and may therefore not be completely reversible. Lossy compression is, however, capable of achieving a compression ratio higher than that attainable with lossless methods.

In many cases, lossy compression is the preferred method, if it does not significantly alter the perceptual quality of the source image. For example, in the recent proliferation of music trading over the Internet, the preferred format is the mp3 (mpeg audio layer three) compression scheme, which attains a level of compression of approximately 11:1. In today's electronic communication systems, cost of bandwidth is at a premium, and therefore a great savings is obtained with the use of compression schemes.

Wavelets provide a powerful linear method for lossy image compression, because the coefficients of the wavelet transform are localized in both space and frequency. As an example, consider the Daubechies wavelet shown in Fig. 10.7(b). The goal is to perform image compression of Fig. 10.10(a), showing a woman holding flowers. Figure 10.10(b) shows the same image after it has been compressed to 68% of its original size. Note that it is difficult to observe any difference between the two images, despite the fact that some information has been lost. If we continue to compress the image to 85% and 97% of its

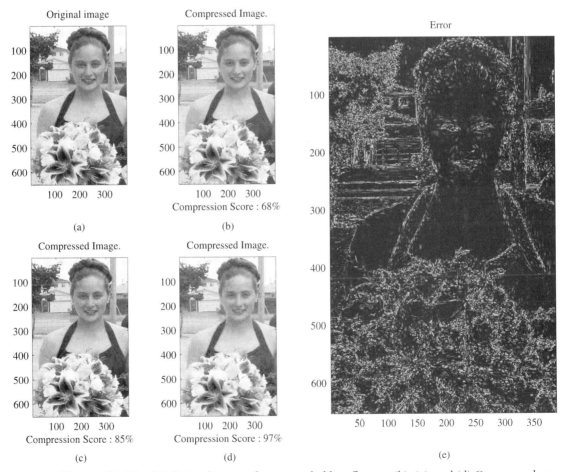

FIGURE 10.10 (a) Original image of a woman holding flowers. (b), (c), and (d) Compressed versions of the image using the Daubechies wavelet of Fig. 10.7(b) with compression scores of 68%, 85%, and 97%, respectively. (e) Difference image obtained by subtracting the compressed image (d) from the original image (a). (See Note 6 under Further Reading).

original size [Figs. 10.10(c) and (d)], visual imperfections become increasingly apparent. The reason for the blurring is that, due to the high compression rate, not enough original information is kept; hence, it is not possible to reconstruct the original image perfectly.

The amount of information lost is shown in the difference image of Fig. 10.10(e), which is obtained by subtracting the compressed image [Fig. 10.10(d)] from the original [Fig. 10.10(a)]. It is interesting to note that the areas of "high activity," or equivalently, areas of high information loss, shown on the difference image correspond to blurred regions in Fig. 10.10(d). These high-activity areas correspond to regions on the original image that contain high-frequency content (spatially, the pixels change rapidly; see the flowers); hence, the difficulty in compression is because there is little redundancy, (i.e., similarity among pixels) in that region of the picture.

10.4 *Nonlinear Systems*

For the linearity assumption to be satisfied, the amplitudes of the signals encountered in a system (e.g., a control system) would have to be restricted to lie inside a range small enough for all components of the system to operate in their "linear region." This restriction ensures that the principle of superposition is essentially satisfied, so that, for all practical purposes, the system can be viewed as linear. But when the amplitudes of the signals are permitted to lie outside the range of linear operation, the system can no longer be considered linear. For example, a transistor amplifier used in a control system exhibits an input–output characteristic that runs into saturation when the amplitude of the signal applied to the amplifier input is large. Other sources of nonlinearity in a control system include backlash between coupled gears and friction between moving parts. In any event, when the deviation from linearity is relatively small, some form of distortion is introduced into the characterization of the system. The effect of this distortion can be reduced by applying feedback around the system, as discussed in Chapter 9.

However, what if the operating amplitude range of a system is required to be large? The answer to this question depends on the intended application and how the system is designed. For example, in a control application, a linear system is likely to perform poorly or to become unstable, because a linear design procedure is incapable of properly compensating for the effects of large deviations from linearity. On the other hand, a nonlinear control system may perform better by directly incorporating nonlinearities into the system design. This point may be demonstrated in the motion control of a robot, wherein many of the dynamic forces experienced vary as the square of the speed. When a linear control system is used for such an application, nonlinear forces associated with the motion of robot links are neglected, with the result that the control accuracy degrades rapidly as the speed of motion is increased. Accordingly, in a robot task such as "pick and place," the speed of motion has to be kept relatively slow so as to realize a prescribed control accuracy. In contrast, a highly accurate control for a large range of robot speeds in a large workplace can be attained by employing a nonlinear control system that compensates for the nonlinear forces experienced in robot motion. The benefit so gained is improved productivity.

Notwithstanding the mathematical difficulty of analyzing nonlinear systems, serious efforts have been made on several fronts to develop theoretical tools for the study of such systems. In this context, four approaches deserve special mention.

▪ 10.4.1 PHASE-SPACE ANALYSIS

The basic idea of this method is to use a graphical approach to study a nonlinear system of first-order differential equations written as

$$\frac{d}{dt}x_j(t) = f_j(\mathbf{x}(t)), \qquad j = 1, 2, \ldots, p, \tag{10.23}$$

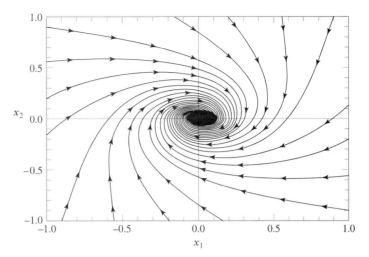

FIGURE 10.11   Phase portrait of a two-dimensional nonlinear dynamical system described by the pair of state equations.

$$\frac{d}{dt}x_1(t) = x_2(t) - x_1(t)(x_1^2(t) + x_2^2(t) - c)$$

and

$$\frac{d}{dt}x_2(t) = -x_1(t) - x_2(t)(x_1^2(t) + x_2^2(t) - c)$$

for the control parameter $c = -0.2$. (Reproduced from T. S. Parker and L. O. Chua, *Practical Numerical Algorithms for Chaotic Systems*, Springer-Verlag, 1989; courtesy of Springer-Verlag.)

where the elements $x_1(t), x_2(t), \ldots, x_p(t)$ define a p-dimensional *state vector* $\mathbf{x}(t)$ and the corresponding $f_1, f_2, \ldots, f_p$ are nonlinear functions; p is referred to as the *order* of the system. Equation (10.23) may be viewed as describing the motion of a point in a p-dimensional space, commonly referred to as the *phase space* of the system; the terminology is borrowed from physics. The phase space is important because it provides us with a visual conceptual tool for analyzing the dynamics of a nonlinear system described by Eq. (10.23). It does so by focusing our attention on the global characteristics of the motion, rather than the detailed aspects of analytic or numeric solutions of the equations.

Starting from a set of initial conditions, Eq. (10.23) defines a solution represented by $x_1(t), x_2(t), \ldots, x_p(t)$. As time t is varied from zero to infinity, this solution traces a curve in the phase space. Such a curve is called a *trajectory*. A family of trajectories, corresponding to different initial conditions, is called a *phase portrait*. Figure 10.11 illustrates the phase portrait of a two-dimensional nonlinear dynamical system.

Largely due to the limitations of our visual capability, the graphical power of phase-space analysis is limited to second-order systems (i.e., with $p = 2$) or systems that can be approximated by second-order dynamics.

■ 10.4.2 DESCRIBING-FUNCTION ANALYSIS

When a nonlinear element is subjected to a sinusoidal input, the *describing function* of the element is defined as the complex ratio of the fundamental component of the output to the sinusoidal input. Thus, the essence of the describing-function method is to approximate the nonlinear elements of a nonlinear system with quasilinear equivalents and then exploit the power of frequency-domain techniques to analyze the approximating system.

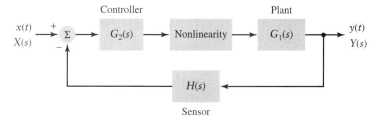

FIGURE 10.12   Feedback control system containing a nonlinear element in its feedback loop.

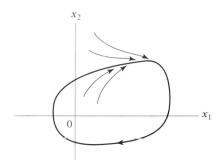

FIGURE 10.13   Limit cycle in a two-dimensional phase space.

Consider, for example, the nonlinear control system described in Fig. 10.12. The feedback loop of the system includes a nonlinearity and three linear elements represented by the transfer functions $G_1(s)$, $G_2(s)$, and $H(s)$. For example, the nonlinear element may be a relay. For a sinusoidal input $x(t)$, the input to the nonlinearity will not be sinusoidal, due to harmonics of the input signal generated by the nonlinearity and fed back to the input via the sensor $H(s)$. If, however, the linear elements $G_1(s)$, $G_2(s)$, and $H(s)$ are of a low-pass type such that the harmonics so generated are attenuated to insignificant levels, we then would be justified in assuming that the input to the nonlinearity is essentially sinusoidal. In a situation of this kind, applying the describing-function method would produce accurate results.

The describing-function method is used mainly to predict the occurrence of *limit cycles* in nonlinear feedback systems. By a "limit cycle," we mean a closed trajectory in the phase space onto which other trajectories converge asymptotically, from both inside and outside, as time approaches infinity. This form of convergence in the phase space is illustrated in Fig. 10.13. A limit cycle is a periodic motion peculiar to nonlinear feedback systems.

■ 10.4.3 LYAPUNOV'S INDIRECT METHOD: STABILITY OF EQUILIBRIUM POINTS

The third approach to the stability analysis of a nonlinear system is based on Lyapunov's indirect method, which states that the stability properties of the system in the neighborhood of an equilibrium point are essentially the same as those obtained through a linearized approximation to the system. By an *equilibrium point*, we mean a point in the phase space at which the state vector of the system can reside forever. Let $\bar{x}_j$ denote the jth element of the equilibrium points, which themselves are denoted by the vector $\bar{\mathbf{x}}$. From the definition of the equilibrium point just given, the derivative $d\bar{x}_j/dt$ vanishes at the equilibrium point for all j, in which case we may write

$$f_j(\bar{\mathbf{x}}) = 0 \qquad \text{for } j = 1, 2, \ldots, p. \tag{10.24}$$

The equilibrium point is also referred to as a *singular point*, signifying the fact that, when it occurs, the trajectory of the system degenerates into the point itself.

To develop a deeper understanding of the equilibrium condition, suppose that the set of nonlinear functions $f_j(\mathbf{x}(t))$ is smooth enough for Eq. (10.23) to be linearized in the neighborhood of $\bar{x}_j$ for $j = 1, 2, \ldots, p$. Specifically, let

$$x_j(t) = \bar{x}_j + \Delta x_j(t), \qquad j = 1, 2, \ldots, p, \tag{10.25}$$

where $\Delta x_j(t)$ is a small deviation from $\bar{x}_j$ for all j at time t. Then, retaining the first two terms in the Taylor series expansion of the nonlinear function $f_j(\mathbf{x}(t))$, we may approximate the function by

$$f_j(\mathbf{x}(t)) \approx x_j + \sum_{k=1}^{p} a_{jk}\,\Delta x_j(t), \qquad j = 1, 2, \ldots, p, \tag{10.26}$$

where the element

$$a_{jk} = \frac{\partial}{\partial x_k} f_j(\mathbf{x})\big|_{x_j = \bar{x}_j} \qquad \text{for } j, k = 1, 2, \ldots, p. \tag{10.27}$$

Hence, substituting Eqs. (10.25) and (10.26) into Eq. (10.23) and then invoking the definition of an equilibrium point, we get

$$\frac{d}{dt}\Delta x_j(t) \simeq \sum_{k=1}^{p} a_{jk}\,\Delta x_j(t), \qquad j = 1, 2, \ldots, p. \tag{10.28}$$

The set of elements $\{a_{jk}\}_{j,k=1}^{p}$ constitutes a $p \times p$ matrix, denoted by $\mathbf{A}$. Provided that $\mathbf{A}$ is nonsingular (i.e., if the inverse matrix $\mathbf{A}^{-1}$ exists), then the approximation described in Eq. (10.28) is sufficient to determine the *local behavior* of the system in the neighborhood of the equilibrium point.

The derivative $\frac{d}{dt}\Delta x_j(t)$ may be viewed as the jth element of a $p \times 1$ *velocity vector*. At an equilibrium point the velocity vector is zero. According to Eq. (10.28), the nature of the equilibrium point is essentially determined by the *eigenvalues* of matrix $\mathbf{A}$ and may therefore be classified in a corresponding fashion. The eigenvalues of $\mathbf{A}$ are the roots of the characteristic equation

$$\det(\mathbf{A} - \lambda\mathbf{I}) = 0, \tag{10.29}$$

where $\mathbf{I}$ is the $p \times p$ identity matrix and λ is an eigenvalue. When the matrix $\mathbf{A}$ has m eigenvalues, the equilibrium point is said to be of *type m*. For the special case of a second-order system, we may classify the equilibrium point as summarized in Table 10.1 and illustrated in Fig. 10.14. Note that in the case of a *saddle point*, trajectories going to the saddle point indicate that the node is stable, whereas trajectories coming from the saddle point indicate that the node is unstable.

> **TABLE 10.1** *Classification of the Equilibrium State of a Second-Order System.*

Type of Equilibrium State $\bar{\mathbf{x}}$	Eigenvalues of Matrix $\mathbf{A}$
Stable node	Real and negative
Stable focus	Complex conjugate with negative real parts
Unstable node	Real and positive
Unstable focus	Complex conjugate with positive real parts
Saddle point	Real with opposite signs
Center	Conjugate purely imaginary

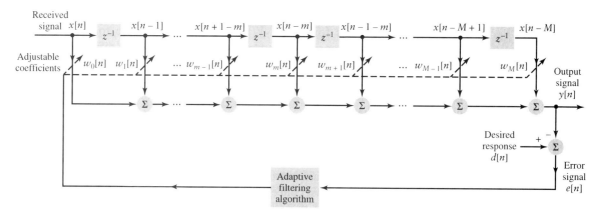

FIGURE 10.16   Adaptive equalizer built around an FIR digital filter of order M.

own parameters (i.e., FIR filter coefficients) continuously and automatically by operating on a pair of signals:

▶ The *received signal* $x[n]$ representing a distorted version of the signal transmitted over the channel

▶ The *desired response* $d[n]$ representing a replica of the transmitted signal

One's first reaction to the availability of a replica of the transmitted signal is likely to be "If such a signal is available at the receiver, why do we need adaptive equalization?" To answer this question, we note that a typical telephone channel changes little during an average data call. Accordingly, prior to data transmission, the equalizer is adjusted with the guidance of a binary *training sequence* transmitted through the channel. A synchronized version of this training sequence is generated at the receiver, where (after a time shift equal to the transmission delay through the channel) it is applied to the equalizer as the desired response. A training sequence commonly used in practice is the *pseudonoise (PN) sequence*—a deterministic periodic sequence with noiselike characteristics. Two identical PN sequence generators are used, one at the transmitter and the other at the receiver. When the training process is completed, the adaptive equalizer is ready for normal operation. The training sequence is switched off, and information-bearing binary data are then transmitted over the channel. The equalizer output is passed through a threshold device, and a decision is made on whether the transmitted binary data symbol is a "1" or a "0." In normal operation, the decisions made by the receiver are correct most of the time. This means that the sequence of symbols produced at the output of the threshold device represents a fairly reliable estimate of the transmitted data sequence and may therefore be used as a substitute for the desired response, as indicated in Fig. 10.17. This second mode of operation is called a *decision-directed mode*, the purpose of which is to *track* relatively slow variations in channel characteristics that may take place during the course of normal data transmission. The adjustment of filter coefficients in the equalizer is thus performed with the use of an *adaptive filtering algorithm* that proceeds as follows:

1. *Training mode*

   (i) Given the FIR filter coefficients at iteration n, the corresponding value $y[n]$ of the actual equalizer output is computed in response to the received signal $x[n]$.

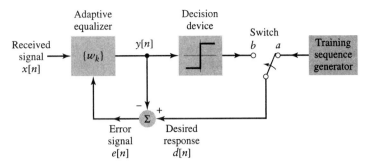

FIGURE 10.17   The two modes of operating an adaptive equalizer. When the switch is in position *a*, the equalizer operates in its training mode. When the switch is moved to position *b*, the equalizer operates in its decision-directed mode.

(ii) The difference between the desired response $d[n]$ and the equalizer output $y[n]$ is computed; this difference constitutes the *error signal*, denoted by $e[n]$.

(iii) The error signal $e[n]$ is used to apply corrections to the FIR filter coefficients.

(iv) Using the updated filter coefficients of the equalizer, the algorithm repeats steps (i) through (iii) until the equalizer reaches a steady state, after which no noticeable changes in the filter coefficients are observed.

To initiate this sequence of iterations, the filter coefficients are set equal to some suitable values (e.g., zero for all of them) at $n = 0$. The details of the corrections applied to the filter coefficients from one iteration to the next are determined by the type of adaptive filtering algorithm employed.

2. *Decision-directed mode.* This second mode of operation starts where the training mode finishes and uses the same set of steps, except for two modifications:

▸ The output of the threshold device is substituted for the desired response.

▸ The adjustments to filter coefficients of the equalizer are continued throughout the transmission of data.

Another useful application of adaptive filtering is in *system identification*. In this application, we are given an unknown dynamic plant, the operation of which cannot be interrupted, and the requirement is to build a model of the plant and its operation. Figure 10.18 shows the block diagram of such a model, which consists of an FIR filter of order *M*. The input signal $x[n]$ is applied simultaneously to the plant and the model. Let $d[n]$ and $y[n]$ denote the corresponding values of the output from the plant and the model, respectively. The plant output $d[n]$ is the desired response in this application. The difference between $d[n]$ and $y[n]$ thus defines the error signal $e[n]$, which is to be minimized according to some specified criterion. This minimization is attained by using an adaptive filtering algorithm that adjusts the model's parameters (i.e., FIR filter coefficients) in a step-by-step fashion, following a procedure similar to that described for the training mode of the adaptive equalizer.

Adaptive equalization and system identification are just two of the many applications of adaptive filters, which span such diverse areas as communications, control, radar, sonar, seismology, radio astronomy, and biomedicine.

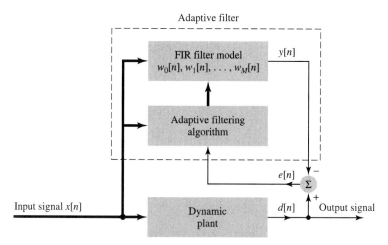

FIGURE 10.18 Block diagram of an FIR model whose coefficients are adjusted by an adaptive filtering algorithm for the identification of an unknown dynamic plant.

10.6 *Concluding Remarks*

The material presented in the previous nine chapters provides a theoretical treatment of signals and systems that paves the way for detailed studies of digital signal processing, communication systems, and control systems. The theory presented therein rests on the following idealizations:

▶ Stationary signals

▶ Linear time-invariant systems

In this chapter we highlighted limitations of that theory, viewed from the perspective of real-life signals and systems. In so doing, we also briefly touched on the topics of nonstationarity, time–frequency analysis, nonlinear systems, and adaptive filters. These topics and those presented in previous chapters emphasize the enormous breadth of what is generally encompassed by the subject of signals and systems.

Another noteworthy point is that throughout this book we have focused on time as the independent variable. We may therefore refer to the material covered herein as *temporal processing*. In *spatial processing*, spatial coordinates play the role of independent variables. Examples of spatial processing are encountered in continuous-aperture antennas, (discrete) antenna arrays, and processing of images. Much of the material presented in this book applies equally well to spatial processing, pointing further to its fundamental nature.

FURTHER READING

1. The classic approach to the characterization and processing of speech signals is discussed in the following books:

   ▶ Flanagan, J. L., *Speech Analysis: Synthesis and Perception* (Springer-Verlag, 1972)

   ▶ Rabiner, L. R., and R. W. Schafer, *Digital Processing of Speech Signals* (Prentice-Hall, 1978)

For a more complete treatment of the subject, see

- Deller, J. R., Jr., J. G. Proakis, and J. H. L. Hanson, *Discrete Time Processing of Speech Signals* (Prentice Hall, 1993)
- Quatieri, T. F., *Discrete Time Speech Signal Processing: Principles and Practice* (Prentice Hall, 2001)

2. The subject of time-frequency analysis is covered in

- Cohen, L., *Time–Frequency Analysis* (Prentice Hall, 1995)

The STFT and its properties are discussed on pages 93–112 of Cohen's book, which also presents a detailed exposition of the orthonormal expansion of signals (pp. 204–209).
Work on time–frequency analysis may be traced to Gabor's classic paper,

- Gabor, D., "Theory of communication," *Journal IEE (London)*, vol. 93, pp. 429–457, 1946

3. For discussions of wavelets and wavelet transforms, their theory, and applications, see

- Strang, G., and T. Q. Nguyen, *Wavelets and Filter Banks* (Wellesley-Cambridge Press, 1996)
- Burrus, C. S., R. A. Gopinath, and H. Guo, *Introduction to Wavelets and Wavelet Transforms—A Primer* (Prentice Hall, 1998)
- Daubechies, I., ed., *Different Perspectives on Wavelets, Proceedings of Symposia in Applied Mathematics*, vol. 47 (American Mathematical Society, 1993)
- Meyer, Y., *Wavelets: Algorithms and Applications* (SIAM, 1993), translated from the French by R. D. Ryan
- Vetterli, M., and J. Kovacevic, *Wavelets and Subband Coding* (Prentice Hall, 1995)
- Qian, S., *Introduction to Time-Frequency and Wavelet Transforms* (Prentice Hall, 2002)

For a proof of the synthesis formula of Eq. (10.15), see Vetterli and Kovacevic, *op. cit.*, pp. 302–304. Properties of the WT are discussed on pages 304–311 of this book.
The Daubechies wavelet is named in honor of Ingrid Daubechies for her pioneering works:

- Daubechies, I. "Time-frequency localization operators: A geometric phase space approach," *IEEE Transactions on Information Theory*, vol. 34, pp. 605–612, 1988
- Daubechies, I., *Ten Lectures on Wavelets*, CBMS Lecture Notes, no. 61 (SIAM, 1992)

The Haar wavelet is named in honor of Alfred Haar for his classic paper,

- Haar, A., "Zur Theorie der Orthogonalen Functionen-Systeme," *Math. Annal.*, vol. 69, pp. 331–371, 1910

For a detailed historical perspective on Wavelets, see pp. 13–31 of Meyer, *op. cit.*

Note however, that the first broad definition of a wavelet is due to A. Grossman and J. Morlet, a physicist and an engineer, respectively. The definition appeared in

- Grossman, A., and J. Morlet, "Decomposition of Hardy functions into square integrable wavelets of constant shape," *SIAM J. Math. Anal.*, vol. 15, pp. 723–736, 1984

4. For a study of describing-function analysis, see

- Atherton, D. P., *Nonlinear-Control Engineering* (Van Nostrand-Reinhold, 1975)

Lyapunov theory is covered in the following books:

- Slotine, J.-J.e., and W. Li, *Applied Nonlinear Control* (Prentice Hall, 1991)
- Khalil, H. K., *Nonlinear Systems* (Macmillan, 1992)
- Vidyasagar, M., *Nonlinear Systems Analysis*, 2d ed. (Prentice Hall, 1993)

These books also discuss converse theorems asserting the existence of Lyapunov functions. The discussion of advanced stability theory includes extensions of the Lyapunov stability theorem to *nonautonomous systems*, by which we mean nonlinear dynamical systems described by equations that depend explicitly on time t—that is,

$$\frac{d}{dt}\mathbf{x}(t) = \mathbf{f}(\mathbf{x}(t), t).$$

Correspondingly, the equilibrium points, $\bar{\mathbf{x}}$, are defined by

$$\mathbf{f}(\bar{\mathbf{x}}, t) = \mathbf{0} \qquad \text{for all } t > t_0.$$

This implies that the system should stay at the point $\bar{\mathbf{x}}$ all the time, thereby making a formulation of the Lyapunov stability theorem much more challenging than that described in Subsection 10.4.4 for autonomous systems.

Alexander M. Lyapunov (1857–1918), a distinguished Russian mathematician and engineer, laid down the foundation of the stability theory of nonlinear dynamical systems, which bears his name. Lyapunov's classic work, *The General Problem of Motion Stability*, was first published in 1892.

5. The theory of adaptive filters and their applications are covered in the following books:
   ▸ Haykin, S., *Adaptive Filter Theory*, 4th ed. (Prentice Hall, 2002)
   ▸ Widrow, B., and S. D. Stearns, *Adaptive Signal Processing* (Prentice-Hall, 1985)

6. The MATLAB code for generating the pictures presented in Figs. 10.6 and 10.10 can be found at the web site www.wiley.com/college/haykin

A

Selected Mathematical Identities

A.1 *Trigonometry*

Consider the right triangle depicted in Fig. A.1. The following relationships hold:

$$\sin \theta = \frac{y}{r}$$

$$\cos \theta = \frac{x}{r}$$

$$\tan \theta = \frac{y}{x} = \frac{\sin \theta}{\cos \theta}$$

$$\cos^2 \theta + \sin^2 \theta = 1$$

$$\cos^2 \theta = \tfrac{1}{2}(1 + \cos 2\theta)$$

$$\sin^2 \theta = \tfrac{1}{2}(1 - \cos 2\theta)$$

$$\cos 2\theta = 2\cos^2 \theta - 1$$

$$= 1 - 2\sin^2 \theta$$

Other identities include the following:

$$\sin(\theta \pm \phi) = \sin \theta \cos \phi \pm \cos \theta \sin \phi$$

$$\cos(\theta \pm \phi) = \cos \theta \cos \phi \mp \sin \theta \sin \phi$$

$$\sin \theta \sin \phi = \tfrac{1}{2}[\cos(\theta - \phi) - \cos(\theta + \phi)]$$

$$\cos \theta \cos \phi = \tfrac{1}{2}[\cos(\theta - \phi) + \cos(\theta + \phi)]$$

$$\sin \theta \cos \phi = \tfrac{1}{2}[\sin(\theta - \phi) + \sin(\theta + \phi)]$$

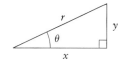

FIGURE A.1   Right triangle.

A.2 Complex Numbers

Let w be a complex number expressed in rectangular coordinates as $w = x + jy$, where $j = \sqrt{-1}$, $x = \text{Re}\{w\}$ is the real part of w, and $y = \text{Im}\{w\}$ is the imaginary part. We express w in polar coordinates as $w = re^{j\theta}$, where $r = |w|$ is the magnitude of w and $\theta = \arg\{w\}$ is the phase of w. The rectangular and polar representations of the number w are depicted in the complex plane of Fig. A.2.

■ A.2.1 CONVERTING FROM RECTANGULAR TO POLAR COORDINATES

$$r = \sqrt{x^2 + y^2}$$
$$\theta = \arctan\left(\frac{y}{x}\right)$$

■ A.2.2 CONVERTING FROM POLAR TO RECTANGULAR COORDINATES

$$x = r\cos\theta$$
$$y - r\sin\theta$$

■ A.2.3 COMPLEX CONJUGATE

If $w = x + jy = re^{j\theta}$, then, using the asterisk to denote complex conjugation, we have

$$w^* = x - jy = re^{-j\theta}$$

■ A.2.4 EULER'S FORMULA

$$e^{j\theta} = \cos\theta + j\sin\theta$$

■ A.2.5 OTHER IDENTITIES

$$ww^* = r^2$$
$$x = \text{Re}(w) = \frac{w + w^*}{2}$$
$$y = \text{Im}(w) = \frac{w - w^*}{2j}$$
$$\cos\theta = \frac{e^{j\theta} + e^{-j\theta}}{2}$$
$$\sin\theta = \frac{e^{j\theta} - e^{-j\theta}}{2j}$$

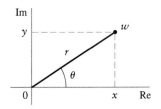

FIGURE A.2   The complex plane.

A.3 Geometric Series

If β is a complex number, then the following relationships hold:

$$\sum_{n=0}^{M-1} \beta^n = \begin{cases} \dfrac{1-\beta^M}{1-\beta}, & \beta \neq 1 \\ M, & \beta = 1 \end{cases}$$

$$\sum_{n=k}^{l} \beta^n = \begin{cases} \dfrac{\beta^k - \beta^{l+1}}{1-\beta}, & \beta \neq 1 \\ l-k+1, & \beta = 1 \end{cases}$$

$$\sum_{n=0}^{\infty} \beta^n = \frac{1}{1-\beta}, \quad |\beta| < 1$$

$$\sum_{n=k}^{\infty} \beta^n = \frac{\beta^k}{1-\beta}, \quad |\beta| < 1$$

$$\sum_{n=-k}^{-\infty} \beta^n = \beta^{-k}\left(\frac{\beta}{\beta-1}\right), \quad |\beta| > 1$$

$$\sum_{n=0}^{\infty} n\beta^n = \frac{\beta}{(1-\beta)^2}, \quad |\beta| < 1$$

A.4 Definite Integrals

$$\int_a^b x^n \, dx = \left. \frac{1}{n+1} x^{n+1} \right|_a^b, \qquad n \neq -1$$

$$\int_a^b e^{cx} \, dx = \left. \frac{1}{c} e^{cx} \right|_a^b$$

$$\int_a^b x e^{cx} \, dx = \left. \frac{1}{c^2} e^{cx}(cx-1) \right|_a^b$$

$$\int_a^b \cos(cx) \, dx = \left. \frac{1}{c} \sin(cx) \right|_a^b$$

$$\int_a^b \sin(cx) \, dx = \left. -\frac{1}{c} \cos(cx) \right|_a^b$$

$$\int_a^b x \cos(cx) \, dx = \left. \frac{1}{c^2}(\cos(cx) + cx\sin(cx)) \right|_a^b$$

$$\int_a^b x \sin(cx) \, dx = \left. \frac{1}{c^2}(\sin(cx) - cx\cos(cx)) \right|_a^b$$

$$\int_a^b e^{gx} \cos(cx) \, dx = \left. \frac{e^{gx}}{g^2+c^2}(g\cos(cx) + c\sin(cx)) \right|_a^b$$

$$\int_a^b e^{gx} \sin(cx) \, dx = \left. \frac{e^{gx}}{g^2+c^2}(g\sin(cx) - c\cos(cx)) \right|_a^b$$

■ A.4.1 GAUSSIAN PULSES

$$\int_{-\infty}^{\infty} e^{-x^2/2\sigma^2}\, dx = \sigma\sqrt{2\pi}, \qquad \sigma > 0$$

$$\int_{-\infty}^{\infty} x^2 e^{-x^2/2\sigma^2}\, dx = \sigma^3\sqrt{2\pi}, \qquad \sigma > 0$$

■ A.4.2 INTEGRATION BY PARTS

$$\int_a^b u(x)\, dv(x) = u(x)v(x)\big|_a^b - \int_a^b v(x)\, du(x)$$

A.5 Matrices

A *matrix* is a set of numbers arranged in a rectangular array. For example,

$$\mathbf{A} = \begin{bmatrix} 2 & 3 \\ -1 & 4 \end{bmatrix}$$

is a matrix with two columns and two rows. We thus say that $\mathbf{A}$ is a two by two matrix. The first and second rows of $\mathbf{A}$ are given by $[2 \quad 3]$ and $[-1 \quad 4]$, respectively. We index the elements of the matrix in terms of their location, which is measured by the row and column in which the element lies. For example, the element in the first row and second column of $\mathbf{A}$ is 3. We refer to a matrix with N rows and M columns as either an N by M matrix or an $N \times M$ matrix. Boldface uppercase symbols are used to denote matrix quantities in this text.

A *vector* is a matrix containing a single column or a single row. A *column vector* is an N by 1 matrix, that is, a single column. For example,

$$\mathbf{b} = \begin{bmatrix} 3 \\ -2 \end{bmatrix}$$

is a two-dimensional column vector. A *row vector* is a M by 1 matrix, that is, a single row. For example,

$$\mathbf{c} = [2 \quad -1]$$

is a two-dimensional row vector. Vectors are denoted with lowercase boldface symbols.

■ A.5.1 ADDITION

If a_{ij} and b_{ij} are the elements in the ith row and jth column of matrices $\mathbf{A}$ and $\mathbf{B}$, respectively, then the matrix $\mathbf{C} = \mathbf{A} + \mathbf{B}$ has elements $c_{ij} = a_{ij} + b_{ij}$.

■ A.5.2 MULTIPLICATION

If a_{ik} is the element in the ith row and kth column of an M-by-N matrix $\mathbf{A}$ and b_{kj} is the element in the kth row and jth column of an N-by-L matrix $\mathbf{B}$, then the M-by-L matrix $\mathbf{C} = \mathbf{AB}$ has elements $c_{ij} = \sum_{k=1}^{N} a_{ik} b_{kj}$.

■ A.5.3 INVERSION

The inverse of an N by N matrix $\mathbf{A}$ is denoted as $\mathbf{A}^{-1}$ and satisfies $\mathbf{A}\mathbf{A}^{-1} = \mathbf{A}^{-1}\mathbf{A} = \mathbf{I}$ where $\mathbf{I}$ is the N by N identity matrix containing unity entries on the diagonal and zeros elsewhere.

Inverse of Two-by-Two Matrix

$$\begin{bmatrix} a & b \\ c & d \end{bmatrix}^{-1} = \frac{1}{ad - bc}\begin{bmatrix} d & -b \\ -c & a \end{bmatrix}$$

Inverse of Product of Matrices

If $\mathbf{A}$ and $\mathbf{B}$ are invertible, then

$$(\mathbf{AB})^{-1} = \mathbf{B}^{-1}\mathbf{A}^{-1}$$

B | Partial-Fraction Expansions

Partial-fraction expansions are used to express a ratio of polynomials as a sum of ratios of lower order polynomials. In essence, a partial-fraction expansion is the inverse of placing a sum of fractions over a common denominator. Partial-fraction expansions are used in analyzing signals and systems to determine inverse Fourier, Laplace, and z-transforms. In that context, we use a partial-fraction expansion to express an arbitrary ratio of polynomials as a sum of terms for which the inverse transform is known.

There are two different standard forms for ratios of polynomials that occur frequently in the study of signals and systems. One arises in the context of representing continuous-time signals and systems, while the other arises in the context of representing discrete-time signals and systems. We shall treat these separately, since the method for performing the partial-fraction expansion differs slightly in each case.

B.1 *Partial-Fraction Expansions of Continuous-Time Representations*

In the study of continuous-time signals and systems, we generally encounter ratios of polynomials of the form

$$W(u) = \frac{B(u)}{A(u)}$$

$$= \frac{b_M u^M + b_{M-1} u^{M-1} + \cdots + b_1 u + b_0}{u^N + a_{N-1} u^{N-1} + \cdots + a_1 u + a_0}. \tag{B.1}$$

We employ the symbol u as a generic variable in this appendix; it should not be confused with the unit step function notation used elsewhere. In a Fourier transform problem, the variable u represents $j\omega$, and in a Laplace transform problem, u represents the complex variable s. Note that the coefficient of u^N in $A(u)$ is unity. We assume that $W(u)$ is a proper rational function; that is, the order of $B(u)$ is less than that of $A(u)$ ($M < N$). If this condition is not satisfied, then long division of $A(u)$ into $B(u)$ is used to write $W(u)$ as the sum of a polynomial in u and a proper rational function representing the remainder of the division. The partial-fraction expansion is then applied to the remainder.

The first step in performing a partial-fraction expansion is to factor the denominator polynomial. If the N roots d_i are distinct, then we may rewrite $W(u)$ as

$$W(u) = \frac{B(u)}{(u - d_1)(u - d_2) \cdots (u - d_N)}$$

In this case, the partial-fraction expansion of $W(u)$ takes the form

$$W(u) = \frac{C_1}{u - d_1} + \frac{C_2}{u - d_2} + \cdots + \frac{C_N}{u - d_N}. \tag{B.2}$$

If a root $u = r$ occurs with multiplicity L, then

$$W(u) = \frac{B(u)}{(u - r)^L (u - d_1)(u - d_2) \cdots (u - d_{N-L})},$$

and the partial-fraction expansion of $W(u)$ is

$$W(u) = \frac{C_1}{u - d_1} + \frac{C_2}{u - d_2} + \cdots + \frac{C_{N-L}}{u - d_{N-L}} + \frac{K_{L-1}}{u - r}$$
$$+ \frac{K_{L-2}}{(u - r)^2} + \cdots + \frac{K_0}{(u - r)^L}. \tag{B.3}$$

Note that as the power to which the denominator terms $(u - r)$ are raised increases, the indices i of the corresponding coefficients K_i decrease.

The constants C_i and K_i are called *residues*. We may obtain the residues by using either of two different approaches. In the method of linear equations, we place all the terms in the partial-fraction expansion of $W(u)$ over a common denominator and equate the coefficient of each power of u to the corresponding coefficient in $B(u)$. This gives a system of N linear equations that may be solved to obtain the residues, as illustrated in the next example. (For hand calculations, this approach is generally limited to $N = 2$ or $N = 3$.)

EXAMPLE B.1 OBTAINING RESIDUES BY SOLVING LINEAR EQUATIONS   Determine the partial-fraction expansion of the function

$$W(u) = \frac{3u + 5}{u^3 + 4u^2 + 5u + 2}.$$

Solution: The roots of the denominator polynomial are $u = -2$ and $u = -1$, the latter with multiplicity two. Hence, the partial-fraction expansion of $W(u)$ is of the form

$$W(u) = \frac{K_1}{u + 1} + \frac{K_0}{(u + 1)^2} + \frac{C_1}{u + 2}.$$

The residues $K_1, K_0,$ and C_1 may be determined by placing the terms in the partial-fraction expansion over a common denominator:

$$W(u) = \frac{K_1(u + 1)(u + 2)}{(u + 1)^2(u + 2)} + \frac{K_0(u + 2)}{(u + 1)^2(u + 2)} + \frac{C_1(u + 1)^2}{(u + 1)^2(u + 2)}$$
$$= \frac{(K_1 + C_1)u^2 + (3K_1 + K_0 + 2C_1)u + (2K_1 + 2K_0 + C_1)}{u^3 + 4u^2 + 5u + 2}.$$

Equating the coefficient of each power of u in the numerator on the right-hand side of this equation to those in $B(u)$ gives the system of three equations in the three unknowns K_1, K_0, and C_1:

$$0 = K_1 + C_1;$$
$$3 = 3K_1 + K_0 + 2C_1;$$
$$5 = 2K_1 + 2K_0 + C_1.$$

Solving these equations, we obtain $K_1 = 1$, $K_0 = 2$, and $C_1 = -1$, so the partial-fraction expansion of $W(u)$ is given by

$$W(u) = \frac{1}{u+1} + \frac{2}{(u+1)^2} - \frac{1}{u+2}.$$

∎

The *method of residues* is based on manipulating the partial-fraction expansion so as to isolate each residue. Hence, this method is usually easier to use than solving linear equations. To apply the method, we multiply each side of Eq. (B.3) by $(u - d_i)$:

$$(u - d_i)W(u) = \frac{C_1(u - d_i)}{u - d_1} + \frac{C_2(u - d_i)}{u - d_2} + \cdots + C_i + \cdots + \frac{C_{N-L}(u - d_i)}{u - d_{N-L}}$$

$$+ \frac{K_{L-1}(u - d_i)}{u - r} + \frac{K_{L-2}(u - d_i)}{(u - r)^2} + \cdots + \frac{K_0(u - d_i)}{(u - r)^L}.$$

On the left-hand side, the multiplication by $(u - d_i)$ cancels the $(u - d_i)$ term in the denominator of $W(u)$. If we now evaluate the resulting expression at $u = d_i$, then all the terms on the right-hand side are zero except for C_i, and we obtain the expression

$$C_i = (u - d_i)W(u)\big|_{u=d_i}. \tag{B.4}$$

Isolation of the residues associated with the repeated root $u = r$ requires multiplying both sides of Eq. (B.3) by $(u - r)^L$ and differentiating. We have

$$K_i = \frac{1}{i!}\frac{d^i}{du^i}\{(u - r)^L W(u)\}\bigg|_{u=r}. \tag{B.5}$$

The next example uses Eqs. (B.4) and (B.5) to obtain the residues.

EXAMPLE B.2 REPEATED ROOTS   Find the partial-fraction expansion of

$$W(u) = \frac{3u^3 + 15u^2 + 29u + 21}{(u+1)^2(u+2)(u+3)}.$$

Solution: Here, we have a root of multiplicity two at $u = -1$ and distinct roots at $u = -2$ and $u = -3$. Hence, the partial-fraction expansion for $W(u)$ is of the form

$$W(u) = \frac{K_1}{u+1} + \frac{K_0}{(u+1)^2} + \frac{C_1}{u+2} + \frac{C_2}{u+3}.$$

We obtain C_1 and C_2 using Eq. (B.4):

$$C_1 = (u+2)\frac{3u^3 + 15u^2 + 29u + 21}{(u+1)^2(u+2)(u+3)}\bigg|_{u=-2}$$

$$= -1;$$

$$C_2 = (u+3)\frac{3u^3 + 15u^2 + 29u + 21}{(u+1)^2(u+2)(u+3)}\bigg|_{u=-3}$$

$$= 3.$$

Now we may obtain K_1 and K_0, using Eq. (B.5):

$$
K_0 = (u + 1)^2 \frac{3u^3 + 15u^2 + 29u + 21}{(u + 1)^2(u + 2)(u + 3)} \bigg|_{u=-1}
$$

$$
= 2;
$$

$$
K_1 = \frac{1}{1!} \frac{d}{du} \left\{ (u + 1)^2 \frac{3u^3 + 15u^2 + 29u + 21}{(u + 1)^2(u + 2)(u + 3)} \right\} \bigg|_{u=-1}
$$

$$
= \frac{(9u^2 + 30u + 29)(u^2 + 5u + 6) - (3u^3 + 15u^2 + 29u + 21)(2u + 5)}{(u^2 + 5u + 6)^2} \bigg|_{u=-1}
$$

$$
= 1.
$$

Hence, the partial-fraction expansion of $W(u)$ is

$$
W(u) = \frac{1}{u + 1} + \frac{2}{(u + 1)^2} - \frac{1}{u + 2} + \frac{3}{u + 3}
$$

  ■

From Eqs. (B.4) and (B.5), we may draw the following conclusions about the residues assuming the coefficients of the numerator and denominator polynomials in $W(u)$ are real valued:

▸ The residue associated with a real root is real.

▸ The residues associated with a pair of complex-conjugate roots are the complex conjugate of each other; thus, only one residue needs to be computed.

B.2 *Partial-Fraction Expansions of Discrete-Time Representation*

In the study of discrete-time signals and systems, we frequently encounter ratios of polynomials having the form

$$
\begin{aligned}
W(u) &= \frac{B(u)}{A(u)} \\
&= \frac{b_M u^M + b_{M-1} u^{M-1} + \cdots + b_1 u + b_0}{a_N u^N + a_{N-1} u^{N-1} + \cdots + a_1 u + 1}.
\end{aligned} \tag{B.6}
$$

In a discrete-time Fourier transform problem, the variable u represents $e^{-j\Omega}$, while in a z-transform problem, u represents z^{-1}. Note that the coefficient of the zeroth power of u in $A(u)$ is unity here. We again assume that $W(u)$ is a proper rational function; that is, the order of $B(u)$ is less than that of $A(u)$ ($M < N$). If this condition is not satisfied, then long division of $A(u)$ into $B(u)$ is used to write $W(u)$ as the sum of a polynomial in u and a proper rational function representing the remainder of the division. The partial-fraction expansion is then applied to the remainder.

Here, we write the denominator polynomial as a product of first-order terms, namely,

$$
A(u) = (1 - d_1 u)(1 - d_2 u) \cdots (1 - d_N u), \tag{B.7}
$$

where d_i^{-1} is a root of $A(u)$. Equivalently, d_i is a root of the polynomial $\widetilde{A}(u)$ constructed by reversing the order of coefficients in $A(u)$. That is, d_i is a root of

$$\widetilde{A}(u) = u^N + a_1 u^{N-1} + \cdots + a_{N-1} u + a_N.$$

If all the d_i are distinct, then the partial-fraction expansion is given by

$$W(u) = \frac{C_1}{1 - d_1 u} + \frac{C_2}{1 - d_2 u} + \cdots + \frac{C_N}{1 - d_N u}. \tag{B.8}$$

If a term $1 - ru$ occurs with multiplicity L in Eq. (B.7), then the partial-fraction expansion has the form

$$\begin{aligned} W(u) = & \frac{C_1}{1 - d_1 u} + \frac{C_2}{1 - d_2 u} + \cdots + \frac{C_{N-L}}{1 - d_{N-L} u} \\ & + \frac{K_{L-1}}{(1 - ru)} + \frac{K_{L-2}}{(1 - ru)^2} + \cdots + \frac{K_0}{(1 - ru)^L}. \end{aligned} \tag{B.9}$$

The residues C_i and K_i may be determined analogously to the continuous-time case. We may place the right-hand side of Eq. (B.8) or Eq. (B.9) over a common denominator and obtain a system of N linear equations by equating coefficients of like powers of u in the numerator polynomials. Alternatively, we may solve for the residues directly by manipulating the partial-fraction expansion in such a way as to isolate each coefficient. This yields the following two relationships:

$$C_i = (1 - d_i u) W(u)\big|_{u = d_i^{-1}}; \tag{B.10}$$

$$K_i = \frac{1}{i!} (-r^{-1})^i \frac{d^i}{du^i} \big\{ (1 - ru)^L W(u) \big\} \Big|_{u = r^{-1}}. \tag{B.11}$$

EXAMPLE B.3 DISCRETE-TIME PARTIAL-FRACTION EXPANSION   Find the partial-fraction expansion of the discrete-time function

$$W(u) = \frac{-14u - 4}{8u^3 - 6u - 2}.$$

Solution: The constant term is not unity in the denominator, so we first divide the denominator and numerator by -2 to express $W(u)$ in standard form. We have

$$W(u) = \frac{7u + 2}{-4u^3 + 3u + 1}.$$

The denominator polynomial $A(u)$ is factored by finding the roots of the related polynomial

$$\widetilde{A}(u) = u^3 + 3u^2 - 4.$$

This polynomial has a single root at $u = 1$ and a root of multiplicity two at $u = -2$. Hence, $W(u)$ can be expressed as

$$W(u) = \frac{7u + 2}{(1 - u)(1 + 2u)^2},$$

and the partial-fraction expansion has the form

$$W(u) = \frac{C_1}{1 - u} + \frac{K_1}{1 + 2u} + \frac{K_0}{(1 + 2u)^2}.$$

The residues are evaluated by using Eqs. (B.10) and (B.11) as follows:

$$C_1 = (1 - u)\,W(u)\big|_{u=1}$$
$$= 1;$$
$$K_0 = (1 + 2u)^2\,W(u)\big|_{u=-1/2}$$
$$= -1;$$
$$K_1 = \frac{1}{1!}\left(\frac{1}{2}\right)\frac{d}{du}\{(1 + 2u)^2\,W(u)\}\bigg|_{u=-1/2}$$
$$= \frac{7(1 - u) + (7u + 2)}{2(1 - u)^2}\bigg|_{u=-1/2}$$
$$= 2.$$

We conclude that the partial-fraction expansion is

$$W(u) = \frac{1}{1 - u} + \frac{2}{1 + 2u} - \frac{1}{(1 + 2u)^2}.$$

■

C Tables of Fourier Representat and Properties

C.1 Basic Discrete-Time Fourier Series Pairs

Time Domain	Frequency Domain
$$x[n] = \sum_{k=0}^{N-1} X[k]e^{ikn\Omega_o}$$ $$Period = N$$	$$X[k] = \frac{1}{N}\sum_{n=0}^{N-1} x[n]e^{-ikn\Omega_o}$$ $$\Omega_o = \frac{2\pi}{N}$$
$$x[n] = \begin{cases} 1, & |n| \le M \\ 0, & M < |n| \le N/2 \end{cases}$$ $$x[n] = x[n+N]$$	$$X[k] = \frac{\sin\left(k\dfrac{\Omega_o}{2}(2M+1)\right)}{N\sin\left(k\dfrac{\Omega_o}{2}\right)}$$
$$x[n] = e^{ip\Omega_o n}$$	$$X[k] = \begin{cases} 1, & k = p, p \pm N, p \pm 2N, \ldots \\ 0, & \text{otherwise} \end{cases}$$
$$x[n] = \cos(p\Omega_o n)$$	$$X[k] = \begin{cases} \frac{1}{2}, & k = \pm p, \pm p \pm N, \pm p \pm 2N, \ldots \\ 0, & \text{otherwise} \end{cases}$$
$$x[n] = \sin(p\Omega_o n)$$	$$X[k] = \begin{cases} \dfrac{1}{2j}, & k = p, p \pm N, p \pm 2N, \ldots \\ \dfrac{-1}{2j}, & k = -p, -p \pm N, -p \pm 2N, \ldots \\ 0, & \text{otherwise} \end{cases}$$
$$x[n] = 1$$	$$X[k] = \begin{cases} 1, & k = 0, \pm N, \pm 2N, \ldots \\ 0, & \text{otherwise} \end{cases}$$
$$x[n] = \sum_{p=-\infty}^{\infty} \delta[n - pN]$$	$$X[k] = \frac{1}{N}$$

2 Basic Fourier Series Pairs

Time Domain	Frequency Domain
$x(t) = \sum_{k=-\infty}^{\infty} X[k]e^{jk\omega_o t}$ $Period = T$	$X[k] = \dfrac{1}{T}\displaystyle\int_0^T x(t)e^{-jk\omega_o t}\,dt$ $\omega_o = \dfrac{2\pi}{T}$
$x(t) = \begin{cases} 1, & \|t\| \le T_o \\ 0, & T_o < \|t\| \le T/2 \end{cases}$	$X[k] = \dfrac{\sin(k\omega_o T_o)}{k\pi}$
$x(t) = e^{jp\omega_o t}$	$X[k] = \delta[k - p]$
$x(t) = \cos(p\omega_o t)$	$X[k] = \tfrac{1}{2}\delta[k - p] + \tfrac{1}{2}\delta[k + p]$
$x(t) = \sin(p\omega_o t)$	$X[k] = \dfrac{1}{2j}\delta[k - p] - \dfrac{1}{2j}\delta[k + p]$
$x(t) = \sum_{p=-\infty}^{\infty}\delta(t - pT)$	$X[k] = \dfrac{1}{T}$

C.3 Basic Discrete-Time Fourier Transform Pairs

Time Domain	Frequency Domain
$x[n] = \dfrac{1}{2\pi}\displaystyle\int_{-\pi}^{\pi} X(e^{j\Omega})e^{j\Omega n}\,d\Omega$	$X(e^{j\Omega}) = \sum_{n=-\infty}^{\infty} x[n]e^{-j\Omega n}$
$x[n] = \begin{cases} 1, & \|n\| \le M \\ 0, & \text{otherwise} \end{cases}$	$X(e^{j\Omega}) = \dfrac{\sin\left[\Omega\left(\dfrac{2M+1}{2}\right)\right]}{\sin\left(\dfrac{\Omega}{2}\right)}$
$x[n] = \alpha^n u[n], \quad \|\alpha\| < 1$	$X(e^{j\Omega}) = \dfrac{1}{1 - \alpha e^{-j\Omega}}$
$x[n] = \delta[n]$	$X(e^{j\Omega}) = 1$
$x[n] = u[n]$	$X(e^{j\Omega}) = \dfrac{1}{1 - e^{-j\Omega}} + \pi \sum_{p=-\infty}^{\infty}\delta(\Omega - 2\pi p)$
$x[n] = \dfrac{1}{\pi n}\sin(Wn), \quad 0 < W \le \pi$	$X(e^{j\Omega}) = \begin{cases} 1, & \|\Omega\| \le W \\ 0, & W < \|\Omega\| \le \pi \end{cases} \quad X(e^{j\Omega})\text{ is }2\pi\text{ periodic}$
$x[n] = (n + 1)\alpha^n u[n]$	$X(e^{j\Omega}) = \dfrac{1}{(1 - \alpha e^{-j\Omega})^2}$

C.4 *Basic Fourier Transform Pairs*

Time Domain	Frequency Domain
$x(t) = \dfrac{1}{2\pi} \displaystyle\int_{-\infty}^{\infty} X(j\omega)e^{j\omega t}\, d\omega$	$X(j\omega) = \displaystyle\int_{-\infty}^{\infty} x(t)e^{-j\omega t}\, dt$
$x(t) = \begin{cases} 1, & \lvert t \rvert \le T_o \\ 0, & \text{otherwise} \end{cases}$	$X(j\omega) = \dfrac{2\sin(\omega T_o)}{\omega}$
$x(t) = \dfrac{1}{\pi t}\sin(Wt)$	$X(j\omega) = \begin{cases} 1, & \lvert \omega \rvert \le W \\ 0, & \text{otherwise} \end{cases}$
$x(t) = \delta(t)$	$X(j\omega) = 1$
$x(t) = 1$	$X(j\omega) = 2\pi\delta(\omega)$
$x(t) = u(t)$	$X(j\omega) = \dfrac{1}{j\omega} + \pi\delta(\omega)$
$x(t) = e^{-at}u(t), \quad \text{Re}\{a\} > 0$	$X(j\omega) = \dfrac{1}{a + j\omega}$
$x(t) = te^{-at}u(t), \quad \text{Re}\{a\} > 0$	$X(j\omega) = \dfrac{1}{(a + j\omega)^2}$
$x(t) = e^{-a\lvert t \rvert}, \quad a > 0$	$X(j\omega) = \dfrac{2a}{a^2 + \omega^2}$
$x(t) = \dfrac{1}{\sqrt{2\pi}}e^{-t^2/2}$	$X(j\omega) = e^{-\omega^2/2}$

C.5 *Fourier Transform Pairs for Periodic Signals*

Periodic Time-Domain Signal	Fourier Transform
$x(t) = \displaystyle\sum_{k=-\infty}^{\infty} X[k]e^{jk\omega_o t}$	$X(j\omega) = 2\pi \displaystyle\sum_{k=-\infty}^{\infty} X[k]\delta(\omega - k\omega_o)$
$x(t) = \cos(\omega_o t)$	$X(j\omega) = \pi\delta(\omega - \omega_o) + \pi\delta(\omega + \omega_o)$
$x(t) = \sin(\omega_o t)$	$X(j\omega) = \dfrac{\pi}{j}\delta(\omega - \omega_o) - \dfrac{\pi}{j}\delta(\omega + \omega_o)$
$x(t) = e^{j\omega_o t}$	$X(j\omega) = 2\pi\delta(\omega - \omega_o)$
$x(t) = \sum_{n=-\infty}^{\infty}\delta(t - nT_s)$	$X(j\omega) = \dfrac{2\pi}{T_s} \displaystyle\sum_{k=-\infty}^{\infty}\delta\!\left(\omega - k\dfrac{2\pi}{T_s}\right)$
$x(t) = \begin{cases} 1, & \lvert t \rvert \le T_o \\ 0, & T_o < \lvert t \rvert < T/2 \end{cases}$ $x(t + T) = x(t)$	$X(j\omega) = \displaystyle\sum_{k=-\infty}^{\infty} \dfrac{2\sin(k\omega_o T_o)}{k}\delta(\omega - k\omega_o)$

C.6 Discrete-Time Fourier Transform Pairs for Periodic Signals

Periodic Time-Domain Signal	Discrete-Time Fourier Transform
$x[n] = \sum\limits_{k=0}^{N-1} X[k]e^{jk\Omega_o n}$	$X(e^{j\Omega}) = 2\pi \sum\limits_{k=-\infty}^{\infty} X[k]\delta(\Omega - k\Omega_o)$
$x[n] = \cos(\Omega_1 n)$	$X(e^{j\Omega}) = \pi \sum\limits_{k=-\infty}^{\infty} \delta(\Omega - \Omega_1 - k2\pi) + \delta(\Omega + \Omega_1 - k2\pi)$
$x[n] = \sin(\Omega_1 n)$	$X(e^{j\Omega}) = \dfrac{\pi}{j} \sum\limits_{k=-\infty}^{\infty} \delta(\Omega - \Omega_1 - k2\pi) - \delta(\Omega + \Omega_1 - k2\pi)$
$x[n] = e^{j\Omega_1 n}$	$X(e^{j\Omega}) = 2\pi \sum\limits_{k=-\infty}^{\infty} \delta(\Omega - \Omega_1 - k2\pi)$
$x[n] = \sum\limits_{k=-\infty}^{\infty} \delta(n - kN)$	$X(e^{j\Omega}) = \dfrac{2\pi}{N} \sum\limits_{k=-\infty}^{\infty} \delta\left(\Omega - \dfrac{k2\pi}{N}\right)$

C.7 *Properties of Fourier Representations*

Property	Fourier Transform $x(t) \xleftrightarrow{FT} X(j\omega)$ $y(t) \xleftrightarrow{FT} Y(j\omega)$	Fourier Series $x(t) \xleftrightarrow{FS;\omega_o} X[k]$ $y(t) \xleftrightarrow{FS;\omega_o} Y[k]$ Period $= T$
Linearity	$ax(t) + by(t) \xleftrightarrow{FT} aX(j\omega) + bY(j\omega)$	$ax(t) + by(t) \xleftrightarrow{FS;\omega_o} aX[k] + bY[k]$
Time shift	$x(t - t_o) \xleftrightarrow{FT} e^{-j\omega t_o}X(j\omega)$	$x(t - t_o) \xleftrightarrow{FS;\omega_o} e^{-jk\omega_o t_o}X[k]$
Frequency shift	$e^{j\gamma t}x(t) \xleftrightarrow{FT} X(j(\omega - \gamma))$	$e^{jk_o\omega_o t}x(t) \xleftrightarrow{FS;\omega_o} X[k - k_o]$
Scaling	$x(at) \xleftrightarrow{FT} \frac{1}{|a|}X\left(\frac{j\omega}{a}\right)$	$x(at) \xleftrightarrow{FS;a\omega_o} X[k]$
Differentiation in time	$\frac{d}{dt}x(t) \xleftrightarrow{FT} j\omega X(j\omega)$	$\frac{d}{dt}x(t) \xleftrightarrow{FS;\omega_o} jk\omega_o X[k]$
Differentiation in frequency	$-jtx(t) \xleftrightarrow{FT} \frac{d}{d\omega}X(j\omega)$	—
Integration/ Summation	$\int_{-\infty}^{t} x(\tau)\,d\tau \xleftrightarrow{FT} \frac{X(j\omega)}{j\omega} + \pi X(j0)\delta(\omega)$	—
Convolution	$\int_{-\infty}^{\infty} x(\tau)y(t - \tau)\,d\tau \xleftrightarrow{FT} X(j\omega)Y(j\omega)$	$\int_{0}^{T} x(\tau)y(t - \tau)\,d\tau \xleftrightarrow{FS;\omega_o} TX[k]Y[k]$
Multiplication	$x(t)y(t) \xleftrightarrow{FT} \frac{1}{2\pi}\int_{-\infty}^{\infty} X(j\nu)Y(j(\omega - \nu))\,d\nu$	$x(t)y(t) \xleftrightarrow{FS;\omega_o} \sum_{l=-\infty}^{\infty} X[l]Y[k - l]$
Parseval's Theorem	$\int_{-\infty}^{\infty} |x(t)|^2\,dt = \frac{1}{2\pi}\int_{-\infty}^{\infty} |X(j\omega)|^2\,d\omega$	$\frac{1}{T}\int_{0}^{T} |x(t)|^2\,dt = \sum_{k=-\infty}^{\infty} |X[k]|^2$
Duality	$X(jt) \xleftrightarrow{FT} 2\pi x(-\omega)$	$x[n] \xleftrightarrow{DTFT} X(e^{j\Omega})$ $X(e^{jt}) \xleftrightarrow{FS;1} x[-k]$
Symmetry	$x(t)\text{ real} \xleftrightarrow{FT} X^*(j\omega) = X(-j\omega)$ $x(t)\text{ imaginary} \xleftrightarrow{FT} X^*(j\omega) = -X(-j\omega)$ $x(t)\text{ real and even} \xleftrightarrow{FT} \text{Im}\{X(j\omega)\} = 0$ $x(t)\text{ real and odd} \xleftrightarrow{FT} \text{Re}\{X(j\omega)\} = 0$	$x(t)\text{ real} \xleftrightarrow{FS;\omega_o} X^*[k] = X[-k]$ $x(t)\text{ imaginary} \xleftrightarrow{FS;\omega_o} X^*[k] = -X[-k]$ $x(t)\text{ real and even} \xleftrightarrow{FS;\omega_o} \text{Im}\{X[k]\} = 0$ $x(t)\text{ real and odd} \xleftrightarrow{FS;\omega_o} \text{Re}\{X[k]\} = 0$

(*continues on next page*)

C.7 (continued)

Property	Discrete-Time FT $x[n] \xleftrightarrow{\;DTFT\;} X(e^{j\Omega})$ $y[n] \xleftrightarrow{\;DTFT\;} Y(e^{j\Omega})$	Discrete-Time FS $x[n] \xleftrightarrow{\;DTFS;\,\Omega_o\;} X[k]$ $y[n] \xleftrightarrow{\;DTFS;\,\Omega_o\;} Y[k]$ Period $= N$
Linearity	$ax[n] + by[n] \xleftrightarrow{\;DTFT\;} aX(e^{j\Omega}) + bY(e^{j\Omega})$	$ax[n] + by[n] \xleftrightarrow{\;DTFS;\,\Omega_o\;} aX[k] + bY[k]$
Time shift	$x[n - n_o] \xleftrightarrow{\;DTFT\;} e^{-j\Omega n_o}X(e^{j\Omega})$	$x[n - n_o] \xleftrightarrow{\;DTFS;\,\Omega_o\;} e^{-jk\Omega_o n_o}X[k]$
Frequency shift	$e^{j\Gamma n}x[n] \xleftrightarrow{\;DTFT\;} X(e^{j(\Omega-\Gamma)})$	$e^{jk_o\Omega_o n}x[n] \xleftrightarrow{\;DTFS;\,\Omega_o\;} X[k - k_o]$
Scaling	$x_z[n] = 0, \quad n \neq 0, \pm p, \pm 2p, \pm 3p, \ldots$ $x_z[pn] \xleftrightarrow{\;DTFT\;} X_z(e^{j\Omega/p})$	$x_z[n] = 0, \quad n \neq 0, \pm p, \pm 2p, \pm 3p, \ldots$ $x_z[pn] \xleftrightarrow{\;DTFS;\,p\Omega_o\;} pX_z[k]$
Differentiation in time	—	—
Differentiation in frequency	$-jnx[n] \xleftrightarrow{\;DTFT\;} \dfrac{d}{d\Omega}X(e^{j\Omega})$	—
Integration/ Summation	$\displaystyle\sum_{k=-\infty}^{n} x[k] \xleftrightarrow{\;DTFT\;} \dfrac{X(e^{j\Omega})}{1 - e^{-j\Omega}}$ $+ \pi X(e^{j0}) \displaystyle\sum_{k=-\infty}^{\infty} \delta(\Omega - k2\pi)$	—
Convolution	$\displaystyle\sum_{l=-\infty}^{\infty} x[l]y[n - l] \xleftrightarrow{\;DTFT\;} X(e^{j\Omega})Y(e^{j\Omega})$	$\displaystyle\sum_{l=0}^{N-1} x[l]y[n - l] \xleftrightarrow{\;DTFS;\,\Omega_o\;} NX[k]Y[k]$
Multiplication	$x[n]y[n] \xleftrightarrow{\;DTFT\;} \dfrac{1}{2\pi}\displaystyle\int_{-\pi}^{\pi} X(e^{j\Gamma})Y(e^{j(\Omega-\Gamma)})\,d\Gamma$	$x[n]y[n] \xleftrightarrow{\;DTFS;\,\Omega_o\;} \displaystyle\sum_{l=0}^{N-1} X[l]Y[k - l]$
Parseval's Theorem	$\displaystyle\sum_{n=-\infty}^{\infty} |x[n]|^2 = \dfrac{1}{2\pi}\displaystyle\int_{-\pi}^{\pi} |X(e^{j\Omega})|^2\,d\Omega$	$\dfrac{1}{N}\displaystyle\sum_{n=0}^{N-1} |x[n]|^2 = \displaystyle\sum_{k=0}^{N-1} |X[k]|^2$
Duality	$x[n] \xleftrightarrow{\;DTFT\;} X(e^{j\Omega})$ $X(e^{jt}) \xleftrightarrow{\;FS;\,1\;} x[-k]$	$X[n] \xleftrightarrow{\;DTFS;\,\Omega_o\;} \dfrac{1}{N}x[-k]$
Symmetry	$x[n] \text{ real} \xleftrightarrow{\;DTFT\;} X^*(e^{j\Omega}) = X(e^{-j\Omega})$ $x[n] \text{ imaginary} \xleftrightarrow{\;DTFT\;} X^*(e^{j\Omega}) = -X(e^{-j\Omega})$ $x[n] \text{ real and even} \xleftrightarrow{\;DTFT\;} \mathrm{Im}\{X(e^{j\Omega})\} = 0$ $x[n] \text{ real and odd} \xleftrightarrow{\;DTFT\;} \mathrm{Re}\{X(e^{j\Omega})\} = 0$	$x[n] \text{ real} \xleftrightarrow{\;DTFS;\,\Omega_o\;} X^*[k] = X[-k]$ $x[n] \text{ imaginary} \xleftrightarrow{\;DTFS;\,\Omega_o\;} X^*[k] = -X[-k]$ $x[n] \text{ real and even} \xleftrightarrow{\;DTFS;\,\Omega_o\;} \mathrm{Im}\{X[k]\} = 0$ $x[n] \text{ real and odd} \xleftrightarrow{\;DTFS;\,\Omega_o\;} \mathrm{Re}\{X[k]\} = 0$

C.8 *Relating the Four Fourier Representations*

Let

$$g(t) \xleftrightarrow{\quad FS;\, \omega_o = 2\pi/T \quad} G[k]$$

$$v[n] \xleftrightarrow{\quad DTFT \quad} V(e^{j\Omega})$$

$$w[n] \xleftrightarrow{\quad DTFS;\, \Omega_o = 2\pi/N \quad} W[k]$$

■ C.8.1 FT REPRESENTATION FOR A CONTINUOUS-TIME PERIODIC SIGNAL

$$g(t) \xleftrightarrow{\quad FT \quad} G(j\omega) = 2\pi \sum_{k=-\infty}^{\infty} G[k]\delta(\omega - k\omega_o)$$

■ C.8.2 DTFT REPRESENTATION FOR A DISCRETE-TIME PERIODIC-SIGNAL

$$w[n] \xleftrightarrow{\quad DTFT \quad} W(e^{j\Omega}) = 2\pi \sum_{k=-\infty}^{\infty} W[k]\delta(\Omega - k\Omega_o)$$

■ C.8.3 FT REPRESENTATION FOR A DISCRETE-TIME NONPERIODIC SIGNAL

$$v_\delta(t) = \sum_{n=-\infty}^{\infty} v[n]\delta(t - nT_s) \xleftrightarrow{\quad FT \quad} V_\delta(j\omega) = V(e^{j\Omega})\Big|_{\Omega = \omega T_s}$$

■ C.8.4 FT REPRESENTATION FOR A DISCRETE-TIME NONPERIODIC SIGNAL

$$w_\delta(t) = \sum_{n=-\infty}^{\infty} w[n]\delta(t - nT_s) \xleftrightarrow{\quad FT \quad} W_\delta(j\omega) = \frac{2\pi}{T_s} \sum_{k=-\infty}^{\infty} W[k]\delta\left(\omega - \frac{k\Omega_o}{T_s}\right)$$

C.9 *Sampling and Aliasing Relationships*

Let

$$x(t) \xleftrightarrow{\quad FT \quad} X(j\omega)$$

$$v[n] \xleftrightarrow{\quad DTFT \quad} V(e^{j\Omega})$$

■ C.9.1 IMPULSE SAMPLING FOR CONTINUOUS-TIME SIGNALS

$$x_\delta(t) = \sum_{n=-\infty}^{\infty} x(nT_s)\delta(t - nT_s) \xleftrightarrow{\quad FT \quad} X_\delta(j\omega) = \frac{1}{T_s} \sum_{k=-\infty}^{\infty} X\left(j\left(\omega - k\frac{2\pi}{T_s}\right)\right)$$

Sampling interval T_s, $X_\delta(j\omega)$ is $2\pi/T_s$ periodic.

■ C.9.2 SAMPLING A DISCRETE-TIME SIGNAL

$$y[n] = v[qn] \xleftrightarrow{\quad DTFT \quad} Y(e^{j\Omega}) = \frac{1}{q} \sum_{m=0}^{q-1} V(e^{j(\Omega - m2\pi)/q})$$

$Y(e^{j\Omega})$ is 2π periodic.

■ C.9.3 SAMPLING THE DTFT IN FREQUENCY

$$w[n] = \sum_{m=-\infty}^{\infty} v[n + mN] \xleftrightarrow{\quad DTFS; \Omega_o = 2\pi/N \quad} W[k] = \frac{1}{N} V(e^{jk\Omega_o})$$

$w[n]$ is N periodic.

■ C.9.4 SAMPLING THE FT IN FREQUENCY

$$g(t) = \sum_{m=-\infty}^{\infty} x(t + mT) \xleftrightarrow{\quad FS; \omega_o = 2\pi/T \quad} G[k] = \frac{1}{T} X(jk\omega_o)$$

$g(t)$ is T periodic.

D Tables of Laplace Transforms and Properties

D.1 Basic Laplace Transforms

Signal $$x(t) = \frac{1}{2\pi j} \int_{\sigma-j\infty}^{\sigma+j\infty} X(s)e^{st}\,ds$$	Transform $$X(s) = \int_{-\infty}^{\infty} x(t)e^{-st}\,dt$$	ROC
$u(t)$	$\dfrac{1}{s}$	$\mathrm{Re}\{s\} > 0$
$tu(t)$	$\dfrac{1}{s^2}$	$\mathrm{Re}\{s\} > 0$
$\delta(t-\tau), \quad \tau \geq 0$	$e^{-s\tau}$	for all s
$e^{-at}u(t)$	$\dfrac{1}{s+a}$	$\mathrm{Re}\{s\} > -a$
$te^{-at}u(t)$	$\dfrac{1}{(s+a)^2}$	$\mathrm{Re}\{s\} > -a$
$[\cos(\omega_1 t)]u(t)$	$\dfrac{s}{s^2+\omega_1^2}$	$\mathrm{Re}\{s\} > 0$
$[\sin(\omega_1 t)]u(t)$	$\dfrac{\omega_1}{s^2+\omega_1^2}$	$\mathrm{Re}\{s\} > 0$
$[e^{-at}\cos(\omega_1 t)]u(t)$	$\dfrac{s+a}{(s+a)^2+\omega_1^2}$	$\mathrm{Re}\{s\} > -a$
$[e^{-at}\sin(\omega_1 t)]u(t)$	$\dfrac{\omega_1}{(s+a)^2+\omega_1^2}$	$\mathrm{Re}\{s\} > -a$

■ D.1.1 BILATERAL LAPLACE TRANSFORMS FOR SIGNALS THAT ARE NONZERO FOR $t < 0$

Signal	Bilateral Transform	ROC
$\delta(t-\tau), \tau < 0$	$e^{-s\tau}$	for all s
$-u(-t)$	$\dfrac{1}{s}$	$\mathrm{Re}\{s\} < 0$
$-tu(-t)$	$\dfrac{1}{s^2}$	$\mathrm{Re}\{s\} < 0$
$-e^{-at}u(-t)$	$\dfrac{1}{s+a}$	$\mathrm{Re}\{s\} < -a$
$-te^{-at}u(-t)$	$\dfrac{1}{(s+a)^2}$	$\mathrm{Re}\{s\} < -a$

D.2 Laplace Transform Properties

Signal	Unilateral Transform $x(t) \xleftrightarrow{\mathcal{L}_u} X(s)$ $y(t) \xleftrightarrow{\mathcal{L}_u} Y(s)$	Bilateral Transform $x(t) \xleftrightarrow{\mathcal{L}} X(s)$ $y(t) \xleftrightarrow{\mathcal{L}} Y(s)$	ROC $s \in R_x$ $s \in R_y$
$ax(t)+by(t)$	$aX(s)+bY(s)$	$aX(s)+bY(s)$	At least $R_x \cap R_y$
$x(t-\tau)$	$e^{-s\tau}X(s)$ if $x(t-\tau)u(t)=x(t-\tau)u(t-\tau)$	$e^{-s\tau}X(s)$	R_x
$e^{s_o t}x(t)$	$X(s-s_o)$	$X(s-s_o)$	$R_x + \mathrm{Re}\{s_o\}$
$x(at)$	$\dfrac{1}{a}X\left(\dfrac{s}{a}\right),\ a>0$	$\dfrac{1}{|a|}X\left(\dfrac{s}{a}\right)$	$\dfrac{R_x}{|a|}$
$x(t)*y(t)$	$X(s)Y(s)$ if $x(t)=y(t)=0$ for $t<0$	$X(s)Y(s)$	At least $R_x \cap R_y$
$-tx(t)$	$\dfrac{d}{ds}X(s)$	$\dfrac{d}{ds}X(s)$	R_x
$\dfrac{d}{dt}x(t)$	$sX(s)-x(0^-)$	$sX(s)$	At least R_x
$\displaystyle\int_{-\infty}^{t}x(\tau)\,d\tau$	$\dfrac{1}{s}\displaystyle\int_{-\infty}^{0^-}x(\tau)\,d\tau+\dfrac{X(s)}{s}$	$\dfrac{X(s)}{s}$	At least $R_x \cap \{\mathrm{Re}\{s\}>0\}$

■ D.2.1 INITIAL-VALUE THEOREM

$$\lim_{s\to\infty} sX(s) = x(0^+)$$

This result does not apply to rational functions $X(s)$ in which the order of the numerator polynomial is equal to or greater than the order of the denominator polynomial. In that case,

$X(s)$ would contain terms of the form cs^k, $k \geq 0$. Such terms correspond to the impulses and their derivatives located at time $t = 0$.

◾ D.2.2 FINAL-VALUE THEOREM

$$\lim_{s \to 0} sX(s) = \lim_{t \to \infty} x(t)$$

This result requires that all the poles of $sX(s)$ be in the left half of the s-plane.

◾ D.2.3 UNILATERAL DIFFERENTIATION PROPERTY, GENERAL FORM

$$\frac{d^n}{dt^n}x(t) \xleftarrow{\quad \mathcal{L}_u \quad} s^n X(s) - \frac{d^{n-1}}{dt^{n-1}}x(t)\Big|_{t=0^-}$$

$$- s\frac{d^{n-2}}{dt^{n-2}}x(t)\Big|_{t=0^-} - \cdots - s^{n-2}\frac{d}{dt}x(t)\Big|_{t=0^-} - s^{n-1}x(0^-)$$

E

Tables of z-Transforms and Properties

E.1 Basic z-Transforms

Signal	Transform	
$x[n] = \dfrac{1}{2\pi j} \oint X(z)\, z^{n-1}\, dz$	$X[z] = \displaystyle\sum_{n=-\infty}^{\infty} x[n]\, z^{-n}$	ROC
$\delta[n]$	1	All z
$u[n]$	$\dfrac{1}{1 - z^{-1}}$	$\|z\| > 1$
$\alpha^n u[n]$	$\dfrac{1}{1 - \alpha z^{-1}}$	$\|z\| > \|\alpha\|$
$n\alpha^n u[n]$	$\dfrac{\alpha z^{-1}}{(1 - \alpha z^{-1})^2}$	$\|z\| > \|\alpha\|$
$[\cos(\Omega_1 n)]u[n]$	$\dfrac{1 - z^{-1}\cos\Omega_1}{1 - z^{-1}2\cos\Omega_1 + z^{-2}}$	$\|z\| > 1$
$[\sin(\Omega_1 n)]u[n]$	$\dfrac{z^{-1}\sin\Omega_1}{1 - z^{-1}2\cos\Omega_1 + z^{-2}}$	$\|z\| > 1$
$[r^n\cos(\Omega_1 n)]u[n]$	$\dfrac{1 - z^{-1}r\cos\Omega_1}{1 - z^{-1}2r\cos\Omega_1 + r^2 z^{-2}}$	$\|z\| > r$
$[r^n\sin(\Omega_1 n)]u[n]$	$\dfrac{z^{-1}r\sin\Omega_1}{1 - z^{-1}2r\cos\Omega_1 + r^2 z^{-2}}$	$\|z\| > r$

■ E.1.1 BILATERAL TRANSFORMS FOR SIGNALS THAT ARE NONZERO FOR $n < 0$

Signal	Bilateral Transform	ROC
$u[-n - 1]$	$\dfrac{1}{1 - z^{-1}}$	$\|z\| < 1$
$-\alpha^n u[-n - 1]$	$\dfrac{1}{1 - \alpha z^{-1}}$	$\|z\| < \|\alpha\|$
$-n\alpha^n u[-n - 1]$	$\dfrac{\alpha z^{-1}}{(1 - \alpha z^{-1})^2}$	$\|z\| < \|\alpha\|$

▌ E.2 *z-Transform Properties*

Signal	Unilateral Transform $x[n] \xleftrightarrow{z_u} X(z)$ $y[n] \xleftrightarrow{z_u} Y(z)$	Bilateral Transform $x[n] \xleftrightarrow{z} X(z)$ $y[n] \xleftrightarrow{z} Y(z)$	ROC $z \in R_x$ $z \in R_y$
$ax[n] + by[n]$	$aX(z) + bY(z)$	$aX(z) + bY(z)$	At least $R_x \cap R_y$
$x[n - k]$	See below	$z^{-k}X(z)$	R_x, except possibly $\|z\| = 0, \infty$
$\alpha^n x[n]$	$X\left(\dfrac{z}{\alpha}\right)$	$X\left(\dfrac{z}{\alpha}\right)$	$\|\alpha\|R_x$
$x[-n]$	—	$X\left(\dfrac{1}{z}\right)$	$\dfrac{1}{R_x}$
$x[n] * y[n]$	$X(z)Y(z)$ if $x[n] = y[n] = 0$ for $n < 0$	$X(z)Y(z)$	At least $R_x \cap R_y$
$nx[n]$	$-z\dfrac{d}{dz}X(z)$	$-z\dfrac{d}{dz}X(z)$	R_x, except possibly addition or deletion of $z = 0$

■ E.2.1 UNILATERAL z-TRANSFORM TIME-SHIFT PROPERTY

$$x[n - k] \xleftrightarrow{z_u} x[-k] + x[-k + 1]z^{-1} + \cdots + x[-1]z^{-k+1} + z^{-k}X(z) \quad \text{for } k > 0$$

$$x[n + k] \xleftrightarrow{z_u} -x[0]z^k - x[1]z^{k-1} - \cdots - x[k - 1]z + z^k X(z) \quad \text{for } k > 0$$

F Introduction to MATLAB

MATLAB (short for Matrix Laboratory) is a matrix processing language that is applicable to scientific and engineering data processing. This short introduction is meant to introduce to the reader the basic tools necessary to understand and follow along with the MATLAB codes presented in this book. As with most learned skills, the best way to learn MATLAB is to sit at a computer to experiment and practice. It is recommended that as the reader learns the theoretical topics presented in this book, they also experiment with the presented code to help solidify the concepts. To further enrich understanding, students are also encouraged to download and experiment with the many supplemental MATLAB files (used to generate many of this book's figures) on this book's website.

F.1 Basic Arithmetic Rules

MATLAB displays values using a standard decimal notation. For very small or large values, you can include a 'power of ten' scale factor denoted by " e " by appending " e " to the original number. Similarly, to make a number complex, append the suffix " i " to the original number. For example,

$$7, \qquad 7e2, \qquad 7i$$

are the numbers 7, 700, and the complex number $7j$, respectively, where $j = \sqrt{-1}$. Using the standard mathematical operations, you can now build expressions:

```
+      addition
-      subtraction
*      multiplication
/      division
^      power
( )    parentheses
```

So to add two complex numbers, we would write

```
>> 3+2i + 2-4i
ans =
  5.0000 - 2.0000i
```

F.2 *Variables and Variable Names*

Variable names must begin with an alphanumeric letter. Following that, any number of letters, digits and underscores can be added, but only the first 19 characters are retained. MATLAB variable names are case sensitive, so "x" and "X" are different variables. The generic MATLAB statement is:

```
>> variable = expression;
```

If you wish to suppress the output of the statement, then append a semicolon ';'. The statement will still be completed but it will not be displayed on screen. This will be helpful for M-files (read on).

You will find it helpful to give the variable name meaning for the value it contains. For example, defining a variable 'rent' is performed by typing

```
>> rent = 650*1.15
rent =
   747.5000
```

Variable names may not start with a number, contain a dash, or contain 'reserved characters'. The following are examples of illegal character names:

```
4home    %x    net-rent    @sum
```

The % character is a special character and means a comment line. MATLAB will ignore any text following the % symbol.

F.3 *Vectors and Matrices*

Perhaps MATLAB's strongest feature is its ability to manipulate vectors and matrices. To create a vector, we use the square brackets "[ ]." For example, setting x=[2 3 1] produces

```
>> x = [2 3 1]
x =
   2   3   1
```

In MATLAB, array and matrix indexing starts at 1 (rather than 0 as is the case for the programming language 'C'). Thus to index the first element in vector x, we would write

```
>> x(1)
ans =
   2
```

Note the difference in the brackets here as well. Use square braces to create the array but use parentheses to index the array. Using square brackets here will result in an error.

Creating matrices is very similar to vectors, as shown here:

```
>> X = [1 2 3; 4 5 6; 7 8 9]
X =
   1   2   3
   4   5   6
   7   8   9
```

Here "X" is a 3-by-3 matrix. In this case, the semicolon is used to signify the end of a row. To index the (1,2) element (i.e., first row, second column), we would write

```
>>X(1,2)
ans =
     2
```

If we wanted to make a new vector "y" out of the second row of "X," then we would write

```
>> y=X(2,:)
y =
   4   5   6
```

In MATLAB, the colon " : " can be read as 'to the end of.' Thus the above expression is read as "let y equal the second row and all columns of matrix X." If we only wanted the first two elements of the third column, we would write

```
>> y=X(1:2,3)
y =
   3
   6
```

In this case, the command is read: "let y equal row 1 to the end of row 2 of X, and take the third column." The special character ' (prime) denotes matrix transposition. For example

```
>> Y=X'
Y =
   1   4   7
   2   5   8
   3   6   9
```

and

```
>> y=x'
y =
   2
   3
   1
```

When performing transposition on a complex matrix, the result is the complex conjugate transpose. For example, defining the complex matrix Z as

```
>> z=[1 2; 3 4] + [4i 2i; -i 5i]
Z =
  1.0000 + 4.0000i 2.0000 + 2.0000i
  3.0000 - 1.0000i 4.0000 + 5.0000i
```

and now taking the conjugate transpose, we obtain

```
>>z'
ans =
  1.0000 - 4.0000i 3.0000 + 1.0000i
  2.0000 - 2.0000i 4.0000 - 5.0000i
```

We add matrices just as we would with scalars.

```
>> A=[1 2; 3 4];
>> B=[2 3; -2 1];
>> C=A+B
C =
   3   5
   1   5
```

Similarly, to multiply A and B,

```
>> D=A*B
D =
  -2    5
  -2   13
```

In addition to the standard matrix multiplication, MATLAB also uses "element by element" multiplication denoted by the symbol `.*`. If A and B have the same dimensions, then `A.*B` denotes the matrix whose elements are the products of the individual elements of A and B, as shown here:

```
>> E=A.*B
E =
   2    6
  -6    4
```

It is important to note the difference between matrices D and E. Just as `.*` denotes element by element multiplication, so does `./` denote element by element division and `.^` denotes element by element powers.

F.4 *Plotting in MATLAB*

MATLAB provides a variety of functions for displaying data as 2-D graphs and for annotating these graphs. The following list summarizes the commonly used functions for plotting:

- ► `plot(X,Y)` – Creates a linear plot of X vs. Y
- ► `loglog(X,Y)` – Creates a plot using logarithmic scales for both axes.
- ► `semilogy(X,Y)` – Creates a plot using a linear scale for the X axis but logarithmic scale for the Y axis.
- ► `title` – Adds a title to the plot.
- ► `xlabel` – Adds a label to the X axis.
- ► `ylabel` – Add a label to the Y axis.
- ► `grid` – Turns on and off grid lines.

For example, to plot a sine curve of frequency three Hz we would write:

```
>> t=0:0.01:1;
>> f=3; %frequency
>> y=sin(2*pi*f*t);
>> plot(t,y,'r-')
>> title('Sine Curve');
>> xlabel('Time (seconds)');
>> ylabel('Amplitude');
```

The command `t=0:0.01:1;` creates a vector t that has first element zero and last element one where each element in the vector is equal to the previous one plus an increment of 0.01.

The command `plot(t,y,'r-')` plots the sine curve y with respect to vector t. The additional commands `r-` indicate to MATLAB to make the color red and the line style solid. You can type `help plot` to see all the possible line styles for plotting.

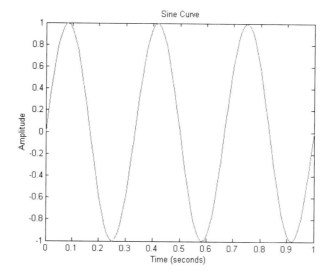

It is often desirable to graph two plots on the same figure. If we wanted to append a cosine curve, we would use the `hold` command to first "hold" the graph and then plot the next curve. For example if we now type

```
>> hold
Current plot held
>> z=cos(2*pi*f*t);
>> plot(t,z,'b--')
```

this appends a blue dashed line representing the cosine curve on the original plot.

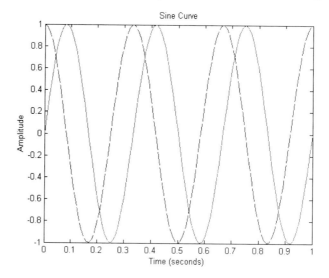

F.5 *M-files*

Normally, when you enter commands in MATLAB you enter them in the command window. However, when you have a large number of commands to write (called a program) it is impractical to keep typing them (especially when you are debugging a program). To solve this problem, we can write and save our commands in script files called M-files. When an M-file is run, MATLAB sequentially executes the commands found in the file. MATLAB even comes with its own editor (invoked by typing `edit`) to write the M-files. Once we have written our file we save the file to disk, and execute it by simply typing its name in the command window. For example, in the previous section we used many commands to produce the figure containing the sine and cosine plots. We could put them all in an M-file by typing

```
>> edit
```

Note how the editor window has popped up. Now in the editor type

```
t=0:0.01:1;
f=3; %frequency
y=sin(2*pi*f*t);
plot(t,y,'r-')
title('Sine Curve');
ylabel('Amplitude');
xlabel('Time (seconds)');
hold
z=cos(2*pi*f*t);
plot(t,z,'b--')
```

Now go to the "file → save as" tab at the top left-hand corner of the editor window and save the file. Now return to the command window and type the name you just saved the file with. MATLAB has now run all the commands in your M-file producing the output plot.

F.6 *Additional Help*

MATLAB contains two extremely useful help commands `help <name>` and `lookfor <criteria>`. The help command will tell you how to use a build in MATLAB function. In the previous section we used the built-in plot function to plot our data for us. Suppose we forgot the proper syntax for the plot function (simply typing plot by itself will produce an error). We can get all the information about the plot function by typing

```
>> help plot
```

MATLAB will then show us the proper syntax for our desired function.

Another useful MATLAB command is `lookfor`. This command will search though all the help files of the various functions and match your search criteria with words from the available help files. For example, if we wanted to obtain the Fourier transform of the vector x but cannot remember the function name to do this, we would type

```
>> lookfor fourier
```

MATLAB will then do a seach and return all the built-in-functions associated with the word "fourier". We can then type `help` to determine the exact syntax needed for a particular file. For example, if we type `lookfor fourier`, MATLAB will return

```
>> lookfor fourier
FFT Discrete Fourier transform.
FFT2 Two-dimensional discrete Fourier Transform.
FFTN N-dimensional discrete Fourier Transform.
IFFT Inverse discrete Fourier transform.
IFFT2 Two-dimensional inverse discrete Fourier
    transform.
IFFTN N-dimensional inverse discrete Fourier transform.
XFOURIER Graphics demo of Fourier series expansion.
DFTMTX Discrete Fourier transform matrix.
INSTDFFT Inverse non-standard 1-D fast Fourier
    transform.
NSTDFFT Non-standard 1-D fast Fourier transform.
FFT Quantized Fast Fourier Transform.
FOURIER Fourier integral transform.
IFOURIER Inverse Fourier integral transform.
```

Note that many internal functions are associated with the word Fourier. Since we are interested in performing the Discrete Fourier transform on vector x, we can now type `help fft` and MATLAB will now tell us the proper syntax to obtain the desired Fourier transform.

In addition to the built-in-help functions of MATLAB, the Mathworks Web site (http://mathworks.com/support/) offers online help as well as paperback books on MATLAB programming for additional help. In addition to this appendix, the following reference books are recommended:

- ▶ *Mastering MATLAB 5, A Comprehensive Tutorial and Reference*
  Authors: Duane Hanselman, Bruce Littlefield
- ▶ *MATLAB Programming for Engineers*
  Authors: Stephen Chapman

Index

UNIVERSITY OF WOLVERHAMPTON
LEARNING & INFORMATION SERVICES